Statistical Thermodynamics for Chemists and Biochemists

Statistical Thermodynamics for Chemists and Biochemists

Arieh Ben-Naim

Department of Physical Chemistry
The Hebrew University of Jerusalem
Jerusalem, Israel

Plenum Press • New York and London

Library of Congress Cataloging-in-Publication Data

Ben-Naim, Arieh.
    Statistical thermodynamics for chemists and biochemists / Arieh
Ben-Naim.
        p.   cm.
    Includes bibliographical references and index.
    ISBN 0-306-43848-8
    1. Statistical thermodynamics.   2. Chemistry, Physical and
theoretical.   I. Title.
QD504.B43   1992
541.3'69--dc20   91-48180
                                                               CIP

ISBN 0-306-43848-8

© 1992 Plenum Press, New York
A Division of Plenum Publishing Corporation
233 Spring Street, New York, N.Y. 10013

Printed in the United States of America

To Merlie

Preface

This book was planned and written with one central goal in mind: to demonstrate that statistical thermodynamics can be used successfully by a broad group of scientists, ranging from chemists through biochemists to biologists, who are not and do not intend to become specialists in statistical thermodynamics. The book is addressed mainly to graduate students and research scientists interested in designing experiments the results of which may be interpreted at the molecular level, or in interpreting such experimental results. It is not addressed to those who intend to practice statistical thermodynamics *per se*.

With this goal in mind, I have expended a great deal of effort to make the book clear, readable, and, I hope, enjoyable. This does not necessarily mean that the book as a whole is easy to read. The first four chapters are very detailed. The last four become progressively more difficult to read, for several reasons. First, presuming that the reader has already acquired familiarity with the methods and arguments presented in the first part, I felt that similar arguments could be skipped later on, leaving the details to be filled in by the reader. Second, the systems themselves become progressively more complicated as we proceed toward the last chapter. Thus, mixtures of liquids are more complicated than pure liquids, aqueous solutions are much more complicated than simple mixtures, and aqueous systems containing biomolecules are far more complicated than simple aqueous solutions. As a rule, I tried to develop in detail only those topics that are new, leaving to the reader any details similar to those encountered in previous chapters.

Chapter 1 presents a brief introduction to statistical thermodynamics. Here the basic rules of the game are summarized and some simple results pertaining to ideal gases are presented. The reader is presumed to be familiar with the basic elements of statistical thermodynamics and classical thermodynamics.

Chapter 2 contains several applications of these tools to very simple systems. Except for section 2.10, the material presented here is contained in most standard introductory textbooks in statistical thermodynamics. Section 2.10 is a detailed treatment of a chemical equilibrium affected by the adsorption of a ligand. The results of this section are applied mainly in Chapter 3, but some more general conclusions also appear in later chapters such as 5, 7, and 8.

Chapter 3 deals with slightly more complex systems, e.g., proteins having a few adsorption sites. The proteins are still considered independent, but the ligands on the sites may be dependent or independent. Here the tools of statistical thermodynamics are applied to investigate the behavior of various adsorption models that are presumed to mimic systems of interest in biology.

In these relatively simple solvable models we also introduce several concepts which normally appear in the context of the theory of liquids, such as the analogue of the solvation process, the pair correlation function and potential of average force, triplet correlations, and the nonadditivity of the triplet potential of average force. All these can be studied exactly within the framework of these models, in terms of the molecular parameters of the models. Particular attention is devoted to elaborating on the various definitions, origins, and manifestations of the cooperativity effects. Gaining familiarity with these concepts within the simplest models should help the reader to understand them when they reappear within the context of the theory of liquids.

Chapter 4 is devoted to one-dimensional (1-D) systems. Although none of the models treated here represents a real system, the study of 1-D models is very useful in gaining insight into various phenomena that do occur in reality, such as the helix–coil transition, phase separation, and the temperature of maximum density of liquid water. These are all real phenomena which may be mimicked by extremely simple and artificial models. The main reward of studying these models comes from their solvability. Section 7.1 treats the helix–coil transition theory as a classic example of the application of the 1-D techniques to solve a problem that arises in physical biochemistry. In reality, however, the helix–coil transition occurs in aqueous solution. The solvent might affect the process to the extent that the "vacuum theory" may not be relevant to the actual process taking place in aqueous solution. (This aspect of the problem is deferred to Chapter 8.)

In Chapter 5 we begin with the theory of the liquid state. Except for section 5.11, we do not survey the various theories suggested for the liquid state. Instead, we focus on the fundamental concepts of molecular distribution functions, their properties, and their relation to thermodynamic quantities.

Chapter 6 is the extension of Chapter 5 to include mixtures of two or more liquids. The most important concepts here are ideal behavior and small deviations from it. Most of the treatment is based on the Kirkwood–Buff theory of solutions. The derivation and a sample application of this powerful theory are presented in detail. We also present the elements of the McMillan–Mayer theory, which is more limited in application. Its main result is the expansion of the osmotic pressure in power series in the solute density. The most useful part of this expansion is the first-order deviation from ideal dilute behavior, a result that may also be obtained from the Kirkwood–Buff theory.

Chapters 5 and 6 may be viewed as introductory to Chapters 7 and 8, which deal with the more complex and more important aqueous solutions. Chapter 7 is devoted to pure liquid water and dilute aqueous solutions of simple solutes. There is a vast literature dealing with theoretical and experimental aspects of these systems. Only the minimum requirements for understanding the outstanding properties of this liquid and its solution are presented here. The emphasis is not on surveying the various theoretical approaches, but on fundamental concepts such as solvation, the structure of water, structural changes induced by a solute, hydrophobic and hydrophilic interactions, and the like. All of these concepts are used to treat the more complicated systems in Chapter 8.

Chapter 8 is the heart of the entire book. Because of the extreme complexity of the systems treated here, we cannot expect any exact solution to any problem. However, statistical thermodynamics can contribute much toward interpreting experimental results, suggesting new experiments, and correlating various aspects of the behavior of these systems.

Some of the "vacuum theories" treated earlier in the book are repeated in Chapter 8. Examples are chemical equilibrium, allosteric phenomena, and helix–coil transition. The general procedure to transform a "vacuum theory" into a "solution theory" is developed. Then we emphasize possible large solvent effects that can significantly alter the "vacuum theory," especially when the solvent is water. A detailed account of the thermodynamics of protein folding and protein association is also presented.

The style of the book is mainly didactic, emphasizing the methods rather than elaborating on specific examples. Occasionally, some numerical examples are presented in the form of a table or a figure. These are included as illustrations only, to give an idea of the order of magnitude of certain quantities. The numerical values are cited in the same units as they appeared in the literature. No attempt was made to use a unified set of units throughout the book.

In covering a wide range of topics, it is inevitable to run out of letters to denote different quantities. Thus, G is used for Gibbs energy throughout the book, but it is also used in the Kirkwood–Buff integrals in the form G_{ij}, as well as being a subscript to denote a gaseous ligand. Appropriate warning comments are made whenever confusion on the part of the reader would otherwise be anticipated.

The criterion used to choose the topics covered in this book was their usefulness in application to problems in chemistry and biochemistry. Thus cluster expansion methods for a real gas, although very useful for the development of the theory of real gases *per se*, was judged not useful except for the second virial coefficient. Similarly, the statistical mechanical extensions of the theory of ionic solutions beyond the Debye–Hückel limiting law were judged not useful in actual applications. Some important topics may have been missed either because of my lack of familiarity with them or because I failed to appreciate their potential usefulness. I would be grateful to receive comments or criticism from readers on this matter or on any other aspect of this book.

Thanks are due to Drs. D. Kramer, G. Haran, R. Mazo, and M. Mezei for reading various parts of the manuscript and offering helpful comments and suggestions. I am particularly grateful to Ms. Merlie Figura for typing the entire manuscript; without her gracious help this book could never have been readied for publication.

Arieh Ben-Naim

Jerusalem, Israel

Contents

1

The Fundamental Tools

1.1. INTRODUCTION

Statistical thermodynamics (ST) is a mathematical tool that bridges the gap between the properties of individual molecules and the macroscopic thermodynamic properties of bulk matter.

A microscopic description of a system of N spherical particles requires the specification of $3N$ coordinates and $3N$ momenta. If N is of the order of 10^{23}, such a detailed description is clearly impractical.

On the other hand, the same system of N particles at equilibrium can be described by only a few thermodynamic parameters, such as volume, temperature, and pressure. Such a drastic reduction in the number of parameters is achieved by averaging over all possible locations and momenta of all the particles involved in the system. The rules employed in averaging are contained in the theory of statistical thermodynamics. This remarkable theory provides a set of relationships between thermodynamic quantities on the one hand and molecular quantities on the other. These relationships are presented in this chapter, and will be referred to as our "rules of the game"; the ultimate proof of their validity is provided *a posteriori* by the success of ST in predicting the thermodynamic quantities of a real system from knowledge of the molecular properties of its constituent particles.

Naturally, one may require a proof of the validity of these rules based on some other, more fundamental postulates. Indeed, such proofs are available in standard textbooks on ST. However, since our aim in this book is the applications of ST rather than its development, we shall not elaborate on the derivation of the rules from more fundamental postulates. The reader should be aware of the fact that no matter how deep one seeks to look for the foundations of the theory, at some point one must accept some postulates which can also be viewed as the "rules of the game." Therefore we start this book with what we believe to be the most convenient set of rules. The proof, or rather the confidence in their validity, will be achieved through their application to systems of interest in chemistry and biochemistry.

In this chapter we also present without derivation some elementary results of ST, e.g., the thermodynamics of ideal gases. These are derived in any elementary textbook on ST. They relate the thermodynamics of an ideal-gas system to the molecular properties of single molecules. For any real system one must use as input not only the properties of single molecules, but also some information on the extent of interactions among a group of a few molecules. These interactions are presumed to be known either from theory or from experiments. In practice, however, the interactions among even simple molecules are not known. Therefore it is a common practice to employ "model potential functions" to describe the interaction between two or more molecules and to use these

1

model functions as input in the theory. In section 1.7 we present a short survey of such model potential functions used in systems of interest in this book.

Both thermodynamics and statistical thermodynamics deal with the same quantities, such as temperature, pressure, energy, and entropy, yet there are some fundamental differences in the way the two theories handle these quantities.

Thermodynamics provides *general* relationships among thermodynamic quantities. For instance, we have the well-known relationships

$$\left(\frac{\partial S}{\partial P}\right)_T = -\left(\frac{\partial V}{\partial T}\right)_P, \qquad \frac{C_V}{T} = \left(\frac{\partial S}{\partial T}\right)_V, \qquad V = \left(\frac{\partial G}{\partial P}\right)_T. \qquad (1.1.1)$$

These are universal relationships in the sense that they apply to any system at equilibrium.

On the other hand ST deals with thermodynamic quantities that are pertinent to *specific* systems. The ultimate goal of ST is to calculate the thermodynamic quantities of a specific system in terms of its molecular properties. For example, the entropy of an ideal gas of simple particles can be computed from the well-known expression

$$S = Nk \ln\left[\left(\frac{2\pi mkT}{h^2}\right)^{3/2} \frac{e^{5/2}}{\rho}\right], \qquad (1.1.2)$$

where N and T are the number of particles and the temperature, respectively, k is the Boltzmann constant, h is the Planck constant, and ρ is the number density ($\rho = N/V$). In Eq. (1.1.2) the entropy is given explicitly in terms of the thermodynamic variables T, V, N as well as in terms of the molecular parameters, in this case the molecular mass m. Thus we have $S = S(T, V, N; m)$. Clearly, for different gases we have different functions $S(T, V, N)$ depending on the parameter m. In more general cases we might obtain a function of the form $S = S(T, V, N; a_1 \cdots a_n)$ where $a_1 \cdots a_n$ are the molecular parameters characterizing the specific system under consideration.

In thermodynamics we often encounter "constants" that are characteristic of a certain equilibrium condition. For instance, for the chemical reaction

$$A + B \rightleftarrows C + D \qquad (1.1.3)$$

at equilibrium in an ideal gas system, we have

$$K(T) = \frac{P_C P_D}{P_A P_D}, \qquad (1.1.4)$$

where P_i is the partial pressure of the species i and K is the equilibrium constant. K depends on T but not on the pressure of the system. Thermodynamics does not offer a method of computing the equilibrium constant $K(T)$. Such a computation is possible for a specific system of A, B, C, and D using the methods of ST.

In thermodynamics we usually do not specify the choice of the independent variables used to describe our system. By writing $(\partial S/\partial T)_{V,N}$ it is implicitly assumed that S is viewed as a function of the independent variables T, V, N. Clearly, if the symbol S is used for the *value* of the entropy of a given system, then S is independent of the choice of the independent variables that describe the system. However, if S denotes the *function* then it becomes essential to specify the choice of the independent variables. Thus, $S(T, V, N)$, $S(T, P, N)$, and $S(T, V, \mu)$ are all different functions of the independent variables. These different functions are not equivalent. For example, having the functions $S(E, V, N)$ or $A(T, V, N)$ or $G(T, P, N)$, one can derive all other thermodynamic quantities of the system using standard thermodynamic relationships. This is not the case if we have the functions $S(T, V, N)$, $S(T, P, N)$ or $G(T, V, N)$. Furthermore, the principle of maximum entropy or minimum Helmholtz or Gibbs energy holds for the functions $S(E, V, N)$ or $A(T, V, N)$ or $G(T, P, N)$ and not for any other choice of the independent variables.

Finally we recall that, within the realm of thermodynamics, each of the thermodynamic quantities is considered to have a well-defined value for each system at equilibrium. In ST we also consider fluctuations of these quantities. The reason is simple. In thermodynamics we deal only with the average properties of the system. In ST we deal with the distribution functions through the use of which average quantities are computed. In addition, various moments of these distributions may be compared or related to other experimental quantities. We shall briefly mention some of these relationships in section 1.4.

1.2. NOTATION

The location of the center of the molecule is denoted by the vector $\mathbf{R} = (x, y, z)$, where x, y, and z are the Cartesian coordinates of a specific point in the molecule, chosen as its center. For instance, for a water molecule, it will be convenient to choose the center of the oxygen atom as the center of the molecule. We shall often use the symbol $\mathbf{R}$ to denote the location of a specific point in the system, not necessarily occupied by the center of a molecule. If there are several molecules, then $\mathbf{R}_i$ denotes the location of the center of the ith molecule.

An infinitesimal element of *volume* is denoted by

$$d\mathbf{R} = dx\, dy\, dz. \qquad (1.2.1)$$

This represents the volume of a small cube defined by the edges dx, dy, and dz, as illustrated in Fig. 1.1. Some texts use the notation d^3R for the element of volume to distinguish it from the infinitesimal vector, denoted by $d\mathbf{R}$. In this book, $d\mathbf{R}$ will always signify an element of volume, except for one case in Chapter 8 when we discuss the force between two particles.

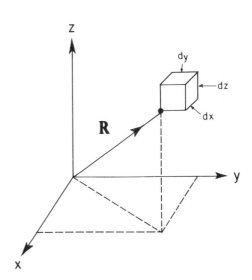

FIGURE 1.1. An infinitesimal element of volume, $d\mathbf{R} = dx\, dy\, dz$, located at point $\mathbf{R}$.

We usually make no distinction between the notation for the *region* (defined, say, by the cube of edges dx, dy, and dz) and its *volume* (given by the product $dx\,dy\,dz$). For instance, we may say that the center of a particle falls within $d\mathbf{R}$, meaning that the center is contained in the *region* of space defined by the cube of edges dx, dy, and dz. On the other hand if $dx = dy = dz = 2$ Å, then $d\mathbf{R}$ is also the volume (8 Å^3) of this region.

The element of volume $d\mathbf{R}$ is understood to be located at the point $\mathbf{R}$. In some cases, it will be convenient to choose an element of volume other than a cubic one. For instance, an infinitesimal spherical shell of radius R and width dR has the volume

$$d\mathbf{R} = 4\pi R^2\, dR. \tag{1.2.2}$$

For simple particles such as hard spheres or argon atoms, the designation of the locations of all the centers of the particles in the system is a sufficient description of the *configuration* of the system. A more detailed description of both location and orientation is needed for more complex molecules. For a rigid, nonspherical molecule, we use $\mathbf{R}_i$ to designate the location of its center and $\mathbf{\Omega}_i$ the orientation of the whole molecule. As an example, consider a water molecule as being a rigid particle.† Figure 1.2 shows one possible set of angles used to describe the orientation of a water molecule. Let $\mathbf{\mu}$ be the vector originating from the center of the oxygen atom and bisecting the H–O–H angle. Two angles, say ϕ and θ, are required to fix the orientation of this vector. In addition, a third angle ψ is needed to describe the rotation of the whole molecule about the axis $\mathbf{\mu}$.

In general, integration over the variable $\mathbf{R}_i$ means integration over the whole volume of the system, i.e.,

$$\int d\mathbf{R}_i = \int_V d\mathbf{R}_i = \int_0^L dx_i \int_0^L dy_i \int_0^L dz_i, \tag{1.2.3}$$

where for simplicity we have assumed that the region of integration is a cube of length

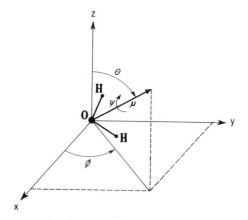

FIGURE 1.2. One possible choice of orientation angles for a water molecule. The vector $\mathbf{\mu}$, originating from the center of the oxygen atom and bisecting the H–O–H angle, can be specified by the two polar angles θ and ϕ. In addition, the angle of rotation of the molecule about the axis $\mathbf{\mu}$ is denoted by ψ.

† The assumption is made that the geometry of the molecule is fixed. This is consistent with the well-known fact that deviations from the equilibrium values for the O–H distances and the H–O–H angle are almost negligible at room temperature (see also Chapter 7).

L. The integration over Ω_i will be understood to be over all possible orientations of the molecule. Using, for instance, the set of Euler angles, we have

$$\int d\Omega_i = \int_0^{2\pi} d\phi_i \int_0^\pi \sin \theta_i \, d\theta_i \int_0^{2\pi} d\psi_i = 8\pi^2. \qquad (1.2.4)$$

Note that for a linear molecule we have one degree of freedom less, hence

$$\int d\Omega_i = \int_0^{2\pi} d\phi_i \int_0^\pi \sin \theta_i \, d\theta_i = 4\pi. \qquad (1.2.5)$$

The configuration of a rigid nonlinear molecule is thus specified by a six-dimensional vector, including both the *location* and the *orientation* of the molecule, namely,

$$\mathbf{X}_i = \mathbf{R}_i, \ \Omega_i = (x_i, y_i, z_i, \phi_i, \theta_i, \psi_i) \qquad (1.2.6)$$

The configuration of a system of N rigid molecules is denoted by

$$\mathbf{X}^N = \mathbf{X}_1, \mathbf{X}_2, \ldots, \mathbf{X}_N. \qquad (1.2.7)$$

Similarly, the infinitesimal element of the configuration of a single molecule is denoted by

$$d\mathbf{X}_i = d\mathbf{R}_i \, d\Omega_i, \qquad (1.2.8)$$

and, for N molecules,

$$d\mathbf{X}^N = d\mathbf{X}_1 \, d\mathbf{X}_2 \cdots d\mathbf{X}_N. \qquad (1.2.9)$$

For a general molecule, the specification of the configuration by six coordinates may be insufficient. A simple example is *n*-butane, a schematic description of which is given in Fig. 1.3. Here, the position of one of the carbon atoms is chosen as the "center" of the molecule. The orientation of the triplet of carbon atoms, say 1, 2, and 3, is denoted by Ω. This specification leaves one degree of freedom, which is referred to as internal rotation and is denoted by ϕ_{23}. This completes our specification of the configuration of the molecule (it is presumed that bond lengths and angles are fixed at their equilibrium values). In the general case, the set of internal rotational angles makes up the vector $\mathbf{P}$ and will be referred to as the *conformation* of the molecule. Thus, the total *configuration*

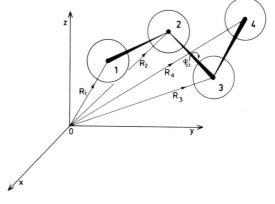

FIGURE 1.3. Schematic description of *n*-butane. The total configuration of the molecule consists of the location of say, the carbon numbered 1 ($\mathbf{R}_1$); the orientation of the (rigid) triplet of carbons, say 1, 2, and 3; and an internal rotational angle, designated by ϕ_{23}.

in the general case includes the coordinates of *location*, *orientation*, and *internal rotation* and is denoted by

$$\mathbf{Y} = \mathbf{R}, \mathbf{\Omega}, \mathbf{P}. \tag{1.2.10}$$

In most of this book, we shall be dealing with rigid molecules, in which case a specification of internal rotations is not required.

1.3. THE FUNDAMENTAL EQUATIONS OF STATISTICAL THERMODYNAMICS

In this section we present the main "rules of the game." These should be adopted before proceeding to any application. Once they are accepted, one can derive other relationships using standard thermodynamic manipulations. The fundamental equations are presented in the following subsections according to the set of independent variables employed in the characterization of the system.

1.3.1. E, V, N Ensemble

We consider first an isolated system having a fixed internal energy E, volume V, and number of particles N. Any real system experiences some degree of interaction with its environment. Therefore a strictly isolated system does not exist. Even if it existed, such a system would be of no interest. No measurements could be performed on such a system. Any measurement performed on an isolated system involves some interaction between the system and its environment. This violates the requirement of isolation.

Nevertheless, an isolated system characterized by the variables E, V, N is a convenient starting point for the theoretical framework of ST. Let $\Omega(E, V, N)$ be the number of quantum-mechanical states of the system characterized by the variables E, V, N. That is also the number of eigenstates of the Hamiltonian of the system having the eigenvalue E. We assume for simplicity that we have a finite number of such eigenstates.

The first relationship that we adopt is between the entropy S of the system and the number of states Ω. This is the famous Boltzmann formula

$$S(E, V, N) = k \ln \Omega(E, V, N), \tag{1.3.1}$$

where $k = 1.38 \times 10^{-23} \, \mathrm{J \, K^{-1}}$ is the Boltzmann constant.

The fundamental thermodynamic relationship for the variation of the entropy in terms of the independent variables E, V, N is

$$dS = \frac{dE}{T} + \frac{P dV}{T} - \frac{\mu \, dN}{T}, \tag{1.3.2}$$

from which one can obtain the temperature T, the pressure P, and the chemical potential μ as partial derivatives of S. Other thermodynamic quantities can be obtained from the standard thermodynamic relationships.

Equations (1.3.1) and (1.3.2) tell us that if we know Ω as a function of the variables E, V, N we can compute the entropy as well as all the thermodynamic quantities of the system at equilibrium. The key connection between thermodynamics and ST is Eq. (1.3.1). In practice there are very few systems for which Ω is known. Therefore this equation, though the cornerstone of the theory, is seldom used in applications.

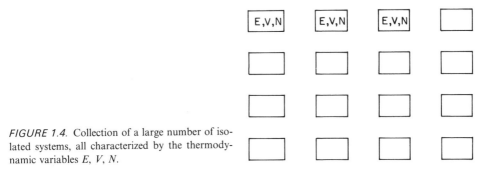

FIGURE 1.4. Collection of a large number of isolated systems, all characterized by the thermodynamic variables E, V, N.

Next we introduce the fundamental distribution function for this system. Suppose
that we have a very large collection of systems, all of which are identical with respect to
the thermodynamic characterization, i.e., all have the same values of E, V and N. Such
a collection of systems is called an E, V, N ensemble (see Fig. 1.4). (It is sometimes
referred to for historical reasons as a microcanonical ensemble.) In such a system the
microscopic or molecular state of each system may be different at any given time. One
of the fundamental postulates of ST is the assertion that the probability of a specific state
i is given by

$$P_i = \frac{1}{\Omega}.$$ (1.3.3)

This is equivalent to the assertion that all states of an E, V, N system have equal probabilities. Since $\sum_{i=1}^{\Omega} P_i = 1$, it follows that each of the P_i is equal to Ω^{-1}.

1.3.2. T, V, N Ensemble

The most useful connection between thermodynamics and ST is the one established
for a system at a given temperature T, volume V, and number of particles N. The
corresponding ensemble is referred to as the isothermal ensemble or the canonical ensemble. To obtain the T, V, N ensemble from the E, V, N ensemble, we replace the boundaries
between the isolated systems of Fig. 1.4 by diathermal (i.e., heat-conducting) boundaries.
The latter permits the flow of heat between systems in the ensemble. The volume and the
number of particles are still kept constant. This ensemble is described schematically in
Fig. 1.5.

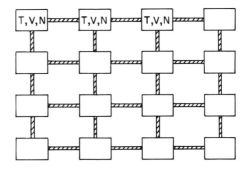

FIGURE 1.5. Same collections of a systems as in
Fig. 1.4, but all the systems are connected by heat-
conducting (diathermal) rods. The systems are
characterized by the thermodynamic variables T,
V, N.

We know from thermodynamics that any two systems at thermal equilibrium (i.e., when heat can be exchanged through their boundaries) have the same temperature. Thus when we replaced the athermal boundaries by diathermal boundaries, the fixed value of the internal energy E is replaced by a fixed value of the temperature. The internal energies of the system can now fluctuate. The probability of finding a system in the ensemble having internal energy E is given by

$$\Pr(E) = \frac{\Omega(E, V, N) \exp(-\beta E)}{Q},\tag{1.3.4}$$

where $\beta = (kT)^{-1}$ and Q is the normalization constant. Note that the probability of finding a specific state having energy E is $\exp(-\beta E)/Q$. Since there are Ω such states, the probability of finding an energy level E is given by (1.3.4). The normalization condition is

$$\sum_E \Pr(E) = 1,\tag{1.3.5}$$

the summation being over all possible energies E. From (1.3.4) and (1.3.5), we have

$$Q(T, V, N) = \sum_E \Omega(E, V, N) \exp(-\beta E).\tag{1.3.6}$$

Note that each term in this sum is a function of the four variables E, V, N, and T, but the quantity obtained after summation over E is a function of only three variables: T, V, N. Formally, we have transformed one of our independent variables E into T. Although Q has been defined as the normalization constant of $\Pr(E)$, it can be viewed as the result of the transformation of variables performed in (1.3.6).

The fundamental connection between $Q(T, V, N)$ as defined in (1.3.6) and thermodynamics is given by

$$A(T, V, N) = -kT \ln Q(T, V, N),\tag{1.3.7}$$

where A is the Helmholtz energy of the system at T, V, N. In the present book, Eq. (1.3.7) is adopted as one of the postulates of the theory, much as we have adopted Eq. (1.3.1). In fact, Eq. (1.3.7) may be derived from (1.3.1) and does not consist of an independent postulate. We can see the connection between A and Q in a qualitative manner as follows. Starting from definition (1.3.6) and using Eq. (1.3.1), we rewrite (1.3.6) as

$$Q(T, V, N) = \sum_E \exp[S(E, V, N)/k] \exp(-\beta E).\tag{1.3.8}$$

We now assume that we can replace the sum over all the energy values by the maximal term, i.e., the term for which

$$\frac{\partial}{\partial E} [\exp\{-\beta[E - TS(E, V, N)]\}] = 0,$$

or equivalently

$$\frac{\partial S(E, V, N)}{\partial E} = \frac{1}{T}.\tag{1.3.9}$$

The value of E that satisfies the condition (1.3.9) is denoted by E^*, which is now a function of the variables T, V, N. Thus we replace the sum in (1.3.8) by the term corresponding to $E = E^*$ to obtain

$$Q(T, V, N) = \exp\{-\beta[E^* - TS(E^*, V, N)]\}$$
$$= \exp[-\beta A(T, V, N)] \qquad (1.3.10)$$

In the last equality on the rhs of (1.3.10) we used the thermodynamic definition of the Helmholtz energy. Since E^* is a function of T, V, N, the entire expression on the rhs of (1.3.10) is also a function of the independent variables T, V, N.

The passage from (1.3.8) to (1.3.10) is not trivial. We replaced a sum of an enormous number of terms, all positive, by only one of its members. The qualitative reason that justifies such a replacement is that, for macroscopic systems, $\Omega(E, V, N)$ steeply increases with E, whereas $\exp(-\beta E)$ is a steeply decreasing function of E. The product of the two produces a function that sharply peaks at $E = E^*$. Since our qualitative argument does not consist of a proof of Eq. (1.3.10), we suggest the adoption of this relationship—as well as similar ones discussed below—as our "rules of the game." We shall elaborate further in section 2.4 on the question of the replacement of a whole sum by a single term.

The quantity $Q(T, V, N)$ defined in Eq. (1.3.6) is called the T, V, N partition function (PF) or the canonical or isothermal–isochoric PF. In the following subsections we shall encounter other partition functions pertaining to systems characterized by the independent variables T, P, N or T, V, μ. The canonical PF is by far the one most used in theoretical work, although other PFs are also useful in some specific applications.

Once the partition function $Q(T, V, N)$ is evaluated, either exactly or approximately, then relation (1.3.7) may be used to obtain the Helmholtz energy. This relation is fundamental in the sense that all the thermodynamic information on the system can be extracted from it by the application of standard thermodynamic relations, i.e., from

$$dA = -S\,dT - P\,dV + \mu\,dN, \qquad (1.3.11)$$

with S the entropy, P the pressure, and μ the chemical potential [for a multicomponent system, the last term on the rhs of (1.3.11) should be interpreted as a scalar product $\boldsymbol{\mu} \cdot d\mathbf{N} = \sum_{i=1}^{c} \mu_i\,dN_i$] we get

$$S = -\left(\frac{\partial A}{\partial T}\right)_{V,N} = k \ln Q + kT\left(\frac{\partial \ln Q}{\partial T}\right)_{V,N} \qquad (1.3.12)$$

$$P = -\left(\frac{\partial A}{\partial V}\right)_{T,N} = kT\left(\frac{\partial \ln Q}{\partial V}\right)_{T,N} \qquad (1.3.13)$$

$$\mu = \left(\frac{\partial A}{\partial N}\right)_{T,V} = -kT\left(\frac{\partial \ln Q}{\partial N}\right)_{T,V}. \qquad (1.3.14)$$

Other relations can be obtained readily; for instance, the internal energy E is obtained from (1.3.7) and (1.3.12):

$$E = A + TS = kT^2\left(\frac{\partial \ln Q}{\partial T}\right)_{V,N}. \qquad (1.3.15)$$

The constant-volume heat capacity can be obtained by differentiating (1.3.15) with respect to temperature. Note that in (1.3.15) E is the *average* internal energy of the system. It should be distinguished from the running index E in Eq. (1.3.6).

1.3.3. T, P, N Ensemble

In the passage from the E, V, N (Fig. 1.4) to the T, V, N (Fig. 1.5) ensemble, we have removed the constraint of a constant energy by allowing exchange of energy between the systems. In this way constant energy has been replaced by constant temperature. In a similar fashion we can remove the constraint of a constant volume by replacing the rigid boundaries between the systems by flexible boundaries, as in Fig. 1.6. In the new ensemble, referred to as the isothermal–isobaric ensemble, the volume of each system may fluctuate. We know from thermodynamics that when two systems are allowed to reach a mechanical equilibrium, they have the same pressure. The volume of each system can attain any value, with its probability distribution given by

$$\Pr(V) = \frac{Q(T, V, N) \exp(-\beta PV)}{\Delta(T, P, N)}, \tag{1.3.16}$$

where P is the pressure of the system at equilibrium. The normalization constant $\Delta(T, P, N)$ defined by

$$\Delta(T, P, N) = \sum_V Q(T, V, N) \exp(-\beta PV)$$

$$= \sum_V \sum_E \Omega(E, V, N) \exp(-\beta E - \beta PV) \tag{1.3.17}$$

is called the isothermal–isobaric PF or simply the T, P, N partition function. Note that in (1.3.17) we have summed over all possible volumes, treating the volume as a discreet variable. In actual application for classical systems this sum should be interpreted as an integral over all possible volumes, namely,

$$\Delta(T, P, N) = C \int_0^\infty dV Q(T, V, N) \exp(-\beta PV), \tag{1.3.18}$$

where C has the dimension of V^{-1}, to render the rhs of (1.3.18) dimensionless. The PF $\Delta(T, P, N)$, though less convenient than $Q(T, V, N)$, is sometimes very useful, especially when connection with experimental quantities measured at constant T and P is required.

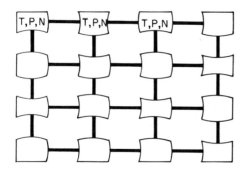

FIGURE 1.6. Same collection of systems as in Fig. 1.4, but now the systems are connected by heat-conducting and "volume-conducting" rods. The systems are characterized by the thermodynamic variables T, P, N.

The fundamental connection between $\Delta(T, P, N)$ and thermodynamics is

$$G(T, P, N) = -kT \ln \Delta(T, P, N), \tag{1.3.19}$$

where G is the Gibbs energy of the system.

As in subsection 1.3.2, we provide a qualitative "derivation" of (1.3.19) based on relation (1.3.7); i.e., we start from the definition (1.3.17) and substitute $Q(T, V, N)$ from (1.3.7):

$$\Delta(T, P, N) = \sum_V \exp[-\beta A(T, V, N)] \exp(-\beta PV). \tag{1.3.20}$$

Again we replace the entire sum on the rhs of (1.3.20) by the maximal term, i.e., the term that fulfills the condition

$$\frac{\partial}{\partial V}[\exp\{-\beta[A(T, V, N) + PV]\}] = 0 \tag{1.3.21}$$

or, equivalently,

$$\frac{\partial A(T, V, N)}{\partial V} = -P. \tag{1.3.22}$$

We see that the volume $V = V^*$ that fulfills the condition (1.3.21) or (1.3.22) is the volume for which the thermodynamic relation (1.3.22) defining the pressure of a T, V, N system is fulfilled.

We now replace the sum on the rhs of (1.3.20) by the maximal term $V = V^*$ to obtain

$$\Delta(T, P, N) = \exp[-\beta A(T, V^*, N) - \beta PV^*]$$

$$= \exp[-\beta G(T, P, N)]. \tag{1.3.23}$$

Note that since V^* is a function of T, P, N, the entire rhs of (1.3.23) is a function of T, P, N.

The relation (1.3.19) is the fundamental equation for the T, P, N ensemble. Once we have the function $\Delta(T, P, N)$, all other thermodynamic quantities may be obtained by standard relations, i.e.,

$$dG = -S\,dT + V\,dP + \mu\,dN. \tag{1.3.24}$$

Hence

$$S = -\left(\frac{\partial G}{\partial T}\right)_{P,N} = k \ln \Delta + kT\left(\frac{\partial \ln \Delta}{\partial T}\right)_{P,N} \tag{1.3.25}$$

$$V = \left(\frac{\partial G}{\partial P}\right)_{T,N} = -kT\left(\frac{\partial \ln \Delta}{\partial P}\right)_{T,N} \tag{1.3.26}$$

$$\mu = \left(\frac{\partial G}{\partial N}\right)_{T,P} = -kT\left(\frac{\partial \ln \Delta}{\partial N}\right)_{T,P}. \tag{1.3.27}$$

The enthalpy can be obtained from (1.3.19) and (1.3.25):

$$H = G + TS = kT^2\left(\frac{\partial \ln \Delta}{\partial T}\right)_{P,N}. \tag{1.3.28}$$

The constant-pressure heat capacity can be obtained by differentiating (1.3.28) with respect to temperature. The isothermal compressibility is obtained by differentiating (1.3.26) with respect to pressure, and so on.

1.3.4. T, V, μ Ensemble

In subsection 1.3.3 we derived the T, P, N PF from the T, V, N PF by replacing the constant variable V by the pressure P. Here we derive another PF by starting from $Q(T, V, N)$ and replacing the constant variable N by μ. To do that we start with the canonical ensemble (Fig. 1.5) and replace the impermeable boundaries by permeable boundaries. The new ensemble is referred to as the T, V, μ ensemble, shown in Fig. 1.7. Note that the volume of each system is still constant. However, by removing the constraint on constant N, we permit fluctuations in the number of particles. We know from thermodynamics that a pair of systems between which there exists a free exchange of particles at equilibrium with respect to material flow, is characterized by a constant chemical potential μ. The variable N can now attain any value with the probability distribution

$$\Pr(N) = \frac{Q(T, V, N) \exp(\beta\mu N)}{\Xi(T, V, \mu)}, \tag{1.3.29}$$

where $\Xi(T, V, \mu)$, the normalization constant, is defined by

$$\Xi(T, V, \mu) = \sum_{N} Q(T, V, N) \exp(\beta\mu N), \tag{1.3.30}$$

where the summation in (1.3.30) is over all possible values of N. The new partition function $\Xi(T, V, \mu)$ is referred to as the grand partition function, the open-system PF or simply the T, V, μ PF.

In Eq. (1.3.30) we have defined the T, V, μ partition function for a one-component system. In a straightforward manner we may generalize the definition for a multicomponent system. Let $\mathbf{N} = N_1, \ldots, N_c$ be the vector representing the composition of the system, where N_i is the number of molecules of species i. The corresponding vector $\boldsymbol{\mu} = \mu_1, \ldots, \mu_c$ includes the chemical potential of each of the species. For an open system with respect to *all* components we have the generalization of definition (1.3.30),

$$\Xi(T, V, \boldsymbol{\mu}) = \sum_{N_1} \cdots \sum_{N_c} Q(T, V, \mathbf{N}) \exp[\beta\boldsymbol{\mu} \cdot \mathbf{N}], \tag{1.3.31}$$

where $\boldsymbol{\mu} \cdot \mathbf{N} = \sum_i \mu_i N_i$ is the scalar product of the two vectors $\boldsymbol{\mu}$ and $\mathbf{N}$.

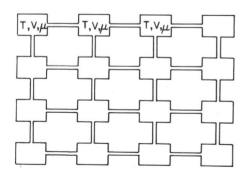

FIGURE 1.7. Same collection of systems as in Fig. 1.4, but now the systems are connected by heat-conducting and material-conducting (permeable) rods. The systems are characterized by the thermodynamic variables T, V, μ.

An important case is a system open with respect to some of the species but closed to the others. For instance, in a two-component system of A and B we may define two partial grand partition functions as follows:

$$\Xi(T, V, N_A, \mu_B) = \sum_{N_B} Q(T, V, N_A, N_B) \exp[\beta \mu_B N_B] \tag{1.3.32}$$

$$\Xi(T, V, \mu_A, N_B) = \sum_{N_A} Q(T, V, N_A, N_B) \exp[\beta \mu_A N_A]. \tag{1.3.33}$$

Equation (1.3.32) corresponds to a system closed with respect to A but open with respect to B. Equation (1.3.33) corresponds to a system closed to B but open to A. A typical experimental system of this kind is an osmotic system, where the solvent but not the solute (or solutes) may be exchanged with environment.

The fundamental connection between the PF defined by (1.3.30) and thermodynamics is

$$P(T, V, \mu)V = kT \ln \Xi(T, V, \mu), \tag{1.3.34}$$

where $P(T, V, \mu)$ is the pressure of a system characterized by the independent variable T, V, μ.

Again we show in a qualitative manner how relation (1.3.34) is obtained from definition (1.3.30). Using the fundamental relation (1.3.7), we have

$$\Xi(T, V, \mu) \Rightarrow \sum_N \exp[-\beta A(T, V, N) + \beta \mu N]. \tag{1.3.35}$$

Now we take the maximal term in the sum on the rhs of (1.3.35), i.e., the term $N = N^*$ for which

$$\frac{\partial}{\partial N} \{\exp[-\beta A(T, V, N) + \beta \mu N]\} = 0, \tag{1.3.36}$$

or equivalently

$$\frac{\partial A(T, V, N)}{\partial N} = \mu. \tag{1.3.37}$$

Note that the value of $N = N^*$ that satisfies the condition (1.3.36) or (1.3.37) is the value for which we obtain the thermodynamic definition of the chemical potential for a system characterized by the variables T, V, N. Replacing the sum in (1.3.35) by its maximal term $N = N^*$, we obtain

$$\Xi(T, V, \mu) = \exp[-\beta A(T, V, N^*) + \beta \mu N^*] = \exp[+\beta P(T, V, \mu)V], \tag{1.3.38}$$

where we used the thermodynamic relation

$$G = \mu N = A + PV. \tag{1.3.39}$$

Note that N^* is the solution of Eq. (1.3.37); therefore it is a function of T, V, μ. Hence the entire rhs of (1.3.38) is also a function of the variables T, V, μ.

The fundamental relation (1.3.34) may be used to obtain all relevant thermodynamic quantities. Thus, using the general differential of PV we obtain,

$$d(PV) = S\,dT + P\,dV + N\,d\mu \tag{1.3.40}$$

$$S = \left(\frac{\partial(PV)}{\partial T}\right)_{V,\mu} = k\ln\Xi + kT\left(\frac{\partial\ln\Xi}{\partial T}\right)_{V,\mu} \tag{1.3.41}$$

$$P = \left(\frac{\partial(PV)}{\partial V}\right)_{T,\mu} = kT\left(\frac{\partial\ln\Xi}{\partial V}\right)_{T,\mu} = kT\frac{\ln\Xi}{V} \tag{1.3.42}$$

$$N = \left(\frac{\partial(PV)}{\partial\mu}\right)_{T,V} = kT\left(\frac{\partial\ln\Xi}{\partial\mu}\right)_{T,V}. \tag{1.3.43}$$

Other quantities such as the Gibbs energy or the energy of the system may be obtained from the standard relations

$$G = \mu N, \tag{1.3.44}$$

$$E = G + TS - TV. \tag{1.3.45}$$

The general structure of the theoretical tool of ST should now be quite clear. For each set of independent variables we define a partition function. This partition function is related to a thermodynamic quantity through one of the fundamental relationships. On the other hand, each of the summands in the PF is proportional to the probability of realizing the specific value of the variable on which the summation is carried out. Having the probability distribution, for each set of independent variables, one can write down various averages over that distribution function. The calculation of such averages consists of the main outcome of ST.

In this section we have listed the most useful partition functions, $Q(T, V, N)$, $\Delta(T, P, N)$, and $\Xi(T, V, \mu)$. Clearly, other PFs can be constructed, for instance for multicomponent systems or for a system in an external field.

The general procedure of transforming one set of independent variables into another set is essentially a Laplace transformation (over either a continuous or a discrete variable). The corresponding procedure within the realm of thermodynamics is known as the Legendre transformation. Note that all the PFs are expressed in terms of independent variables, at least one of which is an extensive variable, e.g., N in $\Delta(T, P, N)$ or V in $\Xi(T, V, \mu)$.

In principle we could have proceeded one step further and transformed the last extensive variable. For instance, we can define

$$\Gamma(T, P, \mu) = C\int_0^\infty \exp(-\beta PV)\Xi(T, V, \mu)\,dV. \tag{1.3.46}$$

However, if we examine the thermodynamic potential corresponding to Γ, we find that

$$\Gamma(T, P, \mu) = C\int_0^\infty \exp(-\beta PV)\exp(+\beta PV)\,dV. \tag{1.3.47}$$

Since the integrand is unity, we cannot proceed to take the maximum term as we did in previous cases. In fact, for a macroscopic system this integral diverges. The physical reason is that there exists no physical ensemble characterized by the independent intensive variables T, P, μ. We know that these variables are connected by the Gibbs–Duhem relation

$$S\,dT - V\,dP + N\,d\mu = 0, \tag{1.3.48}$$

which means that if we had a potential function of the form $f(T, P, \mu)$, it would be identically zero.

The quantity $\Gamma(T, P, \mu)$ is referred to as the generalized PF. It has found applications, although not in the conventional way we use other partition functions. One application is described in section 4.5.5.

1.4. SOME AVERAGE QUANTITIES

As we have already noted, the most common outcomes of ST calculations are average quantities. We present here some simple average quantities in the various ensembles.

In the T, V, N ensemble, the average energy of the system is defined by

$$\langle E \rangle = \sum_E E Pr(E) = \frac{\sum\limits_E E\Omega(E, V, N)\exp(-\beta E)}{Q(T, V, N)}. \tag{1.4.1}$$

Using the definition of $Q(T, V, N)$ in (1.3.6), we find that

$$\langle E \rangle = kT^2 \frac{\partial \ln Q(T, V, N)}{\partial T}. \tag{1.4.2}$$

Note that the average energy of the system denoted here by $\langle E \rangle$ is the same as the internal energy denoted, in thermodynamics, by U. In this book we shall reserve the notation U for potential energy and use $\langle E \rangle$ for the total (potential and kinetic) energy. Sometimes we also use E instead of $\langle E \rangle$. But care must be exercised to make the distinction between a specific energy level E as in (1.3.6), and the average energy of the system, for instance in (1.3.15). When confusion could arise it is better to use the notation $\langle \; \rangle$ for an average quantity.

An important average quantity in the T, V, N ensemble is the average fluctuation in the internal energy, defined by

$$\sigma_E^2 = \langle (E - \langle E \rangle)^2 \rangle. \tag{1.4.3}$$

Using the probability distribution (1.3.4) of section 1.3 we can express σ_E^2 in terms of the constant-volume heat capacity, i.e.,

$$\langle (E - \langle E \rangle)^2 \rangle = \sum_E (E - \langle E \rangle)^2 Pr(E)$$

$$= \sum_E [E^2 Pr(E) - 2E\langle E \rangle Pr(E) + \langle E \rangle^2 Pr(E)]$$

$$= \langle E^2 \rangle - \langle E \rangle^2. \tag{1.4.4}$$

On the other hand, by differentiation of $\langle E \rangle$ in (1.4.1) with respect to T we obtain

$$C_V = \left(\frac{\partial \langle E \rangle}{\partial T} \right)_{V,N} = \frac{\langle E^2 \rangle - \langle E \rangle^2}{kT^2}. \tag{1.4.5}$$

Thus the heat capacity C_V is also a measure of the fluctuation in the energy of the T, V, N system.

Similar relationships hold for the enthalpy in the T, P, N ensemble; namely, using (1.3.17) and (1.3.28), we obtain

$$\langle H \rangle = kT^2 \left(\frac{\partial \ln \Delta}{\partial T} \right)_{P,N} = \langle E \rangle - P\langle V \rangle. \tag{1.4.6}$$

Here, $\langle \ \rangle$ denotes averages in the T, P, N ensemble, using the probability distribution function

$$\text{Pr}(E, V) = \frac{\Omega(E, V, N) \exp(-\beta E - \beta PV)}{\Delta(T, P, N)}, \tag{1.4.7}$$

which is the probability of finding a system in the T, P, N ensemble having energy E and volume V (the latter is treated here as a discrete variable).

The constant-pressure heat capacity is obtained from (1.4.6), and from the definition of Δ in section 1.3 the result is

$$C_P = \left(\frac{\partial \langle H \rangle}{\partial T} \right)_{P,N} = \frac{\langle H^2 \rangle - \langle H \rangle^2}{kT^2}, \tag{1.4.8}$$

where the average quantities in (1.4.8) are taken with the probability distribution (1.4.7).

One of the two other relations of importance in the T, P, N ensemble is the fluctuations in the volume of the system

$$\langle (V - \langle V \rangle)^2 \rangle = \langle V^2 \rangle - \langle V \rangle^2 = kT\kappa_T \langle V \rangle, \tag{1.4.9}$$

where the isothermal compressibility is defined by

$$\kappa_T = -\frac{1}{\langle V \rangle} \left(\frac{\partial \langle V \rangle}{\partial P} \right)_{T,N}. \tag{1.4.10}$$

The second involves the cross-fluctuations of volume and enthalpy, which are related to the thermal expansivity

$$\langle (V - \langle V \rangle)(H - \langle H \rangle) \rangle = \langle VH \rangle - \langle V \rangle \langle H \rangle = kT^2 \alpha \langle V \rangle, \tag{1.4.11}$$

where

$$\alpha = \frac{1}{\langle V \rangle} \left(\frac{\partial \langle V \rangle}{\partial T} \right)_{P,N}. \tag{1.4.12}$$

In the T, V, μ ensemble, of foremost importance is the fluctuation in the number of particles, which, for a one-component system, is given by

$$\langle (N - \langle N \rangle)^2 \rangle = \langle N^2 \rangle - \langle N \rangle^2 = kT \left(\frac{\partial \langle N \rangle}{\partial \mu} \right)_{T,V} = kTV \left(\frac{\partial \rho}{\partial \mu} \right)_T. \tag{1.4.13}$$

In (1.4.13), all average quantities are taken with the probability distribution $\text{Pr}(N)$ given in (1.3.29). The fluctuations in the number of particles in the T, V, μ ensemble can be expressed in terms of the isothermal compressibility, as follows.

We start with the Gibbs–Duhem relation

$$-S\,dT + V\,dP = N\,d\mu \tag{1.4.14}$$

to obtain

$$\left(\frac{\partial P}{\partial \mu}\right)_T = \frac{N}{V} = \rho \qquad (1.4.15)$$

and

$$\left(\frac{\partial \rho}{\partial \mu}\right)_T = \left(\frac{\partial \rho}{\partial P}\right)_T \left(\frac{\partial P}{\partial \mu}\right)_T = \kappa_T \rho^2. \qquad (1.4.16)$$

Combining (1.4.13) and (1.4.16), we obtain the final result

$$\langle N^2 \rangle - \langle N \rangle^2 = kTV\rho^2 \kappa_T. \qquad (1.4.17)$$

Further relations involving cross-fluctuations in the number of particles in a multicomponent system are discussed in Chapter 6. Note that in Eqs. (1.4.14)–(1.4.16) we used the thermodynamic notations for V, N, etc. In applying these relations in the T, V, μ ensemble, the density ρ in (1.4.17) should be understood as

$$\rho = \frac{\langle N \rangle}{V}, \qquad (1.4.18)$$

where the average is taken in the T, V, μ ensemble.

1.5. CLASSICAL STATISTICAL THERMODYNAMICS

In section 1.3 we introduced the quantum-mechanical partition function in the T, V, N ensemble. In most applications of ST to problems in chemistry and biochemistry the classical limit of the quantum-mechanical partition function is used. In this section we present the so-called classical canonical partition function.

The canonical PF introduced in section 1.3 is defined by

$$Q(T, V, N) = \sum_i \exp(-\beta E_i) = \sum_E \Omega(E, V, N) \exp(-\beta E), \qquad (1.5.1)$$

where the first sum is over all possible *states* of the T, V, N system. In the second sum all states having the same energy E are grouped first, and then we sum over all different energy levels. $\Omega(E, V, N)$ is simply the degeneracy of the energy level E (given V and N), i.e., the number of states having the same energy E.

The classical analogue of $Q(T, V, N)$ for a system of N simple particles (i.e., spherical particles having no internal structure) is

$$Q(T, V, N) = (1/h^{3N}N!) \int \cdots \int d\mathbf{p}^N \, d\mathbf{R}^N \exp(-\beta H). \qquad (1.5.2)$$

Here, h is the Planck constant ($h = 6.625 \times 10^{-27}$ erg sec) and H is the classical Hamiltonian of the system, given by

$$H(\mathbf{p}^N, \mathbf{R}^N) = \sum_{i=1}^{N} (\mathbf{p}_i^2/2m) + U_N(\mathbf{R}^N), \qquad (1.5.3)$$

with $\mathbf{p}_i$ the momentum vector of the ith particle (presumed to possess only translational degrees of freedom) and m the mass of each particle. The total potential energy of the system at the specified configuration $\mathbf{R}^N$ is denoted by $U_N(\mathbf{R}^N)$.

It should be noted that expression (1.5.2) is not purely classical, since it contains two remnants of quantum-mechanical origin: the Planck constant h and the N factorial. Therefore, Q defined in (1.5.2) is actually the classical limit of the quantum-mechanical PF in (1.5.1). The purely classical PF consists of the integral expression on the rhs of (1.5.2) without the factor $(h^{3N}N!)^{-1}$. This PF failed to produce the correct form of the chemical potential or the entropy of the system.

The integration over the momenta in (1.5.2) can be performed to obtain

$$h^{-3N} \int_{-\infty}^{\infty} d\mathbf{p}^N \exp\left[-\beta \sum_{i=1}^{N} (\mathbf{p}_i^2/2m)\right] = \left[h^{-1} \int_{-\infty}^{\infty} dp \exp(-\beta p^2/2m)\right]^{3N}$$

$$= \left[h^{-1}(2m/\beta)^{1/2} \int_{-\infty}^{\infty} \exp(-x^2)\, dx\right]^{3N}$$

$$= [(2\pi mkT)^{3/2}/h^3]^N = \Lambda^{-3N}, \tag{1.5.4}$$

where we have introduced the momentum partition function

$$\Lambda = h/(2\pi mkT)^{1/2}, \tag{1.5.5}$$

often referred to as the *thermal de Broglie wavelength* of the particles at temperature T.

Another important quantity is the configurational partition function, defined by

$$Z_N = \int \cdots \int d\mathbf{R}^N \exp[-\beta U_N(\mathbf{R}^N)], \tag{1.5.6}$$

which can be used to rewrite the canonical PF in (1.5.2) as

$$Q(T, V, N) = Z_N/(N!\Lambda^{3N}). \tag{1.5.7}$$

The condition required for the applicability of the PF as given in (1.5.2) is

$$\rho\Lambda^3 \ll 1, \tag{1.5.8}$$

i.e., when the density is low or the mass is large or the temperature is high. Indeed, for most systems of interest in this book, we shall assume the validity of condition (1.5.8). For a system of N nonspherical particles the PF is modified as follows

$$Q(T, V, N) = \frac{q^N}{(8\pi^2)^N \Lambda^{3N} N!} \int \cdots \int d\mathbf{X}^N \exp[-\beta U_N(\mathbf{X}^N)].$$

Here, Λ has the same significance as in (1.5.5) and (1.5.7). The integration on the rhs of (1.5.9) extends over all possible locations and orientations of the N particles. We shall refer to the vector $\mathbf{X}^N = \mathbf{X}_1 \cdots \mathbf{X}_N$ as the *configuration* of the N particles. The factor q includes the rotational vibrational, electronic, or nuclear PFs of a single molecule. These may be treated either classically or quantum mechanically. We shall always assume in this book that the rotational-vibrational and similar partition functions are separable from the configuration partition function. Such an assumption cannot always be granted, especially when strong interaction between the particles can perturb the internal degrees of freedom of the particles involved.

The classical analogue of the T, P, N and the T, V, μ PFs is obtained by using the classical form of Q as in (1.5.9) in the expressions

$$\Delta(T, P, N) = C \int_0^\infty dV\, Q(T, V, N) \exp(-\beta PV) \qquad (1.5.10)$$

and

$$\Xi(T, V, \mu) = \sum_{N=0}^\infty Q(T, V, N) \exp(\beta\mu N). \qquad (1.5.11)$$

Note that in (1.5.10) we have an integration over all possible volumes. The constant C, having the dimensions of a reciprocal volume, will be of no concern to us, since we shall always deal with ratio of Δs, in which case it will be cancelled out.

Most of statistical thermodynamics is concerned with the evaluation of average quantities. The averaging process differs according to the choice of the set of thermodynamic variables which characterize the system under consideration.

Let $Q(\mathbf{X}^N)$ be any function of the configuration $\mathbf{X}^N$. The average of this function is then defined by

$$\langle Q \rangle = \int \cdots \int d\mathbf{X}^N\, P(\mathbf{X}^N) Q(\mathbf{X}^N), \qquad (1.5.12)$$

where $P(\mathbf{X}^N)$ is the distribution function,† i.e., $P(\mathbf{X}^N)\, d\mathbf{X}^N$ is the probability of finding particle 1 in the element $d\mathbf{X}_1, \dots$, particle N in the element $d\mathbf{X}_N$.

We shall often say that $P(\mathbf{X}^N)$ is the probability density of observing the *event* $\mathbf{X}^N$. One should remember, however, that the *probability* of an exact event $\mathbf{X}^N$ is always zero, and that the probabilistic meaning is assigned only to an element of "volume" $d\mathbf{X}^N$ at $\mathbf{X}^N$.

In the classical T, V, N ensemble, the basic distribution function is the probability density for observing the configuration $\mathbf{X}^N$,

$$P(\mathbf{X}^N) = \frac{\exp[-\beta U_N(\mathbf{X}^N)]}{\int \cdots \int d\mathbf{X}^N \exp[-\beta U_N(\mathbf{X}^N)]}. \qquad (1.5.13)$$

Clearly, the normalization condition for $P(\mathbf{X}^N)$ is

$$\int \cdots \int d\mathbf{X}^N P(\mathbf{X}^N) = 1, \qquad (1.5.14)$$

which is consistent with the probabilistic meaning of $P(\mathbf{X}^N)\, d\mathbf{X}^N$. In other words, the probability of finding the system in any one of the possible configurations is unity.

In the classical T, P, N ensemble, the basic distribution function is the probability density of finding a system with a volume V and a configuration $\mathbf{X}^N$:

$$P(\mathbf{X}^N, V) = \frac{\exp[-\beta PV - \beta U_N(\mathbf{X}^N)]}{\int dV \int \cdots \int d\mathbf{X}^N \exp[-\beta PV - \beta U_N(\mathbf{X}^N)]}. \qquad (1.5.15)$$

---

† The concept of "distribution function" as used in statistical mechanics differs from that employed in the mathematical theory of probability. For instance, in a one-dimensional system, the distribution function $F(x)$ is defined as $F(x) = \int_{-\infty}^x f(y)\, dy$ where $f(y)$ is referred to as the density function. Most of the quantities we shall be using in this book are actually density functions and not distribution functions, in the mathematical sense.

The integration over V extends from zero to infinity. The probability density of observing a system with volume V independently of the configuration is obtained from (1.5.15) by integrating over all configurations, i.e.,

$$P(V) = \int \cdots \int d\mathbf{X}^N P(\mathbf{X}^N, V). \tag{1.5.16}$$

Finally, the conditional distribution function defined by

$$P(\mathbf{X}^N/V) = \frac{P(\mathbf{X}^N, V)}{P(V)} = \frac{\exp[-\beta PV - \beta U_N(\mathbf{X}^N)]}{\int \cdots \int d\mathbf{X}^N \exp[-\beta PV - \beta U_N(\mathbf{X}^N)]}$$

$$= \frac{\exp[-\beta U_N(\mathbf{X}^N)]}{\int \cdots \int d\mathbf{X}^N \exp[-\beta U_N(\mathbf{X}^N)]} \tag{1.5.17}$$

is the probability density of finding a system in the configuration $\mathbf{X}^N$, given that the system has the volume V.

In the classical T, V, μ ensemble, the basic distribution function defined by

$$P(\mathbf{X}^N, N) = \frac{(q^N/N!)\exp[\beta\mu N - \beta U_N(\mathbf{X}^N)]}{\sum_{N=0}^{\infty} (q^N/N!)[\exp(\beta\mu N)] \int \cdots \int d\mathbf{X}^N \exp[-\beta U_N(\mathbf{X}^N)]} \tag{1.5.18}$$

is the probability density of observing a system with precisely N particles and the configuration $\mathbf{X}^N$.

The probability of finding a system in the T, V, μ ensemble with exactly N particles is obtained from (1.5.18) by integrating over all possible configurations, namely,

$$P(N) = \int \cdots \int d\mathbf{X}^N P(\mathbf{X}^N, N), \tag{1.5.19}$$

which can be written using the notation of section 1.3 as

$$P(N) = \frac{Q(T, V, N)[\exp(\beta\mu N)]}{\Xi(T, V, \mu)}. \tag{1.5.20}$$

The conditional distribution function defined by

$$P(\mathbf{X}^N/N) = \frac{P(\mathbf{X}^N, N)}{P(N)} = \frac{\exp[-\beta U_N(\mathbf{X}^N)]}{\int \cdots \int d\mathbf{X}^N \exp[-\beta U_N(\mathbf{X}^N)]} \tag{1.5.21}$$

is the probability density of observing a system in the configuration $\mathbf{X}^N$, given that the system contains precisely N particles.

The above expressions are the most fundamental distribution functions; from these we shall compute average quantities in the various ensembles. In Chapter 5 we introduce some additional distribution functions, referred to as molecular distribution functions. The latter are derived from the basic distributions in a rather straightforward manner.

1.6. THE IDEAL GAS

Theoretically, the ideal gas is characterized by the absence of intermolecular forces between the molecules, i.e.,

$$U_N(\mathbf{X}^N) \equiv 0 \tag{1.6.1}$$

for any configuration $\mathbf{X}^N$. Of course, no real system obeys equation (1.6.1). Nevertheless, the results obtained for ideal gases using assumption (1.6.1) are close to those attained by the real gases at very low densities when the occurrence of pairs, triplets, etc., of particles at short interparticle separation becomes exceedingly rare.

Using (1.6.1) in the classical canonical partition function (1.5.9), we immediately obtain

$$
\begin{aligned}
Q(T, V, N) &= \frac{q^N}{(8\pi^2)^N \Lambda^{3N} N!} \int \cdots \int d\mathbf{X}^N \\
&= \frac{q^N}{(8\pi^2)^N \Lambda^{3N} N!} \left[\int_V d\mathbf{R} \int_0^{2\pi} d\phi \int_0^\pi \sin\theta\, d\theta \int_0^{2\pi} d\psi \right]^N \\
&= \frac{q^N V^N}{\Lambda^{3N} N!}.
\end{aligned}
\tag{1.6.2}
$$

For simple spherical particles, sometimes referred to as "structureless" particles, Eq. (1.6.2) reduces to

$$Q(T, V, N) = V^N / \Lambda^{3N} N! \tag{1.6.3}$$

Note that q and Λ depend on the temperature and not on the volume V or on N. An important consequence is that the equation of state of an ideal gas is independent of the particular molecules constituting the system. To see this, we derive the expression for the pressure [see (1.3.13)], differentiating (1.6.2) with respect to volume:

$$P = kT \left(\frac{\partial \ln Q}{\partial V} \right)_{T,N} = \frac{kTN}{V} = \rho kT. \tag{1.6.4}$$

Evidently, this equation of state is universal, it does not depend on the properties of the molecules. This behavior is not shared by other thermodynamic relations of the ideal gas.

For instance, the chemical potential obtained from (1.3.14) and (1.6.2) and using the Stirling approximation† is

$$\mu = -kT\left(\frac{\partial \ln Q}{\partial N}\right)_{T,V} = kT \ln(\Lambda^3 q^{-1}) + kT \ln \rho$$

$$= \mu^{0g}(T) + kT \ln \rho, \qquad (1.6.5)$$

where $\rho = N/V$ is the number density and $\mu^{0g}(T)$ is the standard chemical potential. The latter conveys the properties of the individual molecules in the system. Note that the value of $\mu^{0g}(T)$ depends on the choice of units of ρ. The quantity $\rho\Lambda^3$, however, is dimensionless. Hence μ is independent of the choice of concentration units.

Another useful expression is that for the entropy of an ideal gas, which can be obtained from (1.3.12) and (1.6.2):

$$S = k \ln Q + kT\left(\frac{\partial \ln Q}{\partial T}\right)_{V,N} = \tfrac{5}{2}kN - Nk \ln(\rho\Lambda^3 q^{-1}) + kTN\frac{\partial \ln q}{\partial T}. \qquad (1.6.6)$$

For simple particles this reduces to the well-known relation

$$S = \tfrac{5}{2}kN - Nk \ln \rho\Lambda^3. \qquad (1.6.7)$$

The dependence of both μ and S on the density ρ through $\ln \rho$ is confirmed by experiment. We note here that had we used the purely classical PF [i.e., the integral excluding the factors $h^{3N}N!$ in (1.5.2)], we would not have obtained such a dependence on the density. This demonstrates the necessity of using the quantum-mechanical correction factors $h^{3N}N!$ even in the classical limit of the quantum mechanical PF.

Similarly, the energy of a system of simple particles is obtained from (1.6.3) and (1.6.7) as

$$E = A + TS = kTN \ln \rho\Lambda^3 - kTN + T(\tfrac{5}{2}kN - Nk \ln \rho\Lambda^3) = \tfrac{3}{2}kTN, \qquad (1.6.8)$$

which in this case is entirely due to the kinetic energy of the particles.

The heat capacity for a system of simple particles is obtained directly from (1.6.8) as

$$C_V = (\partial E/\partial T)_V = \tfrac{3}{2}kN, \qquad (1.6.9)$$

which may be viewed as originating from the accumulation of $\tfrac{1}{2}k$ per translational degree of freedom of a particle. For molecules also having rotational degrees of freedom, we get

$$C_V = 3kN, \qquad (1.6.10)$$

which is built up of $\tfrac{3}{2}kN$ from the translational and $\tfrac{3}{2}kN$ from the rotational degrees of

---

† In this book we always use the Stirling approximation in the form

$$\ln N! = N \ln N - N.$$

A better approximation for small values of N is

$$\ln N! = N \ln N - N + \tfrac{1}{2} \ln(2\pi N).$$

However, for N of the order 10^{23}, the first approximation is sufficient. It is also useful to note that, within the Stirling approximation, we have

$$(\partial/\partial N)(\ln N!) = \ln N.$$

freedom. If other internal degrees of freedom are present, there are additional contributions to C_V.

In all of the above discussions, we left unspecified the internal partition function of a single molecule. This, in general, includes contributions from the rotational, vibrational, and electronic states of the molecule. Assuming that these degrees of freedom are independent, the corresponding internal partition function may be factored into a product of the PFs for each degree of freedom, namely,

$$q(T) = q_r(T)q_v(T)q_e(T).$$ (1.6.11)

We shall never need to use the explicit form of the internal PF in this book. Such knowledge is needed for the actual calculation, for instance, of the equilibrium constant of a chemical reaction. We cite here the form of each of the factors in (1.6.11).

Each of these factors has the same form as the PF of the entire system, e.g., for the electronic PF we write

$$q_e(T) = \sum_i \omega_i^{(e)} \exp(-\beta \varepsilon_i^e)$$ (1.6.12)

where ε_i^e is the ith electronic energy level and $\omega_i^{(e)}$ is the corresponding degeneracy. In some cases the electronic excitation energy from the ground state is very large compared with kT. In such a case it is sufficient to retain only the first term on the rhs of (1.6.12), namely,

$$q_e(T) = \omega_0^{(e)} \exp(-\beta \varepsilon_0^e)\left\{1 + \frac{\omega_1^{(e)}}{\omega_0^{(e)}} \exp[-\beta(\varepsilon_1^e - \varepsilon_0^e)] + \cdots\right\}.$$ (1.6.13)

Since $|\beta(\varepsilon_i^e - \varepsilon_0^e)|$ are very large quantities at room temperature, we can neglect all the terms except the first one on the rhs of (1.6.13) to obtain

$$q_e(T) = w_0^{(e)} \exp(-\beta \varepsilon_0^e).$$ (1.6.14)

For the vibrational degree of freedom, the energy levels are given by

$$\varepsilon_n = (n + \tfrac{1}{2})hv \qquad n = 0, 1, 2, 3 \ldots,$$ (1.6.15)

where v is the classical frequency of a one-dimensional harmonic oscillator with a reduced mass μ and force constant f:

$$v = \frac{1}{2\pi} \sqrt{f/\mu}.$$ (1.6.16)

The vibrational partition function is thus

$$q_v(T) = \sum_{n=0} \exp(-\beta \varepsilon_n) = \exp(-\beta hv/2) \sum_{n=0} \exp(-\beta nhv)$$

$$= \frac{\exp(-\beta hv/2)}{1 - \exp(-\beta hv)}.$$ (1.6.17)

The rotational partition function of an unsymmetrical diatomic molecule is derived from the energy levels of a linear rigid rotator. These are

$$\varepsilon_n = \frac{n(n+1)h^2}{8\pi^2 I} \qquad n = 0, 1, 2, 3 \ldots,$$ (1.6.18)

with the degeneracy of the nth energy level

$$\omega_n = 2n + 1, \tag{1.6.19}$$

where I is the moment of inertia about the center of mass of the molecule. Thus,

$$q_r(T) = \sum_n (2n + 1) \exp[-\beta n(n + 1)h^2/8\pi^2 I]$$

$$\approx \int_0^\infty (2n + 1) \exp[-\beta n(n + 1)h^2/8\pi^2 I] \, dn$$

$$= \frac{8\pi^2 IkT}{h^2}, \tag{1.6.20}$$

where the passage from the sum to the integral is valid whenever

$$\frac{h^2}{8\pi^2 Ik} \ll T. \tag{1.6.21}$$

For a symmetrical diatomic molecule, the rotational PF is

$$q_r(T) = \frac{8\pi^2 IkT}{2h^2}. \tag{1.6.22}$$

The factor 2 in the denominator arises because each pair of configurations resulting from the exchange of the two nuclei is considered as a single configuration. In an unsymmetrical molecule such a pair produces two distinguishable configurations.

For a general molecule—the unsymmetrical top—the rotational PF is

$$q_r(T) = \frac{\pi^{1/2}}{\sigma} \left(\frac{8\pi^2 I_A kT}{h^2}\right)^{1/2} \left(\frac{8\pi^2 I_B kT}{h^2}\right)^{1/2} \left(\frac{8\pi^2 I_C kT}{h^2}\right)^{1/2}, \tag{1.6.23}$$

where I_A, I_B, and I_C are the three principal moments of inertia of the molecule and σ is the symmetry number. This is introduced, as in the case of the symmetrical linear molecule, to account for the number of indistinguishable configurations obtained by rotation of the molecule.

1.7. PAIR POTENTIAL AND PAIRWISE ADDITIVITY

The total potential energy of interaction $U_N(\mathbf{X}^N)$ appearing in the classical partition function (1.5.9) is defined as the work required to bring N molecules from infinite separation to the configuration $\mathbf{X}^N$. This work is in general a very complicated function of the configuration $\mathbf{X}^N$. A most fruitful simplifying assumption is the so-called pairwise additivity assumption, which states that the total work required to bring N molecules from infinite separation to the configuration $\mathbf{X}^N$ is equal to the sum of the work required to bring each pair of molecules, say i and j, from infinite separation to the final configuration $\mathbf{X}_i$, $\mathbf{X}_j$. Thus one writes for the total potential energy

$$U_N(\mathbf{X}^N) = \sum_{1 \leq i < j \leq N} U(\mathbf{X}_i, \mathbf{X}_j), \tag{1.7.1}$$

where $U(\mathbf{X}_i, \mathbf{X}_j)$ is called the pair potential and the summation in (1.7.1) extends over all of the $N(N - 1)/2$ pairs of different molecules in the system.

Most of the progress in the molecular theory of liquids has been achieved for systems obeying the pairwise additivity assumption. One system for which (1.7.1) is obeyed is composed of hard spheres. Hard spheres (HS) are idealized particles defined by a pair potential which is zero when the separation is equal or greater than their diameter σ and infinity when the separation is less than σ, i.e.,

$$U^{HS}(\mathbf{R}_i, \mathbf{R}_j) = \begin{cases} 0 & \text{for } |\mathbf{R}_i - \mathbf{R}_j| \geqslant \sigma \\ \infty & \text{for } |\mathbf{R}_i - \mathbf{R}_j| < \sigma \end{cases}. \tag{1.7.2}$$

An example of the hard-sphere pair potential is depicted in Fig. 1.8.

Hard spheres are characterized by the single parameter σ, their diameter. Of course, the definition of the *pair* potential (1.7.2) does not automatically provide a definition of $U_N(\mathbf{R}^N)$ for a system of hard spheres. The most reasonable definition for this would be the following:

$$U_N(\mathbf{R}^N) = \begin{cases} 0 & \text{if } |\mathbf{R}_i - \mathbf{R}_j| \geqslant \sigma \text{ for } \textit{all } i, j = 1, 2, \ldots, N; i \neq j \\ \infty & \text{if at least one } |\mathbf{R}_i - \mathbf{R}_j| \text{ is less than } \sigma. \end{cases} \tag{1.7.3}$$

The total energy is zero if no two particles are closer to each other than σ, and is infinity once at least a single pair is at a closer distance than σ. Clearly, using definition (1.7.2) we obtain

$$\sum_{1 \leqslant i < j \leqslant N} U^{HS}(\mathbf{R}_i, \mathbf{R}_j) = \begin{cases} 0 & \text{if } |\mathbf{R}_i - \mathbf{R}_j| \geqslant \sigma \text{ for all } i, j = 1, 2, \ldots, N; i \neq j \\ \infty & \text{if at least one } |\mathbf{R}_i - \mathbf{R}_j| \text{ is less than } \sigma \end{cases}. \tag{1.7.4}$$

Comparing (1.7.3) and (1.7.4), we get the equality

$$U_N(\mathbf{R}^N) = \sum_{1 \leqslant i < j \leqslant N} U^{HS}(\mathbf{R}_i, \mathbf{R}_j) \tag{1.7.5}$$

It should be remembered that hard spheres are not real particles, and (1.7.5) is valid by virtue of definitions (1.7.2) and (1.7.3). Therefore, the pairwise additivity assumption must be viewed as being a built-in feature of the *definition* of a system of hard spheres. By simple generalization, one can define nonspherical hard particles for which (1.7.1) is fulfilled. Other systems for which the pairwise additivity assumption is presumed to hold are systems of idealized point charges, point dipoles, point quadruples, and the like. A system of real particles such as argon atoms is believed to obey relation (1.7.1) approximately. Although it is now well known that even the simplest molecules do not obey (1.7.1) exactly, it is still considered a useful approximation without which little progress in the theory of liquids, if any, could have been achieved.

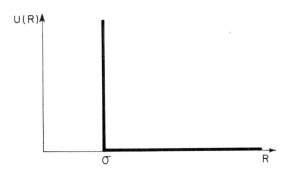

FIGURE 1.8. Pair potential for hard spheres of diameter σ.

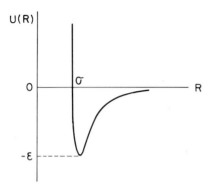

FIGURE 1.9. General form of the pair potential function
for simple and spherical molecules.

The general form of the pair potential for two simple atoms such as argon is depicted
in Fig. 1.9. Note that since the energy of the system is defined up to an arbitrary additive
constant, we can make the choice of fixing $U(R) = 0$ at $R = \infty$. The more important
features of the pair potential are the slopes, rather than the values, at each point. The
most common analytical form of a pair potential is the so-called Lennard–Jones (LJ)
pair potential, which reads

$$U_{LJ}(R) = 4\varepsilon[(\sigma/R)^{12} - (\sigma/R)^6]. \qquad (1.7.6)$$

This is a two-parameter function which conveys the general form of the expected "real"
pair potential of a pair of simple atoms. The parameter σ can be conveniently assigned
the meaning of an effective diameter of the particles, whereas ε can serve as a measure
of the strength of the interaction between the two particles.

Let us examine a few features of the LJ function (1.7.6). Clearly,

$$U_{LJ}(R = \sigma) = 0. \qquad (1.7.7)$$

For $R < \sigma$, the two particles exert strong repulsive forces on each other, and therefore
this region is effectively impenetrable, which justifies the meaning of σ as an effective
diameter. The minimum of $U_{LJ}(R)$ occurs at $R = 2^{1/6}\sigma$ and its value is

$$U_{LJ}(R = 2^{1/6}\sigma) = -\varepsilon. \qquad (1.7.8)$$

Two LJ curves with parameters for neon and argon are shown in Fig. 1.10. The force
exerted on one particle by the other is given by

$$F(R) = -\partial U_{LJ}(R)/\partial R. \qquad (1.7.9)$$

This is negative (indicating attraction) for $R > 2^{1/6}\sigma$ and positive (repulsion) for
$R < 2^{1/6}\sigma$. The force is zero at the point $R = 2^{1/6}\sigma$.

The leading, long-range, $\sim R^{-6}$ behavior of the pair potential has some theoretical
basis. On the other hand, the $\sim R^{-12}$ behavior at short distances is simply a convenient
analytical way of expressing the strong repulsive forces. Therefore the overall LJ function
should be viewed basically as a model for the pair potential operating between two real
particles. We can now define a system of LJ particles as a system of imaginary particles
for which the pair potential is the LJ function (1.7.6) and for which the total potential
function for a system of N such particles obeys the assumption of pairwise additivity.

The values of σ and ε do not have any absolute significance. First, because of the
arbitrariness of the LJ function, one could have chosen an analytical function quite

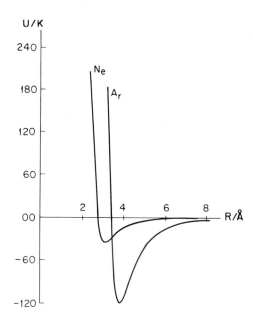

FIGURE 1.10. The form of Lennard–Jones curves for neon (with $\varepsilon/k = 34.9$ K and $\sigma = 2.78$ Å) and for argon (with $\varepsilon/k = 119.8$ K and $\sigma = 3.405$ Å). The parameters are obtained from data on the second virial coefficients for these gases.

different from (1.7.6) and characterized by different parameters. Second, even if the LJ function were the best one to represent the pair potential of the real molecules, there is as yet no experimental method of determining the parameters σ and ε uniquely.

We now turn to a brief consideration of the origin of nonadditivity effects of the potential function. We demonstrate the idea with a simple example. Consider a system of three particles (Fig. 1.11) whose pair potential is a superposition of a hard-sphere and a dipole–dipole interaction:

$$U_2(\mathbf{R}_1, \mathbf{\Omega}_1, \mathbf{R}_2, \mathbf{\Omega}_2) = U^{\text{HS}}(|\mathbf{R}_2 - \mathbf{R}_1|) + U^{\text{DD}}(\mathbf{R}_1, \mathbf{\Omega}_1, \mathbf{R}_2, \mathbf{\Omega}_2), \qquad (1.7.10)$$

where $U^{\text{HS}}(R)$ is defined in (1.7.2). (Note that U^{HS} is a function of the scalar distance $|\mathbf{R}_2 - \mathbf{R}_1|$ only.) The dipole–dipole interaction is given by

$$U^{\text{DD}}(\mathbf{R}_1, \mathbf{\Omega}_1, \mathbf{R}_2, \mathbf{\Omega}_2) = \frac{\boldsymbol{\mu}_1 \cdot \boldsymbol{\mu}_2 - 3(\boldsymbol{\mu}_1 \cdot \mathbf{u}_{12})(\boldsymbol{\mu}_2 \cdot \mathbf{u}_{12})}{|\mathbf{R}_2 - \mathbf{R}_1|^3}. \qquad (1.7.11)$$

Here, $\boldsymbol{\mu}_1$ is the dipole moment vector of the ith particle and $\mathbf{u}_{12}$ is a unit vector along

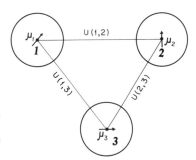

FIGURE 1.11. A configuration of three hard spheres with a point dipole embedded at each of their centers. The dipole vectors are denoted by $\boldsymbol{\mu}_1$, $\boldsymbol{\mu}_2$, and $\boldsymbol{\mu}_3$. The pair potentials are $U(1, 2)$, $U(1, 3)$ and $U(2, 3)$.

$\mathbf{R}_2 - \mathbf{R}_1$. For any given configuration of the three particles, we have in this case the relation of additivity

$$U_3(\mathbf{X}_1, \mathbf{X}_2, \mathbf{X}_3) = U_2(\mathbf{X}_1, \mathbf{X}_2) + U_2(\mathbf{X}_1, \mathbf{X}_3) + U_2(\mathbf{X}_2, \mathbf{X}_3), \qquad (1.7.12)$$

where $\mathbf{X}_i$ stands for $\mathbf{R}_i, \mathbf{\Omega}_i$. Next, suppose that we choose to characterize the configuration of the system by the locations of the centers of the particles only, ignoring their orientations. For two such particles we introduce the average interaction energy at $\mathbf{R}_i, \mathbf{R}_j$, defined by

$$V_2(\mathbf{R}_i, \mathbf{R}_j) \equiv \langle U_2(\mathbf{X}_i, \mathbf{X}_j) \rangle_2$$

$$= \frac{\displaystyle\iint d\mathbf{\Omega}_i \, d\mathbf{\Omega}_j \, U_2(\mathbf{X}_i, \mathbf{X}_j) \exp[-\beta U_2(\mathbf{X}_i, \mathbf{X}_j)]}{\displaystyle\iint d\mathbf{\Omega}_i \, d\mathbf{\Omega}_j \exp[-\beta U_2(\mathbf{X}_i, \mathbf{X}_j)]}. \qquad (1.7.13)$$

The symbol $\langle \ \rangle_2$ stands for an average over all the orientations of the pair of particles. The average interaction energy for the triplet of particles at positions $\mathbf{R}_1, \mathbf{R}_2, \mathbf{R}_3$ is defined similarly as

$$V_3(\mathbf{R}_1, \mathbf{R}_2, \mathbf{R}_3) = \langle U_3(\mathbf{X}_1, \mathbf{X}_2, \mathbf{X}_3) \rangle_3, \qquad (1.7.14)$$

where here the symbol $\langle \ \rangle_3$ means an average over all the orientations of the *three* particles, namely,

$$\langle U_3(\mathbf{X}_1, \mathbf{X}_2, \mathbf{X}_3) \rangle_3 = \frac{\displaystyle\iiint d\mathbf{\Omega}_1 \, d\mathbf{\Omega}_2 \, d\mathbf{\Omega}_3 \, U_3(\mathbf{X}_1, \mathbf{X}_2, \mathbf{X}_3) \exp[-\beta U_3(\mathbf{X}_1, \mathbf{X}_2, \mathbf{X}_3)]}{\displaystyle\iiint d\mathbf{\Omega}_1 \, d\mathbf{\Omega}_2 \, d\mathbf{\Omega}_3 \exp[-\beta U_3(\mathbf{X}_1, \mathbf{X}_2, \mathbf{X}_3)]}$$

$$(1.7.15)$$

We recall that if we record all of the positions and orientations of the particles, we have a perfect additivity as given in (1.7.12). However, if we disregard the orientation, we have a new average pair potential (1.7.13) and a new triplet potential (1.7.15), which in this case does not necessarily obey the additivity assumption. To see this more clearly, let us substitute (1.7.12) in (1.7.14) to obtain

$$V_3(\mathbf{R}_1, \mathbf{R}_2, \mathbf{R}_3) = \langle U_3(\mathbf{X}_1, \mathbf{X}_2, \mathbf{X}_3) \rangle_3$$

$$= \langle U_2(\mathbf{X}_1, \mathbf{X}_2) \rangle_3 + \langle U_2(\mathbf{X}_1, \mathbf{X}_3) \rangle_3 + \langle U_2(\mathbf{X}_2, \mathbf{X}_3) \rangle_3$$

$$\neq \langle U_2(\mathbf{X}_1, \mathbf{X}_2) \rangle_2 + \langle U_2(\mathbf{X}_1, \mathbf{X}_3) \rangle_2 + \langle U_2(\mathbf{X}_2, \mathbf{X}_3) \rangle_2$$

$$= V_2(\mathbf{R}_1, \mathbf{R}_2) + V_2(\mathbf{R}_1, \mathbf{R}_3) + V_2(\mathbf{R}_2, \mathbf{R}_3) \qquad (1.7.16)$$

The inequality in (1.7.16) arises because, in general, the averages $\langle \ \rangle_2$ and $\langle \ \rangle_3$ do not produce the same results. In other words, the average interaction for a pair of particles may be different if, in the process of averaging, a third particle is present in the system.

The above example illustrates a general principle. If we have a very detailed specification of our system, then we hope to get complete additivity. Once we average over part of the degrees of freedom of the system, we may get a simpler description of the system, but with the sacrifice of additivity.

It should be noted that the work required to bring two particles from infinite separation $|\mathbf{R}_1 - \mathbf{R}_2| \to \infty$, to close configuration, $\mathbf{R}_1$, $\mathbf{R}_2$, ignoring their orientations is

$$\Delta A(\mathbf{R}_1, \mathbf{R}_2) = A(\mathbf{R}_1, \mathbf{R}_2) - A(|\mathbf{R}_1 - \mathbf{R}_2| \to \infty)$$

$$= -kT \ln \left\{ \frac{\displaystyle\int\int \exp[-\beta U_2(\mathbf{X}_1, \mathbf{X}_2)] \, d\mathbf{\Omega}_1 \, d\mathbf{\Omega}_2}{\displaystyle\int d\mathbf{\Omega}_1 \, d\mathbf{\Omega}_2} \right\}$$

$$= -kT \ln \langle \exp(-\beta U_2) \rangle_0. \tag{1.7.17}$$

Note that $\langle\ \rangle_0$ is an average of the quantity $\exp(-\beta U_2)$ giving the same weight to all orientations of the pair of particles. This is different from the average in (1.7.13). ΔA in (1.7.17) can be split into two components; the entropy and the energy changes corresponding to the same process, i.e.,

$$\Delta S(\mathbf{R}_1, \mathbf{R}_2) = -\frac{\partial \Delta A(\mathbf{R}_1, \mathbf{R}_2)}{\partial T}$$

$$= k \ln \langle \exp(-\beta U_2) \rangle_0 + \frac{1}{T} \langle U_2 \rangle_2 \tag{1.7.18}$$

$$\Delta E(\mathbf{R}_1, \mathbf{R}_2) = \Delta A(\mathbf{R}_1, \mathbf{R}_2) + T\Delta S(\mathbf{R}_1, \mathbf{R}_2)$$

$$= \langle U_2 \rangle_2 = V_2(\mathbf{R}_1, \mathbf{R}_2). \tag{1.7.19}$$

Thus $V_2(\mathbf{R}_1, \mathbf{R}_2)$ as defined in (1.7.13) is equal to the *energy* change for the process described above.

The force acting on, say, particle 1, given the configuration $\mathbf{R}_1$, $\mathbf{R}_2$ and ignoring the orientations, is obtained from the gradient of $\Delta A(\mathbf{R}_1, \mathbf{R}_2)$ with respect to $\mathbf{R}_1$, i.e.,

$$\mathbf{F}_1(\mathbf{R}_1, \mathbf{R}_2) = -\nabla_1 \Delta A(\mathbf{R}_1, \mathbf{R}_2)$$

$$= \frac{-\displaystyle\int \exp[-\beta U_2(\mathbf{X}_1, \mathbf{X}_2)] \nabla_1 U_2(\mathbf{X}_1, \mathbf{X}_2) \, d\mathbf{\Omega}_1 \, d\mathbf{\Omega}_2}{\displaystyle\int \exp[-\beta U_2(\mathbf{X}_1, \mathbf{X}_2)] \, d\mathbf{\Omega}_1 \, d\mathbf{\Omega}_2}$$

$$= \langle -\nabla_1 U_2(\mathbf{X}_1, \mathbf{X}_2) \rangle_2. \tag{1.7.20}$$

Here $\langle\ \rangle_2$ is an average of the same kind as in (1.7.13), and $-\nabla_1 U_2(\mathbf{X}_1, \mathbf{X}_2)$ is the force acting on 1, given the configuration $\mathbf{X}_1$, $\mathbf{X}_2$. In (1.7.20) we averaged this force over all orientations with the same distribution function used in (1.7.13).

Throughout the book we shall encounter many other examples where integration over part of the degrees of freedom of the particles produces average interactions and forces. These, in general, have different properties from the original interactions, i.e., the interaction before the averaging process has been performed.

1.8. VIRIAL EXPANSION AND VAN DER WAALS EQUATION

In section 1.6, we noted that the equation of state of an ideal gas is "universal," i.e., it does not depend on any particular property of the gas under observation. For any gas at sufficiently low density, we have

$$\frac{P}{kT} = \frac{N}{V} = \rho. \tag{1.8.1}$$

The universal character of the equation of state ceases to hold for nonideal gases. This fact is already revealed by the well known van der Waals equation, which may be written as

$$(P + \rho^2 a)(1 - \rho b) = \rho k T, \tag{1.8.2}$$

where the constants a and b were first introduced by van der Waals to account for the attraction between the particles and the finite volume of the particles, respectively. As we shall soon see, these two constants are actually two features of the intermolecular potential operating between the particles, and any division between attraction and volume (or size) of the particles is quite arbitrary. Let us rewrite (1.8.2) as an expansion of P/kT in the density ρ:

$$\frac{P}{kT} = \rho - \rho^2 \frac{a}{kT} + \frac{P\rho b}{kT} + \rho^3 \frac{ab}{kT}. \tag{1.8.3}$$

It is now known that this expansion is valid up to terms of order ρ^2. Since at low densities $P/kT \approx \rho$, we see that the term $P\rho b/kT$ is of order ρ^2. Therefore, if we are going to retain the expansion up to the second order in the density, we have

$$\frac{P}{kT} = \rho + \left(b - \frac{a}{kT}\right)\rho^2 + \cdots. \tag{1.8.4}$$

The term of order ρ^3 has been dropped in (1.8.4). We already see here that a and b appear in a certain combination as a coefficient of the second power of the density. Statistical mechanics gives a general procedure for identifying this coefficient, as well as higher ones in the density expansion of the pressure. This is shown in the rest of this chapter.

In section 1.6 we noted that the equation of state of an ideal gas (1.8.1) may be obtained either from a model system for which $U(\mathbf{X}^N) = 0$, which in reality does not exist, or for a real system at very low density. Equation (1.8.1) is obeyed for a real gas, provided the density is so low that interaction between two or more particles can be neglected. At higher densities we must take into account the effect of interactions among the particles on the equation of state of the system.

Formally we try an expansion βP in power series in the density, presuming we know the limiting behavior (1.8.1); we write

$$\beta P = \rho \left(\frac{\partial(\beta P)}{\partial \rho}\right)_{T, \rho = 0} + \frac{1}{2}\rho^2 \left(\frac{\partial^2(\beta P)}{\partial \rho^2}\right)_{T, \rho = 0} + \cdots$$

$$= \rho + B_2(T)\rho^2 + B_3(T)\rho^3 + \cdots, \tag{1.8.5}$$

where the coefficients $B_k(T)$ are evaluated at $\rho = 0$, and hence are functions of the temperature only.

Experimentally these coefficients may be determined as limiting slopes of experimental data. For instance, by plotting $(\beta P - \rho)/\rho$ as a function of ρ, we can evaluate $B_2(T)$ from the limiting slope

$$B_2(T) = \lim_{\rho \to 0} \frac{\partial}{\partial \rho}\left[\frac{\beta P - \rho}{\rho}\right].$$ (1.8.6)

One of the most remarkable achievements of ST is that it provides explicit expressions for the virial coefficients B_k in terms of the intermolecular interaction energy of a group of k particles. In particular the second virial coefficient B_2 is determined by an integral over a function of the interaction potential between two particles.

We now derive the expression for the second virial coefficient. A more general procedure is available to obtain expressions for the higher virial coefficients, but these are rarely used in practical applications. We start with the GPF of a one-component system, Eq. (1.3.34), characterized by the variables T, V, μ, or T, V, λ, where λ is the absolute activity $\lambda = \exp[\beta\mu]$.

$$\exp[\beta PV] = \Xi(T, V, \lambda) = \sum_{N \geq 0} Q(T, V, N)\lambda^N = 1 + \sum_{N \geq 1} Q_N \lambda^N,$$ (1.8.7)

where $Q(T, V, 0) = 1$; i.e., the empty system has one state with zero energy. We also use the shorthand notation Q_N for $Q(T, V, N)$ since T, V will be constant throughout the following treatment.

Equation (1.8.7) is a power series in λ. We wish to reexpress the pressure as a power series in the density. The average density, in an open system, is obtained from

$$\rho = \bar{N}/V = \frac{\lambda}{V}\frac{\partial \ln \Xi}{\partial \lambda} = \lambda \frac{\partial(\beta P)}{\partial \lambda}.$$ (1.8.8)

From now on we shall write all power series up to second order only; thus, expanding βPV from (1.8.7), we obtain

$$\beta PV = \ln\left(1 + \sum_{N \geq 1} Q_N \lambda^N\right) = \sum_{N \geq 1} Q_N \lambda^N - \frac{1}{2}\left[\sum_{N \geq 1} Q_N \lambda^N\right]^2 + \cdots$$

$$= Q_1 \lambda + (Q_2 - \tfrac{1}{2}Q_1^2)\lambda^2 + \cdots.$$ (1.8.9)

To obtain the density ρ from (1.8.8), we differentiate βP in (1.8.9) term by term to obtain

$$\rho = \lambda \frac{\partial(\beta P)}{\partial \lambda} = \frac{Q_1}{V}\lambda + \frac{2(Q_2 - \tfrac{1}{2}Q_1^2)}{V}\lambda^2 + \cdots.$$ (1.8.10)

Next we invert the expansion (1.8.10) to obtain λ as a power series in ρ. We first write

$$\lambda = a_0 + a_1\rho + a_2\rho^2 + \cdots.$$ (1.8.11)

To determine the coefficient a_i in (1.8.11), we substitute λ from (1.8.11) into (1.8.10) to obtain an identity in ρ

$$\rho = \frac{Q_1}{V}(\sum a_i\rho^i) + \frac{2(Q_2 - \tfrac{1}{2}Q_1^2)}{V}(\sum a_i\rho^i)^2 + \cdots.$$ (1.8.12)

The first two coefficients in (1.8.11) may be determined from the fact that as $\rho \to 0$, the system behaves as an ideal gas (section 1.6), so that

$$\lambda = \exp[\beta\mu] = \exp[\ln \rho\Lambda^3 q^{-1}] = \frac{\Lambda^3}{q}\rho. \qquad (1.8.13)$$

Hence $a_0 = 0$ and $a_1 = \Lambda^3/q$ in (1.8.11); q is the internal PF of a single molecule.

Equating coefficients of the same power in ρ in (1.8.12), we obtain

$$\frac{Q_1}{V}a_1 = 1,$$

or

$$a_1 = \frac{V}{Q_1} = \frac{\Lambda^3}{q}, \qquad (1.8.14)$$

which we already had from (1.8.13), and

$$a_2 = -2(Q_2 - \tfrac{1}{2}Q_1^2)\frac{V^2}{Q_1^3}. \qquad (1.8.15)$$

We now substitute (1.8.12) in (1.8.9) to obtain the required expansion of the pressure in power series of the density, i.e.,

$$\beta P = \rho - \frac{V}{Q_1^2}(Q_2 - \tfrac{1}{2}Q_1^2)\rho^2 + \cdots$$

$$= \rho + B_2(T)\rho^2 + \cdots, \qquad (1.8.16)$$

where we have identified the second virial coefficient $B_2(T)$ by comparison with the formal Taylor series (1.8.5)

$$B_2(T) = -\frac{V}{Q_1^2}(Q_2 - \tfrac{1}{2}Q_1^2). \qquad (1.8.17)$$

We now express $B_2(T)$ in terms of the intermolecular potential operating between two particles and show that B_2 is a function of T only, not of V, as might be inferred from (1.8.17).

For rigid, nonspherical molecules we have

$$Q_1 = \frac{q}{8\pi^2\Lambda^3}\int d\mathbf{X}_1 = \frac{qV}{\Lambda^3} \qquad (1.8.18)$$

and

$$Q_2 = \frac{1}{2}\left(\frac{q}{8\pi^2\Lambda^3}\right)^2\int \exp[-\beta U(\mathbf{X}_1, \mathbf{X}_2)]\, d\mathbf{X}_1\, d\mathbf{X}_2. \qquad (1.8.19)$$

Hence, substituting (1.8.18) and (1.8.19) in (1.8.17) we obtain

$$B_2(T) = -\frac{1}{2V(8\pi^2)^2}\int \{\exp[-\beta U(\mathbf{X}_1, \mathbf{X}_2)] - 1\}\, d\mathbf{X}_1\, d\mathbf{X}_2$$

$$= -\frac{1}{2(8\pi^2)}\int \{\exp[-\beta U(\mathbf{X})] - 1\}\, d\mathbf{X}. \qquad (1.8.20)$$

In the last step on the rhs of (1.8.20) we exploit the fact that $U(\mathbf{X}_1, \mathbf{X}_2)$ is actually a function of six coordinates, not twelve as implied in $\mathbf{X}_1, \mathbf{X}_2$; i.e., we can hold $\mathbf{X}_1$ fixed, say at the origin, and view the potential function $U(\mathbf{X}_1, \mathbf{X}_2)$ as depending on the relative locations and orientations of the second particle, which we denote by $\mathbf{X}$. Thus, integrating over $\mathbf{X}_1$ produces a factor $V8\pi^2$, and the final form of $B_2(T)$ is obtained.

The integration over $\mathbf{X} = \mathbf{R}, \mathbf{\Omega}$ is still over the entire volume of the system. However, assuming that the potential function $U(\mathbf{X})$ has a short range, say of a few molecular diameters, the integral over the entire volume is actually over only a very short distance from the particle that we held fixed at the origin. This is the reason why $B_2(T)$ is not a function of the volume. In section 6.12, when discussing ionic solutions, we shall see that if the particles are charged, then the fact that the coulombic interaction is of long range causes difficulties in the theory of ionic solutions.

Expression (1.8.20) can be further simplified if the pair potential is a function of the scalar distance $R = |\mathbf{R}_2 - \mathbf{R}_1|$. In that case, the integration over the orientations produces the factor $8\pi^2$ and the integration over the volume can be performed using polar coordinates to obtain

$$B_2(T) = -\frac{1}{2} \int_0^\infty \{\exp[(-\beta U(R))] - 1\} 4\pi R^2 \, dR. \tag{1.8.21}$$

Note that although we took infinity as the upper limit of the integral, in practice the integration extends to a finite distance of the order of a few molecular diameters, i.e., the effective range of the interaction potential. Beyond this limit $U(R)$ is zero and therefore the integrand becomes zero as well. Hence the extension of the range of integration does not affect the value of $B_2(T)$.

$B_2(T)$ is the most useful of the virial coefficients. As we see, for a given function $U(R)$, $B_2(T)$ is uniquely determined. The inverse of the last statement is in general incorrect; i.e., one cannot determine uniquely the potential from a knowledge of $B_2(T)$. Nevertheless, this relation is still useful for characterization of the potential. One usually makes a reasonable choice of a functional form of the potential, for instance the Lennard–Jones function, and then determines the Lennard–Jones parameters σ and ε (see section 1.7) that give the best fit to the experimental values of $B_2(T)$, as well as its temperature dependence.

In our derivation of the expansion (1.8.16), we retained in each step only first-order corrections to the ideal-gas limit. Had we retained second-order terms, we could have obtained an explicit expression for the third virial coefficient in (1.8.5). This expression as well as the higher coefficients become increasingly more complicated and therefore less useful. To demonstrate the nature of the difficulty, we cite here the expression for $B_3(T)$, which reads

$$B_3(T) = -\frac{1}{3(8\pi^2)^2} \int \{\exp[-\beta U_3(\mathbf{X}_1, \mathbf{X}_2, \mathbf{X}_3)] - \exp[-\beta U(\mathbf{X}_1, \mathbf{X}_2) - \beta U(\mathbf{X}_2, \mathbf{X}_3)]$$

$$- \exp[-\beta U(\mathbf{X}_1, \mathbf{X}_2) - \beta U(\mathbf{X}_1, \mathbf{X}_3)]$$

$$- \exp[-\beta U(\mathbf{X}_1, \mathbf{X}_3) - \beta U(\mathbf{X}_2, \mathbf{X}_3)]$$

$$+ \exp[-\beta U(\mathbf{X}_1, \mathbf{X}_2)] + \exp[-\beta U(\mathbf{X}_1, \mathbf{X}_3)]$$

$$+ \exp[-\beta U(\mathbf{X}_2, \mathbf{X}_3)] - 1\} \, d\mathbf{X}_2 \, d\mathbf{X}_3. \tag{1.8.22}$$

We see that this expression is fairly complicated. If the total potential energy is pairwise additive, in the sense that

$$U_3(\mathbf{X}_1, \mathbf{X}_2, \mathbf{X}_3) = U(\mathbf{X}_1, \mathbf{X}_2) + U(\mathbf{X}_1, \mathbf{X}_3) + U(\mathbf{X}_2, \mathbf{X}_3), \qquad (1.8.23)$$

the integrand in (1.8.22) simplifies to

$$B_3(T) = -\frac{1}{3(8\pi^2)^2} \int f(\mathbf{X}_1, \mathbf{X}_2) f(\mathbf{X}_1, \mathbf{X}_3) f(\mathbf{X}_2, \mathbf{X}_3) \, d\mathbf{X}_2 \, d\mathbf{X}_3 \qquad (1.8.24)$$

where f, the so-called *Mayer f-function*, is defined by

$$f(\mathbf{X}_i, \mathbf{X}_j) = \exp[-\beta U(\mathbf{X}_i, \mathbf{X}_j)] - 1. \qquad (1.8.25)$$

We see that even when we assume pairwise additivity of the potential we still have too complicated a function to be useful either for computing a theoretical value of $B_3(T)$ or for determining the interaction potential.

Extending the same procedure for mixtures will give us the second virial coefficient for a mixture of two components, say A and B. The first-order correction to the ideal gas behavior is

$$\beta P = \rho_A + \rho_B + B_{AA}\rho_A^2 + B_{BB}\rho_B^2 + B_{AB}\rho_A\rho_B + \cdots. \qquad (1.8.26)$$

In terms of the total density $\rho_T = \rho_A + \rho_B$ and the mole fraction $x_A = \rho_A/\rho_T$, (1.8.26) can be rewritten as

$$\beta P = \rho_T + [B_{AA}x_A^2 + B_{BB}x_B^2 + B_{AB}x_Ax_B]\rho_T^2 + \cdots, \qquad (1.8.27)$$

where the term in the square brackets may be interpreted as the average second virial coefficient of the mixture. $B_{\alpha\beta}$ is related to $U_{\alpha\beta}$ by the same relation as B_2 to U in (1.8.20) or (1.8.21).

Having obtained expressions for $B_2(T)$ we now return to identify the parameters a and b in Eq. (1.8.4). We assume for concreteness that $U(R)$ in (1.8.20) is a Lennard–Jones potential. We know that for $R \leq \sigma$, the potential function is positive and increases steeply to infinity as we reduce the distance. Therefore, we can approximate the integral in (1.8.21) by

$$B_2(T) = -\frac{1}{2}\left[\int_0^\sigma \{\exp[-\beta U(R)] - 1\}4\pi R^2 \, dR \right.$$

$$\left. + \int_\sigma^\infty \{\exp[-\beta U(R)] - 1\}4\pi R^2 \, dR\right]$$

$$\approx \frac{1}{2}\left[\int_0^\sigma 4\pi R^2 \, dR + \beta \int_\sigma^\infty U(R)4\pi R^2 \, dR\right]; \qquad (1.8.28)$$

i.e., we put $\exp[-\beta U(R)] = 0$ for $R \leq \sigma$ and expand to the first order the exponent $\exp[-\beta U(R)] \approx 1 - \beta U(R)$ for $R > \sigma$. If we now make the identifications

$$b = \frac{1}{2}\int_0^\sigma 4\pi R^2 \, dR = \frac{1}{2}\frac{4\pi\sigma^3}{3} \qquad (1.8.29)$$

and

$$\frac{a}{kT} = \frac{-\displaystyle\int_{\sigma}^{\infty} U(R)4\pi R^2 \, dR}{2kT}, \qquad (1.8.30)$$

we see that the constant b is related to the "volume" of the particles, whereas the constant a is roughly an average "attraction" between them. These two features appear in (1.8.21) as parts of the integral over the full pair potential $U(R)$. The splitting into two terms is of course arbitrary; we can choose σ at any point for which $\exp[-\beta U(R)] \approx 0$ for all $R \leq \sigma$.

SUGGESTED READINGS

The best introductions to statistical thermodynamics for chemists are:

T. L. Hill, *Introduction to Statistical Thermodynamics* (Addison-Wesley, Reading, MA, 1960).
D. A. McQuarrie, *Statistical Mechanics* (Harper and Row, New York, 1976).

For a thorough discussion of the fundamental principles, see:

R. C. Tolman, *The Principles of Statistical Mechanics* (Clarendon Press, Oxford, 1938).
A. Munster, *Statistical Thermodynamics* (Springer-Verlag, Berlin, 1969).

2

Simple Systems without Interactions

2.1. INTRODUCTION

In this chapter we present some applications of ST to very simple systems. The simplicity here arises from either negligible or total absence of interparticle interactions. Lack of interaction usually implies independence of the particles. This, in turn, leads to a relatively easy solution for the PF of the system. A total lack of interactions never exists in real systems. Nevertheless, such idealized systems are interesting for two reasons. First, some systems behave, to a good approximation, as if there are no interactions (e.g., a real gas at very low densities, adsorption of molecules on sites that are far apart). Second, real systems with interactions can be viewed and treated as extensions of idealized simple systems. For instance, the theory of real gases is based on corrections due to interactions between pairs, triplets, etc. Even in the very simple systems, some interactions between particles or between particles and an external field are essential to the maintenance of equilibrium. Lack of interactions usually leads to solvability of the PF, but this is not always so. In Chapter 3 we shall study systems with interactions among a small number of particles for which a PF can be written explicitly. Likewise, the inherent simplicity of the one-dimensional systems studied in Chapter 4 also leads to solvability of the PF.

2.2. THE CHEMICAL POTENTIAL OF AN IDEAL GAS

In section 1.6 we derived the thermodynamic quantities of an ideal gas. Of particular importance is the chemical potential, which is now written as

$$\mu = kT \ln q^{-1} + kT \ln \frac{N}{V} \Lambda^3. \tag{2.2.1}$$

Here, q is the internal partition function of a single molecule, V the volume of the system, N the number of particles, and Λ^3 the momentum partition function.

It is convenient to assign meaning to the two terms in (2.2.1) as follows: the chemical potential is defined, in the T, V, N ensemble, as

$$\mu = \left(\frac{\partial A}{\partial N} \right)_{T,V} = A(T, V, N+1) - A(T, V, N), \tag{2.2.2}$$

where the second equality follows from the extensive property of the Helmholtz energy and the fact that the addition of one particle to a macroscopic system can be viewed as an infinitesimal change in the variable N.

Suppose that instead of just adding one particle to the system, we introduce the particle at a fixed position in the system, say, at $\mathbf{R}_0$. The corresponding change in the Helmholtz energy is defined by

$$\mu^* = A(T, V, N + 1, \mathbf{R}_0) - A(T, V, N)$$

$$= -kT \ln\left[\frac{q^{N+1}V^N}{\Lambda^{3N}N!}\right] + kT \ln\left[\frac{q^N V^N}{\Lambda^{3N}N!}\right] = kT \ln q^{-1}. \qquad (2.2.3)$$

We see that this quantity is equal to the first term on the rhs of (2.2.1). We shall refer to μ^* as the *pseudo chemical potential*.

Combining (2.2.1) with (2.2.3), we obtain

$$\mu = \mu^* + kT \ln \frac{N}{V} \Lambda^3. \qquad (2.2.4)$$

Thus the work required to introduce one particle into the system is divided into two parts, of which the first is the work required to place the particle at a fixed position in the system. Next the particle is freed; the corresponding work is $kT \ln \rho\Lambda^3$. We shall refer to this term as the *liberation Helmholtz energy*. It is easy to show that the corresponding liberation Gibbs energy has the same form as in (2.2.4) with the replacement of V by the average volume in the T, P, N ensemble (see Appendix E).

When the particle is released from its fixed position, there are three sources of the change in the Helmholtz energy. First, it can acquire translational kinetic energy, the corresponding contribution being $kT \ln \Lambda^3$. Second, the particle can now wander in the entire volume V, the corresponding contribution being $-kT \ln V$. Finally, when the particle is at a fixed position, it is distinguishable from all the N indistinguishable particles. This is the reason for having $N!$ in both terms on the rhs of (2.2.3). Once the particle is released, it is no longer distinguishable from other members of the same species. We shall refer to the term $kT \ln N$ as being due to the *assimilation* of one particle by N indistinguishable particles. Together the three contributions give rise to the liberation term $kT \ln \rho\Lambda^3$, where $\rho\Lambda^3$ is a dimensionless quantity. Note that since we are using classical ST, $\rho\Lambda^3 < 1$; therefore the liberation Helmholtz energy is always negative.

Each of the factors mentioned above can be changed independently. If we change the temperature (at V, N constant), the change in the chemical potential is

$$\mu(T_2) - \mu(T_1) = \mu^*(T_2) - \mu^*(T_1) + kT_2 \ln \Lambda^3(T_2) - kT_1 \ln \Lambda^3(T_1). \qquad (2.2.5)$$

On the other hand, at a fixed temperature we may change the volume, say from V to $2V$; the corresponding change of the chemical potential is

$$\mu(2V) - \mu(V) = kT \ln \frac{V}{2V} = kT \ln \tfrac{1}{2}. \qquad (2.2.6)$$

The change in the chemical potential for the expansion process is negative. Finally, we can change N and keep T and V constant, for instance, by eliminating $\frac{1}{2}N$ of the particles. The corresponding change in the chemical potential is

$$\mu(\tfrac{1}{2}N) - \mu(N) = kT \ln \frac{N}{2N} = kT \ln \tfrac{1}{2}. \qquad (2.2.7)$$

Formally, in both (2.2.6) and (2.2.7), the density $\rho = N/V$ has been cut to half of its initial value. However, the change in density in (2.2.6) is due to expansion, whereas in (2.2.7) it is due to a change in the number of particles.

In Chapter 5 we shall generalize Eq. (2.2.1) to include also the interactions among the particles. We shall find that μ^* in (2.2.4) will change, but the liberation term will not be affected (provided we remain in the realm of the classical limit of ST).

2.3. MIXTURE OF IDEAL GASES

The study of mixtures of ideal gases is important for two reasons: first, these systems serve as a reference system for the study of real mixtures and solutions; and second, these systems are special cases of systems in which a chemical reaction occurs. As we shall see in section 2.4, a system of reacting species in an ideal-gas mixture reduces to a nonreacting ideal-gas mixture whenever we block the flow of material from one species to another— e.g., by adding an inhibitor or by removing a catalyst.

In this section we consider a system consisting of two components of structureless particles of species A and B. Let N_A and N_B be the number of particles of species A and B, respectively, contained in a volume V at temperature T. The T, V, N_A, N_B PF of this system is

$$Q(T, V, N_A, N_B) = \frac{V^{N_A + N_B}}{\Lambda_A^{3N_A}\Lambda_B^{3N_B}N_A!N_B!}$$

$$= Q(T, V, N_A)Q(T, V, N_B). \tag{2.3.1}$$

As expected, the PF of the system is simply the product of the PFs of the two pure systems of A and B each in the same volume V and at temperature T.

From (2.3.1), it follows that the Helmholtz energy of the system is

$$A(T, V, N_A, N_B) = -kT \ln Q(T, V, N_A, N_B)$$

$$= -kT \ln Q(T, V, N_A) - kT \ln Q(T, V, N_B)$$

$$= A(T, V, N_A) + A(T, V, N_B). \tag{2.3.2}$$

The Helmholtz energy of the system is simply the sum of the Helmholtz energies of the two pure components, each occupying separately a volume V.

From (2.3.2) we can derive all the other thermodynamic quantities of the system. For instance, the entropy of the system is

$$S(T, V, N_A, N_B) = -\left(\frac{\partial A(T, V, N_A, N_B)}{\partial T}\right)_{V, N_A, N_B}$$

$$= S(T, V, N_A) + S(T, V, N_B), \tag{2.3.3}$$

and similarly the energy of the system is

$$E(T, V, N_A, N_B) = A(T, V, N_A, N_B) + TS(T, V, N_A, N_B)$$

$$= E(T, V, N_A) + E(T, V, N_B). \tag{2.3.4}$$

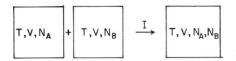

FIGURE 2.1. Mixing of N_A particles of species A in T, V with N_B particles of species B in T, V to form a mixture of $N_A + N_B$ particles at the same T and V.

From these results we can conclude that a system of two (noninteracting) components characterized by the variables T, V, N_A, N_B is thermodynamically equivalent to two separate systems, each containing one component at the same volume V and temperature T. In other words, when we combine two systems, characterized by T, V, N_A and T, V, N_B into a single system characterized by T, V, N_A, N_B, there is no change in any of the thermodynamic quantities. This conclusion is not valid for systems of interacting or reacting particles. These cases will be discussed in section 2.4 and Chapter 5.

In particular, we note that in the process of combining the two systems (Fig. 2.1), the species A and B do mix in the process, yet no entropy change is observed. We can therefore conclude that the process of *mixing* (of ideal gases) by itself has no effect on the entropy of the system.

There is another process in which mixing is involved and in which both the Helmholtz energy and the entropy do change. Consider process II in Fig. 2.2. We start with two pure components A and B as in the previous experiment. We let the two components mix in a volume $2V$, instead of V as in the previous experiment. This can be achieved by removing a partition separating the two compartments.

The partition function in the initial state is exactly the same as in (2.3.1). In the final state, however, we have

$$Q(T, 2V, N_A, N_B) = \frac{(2V)^{N_A + N_B}}{\Lambda_A^{3N_A}\Lambda_B^{3N_B}N_A!N_B!}. \tag{2.3.5}$$

The change in the Helmholtz energy is thus

$$\Delta A = A(T, 2V, N_A, N_B) - A(T, V, N_A) - A(T, V, N_B)$$
$$= -kT(N_A + N_B)\ln(2V) + kTN_A \ln V + kTN_B \ln V$$
$$= kTN_A \ln \tfrac{1}{2} + kTN_B \ln \tfrac{1}{2}. \tag{2.3.6}$$

This is sometimes referred to as the "free energy of mixing."

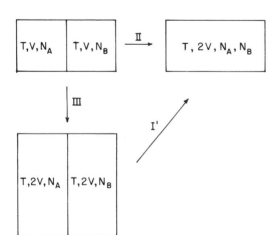

FIGURE 2.2. Process II shows the removal of a partition that separates two compartments T, V, N_A and T, V, N_B to form a new system T, $2V$, $N_A + N_B$. Process III shows the expansion of each compartment from V to $2V$. Process I' resembles that of Fig. 2.1, but with $2V$ instead of V.

The corresponding entropy change is

$$\Delta S = -kN_A \ln \tfrac{1}{2} - kN_B \ln \tfrac{1}{2}, \qquad (2.3.7)$$

which is often referred to as the "entropy of mixing." However, a glance at Eq. (2.3.6) shows that this term originates not in the *mixing* process but in the *expansion* process— each gas expands from V into $2V$. The same entropy change would have been obtained in the process of expanding each component *separately* from V to $2V$, process III in Fig. 2.2.

From the thermodynamic point of view, the two processes II and III depicted in Fig. 2.2 are equivalent, producing the same entropy change (2.3.7). Nevertheless, ΔS of process II is referred to as the entropy of mixing, whereas the same entropy change in process III is referred to as the entropy of expansion. It is true that in process II we observe mixing, as we observed "mixing" in process I. The additional complexity of process II is that we have *both* mixing *and* expansion. In III, on the other hand, we have only expansion. But the final product of processes II and III leads to two thermodynamically equivalent systems, as can be verified by performing process I', which involves only mixing. This is exactly the same as process I (except for replacing V by $2V$). Therefore the conclusion reached from process I holds also for II. The thermodynamic changes (2.3.6) and (2.3.7) are the result of expanding from V to $2V$ and have nothing to do with the mixing process.

Consider now processes II and II' in Fig. 2.3. In process II (as in Fig. 2.2) we mix two components, A and B. For simplicity, we take $N = N_A = N_B$. In II' the two compartments contain only A particles. We remove the partition separating the two compartments. In process II we find

$$\Delta S = -kN_A \ln \tfrac{1}{2} - kN_B \ln \tfrac{1}{2} = -2Nk \ln \tfrac{1}{2} > 0. \qquad (2.3.8)$$

In II', on the other hand, the entropy change is zero.

$$\Delta S = 0. \qquad (2.3.9)$$

The thermodynamic view is that the two processes II and II' are the same except for the mixing that takes place in II. Therefore, $\Delta S > 0$ in (2.3.8) is attributed to the mixing process. In II', on the other hand, nothing is observed to happen. On the molecular level the picture is quite different. In II, each of the components has expanded from V into $2V$. As discussed above, the positive entropy change in (2.3.8) is due to this expansion process. On the other hand, in II' two processes occur. First, the volume accessible to each particle has increased from V to $2V$, giving rise to an entropy change of $\Delta S(\text{expansion}) = -2Nk \ln \tfrac{1}{2}$. In addition, the assimilation term has also changed. Initially, each particle is assimilated by N particles, but in the final state by $2N$ particles. The resulting change in

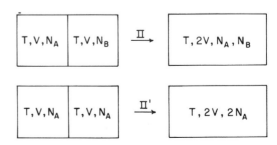

FIGURE 2.3. Same process as II as in Fig. 2.2. A is different from B. Process II' shows the same process as II, but A molecules are in the two compartments.

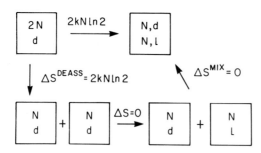

FIGURE 2.4. A spontaneous process of racemization is performed along a different route in three steps. First, the system of $2N$ identical particles is deassimilated into two groups, each of which contains N particles. One group of d particles is then transformed into l particles. Finally, the two distinguishable components d and l are mixed. The entropy changes only in the deassimilation process in the first step.

entropy is $\Delta S(\text{assimilation}) = -k \ln (2N)!/(N!)^2$ which for large N is

$$\Delta S(\text{assimilation}) = 2Nk \ln \tfrac{1}{2}. \qquad (2.3.10)$$

Therefore the combined change in entropy for process II′ is

$$\Delta S(\text{in II}') = \Delta S(\text{expansion}) + \Delta S(\text{assimilation})$$
$$= -2Nk \ln \tfrac{1}{2} + 2Nk \ln \tfrac{1}{2} = 0. \qquad (2.3.11)$$

The net change in entropy is zero. Note, however, that the cancellation in (2.3.11) is between two quantities that are of equal magnitude but of different origin.

We have already found that for a *pure mixing* process $\Delta S = 0$ (Fig. 2.1). On the other hand, the contribution due to assimilation in (2.3.10) is negative. We should expect, therefore, that the reversal of the assimilation process, to which we shall refer as a *deassimilation* process, will result in a positive change in entropy. A process for which $\Delta S > 0$ in an isolated system should be spontaneous. Is there a spontaneous process which is driven by pure deassimilation?

Figure 2.4 shows such a process. Consider an ideal gas of $2N$ molecules that contains one chiralic center. Initially, we prepare the system in such a way that all the molecules are in one of the enantiomeric forms, say the d enantiomer. We then introduce a catalyst that induces a racemization process in adiabatic conditions. At equilibrium, we obtain N molecules of the d enantiomer and N molecules of the l enantiomer. The entropy change in this spontaneous process is

$$\Delta S = 2kN \ln 2 > 0. \qquad (2.3.12)$$

If we analyze carefully the various factors involved in the expression of the chemical potential of the d and l molecules, we find that the momentum partition function as well as any other internal partition function of each molecule, as well as the volume accessible to each molecule, are unchanged during the entire process. The only change that does take place is in the assimilation terms of d and l; hence the entropy change in (2.3.12) is due to the deassimilation of $2N$ identical molecules into two subgroups of distinguishable molecules, N of one kind and N of a second kind. This process, along with an alternative route for its realization, is described in Fig. 2.4.

2.4. CHEMICAL EQUILIBRIUM IN AN IDEAL-GAS MIXTURE

One of the most important aspects of the application of ST is that it provides a method of explicit calculation of the equilibrium constant of a chemical reaction. From the general ST expression for the chemical potential (section 2.2), we can obtain a general expression for the equilibrium constant. In some simple cases, a numerical value of the equilibrium constant may be computed, and this value may be compared with the corresponding experimental value. In more complex systems, such calculations are not feasible. Nevertheless, the general expressions are still useful for qualitative predictions of expected trends of changes in the equilibrium constant upon changing some parameters of the system.

Because of their central importance to almost all applications of ST to chemistry and biochemistry, we shall now study in detail some aspects of systems in chemical equilibrium.

2.4.1. Simple Isomerization Equilibrium

We start with the simplest case of a chemical reaction. Consider a system of N particles in an ideal-gas phase of volume V and temperature T. We assume that each particle has only two internal (say, electronic or vibrational) energy levels, which we denote by E_A and E_B, with degeneracies ω_A and ω_B, respectively. No other internal degrees of freedom are ascribed to the particles.

The partition function of such a system is simply

$$Q(T, V, N) = \frac{q^N V^N}{N! \Lambda^{3N}}, \tag{2.4.1}$$

where

$$q = \omega_A e^{-\beta E_A} + \omega_B e^{-\beta E_B}$$
$$= q_A + q_B. \tag{2.4.2}$$

Here, q is the internal PF of a single particle. Thus, if we know the molecular parameters m, ω_A, ω_B, E_A, E_B, we can compute all the relevant thermodynamic quantities of our system.

Until this point we have considered a system of *one* component. No chemical equilibrium was mentioned. Now we choose to view the *same* system from a *different* point of view. Each molecule of the system can be in one of the two states. We call a molecule occupying the state A an A molecule, and a molecule occupying the state B a B molecule. We stress that this classification of molecules into two groups is purely theoretical. It does not depend on whether or not we can detect or isolate the A molecules or the B molecules.

We now denote

$$Y_A = \frac{q_A V}{\Lambda^3}, \qquad Y_B = \frac{q_B V}{\Lambda^3} \tag{2.4.3}$$

and

$$Y = Y_A + Y_B, \tag{2.4.4}$$

and we can rewrite the PF (2.4.1) as

$$Q(T, V, N) = \frac{Y^N}{N!} = \frac{(Y_A + Y_B)^N}{N!}$$

$$= \frac{1}{N!} \sum_{N_A = 0}^{N} \frac{N!}{N_A! N_B!} Y_A^{N_A} Y_B^{N_B}, \tag{2.4.5}$$

where we used the binomial expansion. The summation is over all possible values of N_A (where N_B is defined by $N_B = N - N_A$).

We now define the quantity

$$Q(T, V, N_A, N_B) = \frac{Y_A^{N_A} Y_B^{N_B}}{N_A! N_B!}, \tag{2.4.6}$$

and rewrite (2.4.5) as

$$Q(T, V, N) = \sum_{N_A = 0} Q(T, V, N_A, N_B), \tag{2.4.7}$$

with $N_B = N - N_A$.

We identify $Q(T, V, N_A, N_B)$ in (2.4.6) as the PF of a *two-component mixture* of A and B in an ideal gas (section 2.3). How did we get a two-component system from a one-component system? We did not! The partition function of our system is (2.4.7), not (2.4.6). Equation (2.4.6) is the PF of a two-component system having any arbitrary but fixed values of N_A and N_B. In our case, the summation over N_A makes the sum (2.4.7) independent of N_A, so that the resulting sum on the rhs of (2.4.7) is a function of N only. However, there exists a particular value of N_A (and N_B) for which $Q(T, V, N)$ may be approximated by $Q(T, V, N_A, N_B)$. This is an important property of the PF of a macroscopic system at equilibrium. We shall examine this aspect shortly, after we have introduced a few useful concepts.

According to the general rule of ST, each term in the PF is proportional to the probability of the event characterized by that term. Here $Q(T, V, N_A, N_B)$ is proportional to the probability of finding our (one-component) system in a state such that precisely N_A molecules are in the state A and the rest, N_B molecules, in the state B. Thus

$$P(N_A) = \frac{Q(T, V, N_A, N_B)}{Q(T, V, N)}. \tag{2.4.8}$$

With the help of this distribution function we can calculate three specific values of N_A:

1. The most probable value of N_A, denoted N_A^*;
2. The average value of N_A, denoted $\langle N_A \rangle$; and
3. The equilibrium value of N_A, denoted N_A^e.

These are three conceptually different quantities. As we shall soon see, their numerical values are, from the macroscopic point of view, identical.

The most probable value of N_A is defined as the value of N_A for which

$$\frac{\partial P(N_A)}{\partial N_A} = 0. \tag{2.4.9}$$

It is easy to see that $P(N_A)$ has one extremum, which is a maximum. From (2.4.6) and (2.4.9) we obtain, using the Stirling approximation,

$$\frac{\partial \ln P(N_A)}{\partial N_A} = \ln Y_A - \ln Y_B - \ln N_A + \ln(N - N_A) = 0. \tag{2.4.10}$$

Solving (2.4.10) for N_A^*, we obtain

$$N_A^* = N \frac{Y_A}{Y_A + Y_B}. \tag{2.4.11}$$

For $N_A = N_A^*$, we also have

$$\frac{\partial^2 \ln P(N_A)}{\partial N_A^2} = -\left[\frac{1}{N_A} + \frac{1}{N - N_A}\right]_{N_A = N_A^*} < 0, \tag{2.4.12}$$

which means that $\ln P(N_A)$, as well as $P(N_A)$, has a maximum at $N_A = N_A^*$.

The average value of N_A is defined by

$$\langle N_A \rangle = \sum_{N_A = 0}^{N} N_A P(N_A)$$

$$= \frac{1}{Q(T, V, N)} \sum_{N_A} N_A \frac{Y_A^{N_A} Y_B^{N_B}}{N_A! N_B!}. \tag{2.4.13}$$

To evaluate the sum on the rhs of (2.4.13), we note the identity

$$Y_A \frac{\partial}{\partial Y_A} \left[\sum_{N_A} \frac{Y_A^{N_A} Y_B^{N_B}}{N_A! N_B!}\right] = \sum_{N_A} N_A \frac{Y_A^{N_A} Y_B^{N_B}}{N_A! N_B!}. \tag{2.4.14}$$

Hence using (2.4.5) we obtain

$$\langle N_A \rangle = \frac{Y_A}{Q} \frac{\partial}{\partial Y_A} \left[\sum_{N_A} \frac{Y_A^{N_A} Y_B^{N_B}}{N_A! N_B!}\right]$$

$$= \frac{Y_A}{Q} \frac{\partial}{\partial Y_A} \left[\frac{(Y_A + Y_B)^N}{N!}\right]$$

$$= \frac{Y_A}{Q} \left[\frac{N(Y_A + Y_B)^{N-1}}{N!}\right]$$

$$= N \frac{Y_A}{Y_A + Y_B}. \tag{2.4.15}$$

We find that the value of $\langle N_A \rangle$ in (2.4.15) is equal to the value of N_A^* in (2.4.11).

The third quantity is the equilibrium value of N_A. This is defined as the solution of the equation

$$\mu_A(N_A^e) = \mu_B(N_A^e), \tag{2.4.16}$$

where μ_A and μ_B are the chemical potentials of A and B, respectively. How do we get the chemical potentials of A and B if the system is a one-component system? Indeed, if we insist on viewing our system as a one-component system we cannot define μ_A and μ_B. The way to achieve such a definition is to "freeze" the chemical equilibrium. In other words, suppose we can prepare a system of N_A molecules in state A and N_B molecules in state B, where N_A and N_B are *fixed*. This is a different system from the one we started with. In practice, we may think of a catalyst the removal of which precludes the conversion between the A and B states. Once the catalyst is removed, the system becomes a two-component system of A and B. In this case the chemical potential of each species may be defined, e.g., as

$$\mu_A = \left(\frac{\partial A}{\partial N_A}\right)_{T,V,N_B} = -kT\left(\frac{\partial \ln Q(T, V, N_A, N_B)}{\partial N_A}\right)_{T,V,N_B}. \tag{2.4.17}$$

It is important to realize that μ_A is defined as the derivative of $Q(T, V, N_A, N_B)$ (see 2.4.6)—i.e., for a two-component system of A and B. It cannot be defined using the function $Q(T, V, N)$ as appears in (2.4.7). We shall soon see that the removal of a catalyst is equivalent to placing a partition between two compartments in a system in which the components A and B are defined in terms of their locations in different compartments.

Using (2.4.6) as the PF of our new "frozen-in" system, we obtain

$$\mu_A = kT \ln \frac{N_A}{Y_A} \tag{2.4.18}$$

$$\mu_B = kT \ln \frac{N_B}{Y_B}. \tag{2.4.19}$$

Hence the solution of (2.4.16) for N_A^e is

$$kT \ln \frac{N_A^e}{Y_A} = kT \ln \frac{(N - N_A^e)}{Y_B} \tag{2.4.20}$$

or, equivalently,

$$N_A^e = N \frac{Y_A}{Y_A + Y_B}, \tag{2.4.21}$$

which again has the same value as N_A^* and $\langle N_A \rangle$.

The equilibrium constant for the reaction

$$A \rightleftarrows B \tag{2.4.22}$$

is defined by

$$K = \frac{N_B^e}{N_A^e} = \frac{Y_B}{Y_A} = \frac{q_B}{q_A}. \tag{2.4.23}$$

Thus, knowing the molecular parameters ω_A, ω_B, E_A, and E_B permits the calculation of the equilibrium constant for this particular reaction. Note that μ_A and μ_B are definable for any N_A and N_B. The condition of chemical equilibrium $\mu_A = \mu_B$ is attained only for a specific value of N_A (and N_B), which is the solution of Eq. (2.4.20).

2.4.2. An Analogue of the Isomerization Equilibrium

We now discuss a different system, the treatment of which is equivalent to the treatment of the isomerization reaction discussed above. Consider again a one-component system at T, V, N. The PF of this system is, as before,

$$Q(T, V, N) = \frac{q^N V^N}{N! \Lambda^{3N}}, \tag{2.4.24}$$

where q is the internal PF of a single molecule.

Suppose we now place a partition that divides the volume V into two compartments of volumes V_A and V_B, with N_A and N_B in V_A and V_B, respectively (Fig. 2.5). Note that the molecules in the two compartments are the *same* molecules. However, we may refer to the molecule in the volume V_A as the A molecules and in V_B as the B molecules. Although the molecules are of the same species, they are distinguishable by the very fact that they occupy different compartments.

The partition function for the system with the inserted partition is

$$Q(T, V, N_A, N_B) = Q(T, V_A, N_A) Q(T, V_B, N_B)$$

$$= \frac{V_A^{N_A} V_B^{N_B} q^N}{N_A! N_B! \Lambda^{3N}}, \tag{2.4.25}$$

where $N = N_A + N_B$ and $q = q_A = q_B$, $\Lambda = \Lambda_A = \Lambda_B$. Clearly we can prepare such a system with any number N_A in V_A and any number N_B in V_B (subject to the restriction that the densities $\rho_A = N_A/V_A$ and $\rho_B = N_B/V_B$ are low enough so the two subsystems behave as ideal gases).

The relation between the PF in (2.4.24) and the PF in (2.4.25) is

$$Q(T, V, N) = \frac{q^N (V_A + V_B)^N}{N! \Lambda^{3N}}$$

$$= \frac{q^N}{N! \Lambda^{3N}} \sum_{N_A=0}^{N} \frac{N!}{N_A! N_B!} V_A^{N_A} V_B^{N_B}$$

$$= \sum_{N_A=0}^{N} \frac{Y_A^{N_A}}{N_A!} \frac{Y_B^{N_B}}{N_B!}, \tag{2.4.26}$$

with $N_B = N - N_A$ and $Y_A = qV_A/\Lambda^3$, $Y_B = qV_B/\Lambda^3$. The analogy with (2.4.5) is evident. In (2.4.5) we classified our system of N molecules into two groups according to their internal energy levels. Here we classified the system into two groups according to the location of the molecules in V_A or in V_B. Each term in (2.4.26) corresponds to a specific

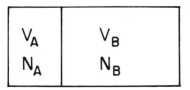

FIGURE 2.5. N_A and N_B molecules of the same species in V_A and V_B, respectively.

partition of the system into N_A molecules in V_A and N_B molecules in V_B. Placing the partition in our system is the analogue of inserting an inhibitor in the previous system.

The chemical potentials of the A and B components in this system are

$$\mu_A = kT \ln \rho_A \Lambda^3 q^{-1} \tag{2.4.27}$$

$$\mu_B = kT \ln \rho_B \Lambda^3 q^{-1}. \tag{2.4.28}$$

In general, we can choose arbitrary densities, and hence $\mu_A \neq \mu_B$. However, there is one density for which we obtain the condition of chemical equilibrium

$$\mu_A = \mu_B, \tag{2.4.29}$$

i.e.,

$$\frac{N_A^e}{N_B^e} = \frac{V_A}{V_B}. \tag{2.4.30}$$

Thus, at equilibrium, the number of particles in each compartment is proportional to the volume of that compartment. The last relation is the analogue of the equilibrium constant which in the previous case is given by (2.4.23). In both cases we have shown that the classification into two (or more) components is a matter of choice. There are infinitely many ways of achieving such a classification.

We now turn to reinterpret the equilibrium value N_A^e in a somewhat different way.

Consider the system with the inserted partition (or the inhibitor) for which the PF is (2.4.25) or (2.4.6). The Helmholtz energy of such a system is

$$A(T, V, N_A, N_B) = -kT \ln Q(T, V, N_A, N_B). \tag{2.4.31}$$

We have seen that $Q(T, V, N_A, N_B)$ is proportional to the probability $P(N_A)$ of finding N_A molecules in state A. We found also that $P(N_A)$ has a maximum value for N_A^* (or N_A^e) for which

$$N_A^* = N \frac{Y_A}{Y_A + Y_B}. \tag{2.4.32}$$

From (2.4.31) we see that a maximum of $P(N_A)$ corresponds to a minimum of $A(T, V, N_A, N_B)$ at the point N_A^* (and $N_B^* = N - N_A^*$). It is also clear that we always have the inequality

$$A(T, V, N) < A(T, V, N_A, N_B), \tag{2.4.33}$$

simply because Q in (2.4.26) is a sum of many positive terms, whereas Q in (2.4.25) is only one of these terms. Thus for any process for which we remove the partition in Fig. 2.5 (or introduce a catalyst), the change in the Helmholtz energy is always negative:

$$\Delta A = A(T, V, N) - A(T, V, N_A, N_B) < 0. \tag{2.4.34}$$

It is never zero. However, for the particular number N_A^e (and $N_B^e = N - N_A^e$) this change is so small that from the macroscopic point of view we can simply put $\Delta A = 0$. The exact value of ΔA at N_A^e is

$$\Delta A = -kT \ln \frac{Q(T, V, N)}{Q(T, V, N_A, N_B)}$$

$$= kT \ln \left(\frac{V_A}{V_A + V_B}\right)^{N_A} + kT \ln \left(\frac{V_B}{V_A + V_B}\right)^{N_B} - kT \ln \frac{N_A! N_B!}{N!}$$

$$= kT \ln \frac{N_A^{N_A} N_B^{N_B}}{N_A! N_B!} \frac{N!}{N^N} < 0. \tag{2.4.35}$$

In the last equality we substitute the values of N_A^e and N_B^e from (2.4.30).

For N_A, N_B large enough, we can use the Stirling approximation in (2.4.35) to obtain

$$\Delta A = kT \ln \frac{e^{N_A} e^{N_B}}{e^N} = 0. \tag{2.4.36}$$

To conclude, the removal of a partition (or an inhibitor) separating two compartments always involves a negative change in the Helmholtz energy. This is true even when the values of N_A and N_B were chosen to be the equilibrium values of N_A^e and N_B^e, respectively. However, within the Stirling approximation, the last change is negligibly small, and practically we can say that no change in the Helmholtz energy has occurred.

We now show that the last conclusion is equivalent to replacing the sum in Eq. (2.4.7) (or 2.4.26) by a single term, i.e.,

$$\sum_{N_A = 0}^{N} Q(T, V, N_A, N_B) \approx Q(T, V, N_A^e, N_B^e), \tag{2.4.37}$$

when N_A^e fulfills condition (2.4.21). This is quite a remarkable result. The sum on the left-hand side is over a great number of terms (N on the order of 10^{23}), including the term on the rhs of (2.4.37). How can this equality hold? The answer is that (2.4.37) is not an exact mathematical equality—it can never be an equality since all the terms in the sum are positive. We shall now see in what sense we claim that (2.4.37) is an equality. To do that, we start from the rhs of (2.4.37) and proceed to obtain the lhs.

$$Q(T, V, N_A^e, N_B^e) = \frac{Y_A^{N_A} Y_B^{N_B}}{N_A! N_B!} = \left[\frac{Y_A e}{N_A^e}\right]^{N_A^e} \left[\frac{Y_B e}{N_B^e}\right]^{N_B^e}$$

$$= e^N \left[\frac{Y_A}{N_A^e}\right]^{N_A^e} \left[\frac{Y_B}{N_B^e}\right]^{N_B^e} = e^N \left[\frac{Y_A}{N_A^e}\right]^{N_A^e} \left[\frac{Y_A}{N_A^e}\right]^{N_A^e}$$

$$= e^N \left[\frac{Y_A}{N_A^e}\right]^{N} = e^N \left[\frac{Y_A + Y_B}{N_A^e + N_B^e}\right]^{N}$$

$$= e^N \frac{(Y_A + Y_B)^N}{N^N} = \frac{(Y_A + Y_B)^N}{N!} = Q(T, V, N), \tag{2.4.38}$$

where we have used the Stirling approximation in the form

$$\ln N! = N \ln N - N \tag{2.4.39}$$

for N_A, N_B, and N, as well as the identity

$$\frac{Y_A}{N_A^e} = \frac{Y_B}{N_B^e} = \frac{Y_A + Y_B}{N_A^e + N_B^e}. \tag{2.4.40}$$

We see that, within the Stirling approximation, (2.4.27) turns into an equality. In thermo-dynamic terms this conclusion is equivalent to the statement that placing or removing a partition, when the two compartments contain the quantities N_A^e and N_B^e, respectively, does not result in any measurable change in the Helmholtz energy. The same holds true when placing an inhibitor after a chemical equilibrium has been reached.

In section 1.3 we have also replaced sums of terms by a single term. The reasons for doing so are essentially the same as in the example discussed above. In all cases we have a sum over a large number of positive terms. These terms are sharply peaked at some value of the summation index (E^*, V^*, or N^* of section 1.3). For a large system, the thermodynamic quantities computed from the entire sum and from the maximal terms are practically identical. This is also the reason why we can choose any one of the partition functions that we find convenient to treat our macroscopic system.

2.4.3. Standard Thermodynamic Quantities of a Chemical Reaction

The standard Helmholtz energy is defined as follows: First we write the chemical potentials of A and B, in their thermodynamic representation, as

$$\mu_A = kT \ln \frac{N_A}{Y_A} = kT \ln \Lambda_A^3 q_A^{-1} + kT \ln \rho_A$$

$$= \mu_A^0 + kT \ln \rho_A, \tag{2.4.41}$$

$$\mu_B = kT \ln \frac{N_B}{Y_B} = kT \ln \Lambda_B^3 q_B^{-1} + kT \ln \rho_B$$

$$= \mu_B^0 + kT \ln \rho_B. \tag{2.4.42}$$

The standard Helmholtz energy of the reaction $A \rightarrow B$ is defined by

$$\Delta A^0 = \mu_B^0 - \mu_A^0 = kT \ln \frac{q_A}{q_B} = -kT \ln K. \tag{2.4.43}$$

This is the well-known relation between the chemical equilibrium K and the standard Helmholtz energy of the reaction.

Note that K may be *measured* through the relation

$$K = \frac{N_B^e}{N_A^e} \tag{2.4.44}$$

at equilibrium. This is possible only if we have a means of measuring the relative concentrations of A and B at equilibrium. On the other hand, K can be *calculated*, theoretically, through

$$K = \frac{q_B}{q_A} = \frac{\omega_B}{\omega_A} e^{-\beta[E_B - E_A]},\qquad (2.4.45)$$

whenever the molecular information on A and B is available. In our example of the two energy levels discussed above, all we need to know is the quantities ω_A, ω_B, E_A, and E_B.

The standard entropy and energy of this reaction are obtained from the relations

$$\Delta S^0 = -\frac{\partial \Delta A^0}{\partial T} = k \ln \frac{\omega_B}{\omega_A}\qquad (2.4.46)$$

and

$$\Delta E^0 = \Delta A^0 + T\Delta S^0 = E_B - E_A.\qquad (2.4.47)$$

These results are quite instructive, though they were obtained for this particular simple system. The standard entropy of the reaction depends only on the ratio of the degeneracies of the two levels. The standard energy depends only on the difference in the energies of the two levels. In this example the two quantities ΔS^0 and ΔE^0 are completely independent of each other, in the sense that each depends on a different set of parameters of the molecules. In more complicated examples discussed in Chapter 8, such a neat separation of parameters does not occur.

Two limiting cases of K are of interest. We rewrite the chemical equilibrium (2.4.45) as

$$K = \exp\left(-\frac{\Delta E^0 - T\Delta S^0}{kT}\right).\qquad (2.4.48)$$

At very low temperature, the term $T\Delta S^0$ will become very small compared with ΔE^0. In this limit only the difference in the energy levels will determine the equilibrium constant. If we let $T \to 0$, we eventually get $K \to 0$; this means that the concentration of B is zero, i.e., all molecules are in the ground state A.

At very high temperature, $T\Delta S^0$ is the dominating term. Hence, as $T \to \infty$ we have

$$K = e^{\Delta S^0/k} = \omega_B/\omega_A,\qquad (2.4.49)$$

i.e., the equilibrium constant will be equal to the ratio of the degeneracies, independent of the difference in the energy levels.

Figure 2.6 shows the dependence of the mole fraction $x_A = N_A^e/(N_A^e + N_B^e) = (1 + K)^{-1}$ on the temperature for different values of ω_B/ω_A. All the curves start with

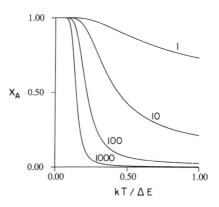

FIGURE 2.6. Dependence of the mole fraction x_A on the temperature (in units of $\Delta E/k$) for different ratios of the degeneracies ω_B/ω_A (as indicated next to each curve).

$x_A = 1$ at $T \to 0$. At $T \to \infty$, the limiting value depends on ω_B/ω_A. For $\omega_B/\omega_A = 1$, $x_A \to \frac{1}{2}$, and for $\omega_B/\omega_A = 1000$, $x_A \to (1001)^{-1}$ at this limit. These curves are very typical for any system at chemical equilibrium. We shall see several variations of this behavior in Chapters 4 and 8.

2.4.4. Generalizations

We will discuss here two generalizations of the simple example treated in the preceding section. First, consider again an isomerization reaction. This could be a *cis–trans* conversion in dichloroethylene or a helix–coil transition in a polypeptide, symbolically

$$A \rightleftarrows B, \tag{2.4.50}$$

where now, instead of two internal energy levels E_A and E_B as in (2.4.2), each molecule has a series of energy levels, depicted schematically in Fig. 2.7.

In some practical cases, the set of all states of a single molecule may be split into two groups in such a way that transitions between states in one group are much faster than transitions between states belonging to different groups. We denote by $i \in A$ all the internal states that belong to one group, A, and by $i \in B$ the states belonging to group B. In this way we have chosen to view the one-component system of N molecules as a mixture of two components A and B with the corresponding internal partition functions q_A and q_B defined by

$$q = \sum_i \exp(-\beta E_i) = \sum_{i \in A} \exp(-\beta E_i) + \sum_{i \in B} \exp(-\beta E_i) = q_A + q_B. \tag{2.4.51}$$

If the transitions between states belonging to different groups is extremely slow, then the system behaves like a mixture of two independent components. In principle we can even separate the system into two pure components or prepare a mixture with any composition of N_A and N_B. However, if the transitions are fast, then the system is essentially a one-component system and each molecule has an internal partition function q. From the theoretical point of view, any division of the set of the states into two groups defines a mixture-model point of view. Such a division is independent of the rate of transitions between the states and of the distribution of the states, like the one in Fig. 2.7. (For instance, one group could be chosen to consist of all states labeled by even numbers, the other with odd numbers, or A could be chosen as the ground state only

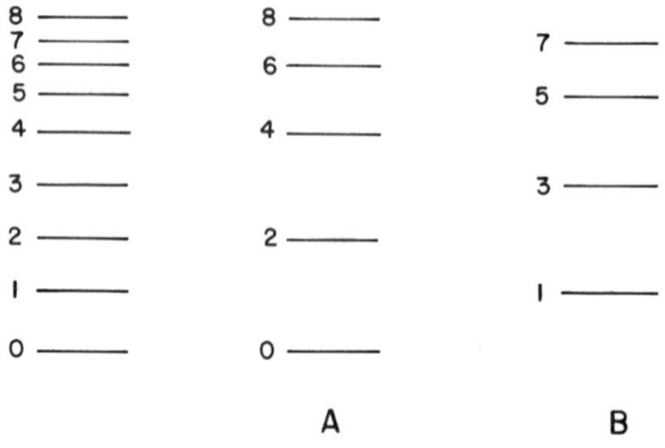

FIGURE 2.7. All the energy levels of a molecule are divided into two groups, A and B. This division defines the two species A and B.

and B all the rest.) The choice of grouping is arbitrary. We define the corresponding canonical PFs

$$Q(T, V, N) = \frac{q^N V^N}{N! \Lambda^{3N}} = \frac{V^N}{N! \Lambda^{3N}} (q_A + q_B)^N = \frac{V^N}{N! \Lambda^{3N}} \sum_{N_A} \frac{N!}{N_A! N_B!} q_A^{N_A} q_B^{N_B}$$

$$= \sum_{N_A} Q(T, V, N_A, N_B), \tag{2.4.52}$$

where, as in section 2.4.1, $Q(T, V, N_A, N_B)$ is the partition function of a mixture of N_A molecules of A and N_B molecules of B in the presence of an inhibitor that prevents the transitions between the groups A and B. The sum over all possible N_A brings us back to the one-component system. Note again that the mixture-model point of view is independent of whether or not such an inhibitor actually exists.

We see that (2.4.52) is formally the same as (2.4.7) except for the interpretation of q_A and q_B in (2.4.51). Therefore we can proceed to obtain all the thermodynamic results in the same formal manner as in the simpler case, provided we keep in mind the new interpretation of q_A and q_B. Thus the equilibrium values of N_A and N_B are

$$N_A^e = N \frac{Y_A}{Y_A + Y_B}, \qquad N_B^e = N \frac{Y_B}{Y_A + Y_B}, \tag{2.4.53}$$

where

$$Y_A = \frac{q_A V}{\Lambda_A^3}, \qquad Y_B = \frac{q_B V}{\Lambda_B^3}, \tag{2.4.54}$$

and the equilibrium constant is

$$K = \frac{N_B^e}{N_A^e} = \frac{q_B}{q_A}, \tag{2.4.55}$$

which is formally the same as (2.4.23). In contrast to the simple system with two energy levels treated in the previous subsections, the standard thermodynamic quantities are not so simple as the ones obtained in section 2.4.3. The standard Helmholtz energy is defined by

$$\Delta A^0 = -kT \ln K = kT \ln \frac{q_A}{q_B}. \tag{2.4.56}$$

The standard entropy and energy are

$$\Delta S^0 = k \ln K + \frac{1}{T} (\bar{E}_B - \bar{E}_A), \tag{2.4.57}$$

$$\Delta E^0 = \bar{E}_B - \bar{E}_A. \tag{2.4.58}$$

Here $\bar{E}_A$ and $\bar{E}_B$ are the average "energy level" of groups A and B, respectively. These are defined by

$$\bar{E}_A = \sum_{i \in A} E_i \exp[-\beta E_i]/q_A = \sum_{i \in A} E_i P_i(A), \tag{2.4.59}$$

$$\bar{E}_B = \sum_{i \in B} E_i \exp[-\beta E_i]/q_B = \sum_{i \in B} E_i P_i(B), \tag{2.4.60}$$

where $P_i(A)$ is the probability of finding the state i within the group A. This is different from the probability of finding the state i, which is

$$P_i = \frac{\exp(-\beta E_i)}{q}. \tag{2.4.61}$$

Equations (2.4.57) and (2.4.58) should be compared with (2.4.46) and (2.4.47). Here, we do not have a neat separation into the ratio of the degeneracies and the difference in the energy levels of the two states.

It is sometimes convenient to rewrite the equilibrium constant K in terms of the energy differences, rather than energy levels. Let E_i^A and E_i^B be the energy levels in each group. Then

$$K = \frac{q_B}{q_A}$$

$$= \frac{\sum\limits_{i=0} \exp[-\beta E_i^B]}{\sum\limits_{i=0} \exp[-\beta E_i^A]}$$

$$= \exp[-\beta(E_0^B - E_0^A)] \frac{\sum\limits_{i=1} \exp[-\beta(E_i^B - E_0^B)]}{\sum\limits_{i=1} \exp[-\beta(E_i^A - E_0^A)]}$$

$$= \exp[-\beta(E_0^B - E_0^A)] \frac{q_B'}{q_A'}, \tag{2.4.62}$$

where E_0^B and E_0^A are the energies of the ground states in each group (here assumed nondegenerate), and q_B' and q_A' are the modified partition functions, expressed in terms of the energy levels above the ground states.

In biochemistry, when dealing with huge molecules such as proteins or nucleic acids, it is often convenient to treat the system as being a quasi-two-state system. For instance, a protein may be viewed as having two states; the folded **F** (or native) and the unfolded **U** (or denatured) states. Clearly, each of these states consists of many quantum-mechanical states. It is convenient to define the Helmholtz energy of each of these combined states as follows:

$$A_U = -kT \ln q_U = -kT \ln \sum_{i \in U} \exp[-\beta E_i] \tag{2.4.63}$$

$$A_F = -kT \ln q_F = -kT \ln \sum_{i \in F} \exp[-\beta E_i]. \tag{2.4.64}$$

These Helmholtz energies are then used as the "energy levels" of the protein. The equilibrium constant is thus

$$K = \frac{q_U}{q_F} = \exp[-\beta(A_U - A_F)]. \tag{2.4.65}$$

The standard Helmholtz energy of this reaction is simply

$$\Delta A^0 = -kT \ln K = A_U - A_F. \tag{2.4.66}$$

In all of the examples treated in the next chapter, we shall treat A_U and A_F as if they were two energy levels. This is a necessary approximation in treating very complex systems such as proteins. However, one should bear in mind that A_U and A_F are free energies and that in general one should take into account their temperature dependence.

The generalization to more complicated chemical reactions is quite straightforward. We shall briefly discuss the case of a reaction of the type

$$v_A A + v_B B \rightleftarrows v_C C + v_D D, \tag{2.4.67}$$

where v_i is the stoichiometric coefficient of the ith component.

The PF of a system containing A, B, C, and D at equilibrium is quite complicated. However, we already know that we can "freeze in" the equilibrium and treat the system as if it were a mixture of ideal gases. Thus for any composition of N_A, N_B, N_C, and N_D the PF of the frozen system is

$$Q(T, V, N_A, N_B, N_C, N_D) = \prod_i \frac{(q_i V)^{N_i}}{\Lambda_i^{3N_i} N_i!}, \tag{2.4.68}$$

where the product over i is over all species $i = A, B, C, D$.

In order to obtain the equilibrium constant, we use the thermodynamic condition of chemical equilibrium. First we write the chemical potential of each species i as

$$\mu_i = kT \ln \rho_i \Lambda_i^3 q_i^{-1} = \mu_i^0 + kT \ln \rho_i, \tag{2.4.69}$$

where q_i is the internal PF of the molecule of species i. The energy levels for all species are measured relative to some common zero-energy level. The thermodynamic equilibrium condition is

$$v_A \mu_A + v_B \mu_B = v_C \mu_C + v_D \mu_D, \tag{2.4.70}$$

which leads to the equilibrium constant

$$K_\rho = \exp\{-\beta[v_C \mu_C^0 + v_D \mu_D^0 - v_A \mu_A^0 - v_B \mu_B^0]\}, \tag{2.4.71}$$

or

$$K_\rho = \prod_i \rho_i^{v_i} = \prod_i \left(\frac{q_i}{\Lambda_i^3}\right)^{v_i}. \tag{2.4.72}$$

Here we have expressed the equilibrium constant in terms of the densities ρ_i. A more common expression for the equilibrium constant in an ideal-gas mixture is in terms of the partial pressures of the species. Thus, substituting $P_i = kT\rho_i$ in (2.4.72), we obtain

$$K_\rho = \prod_i \rho_i^{v_i} = \prod_i (P_i/kT)^{v_i} = (kT)^{-\Sigma v_i} \prod_i P_i^{v_i} = (kT)^{-\Sigma v_i} K_p. \tag{2.4.73}$$

Note that in (2.4.72) and (2.4.73) we use the convention that v_i is positive for a product species (C and D) and negative for a reactant species (A and B).

2.4.5. Heat Capacity of a System in Chemical Equilibrium

We now return to the case of isomerization discussed in the preceding subsection. Again we have N molecules in an ideal-gas phase, and each molecule has an internal PF given by

$$q = \sum_i \exp(-\beta E_i). \tag{2.4.74}$$

The summation in (2.4.74) is over *all* states of a single molecule (excluding the momentum PF). The PF of the system is

$$Q(T, V, N) = \frac{q^N V^N}{N! \Lambda^{3N}}. \tag{2.4.75}$$

The internal energy is

$$E = -T^2 \frac{\partial(A/T)}{\partial T} = \tfrac{3}{2} NkT + N \sum_i P_i E_i. \tag{2.4.76}$$

Here each molecule has a kinetic energy of $\tfrac{3}{2}kT$ and an average internal energy of $\Sigma P_i E_i$, where the sum is over *all* internal states of a single molecule.

The constant-volume heat capacity is given by

$$C_V = \left(\frac{\partial E}{\partial T}\right)_{V,N}$$

$$= \tfrac{3}{2} Nk + \frac{N}{kT^2}\left[\sum_i P_i E_i^2 - \left(\sum_i P_i E_i\right)^2\right]$$

$$= \tfrac{3}{2} Nk + \frac{N}{kT^2}[\overline{(E_i^2)} - (\bar{E}_i)^2]. \tag{2.4.77}$$

This is the general expression for the heat capacity.

Now we choose to view our system as a two-component system. We follow the same grouping of all states into two groups as in section 2.4.4 and rewrite (2.4.76) as

$$E = \tfrac{3}{2} NkT + N \sum_i P_i E_i$$

$$= \tfrac{3}{2} NkT + \sum_i N_i E_i$$

$$= \tfrac{3}{2} NkT + N\left\{\sum_{i \in A} \frac{E_i \exp[-\beta E_i]}{q} + \sum_{i \in B} \frac{E_i \exp[-\beta E_i]}{q}\right\}$$

$$= \tfrac{3}{2} NkT + \bar{N}_A \bar{E}_A + \bar{N}_B \bar{E}_B, \tag{2.4.78}$$

where $\bar{E}_A$ and $\bar{E}_B$ are the average energy levels of each group, as defined in (2.4.59) and (2.4.60), and $\bar{N}_A$ and $\bar{N}_B$ are the average number of molecules in each group defined by $\bar{N}_A = Nq_A/q$ and $\bar{N}_B = Nq_B/q$, respectively. The heat capacity in this representation is now

$$C_V = \tfrac{3}{2} Nk + \bar{N}_A\left(\frac{\partial \bar{E}_A}{\partial T}\right) + \bar{N}_B\left(\frac{\partial \bar{E}_B}{\partial T}\right) + (\bar{E}_B - \bar{E}_A)\left(\frac{\partial \bar{N}_B}{\partial T}\right). \tag{2.4.79}$$

The first term on the rhs of (2.4.79) is as in (2.4.77) due to the translational degrees of freedom. The significance of the other two terms is as follows: Suppose that transition between states belonging to the two groups is forbidden, say by introducing an inhibitor that prevents such transitions; then $\bar{N}_A$ and $\bar{N}_B$ are fixed quantities and the last term on the rhs of (2.4.79) is zero. If we allow transitions between the two groups (say by removing the inhibitor, or introducing a catalyst), then $\bar{N}_A$ and $\bar{N}_B$ are the equilibrium values of N_A and N_B, and as such they are temperature dependent. The last term on the rhs of (2.4.79) is the change in the heat capacity of the system caused by introducing the catalyst. We denote this contribution by

$$\Delta C^r = (\bar{E}_B - \bar{E}_A)\left(\frac{\partial \bar{N}_B}{\partial T}\right). \tag{2.4.80}$$

We now show that this term is always positive; i.e., introducing the catalyst to the system always *increases* the heat capacity of the system.

In the case of an ideal gas it is easy to calculate the temperature derivative of $\bar{N}_B$, for instance from

$$\bar{N}_B = \frac{q_B}{q} N \tag{2.4.81}$$

we obtain

$$\frac{\partial \bar{N}_B}{\partial T} = \frac{N x_A x_B}{kT^2} (\bar{E}_B - \bar{E}_A). \tag{2.4.82}$$

Hence

$$\Delta C^r = \frac{N x_A x_B}{kT^2} (\bar{E}_B - \bar{E}_A)^2 \geq 0, \tag{2.4.83}$$

where x_A and x_B are the mole fractions of A and B at equilibrium.

This result was obtained here for a special case of an ideal-gas mixture. We shall encounter the same results for a more general case in Chapters 5 and 7.

2.5. IDEAL GAS IN AN EXTERNAL ELECTRIC FIELD

We consider here a system consisting of N molecules in a volume V at temperature T placed in a uniform electric field $\mathbf{D}$. $\mathbf{D}$ is actually the electric displacement field, which is controllable by external charges (say on the surfaces of a condenser).

The canonical partition function is written as

$$Q(T, V, N, D) = \frac{q^N}{(8\pi^2)^N N! \Lambda^{3N}} \int \cdots \int \exp\left[-\beta U(\mathbf{X}^N) - \beta \sum_{i=1}^{N} U_D(\mathbf{X}_i)\right] d\mathbf{X}^N, \tag{2.5.1}$$

where $U(\mathbf{X}^N)$ is the total interaction energy among the particles and $U_D(\mathbf{X}_i)$ is the interaction of a single molecule with the external field.

In the present treatment we assume that interactions among the molecules are negligible; therefore we put $U(\mathbf{X}^N) \equiv 0$ as in the case of an ideal gas. The remaining integral may now be factored into a product of N factors

$$\int \cdots \int \exp\left[-\beta \sum_{i=1}^{N} U_D(\mathbf{X}_i)\right] d\mathbf{X}^N = \left[\int \exp[-\beta U_D(\mathbf{X})] d\mathbf{X}\right]^N. \tag{2.5.2}$$

We denoted by q_D the contribution of a single molecule to the total PF of the system due to the external field. This is sometimes included in the rotational PF of a single molecule. Here, the rotational partition function is included in q. Thus we rewrite (2.5.1) as

$$Q(T, V, N, D) = \frac{q^N V^N}{N! \Lambda^{3N}} q_D^N,$$ (2.5.3)

where

$$q_D = \frac{1}{8\pi^2} \int \exp[-\beta U_D(\mathbf{X})] \, d\mathbf{\Omega},$$ (2.5.4)

where the integration in (2.5.4) is only on the orientations of a single molecule.

We now assume that the molecules interact with the external field in two ways. First, if it has a permanent dipole moment μ_0, its interaction with the field is

$$-(\mathbf{\mu_0} \cdot \mathbf{D}) = -\mu_0 D \cos \theta,$$ (2.5.5)

where θ is the angle between the direction of the dipole and the direction of the field $\mathbf{D}$ (see Fig. 2.8). In (2.5.5), μ_0 and D are the absolute magnitudes of the vectors $\mathbf{\mu_0}$ and $\mathbf{D}$. In writing (2.5.5) we choose the zero energy at the orientation $\theta = \pi/2$.

The second interaction is between an induced dipole and the external field. We assume for simplicity that the induced dipole μ_{ind} is in the direction of the field and is linear in D, i.e.,

$$\mu_{\text{ind}} = \alpha D,$$ (2.5.6)

where α is the polarizability of the molecule. (In general, the polarizability is a tensor, and the induced dipole moment is not necessarily in the direction of $\mathbf{D}$.)

Since the induced dipole moment is proportional to D, the interaction energy between the molecule and the field D is

$$-\int \alpha D \, dD = -\frac{\alpha D^2}{2}.$$ (2.5.7)

With (2.5.5) and (2.5.7), we have the total interaction between a single molecule and the field

$$U(\mathbf{\Omega}) = -\frac{\alpha D^2}{2} - \mu_0 D \cos \theta.$$ (2.5.8)

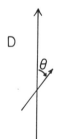

D

FIGURE 2.8. The direction of the dipole moment relative to the direction of the electric field.

Hence for q_D we have

$$q_D = \frac{1}{8\pi^2} \int \exp[-\beta U_D(\Omega)] \, d\Omega$$

$$= \frac{1}{8\pi^2} \int \exp[-\beta(-\tfrac{1}{2}\alpha D^2 - \mu_0 D \cos \theta)] \, d\Omega. \qquad (2.5.9)$$

Since the integrand depends on the angle θ only, we may choose the three angles $\Omega = \theta, \psi, \phi$ such that θ and ψ are the polar angles describing the orientation of the permanent dipole (D being directed along the z axis) and ϕ is the angle of rotation about the vector μ_0.

Thus

$$q_D = \frac{1}{8\pi^2} \int_0^\pi \exp\left[\frac{\beta\alpha D^2}{2} + \beta\mu_0 D \cos \theta\right] \sin \theta \, d\theta \int_0^{2\pi} d\psi \int_0^{2\pi} d\phi$$

$$= \frac{1}{2} \exp\left[\frac{\beta\alpha D^2}{2}\right] \int_0^\pi \exp[\beta\mu_0 D \cos \theta] \sin \theta \, d\theta$$

$$= q_{\text{ind}} q_\mu. \qquad (2.5.10)$$

The two factors q_{ind} and q_μ are the contributions due to the polarizability and the permanent dipole of the molecule, respectively. The integral defining q_μ may be computed as follows:

$$q_\mu = \tfrac{1}{2} \int_0^\pi \exp[\beta\mu_0 D \cos \theta] \sin \theta \, d\theta. \qquad (2.5.11)$$

Define $y = \beta\mu_0 D$ and $x = \cos \theta$; $dx = -\sin \theta \, d\theta$. Transforming variables, we obtain

$$q_\mu = \tfrac{1}{2} \int_1^{-1} \exp(yx)(-dx) = \tfrac{1}{2} \int_{-1}^1 \exp(yx) \, dx = \frac{e^y - e^{-y}}{2y} = \frac{\sinh y}{y}. \qquad (2.5.12)$$

We can now write the total partition function of the system as

$$Q(T, V, N, D) = \frac{q^N V^N}{\Lambda^{3N} N!} \left[\exp(\tfrac{1}{2}\beta\alpha D^2) \frac{\sinh y}{y}\right]^N, \qquad (2.5.13)$$

where $y = \beta\mu_0 D$.

Next we apply the total differential of the Helmholtz energy in our system

$$dA = -S \, dT - P \, dV - M \, dD + \mu \, dN. \qquad (2.5.14)$$

We first compute M, which is the total average dipole moment of the system.

$$M = -\frac{\partial A}{\partial D} = kT \frac{\partial \ln Q}{\partial D} = kTN \frac{\partial \ln q_D}{\partial D}$$

$$= N\alpha D + \frac{N \int_0^\pi \mu_0 \cos \theta \exp[y \cos \theta] \sin \theta \, d\theta}{\int_0^\pi \exp[y \cos \theta] \sin \theta \, d\theta}$$

$$= N(\alpha D + \mu_0 \langle \cos \theta \rangle). \qquad (2.5.15)$$

We see that M/N consist of two terms, the induced dipole moment of a single molecule (which here is always in the direction of **D**) and the average component of the permanent dipole in the direction of **D**. The second term can be written more explicitly by evaluating the integral in (2.5.15), i.e.,

$$\mu_0\langle\cos\theta\rangle = \mu_0 \frac{-\int_{-1}^{1} x\exp(yx)\,dx}{\int_{-1}^{1}\exp(yx)\,dx}$$

$$= \mu_0 \frac{\dfrac{d}{dy}\int_{-1}^{1}\exp(yx)\,dx}{\int_{-1}^{1}\exp(yx)\,dx}$$

$$= \mu_0 \frac{\dfrac{d}{dy}\left(\dfrac{\sinh y}{y}\right)}{\dfrac{\sinh y}{y}}$$

$$= \mu_0\left(\coth y - \frac{1}{y}\right) = \mu_0 L(y), \tag{2.5.16}$$

where the function $L(y)$ defined in (2.5.16) is called the Langevin function (Fig. 2.9). Note that for $y\to\infty$ (e.g., $D\to\infty$ or $T\to 0$) we have $L(y)\to 1$. This is the saturation limit where all the dipoles are oriented exactly in the direction of **D**. Ordinarily, the value of y is quite small, so that only the first-order term in the expansion of $L(y)$ in y is taken into consideration; thus

$$L(y) = L(0) + yL'(0) + \cdots = \tfrac{1}{3}y - \tfrac{1}{45}y^3 + \cdots. \tag{2.5.17}$$

Hence the total dipole moment of the system is, to the first approximation in y,

$$M = NaD + N\mu_0\frac{1}{3}\frac{\mu_0 D}{kT} = ND\left(\alpha + \frac{\mu_0^2}{3kT}\right). \tag{2.5.18}$$

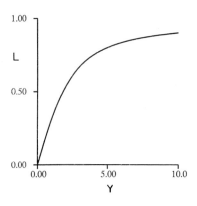

FIGURE 2.9. The Langevin function $L(y)$ defined in Eq. (2.5.16).

Using the general relation between the electric field strength $\mathscr{E}$, the electric induction D, and the polarization P (not to be confused with the pressure P),

$$D = \mathscr{E} + 4\pi P \tag{2.5.19}$$

and given that the definition of the dielectric constant ε is

$$D = \varepsilon \mathscr{E} \tag{2.5.20}$$

where the polarization P is simply the total dipole moment per unit volume, i.e.,

$$P = \frac{M}{V}, \tag{2.5.21}$$

we obtain

$$D = \frac{D}{\varepsilon} + \frac{4\pi M}{V} = \frac{D}{\varepsilon} + \frac{4\pi ND}{V}\left(\alpha + \frac{\mu_0^2}{3kT}\right). \tag{2.5.22}$$

Eliminating ε from (2.5.22) we obtain

$$\frac{\varepsilon - 1}{\varepsilon} = 4\pi\rho\left(\alpha + \frac{\mu_0^2}{3kT}\right). \tag{2.5.23}$$

For gases at very low density $\rho \to 0$, ε is very close to unity. Hence, in this limit we have

$$\frac{\varepsilon - 1}{4\pi\rho} = \alpha + \frac{\mu_0^2}{3kT}. \tag{2.5.24}$$

Thus plotting the experimental quantity $(\varepsilon - 1)/4\pi\rho$ as a function of $1/T$ should give the polarizability α from the intercept and the dipole moment μ_0 from the slope.

We now turn to compute some of the thermodynamic quantities of the gas. The chemical potential μ (not to be confused with μ_0) is given by

$$\mu = \frac{\partial A}{\partial N} = kT\ln(\rho\Lambda^3 q^{-1}q_D^{-1})$$

$$= \mu(D = 0) - kT\ln q_D, \tag{2.5.25}$$

where $\mu(D = 0)$ is the chemical potential of the gas at the same T and ρ in the absence of the electric field $(D = 0)$. The second term is obtained from (2.5.13) for small y,

$$-kT\ln q_D = \frac{-\alpha D^2}{2} - \frac{D^2\mu_0^2}{6kT} = \frac{-D^2}{2}\left(\alpha + \frac{\mu_0^2}{3kT}\right). \tag{2.5.26}$$

Clearly this quantity is always negative; the chemical potential of a gas is always reduced by turning on the electric field, i.e.,

$$\mu(D) < \mu(D = 0). \tag{2.5.27}$$

Another way of interpreting (2.5.27) is the following: Suppose we have two subsystems consisting of the same molecules, one subjected to an external field and the second not. At equilibrium we have the equality

$$\mu(D) = \mu(D = 0), \tag{2.5.28}$$

or

$$kT \ln \left[\frac{\rho(D)}{\rho(D = 0)} \right]_{eq} = kT \ln q_D > 0. \tag{2.5.29}$$

The last result means that at equilibrium, the molecules will be distributed between the two phases in favor of the phase at which the electric field is applied; i.e., the electric field attracts the particles.

The entropy of the system, in the limit $y \to 0$ is given by

$$S(D) = -\frac{\partial A}{\partial T} = S(D = 0) - \frac{Nk}{6} \left(\frac{\mu_0 D}{kT} \right)^2, \tag{2.5.30}$$

which states that the entropy decreases when the electric field is applied. Note that the difference $S(D) - S(D = 0)$ is independent of the polarizability and depends only on the average orientation of the permanent dipoles. The reduction in the entropy is due to partial ordering of the permanent dipoles in the direction of the field.

The energy of the gas, in the same limit $y \to 0$, is obtained from

$$E = A + TS$$

$$E(D) = E(0) - D^2 N \left(\frac{\alpha}{2} + \frac{1}{3} \frac{\mu_0^2}{kT} \right). \tag{2.5.31}$$

The additional term is negative and depends on the work required to create an induced dipole as well as that needed to orient the permanent dipoles.

2.6. IDEAL GAS IN A GRAVITATIONAL OR CENTRIFUGAL FIELD

We consider an ideal gas (i.e., no interactions among the particles) consisting of N molecules at a temperature T contained in a vertical column of height H and cross-section a (Fig. 2.10). The system is subjected to an external field of force (either gravitational or centrifugal) such that at each height z, the interaction between the particle of mass m and the external field is

$$U(z) = mgz \text{ (in a gravitational field)} \tag{2.6.1}$$

$$U(z) = -\frac{m}{2} w^2 z^2 \text{ (in a centrifugal field).} \tag{2.6.2}$$

Here, g is the gravitational acceleration, z is the height of the particle measured relative to the bottom of the column in the case of the gravitational field or the distance from the axis of rotation in the case of a centrifugal field, and w is the angular velocity of rotation of the centrifuge. In general, g depends on the distance from the center of the earth. However, since H is very small compared to the radius of the earth, one can assume

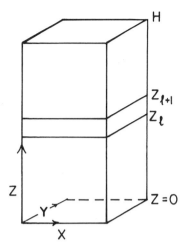

FIGURE 2.10. A column of height H in a gravitational field.

that g is almost constant within the system. Also we can neglect gravitational forces due to other particles in the system. In writing (2.6.1), we have set arbitrarily the zero of the potential energy at $z = 0$. Since in both cases (2.6.1) and (2.6.2) the external field operates on each particle separately, the treatment of the two cases is similar. We pursue only the first case.

The canonical partition function for the system in the gravitational field is

$$Q(T, V, N; g) = \frac{1}{N! \Lambda^{3N}} \int \exp\left[-\sum_{i=1}^{N} mgz_i \right] d\mathbf{R}^N, \qquad (2.6.3)$$

where we assumed that the particles are spherical and structureless (i.e., have no internal degrees of freedom). Since the integrand is a product of functions each of which depends only on the coordinates of one particle, the entire integral may be performed to obtain

$$Q(T, V, N; g) = \frac{a^N}{N! \Lambda^{3N}} \left[\int_0^H \exp(-\beta mgz) \, dz \right]^N$$

$$= \frac{a^N}{N! \Lambda^{3N}} \left[\frac{1 - \exp(-\beta mgH)}{\beta mg} \right]^N$$

$$= \frac{V^N}{N! \Lambda^{3N}} q_g^N, \qquad (2.6.4)$$

where we defined

$$q_g = \frac{1}{H} \int_0^H \exp(-\beta mgz) \, dz. \qquad (2.6.5)$$

The chemical potential of the gas in the system is defined by

$$\mu = \left(\frac{\partial A}{\partial N} \right)_{T,V,g} = kT \ln \rho \Lambda^3 q_g^{-1}, \qquad (2.6.6)$$

where $\rho = N/V$ is the bulk density of the entire system. Note that when $g \to 0$, $q_g = 1$, $\mu = kT \ln \rho \Lambda^3$, and we return to the chemical potential of an ideal gas in the absence of an external field. For $\beta mgH \ll 1$, the first-order correction to the chemical potential is

$$\mu = kT \ln \rho \Lambda^3 + \tfrac{1}{2}mgH.$$

The new phenomenon of interest in this system is the density distribution along the z axis. To obtain this distribution, we divide the length of the system into M narrow strips or "phases" in such a way that in each strip of volume $\Delta V = a\Delta z$, the intensive variables are almost constant. But Δz should be large enough so that macroscopic thermodynamics applies in the volume $a\Delta z$. With these assumptions we rewrite the integral in (2.6.4) as

$$\left[\int_0^H \exp(-\beta mgz)\, dz \right]^N = \left[\sum_{l=1}^M \int_{z_l}^{z_l + \Delta z} \exp(-\beta mgz)\, dz \right]^N$$

$$= \left(\sum_{l=1}^M q_l \right)^N, \qquad (2.6.7)$$

where

$$q_l = \int_{z_l}^{z_l + \Delta z} \exp(-\beta mgz)\, dz \approx \exp(-\beta mgz_l)\, \Delta z \qquad (2.6.8)$$

and where $z_1 = 0$ and $z_M + \Delta z = H$.

Expanding the product in (2.6.7) by the multinomial theorem,

$$\left(\sum_{l=1}^M q_l \right)^N = \sum_{N_1, N_2, \ldots, N_M} \frac{N!}{\prod N_l!} \prod_{l=1}^M q_l^{N_l}, \qquad (2.6.9)$$

which may be substituted in (2.6.4) to obtain

$$Q(T, V, N; g) = \sum_{\substack{N_1, N_2, \ldots, N_M \\ \sum N_l = N}} \prod_{l=1}^M \frac{(aq_l)^{N_l}}{N_l! \Lambda^{3N_l}}. \qquad (2.6.10)$$

This has the form of an equilibrium distribution of the N particles into M species, the species being distinguished by their different heights z_i in the system. Each term corresponds to a specific set of $N_1 \cdots N_M$, N_l being the number of particles in the lth strip and $\sum N_l = N$ the total number of particles in the system.

In order to define the chemical potential of the lth strip we assume that $\bar{N}_l$ is the average, or the equilibrium, number of particles in the lth strip. Hence we may treat the lth strip as a macroscopic system, for which the PF is

$$Q_l(T, a\Delta z, \bar{N}_l; g) = \frac{(aq_l)^{\bar{N}_l}}{\bar{N}_l! \Lambda^{3\bar{N}_l}} \qquad (2.6.11)$$

and the corresponding chemical potential

$$\mu_l = \frac{-kT \partial \ln Q_l}{\partial \bar{N}_l} = kT \ln \frac{\bar{N}_l \Lambda^3}{aq_l} = kT \ln \rho_l \Lambda^3 + mgz_l, \qquad (2.6.12)$$

where ρ_l is the equilibrium density at the height z_l, or in the lth strip. Since at equilibrium the chemical potential is constant throughout the entire system (otherwise matter will flow from the region of high to that of low chemical potential) we must have

$$\mu_1 = \mu_2 = \mu_3 = \cdots = \mu_M; \qquad (2.6.13)$$

from this we obtain the density distribution; i.e., from $\mu_1 = \mu_l$ we obtain

$$\left(\frac{\rho_l}{\rho_1}\right)_{eq} = \exp[-\beta mg(z_l - z_1)], \qquad (2.6.14)$$

where $z_1 = 0$ was chosen as zero. This is the well-known Boltzmann distribution of densities. The higher the strip, the less dense the gas is. Assuming that the ideal-gas law holds for each of the strips, we can write, for each l, the corresponding pressure

$$P_l = \rho_l kT. \qquad (2.6.15)$$

Hence

$$P_l = P_1 \exp(-\beta mgz_l), \qquad (2.6.16)$$

which gives the pressure as a function of the height z_l, P_1 being the pressure at the bottom of the system.

The probability of finding a particle at a height between z and $z + dz$ is proportional to the density of particles at that height, i.e.,

$$P(z)\, dz = a \exp(-\beta mgz)\, dz, \qquad (2.6.17)$$

where a is the normalization constant, obtained from the condition

$$\int_0^H P(z)\, dz = 1. \qquad (2.6.18)$$

Hence

$$a = \frac{\beta mg}{1 - \exp(-\beta mgH)}. \qquad (2.6.19)$$

Note that the bulk density defined in (2.6.6) is related to the densities ρ_l at each level by

$$\rho = \frac{N}{V} = \frac{\displaystyle\sum_{l=1}^{M} \bar{N}_l}{V} = \frac{a\Delta z}{V}\sum_l \rho_l = \frac{a\Delta z}{V}\sum_l \rho_1 \exp(-\beta mgz_l). \qquad (2.6.20)$$

From (2.6.4), (2.6.7), and (2.6.20) we also have

$$\rho q_g^{-1} = \frac{Ha\Delta z \sum_l \rho_1 \exp(-\beta mgz_l)}{V \sum_l \exp(-\beta mgz_l)\, \Delta z} = \rho_1. \qquad (2.6.21)$$

Substituting in (2.6.6), we obtain

$$\mu = kT \ln \rho \Lambda^3 q_g^{-1} = kT \ln \rho_1 \Lambda^3 = \mu_1, \qquad (2.6.22)$$

where μ is the chemical potential of the bulk gas and μ_1 is the chemical potential of the first ($l = 1$) strip. Because of relation (2.6.13), μ is also equal to any μ_l.

The Helmholtz energy of the entire system is

$$A = -kT \ln Q = -kT \ln \frac{V^N q_g^N}{N! \Lambda^{3N}} = A(g = 0) - kT \ln q_g, \tag{2.6.23}$$

where $A(g = 0)$ is the Helmholtz energy of the system in the absence of the gravitational field, $g = 0$.

For a very weak field, $\beta mgH \ll 1$, we have

$$A = A(g = 0) - \tfrac{1}{2} NmgH.$$

The entropy of the system is obtained from (2.6.23) as

$$S = -\frac{\partial A}{\partial T} = S(g = 0) + kN \ln q_g + \frac{Nmg\bar{z}}{T}, \tag{2.6.24}$$

where $\bar{z}$ is defined by

$$\bar{z} = \int_0^H P(z)z \, dz. \tag{2.6.25}$$

The energy of the system is

$$E = A + TS = E(g = 0) + Nmg\bar{z} = \tfrac{3}{2} NkT + Nmg\bar{z}. \tag{2.6.26}$$

The first term on the rhs of (2.6.26) is the average kinetic energy of the particles. The second term is the average potential energy of the particles in the gravitational field.

The heat capacity of the system is

$$
\begin{aligned}
C_V = \left(\frac{\partial E}{\partial T}\right)_{N,V} &= \tfrac{3}{2} Nk + Nmg \frac{\partial \bar{z}}{\partial T} \\
&= \tfrac{3}{2} Nk + \frac{N}{kT^2} (mg)^2 \overline{(z - \bar{z})^2} \\
&= C_V(g = 0) + \Delta C_V.
\end{aligned}
\tag{2.6.27}
$$

Since $\overline{(z - \bar{z})^2}$ is always positive, the heat capacity always increases upon turning on the gravitational field, i.e., $\Delta C_V > 0$. This is so because the field adds another degree of freedom that may absorb heat. If the total heat supplied to the system is ΔQ, some of it, say ΔQ_k, is spent to increase the average kinetic energy of the particles and another part ΔQ_p is spent in raising the average potential energy of the particles (or in increasing the average height $\bar{z}$). Thus

$$C_V = \frac{\Delta Q}{\Delta T} = \frac{\Delta Q_k + \Delta Q_p}{\alpha \Delta Q_k} = \frac{1}{\alpha} + \frac{\Delta Q_p}{\Delta Q_k}, \tag{2.6.28}$$

where the change in temperature ΔT is assumed to be proportional to ΔQ_k only. If $g = 0$, $\Delta Q_p = 0$, i.e., no heat is taken to increase $\bar{z}$, hence $C_V = 1/\alpha = \tfrac{3}{2} Nk$. However, for $g \neq 0$, $\Delta Q_p \neq 0$ and the second term on the rhs of (2.6.28) will always be positive due to the redistribution of particles along the z-axis in favor of higher $\bar{z}$. This argument is similar to the one we encountered in connection with the heat capacity of a system in chemical equilibrium (section 2.4).

2.7. NONINTERACTING MAGNETIC DIPOLES IN A MAGNETIC FIELD

This system consists of M lattice points at temperature T subjected to a uniform external magnetic field $\mathbf{H}$. At each lattice point we have a magnetic dipole, which can attain two orientations: one in the direction of the magnetic field, which we denote by $+$, and the other against the direction of the field, which we denote by $-$.

Let m_0 be the magnetic moment of each particle (nucleus or electron); the potential energies of interaction are $-m_0 H$ and $+m_0 H$ for the dipoles in the $+$ and $-$ orientations, respectively. No interaction between the dipoles are assumed.

We denote by N_+ and N_- the number of dipoles in the $+$ and $-$ orientations at any particular configuration of the system. The canonical partition function for such a system is

$$Q(T, M, H) = \sum_{N_+ = 0}^{M} \frac{M!}{N_+!N_-!} q_+^{N_+} q_-^{N_-}, \tag{2.7.1}$$

where

$$q_\pm = \exp(\pm \beta m H). \tag{2.7.2}$$

No other internal degrees of freedom are ascribed to the particles.

It is clear from the form of the PF in (2.7.1) that the system corresponds to a two-state equilibrium system (section 2.4). The partition function may be summed to obtain

$$Q(T, M, H) = (q_+ + q_-)^M = (e^{\beta mh} + e^{-\beta mH})^M = [2\cosh(\beta mH)]^M, \tag{2.7.3}$$

where $N_+ + N_- = M$.

We can obtain the average values of N_+ and N_- (or the most probable values) by exactly the same procedure as section 2.4. The result can be put in the form of an equilibrium constant, i.e.,

$$\frac{\bar{N}_+}{\bar{N}_-} = \frac{q_+}{q_-} = \exp(2\beta m_0 H) = K. \tag{2.7.4}$$

Clearly, if $H = 0$, then $\bar{N}_+ = \bar{N}_-$. If $H \to \infty$, then all dipoles are oriented in the $+$ direction; hence $\bar{N}_+ \to M$, $\bar{N}_- \to 0$. The converse hold for $H \to -\infty$.

We define the average magnetization of the system at equilibrium as the excess of the number of dipoles in the H direction times m_0, i.e.,

$$I = (\bar{N}_+ - \bar{N}_-)m_0, \tag{2.7.5}$$

which can be obtained from the equilibrium condition (2.7.4):

$$\bar{N}_+ = \frac{KM}{1 + K}, \qquad \bar{N}_- = \frac{M}{1 + K}, \tag{2.7.6}$$

hence

$$I = \frac{m(KM - M)}{1 + K} = mM \frac{q_+ - q_-}{q_+ + q_-} = m_0 M \tanh(\beta m_0 H). \tag{2.7.7}$$

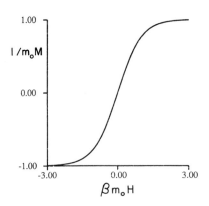

FIGURE 2.11. The magnetization as a function of the magnetic field, Eq. (2.7.9).

Alternatively we may use the fundamental differential of dA for this system:

$$dA = -S\,dT - I\,dH + \mu\,dM \tag{2.7.8}$$

to obtain I from the PF (2.7.3); i.e.,

$$I = -\left(\frac{\partial A}{\partial H}\right)_{T,M} = kT\left(\frac{\partial \ln Q}{\partial H}\right)_{T,M} = m_0 M \frac{\sinh(\beta m_0 H)}{\cosh(\beta m_0 H)}$$

$$= m_0 M \tanh(\beta m_0 H). \tag{2.7.9}$$

Figure 2.11 shows $I/m_0 M$ as a function of the parameter $\beta m_0 H$. Note the limiting behavior at $H \to 0$ and at $H \to \infty$.

2.8. SIMPLE ADSORPTION ISOTHERMS

In chemistry and biochemistry we encounter many examples of the adsorption of small molecules or ligands on a surface of a large molecule or on a solid. Many cases of adsorption phenomena have been studied: noninteracting or interacting ligands, adsorption from a gaseous or from a liquid phase, single or multiple occupancy, binding with or without perturbation to the adsorbent surface, and so on.

In this section we derive the theory of the simplest case, known as the Langmuir adsorption isotherm. The experimental observation is the following: We have an adsorbing system which could be a piece of a solid or a solution of polymer molecules. The surface of this material can adsorb ligands in equilibrium with a gaseous phase at a given temperature T and pressure P. If we measure the quantity of adsorbed ligand on the surface, say by weighing the solid, we find that the weight, or the quantity of ligand adsorbed on the solid has the following functional form

$$N = M\frac{KP}{1 + KP}. \tag{2.8.1}$$

N is the number of ligand molecules (or moles) adsorbed and P is the partial pressure

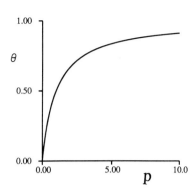

FIGURE 2.12. A simple Langmuir isotherm.

of the gas at equilibrium with the adsorbed molecules. M and K are constants. Figure 2.12 presents a typical Langmuir isotherm where $\theta = N/M$ is plotted as a function of P.

Our goal is to find a molecular model for the adsorption process that reproduces the same functional dependence of N on P as in (2.8.1) and thereby to provide a molecular interpretation of the quantities M and K in this equation. Later, we shall generalize this simple isotherm in various directions.

2.8.1. The Molecular Model and Its Solution

Consider first an adsorbing crystal, the surface of which provides M adsorbing sites for molecules of type A. We make the following assumptions:

1. Each site can adsorb only one A molecule.
2. The interaction between the site and the adsorbed molecule is characterized by a single parameter U. This is defined as the energy change for bringing A from a fixed position at infinite separation from the crystal to a specific site on the surface of the crystal.
3. The sites are identical in the sense that they are all characterized by the same parameter, U.
4. Neither the internal degrees of freedom of the crystal nor those of A are affected by the adsorption process.
5. The sites are distinguishable; i.e., different configurations of the molecules on the M sites are different quantum-mechanical states.
6. The adsorbed molecules are indistinguishable, i.e., exchanging any two A molecules on two sites gives the same state of the system.
7. The sites are independent. In general this assumption requires that the adsorption of A on one site does not affect the probability of adsorption on any other site. Dependence between sites can occur either because of interactions between adsorbed molecules on different (usually nearest-neighbors) sites or by indirect interaction transmitted by the sites. Since we have already made assumption 4, it is sufficient to state here that there are no direct interactions between adsorbed molecules on different sites. Note that assumption 1 is equivalent to infinite repulsive interaction between two ligands brought to the same site.
8. The adsorbed molecules A are in equilibrium with an ideal gas phase containing A molecules.

This long list of assumptions defines the simplest adsorption model. In later sections we shall discuss more complex cases where each one of the above assumptions will be removed. For instance, multiple occupancy is discussed in section 2.9. The case with different types of sites is discussed in section 2.8.7, indistinguishable adsorbing sites in section 2.9.2, dependence between sites in Chapter 3, and equilibrium with a liquid phase in Chapter 8.

We shall see that many quite complex phenomena can be viewed as generalizations of this particular model—each generalization consists of relaxing one or more of the assumptions made above. In this section we shall elaborate in great detail on some fundamental concepts, far more than is necessary for solving the particular model treated here. However, acquiring familiarity with these concepts will be useful for the study of the more complicated cases treated in subsequent sections and chapters.

With the above description of the model we now turn to construct the partition function of our system. The system consists of a crystal having M sites on its surface and N ligands occupying these sites ($N \leq M$). Since the crystal is presumed to be unaffected by the adsorption process, we may write the PF of the entire system (crystal and adsorbents) as

$$Q_{\text{system}} = Q_{\text{cryst}} Q_{\text{ad}}, \tag{2.8.2}$$

where Q_{cryst} is the PF of the empty crystal and Q_{ad} includes both the properties of the ligands and the interaction energy between the ligands and the crystal.

From now on we ignore the factor Q_{cryst}. This is the same as treating the ratio $Q_{\text{system}}/Q_{\text{cryst}}$ or the excess Helmholtz energy $A_{\text{system}} - A_{\text{cryst}}$. Furthermore, the molecules adsorbed on different sites are presumed to be independent. We shall therefore write

$$Q_{\text{ad}} = q_A^N \sum_E g(E) \exp(-\beta E), \tag{2.8.3}$$

where q_A is the internal PF of a single A molecule and the sum is over all energy levels associated with different configurations of the A molecules on the surface of the crystal.

In this particularly simple model, there is only one energy level in the sum over E in (2.8.3). Once we have fixed the number of adsorbed molecules N, the energy level is NU, the total interaction energy between all the ligands and all the sites. The only quantity that we need to compute is the configurational degeneracy $g(E)$, i.e., the number of distinguishable quantum-mechanical states that correspond to the single energy level $E = NU$. This number is simply the number of ways that N indistinguishable molecules can be distributed among M distinguishable sites, i.e.,

$$g(E) = \frac{M!}{N!(M-N)!} = \binom{M}{N}. \tag{2.8.4}$$

This number is obtained as follows: There are M different sites to place the first particle, $M-1$ different sites for the second particle, $M-(N-1)$ different sites for the Nth particle. Therefore, the total number of ways of placing N particles on M sites is

$$M(M-1)(M-2) \cdots (M-N+1) = \frac{M!}{(M-N)!}. \tag{2.8.5}$$

In this calculation we have also counted configurations in which adsorbed molecules are permuted on different sites; these are considered indistinguishable configurations. Therefore we have to divide by $N!$ to obtain $g(E)$, the number of distinguishable configurations.

The PF of the adsorbed molecules is thus

$$Q_{ad}(T, M, N) = \frac{q_A^N M!}{N!(M-N)!} \exp[-\beta N U]. \tag{2.8.6}$$

This is an explicit function of T, M, N and the molecular parameters, U and those included in q_A. By writing the PF we have solved the model, in the sense that all relevant thermodynamic quantities can be computed from the partition function.

2.8.2. Thermodynamics and the Langmuir Isotherm

The Helmholtz energy is

$$A_{ad} = -kT \ln Q_{ad} = -kTN \ln q_A + NU - kTM \ln M$$
$$+ kTN \ln N + kT(M-N)\ln(M-N). \tag{2.8.7}$$

The Helmholtz energy per site is

$$a_{ad} = \frac{A_{ad}}{M} = -kT\theta \ln q_A + \theta U + kT[\theta \ln \theta + (1-\theta)\ln(1-\theta)], \tag{2.8.8}$$

where $\theta = N/M$ is the fraction of occupied sites, $(1-\theta)$ being the fraction of empty sites.

The entropy per site is

$$s_{ad} = \frac{S_{ad}}{M} = \frac{-1}{M}\left(\frac{\partial A_{ad}}{\partial T}\right)_{N,M} = \frac{\partial}{\partial T}[kT\theta \ln q_A] - k[\theta \ln \theta + (1-\theta)\ln(1-\theta)]. \tag{2.8.9}$$

Note that besides the entropy associated with the internal degrees of freedom, we have the so-called configuration entropy.

It is easy to see (see section 2.8.4) that θ is the probability of finding a specific site occupied and $(1-\theta)$ the probability of finding it empty, i.e.,

$$\Pr(\text{empty}) = (1-\theta), \qquad \Pr(\text{occupied}) = \theta. \tag{2.8.10}$$

Hence the configurational entropy has the general form

$$s_{conf} = -k \sum_i P_i \ln P_i, \tag{2.8.11}$$

where the sum is over all possible states of a single site—i.e., occupied and empty.

The chemical potential of the adsorbed molecules is

$$\mu_A^{ad} = \left(\frac{\partial A}{\partial N}\right)_{T,M} = -kT \ln q_A + U + kT \ln\left[\frac{\theta}{1-\theta}\right]. \tag{2.8.12}$$

It is of interest to examine the two limiting cases: For $\theta \ll 1$, we have

$$\mu_A^{ad} = \mu_A^{0,ad} + kT \ln \theta. \tag{2.8.13}$$

This has the general form of a chemical potential in an ideal dilute solution, $\mu_A^{ad} \to -\infty$ when $\theta \to 0$. The other extreme case is when $\theta \to 1$; then $\mu_A^{ad} \to +\infty$. The form

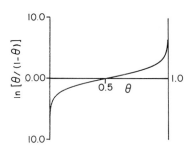

FIGURE 2.13. The function $\ln[\theta/(1 - \theta)]$ appearing in Eq. (2.8.12).

of the function $\ln[\theta/(1 - \theta)]$ is depicted in Fig. 2.13. At the limit $\theta \to 0$ there is an infinite driving force to attract A molecules from some reference state. On the other hand, at the limit $\theta \to 1$, the sites are all occupied and there is an infinite repulsive force or resistance for accepting further molecules on the surface.

Next we assume that the adsorbed molecules are at equilibrium with an ideal gas consisting of A molecules. The chemical potential of the molecules in the gaseous phase is

$$\mu_A^g = kT \ln \rho_A \Lambda_A^3 q_A^{-1}. \tag{2.8.14}$$

Note that, following the assumption of the model, q_A in (2.8.14) is the same as q_A in (2.8.12); i.e., the internal degrees of freedom of A do not change in the process of adsorption. In some more realistic models we must include in q_A some new degrees of freedom, such as vibrations of A on the site, that do not occur in the gaseous phase.

At equilibrium we have

$$\mu_A^{\text{ad}} = kT \ln \left(q_A^{-1} \frac{\theta}{1 - \theta} \right) + U = \mu_A^g = kT \ln \rho_A \Lambda_A^3 q_A^{-1}, \tag{2.8.15}$$

which can be solved for θ to obtain

$$\theta = \frac{K' \rho_A}{1 + K' \rho_A}, \qquad K' = \Lambda_A^3 \exp(-\beta U). \tag{2.8.16}$$

This is essentially the adsorption isotherm. We may rewrite it in terms of pressure instead of density by using the ideal-gas equation of state $P = kT\rho_A$, i.e.,

$$N = M \frac{KP}{1 + KP}, \qquad \text{where } K = \frac{K'}{kT}. \tag{2.8.17}$$

We recall our starting empirical relation (2.8.1). What we have obtained is a molecular model that reproduces the same functional dependence of N on P. Here M is identified with the number of sites, and we have an explicit expression for the constant K in terms of the molecular parameters of the system; these are the mass of each of the molecules, contained in Λ_A^3, and the adsorption energy U.

2.8.3. Rederivation of the Langmuir Isotherm Using the Grand Partition Function

We now apply the grand PF for the same model. Since we have already solved the model, the present treatment simply demonstrates an alternative, but equivalent, method of obtaining the solution of the same problem. The application of the grand PF is, however, essential for the treatment of more complicated models, such as the one discussed in section 2.8.7, or most of the models of Chapter 3. Therefore familiarity with the grand PF is useful even when applied to the simplest Langmuir model.

The grand PF is defined for our system by

$$\Xi(T, M, \mu) = \sum_{N=0}^{M} \lambda_A^N Q(T, M, N), \tag{2.8.18}$$

where $\lambda_A = \exp(\beta\mu)$ is the absolute activity of A; i.e., we open the system with respect to A. This can be done, for instance, by letting the gaseous phase be at equilibrium with a reservoir of A molecules at a constant chemical potential μ. In (2.8.18) we sum over all possible values of N. Each term is proportional to the probability of finding precisely N molecules on the adsorbing surface.

Introducing (2.8.6) into (2.8.18), we can sum over N to obtain

$$\Xi(T, M, \mu) = \sum_{N=0}^{M} \lambda_A^N q_A^N \frac{M!}{N!(M-N)!} \exp(-\beta NU) = (1 + Y_A\lambda_A)^M = \xi^M, \tag{2.8.19}$$

where

$$Y_A = q_A \exp(-\beta U) \tag{2.8.20}$$

and

$$\xi = 1 + Y_A\lambda_A. \tag{2.8.21}$$

ξ is referred to as the grand PF of one site. It has the same formal structure as the grand PF of the entire system, namely, the sum over all possible states of a single site, empty and occupied. The most important feature of Ξ is its factorization into M equal factors. This feature follows from the assumption of the independence of the sites.

Using the thermodynamic relation (1.3.43) we obtain the average number of adsorbed molecules:

$$\bar{N} = kT \frac{\partial \ln \Xi}{\partial \mu} = \lambda_A \frac{\partial \ln \Xi}{\partial \lambda_A} = \lambda_A M \frac{\partial \ln \xi}{\partial \lambda_A}, \tag{2.8.22}$$

or

$$\theta = \frac{\bar{N}}{M} = \frac{Y_A\lambda_A}{1 + Y_A\lambda_A}, \tag{2.8.23}$$

which is essentially the Langmuir isotherm. However, in contrast to Eqs. (2.8.16) or (2.8.17), this relation is more general, since it applies to any state of the A molecules at a given chemical potential μ (or, equivalently, λ_A). No assumption of ideality of the gas at equilibrium with the surface has been used.

If A is an ideal gas, then

$$\mu_A^g = kT \ln \rho_A \Lambda_A^3 q_A^{-1}, \tag{2.8.24}$$

or

$$\lambda_A = \rho_A \Lambda_A^3 q_A^{-1}. \tag{2.8.25}$$

Hence

$$Y_A \lambda_A = \rho_A \Lambda_A^3 \exp(-\beta U) = K' \rho_A, \tag{2.8.26}$$

and (2.8.23) reduces to (2.8.16). Note that in (2.8.23) we have the average number $\bar{N}$, whereas in (2.8.16) we had a fixed value of N. It is also useful to examine the average quantity $\bar{N}$ in relation to ξ. It follows from the general expression for $\bar{N}$ (section 1.3.4) that

$$\bar{N} = kT \frac{\partial \ln \Xi}{\partial \mu} = \sum_{N=0}^{M} N \Pr(N), \tag{2.8.27}$$

where $\Pr(N)$ is the probability of finding N particles on the surface. Likewise, $\bar{N}/M$ may be interpreted as the average number of ligands per *single* site, which is the same as the probability of finding a specific site occupied (see also section 2.8.4). From (2.8.23), we have

$$\theta = \frac{\bar{N}}{M} = \frac{Y_A \lambda_A}{\xi} = 0 \frac{1}{\xi} + 1 \frac{Y_A \lambda_A}{\xi} = 0 \times p_r(0) + 1 \times p_r(1); \tag{2.8.28}$$

thus the average occupation number of a single site is written in the same form as (2.8.27), but for a single site instead of the whole system. The probabilities of finding a specific site empty or occupied are read from the grand PF of a single site (2.8.21), i.e.,

$$p_r(0) = \frac{1}{\xi} \tag{2.8.29}$$

$$p_r(1) = Y_A \frac{\lambda_A}{\xi} = \frac{\bar{N}}{M}. \tag{2.8.30}$$

We shall further elaborate on these probabilities in the next subsection. In this section we made the distinction between the probabilities Pr and p_r pertaining to the entire system and to a single site, respectively. In the next section we shall make no such distinction. We shall always use Pr for probability. The system to which it pertains should be clear from the context.

2.8.4. Occupation Probabilities and Free Energy of "Cavity" Formation

In this section we further elaborate on the definitions of the probabilities $\Pr(0)$ and $\Pr(1)$ and their relation to the thermodynamics of adsorption. Although the relation $\Pr(1) = \bar{N}/M = \theta$ is intuitively clear, we rederive it using a detailed argument. In this section we return to the canonical ensemble; i.e., N is fixed. The quantity $\theta = N/M$ is the mole fraction of the occupied sites (in the open system the corresponding quantity $\bar{N}/M$ is the average mole fraction of the occupied sites). Clearly, this quantity pertains to the entire system. In this system, the probability of finding *any* site occupied is unity. If $N \neq 0$, there is always one site occupied. However, if we select a *specific site*, say the ith site, we can ask what the probability is of finding this specific site occupied? In our particular model, the answer can be given, using the so-called classical definition of probability; the probability Pr (ith site occupied) is equal to the total number of configurations that are consistent with the event "ith site occupied" divided by the total number of configurations of the system, i.e.,

$$\Pr(i\text{th site occupied}) = \frac{\binom{M-1}{N-1}}{\binom{M}{N}} = \frac{N}{M} = \theta. \qquad (2.8.31)$$

The total number of configurations consistent with the requirement that the ith site be occupied is calculated by placing one ligand on the ith site and then counting all the possible configurations of the remaining $N-1$ particles on the $M-1$ sites. The argument used here is the same as the one used in calculating the probability of finding an even result by tossing a die, which is

$$\Pr(\text{even}) = \tfrac{3}{6} = \tfrac{1}{2}. \qquad (2.8.32)$$

The implicit assumptions made in these calculations are that each of the possible outcomes is an elementary event and that all elementary events have equal probabilities: i.e., each possible outcome in tossing a die has probability $\tfrac{1}{6}$, likewise each specific configuration of the system has probability of $\binom{M}{N}^{-1}$. These implicit assumptions are valid for systems for which all the elementary events have equal probabilities. In our system, each of the $\binom{M}{N}$ configurations has the same probability, and therefore the method used in (2.8.31) is applicable.

This method becomes invalid when there are interactions among the ligands, in which case different configurations do not have the same probability. Nevertheless, the result (2.8.31) is still valid even when there are interactions. To prove that, we need a different argument. This is presented below. A similar argument applies in the theory of homogeneous fluids, to which we return in Chapter 5.

Suppose that we have labeled all the N ligands. Let $\Pr^k(i\text{th})$ be the probability of finding a specific ligand say, the kth, on a specific site, the ith site. Since the kth ligand must be found somewhere on one of the M sites, we must have the normalization condition

$$\sum_{i=1}^{M} \Pr^k(i\text{th}) = 1. \qquad (2.8.33)$$

Since the M sites are equivalent, the probabilities $\Pr^k(i\text{th})$ are independent of the index i. Therefore the sum over i in (2.8.33) produces M equal terms.

$$\sum_{i=1}^{M} \Pr^k(i\text{th}) = M\Pr^k(i\text{th}) = 1. \qquad (2.8.34)$$

From (2.8.34) we obtain

$$\Pr^k(i\text{th}) = \frac{1}{M}. \qquad (2.8.35)$$

Note that although this quantity is independent of the specific site, we still leave the index i, to stress that we are considering a *specific* site, the ith site.

Since the ligands are also equivalent, the probabilities $\text{P}_\text{r}^k(i\text{th})$ are also independent of the index k. In addition, the N events "particle k occupies a specific site" are disjoint events $(k = 1, \ldots, N)$; therefore, the probability of finding *any* of the N ligands on the specific site is the sum over k, i.e.,

$$\text{Pr}(1) = \text{Pr}(\text{any ligand on } i) = \sum_{k=1}^{N} \text{Pr}^k(i\text{th}) = \frac{N}{M}, \qquad (2.8.36)$$

which is the same result as (2.8.31). We have returned to the notation $\text{Pr}(1)$ as in (2.8.30), eliminating the index i. It should be remembered, however, that we always refer to a specific site. To conclude, the probability of finding *any* site empty or occupied is unity (if $0 < N < M$). The probabilities of finding a specific site empty or occupied are $(1 - \theta)$ and θ, respectively. These quantities are also related to the free energy change for the following process. Consider again the same system of N ligands on M sites. The Helmholtz energy of this system is (2.8.7)

$$A_{\text{ad}} = -kT \ln Q_{\text{ad}}. \qquad (2.8.37)$$

Next we form "a cavity at i". By this we mean that we constrain the system in such a way that the ith site must always remain empty. We denote by $A_{\text{ad}}(\text{cav})$ and $Q_{\text{ad}}(\text{cav})$ the Helmholtz energy and the PF of a system of N ligands on M sites with the condition that the ith site is empty. The corresponding relation is

$$A_{\text{ad}}(\text{cav}) = -kT \ln Q_{\text{ad}}(\text{cav})$$

$$= -kT \ln \left[\frac{q_A^N (M - 1)!}{N!(M - N - 1)!} \exp(-\beta NU) \right]. \qquad (2.8.38)$$

Note that the energy and the number of ligands in (2.8.38) are the same as in (2.8.6). The only difference is that the N ligands are now distributed over $M - 1$ sites. Therefore the change in the Helmholtz energy for the process of forming a "cavity" at a specific site is

$$\Delta A(\text{cav at } i) = A_{\text{ad}}(\text{cav}) - A_{\text{ad}} = -kT \ln \frac{(M - 1)!(M - N)!}{(M - N - 1)!M!} = -kT \ln(1 - \theta). \qquad (2.8.39)$$

This is an important relation. The Helmholtz energy change for creating "a cavity at i" is related to the probability of finding a specific site empty. We shall see the analogue of this quantity in Chapter 5 when we discuss cavity formation in liquids. We note that in this particular model the Helmholtz energy change is purely an entropy effect, i.e.,

$$\Delta S(\text{cav at } i) = -\frac{\partial \Delta A(\text{cav at } i)}{\partial T} = k \ln(1 - \theta) \qquad (2.8.40)$$

and

$$\Delta E(\text{cav at } i) = 0. \qquad (2.8.41)$$

There are always fewer possibilities of distributing N particles on $M - 1$ sites than on M sites. This is the reason for the negative entropy change (2.8.40). The corresponding energy change is obviously zero.

   If there are interactions among the ligands, the method of calculation of $\Delta A(\text{cav at } i)$ as in (2.8.38) and (2.8.39) does not apply. In this case we do not have an explicit expression for either Q_{ad} or $Q_{\text{ad}}(\text{cav})$. Nevertheless, the result $\Delta A(\text{cav at } i) = -kT \ln(1 - \theta)$ obtained in (2.8.39) is still valid. The argument leading to this result, when interactions exist, is different from the one given above. In the most general case, the PFs of two systems with and without a cavity are defined by

$$Q_{ad}(cav) = \sum_{cav} \exp(-\beta E_j) \tag{2.8.42}$$

$$Q_{ad} = \sum_{all\ j} \exp(-\beta E_j). \tag{2.8.43}$$

In (2.8.42), the sum is over all states that are consistent with the presence of a cavity in the system. In (2.8.43), the sum is over all possible states of the system. Since the probability of finding any particular state j is $\exp(-\beta E_j)/Q_{ad}$, the probability of finding a "cavity" is the sum over the probabilities of all the (disjoint) events consistent with the requirement that such a cavity exists,

$$Pr(cav) = \sum_{cav} P_j = \frac{Q_{ad}(cav)}{Q_{ad}}. \tag{2.8.44}$$

This is a very general result, valid for a system with interactions, and also for any size of a cavity. In particular, if the size of the cavity is equal to one site only, say the ith site, then we already have relation (2.8.36), from which we infer that the probability of finding a site i empty is

$$Pr(0) = Pr(cav\ at\ i) = 1 - \frac{N}{M} = 1 - \theta. \tag{2.8.45}$$

Combining (2.8.44) and (2.8.45), we obtain

$$\Delta A(cav\ at\ i) = -kT \ln(1 - \theta), \tag{2.8.46}$$

which is the same as (2.3.39). We stress again that this relation is valid for a cavity having the size of one site only. If there are no interactions, it is easy to calculate the work required to create a cavity of any size. For instance, a cavity of two sites is discussed below. The situation is far more complicated in the case of interacting ligands. We shall see the analogy of these cases in the theory of fluids in Chapter 5.

Returning to our model, let i and j be two specific sites. These could be either adjacent or far apart. The probability of finding these two sites occupied is denoted by $Pr(1, 1)$. This can be easily computed as follows

$$Pr(1, 1) = \frac{\binom{M-2}{N-2}}{\binom{M}{N}} = \frac{N(N-1)}{M(M-1)} = \theta \frac{N-1}{M-1}. \tag{2.8.47}$$

We place two particles at the specific sites, and distribute the $N - 2$ remaining particles over the remaining $M - 2$ sites.

Similarly, the probability of finding the two sites empty—i.e., a cavity of size two, is

$$Pr(0, 0) = \frac{\binom{M-2}{N}}{\binom{M}{N}} = \frac{(M-N)(M-N-1)}{M(M-1)} = (1-\theta)\left(1 - \frac{N}{M-1}\right), \quad (2.8.48)$$

where we distribute the N particles over the $M - 2$ available sites (i.e., excluding the ith and the jth sites that are left empty).

Recall that θ and $(1 - \theta)$ are the probabilities of finding a specific site occupied and empty, respectively. From (2.8.47) and (2.8.48), we see that even when there are no interactions the probability $Pr(1, 1)$ is not exactly equal to the product of the probabilities of finding each of the sites occupied.

It is only in the limit of the macroscopic system, i.e., when $M \to \infty$ and $N \to \infty$, but N/M is constant, that we obtain complete independence of the two sites, i.e.,

$$Pr(1, 1) = \theta^2 = (Pr(1))^2 \quad (2.8.49)$$

$$Pr(0, 0) = (1 - \theta)^2 = (Pr(0))^2 \quad (2.8.50)$$

$$Pr(1, 0) = \theta(1 - \theta) = Pr(0)Pr(1). \quad (2.8.51)$$

The difference between, say, (2.8.47) and (2.8.49) arises from the finite size of the system.

We can introduce various conditional probabilities; for instance, the probability of finding site j occupied given that site i is occupied, is

$$Pr(1/1) = \frac{Pr(1, 1)}{Pr(1)} = \frac{N-1}{M-1} \to \theta. \quad (2.8.52)$$

Clearly, if we place a particle at i, the probability of finding site j occupied is not exactly equal to θ, but to the new mole fraction of $(N - 1)/(M - 1)$, which, in the limit of macroscopic system, approaches θ.

In the next chapter, we shall discuss pair distributions of the type $Pr(1, 1)$, $Pr(0, 0)$, or $Pr(1, 0)$ for which no factorization of the type (2.8.49) to (2.8.51) is possible. There, the dependence between the sites is due to (direct or indirect) interactions between the sites. We shall always assume that the system is macroscopically large. Here we note that deviations from independence due to the finite size of the system could be quite large. For instance, with $M = 2$ and $N = 2$, we have

$$Pr(1) = \tfrac{2}{4} = \tfrac{1}{2} \quad (2.8.53)$$

but

$$Pr(1/1) = \tfrac{1}{3}. \quad (2.8.54)$$

2.8.5. The Analogue of the Pseudochemical Potential

In section 2.2 we introduced the concept of the pseudochemical potential of an ideal gas. A similar concept is defined here for the adsorbed molecules. The pseudochemical potential is defined as the change in the Helmholtz energy for the process of introducing a ligand at a *fixed* site, say the ith.

$$\mu_A^*(\text{at } j) = -kT \ln \frac{Q(T, M, N+1, \text{at } j)}{Q(T, M, N)}$$

$$= -kT \ln \left[\frac{(M-1)!(M-N)!}{(M-N-1)!M!} q_A \exp(-\beta U) \right]$$

$$= -kT \ln q_A + U + kT \ln(1 - \theta). \tag{2.8.55}$$

Combining this with (2.8.12), we write the chemical potential of the adsorbed molecules

$$\mu_A^{ad} = \mu_A^*(\text{at } j) + kT \ln \theta. \tag{2.8.56}$$

This is the analogue of Eq. (2.2.4). Here the addition of one molecule to the system is split into two steps. First, we placed the molecule at a specific site, say j. The corresponding work is μ_A^*, given in (2.8.55). Then we liberate the particle, in the sense that the constraint on the fixed site is released. The corresponding change in the Helmholtz energy is $kT \ln \theta = kT \ln(N/M)$. In contrast to the case of an ideal gas (Eq. 2.2.4) where we recovered the momentum PF, here there are no translational degrees of freedom. The analogue of $-kT \ln V$ in (2.2.4) is $-kT \ln M$ and results from the accessibility of all the M sites. As in Eq. (2.2.4), the term $kT \ln N$ arises from the assimilation of the released particles among the N indistinguishable particles. The term $kT \ln \theta$ is therefore the liberation Helmholtz energy of the adsorbed molecules.

Some related quantities that will be found useful in the analysis of the more complicated systems treated in Chapter 3 are the following: The Helmholtz energy change for the process of bringing a molecule from a fixed position, say $\mathbf{R}_0$, in an ideal-gas phase, to a fixed site, say j, is

$$\Delta A^*(\mathbf{R}_0 \to \text{site } j) = \mu_A^*(\text{at } j) - \mu_A^* = U - kT \ln(1 - \theta). \tag{2.8.57}$$

The internal PF, q_A, does not appear in (2.8.57).

The Helmholtz energy change for the same process as above but *given* that the jth site is empty is

$$\Delta A^*(\mathbf{R}_0 \to \varnothing \text{ site } j) = -kT \ln \left\{ \frac{Q(T, M, N+1, \text{at } j)}{Q(T, M, N, \varnothing \text{ site } j)} \right\} = U. \tag{2.8.58}$$

The symbol $\varnothing$ is used for empty. Similarly, if the jth site is already occupied, we have $\Delta A^*(\mathbf{R}_0 \to \text{occupied site } j) = \infty$, which results from the infinite repulsion felt if we attempt to place another particle on an occupied site.

The last three quantities may be combined, using the probabilities $\Pr(0) = 1 - \theta$ and $\Pr(1) = \theta$, to obtain

$$\exp[-\beta \Delta A^*(\mathbf{R}_0 \to \text{site } j)]$$

$$= \Pr(0) \exp[-\beta \Delta A^*(\mathbf{R}_0^* \to \varnothing \text{ site } j)]$$

$$+ \Pr(1) \exp[-\beta \Delta A^*(\boldsymbol{\rho}_0 \to \text{occupied site } j)]$$

$$= (1 - \theta) \exp[-\beta \Delta A^*(\mathbf{R}_0 \to \varnothing \text{ site } j)] + \theta \times 0. \tag{2.8.59}$$

This is simply another way of rewriting Eq. (2.8.57) as an average quantity. We shall encounter many other averages of similar form throughout the book. The last form on the rhs of (2.8.59) is the analogue of the solvation Helmholtz energy of a hard sphere, or the work required to create a small cavity in a liquid, section 6.13.

2.8.6. Mixture of Ligands

We generalize the simple model to the case that each site may adsorb either a molecule of type A or a molecule of type B, at equilibrium with a reservoir at constant μ_A and μ_B, respectively. This model is relevant to systems where two ligands compete for the same site. Since the sites are still independent, we have the general result for the grand PF which is a straightforward generalization of Eq. (2.8.19).

$$\Xi(T, M, \mu_A, \mu_B) = \xi^M, \tag{2.8.60}$$

where ξ is the grand PF of a single site. For the case of a mixture of A and B, we have

$$\xi = q(0) + q(A)\lambda_A + q(B)\lambda_B, \tag{2.8.61}$$

where the three terms correspond to the three states of the sites: empty, occupied by A, and occupied by B, and where the PF of a single site in the states i is $q(i)$. The average numbers of A and B molecules are

$$\bar{N}_A = \lambda_A M \frac{\partial \ln \xi}{\partial \lambda_A} = M \frac{q(A)\lambda_A}{1 + q(A)\lambda_A + q(B)\lambda_B} \tag{2.8.62}$$

$$\bar{N}_B = \lambda_B M \frac{\partial \ln \xi}{\partial \lambda_B} = M \frac{q(B)\lambda_B}{1 + q(A)\lambda_A + q(B)\lambda_B}, \tag{2.8.63}$$

where we put $q(0) = 1$.

Note that in (2.8.61) we have

$$q(A) = q_A \exp(-\beta U_A) \tag{2.8.64}$$

$$q(B) = q_B \exp(-\beta U_B), \tag{2.8.65}$$

where U_A and U_B are the binding energies of A and B to the sites. If either λ_A or λ_B is zero, then this case reduces to the simple Langmuir model. Generalization to any number of ligands is straightforward.

2.8.7. Two Kinds of Site

Another simple generalization of the simple model occurs when we have two kinds of site, say M_α sites of type α and M_β sites of type β, but only one type of ligand molecules, A. The generalization of (2.8.19) in this case is

$$\Xi(T, M_\alpha, M_\beta, \mu_A) = \xi_\alpha^{M_\alpha} \xi_\beta^{M_\beta}, \tag{2.8.66}$$

where

$$\xi_\alpha = q_\alpha(0) + q_\alpha(1)\lambda_A \tag{2.8.67}$$

$$q_\alpha(1) = q_\alpha(0)q_A \exp(-\beta U_\alpha), \tag{2.8.68}$$

where $q_\alpha(1)$ is the PF of a site of type α occupied by an A molecule with binding energy U_α and $q_\alpha(0)$ is the PF of an empty site of type α. Similar expressions apply for β.

The average number of adsorbed molecules in this case is

$$\bar{N}_A = \lambda_A \frac{\partial \ln \Xi}{\partial \lambda_A} = \lambda_A M_\alpha \frac{\partial \ln \xi_\alpha}{\partial \lambda_A} + \lambda_A M_\beta \frac{\partial \ln \xi_\beta}{\partial \lambda_A}$$

$$= M_\alpha \frac{q_\alpha(1)\lambda_A}{1 + q_\alpha(1)\lambda_A} + M_\beta \frac{q_\beta(1)\lambda_A}{1 + q_\beta(1)\lambda_A}$$

$$= \bar{N}_\alpha + \bar{N}_\beta. \tag{2.8.69}$$

In this case, $\bar{N}_A$ is the sum of the average number of molecules adsorbed on sites of type α and on sites of type β. If either M_α or M_β is zero, this case reduces to the simple Langmuir model.

The above models can easily be generalized to the cases of multicomponent ligands or any number of kinds of sites.

2.9. MULTIPLE OCCUPANCY OF THE SITES

We generalize here the simple model of section 2.8.1 with respect to condition (1). Instead of single occupancy of each site we allow multiple occupancy. The sites are still independent in the sense that what happens on one site does not affect any other site. Until now we have assumed that there are no interactions (direct or indirect) between adsorbed molecules. Here, the lack of interaction still applies between any pair of sites or any pair of molecules on different sites. However, ligands within the same site can interact with each other. These interactions will be taken into account explicitly in Chapter 3.

Let m be the maximum occupancy of any single site. We denote by $q(i)$ the PF of a site containing i adsorbed particles; $q(i)$ contains both the internal PFs of the i ligands and any interactions between the ligands and the site and among the ligands themselves.

As in the simple model, we assume again that the M sites are identical, independent, and distinguishable, and the total number of adsorbed ligands is N. For any configuration of the system, let $a_i(i = 0, 1, \ldots, m)$ be the number of sites containing exactly i molecules. By configuration of the system we mean a particular distribution of N molecules on the M sites characterized by the vector $\mathbf{a} = (a_0, a_1, \ldots, a_m)$. (Note that we can also talk about the various configurations within each site. This should be taken into account in each of $q(i)$, as we shall indeed do explicitly in Chapter 3.)

For the present treatment we can put $q(0) = 1$. If $q(0) \neq 1$, we can write $q'(i) = q(0)q(i)$ and factor out $q(0)^M$. This factor will not affect the thermodynamics of adsorption and therefore can be ignored.

For a specific configuration $\mathbf{a}$ the number of ways we can distribute N particles on M sites is

$$\frac{M!}{\prod_{i=0}^{m} a_i!}. \tag{2.9.1}$$

This is a straightforward generalization of Eq. (2.8.4) for the case $a_0 = M - N$ and $a_1 = N$. Note that the factorials $a_i!$ arise from the possibility of exchanging the content of any pair of sites having the same i, without changing the configuration of the system as a whole. Note that this number does not include exchange of particles within a given site. The latter is included in $q(i)$ and will be treated explicitly in Chapter 3.

The partition function for our system is thus

$$Q(T, M, N) = \sum_{\mathbf{a}} \frac{M!}{\prod\limits_{i=0}^{m} a_i!} \prod_{i=0}^{m} q(i)^{a_i}, \qquad (2.9.2)$$

where the sum is over all the vectors $\mathbf{a}$ subject to the two restrictions

$$\sum_{i=0}^{m} a_i = M, \qquad \sum_{i=0}^{m} ia_i = N. \qquad (2.9.3)$$

Equation (2.9.2) may be viewed as a PF of a system of $m + 1$ species at chemical equilibrium. The species are characterized by the index i.

The sum in (2.9.2) is very complicated, because of the additional restrictions (2.9.3). It is here that the passage to the grand PF proves most useful. We thus open the system with respect to the ligand, and the corresponding grand PF is

$$\Xi(T, M, \mu_A) = \sum_{N=0}^{mM} \lambda_A^N Q(T, M, N). \qquad (2.9.4)$$

We now use the identity

$$\sum_{N=0}^{mM} \sum_{\substack{\mathbf{a} \\ \left[\begin{array}{c} \Sigma a_i = M \\ \Sigma ia_i = N \end{array} \right]}} = \sum_{\substack{\mathbf{b} \\ [\Sigma a_i = M]}} \qquad (2.9.5)$$

The summation over all possible N values as required in the grand PF is the same as the removal of the condition of constant N in the inner summation. Since this simplification is fundamental to the application of the grand PF, it is helpful to consider a simple case. Suppose we have a matrix with elements a_{ij}. The sum of the form

$$\sum_{i+j=N} a_{ij}$$

is over all elements along a diagonal line in the matrix, for which the sum of the indices $i + j$ is constant. If we now sum over all possible N, we obtain the identity

$$\sum_{N} \sum_{i+j=N} a_{ij} = \sum_{i} \sum_{j} a_{ij}.$$

On both sides we sum over all the elements of the matrix. On the rhs we sum over each row first and then over all rows (or vice versa); on the lhs we sum first over each diagonal and then over all diagonals. The result is of course the same in the two sums. In (2.9.5) we have the same situation with respect to the condition $\sum ia_i = N$. Summation over all N removes this condition. The condition of constant M is still left. Thus we have, from (2.9.4) and (2.9.2),

$$\Xi(T, M, \mu_A) = \sum_{\substack{\mathbf{a} \\ [\Sigma a_i = M]}} M! \prod_{i=0}^{m} \frac{q(i)^{a_i} \lambda_A^{ia_i}}{a_i!}. \qquad (2.9.6)$$

Note also that since N is not a variable in (2.9.6), λ_A^N was written in an equivalent way as $\lambda_A^N = \prod_{i=1}^{m} \lambda_A^{ia_i}$.

The removal of the condition $\sum ia_i = N$ makes the summation (2.9.6) feasible. We use the multinomial theorem:

$$(X_1 + X_2 + \cdots + X_m)^M = \sum_{[\sum_i a_i = M]}^{\mathbf{a}} M! \prod_{i=1}^{m} \frac{X_i^{a_i}}{a_i!}, \tag{2.9.7}$$

which is a generalization of the binomial theorem ($M = 2$). Thus we have

$$\Xi(T, M, \mu_A) = \left[\sum_{i=0}^{m} \lambda_A^i q(i) \right]^M = \xi^M, \tag{2.9.8}$$

where

$$\xi = \sum_{i=0}^{m} \lambda_A^i q(i). \tag{2.9.9}$$

What we have achieved is a factorization of the grand PF into M identical factors ξ. Each factor has the same form as a PF of a single site. This results from the assumption of independence of the sites. The equality of all the M factors ξ arises from the identity of all the sites. If, for example, we had M_1 sites of one kind and M_2 sites of another kind, but still all sites are presumed independent, the corresponding grand PF is

$$\Xi(T, M_1, M_2, \mu_A) = \xi_1^{M_1} \xi_2^{M_2}. \tag{2.9.10}$$

The average number $\bar{N}$ is given by

$$\bar{N} = \lambda_A \left(\frac{\partial \ln \Xi}{\partial \lambda_A} \right)_{T,M} = \lambda_A M \frac{\partial \ln \xi}{\partial \lambda_A}. \tag{2.9.11}$$

Hence the average number of particles per site is

$$\bar{i} = \frac{\bar{N}}{M} = \lambda_A \frac{\partial \ln \xi}{\partial \lambda_A} = \frac{\sum\limits_{i=0}^{m} iq(i)\lambda_A^i}{\sum\limits_{i=0}^{m} q(i)\lambda_A^i}. \tag{2.9.12}$$

The probability of finding i molecules at a given site is

$$\Pr(i) = \frac{q(i)\lambda_A^i}{\sum\limits_{i=0}^{m} q(i)\lambda_A^i}. \tag{2.9.13}$$

Therefore the average quantity $\bar{i}$ in (2.9.12) is rewritten as

$$\bar{i} = \sum_{i=1}^{m} i \Pr(i). \tag{2.9.14}$$

This should be compared with (2.8.27). This quantity will be the central one in the analysis of several specific examples treated in the next chapter.

2.9.1. Thermodynamic Derivation

An equivalent form of Eq. (2.8.12) or (2.8.14) in terms of equilibrium constants can be obtained as follows: To establish the relationship between the two notations we rederive Eq. (2.8.12) using purely thermodynamic arguments. We denote by S_i a site containing i molecules of A. A set of equilibrium reactions can be written

$$S_0 + A \rightleftarrows S_1$$

$$S_1 + A \rightleftarrows S_2 \qquad (2.9.15)$$

$$S_{m-1} + A \rightleftarrows S_m.$$

Assuming that A is in equilibrium with an ideal-gas phase, we can define the equilibrium constant for the addition of one A to S_{i-1} by

$$\bar{K}_i = \frac{[S_i]}{\rho_A[S_{i-1}]}, \text{ for } i = 1, 2, \ldots, m \qquad (2.9.16)$$

where $[S_i]$ is the mole fraction of sites of type S_i. Thus, for $i \neq 0$

$$[S_i] = \bar{K}_i\rho_A[S_{i-1}] = \bar{K}_i\bar{K}_{i-1}\rho_A^2[S_{i-2}] = \cdots = \bar{K}_1\bar{K}_2 \cdots \bar{K}_i\rho_A^i[S_0]. \qquad (2.9.17)$$

Since

$$1 = \sum_{i=0}^{m} [S_i] = \sum_{i=1}^{m} \rho_A^i \prod_{j=1}^{i} \bar{K}_j[S_0] + [S_0] \qquad (2.9.18)$$

we can eliminate $[S_0]$

$$[S_0] = \left[1 + \sum_{i=1}^{m} \rho_A^i \prod_{j=1}^{i} \bar{K}_j\right]^{-1}. \qquad (2.9.19)$$

The average number of molecules adsorbed per site is therefore

$$\bar{i} = \sum_{i=1}^{m} i[S_i] = \sum_{i=1}^{m} i[S_0]\rho_A^i \prod_{j=1}^{i} \bar{K}_j$$

$$= \frac{\displaystyle\sum_{i=1}^{m} i\rho_A^i \prod_{j=1}^{i} \bar{K}_j}{1 + \displaystyle\sum_{i=1}^{m} \rho_A^i \prod_{j=1}^{i} \bar{K}_j}. \qquad (2.9.20)$$

Note that $\prod_{j=1}^{i} \bar{K}_j$ may be interpreted as the equilibrium constant for the reaction

$$S_0 + iA \rightleftarrows S_i. \qquad (2.9.21)$$

From (2.9.17) we have, for $i \neq 0$

$$K^{(i)} \equiv \frac{[S_i]}{[S_0]\rho_A^i} = \prod_{j=1}^{i} \bar{K}_j. \qquad (2.9.22)$$

Therefore we can rewrite (2.9.20) as

$$\bar{i} = \frac{\displaystyle\sum_{i=1}^{m} iK^{(i)}\rho_A^i}{1 + \displaystyle\sum_{i=1}^{m} K^{(i)}\rho_A^i}. \qquad (2.9.23)$$

This should be compared with Eq. (2.7.12). Of course, (2.9.12) is more general in the sense that it is not restricted by the assumption of ideality as used in (2.9.23). To establish the connection between $K^{(i)}$ and the molecular properties of the system, we take the special case of (2.7.12) for an ideal gas, for which

$$\lambda_A = \exp(\beta\mu_A) = \Lambda_A^3 q_A^{-1}\rho_A. \tag{2.9.24}$$

For this special case, Eq. (2.9.12) reduces to

$$\bar{i} = \frac{\sum_i iq(i)(\Lambda_A^3 q_A^{-1}\rho_A)^i}{\sum_i q(i)(\Lambda_A^3 q_A^{-1}\rho_A)^i}. \tag{2.9.25}$$

Comparing with (2.9.23), we obtain

$$K^{(i)} = cq(i)(\Lambda_A^3 q_A^{-1})^i; \tag{2.9.26}$$

where c may be easily determined by writing $K^{(i)}$ as the equilibrium constant of the reaction (2.9.21) (using an argument similar to that in section 2.4.4):

$$K^{(i)} = \frac{q(i)\Lambda_A^{3i}}{q(0)q_A^i}. \tag{2.9.27}$$

2.9.2. Indistinguishable Sites

Up to now we have always assumed that the sites are distinguishable [assumption (5) of section 2.8.1], e.g., the sites are localized on a solid surface. A very common system in biochemistry is one consisting of polymer units each adsorbing one or more ligands. The only difference between the two cases is that the sites in the latter case are indistinguishable (presuming that the polymers are identical, say hemoglobin).

In actual experimental cases, the polymers as well as the ligands are in a solvent, but in this section we assume that the system contains M adsorbing polymers in an ideal-gas phase; i.e., there are no significant interactions between the sites. All other features of the model are identical with the model treated earlier in this section. The canonical PF is

$$Q(T, V, M, N) = \sum_{\mathbf{a}} \prod_{i=0}^{m} \frac{q'(i)^{a_i}}{a_i!}, \tag{2.9.28}$$

which is almost the same as the PF in (2.9.2). There are two differences between (2.9.28) and (2.9.2). First, we have replaced $q(i)$ of (2.9.2) by $q'(i)$. The reason is that now, each polymer has translational and rotational degrees of freedom which depend on i. Thus we can write $q'(i) = q_{\text{vib}}(0)q_{\text{rot}}(i)q_{\text{tr}}(i)q_{\text{ad}}(i)$. The vibrational degrees of freedom of the polymer are presumed to be independent of i. The rotational and the translational degrees of freedom of each polymer depend on i even under assumption (4) of section 2.8. Polymers with i ligands have different mass and different moments of inertia. However, in actual applications we shall assume that the polymer is large enough compared with the ligand (say hemoglobin and oxygen) that we may approximate $q'(i)$ by

$$q'(i) = q(0)q_{\text{ad}}(i),$$

and $q(0)^M$ will be factored out of the sum (2.9.28). This, as in section 2.8, will not affect the binding thermodynamics of our system and therefore may be ignored. The second difference is the division by $M!$ made in (2.9.28) to account for the indistinguishability of the M adsorbing units. The same two restrictions (2.9.3) also apply here.

To obtain a more convenient form of the PF we pass to a system open with respect to the ligand, but still closed with respect to the polymers. The corresponding (semi-) grand PF is

$$\Xi(T, V, M, \mu_A) = \sum_{N=0}^{mM} \lambda_A^N Q(T, V, M, N)$$

$$= \xi^M/M!. \tag{2.9.29}$$

Thus the only difference between this PF and the previous one, Eq. (2.9.8), is the factor $M!$.

The two systems are of course different. For instance, the pressure in the present system is determined by both the polymer and the ligand molecules. However, if we are interested only in the adsorption isotherms, the two systems behave identically. In particular the average number of ligands adsorbed on the M sites is

$$\bar{N} = \lambda_A \left(\frac{\partial \ln \Xi}{\partial \lambda_A} \right)_{T,V,M} = M \frac{\sum iq(i)\lambda_A^i}{\sum q(i)\lambda_A^i}, \tag{2.9.30}$$

which is the same as (2.9.12).

Because of this property we shall assume, in the next chapter, that the sites are localized, although in an actual system they are not. A more difficult problem is the effect of the solvent on the adsorption isotherm. Discussion of this will be postponed to Chapter 8, after we have acquired familiarity with the concept of solvation.

2.10. ADSORPTION WITH CONFORMATION CHANGES IN THE ADSORBENT MOLECULES

We extend here the model of section 2.8.1 by relaxing one of the assumptions, namely that the adsorbed ligand does not affect the adsorbent molecules. In many cases in biochemistry, the binding of the ligands causes a conformational change. This effect will be studied in great detail in Chapter 3, where the conformational changes in the polymer will be shown to be connected with the cooperativity of the binding process. Some of the concepts introduced in this section will also be generalized in Chapters 7 and 8, where we shall study the structural changes in the solvent induced by a solute, an important phenomenon in aqueous solutions.

In the present section we deal with the simplest effect of a ligand on the conformational equilibrium of the adsorbent molecules. We shall study in great detail the thermodynamics of such systems. Some of the results will be used in Chapter 3 where more complicated systems will be treated.

2.10.1. The Model and Its Solution

As in the simplest Langmuir model we start with M independent, identical and, for simplicity, localized sites. As we have seen in section 2.9, a factor of $M!$ is introduced in the case of free adsorbent molecules, but this will have no effect on the properties in which we are interested in this section. Also, the polymer is assumed to be large compared with the ligands so that changes in mass or in moment of inertia caused by adsorption can be neglected. Each site can adsorb only one ligand, which we assume to be a simple gas molecule G.

The new feature of the present model is that each site or polymer can be in one of two states, low energy (L) or high energy (H). These two states are in chemical equilibrium

$$L \rightleftarrows H. \tag{2.10.1}$$

The adsorbed molecule G has different adsorption energies U_L and U_H, according to whether it is adsorbed on the L or the H form. The L and H forms can be thought of as two conformations of a polymer, which we symbolically denote by a square and a circle.

$$\square \rightleftarrows \bigcirc. \tag{2.10.2}$$

Thus the generalization of the simple Langmuir model applies to the two assumptions made in the introduction to section 2.8. First, there are two adsorption parameters U_L and U_H instead of one. Second, the process of adsorption might perturb the sites, in the sense that the equilibrium composition of L and H is in general affected by the adsorption of G.

In order to minimize the parameters needed to describe the model, we assume that L and H are characterized only by their *energy levels* E_L and E_H (with $E_L < E_H$). The PFs of a site in either state, Q_L and Q_H, are given by

$$Q_L = \exp(-\beta E_L), \qquad Q_H = \exp(-\beta E_H). \tag{2.10.3}$$

Note that, in general, E_L and E_H should be interpreted as free energies defined by

$$Q_L = \sum_{i \in L} \exp(-\beta E_i) = \exp(-\beta A_L) \tag{2.10.4}$$

(see section 2.4), where the sum $i \in L$ is over all states of the adsorbent molecules which we recognize as belonging to the L states. The L states include many microstates, which in this particular treatment we do not care to take into consideration. From now on we refer to E_L and E_H as the only molecular parameters of the adsorbing molecules, and we suppress any other parameters that might enter into Q_L and Q_H (say, vibrations, rotations, and so on).

From the point of view of chemical equilibrium, we write the canonical PF of the system of M empty sites at temperature T as

$$Q(T, M) = (Q_L + Q_H)^M = \sum_{M_L = 0}^{M} \frac{M!}{M_L! M_H!} Q_L^{M_L} Q_H^{M_H}$$

$$= \sum_{M_L} Q^*(T, M_L, M_H). \tag{2.10.5}$$

Note that if the molecules are in a gaseous phase, then Q is also a function of the volume of the system. In such a case we must include at least the translational PF of the molecules; i.e., instead of (2.10.5) we must write

$$Q(T, V, M) = \frac{V^M}{\Lambda^{3M} M!} (Q_L + Q_H)^M, \tag{2.10.6}$$

where V is the volume of the system and Λ^{3M} the momentum PF of the adsorbent molecules. Since the factor $V^M/\Lambda^{3M} M!$ is constant in the entire treatment that follows, we can ignore it. (This would be the same as freezing in the translation of the molecules, i.e., considering localized sites.)

Following the discussion of chemical equilibrium in section 2.4, we view Q^* defined in (2.10.5) as the PF of a "frozen" system with fixed values of M_L and M_H.

The equilibrium (or average) values of M_L and M_H are obtained from the condition

$$\frac{\partial \ln Q^*}{\partial M_L} = 0, \tag{2.10.7}$$

which leads to the equilibrium condition (see section 2.4)

$$\frac{\bar{M}_H}{\bar{M}_L} = \frac{Q_H}{Q_L}. \tag{2.10.8}$$

We define the constant

$$K = \frac{Q_H}{Q_L} = \exp[-\beta(E_H - E_L)] \tag{2.10.9}$$

by our assumptions $E_H - E_L > 0$, and hence $0 \leq K \leq 1$. From (2.10.8) and (2.10.9) we have

$$\frac{\bar{M}_H}{\bar{M}_L} = K. \tag{2.10.10}$$

It should be noted that K is *defined* in (2.10.9) in terms of the *molecular* parameters E_H and E_L of the adsorbing molecules. By virtue of the equilibrium condition (2.10.8), it is also equal to the ratio $\bar{M}_H/\bar{M}_L$. However, in the following, we shall study the case where the equilibrium ratio $\bar{M}_H/\bar{M}_L$ changes (upon adsorption). In that case, the equality (2.10.10) does not hold. On the other hand, the meaning of K as a constant defined by (2.10.9) remains unchanged.

We also introduce the mole fractions of sites in the two states by

$$x_L^0 = \frac{\bar{M}_L}{M} = \frac{Q_L}{Q_L + Q_H} = \frac{1}{1 + K}, \tag{2.10.11}$$

$$x_H^0 = \frac{\bar{M}_H}{M} = \frac{Q_H}{Q_L + Q_H} = \frac{K}{1 + K}. \tag{2.10.12}$$

These are also the probabilities of finding a specific site in the L or H state for the empty system. (The term empty refers here to the system of absorbent molecules in the absence of ligands.)

We now turn to the case where, in addition to the M sites, we also have N adsorbed molecules distributed over the sites. For simplicity, we assign no internal degrees of freedom to these molecules: they are characterized solely by their adsorption energies U_L and U_H.

The canonical partition function for such a system is

$$Q(T, M, N) = \sum_{\substack{M_L + M_H = M \\ N_L + N_H = N}} \binom{M}{M_L}\binom{M_L}{N_L}\binom{M_H}{N_H} Q_L^{M_L} Q_H^{M_H} q_L^{N_L} q_H^{N_H}, \tag{2.10.13}$$

where

$$q_L = \exp(-\beta U_L), \qquad q_H = \exp(-\beta U_H). \tag{2.10.14}$$

N_L and N_H are the number of molecules adsorbed on L and H sites, respectively. The summation in (2.10.13) extends over all possible values of M_L, M_H, N_L, N_H with the conditions

$$M_L + M_H = M, \qquad N_L + N_H = N, \qquad N_L \leq M_L, \qquad N_H \leq M_H. \qquad (2.10.15)$$

Clearly the auxiliary parameters N_L, N_H, M_L, M_H serve as intermediate quantities which determine the energy levels of the composed system, namely,

$$E(N_L, N_H, M_L, M_H) = M_L E_L + M_H E_H + N_L U_L + N_H U_H. \qquad (2.10.16)$$

The degeneracy of these energy levels is given by the product of the three combinatorial factors in (2.10.13).

The summation in (2.10.13) cannot be carried out to obtain a closed form of the partition function. This is easily achieved, however, by transforming to an open system with respect to the gas molecules; i.e., we define the grand partition function by

$$\Xi(T, M, \lambda) = \sum_{N=0}^{M} \lambda^N Q(T, M, N)$$

$$= \sum_{M_L=0}^{M} \sum_{N_L=0}^{M_L} \sum_{N_H=0}^{M_H} \lambda^{N_L} \lambda^{N_H} \binom{M}{M_L}\binom{M_L}{N_L}\binom{M_H}{N_H} Q_L^{M_L} Q_H^{M_H} q_L^{N_L} q_H^{N_H}$$

$$= \sum_{M_L=0}^{M} \binom{M}{M_L} Q_L^{M_L} (1 + \lambda q_L)^{M_L} Q_H^{(M-M_L)} (1 + \lambda q_H)^{(M-M_L)}$$

$$= (Q_L + Q_H + \lambda q_L Q_L + \lambda q_H Q_H)^M$$

$$= \xi^M, \qquad (2.10.17)$$

where $\lambda = \exp(\beta\mu)$ is the absolute activity of the gas and μ is its chemical potential.

Note that $\Xi(T, M, \lambda)$ is the semigrand PF of a system of M adsorbent molecules and any number of ligands $N \leq M$. No free ligands are taken into account in our definition of the system. Likewise $Q(T, M, N)$ pertain to a system of M adsorbent and N ligand molecules. Of course, we can include free ligands in the definition of our system, for instance by including a factor $q_G^{N_G}/N_G!$ where N_G is the number of free G molecules. For the purpose of this section, the restricted definition of our system is more convenient.

As we have seen in section 2.8, the last form of the PF is typical of a system of independent sites; ξ, defined in (2.10.17), may be viewed as the grand PF of a single site.

The four terms in ξ correspond to the four possible states of each site, i.e., empty L, empty H, occupied L, and occupied H. Symbolically, these states are denoted as

$$\square, \bigcirc, \boxed{G}, \textcircled{G}. \qquad (2.10.18)$$

The terms in (2.10.17) correspond to the probabilities of finding a specific site in each of the states indicated in (2.10.18), namely

$$\Pr(L, 0) = Q_L/\xi, \qquad \Pr(H, 0) = Q_H/\xi,$$
$$\Pr(L, G) = \lambda q_L Q_L/\xi, \qquad \Pr(H, G) = \lambda q_H Q_H/\xi. \qquad (2.10.19)$$

The probabilities of finding a site empty or occupied, independently of its state, are

$$Pr(0) = Pr(L, 0) + Pr(H, 0) = \frac{Q_L + Q_H}{\xi} \tag{2.10.20}$$

$$Pr(G) = Pr(L, G) + Pr(H, G) = \frac{\lambda q_L Q_L + \lambda q_H Q_H}{\xi}. \tag{2.10.21}$$

Note that as $\lambda \to 0$, $Pr(0) \to 1$, while $Pr(G) \to 0$. On the other hand, as $\lambda \to \infty$, $Pr(0) \to 0$, while $Pr(G) \to 1$, as expected. For later applications it will also be useful to introduce the conditional probability $Pr(L/G)$ of finding a site in state L when it is known to be occupied by a gas molecule, i.e.,

$$Pr(L/G) = \frac{Pr(L, G)}{Pr(G)} = \frac{\lambda q_L Q_L / \xi}{(\lambda q_L Q_L + \lambda q_H Q_H)/\xi} = \frac{q_L Q_L}{q_L Q_L + q_H Q_H}. \tag{2.10.22}$$

Similarly, the conditional probability of finding a site in state H, given that it is occupied, is

$$Pr(H/G) = \frac{q_H Q_H}{q_L Q_L + q_H Q_H}. \tag{2.10.23}$$

The average number of gas molecules in the system is given by the standard relation (section 1.3.4).

$$\bar{N} = \lambda \frac{\partial \ln \Xi}{\partial \lambda} = \frac{\lambda M}{\xi} (q_L Q_L + q_H Q_H). \tag{2.10.24}$$

Let θ be the average fraction of sites occupied. This is also equal to the probability of finding a specific site being occupied (see section 2.8). From (2.10.19) and (2.10.24) we obtain

$$x = Pr(L, G) + Pr(H, G) = Pr(G). \tag{2.10.25}$$

Elimination of λ from (2.10.24) gives

$$\lambda = \left(\frac{\theta}{1 - \theta}\right) \frac{Q_L + Q_H}{q_L Q_L + q_H Q_H}, \tag{2.10.26}$$

from which we can obtain all the partial thermodynamic quantities of the adsorbed gas. First, the chemical potential is

$$\mu = kT \ln\left(\frac{\theta}{1 - \theta}\right) - kT \ln\left(\frac{q_L Q_L + q_H Q_H}{Q_L + Q_H}\right). \tag{2.10.27}$$

Using the distribution of two states in the empty system given in (2.10.11) and (2.10.12), we may rewrite (2.10.27) as

$$\mu = kT \ln\left(\frac{\theta}{1 - \theta}\right) - kT \ln(q_L x_L^0 + q_H x_H^0)$$

$$= kT \ln\left(\frac{\theta}{1 - \theta}\right) - kT \ln\langle \exp(-\beta B_G)\rangle_0, \tag{2.10.28}$$

where we have introduced the quantity B_G which will be referred to as the *binding energy of G to the site*. More specifically, B_G is a function of the state of the site $\alpha = L, H$ and is defined as follows:

$$B_G(\alpha) = \begin{cases} U_L & \text{if } \alpha = L \\ U_H & \text{if } \alpha = H. \end{cases} \qquad (2.10.29)$$

Thus in (2.10.28) we denoted by $\langle \ \rangle_0$ an average of the function $\exp[-\beta B_G(\alpha)]$. The average is taken over all states of the adsorbent molecules, here L and H, with the probability distribution x_L^0 and x_H^0 as given in (2.10.11) and (2.10.12), i.e., of the empty system.

The partial molar (or molecular) entropy of the adsorbed ligand is obtained from (2.10.28) by differentiation with respect to the temperature:

$$\bar{S} = -\left(\frac{\partial \mu}{\partial T}\right)_{M,N}$$

$$= -k \ln\left(\frac{\theta}{1-\theta}\right) + k \ln\langle \exp(-\beta B_G)\rangle_0$$

$$+ \frac{1}{T}\left[\frac{q_L Q_L(E_L + U_L) + q_H Q_H(E_H + U_H)}{q_L Q_L + q_H Q_H} - \frac{Q_L E_L + Q_H E_H}{Q_L + Q_H}\right]. \qquad (2.10.30)$$

Using the conditional probabilities of (2.10.22) and (2.10.23), relation (2.10.30) may be written as

$$\bar{S} = -k \ln\left(\frac{\theta}{1-\theta}\right) + k \ln\langle \exp(-\beta B_G)\rangle_0 + \frac{1}{T}[\langle E_P + B_G\rangle_G - \langle E_P\rangle_0] \qquad (2.10.31)$$

E_P is the energy of the site or, of the polymer, and is defined similarly to (2.10.29) as

$$E_P(\alpha) = \begin{cases} E_L & \text{for } \alpha = L \\ E_H & \text{for } \alpha = H, \end{cases} \qquad (2.10.32)$$

where the symbol $\langle \ \rangle_G$ signifies a conditional average taken with the distribution given in (2.10.22) and (2.10.23).

The partial molar (or molecular) energy of the gas is given by

$$\bar{E} = \mu + T\bar{S} = \langle E_P + B_G\rangle_G - \langle E_P\rangle_0 = \langle B_G\rangle_G + (\langle E_P\rangle_G - \langle E_P\rangle_0). \qquad (2.10.33)$$

2.10.2. The Adsorption Isotherm

From (2.10.26) we can express θ as a function of λ, i.e.,

$$\theta = \frac{K_{in}\lambda}{1 + K_{in}\lambda}, \qquad (2.10.34)$$

where K_{in} is defined by

$$K_{in} = \frac{q_L Q_L + q_H Q_H}{Q_L + Q_H} = \langle \exp(-\beta B_G)\rangle_0. \qquad (2.10.35)$$

This is the adsorption isotherm in terms of the absolute activity λ of the ligand. To transfer this expression to a more familiar form, we assume that the ligand is in equilibrium with an ideal gas, for which the chemical potential is

$$\mu_G^{\text{i.g.}} = kT \ln(\rho_G \Lambda_G^3) = kT \ln\left(\frac{P\Lambda_G^3}{kT}\right), \tag{2.10.36}$$

where P is the partial pressure of the gas. At equilibrium we have

$$\mu_G^{\text{i.g.}} = \mu, \tag{2.10.37}$$

where μ is the chemical potential of the adsorbed ligand, given in (2.10.28). (In section 2.8.2 this was denoted by μ^{ad}.)

For this special case we have

$$\theta = \frac{K_{\text{ad}}P}{1 + K_{\text{ad}}P}, \tag{2.10.38}$$

where

$$K_{\text{ad}} = \frac{K_{\text{in}}\Lambda_G^3}{kT}. \tag{2.10.39}$$

K_{ad} is the adsorption constant that can be obtained from the experimental curve of θ as a function of the pressure P.

The quantity denoted by K_{in} and defined in (2.10.35) will be referred to as the *intrinsic binding constant*. This quantity is related to the change in the Helmholtz energy for the process of bringing G from a fixed position, say $\mathbf{R}_0$, in an ideal gas to a *fixed* and *empty* specific site, say j. The argument leading to this relation is similar to the one given in section 2.8.5.

$$\Delta A^*(\mathbf{R}_0 \to \text{specific } \varnothing \text{ site } j) = -kT \ln \frac{Q(T, M, N+1; \text{site } j \text{ occupied})}{Q(T, M, N; \varnothing \text{ site } j)}.$$

$$= -kT \ln\left(\frac{q_L Q_L + q_H Q_H}{Q_L + Q_H}\right)$$

$$= -kT \ln\langle\exp(-\beta B_G)\rangle_0. \tag{2.10.40}$$

Comparing (2.10.40) with (2.10.35), we have the relation

$$K_{\text{in}} = \exp[-\beta \Delta A^*(\mathbf{R}_0 \to \text{specific } \varnothing \text{ site } j)]. \tag{2.10.41}$$

Note that in (2.10.40) we returned to the canonical PF. Since the only change that occurs is in the site j, we need to consider only the change in the PF of a single site from empty to occupied.

We also note that ΔA^* defined in (2.10.40) is different from a similar quantity defined for the process of bringing the ligand from $\mathbf{R}_0$ to the specific site j (empty or occupied), for which we have the relation (see section 2.8.5)

$$\Delta A^*(\mathbf{R}_0 \to \text{specific site } j) = \Delta A^*(\mathbf{R}_0 \to \text{specific } \varnothing \text{ site } j) - kT \ln(1 - \theta). \tag{2.10.42}$$

The latter quantity for this particular model (no internal degrees of freedom assigned to the ligand) is equal to the pseudochemical potential.

$$\mu^*(\text{at } j) = \Delta A^*(\mathbf{R}_0 \rightarrow \text{specific site } j). \qquad (2.10.43)$$

This can be combined with expression (2.10.28), but viewed in the T, M, N ensemble, to obtain the familiar relation

$$\mu = \mu^*(\text{at } j) + kT \ln \theta, \qquad (2.10.44)$$

i.e., the chemical potential of the adsorbed ligands is split into the pseudochemical potential and the liberation Helmholtz energy, $kT \ln \theta$.

The two processes indicated in (2.10.40) and (2.10.42) can be made formally indistinguishable if we define a modified binding energy as follows:

$$B'_G(\alpha) = \begin{cases} U_L & \text{if } \alpha = L \text{ and } \varnothing \\ U_H & \text{if } \alpha = H \text{ and } \varnothing \\ \infty & \text{if } \alpha = \text{occupied.} \end{cases}$$

The work required to bring the ligand from $\mathbf{R}_0$ to a specific site (empty or occupied) is the same as in (2.10.12), but in terms of B'_G is written as follows

$$\Delta A^*(\mathbf{R}_0 \rightarrow \text{specific site } j) = -kT \ln \langle \exp(-\beta B_G) \rangle_0 - kT \ln \text{Pr}(0)$$
$$= -kT \ln[x_L \text{Pr}(0)q_L + x_H \text{Pr}(0)q_H + 0 \times \text{Pr}(1)]$$
$$= -kT \ln \langle \exp(-\beta B'_G) \rangle_0,$$

where $\text{Pr}(0)$, $\text{Pr}(1)$ are probabilities of the empty and the occupied sites.

We see that with the new definition of B'_G we can absorb the term $-kT \ln(1 - \theta)$ into a new average which has the same form as (2.10.40). The last form of ΔA^* brings this quantity to a complete analogy with the solvation Helmholtz energy, which is discussed in Chapter 6. In general, in the theory of solvation, the binding energy of a solute is always defined in generalization of B'_G to include the repulsive interaction with its neighboring molecules.

In the next chapter, the quantity K_{in} defined in (2.10.35) will serve as a central tool to study induced (or allosteric) conformational changes. It is called intrinsic since it involves only the energetics of the binding to the site and not the factor $-kT \ln(1 - \theta)$ (or other combinatorial factors that will appear later), which has nothing to do with the energetics of the binding.

If $U_L = U_H = U$, then K_{in} defined in (2.10.35) reduces to

$$K_{in} = \langle \exp(-\beta B_G) \rangle_0 = \exp(-\beta U), \qquad (2.10.45)$$

which is the intrinsic binding constant for the simple Langmuir isotherm (see for instance Eq. (2.8.3) for the case $q_A = 1$).

Another simple limiting case occurs when $E_L = E_H = E$, but $U_L \neq U_H$. In this case

$$K_{in} = \tfrac{1}{2}[\exp(-\beta U_L) + \exp(-\beta U_H)], \qquad (2.10.46)$$

which is the average of two intrinsic binding energies in a system of equal numbers of L and H sites.

It should be noted that the average quantity that appears in the definition of K_{in} in (2.10.35) is an average with the probability distribution given in (2.10.11) and (2.10.12) of the *unperturbed* adsorbent molecules. As we shall soon see, the process of adsorption will, in general, affect the distribution of adsorbent molecules in the two states L and H. However, the average $\langle \exp(-\beta B_G) \rangle_0$ is computed in the absence of the ligand molecules, using the distributions (2.10.11) and (2.10.12). From (2.10.19) we may write the mole fractions of sites in the L and in the H states at any λ as follows:

$$x_L = \Pr(L) = \Pr(L, O) + \Pr(L, G) = \frac{Q_L + \lambda q_L Q_L}{\xi} \tag{2.10.47}$$

$$x_H = \Pr(H) = \Pr(H, O) + \Pr(H, G) = \frac{Q_H + \lambda q_H Q_H}{\xi}. \tag{2.10.48}$$

These are functions of λ and reduce to (2.10.11) and (2.10.12) for $\lambda \to 0$. $\Pr(L)$ is the probability of finding a specific site in state L, independently of its occupation by a ligand. Similar meaning is ascribed to $\Pr(H)$.

2.10.3. Thermodynamics of the Adsorption Process

Choosing the reference state of the ligand as the state of G at some fixed position in vacuum, we define the adsorption process as the process of transferring G from a fixed position in vacuum into a *fixed*, *empty* site. The corresponding thermodynamic quantities are

$$\Delta A_G^* = -kT \ln \langle \exp(-\beta B_G) \rangle_0 \tag{2.10.49}$$

$$\Delta S_G^* = k \ln \langle \exp(-\beta B_G) \rangle_0 + \frac{1}{T} (\langle B_G \rangle_G + \langle E_P \rangle_G - \langle E_P \rangle_0) \tag{2.10.50}$$

$$\Delta E_G^* = \langle B_G \rangle_G + \langle E_P \rangle_G - \langle E_P \rangle_0. \tag{2.10.51}$$

Note that conditional averages appear in ΔE_G^*, but not in ΔA_G^*. We shall now reinterpret this observation from a chemical equilibrium point of view; i.e., the quantity $\langle E_P \rangle_G - \langle E_P \rangle_0$ will be related to the shift in the chemical equilibrium $L \rightleftarrows H$ induced by the adsorption of G. To do that we adopt a mixture-model point of view of the same system. This approach will be generalized in Chapter 5 to treat any liquid and in particular aqueous solutions. In our model we have two states of the adsorbent molecules. The mixture-model approach follows from the classification of molecules in state L as L-molecule, and likewise molecules in state H as an H-molecule. This is the same procedure we have discussed in section 2.4. We now assign partial molecular quantities to the species L and H, viewing M_L and M_H as independent variables. Theoretically these are defined as partial derivatives of the corresponding thermodynamic functions (see below). From the physical point of view, these can be defined only if we have a means of varying M_L and M_H independently, e.g., by placing an inhibitor that prevents the conversion between the two species. We now define three PFs, differing in the degree of detail we want to look at our system. These are

$$Q^{**}(T, M_L, M_H, N_L, N_H) = \binom{M}{M_L}\binom{M_L}{N_L}\binom{M_H}{N_H} Q_L^{M_L} Q_H^{M_H} q_L^{N_L} q_H^{N_H} \qquad (2.10.52)$$

$$Q^*(T, M_L, M_H, N) = \sum_{N_L + N_H = N} Q^{**}(T, M_L, M_H, N_L, N_H) \qquad (2.10.53)$$

$$Q(T, M, N) = \sum_{M_L + M_H = M} Q^*(T, M_L, M_H, N). \qquad (2.10.54)$$

The first PF, Q^{**}, is the most detailed, in the sense that it corresponds to a system with fixed values of N_L, N_H, M_L, M_H. The energy levels are given by (2.10.16) and the degeneracy of each energy level is the number of configurations, or the number of ways of choosing M_L out of M, distributing N_L over M_L, and distributing N_H over M_H. Here the variables N_L, N_H, M_L, M_H are considered as independent variables, the system is viewed as a *four*-component system of N_L occupied L sites, N_H occupied H sites, $M_L - N_L$ empty L sites, and $M_H - N_H$ empty H sites.† This system may be referred to as the *completely frozen-in* system.

The second PF, $Q^*(T, M_L, M_H, N)$ defined in (2.10.53) pertains to a system that may be referred to as a *partially equilibrated system* (PES) or, equivalently, as a *partially frozen-in* system. Here M_L and M_H are fixed, but the ligand molecules are in equilibrium, i.e., N_L and N_H are not fixed. Only the sum $N_L + N_H$ is constant. It is at this intermediate level of details that we shall be working in this section.

The third PF, $Q(T, M, N)$ pertains to a system which may be referred to as the *completely equilibrated system* (CES). The independent variables are M and N; i.e., this is a two-component system, with the least-detailed description.

In the partially equilibrated system, we have an equilibrium with respect to the flow of G molecules on the M_L and M_H sites. The equilibrium values of N_L and N_H, denoted by $\bar{N}_L$ and $\bar{N}_H$, are obtainable from the condition that Q^{**} be maximum with respect to N_L and N_H subject to the restriction $N_L + N_H = N$ (M_L and M_H being constants). This procedure leads to the equilibrium constant for the partially equilibrated system, namely

$$\frac{\bar{N}_H(M_L - \bar{N}_L)}{\bar{N}_L(M_H - N_H)} = \frac{q_H}{q_L} \equiv h. \qquad (2.10.55)$$

Note that h is *defined* as the ratio $q_H/q_L = \exp[-\beta(U_H - U_L)]$. In (2.10.55), it also serves as an equilibrium constant in the partially equilibrated system. The relevant reaction can be symbolically written as

$$\boxed{G} + \bigcirc \rightleftarrows \square + \boxed{G}, \qquad (2.10.56)$$

i.e., the reaction in which G is transferred from an occupied L site to an empty H site to form an occupied H site and an empty L site.

---

† This may become clearer if we write $Q_L^{M_L} Q_H^{M_H} q_L^{N_L} q_H^{N_H} = Q_L^{M_L - N_L} Q_H^{M_N - N_N}(q_L Q_L)^{N_L}(q_H Q_H)^{N_N}$. In the second form each factor corresponds to each of the species.

We can solve (2.10.55) for $\bar{N}_H$ in terms of the other parameters of the system; the result is

$$\bar{N}_H = \frac{-(N - M_L - hN - hM_H) \pm [(N - M_L - hN - hM_H)^2 - 4(h-1)hNM_H]^{1/2}}{2(h-1)}.$$

(2.10.57)

Clearly there is only one solution which is physically plausible. To determine the physical solution we may examine some special cases. For instance, if we choose $h = 0$, i.e., $U_H - U_L = \infty$, in which case adsorption of G on the L site is absolutely preferable to adsorption on the H site, (2.10.57) reduces to

$$\bar{N}_H = \frac{-(N - M_L) \pm [(N - M_L)^2]^{1/2}}{-2}.$$

(2.10.58)

We now distinguish between two cases. If $N < M_L$, then all the gas molecules G will preferentially be adsorbed on L, and hence $\bar{N}_H = 0$. If, on the other hand, $N > M_L$, then the M_L sites will first be filled, and the remaining $N - M_L$ gas molecules will fill the H sites; hence $\bar{N}_H = N - M_L$. Both of these cases are obtained only if we choose the *minus* sign in (2.10.58). If we now increase h from zero to any value $0 \le h \le \infty$, then by continuity of the solution of (2.10.57), the minus sign should be retained. Incidentally, for $h = 1$, i.e., when $U_L = U_H$, we can solve (2.10.55) directly to obtain the result $\bar{N}_H = NM_H/M$ and $\bar{N}_L = NM_L/M$. Another way to determine the correct sign is to note that for $h < 1$, N_H becomes negative for $+$ and positive for $-$.

For later use, we shall be interested only in the limit of $N \ll M$ (the limit of infinite dilution on the sites). Expanding (2.10.57) to linear order in N, we obtain

$$\bar{N}_L = \frac{NM_L}{M_L + hM_H}, \qquad \bar{N}_H = \frac{hNM_H}{M_L + hM_H}, \qquad (\text{PES, } N \ll M), \qquad (2.10.59)$$

or the equilibrium condition

$$\frac{\bar{N}_H}{\bar{N}_L} = \frac{hM_H}{M_L}, \qquad (\text{PES, } N \ll M), \qquad (2.10.60)$$

where we have stressed in (2.10.59) and (2.10.60) that this result pertains to the partially equilibrated system (PES) at very low dilution.

Next we turn to the completely equilibrated system (CES), which is obtained by letting M_L and M_H reach their equilibrium values, which we denote by $\bar{M}_L$ and $\bar{M}_H$, respectively. These are obtained by taking the maximum of $Q^*(T, M_L, M_H, N)$ with respect to M_L and M_H, subject to the condition $M_L + M_H = M$. This procedure leads to a new equilibrium condition

$$\frac{\bar{M}_H - \bar{N}_H}{\bar{M}_L - \bar{N}_L} = \frac{Q_H}{Q_L} = K, \qquad (\text{CES}). \qquad (2.10.61)$$

This should be compared with (2.10.9) and (1.10.10). From (2.10.55) and (2.10.61), when evaluated at the condition of CES, we can solve for $\bar{N}_L$, $\bar{N}_H$, $\bar{M}_L$, and $\bar{M}_H$ in terms of the molecular parameters h and K and the macroscopic parameters N, M, and T. The results are

$$\bar{N}_L = \frac{N}{1 + hK}, \qquad \bar{N}_H = \frac{hKN}{1 + hK} \qquad (2.10.62)$$

$$\bar{M}_L = M \frac{1 + hK - \theta K(h-1)}{(1+K)(1+hK)}, \qquad \bar{M}_H = M \frac{K + hK^2 + \theta K(h-1)}{(1+K)(1+hK)}, \qquad (2.10.63)$$

where $\theta = N/M$.

From (2.10.61) and (2.10.63), we also have the equilibrium conditions in the completely equilibrated system CES,

$$\frac{\bar{N}_H}{\bar{N}_L} = hK, \qquad \text{(CES)} \qquad (2.10.64)$$

$$\frac{\bar{M}_H}{\bar{M}_L} = K \frac{1 + hK + \theta(h-1)}{1 + hK - \theta K(h-1)}, \qquad \text{(CES)}. \qquad (2.10.65)$$

Note that the equilibrium ratio $\bar{M}_H/\bar{M}_L$ approaches K [see Eq. (2.10.10)] for $\theta \to 0$ (empty sites) or for $h = 1$ (no preferential binding $U_L = U_H$). For any other θ, the equilibrium ratio is modulated by the concentration of the ligand.

The mole fractions of L and H sites can now be written for any θ as

$$x_L = \frac{\bar{M}_L}{M} = x_L^0 + \theta \frac{K(1-h)}{(1+K)(1+hK)} \qquad (2.10.66)$$

$$x_H = \frac{\bar{M}_H}{M} = x_H^0 - \theta \left[\frac{K(1-h)}{(1+K)(1+hK)} \right]. \qquad (2.10.67)$$

Note that for $h < 1$, $x_L > x_L^0$; i.e., the ligand stabilizes the L state. In this representation we view x_L and x_H as functions of θ (Fig. 2.14). We define the differential shift in x_L by

$$d_L = \frac{\partial x_L}{\partial \theta}. \qquad (2.10.68)$$

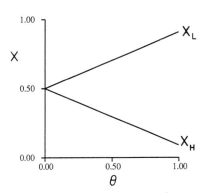

FIGURE 2.14. x_L and x_H as functions of θ for $K = 1$ and $h = 0.1$ [Eqs. (2.10.66) and (2.10.67)].

The derivative $\partial x_L/\partial\theta$ may also be interpreted as the shift of the equilibrium value of $\bar{M}_L$ caused by the addition of one G molecule, i.e.,

$$d_L = \lim_{\Delta N \to 0} \frac{\bar{M}_L(N+\Delta N) - \bar{M}_L(N)}{\Delta N} = \left(\frac{\partial\bar{M}_L}{\partial N}\right)_{T,M}$$

$$= \left(\frac{\partial x_L}{\partial\theta}\right)_T$$

$$= \frac{K(1-h)}{(1+K)(1+hK)}. \tag{2.10.69}$$

The last expression will be used in the next section to express the partial molecular quantities of the absorbed molecules.

Before proceeding to some thermodynamic quantities of this system, it is instructive to check a few limiting cases:

1. For $h \to 0$, i.e., $U_H - U_L \to \infty$, adsorption on L will be overwhelmingly preferable. In this case $\bar{N}_L \to N$, $\bar{N}_H \to 0$, and $d_L \to K/(1+K)$. This is the maximum effect that G can have on the equilibrium $L \rightleftarrows H$ for any given K. The absolute maximum is $d_L = 1$ when $K \to \infty$.
2. For $h = 1$, i.e., $U_H = U_L$, there is no preference in binding on L or H. Hence, $\bar{N}_L = N/(1+K)$, $\bar{N}_H = KN/(1+K)$, $\bar{N}_H/\bar{N}_L = K = \bar{M}_H/\bar{M}_L$, $d_L = 0$.
3. For $K \to 0$, i.e., $E_H - E_L \to \infty$, $\bar{M}_L \to M$, $\bar{M}_H \to 0$, $\bar{N}_L \to N$, $\bar{N}_H \to 0$, $d_L \to 0$.
4. For $\theta \to 0$, $x_H/x_L = x_H^0/x_L^0 = K$.
5. For $\theta \to 1$, $x_H/x_L = Kh = \bar{N}_H/\bar{N}_L$.

Figure 2.15a shows the behavior of the quantity d_L as a function of K for different values of h. Since we have chosen $E_L - E_H < 0$, we can restrict the variation of K to the region $0 \le K \le 1$.

When $K = 0$, the L form is infinitely more stable than the H form. In this case, for any finite value of h, the equilibrium will not be affected by the ligand. We can say that the system is *infinitely resistant* to conformational changes. On the other hand, for $K = 1$, the two energy levels are equal and the system is very responsive to the binding. For

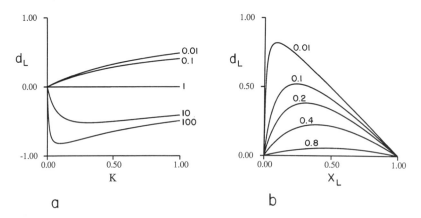

FIGURE 2.15. (a) The dependence of d_L on K for different values of h. (b) The dependence of d_L on x_L for different values of h.

$K = 1$, we have $x_L = x_H = \frac{1}{2}$. If we use $h = 0$, we get $d_L = \frac{1}{2}$, i.e., *total* conversion from $H \rightarrow L$. If $h = \infty$, we obtain $d_L = -\frac{1}{2}$, i.e., *total* conversion from $L \rightarrow H$. At any intermediate K, say $K = 0.1$, we have $x_L = 1/1.1$, $x_H = 0.1/1.1$. If we use $h = 0$, $d_L = 0.1/1.1 = 0.09$, again a total conversion from $H \rightarrow L$, whereas for $h = \infty$, $d_L = -1/1.1 = -0.9$, i.e., a total conversion from $L \rightarrow H$. Clearly, the maximum total conversion is ± 1.

A different presentation of the same effects is shown in Fig. 2.15b, when d_L is now plotted against the mole fraction x_L. Although x_L is not a molecular parameter of the system, it is useful to think of the response of the system to the binding as a function of the concentrations of the two forms L and H. We shall make use of the result (2.10.69) in the next chapter when we analyze various allosteric effects.

2.10.4. Partial Molecular Thermodynamic Quantities in the Mixture Model Formalism

In this subsection we evaluate the chemical potential and the partial molecular entropy and energy of the gas, using the mixture model approach. The final results will be exactly the same as in (2.10.28), (2.10.30), and (2.10.31), thus proving the equivalence of the two methods. However, the more important aspect of this approach is the division of each thermodynamic quantity into two terms: one corresponds to a PES and the second is a relaxation term corresponding to relaxation to the CES. We first calculate the chemical potential in the PES, i.e., when M_L and M_H are fixed; this will be denoted by $\mu^f(M_L, M_H)$. Next we substitute $\bar{M}_L$ and $\bar{M}_H$ for M_L and M_H to get $\mu^f(\bar{M}_L, \bar{M}_H)$; this will turn out to be exactly the same as (2.10.28). Carrying out a similar procedure for S^f and E^f, we shall see that these are only parts of $\bar{S}$ and $\bar{E}$, respectively. For the sake of mathematical simplicity we shall restrict ourselves to the case $N \ll M$, or $\theta \ll 1$.

The chemical potential of the gas in any PES may be obtained by differentiating Q^* with respect to N, i.e.,

$$\mu^f(M_L, M_H) = \left(\frac{\partial A^f}{\partial N}\right)_{T,M_L,M_H} = \left(\frac{-kT\partial \ln Q^*}{\partial N}\right)_{T,M_L,M_H}, \qquad (2.10.70)$$

where

$$A^f = -kT \ln Q^*(T, M_L, M_H, N) \qquad (2.10.71)$$

and $Q^*(T, M_L, M_H, N)$ has been defined in (2.10.53). This PF pertains to the partially equilibrated system (PES) or equivalently to a system in which the conversion of L and H is frozen-in. The superscript f will be used to distinguish partial molecular quantities in the PES from the ordinary partial molecular quantities defined in the CES.

Evaluating μ^f in the PES, taking the limit $\theta \ll 1$, and then substituting $\bar{M}_L$ and $\bar{M}_H$ for M_L and M_H, we get, after some elementary algebra, the final result

$$\mu^f(\bar{M}_L, \bar{M}_H) = kT \ln \theta - kT \ln\left(\frac{q_L Q_L + q_H Q_H}{Q_L + Q_H}\right), \qquad (2.10.72)$$

which is exactly equal to μ in (2.10.27) taken at the limit $\theta \ll 1$.

Note that there is a conceptual difference between $\mu^f(\bar{M}_L, \bar{M}_H)$ in (2.10.72) and μ in (2.10.27), although they are numerically equal: $\mu^f(M_L, M_H)$ is defined as the chemical potential of G for any fixed values of M_L, M_H. This might correspond to a system in which an inhibitor to the reaction (2.10.1) is present. Thus $\mu^f(\bar{M}_L, \bar{M}_H)$ is a specific value of $\mu^f(M_L, M_H)$ evaluated for the specific point $M_L = \bar{M}_L$ and $M_H = \bar{M}_H$, i.e., we let the reaction $L \rightleftarrows H$ attain its equilibrium composition $\bar{M}_L, \bar{M}_H$, then we freeze-in the equilibrium (by introducing the inhibitor).

On the other hand, μ is defined in the completely equilibrated system (2.10.28), i.e., in the absence of the inhibitor. The equality

$$\mu^f(\bar{M}_L, \bar{M}_H) = \mu(\bar{M}_L, \bar{M}_H) \tag{2.10.73}$$

simply means that at the specific composition $\bar{M}_L, \bar{M}_H$, the chemical potential of G is not affected by the presence or absence of the inhibitor. This equality is peculiar to the chemical potential of G. As we shall soon see, no other partial molecular quantity of G shares this property. For instance, the partial molecular entropy and energy are different when defined in the PES or in the CES.

The partial molecular quantities in the CES have already been written in (2.10.31) and (2.10.33). We now compute S^f and E^f for the PES. For any composition M_L, M_H, we define

$$S^f(M_L, M_H) = -\left(\frac{\partial \mu^f(M_L, M_H)}{\partial T}\right)_{N, M_L, M_H}, \tag{2.10.74}$$

differentiating (2.10.72) with respect to T (but *before* substituting $M_L = \bar{M}_L$ and $M_H = \bar{M}_H$), taking the limit $\theta \ll 1$, and then evaluating S^f at the point $M_L = \bar{M}_L$ and $M_H = \bar{M}_H$, we obtain after some lengthy but straightforward algebra the result

$$S^f(\bar{M}_L, \bar{M}_H) = -k \ln \theta + k \ln \langle \exp(-\beta B_G) \rangle_0 + \frac{1}{T} \langle B_G \rangle_G. \tag{2.10.75}$$

Similarly the partial molecular energy of G in the PES, evaluated at the point $M_L = \bar{M}_L$ and $M_H = \bar{M}_H$, is

$$E^f(\bar{M}_L, \bar{M}_H) = \mu^f(\bar{M}_L, \bar{M}_H) + TS^f(\bar{M}_L, \bar{M}_H) = \langle B_G \rangle_G. \tag{2.10.76}$$

The last two results, (2.10.75) and (2.10.76), should be compared with the corresponding values in the CES, (2.10.31) and (2.10.33), respectively. In both cases the difference includes the quantity $\langle E_P \rangle_G - \langle E_P \rangle_0$, which is the shift of the average energy of the polymer induced by G. The difference can be rewritten in an equivalent form using the mixture-model approach. In the PES we have a three-component system with composition N, M_L, M_H. Thus, as for any extensive quantity we write the differential of the Helmholtz energy in the PES as

$$dA = \left(\frac{\partial A}{\partial N}\right)_{M_L, M_H} dN + \left(\frac{\partial A}{\partial M_L}\right)_{N, M_H} dM_L + \left(\frac{\partial A}{\partial M_H}\right)_{N, M_L} dM_H; \tag{2.10.77}$$

dividing by dN and evaluating the derivative at the CES, we have

$$\mu = \left(\frac{\partial A}{\partial N}\right)_M = \left(\frac{\partial A}{\partial N}\right)_{\bar{M}_L, \bar{M}_H} + (\mu_L - \mu_H)\left(\frac{\partial \bar{M}_L}{\partial N}\right)_M \tag{2.10.78}$$

$$\mu = \mu^f + (\mu_L - \mu_H)\left(\frac{\partial \bar{M}_L}{\partial N}\right)_M, \tag{2.10.79}$$

where μ_L and μ_H are the chemical potentials of L and H in the PES. (Note that these quantities are definable *only* in the PES, so there is no need to add the superscript f, as we have done for the partial molecular quantities of G.)

Similarly, for the entropy and the enthalpy of the system, we derive the analogue of (2.10.79), i.e.,

$$\bar{S} = S^f + (\bar{S}_L - \bar{S}_H)\left(\frac{\partial \bar{M}_L}{\partial N}\right)_M \tag{2.10.80}$$

$$\bar{E} = E^f + (\bar{E}_L - \bar{E}_H)\left(\frac{\partial \bar{M}_L}{\partial N}\right)_M. \tag{2.10.81}$$

Note that in (2.10.77) M_L and M_H are independent quantities, but in (2.10.79) as well as in (2.10.80) and (2.10.81) we evaluate the derivative at equilibrium, i.e., we must take $M_L = \bar{M}_L$ and $M_H = \bar{M}_H$. The quantity $(\partial \bar{M}_L/\partial N)_M$ is the differential change in the composition of the adsorbent molecules caused by the ligand. This quantity is in general nonzero [except for some limiting cases such as $h = 1$ or $K = 0$, for which see Eq. (2.10.69)]. By comparing (2.10.79) with (2.10.80) and (2.10.81), we find that there is a fundamental difference between the chemical potential and any other partial molecular quantity. We shall now see that this difference is equivalent to the difference we have observed in the expressions (2.10.28), (2.10.31), and (2.10.33).

Because of the existence of chemical equilibrium, at the point $\bar{M}_L$, $\bar{M}_H$, the chemical potentials μ_L and μ_H are equal, i.e.,

$$\mu_L = \mu_H. \tag{2.10.82}$$

Therefore the second term on the rhs of (2.10.79) is zero, whether $\partial \bar{M}_L/\partial N$ is zero or nonzero. On the other hand, $\bar{S}_L - \bar{S}_H$ and $\bar{E}_L - \bar{E}_H$ are in general nonzero. Therefore both $\bar{S}$ and $\bar{E}$ in (2.10.80) and (2.10.81) have two contributions, one due to adsorption in the frozen-in system S^f and E^f and the second due to the relaxation effect, i.e., the effect induced on the equilibrium composition $\bar{M}_L$, $\bar{M}_H$ by the ligand molecules. We can evaluate all the quantities in (2.10.80) and (2.10.81) for the PES, the results being

$$(\mu_L - \mu_H) = 0 \tag{2.10.83}$$

$$(\bar{S}_L - \bar{S}_H)\left(\frac{\partial \bar{M}_L}{\partial N}\right)_{T,M} = \frac{E_L - E_H}{T}\frac{K(1 - h)}{(1 + K)(1 + hK)} \tag{2.10.84}$$

$$(\bar{E}_L - \bar{E}_H)\left(\frac{\partial \bar{M}_L}{\partial N}\right)_{T,M} = (E_L - E_H)\frac{K(1 - h)}{(1 + K)(1 + hK)}. \tag{2.10.85}$$

Note that $\bar{E}_i$ and $\bar{S}_i$ are partial (molecular) quantities, whereas E_L and E_H are the molecular quantities of the model. We see from the last three relations that if $E_L \neq E_H$, $K \neq 0$, and $h \neq 1$, then both $\bar{S}$ and $\bar{E}$ get finite contributions from the relaxation terms. In any case, the chemical potential does not get such a contribution. These are exact results for our particular model, but the conclusion has a more general validity beyond this model (see also Chapters 3, 7, and 8). We note also that the difference between $\bar{S}$

and S^f can now be rewritten as

$$\frac{1}{T}(\langle E_P \rangle_G - \langle E_P \rangle_0) = (\bar{S}_L - \bar{S}_H)\left(\frac{\partial \bar{M}_L}{\partial N}\right)_{T,M}, \qquad (2.10.86)$$

and similarly

$$\langle E_P \rangle_G - \langle E_P \rangle_0 = (\bar{E}_L - \bar{E}_H)\left(\frac{\partial \bar{M}_L}{\partial N}\right)_{T,M}; \qquad (2.10.87)$$

i.e., the relaxation terms, in the mixture-model approach, are the same as the effect of G on the average energy of a polymer molecule.

In this section we have treated a very simple model where the adsorbent molecules have two states, and we examined the effect of an adsorbed molecule on the distribution of adsorbent molecules in the two states. This problem also arises in other, more complicated situations, e.g., when a solute dissolves in a solvent, particularly in aqueous solutions. We shall discuss some of these problems in Chapter 7. Now that we have all the required quantities we can answer a question that is often raised about aqueous solutions, but can be reformulated in our model as well. The question is: How could it be that a process causes a structural change in the solvent, yet this change does not affect the standard free-energy change of the process? An exact answer may be given to this question, within this specific model. To do this we reformulate our problem in a slightly different way. Consider an empty system (pure solvent) with composition $\bar{M}_L^0$ and $\bar{M}_H^0$. Now add one ligand molecule to this system. The addition of the ligand may cause some "structural change" in the adsorbent molecules. As a matter of fact, we know from (2.10.69) exactly how much structural change is brought about by the addition of one ligand molecule. This quantity is in general nonzero (except for the cases $h = 1$ or $K = 0$). The new composition will be denoted by $\bar{M}_L$ and $\bar{M}_H$, respectively. The chemical potential of the ligand may be obtained as follows (note that $M = \bar{M}_L^0 + \bar{M}_H^0 = \bar{M}_L + \bar{M}_H$).

$$\begin{aligned}
\mu(\bar{M}_L, \bar{M}_H) &= \lim_{M \to \infty} [A(\bar{M}_L, \bar{M}_H, N = 1) - A(\bar{M}_L^0, \bar{M}_H^0)] \\
&= \lim_{M \to \infty} [A(\bar{M}_L^0, \bar{M}_H^0, N = 1) - A(\bar{M}_L^0, \bar{M}_H^0)] \\
&\quad + \lim_{M \to \infty} [A(\bar{M}_L, \bar{M}_H, N = 1) - A(\bar{M}_L^0, \bar{M}_H^0, N = 1)] \\
&= \mu'(\bar{M}_L^0, \bar{M}_H^0) + \Delta\mu^r, \qquad (2.10.88)
\end{aligned}$$

where μ' is the chemical potential of the ligand in the PES and $\Delta\mu^r$ is the contribution to μ from structural changes in the adsorbent molecules. Expanding $A(\bar{M}_L^0, \bar{M}_H^0, N = 1)$ about the equilibrium point $(\bar{M}_L, \bar{M}_H, N = 1)$, we obtain

$$\begin{aligned}
-\Delta\mu^r &= [(\bar{M}_L^0 - \bar{M}_L)\mu_L(\bar{M}_L, \bar{M}_H, N = 1) + (\bar{M}_H^0 - \bar{M}_H)\mu_H(\bar{M}_L, \bar{M}_H, N = 1)] \\
&\quad + \tfrac{1}{2}(\mu_{LL} - 2\mu_{HL} + \mu_{HH})(\bar{M}_L^0 - \bar{M}_L)^2 + \cdots \qquad (2.10.89)
\end{aligned}$$

where $\mu_{ij} = \partial^2 A/(\partial \bar{M}_i, \partial \bar{M}_j)$. (This should be understood as the second derivative of A with respect to M_i and M_j at the PES and then evaluated at $\bar{M}_L$, $\bar{M}_H$.)

Now, since at the point $(\bar{M}_L, \bar{M}_H, N = 1)$ a chemical equilibrium exists, we have $\mu_L = \mu_H$ and obviously $\bar{M}_L^0 - \bar{M}_L = -(\bar{M}_H^0 - \bar{M}_H)$. Therefore the first term on the rhs of (2.10.89) is zero. In addition, the second term will be as small as we wish in the limit of $M \to \infty$ as required in (2.10.88). This is so because $(\bar{M}_L^0 - \bar{M}_L)^2$ is finite but $(\mu_{LL} - 2\mu_{HL} + \mu_{HH})$ tends to zero as M^{-1}. More explicitly, in this model we have

$$(\mu_{LL} - 2\mu_{HL} + \mu_{HH}) = \frac{kT}{Mx_L^0 x_H^0} \xrightarrow[M \to \infty]{} 0. \qquad (2.10.90)$$

Similarly, all higher-order terms in the expansion (2.10.89) will tend to zero even faster than the second term. We therefore conclude that in the limit of a macroscopic system, $\Delta\mu^r \to 0$ and hence $\mu = \mu'$, which is the same result we obtained before. Had we carried out the same procedure for, say, the entropy, we would have obtained instead of (2.10.88) and (2.10.89) the relations

$$\bar{S}(\bar{M}_L, \bar{M}_H) = S^f(\bar{M}_L^0, \bar{M}_H^0) + \Delta S^r, \qquad (2.10.91)$$

where

$$-\Delta S^r = (\bar{S}_L - \bar{S}_H)(\bar{M}_L^0 - \bar{M}_L) + \tfrac{1}{2}(S_{LL} - 2S_{LH} + S_{HH})(\bar{M}_L^0 - \bar{M}_L)^2 + \cdot \cdot (2.10.92)$$

where we use the notation

$$S_{ij} = \frac{\partial^2 S}{\partial \bar{M}_i \, \partial \bar{M}_j}. \qquad (2.10.93)$$

Here again the second term and all higher-order terms in (2.10.92) will tend to zero as $M \to \infty$, but in contrast to (2.10.89), ΔS^r still contains the first-order term in (2.10.92), which is nonzero.

Thus we see that although $\Delta\mu^r$ tends to zero in the limit of a macroscopic system, $(\partial \bar{M}_L / \partial N)$ remains finite. This may be seen in yet another way through the identity (see section 5.13.10)

$$\left(\frac{\partial \bar{M}_L}{\partial N}\right)_{M,T} = -\left[\frac{\partial(\mu_L - \mu_H)}{\partial N}\right]_{T,M_L,M_H} \frac{1}{\mu_{LL} - 2\mu_{LH} + \mu_{HH}}. \qquad (2.10.94)$$

The two factors on the rhs of (2.10.94) tend to zero as M^{-1}. In particular, in our model we have

$$\left(\frac{\partial(\mu_L - \mu_H)}{\partial N}\right)_{T,M_L,N_H} = \frac{-kT(1 - h)}{M(x_L^0 + hx_H^0)}. \qquad (2.10.95)$$

Thus, although the leading term in (2.10.89), which is determined by (2.10.90) tends to zero as $M \to \infty$, the ratio of the two quantities (2.10.95) and (2.10.90) is finite in this limit. This answers the question posed above.

SUGGESTED READINGS

Simple applications of Statistical Thermodynamics may be found in many textbooks, e.g.:

T. L. Hill, *Introduction to Statistical Thermodynamics* (Addison-Wesley, Reading, MA, 1960).

N. Davidson, *Statistical Mechanics* (McGraw-Hill, New York, 1962).

3

Simple Systems with Interactions

In this chapter we introduce interactions among the particles of the system. These interactions are simple enough so that the treatment of the system by ST is still feasible. The models presented in the following sections are devised to mimic the behavior of binding or adsorption phenomena frequently occurring in biochemistry. Examples are the binding of oxygen to hemoglobin, the binding of ligands to multisubunit enzymes, the so-called allosteric effect, and the far more complex binding of proteins, such as repressors, to DNA.

Throughout this chapter, we use very simple models to introduce a number of concepts that are frequently used in the context of the theory of fluids. Examples are the pair-correlation function, direct and indirect correlations, potential of average force, nonadditivity of the triplet correlation function, and so on. All these will be introduced again in Chapter 5. However, it is easier to grasp these concepts within the simple models. This should facilitate understanding them in more complex systems.

In section 2.9 we treated a system of M *sites*, each having room for m ligands. In this chapter we shall refer to M *polymer* molecules, each having m *sites*. Thus the term "site" is reserved for the location of attachment of one ligand only. Usually a polymer has a few binding sites. Our main objective will be to study the effects of the interactions among the ligands occupying the same polymer on the binding isotherms of these systems.

In actual examples the polymers are not localized. They form a dilute solution within the solvent. The consideration of M indistinguishable polymers causes only a minor difference in the PF of the system (see section 2.9.2). A more serious modification should be considered when taking into account the presence of the solvent. This will be deferred to Chapter 8.

For most of the present chapter, we can think of the M polymers, very large compared with the ligands, as forming an ideal gas in which the translational and rotational degrees of freedom have been frozen in. This makes the polymers localized, and therefore distinguishable. The release of the polymers from their fixed positions adds the translational-rotational PF and a factor $M!$. These factors do contribute to the PF of the system, but do not change the thermodynamics of adsorption.

3.1. TWO IDENTICAL SITES ON A POLYMER: DIRECT INTERACTION BETWEEN THE LIGANDS

In this section we examine the simplest case of an adsorption process with ligand–ligand interactions. The model treated here is a special case of the multiple occupancy model discussed in section 2.9. Here, $m = 2$, i.e., there are a maximum of two ligands on a given polymer. The system consists of M identical, independent, and localized polymers, each of which contains two identical sites for binding a ligand G with binding energy U. In addition there is a *direct* interaction between the two ligands occupying the two sites of the same polymer.

3.1.1. The Binding Isotherm

The grand PF of the system (open with respect to G only) is

$$\Xi(T, M, \lambda) = \xi^M, \tag{3.1.1}$$

where

$$\xi(T, \lambda) = Q(0, 0) + [Q(0, 1) + Q(1, 0)]\lambda + Q(1, 1)\lambda^2. \tag{3.1.2}$$

The four terms on the rhs of (3.1.2) correspond to the four states of the polymer (Fig. 3.1). Here, the polymer is unaffected by the binding of ligands; therefore we may factor out the quantity $Q(0, 0)$, the PF of the empty polymer. In this model the two states $(0, 1)$ and $(1, 0)$ are distinguishable, but the corresponding PFs are identical. Thus we write

$$Q(0) = Q(0, 0) = 1 \tag{3.1.3}$$

$$Q(1) = Q(0, 1) = Q(1, 0) = \exp(-\beta U) = q \tag{3.1.4}$$

$$Q(2) = Q(1, 1) = \exp(-\beta 2U - \beta U_{12}) = q^2 S, \tag{3.1.5}$$

where we define

$$q = \exp(-\beta U), \qquad S = \exp(-\beta U_{12}). \tag{3.1.6}$$

The average number of ligands per polymer is

$$\bar{n} = \lambda \frac{\partial \ln \xi}{\partial \lambda} = \frac{2(q\lambda + q^2 S \lambda^2)}{\xi}, \tag{3.1.7}$$

and the average fraction of occupied sites is

$$\theta = \frac{\bar{n}}{2} = \frac{q\lambda (1 + q\lambda S)}{1 + 2q\lambda + q^2 S \lambda^2}. \tag{3.1.8}$$

Clearly, when $S = 1$, i.e., there is no interaction between ligands, Eq. (3.1.8) reduces to the simple Langmuir isotherm.

If the ligands are at equilibrium with an ideal-gas phase, then

$$\lambda = \lambda^{\mathrm{id}} = \Lambda^3 \rho = \frac{\Lambda^3 P}{kT}. \tag{3.1.9}$$

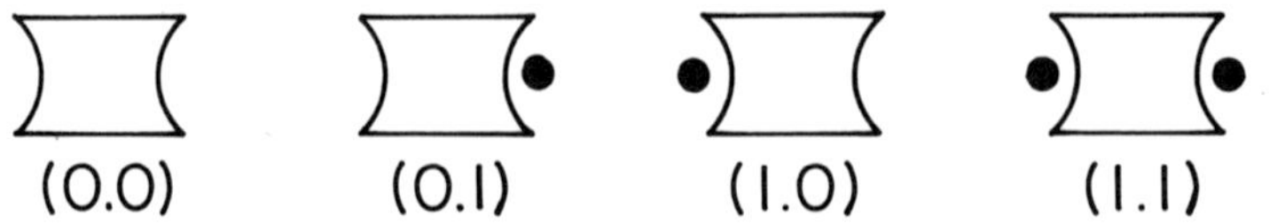

$$(0.0) \qquad (0.1) \qquad (1.0) \qquad (1.1)$$

FIGURE 3.1. Four states of a polymer, corresponding to the four terms in Eq. (3.1.2).

We denote by K the Langmuir binding constant, in the same way as we did in the case of the Langmuir isotherm; i.e., we define

$$K = \Lambda^3 \frac{q}{kT}. \tag{3.1.10}$$

The binding isotherm in our case, (3.1.8), is rewritten as

$$\theta = \frac{KP(1 + KPS)}{1 + 2KP + K^2 P^2 S}. \tag{3.1.11}$$

We may define a modified binding "constant" K^* by

$$K^* = K \frac{1 + KPS}{1 + KP} \tag{3.1.12}$$

to rewrite (3.1.11) as

$$\theta = \frac{K^*P}{1 + K^*P}. \tag{3.1.13}$$

This has the same form as the simple Langmuir isotherm. However, the "constant" K^* depends on P through (3.1.12). It becomes the Langmuir constant K when $S = 1$.

Differentiating θ with respect to λ and evaluating at $\lambda = 0$, we obtain

$$\left(\frac{\partial \theta}{\partial \lambda}\right)_{\lambda = 0} = q. \tag{3.1.14}$$

Thus the initial slope is determined only by $q = \exp(-\beta U)$, as in the simple Langmuir isotherm. This is clear, since initially the binding of a ligand on an empty polymer does not feel the interaction with the second ligand. The interactions enter in the initial curvature of the isotherm, namely

$$\left(\frac{\partial^2 \theta}{\partial \lambda^2}\right)_{\lambda - 0} = 2q^2(S - 2), \tag{3.1.15}$$

which could be positive or negative according to whether $S > 2$ or $S < 2$, respectively.

The explicit expression for the chemical potential of the ligand in terms of θ may be obtained from (3.1.8):

$$q\lambda = \frac{(1 - 2\theta) \pm \sqrt{(1 - 2\theta)^2 - 4\theta(\theta - 1)S}}{2S(\theta - 1)}. \tag{3.1.16}$$

It is easily verified that the solution with the minus sign is the physically acceptable solution for our model. This can be checked by taking some limiting cases; e.g., for $\theta \to 0$ we must have $\lambda \to 0$, or for $\theta \to 1$, λ must tend to $+\infty$, or for $S = 1$, we return to the solution for the Langmuir case.

3.1.2. Distribution Functions

The probabilities of finding the polymer in any one of the possible states (Fig. 3.1) can be read from the PF (3.1.2). These are:

$$\Pr(0, 0) = \frac{1}{\xi}$$

$$\Pr(1, 0) = \Pr(0, 1) = \frac{q\lambda}{\xi}$$

$$\Pr(1, 1) = \frac{q^2\lambda^2 S}{\xi}. \tag{3.1.17}$$

Note that $\Pr(1, 0)$ is the probability of finding a specific site (say the left or right one) occupied and the second site on the same polymer empty. The probability of finding *any* one of the sites, on a given polymer, occupied is the sum of

$$\Pr(1, 0) + \Pr(0, 1) = \frac{2q\lambda}{\xi}. \tag{3.1.18}$$

The probability of finding one specific site occupied independent of the occupation state of the second site on the same polymer is

$$\Pr(1) = P(1, 0) + P(1, 1) = \frac{q\lambda}{\xi} + \frac{q^2\lambda^2 S}{\xi}. \tag{3.1.19}$$

We define the ligand–ligand correlation function by

$$g(1, 1) = \frac{\Pr(1, 1)}{\Pr(1)\,\Pr(1)} = \frac{S\xi}{(1 + \lambda qS)^2}. \tag{3.1.20}$$

When $S = 1$, $g(1, 1) = 1$, and there is no correlation between the sites. Note that as $\lambda \to 0$, both the probabilities $\Pr(1, 1)$ and $\Pr(1)$ in (3.1.20) tend to zero. However, the ratio is finite in this limit. We shall need only this limit of $g(1, 1)$ (see Appendix I)

$$g^0(1, 1) = \lim_{\lambda \to 0} \frac{\Pr(1, 1)}{\Pr(1)\,\Pr(1)} = \frac{Q(1, 1)}{[Q(0, 1)]^2} = S. \tag{3.1.21}$$

This quantity will be referred to as the *pair correlation function at infinite dilution*. It has a simple interpretation in terms of the Helmholtz energy change for the reaction, which we can write symbolically as

$$2(0, 1) \to (1, 1) + (0, 0); \tag{3.1.22}$$

i.e., two singly occupied polymers are converted to one doubly occupied and one empty polymer. The corresponding change in the Helmholtz energy is (note that by assumption (3.1.3), $A(0, 0) = 0$)

$$\Delta A = A(1, 1) - 2A(0, 1), \tag{3.1.23}$$

hence

$$g^0(1, 1) = \exp[-\beta(A(1, 1) - 2A(0, 1))] = S. \tag{3.1.24}$$

If $S = 1$, then $\Delta A = 0$ or $g^0(1, 1) = 1$, and we say that there is no correlation between the two ligands. If $S > 1$, i.e., $U_{12} < 0$, then $g^0(1, 1) > 1$ or $\Delta A < 0$; this is the case of positive correlation. If $S < 1$, i.e., $U_{12} > 0$, then $g^0(1, 1) < 1$ or $\Delta A > 0$; this case is said to be one of negative correlation between the ligands.

There are several ways to introduce the concept of cooperativity. In general, cooperativity between two sites means that an event that occurs on one site affects the properties of the second site. We mention here three possible ways of defining cooperativity.

1. For any λ, the probability of finding a specific site occupied is $Pr(1)$. The conditional probability of finding a specific site occupied, given that the second site on the same polymer is already occupied, is defined by

$$Pr(1/1) = \frac{Pr(1, 1)}{Pr(1)} = \frac{q\lambda S}{1 + q\lambda S} = g(1, 1)Pr(1). \qquad (3.1.25)$$

Thus the correlation function $g(1, 1)$ measures the deviation of the conditional probability from the unconditional probability $Pr(1)$. If $S = 1$, the probability of finding the second site occupied, given that the first site is occupied, is simply the Langmuir probability $\lambda q/(1 + q\lambda)$. When $S > 1$, we say that there is a positive cooperativity; when $S < 1$, negative cooperativity.

2. The Helmholtz energy change for bringing a ligand from a fixed position in a vacuum to a specific site on an empty polymer (see section 2.8) is

$$\Delta A^{*(1)} = A(1, 0) - A(0, 0) = -kT \ln \frac{Q(1, 0)}{Q(0, 0)} = U. \qquad (3.1.26)$$

The Helmholtz energy for bringing a ligand from a fixed position in vacuum to a specific empty site, given that the second site on the same polymer is occupied, is

$$\Delta A^{*(2)} = A(1, 1) - A(1, 0) = -kT \ln \frac{Q(1, 1)}{Q(1, 0)} = U + U_{12}. \qquad (3.1.27)$$

We see that the effect of an occupied site on the quantity ΔA^* is simply to add the interaction energy U_{12}. In this sense a positive cooperativity corresponds to $U_{12} < 0$, i.e., $\Delta A^{*(2)} < \Delta A^{*(1)}$. Negative cooperativity corresponds to $U_{12} > 0$, i.e., $\Delta A^{*(2)} > \Delta A^{*(1)}$. The two Helmholtz energy changes are connected by the correlation function $g^0(1, 1)$ as follows (see (3.1.24) and note that in this particular model $A(0, 0) = 0.$):

$$\exp[-\beta \Delta A^{*(2)}] = g^0(1, 1) \exp[-\beta \Delta A^{*(1)}]. \qquad (3.1.28)$$

3. The energy changes corresponding to the two processes described above can be easily computed, in this case, from

$$\Delta E^{*(1)} = \frac{\partial(\beta \Delta A^{*(1)})}{\partial \beta} = U, \qquad (3.1.29)$$

$$\Delta E^{*(2)} = \frac{\partial(\beta \Delta A^{*(2)})}{\partial \beta} = U + U_{12}. \qquad (3.1.30)$$

These are also the same as the partial molecular energies of the ligand, which may be computed from (3.1.16) and evaluated at $\theta = 0$ and $\theta = 1$ for the first and second ligand.

As expected, the adsorption *energy* on the first site is simply U; the adsorption energy on the second site is modified by the quantity U_{12}. In this model, the correlation arises only from the direct interaction between the two ligands, and therefore the differences between ΔA^*s in (3.1.26) and in (3.1.27) is the same as the difference between the

ΔE^*s in (3.1.29) and in (3.1.30). We shall see that this is not the case in the more general models. Different sources of cooperativity differently affect the free energy of adsorption and the energies of adsorption.

3.1.3. Generalizations

Two simple generalizations of this model are the following:

1. The two sites are different, but only one ligand can bind to either of the sites with binding energies U_1 and U_2. The four possible states of the polymer are the same as in Fig. 3.1. The PF of a single polymer is now

$$\xi = 1 + \exp(-\beta U_1)\lambda + \exp(-\beta U_2)\lambda + \exp[-\beta(U_1 + U_2 + U_{12})]\lambda^2$$
$$= 1 + q_1\lambda + q_2\lambda + q_1q_2S\lambda^2, \tag{3.1.31}$$

and the corresponding binding isotherm is

$$\bar{n} = \lambda \frac{\partial \ln \xi}{\partial \lambda} = \frac{(q_1 + q_2)\lambda + 2q_1q_2S\lambda^2}{\xi}. \tag{3.1.32}$$

Compare now Eq. (3.2.32), with $S = 1$ (no interaction), to Eq. (3.1.7), with $S \neq 1$. Clearly the two isotherms are indistinguishable if

$$2q = q_1 + q_2$$

and

$$2q^2S = 2q_1q_2.$$

This means that for any two-different-sites model with q_1 and q_2, without interaction, one can always find a two-identical-sites model, with interactions, such that $q = (q_1 + q_2)/2$ and $S = 4q_1q_2/(q_1 + q_2)^2$. The two systems will have the same binding isotherm. Note that if $q_1 \neq q_2$, then $(q_1 + q_2)^2 > 4q_1q_2$; therefore S must be less than unity to satisfy this equivalence of the two systems.

2. The two sites are different and each site can bind a different ligand, say G_1 and G_2, with binding energies U_1 and U_2, respectively. The PF is similar to (3.1.31), but in this case we have two different activities, λ_1 and λ_2; hence

$$\xi = 1 + q_1\lambda_1 + q_2\lambda_2 + q_1q_2S\lambda_1\lambda_2. \tag{3.1.33}$$

The average number of G_1's and of G_2's on a single polymer are

$$\bar{n}_1 = \lambda_1 \frac{\partial \ln \xi}{\partial \lambda_1} = \frac{q_1 + q_1q_2S\lambda_2}{\xi} \tag{3.1.34}$$

$$\bar{n}_2 = \lambda_2 \frac{\partial \ln \xi}{\partial \lambda_2} = \frac{q_2 + q_1q_2S\lambda_1}{\xi}. \tag{3.1.35}$$

3.1.4. Some Numerical Examples

We explore here some numerical aspects of the binding isotherm for this model. The examples worked out in the present section will serve as reference cases for the forthcoming sections of this chapter. We assume that U, the binding energy, is given, hence $q = \exp(-\beta U)$ will be constant. We define a new activity of the ligand by

$$x = \lambda q \tag{3.1.36}$$

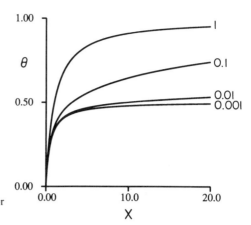

FIGURE 3.2. Binding isotherms [Eq. (3.1.37)] for various values of $S < 1$. Negative cooperativity.

and rewrite the binding isotherm (3.1.8) as

$$\theta = \frac{x + x^2 S}{1 + 2x + x^2 S}.$$ (3.1.37)

When $S = 1$, we have the Langmuir isotherm. Figure 3.2 shows the binding isotherms for several values of $S < 1$, i.e., for negative cooperativity. Note that when $S \ll 1$, there is very strong repulsion between the two ligands on the same polymer. Therefore the ligands will tend to spread first on different polymers until a "saturation" of first sites is achieved. Upon further increase of x, we shall eventually reach a complete saturation of all the sites. In Fig. 3.2 we can see the intermediate "saturation" at $\theta = \frac{1}{2}$ for the case of $S \ll 1$. The eventual saturation cannot be shown on this scale and will be demonstrated on a $\theta(\ln x)$ plot in Fig. 3.3.

On the same scale of Fig. 3.2 it is difficult to demonstrate the case of positive cooperativity, i.e., with $S > 1$. The curves in these cases will be too close to the ordinate.

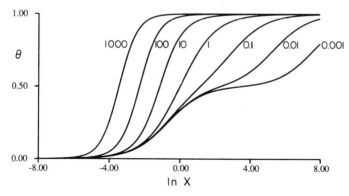

FIGURE 3.3. Plot of $\theta(\ln x)$ for the same values of S as in Figs. 3.2 and 3.4.

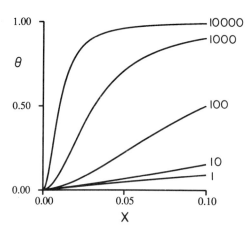

FIGURE 3.4. Binding isotherms [Eq. (3.1.37)] for various values of $S > 1$. Positive cooperativity.

Therefore we plot in Fig. 3.4 some curves with $S > 1$. Note that the upper Langmuir curve ($S = 1$) of Fig. 3.2 is now the lower curve of Fig. 3.4. We have amplified the region $0 \leq x \leq 0.1$ to show how the isotherms change with increasing values of S. Note that for $S = 2$ the initial curvature is zero [Eq. (3.1.5)], but for $S > 2$ the initial curvature is positive in contrast to all the curves of Fig. 3.2, all of which have negative curvature. Furthermore, as S increases, the curves rise steeply and change the sign of their curvature. This is the typical S-shaped curve characterizing positive cooperativity.

It is sometimes convenient to plot the isotherm $\theta(\ln x)$, Fig. 3.3. Here we can demonstrate both positive and negative cooperativity on the same plot. Also, in the case of $S \ll 1$ we can see the intermediate "saturation" at $\theta = \frac{1}{2}$, followed by the eventual saturation when $x \to \infty$.

The slope and the curvature of the $\theta(x)$ function (3.1.37) are

$$\theta'(x) = \frac{\partial \theta}{\partial x} = \frac{1 + 2xS + x^2S}{(1 + 2x + x^2S)^2} \tag{3.1.38}$$

$$\theta''(x) = \frac{\partial^2 \theta}{\partial x^2} = \frac{-2(x^3S^2 + 3x^2S^2 + 3xS - S + 2)}{(1 + 2x + x^2S)^3}. \tag{3.1.39}$$

We define the value of x at which the slope is maximal by

$$\theta''(x_{\max}) = 0. \tag{3.1.40}$$

For $S = 2$, $x_{\max} = 0$. As we increase $S > 2$, the value of $x_{\max}$ initially moves to the right. At about $S \approx 6$, the value of $x_{\max}$ starts to move to the left as we increase S. The maximum value of $x_{\max}$ is about 0.125.

In discussing the binding isotherm of oxygen to hemoglobin in section 3.6 we shall be interested in an S-shaped curve of a certain steepness with the maximal steepness to occur at some specific range of concentrations. For instance, suppose that we require a binding curve with a certain maximal steepness of $\theta'(x_{\max}) = 20$ at $x_{\max} = 10$. These requirements cannot be met in this model for any value of S. For if we increase S, we can increase the steepness as we wish, but we cannot at the same time require that $x_{\max}$ occur at any required value. In fact, the analysis made above shows that $x_{\max}$ changes within a quite narrow range of, say, $0 \leq x_{\max} \lesssim 0.125$. The conclusion is that in order to fulfill any given requirements on x_{max} and $\theta'(x_{max})$, we need a more complicated cooperative mechanism. We shall see that our requirements on the binding isotherm of hemoglobin are even more stringent than the ones specified above. In particular, we shall need a positive cooperative system (hemoglobin) the $\theta(x)$ curve of which falls below that of a noncooperative system (myoglobin). This requirement can never be met in this model. As shown in Fig. 3.4, all the positive cooperative curves $(S > 1)$ fall above the noncooperative curve $(S = 1)$.

3.2. TWO IDENTICAL SITES ON A POLYMER HAVING TWO CONFORMATIONAL STATES: DIRECT AND INDIRECT CORRELATIONS

3.2.1. The Model and Its Solution

The model treated here combines aspects of the models treated in sections 2.10 and 3.1. We still have M polymers that are identical and independent. For simplicity, we can also assume that they are localized so that factors like $M!$ and the translational and rotational PF of the polymer can be omitted. Each polymer is allowed to be in one of two possible conformational states, which we denote by L and H. These two states are characterized only by their energy levels, E_L and E_H, and we assume that $E_L < E_H$, i.e., L has a lower energy than H.

As we have pointed out in section 2.4, each of the states L and H comprises a large number of quantum-mechanical states. Therefore, strictly speaking, we must use free-energy levels in writing the partition functions Q_L and Q_H [see (3.2.3) below]. However, in order to focus on some new aspects of this system, we assume that the states L and H are characterized only by their energies E_L and E_H.

Each polymer has two identical sites. The sites are distinguishable, say the left-hand site and the right-hand site. In this chapter we shall always refer to a site as a location binding one ligand. The polymer as a whole will have multiple sites—in this particular section, there are only two sites per polymer.

In the absence of ligands, there is an equilibrium between the two conformations (Fig. 3.5).

$$L \rightleftarrows H. \tag{3.2.1}$$

The mole fraction of the L and the H states are given by (section 2.10)

$$x_L^0 = \frac{Q_L}{Q_L + Q_H}, \qquad x_H^0 = \frac{Q_H}{Q_L + Q_H}, \tag{3.2.2}$$

where

$$Q_L = \exp(-\beta E_L), \qquad Q_H = \exp(-\beta E_H). \tag{3.2.3}$$

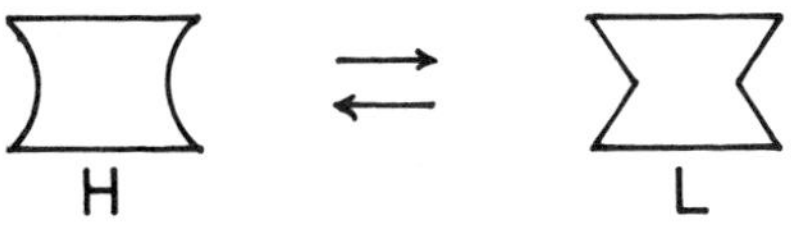

FIGURE 3.5. Two conformations L and H of the polymer.

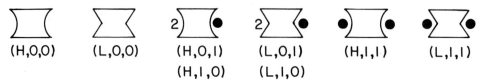

FIGURE 3.6. Eight possible configurations of the polymer with ligands.

We assume that the binding energy of the ligand on a site is either U_L or U_H, according to whether G is bound to the polymer in the state L or H, respectively. The ligand–ligand interactions are U_{12}^L and U_{12}^H, depending on the state of the polymer (e.g., the distance between the ligands is different in states L and H).

The grand PF of a single polymer is

$$\xi = Q(0) + 2Q(1)\lambda + Q(2)\lambda^2$$
$$= Q_L + Q_H + 2[Q_L q_L + Q_H q_H]\lambda + [Q_L q_L^2 S_L + Q_H q_H^2 S_H]\lambda^2. \qquad (3.2.4)$$

Here $Q(i)$ is the PF of a single polymer with i ligands. For $i = 0$ there are two states. For $i = 1$ there are four states, and for $i = 2$, two states. These are shown in Fig. 3.6.

We have defined

$$q_L = \exp(-\beta U_L), \qquad q_H = \exp(-\beta U_H) \qquad (3.2.5)$$
$$S_L = \exp(-\beta U_{12}^L), \qquad S_H = \exp(-\beta U_{12}^H). \qquad (3.2.6)$$

The eight terms in (3.2.4) correspond to the eight possible states of the polymer as shown in Fig. 3.6.

The average number of ligands per polymer is given by

$$\bar{n} = \lambda \frac{\partial \ln \xi}{\partial \lambda} = \frac{2[Q(1)\lambda + Q(2)\lambda^2]}{\xi}, \qquad (3.2.7)$$

and the average fraction of occupied sites is

$$\theta = \frac{\bar{n}}{2} = \frac{(Q_L q_L + Q_H q_H)\lambda + (Q_L q_L^2 S_L + Q_H q_H^2 S_H)\lambda^2}{\xi}. \qquad (3.2.8)$$

Note that $0 \leq \theta \leq 1$, whereas $0 \leq \bar{n} \leq 2$.

3.2.2. Probabilities

The mole fractions of polymers in the state L or H is the same as the probability of finding a particular polymer in state L or H, respectively. These can be read from the grand PF of a single polymer; i.e., we rewrite ξ in (3.2.4) as

$$\xi = (Q_L + 2Q_L q_L \lambda + Q_L q_L^2 S_L \lambda^2) + (Q_H + 2Q_H q_H \lambda + Q_H q_H^2 S_H \lambda^2), \qquad (3.2.9)$$

where each term in (3.2.9) corresponds to one of the states, L or H, of the polymer. Hence

$$\Pr(L) = x_L = \frac{Q_L + 2Q_L q_L \lambda + Q_L q_L^2 S_L \lambda^2}{\xi} \tag{3.2.10}$$

$$\Pr(H) = x_H = \frac{Q_H + 2Q_H q_H \lambda + Q_H q_H^2 S_H \lambda^2}{\xi}. \tag{3.2.11}$$

Two limiting cases can be obtained immediately from (3.2.10) and (3.2.11). For $\lambda \to 0$, the polymers will be empty and x_L and x_H reduce to the distribution of the empty polymers (3.2.2). The other extreme case is $\lambda \to \infty$; here, the polymers are fully occupied and the corresponding mole fractions are

$$\lim_{\lambda \to \infty} x_L = \frac{Q_L q_L^2 S_L}{Q_L q_L^2 S_L + Q_H q_H^2 S_H}, \qquad \lim_{\lambda \to \infty} x_H = \frac{Q_H q_H^2 S_H}{Q_L q_L^2 S_L + Q_H q_H^2 S_H}. \tag{3.2.12}$$

For any λ we can write the probabilities of each of the eight states of the polymer. These are

$$\Pr(L, 0, 0) = \frac{Q_L}{\xi}, \qquad \Pr(H, 0, 0) = \frac{Q_H}{\xi} \tag{3.2.13}$$

$$\Pr(L, 0, 1) = \Pr(L, 1, 0) = \frac{Q_L q_L \lambda}{\xi}, \qquad \Pr(H, 0, 1) = \Pr(H, 1, 0) = \frac{Q_H q_H \lambda}{\xi} \tag{3.2.14}$$

$$\Pr(L, 1, 1) = \frac{Q_L q_L^2 S_L \lambda^2}{\xi}, \qquad \Pr(H, 1, 1) = \frac{Q_H q_H^2 S_H \lambda^2}{\xi}. \tag{3.2.15}$$

The notation of the state of the polymer includes here the conformational state L or H and the occupancy state 0 or 1 for empty or occupied. Note that $\Pr(L, 0, 1)$ is the probability of finding a specific polymer in conformation L, with its lhs site empty and the rhs site occupied. By the assumption of identity of the sites, this is equal to $\Pr(L, 1, 0)$. The probability of finding the conformation L and (any) *one* site occupied is $\Pr(L, 0, 1) + \Pr(L, 1, 0)$, or simply $2Q_L q_L \lambda / \xi$.

A variety of conditional probabilities can be introduced in this system. For instance,

$$\Pr(0, 0/L) = \frac{\Pr(L, 0, 0)}{\Pr(L)} = \frac{Q_L}{Q_L + 2Q_L q_L \lambda + Q_L q_L^2 S_L \lambda^2} \tag{3.2.16}$$

is the probability of finding the two sites empty, given that the conformational state is L. A similar definition applies for H.

The quantity

$$\Pr(1, 1/L) = \frac{\Pr(L, 1, 1)}{\Pr(L)} = \frac{Q_L q_L^2 S_L \lambda^2}{Q_L + 2Q_L q_L \lambda + Q_L q_L^2 S_L \lambda^2} \tag{3.2.17}$$

is the probability of finding the two sites occupied, given that the state is L (and similarly for H).

The probability of finding the two sites occupied, independently of the state of the polymer, is the sum of the two probabilities in (3.2.15), one for L, one for H, i.e.

$$\Pr(1, 1) = \frac{Q_L q_L^2 S_L \lambda^2 + Q_H q_H^2 S_H \lambda^2}{\xi}. \tag{3.2.18}$$

The probability of finding a specific site, say the rhs site, occupied and the lhs site empty, independently of the state of the polymer, L or H, is the sum

$$\Pr(0, 1) = \Pr(L, 0, 1) + \Pr(H, 0, 1) = \frac{(Q_L q_L + Q_H q_H)\lambda}{\xi}. \qquad (3.2.19)$$

On the other hand, the probability of finding the rhs site occupied, independently of the state of the polymer (L or H) and independently of the occupational state of the lhs site is the sum

$$\Pr(1) = \Pr(0, 1) + \Pr(1, 1) = \frac{(Q_L q_L + Q_H q_H)\lambda}{\xi} + \frac{Q_L q_L^2 S_L \lambda^2 + Q_H q_H^2 S_H \lambda^2}{\xi}. \qquad (3.2.20)$$

3.2.3. Correlation Function

We now define the correlation function between two ligands on the same polymer by

$$g(1, 1) = \frac{\Pr(1, 1)}{[\Pr(1)]^2}. \qquad (3.2.21)$$

Note again (see section 3.1) that $\Pr(1, 1)$ and $\Pr(0, 1)$ tend to zero as $\lambda \to 0$; however, the ratio defined in (3.2.21) is finite in this limit, i.e.,

$$g^0(1, 1) = \lim_{\lambda \to 0} g(1, 1) = \frac{(Q_L q_L^2 S_L + Q_H q_H^2 S_H)(Q_L + Q_H)}{(Q_L q_L + Q_H q_H)^2}. \qquad (3.2.22)$$

In (3.2.22), the pair-correlation function includes an average of the two quantities S_L and S_H. We note that even if $S_L = S_H = 1$, i.e., no direct correlation, we still have a nonzero correlation between the two ligands. In order to further explore this indirect effect, it is convenient to separate the direct and the indirect effects. From now on we assume that $S_L = S_H$, i.e., the direct interaction between the two ligands is independent of the states of the polymer. Thus, for any λ, we rewrite (3.2.21) as

$$g(1, 1) = S \frac{\Pr(L, 0, 0)q_L^2 + \Pr(H, 0, 0)q_H^2}{[\Pr(L, 0, 0)q_L + \Pr(H, 0, 0)q_H + \Pr(L, 0, 0)q_L^2\lambda + \Pr(H, 0, 0)q_H^2\lambda]^2}. \qquad (3.2.23)$$

We now define the binding energy of the ligand G to the site as

$$B_G(\alpha) = \begin{cases} U_L & \text{if } \alpha = L \\ U_H & \text{if } \alpha = H \end{cases}. \qquad (3.2.24)$$

Similarly, the binding energy of a pair of ligands to the two sites is defined by

$$B_{GG}(\alpha) = 2B_G(\alpha) = \begin{cases} 2U_L & \text{if } \alpha = L \\ 2U_H & \text{if } \alpha = H \end{cases}, \qquad (3.2.25)$$

and the correlation function in (3.2.23) is rewritten as

$$g(1, 1) = S \frac{\langle \exp(-\beta B_{GG}) \rangle_{0,0}}{[\langle \exp(-\beta B_G) \rangle_{0,0} + \lambda \langle \exp(-\beta B_{GG}) \rangle_{0,0}]^2}. \qquad (3.2.26)$$

Note that the averages in (3.2.26) are taken with respect to the distribution (3.2.13) of the empty polymer. These probabilities depend on λ. We examine two limiting cases: For $\lambda \to 0$, we find

$$\Pr(L, 0, 0) \to \frac{Q_L}{Q_L + Q_H} = x_L^0$$

$$\Pr(H, 0, 0) \to \frac{Q_H}{Q_L + Q_H} = x_H^0, \qquad (3.2.27)$$

but for $\lambda \to \infty$, we have

$$\Pr(L, 0, 0) = \Pr(H, 0, 0) \to 0. \qquad (3.2.28)$$

Clearly, for $\lambda \to \infty$ the probability of finding a polymer empty is always zero.

From now on we restrict ourselves to the limit $\lambda = 0$; hence (3.2.26) still applies with the reinterpretation of the averages as in the empty polymer (see Appendix I).

$$g^0(1, 1) = S \frac{x_L^0 q_L^2 + x_H^0 q_H^2}{[x_L^0 q_L + x_H^0 q_H]^2}. \qquad (3.2.29)$$

Note that x_L^0 and x_H^0 are the probabilities of finding an empty polymer in state L or H in an *empty* system with respect to G, i.e., $\lambda = 0$. For any finite λ, the system is not empty with respect to G, there is still a finite probability of finding an *empty polymer* in state L or H, these are given in Eq. (3.2.13).

Comparing $g^0(1, 1)$ in (3.2.29) with the corresponding result in the model treated in section 3.1 [see Eq. (3.1.21)], we find that in addition to the *direct* correlation resulting from the direct ligand–ligand interaction U_{12}, we have an additional source of correlation originating from the new feature of the present model, i.e., the two conformational states of the polymer.

We define the *indirect* correlation function by

$$y(1, 1) = \frac{g^0(1, 1)}{S} = \frac{x_L^2 q_L^2 + x_H^0 q_H^2}{(x_L^0 q_L + x_H^0 q_H)^2} = \frac{(1 + h^2 K)(1 + K)}{(1 + hK)^2}, \qquad (3.2.30)$$

where $h = q_H/q_L$ and $K = Q_H/Q_L$. For any K and h, the indirect correlation is always greater than unity. This is clearly seen if we rewrite (3.2.30) as

$$y(1, 1) = 1 + \frac{(1 - h)^2 K}{(1 + hK)^2} \geq 1. \qquad (3.2.31)$$

Figure 3.7a shows the indirect correlation function $y(1, 1)$ as a function of the parameter K for a few values of h. In Fig. 3.7b we plot the same quantity as a function of the mole fraction of the L form. Two limiting cases are of interest. If $q_L = q_H$, i.e., $h = 1$, then $y(1, 1) = 1$ independently of the value of K. This brings us back to the model of section 3.1. From (3.2.31) it follows also that if $y(1, 1) = 1$ for any $K \neq 0$, then h must be equal

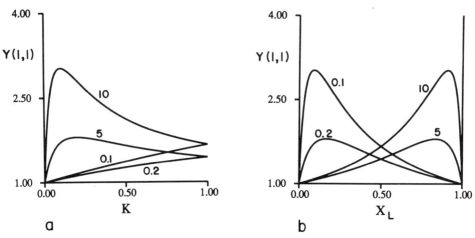

FIGURE 3.7. Indirect correlation function $y(1, 1)$ defined in Eq. (3.2.31). (a) As a function of K for various values of h. (b) As a function of x_L for various values of h.

to unity. Thus $h \neq 1$ is a sufficient and a necessary condition for the existence of indirect correlation.

The second case is $Q_L = Q_H$; i.e., $K = 1$. This is the case where the equilibrium $L \rightleftarrows H$ is maximally responsive (see section 2.10) to the effect of the ligand. In this case we have

$$y(1, 1) = 1 + \left(\frac{1-h}{1+h}\right)^2 = 2\frac{1+h^2}{(1+h)^2} \geq 1. \qquad (3.2.32)$$

This reaches the maximum value of $y = 2$ when either $h = 0$ or $h = \infty$. These results will be reinterpreted from a thermodynamical point of view in subsection 3.2.6. Qualitatively, the reason for having $y(1, 1) \geq 1$ is the following: When the first ligand binds, it shifts the equilibrium $L \rightleftarrows H$ toward that species for which the binding energy is lower. For instance, if $U_L < U_H$, then the first ligand will shift the equilibrium towards L. When the second ligand joins in, it will bind to a site which has a larger probability of being in state L. Therefore the binding free energy will now be lower than the binding free energy of the first ligand. This argument will be made more precise in subsection 3.2.6.

At this stage we can conclude that the ligand–ligand correlation has two independent sources: The *direct* interaction U_{12}, which may be either positive or negative, and an *indirect* correlation, which in this particular model is always positive, i.e., $y > 1$. The latter arises from the effect of the ligand on the equilibrium distribution of the two states L and H. If $U_L = U_H$, the ligand cannot affect the equilibrium distribution, and the indirect correlation disappears.

3.2.4. Binding Helmholtz Energies on First and Second Site, Cooperativity

We define the process of binding on the first site as the process of bringing G from a fixed position in the vacuum onto a specific site of an empty polymer (see also definitions in section 2.8). Symbolically the process is

$$G + (0, 0) \rightarrow (0, 1). \qquad (3.2.33)$$

The corresponding Helmholtz energy change is

$$\Delta A_G^{*(1)} = A(0, 1) - A(0, 0)$$

$$= -kT \ln \frac{Q(0, 1)}{Q(0, 0)} = -kT \ln \frac{Q_L q_L + Q_H q_H}{Q_L + Q_H}$$

$$= -kT \ln \langle \exp(-\beta B_G) \rangle_{0,0}. \tag{3.2.34}$$

Note that since we are considering a structureless ligand, the partition function of the ligand at a fixed position in the vacuum is unity. The result (3.2.24) is, however, also valid if we include internal partition functions of G but assume that the internal degrees of freedom are unchanged in the process (3.2.33). The result (3.2.34) is formally the same as for the solvation Helmholtz energy of a solute G. (This topic is discussed in Chapter 6.)

The average in (3.2.34) is taken with the distribution of the two states L and H of the *empty* polymer, given in Eq. (3.2.2). Here we introduce the subscript $(0, 0)$ to stress the condition that *both* sites are empty to distinguish this average from the conditional average defined below.

The process of binding on the second site is defined as the process of bringing G from a fixed position in the vacuum onto a fixed empty site, given that the other site on the same polymer is already occupied. Symbolically, the process is

$$G + (0, 1) \rightarrow (1, 1), \tag{3.2.35}$$

and the corresponding Helmholtz energy change is

$$\Delta A_G^{*(2)} = A(1, 1) - A(0, 1)$$

$$= -kT \ln \frac{Q(1, 1)}{Q(0, 1)} = -kT \ln \frac{Q_L q_L^2 S + Q_H q_H^2 S}{Q_L q_L + Q_H q_H}$$

$$= U_{12} - kT \ln \langle \exp(-\beta B_G) \rangle_{0,1}. \tag{3.2.36}$$

The quantity $\Delta A_G^{*(2)}$ differs from $\Delta A_G^{*(1)}$ in two respects: First we have U_{12} in (3.2.36), which is absent in (3.2.34). Secondly, the average in (3.2.36) is taken with the distribution probabilities of the two states of the polymer, given that one site is occupied and the second is empty. This is denoted by the subscript $(0, 1)$ in (3.2.36). This distribution is obtained from (3.2.14) and (3.2.19).

$$\Pr(L/0, 1) = \frac{\Pr(L, 0, 1)}{\Pr(0, 1)} = \frac{Q_L q_L}{Q_L q_L + Q_H q_H} \tag{3.2.37}$$

$$\Pr(H/0, 1) = \frac{\Pr(H, 0, 1)}{\Pr(0, 1)} = \frac{Q_H q_H}{Q_L q_L + Q_H q_H}. \tag{3.2.38}$$

Thus $\Pr(L/0, 1)$ is the probability of finding the state L of the polymer given that one site is empty and the second site is occupied. Similar meaning applies to $\Pr(H/0, 1)$.

The connection between $\Delta A_G^{*(1)}$ and $\Delta A_G^{*(2)}$ may be obtained by rewriting $\Delta G_G^{*(2)}$ in (3.2.36) as follows:

$$\Delta A_G^{*(2)} = U_{12} - kT \ln \left(\frac{Q_L q_L^2 + Q_H q_H^2}{Q_L q_L + Q_H q_H} \times \frac{Q_L + Q_H}{Q_L + Q_H} \right)$$

$$= U_{12} - kT \ln \left[\frac{\langle \exp(-\beta B_{GG}) \rangle_{0,0}}{\langle \exp(-\beta B_G) \rangle_{0,0}} \right]. \tag{3.2.39}$$

Hence from (3.2.34), (3.2.39), and the definition of $g^0(1, 1)$ in (3.2.29) we obtain the required connection,

$$\exp(-\beta \Delta A_G^{*(2)}) = g^{(0)}(1, 1) \exp(-\beta \Delta A_G^{*(1)}). \tag{3.2.40}$$

Thus the correlation function transfers from $\Delta A_G^{*(1)}$ to $\Delta A_G^{*(2)}$. The binding Helmholtz energy of the *pair* (GG) of ligands on the polymer is defined by

$$\exp(-\beta \Delta A_{GG}^*) = \langle \exp(-\beta B_{GG}) \rangle_{0,0}. \tag{3.2.41}$$

This corresponds to the process written symbolically as

$$(GG) + (0, 0) \to (1, 1). \tag{3.2.42}$$

Note that U_{12} is not involved in ΔA_{GG}^*. It is included in both (GG) and the final state $(1, 1)$.

We can now reinterpret the correlation function $g^0(1, 1)$ as

$$\begin{aligned}
g^0(1, 1) &= \exp[-\beta(\Delta A_G^{*(2)} - \Delta A_G^{*(1)})] \\
&= \exp[-\beta(A(1, 1) - 2A(1, 0) + A(0, 0))] \\
&= \exp[-W(1, 1)],
\end{aligned} \tag{3.2.43}$$

where the exponent in (3.2.43) contains the Helmholtz energy change for the process

$$2(0, 1) \to (0, 0) + (1, 1); \tag{3.2.44}$$

i.e., we convert two singly occupied polymers into one doubly occupied and one empty polymer. Another way of interpreting the quantity $W(1, 1)$ is

$$\begin{aligned}
W(1, 1) &= \Delta A_G^{*(2)} - \Delta A_G^{*(1)} = U_{12} + \Delta A_{GG}^* - 2\Delta A_G^{*(1)} \\
&= U_{12} + \delta W,
\end{aligned} \tag{3.2.45}$$

where the work required for the process (3.2.44) is rewritten in (3.2.45) as the direct work U_{12} (when the same process takes place in the vacuum) and an indirect part δW which originates from the conformational equilibrium in the polymer. These quantities will reappear in the theory of solutions, and particularly in aqueous solutions, where δW will be referred to as the solvent-induced interaction and $W(1, 1)$ as the analogue of the potential of average force between two particles in a solvent.

We have found that in this particular model $y(1, 1)$ defined in (3.2.30) is always greater than unity. This is equivalent to the statement that

$$\delta W = -kT \ln y(1, 1) < 0. \tag{3.2.46}$$

As noted in section 3.1, the concept of cooperativity may be defined in various ways. The following are four equivalent definitions:

1. Positive cooperativity

$$W(1, 1) < 0 \tag{3.2.47}$$

2. Or [note (3.2.43)]

$$g^0(1, 1) - 1 > 0 \tag{3.2.48}$$

3. Or

$$\Delta A_G^{*(2)} < \Delta A_G^{*(1)} \tag{3.2.49}$$

4. Or

$$\Pr(1/1) > \Pr(1). \tag{3.2.50}$$

The positive character of cooperativity is best expressed by (3.2.50); i.e., the conditional probability of finding the second site occupied, given that the first site is already occupied, is larger than the probability of finding the same polymer with one site occupied (both probabilities are irrespective of the conformational state of the polymer). As for the source of the cooperativity, we have already noted two sources: the direct interaction U_{12}, which might contribute either positive or negative cooperativity, and the indirect part δW, which, in our model, is always negative (i.e., this part produces positive cooperativity).

Negative cooperativity is defined with the reverse signs in (3.2.47)–(3.2.50) and no cooperativity corresponds to equalities in (3.2.47)–(3.2.50). Note for instance that the overall cooperativity might be zero, i.e., $W(1, 1) = 0$ but the system can still be cooperative in the sense that $U_{12} \neq 0$ and $\delta W \neq 0$, but these two exactly cancel each other.

Another way of defining cooperativity which is not equivalent to the definitions given above uses the energy of the binding. This is discussed in the following subsection.

3.2.5. Energy of Binding on First and Second Sites

We define the energy change corresponding to the two processes described in the previous subsection, i.e., processes (3.2.33) and (3.2.35). These are obtainable by differentiation of (3.2.34) and (3.2.36) with respect to $\beta = (kT)^{-1}$.

$$\Delta E_G^{*(1)} = \frac{\partial(\beta \Delta A_G^{*(1)})}{\partial \beta}$$

$$= \frac{Q_L q_L (E_L + U_L) + Q_H q_H (E_H + U_H)}{Q_L q_L + Q_H q_H} - \frac{Q_L E_L + Q_H E_H}{Q_L + Q_H}$$

$$= \langle B_G \rangle_{0,1} + \langle E_p \rangle_{0,1} - \langle E_p \rangle_{0,0} \tag{3.2.51}$$

$$\Delta E_G^{*(2)} = \frac{\partial(\beta \Delta A_G^{*(2)})}{\partial \beta}$$

$$= U_{12} + \frac{Q_L q_L^2 (E_L + 2U_L) + Q_H q_H^2 (E_H + 2U_H)}{Q_L q_L^2 + Q_H q_H^2}$$

$$- \frac{Q_L q_L (E_L + U_L) + Q_H q_H (E_H + U_H)}{Q_L q_L + Q_H q_H}$$

$$= U_{12} + \langle B_{GG} \rangle_{1,1} - \langle B_G \rangle_{0,1} + \langle E_p \rangle_{1,1} - \langle E_p \rangle_{0,1} \tag{3.2.52}$$

$$= U_{12} + \langle B_G \rangle_{1,1} + \langle E_p + B_G \rangle_{1,1} - \langle E_p + B_G \rangle_{0,1}, \tag{3.2.53}$$

where

$$B_G(\alpha) = \begin{cases} U_L & \text{for } \alpha = L \\ U_H & \text{for } \alpha = H, \end{cases} \tag{3.2.54}$$

$$E_p(\alpha) = \begin{cases} E_L & \text{for } \alpha = L \\ E_H & \text{for } \alpha = H, \end{cases} \tag{3.2.55}$$

$$B_{GG}(\alpha) = \begin{cases} 2U_L & \text{for } \alpha = L \\ 2U_H & \text{for } \alpha = H. \end{cases} \tag{3.2.56}$$

Comparing (3.2.51) with (3.2.52), we see that there are three factors that make $\Delta E_G^{*(2)}$ different from $\Delta E_G^{*(1)}$.

1. The direct ligand–ligand interaction.
2. The average binding energy of the ligand to the site, and
3. The change in the average energy of the polymer, induced by the binding of G.

The reason for writing (3.2.53) in the last form will be clear in subsection 3.2.6. Note also that in this section E_L and E_H are treated as if they truly are energy levels, hence temperature independent. In reality these are Helmholtz energy levels and therefore in general are temperature dependent.

The energy change corresponding to the reaction (3.2.44) can be easily calculated from

$$\Delta E = \frac{\partial(\beta W(1, 1))}{\partial \beta} = \Delta E_G^{*(2)} - \Delta E_G^{*(1)}$$

$$= U_{12} + (E_L - E_H)\left(\frac{2Kh}{1 + Kh} - \frac{K}{1 + K} - \frac{Kh^2}{1 + Kh^2}\right)$$

$$+ 2(U_L - U_H)\left(\frac{Kh}{1 + Kh} - \frac{Kh^2}{1 + Kh^2}\right). \tag{3.2.57}$$

In the previous subsection we defined cooperativity with respect to $W(1, 1)$ or equivalently with respect to the difference in the binding free energies $\Delta A_G^{*(2)} - \Delta A_G^{*(1)}$. We could have defined another cooperativity in terms of the sign of ΔE; $\Delta E < 0$ corresponds to the case when the binding energy to the second site is more negative than the binding energy to the first site. The sign of ΔE does not have to bear any relation to the sign of $W(1, 1)$. Of course, if there is only one state of the polymer, or if $h = 1$, then

$$W(1, 1) = \Delta E = U_{12}. \tag{3.2.58}$$

In (3.2.58) the cooperativity is due only to the direct interaction U_{12}. This brings us back to section 3.1, in which case $W(1, 1)$ and ΔE are equal. However, when indirect correlation exists, then the two quantities $W(1, 1)$ and ΔE are in general different, and therefore, in general can have different signs.

It follows also that, in general, a positive cooperativity in the sense of $W(1, 1) < 0$, does not necessarily imply a positive cooperativity in the sense of $\Delta E < 0$. We shall return to this point in section 3.2.6. Here we continue to analyze the components that contribute to ΔE in (3.2.57).

First note that both $\langle B_G \rangle_{0,1}$ and $\langle B_G \rangle_{1,1}$ may vary between the two bounds U_L and U_H. For instance, if $U_L < U_H$, then

$$U_L \le \langle B_G \rangle_{0,1} \le U_H, \quad \text{and} \quad U_L \le \langle B_G \rangle_{1,1} \le U_H. \tag{3.2.59}$$

But the difference $\langle B_G \rangle_{1,1} - \langle B_G \rangle_{0,1}$ must be negative. The reason is the same as the argument given in connection with the positive cooperativity of the indirect part of $W(1, 1)$ (see sections 3.2.3 and 3.2.6 below). To see that, we write this difference as follows

$$\langle B_G \rangle_{1,1} - \langle B_G \rangle_{0,1} = [U_L \Pr(L/1, 1) + U_H \Pr(H/1, 1)]$$
$$- [U_L \Pr(L/0, 1) + U_H \Pr(H/0, 1)]$$
$$= (U_L - U_H)(\Pr(L/1, 1) - \Pr(L/1, 0)) \le 0. \tag{3.2.60}$$

If $U_L - U_H < 0$, then $\Pr(L/1, 1) - \Pr(L/1, 0)$ is positive; i.e., G will cause a shift toward L. If $U_L - U_H > 0$, then $\Pr(L/1, 1) - \Pr(L/0, 1)$ is negative.

Both $\Delta E_G^{*(1)}$ and $\Delta E_G^{*(2)}$ contain a term which may be referred to as the conformational change induced by G on the polymer. In $\Delta E_G^{*(1)}$ we have (see Eq. 3.2.51)

$$\langle E_p \rangle_{0,1} - \langle E_p \rangle_{0,0} = [E_L \Pr(L/0, 1) + E_H \Pr(H/0, 1)]$$
$$- [E_L \Pr(L/0, 0) + E_H \Pr(H/0, 0)]$$
$$= (E_L - E_H)[\Pr(L/0, 1) - \Pr(L/0, 0)]. \tag{3.2.61}$$

Here the sign of the quantity may be either positive or negative, depending on the signs of both $E_L - E_H$ and $\Pr(L/0, 1) - \Pr(L/0, 0)$.

Similarly, in $\Delta E_G^{*(2)}$ we have the term (see Eq. 3.2.53)

$$\langle E_p + B_G \rangle_{1,1} - \langle E_p + B_G \rangle_{0,1} = (E_L + U_L - E_H - U_H)[\Pr(L/1, 1) - \Pr(L/0, 1)] \tag{3.2.62}$$

which again can be either positive or negative. The contributions of both (3.2.61) and (3.2.62) are included in the middle term on the rhs of (3.2.57).

3.2.6. Cooperativity and Induced Conformational Changes

In section 2.10 we made a detailed analysis of the conformational changes induced by the ligand. The system discussed in the present section is essentially the same as the one treated in section 2.10 with the additional feature that there are two sites per polymer instead of one. We shall therefore extend the treatment of section 2.10 to include only this new feature of the present model.

We start with the differential shift in the equilibrium concentration of, say, the L form, which is defined in Eq. (2.10.69).

$$d_L = \left(\frac{\partial \bar{M}_L}{\partial N} \right)_{T,M} = \left(\frac{\partial x_L}{\partial \theta} \right)_T = \frac{K(1 - h)}{(1 + K)(1 + hK)}. \tag{3.2.63}$$

In our model we can ask two questions regarding the equilibrium shift, say of the L form, induced by the binding of either the *first* or the *second* ligand. Clearly the first ligand has exactly the same effect as in (3.2.63). The first ligand that binds to the polymer with the initial equilibrium concentration x_L^0 clearly does not know of the effect of the second site. In the notation of the present section, the appropriate differential shift is

$$d_L^{(1)} = \Pr(L/0, 1) - \Pr(L/0, 0) = \frac{Q_L q_L}{Q_L q_L + Q_H q_H} - \frac{Q_L}{Q_L + Q_H}$$

$$= \frac{1}{1 + Kh} - \frac{1}{1 + K} = \frac{K(1 - h)}{(1 + Kh)(1 + K)}, \quad (3.2.64)$$

which is the same as (3.2.63), as expected. If we choose $U_L < U_H$, i.e., $h = q_H/q_L < 1$, we obtain

$$d_L^{(1)} > 0, \qquad\qquad (3.2.65)$$

which simply means that for any K, the ligand will always shift the equilibrium toward the L form.

The shift caused by the binding of the second ligand is likewise defined by

$$d_L^{(2)} = \Pr(L/1, 1) - \Pr(L/0, 1) = \frac{Q_L q_L^2}{Q_L q_L^2 + Q_H q_H^2} - \frac{Q_L q_L}{Q_L q_L + Q_H q_H}$$

$$= \frac{1}{1 + Kh^2} - \frac{1}{1 + Kh} \qquad (3.2.66)$$

$$= \frac{Kh(1 - h)}{(1 + Kh^2)(1 + Kh)}.$$

Again, for $h < 1$ this will always be positive, for the same reason as above. Note that (3.2.66) is obtained from (3.2.64) by replacing K by Kh. This is understandable since the second ligand approaching the polymer sees a singly occupied polymer with equilibrium constant of Kh, which may be obtained simply by putting $\theta = 1$ in Eq. (2.10.65). Therefore the effect of the second ligand is formally the same as that of the first but with the modified constant Kh instead of K.

Thus, in order to estimate $d_L^{(2)}$, we can look at the $d_L^{(1)}$ curve in Fig. 2.15a but read the values at Kh instead of K. Alternatively, we can read the value of $d_L^{(2)}$ simply at the corresponding value of x_L in Fig. 2.15b. The difference between the two is

$$d_L^{(2)} - d_L^{(1)} = \frac{Kh(1 - h)}{(1 + Kh^2)(1 + Kh)} - \frac{K(1 - h)}{(1 + Kh)(1 + K)}$$

$$= \frac{K(1 - h)^2(Kh - 1)}{(1 + K)(1 + Kh)(1 + Kh^2)}. \qquad (3.2.67)$$

The sign of this quantity depends on whether $Kh > 1$ or $Kh < 1$.

We can now see the physical reason for the result $y(1, 1) \geq 1$ in this model. To see this we rewrite the analogue of (3.2.40) by excluding the direct interaction, i.e.,

$$\langle \exp(-\beta B_G) \rangle_{0,1} = y(1, 1)\langle \exp(-\beta B_G) \rangle_{0,0}, \tag{3.2.68}$$

or, equivalently,

$$
\begin{aligned}
y(1, 1) &= \left[\frac{Q_L q_L^2 + Q_H q_H^2}{Q_L q_L + Q_H q_H} \right] \times \left[\frac{Q_L q_L + Q_H q_H}{Q_L + Q_H} \right]^{-1} \\
&= \frac{q_L \Pr(L/0, 1) + q_H \Pr(H/0, 1)}{q_L \Pr(L/0, 0) + q_H \Pr(H/0, 0)} \\
&= \frac{1 + \Pr(H/0, 1)(h - 1)}{1 + \Pr(H/0, 0)(h - 1)}.
\end{aligned}
\tag{3.2.69}
$$

If $h < 1$, then the L form is stabilized by the ligand; hence $\Pr(H/0, 1) < \Pr(H/0, 0)$, and the numerator is larger than the denominator. If $h > 1$, then the H form is stabilized by the ligand; hence $\Pr(H/0, 1) > \Pr(H/0, 0)$, and again the numerator is larger than the denominator. Thus in any case we have $y(1, 1) \geq 1$, a result that we had already reached in section 3.2.3.

Another way of looking at this result is the following. Both the numerator and the denominator in (3.2.69) are averages of the quantity $\exp(-\beta B_G)$. If $h < 1$, i.e., $q_H < q_L$, this average is bound by

$$q_H \leq \langle \exp(-\beta B_G) \rangle \leq q_L. \tag{3.2.70}$$

But since the distribution in the numerator favors the L form compared to the denominator, it is clear that the average in the numerator will be closer to q_L than to q_H, and hence $y(1, 1) \geq 1$. On the other hand, when $h > 1$, the distribution in the numerator favors the H form compared to the denominator, but now $q_H > q_L$, and again $y(1, 1) \geq 1$. This is equivalent to the statement that

$$\Delta A_G^{*(2)} < \Delta A_G^{*(1)}. \tag{3.2.71}$$

Note, however, that although the difference $\Delta A_G^{*(2)} - \Delta A_G^{*(1)}$ is affected by the amount of conformational change induced by the ligand, each of the quantities $\Delta A_G^{*(i)}$ separately is not affected by the conformational changes caused by the binding of the ith ligand. This is an important observation which will appear in several forms throughout this book. $\Delta A_G^{*(1)}$ in (3.2.34) is an average with the probability distribution of the *empty* polymer, denoted by $(0, 0)$ in (3.2.34). This is so even when the first ligand induces a finite conformational change in the polymer $(d_L^{(1)} \neq 0)$ in (3.2.64). Similarly, $\Delta A_G^{*(2)}$ is expressed as an average over the distribution of the singly occupied polymer, denoted by $(0, 1)$ in (3.2.36). It is independent of the conformational changes induced by the *second* ligand, which could be finite; $d_L^{(2)} \neq 0$ in (3.2.66). It is dependent on the changes caused by the *first* ligand. In other words, when the second ligand lands on a singly occupied polymer, it sees an equilibrium constant Kh for the reaction $L \rightleftarrows H$ which differs from the initial equilibrium constant of the empty polymer, which is K. (Note also that the corresponding equilibrium constant for the doubly occupied polymer is Kh^2, but this does not enter into the above consideration.)

The situation is markedly different for the binding *energy* and the corresponding binding *entropy*. $\Delta E_G^{*(1)}$ as defined in (3.2.51) contains a term due to the conformational changes induced by the *first* ligand.

$$\Delta E_G^{*(1)} = \langle B_G \rangle_{0,1} + (E_L - E_H)[\Pr(L/0, 1) - \Pr(L/0, 0)] \qquad (3.2.72)$$

$$T\Delta S_G^{*(1)} = -\Delta A_G^{*(1)} + \Delta E_G^{*(1)}$$

$$= kT \ln \langle \exp(-\beta B_G) \rangle_{0,0} + \langle B_G \rangle_{0,1} + (E_L - E_H)[\Pr(L/0, 1) - \Pr(L/0, 0)]. \qquad (3.2.73)$$

We have already noted that $\langle B_G \rangle_{0,1}$ is bound by (U_L, U_H), depending on which of the two is the larger. The second term on the rhs of (3.2.72) can be either positive or negative, depending on both signs of $E_L - E_H$ and of $(U_L - U_H)$. Note, however, that this term cancels out when the combination $\Delta E_G^{*(1)} - T\Delta A_G^{*(1)}$ is formed.

Similarly, for the second ligand we have [see Eq. (3.2.53)]

$$\Delta E_G^{*(2)} = U_{12} + \langle B_G \rangle_{1,1} + \langle E_p + B_G \rangle_{1,1} - \langle E_p + B_G \rangle_{0,1}, \qquad (3.2.74)$$

$$T\Delta S_G^{*(2)} = -\Delta A_G^{*(2)} + \Delta E_G^{*(2)}$$

$$= kT \ln \langle \exp(-\beta B_G) \rangle_{0,1} + \langle B_G \rangle_{1,1} + \langle E_p + B_G \rangle_{1,1} - \langle E_P + B_G \rangle_{0,1}. \qquad (3.2.75)$$

Again, both $\Delta E_G^{*(2)}$ and $\Delta S_G^{*(2)}$ depend on the conformational changes induced by the second ligand [see also (3.2.62)]. However, when the combination $\Delta E_G^{*(2)} - T\Delta S_G^{*(2)}$ is formed, these effects cancel out and do not appear in $\Delta A_G^{*(2)}$.

3.2.7. The Binding Isotherm

We define the first and second intrinsic binding constants by

$$K_1 = \exp(-\beta \Delta A_G^{*(1)}) = \frac{Q_L q_L + Q_H q_H}{Q_L + Q_H} \qquad (3.2.76)$$

$$K_2 = \exp(-\beta \Delta A_G^{*(2)}) = S \frac{Q_L q_L^2 + Q_H q_H^2}{Q_L q_L + Q_H q_H}, \qquad (3.2.77)$$

in terms of which the binding isotherm (3.2.8) can be written as

$$\theta = \frac{K_1 \lambda + K_1 K_2 \lambda^2}{1 + 2K_1 \lambda + K_1 K_2 \lambda^2}$$

$$= \frac{K_1 \lambda + K_1^2 S y \lambda^2}{1 + 2K_1 \lambda + K_1^2 S y \lambda^2}. \qquad (3.2.78)$$

In section 3.1, we explored the behavior of this function for the case $y = 1$, no indirect effects. Here, on the other hand, we take the case $S = 1$ to explore the new feature of this system. For $S = 1$, the isotherm (3.2.78) is formally the same as in (3.1.8), with y in (3.2.78) playing the role of S in (3.1.8). In section 3.1 the parameters q and S were independent parameters of the model. In fact, in section 3.1.4 we examined the function $\theta(x)$, for different values of S ($x = q\lambda$ was the modified concentration variable). In contrast, the quantities K_1 and y in (3.2.78) are not independent quantities. Both are determined by the molecular parameters of the system, h and K. Therefore, in exploring the behavior of this isotherm, we cannot choose any pair of independent parameters K_1 and y. This can be made more conspicuous by rewriting (3.2.78) in terms of K and h. Defining the new concentration variable $x = q_L \lambda$, we have

$$\theta = \frac{x\left(\dfrac{1 + Kh}{1 + K}\right) + x^2\left(\dfrac{1 + Kh^2}{1 + K}\right)}{1 + 2x\left(\dfrac{1 + Kh}{1 + K}\right) + x^2\left(\dfrac{1 + Kh^2}{1 + K}\right)}. \tag{3.2.79}$$

This should be compared with (3.1.8). Here, in contrast to the model treated in section 3.1, we have two degrees of freedom h and K (after fixing q_L that was absorbed in x). In contrast, Eq. (3.1.8) had only one degree of freedom, S (after absorbing q into x).

3.3. TWO SUBUNITS EACH HAVING ONE SITE: ALLOSTERIC EFFECT

This model extends one aspect of the model treated in Section 3.2. The adsorbing polymers are still identical, independent, and for convenience also assumed to be localized. Each polymer consists of two subunits that are distinguishable, say the right- and left-hand ones. On each subunit there is one adsorption site, and each subunit can attain one of two conformational states, L and H. This is perhaps the simplest model that manifests the so-called allosteric effect.

Originally the term "allosteric" was used for regulatory enzymes, where an effector binds to a site other than the active site and thereby changes the properties of the active site. Now it is also used whenever two ligands on two different sites of the same protein can communicate through the medium, which is the protein itself. Literally, the model of section 3.2 is also allosteric in the sense that adsorption on one site affects the adsorption on a different site. In the model treated in this section, the communication between the two sites is transmitted through the interaction energies between the two subunits.

We shall examine in this section the details of the mechanism of the communication between the two ligands on the two subunits.

3.3.1. The Empty Polymer

The polymer consists of two subunits, each of which can attain one of the two states, denoted L and H, having energies E_L and E_H, respectively. In addition, we have intersubunits interactions, which we denote by E_{LL}, $E_{LH} = E_{HL}$, and E_{HH}, depending on the state of the two subunits. Note that in general E_{LH} could be different from E_{HL}, Fig. 3.8. But for simplicity we assume that $E_{LH} = E_{HL}$. Denote

$$Q_\alpha = \exp(-\beta E_\alpha), \qquad Q_{\alpha\beta} = \exp(-\beta E_{\alpha\beta}), \tag{3.3.1}$$

where the subscripts α and β can be either L or H.

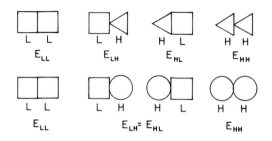

FIGURE 3.8. Two subunits with symmetrical (lower) and unsymmetrical (upper) interactions.

We first examine the equilibrium properties of this system in the absence of ligands—i.e., the empty polymer. The PF of a single polymer is

$$\xi(0) = Q_L^2 Q_{LL} + 2Q_L Q_H Q_{LH} + Q_H^2 Q_{HH}. \tag{3.3.2}$$

Note that this is actually the canonical partition function for a single polymer. We denote it by $\xi(0)$ to stress that this is the limit of the grand PF of a single polymer obtained for $\lambda \to 0$ (see subsection 3.3.3).

The four states of the polymer are LL, LH, HL, and HH (see first row of Fig. 3.9) with corresponding probabilities

$$x_{LL}^0 = \frac{Q_L^2 Q_{LL}}{\xi(0)}, \qquad x_{LH}^0 = x_{HL}^0 = \frac{Q_L Q_H Q_{LH}}{\xi(0)}, \qquad x_{HH}^0 = \frac{Q_H^2 Q_{HH}}{\xi(0)}. \tag{3.3.3}$$

Formally this model can be viewed as an extension of the model treated in section 3.2, but instead of two states, we have four states, with the three energy levels $2E_L + E_{LL}$, $E_L + E_H + E_{LH}$, and $2E_H + E_{HH}$ (the second energy level is doubly degenerated). However, to study the details of the response of this polymer to the adsorption of ligands, it is more convenient to view it as a combination of two subunits, each having two states L and H, and to assume that information is transmitted between the subunits through the interaction parameters $E_{\alpha\beta}$.

In Eq. (3.3.3) the superscript zero indicates the empty system, i.e., the polymer in the absence of ligands. Note that x_{LH}^0 is the probability of finding, say, the rhs subunits in the H state and the lhs subunit in the L state. The mole fraction of the polymers such that any one of its subunits is in the L state and the other in the H state is the sum of x_{LH}^0 and x_{HL}^0, which in our case is simply $2x_{LH}^0$.

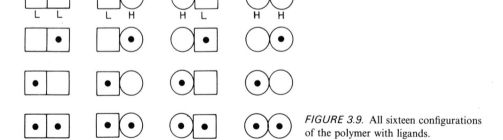

FIGURE 3.9. All sixteen configurations of the polymer with ligands.

We define the probability of finding a specific subunit (say the lhs) in state L, independently of the state of the second subunit on the same polymer, by

$$x_L^0 = x_{LL}^0 + x_{LH}^0 = \frac{Q_L^2 Q_{LL} + Q_L Q_H Q_{LH}}{\xi(0)}, \qquad (3.3.4)$$

and, similarly,

$$x_H^0 = x_{HH}^0 + x_{HL}^0 = \frac{Q_H^2 Q_{HH} + Q_L Q_H Q_{LH}}{\xi(0)}. \qquad (3.3.5)$$

Clearly if there are no interactions between the subunits, i.e., $E_{\alpha\beta} = 0$ for any α,β, then

$$x_L^0 \to \frac{Q_L^2 + Q_L Q_H}{(Q_L + Q_H)^2} = \frac{Q_L}{Q_L + Q_H} = x_L^{0,\infty}, \qquad (3.3.6)$$

$$x_H^0 \to \frac{Q_H^2 + Q_L Q_H}{(Q_L + Q_H)^2} = \frac{Q_H}{Q_L + Q_H} = x_H^{0,\infty}. \qquad (3.3.7)$$

The symbols $x_L^{0,\infty}$, $x_H^{0,\infty}$ stand for the probability of finding a subunit in state L or H in the absence of ligands (0) and in the absence of intersubunit interaction. This may be obtained by separating the two subunits to infinite distance, hence the superscript ∞. Clearly in this limit the present model becomes identical with the model treated in section 2.10.

It follows from the definition (3.3.3) that in the limit of infinite separation between the two subunits,

$$x_{\alpha,\beta}^{0,\infty} = \frac{Q_\alpha Q_\beta}{(Q_L + Q_H)^2} = x_\alpha^{0,\infty} x_\beta^{0,\infty}, \qquad (3.3.8)$$

which is the expected result; that is, the probability of finding one subunit in state α and the second in state β is simply the product of the two probabilities.

There is another important case where there is an effective independence between the two subunits. This occurs when the interaction energies are such that

$$E_{LH} = \tfrac{1}{2}(E_{LL} + E_{HH}), \qquad (3.3.9)$$

or, equivalently, when

$$Q_{LH}^2 = Q_{LL} Q_{HH}. \qquad (3.3.10)$$

In this case the PF can be written as

$$\xi(0) = Q_L^2 Q_{LL} + 2 Q_L Q_H \sqrt{Q_{LL} Q_{HH}} + Q_H^2 Q_{HH}$$
$$= (Q_L \sqrt{Q_{LL}} + Q_H \sqrt{Q_{HH}})^2. \qquad (3.3.11)$$

From (3.3.3) and (3.3.4), we find for this case

$$x_L^0 x_L^0 = \frac{(Q_L^2 Q_{LL} + Q_L Q_H \sqrt{Q_{LL} Q_{HH}})^2}{(Q_L \sqrt{Q_{LL}} + Q_H \sqrt{Q_{HH}})^4} = \frac{Q_L^2 Q_{LL}}{(Q_L \sqrt{Q_{LL}} + Q_H \sqrt{Q_{HH}})^2} = x_{LL}^0,$$

and, similarly, for other combination of indices we have

$$x_{LL}^0 = x_L^0 x_L^0, \qquad x_{LH}^0 = x_L^0 x_H^0, \qquad x_{HH}^0 = x_H^0 x_H^0. \qquad (3.3.12)$$

Thus whenever the condition (3.3.9) or (3.3.10) is fulfilled, the probability distribution of the states of the polymer is the product of the probability distributions of the states of each subunit. This by definition is called independence of the probability distribution. We see that the existence of interaction between the subunits does not imply dependence. Note also that the quantities in (3.3.12) are different from those in (3.3.8).

Consider now the following reaction:

$$(LL) + (HH) \rightarrow 2(LH). \tag{3.3.13}$$

For this reaction the equilibrium constant is (see 3.3.3)

$$\eta = \frac{(x_{LH}^0)^2}{x_{LL}^0 x_{HH}^0} = \frac{(Q_L Q_H Q_{LH})^2}{Q_L^2 Q_{LL} Q_H^2 Q_{HH}} = \frac{Q_{LH}^2}{Q_{LL} Q_{HH}} = \exp[-\beta(2E_{LH} - E_{LL} - E_{HH})]. \tag{3.3.14}$$

Thus in the empty system the equilibrium constant η is determined only by the interaction energies $E_{\alpha\beta}$. Condition (3.3.9) is equivalent to the condition $\eta = 1$. We shall see in section 3.3.6 that the equilibrium constant η is also responsible for transmitting information between the two ligands occupying the two subunits.

It is easy to show that if (3.3.12) holds, i.e., if $x_{\alpha\beta}^0$ is a product of x_α^0 and x_β^0 as defined in (3.3.4) and (3.3.5), then $\eta = 1$. This follows directly from the definitions of $x_{\alpha\beta}^0$ and x_α^0 and x_β^0. Therefore $\eta = 1$ is a necessary and a sufficient condition for independence of the subunits.

3.3.2. Potential of Average Force between the Subunits

Consider now the work required to bring the two subunits from fixed positions at infinite separation $R = \infty$ to the final position $R = d$, where d is the distance between the subunits in the final state of our model. This work is denoted $\Delta A(\infty \rightarrow d)$ and is given by the ratio of the corresponding PFs in the initial and final states.

$$\exp[-\beta\Delta A(\infty \rightarrow d)] = \frac{Q_L^2 Q_{LL} + 2Q_L Q_H Q_{LH} + Q_H^2 Q_{HH}}{(Q_L + Q_H)^2}, \tag{3.3.15}$$

where $(Q_L + Q_H)$ is the PF of each subunit at infinite separation from the other (section 2.10). Recalling the probabilities of the states L and H of each subunit at infinite separation,

$$x_L^{0,\infty} = \frac{Q_L}{Q_L + Q_H}, \qquad x_H^{0,\infty} = \frac{Q_H}{Q_L + Q_H}, \tag{3.3.16}$$

we have

$$\exp[-\beta\Delta A(\infty \rightarrow d)] = \sum_{\alpha\beta} x_\alpha^{0,\infty} x_\beta^{0,\infty} Q_{\alpha\beta} = \langle\exp(-\beta E_{PP})\rangle_{0,\infty}, \tag{3.3.17}$$

where E_{PP} is the polymer–polymer interaction. It is a function of two parameters α, β, defined by

$$E_{pp}(\alpha, \beta) = \begin{cases} E_{LL} & \text{for } \alpha = L, \beta = L \\ E_{LH} & \text{for } \alpha = L, \beta = H, \quad \text{or} \quad \alpha = H, \beta = L \\ E_{HH} & \text{for } \alpha = H, \beta = H. \end{cases} \tag{3.3.18}$$

Note that $E_{PP}(\alpha, \beta)$ is a function of $\alpha\beta$, whereas E_{LL}, E_{LH}, and E_{HH} are constants. The average in (3.3.17) is taken with respect to the probability distribution at infinite separation (3.3.16) and is denoted by $\langle\ \rangle_{0,\infty}$.

In most of this chapter we shall discuss the binding properties of the entire polymer in its final form. In particular, the interaction energies $E_{\alpha\beta}$ will be constants. However, for the sake of the following derivation, we view each of the interaction energies $E_{\alpha\beta}$ as a function of the intersubunit distance R. In this case (3.3.17) is written, for any R, as

$$\exp[-\beta\Delta A(\infty \to R)] = \langle\exp[-\beta E_{PP}(R)]\rangle_{0,\infty}. \tag{3.3.19}$$

Note that $E_{pp}(R)$ is a function of R, but the probability distribution of states used in the average (3.3.19) corresponds to $R = \infty$. Physically we can think of a system for which we have "frozen in" the equilibrium $L \rightleftarrows H$ at $R \to \infty$. We bring the two subunits from infinity to R but keep the mole fraction of states $x_L^{0,\infty}$ and $x_H^{0,\infty}$ that correspond to $R = \infty$.

We now take the derivative of (3.3.19) with respect to R, using (3.3.15) to obtain

$$-\frac{\partial\Delta A(\infty \to R)}{\partial R} = \frac{\displaystyle\sum_{\alpha\beta} Q_\alpha Q_\beta Q_{\alpha\beta}(R)\left[-\dfrac{\partial E_{\alpha\beta}(R)}{\partial R}\right]}{\displaystyle\sum_{\alpha\beta} Q_\alpha Q_\beta Q_{\alpha\beta}(R)}$$

$$= \sum_{\alpha\beta} x_{\alpha\beta}^0(R)\left[-\frac{\partial E_{\alpha\beta}(R)}{\partial R}\right] = \left\langle-\frac{\partial E_{pp}(R)}{\partial R}\right\rangle_0. \tag{3.3.20}$$

Here, $-\partial E_{\alpha\beta}(R)/\partial R$ is the direct force operating between the two subunits in states α, β at a distance R. The sum on the rhs of (3.3.20) is over all possible states of the (empty) subunits, with the distribution $x_{\alpha\beta}^0(R)$ at that particular R (note that each of $x_{\alpha\beta}^0(R)$ as defined in (3.3.3) is a function of R through $Q_{\alpha\beta}$). This justified the reference to $\Delta A(\infty \to R)$ as the potential of average force between the two subunits.

We shall encounter the concept of the potential of average force in various places throughout the book. In general, this concept applies to averages over states (or configurations) of the surrounding particles. The simplest case is discussed in section 3.3.5. Here, the average is over the states of the interacting particles themselves.

It should be noted that (3.3.20) is an average over the various forces; this is different from the force defined by the average interaction energy. The reason, as we shall soon see, is that the averaging and the differentiation processes do not commute.

The energy change for the process of bringing the two subunits from ∞ to R is obtained from (3.3.15) as follows:

$$\Delta E(\infty \to R) = \frac{\partial(\beta\Delta A(\infty \to R))}{\partial\beta}$$

$$= \frac{\displaystyle\sum_{\alpha\beta} Q_\alpha Q_\beta Q_{\alpha\beta}(E_\alpha + E_\beta + E_{\alpha\beta})}{\displaystyle\sum_{\alpha\beta} Q_\alpha Q_\beta Q_{\alpha\beta}} - 2\frac{Q_L E_L + Q_H E_H}{Q_L + Q_H}$$

$$= \sum_{\alpha\beta} x_{\alpha\beta}^0(E_\alpha + E_\beta + E_{\alpha\beta}) - 2(x_L^{0,\infty} E_L + x_H^{0,\infty} E_H)$$

$$= \sum_{\alpha\beta} x_{\alpha\beta}^0 E_{\alpha\beta} + \sum_{\alpha\beta} x_{\alpha\beta}^0(E_\alpha + E_\beta) - 2\sum_\alpha x_\alpha^{0,\infty} E_\alpha$$

$$= \sum_{\alpha\beta} x_{\alpha\beta}^0 E_{\alpha\beta} + \sum_{\alpha\beta} (E_\alpha + E_\beta)(x_{\alpha\beta}^0 - x_{\alpha\beta}^{0,\infty}), \tag{3.3.21}$$

where the energy change in the process is written as the average interaction energy between the subunits at the final position, with the distribution $x_{\alpha\beta}^0$, and the change in the average energies of the two subunits when they are brought from infinity to R. Another way of writing (3.3.21) is obtained by rearranging its terms

$$\Delta E(\infty \to R) = \sum_{\alpha\beta} x_{\alpha\beta}^{0,\infty} E_{\alpha\beta} + \sum_{\alpha\beta} (E_{\alpha\beta} + E_\alpha + E_\beta)(x_{\alpha\beta}^0 - x_{\alpha\beta}^{0,\infty}). \qquad (3.3.22)$$

Here the two terms on the rhs of (3.3.22) have the following interpretation; we start with the two subunits at infinite separation, the distribution of states is $x_{\alpha\beta}^{0,\infty}$. We freeze the equilibrium $L \rightleftarrows H$, i.e., we keep the concentrations of the species $\alpha\beta$ constant, and bring the two subunits to the final position R. The energy change is the first term on the rhs of (3.3.22). The second term on the rhs of (3.3.22) may be interpreted as the relaxation term. This is the energy change due to relaxing the constraint on the fixed concentrations $x_{\alpha\beta}^{0,\infty}$.

Clearly $x_{\alpha\beta}^0 - x_{\alpha\beta}^{0,\infty}$ is the change in the concentrations of each species $\alpha\beta$ in the relaxation process, and the energy change carried in this process is the second term on the rhs of (3.3.22).

The entropy change corresponding to this process is

$$T\Delta S(\infty \to R) = -\Delta A(\infty \to R) + \Delta E(\infty \to R)$$

$$= kT \ln \sum_{\alpha\beta} x_{\alpha\beta}^{0,\infty} Q_{\alpha\beta} + \sum_{\alpha\beta} x_{\alpha\beta}^0 E_{\alpha\beta} + \sum_{\alpha\beta} (E_\alpha + E_\beta)(x_{\alpha\beta}^0 - x_{\alpha\beta}^{0,\infty}).$$

$$(3.3.23)$$

Note that in the above equations $x_{\alpha\beta}^0(R)$ and $E_{\alpha\beta}(R)$ are functions of R, but E_α and $x_{\alpha\beta}^{0,\infty}$ are not. Taking the derivative of (3.3.21) with respect to R, we obtain

$$\frac{\partial \Delta E(\infty \to R)}{\partial R} = \sum_{\alpha\beta} x_{\alpha\beta}^0(R) \frac{\partial E_{\alpha\beta}(R)}{\partial R} + \sum_{\alpha\beta} \frac{\partial x_{\alpha\beta}^0(R)}{\partial R} [E_{\alpha\beta}(R) + E_\alpha + E_\beta]. \quad (3.3.24)$$

The first term on the rhs of (3.3.24) is the same as (3.3.20); the second term includes the change in the distribution of states as we change the distance R. The latter may be identified with the derivative of $T\Delta S(\infty \to R)$ with respect to R. This can be obtained from (3.3.23).

$$T\frac{\partial \Delta S(\infty \to R)}{\partial R} = \sum_{\alpha\beta} \frac{\partial x_{\alpha\beta}^0(R)}{\partial R} [E_{\alpha\beta}(R) + E_\alpha + E_\beta]. \qquad (3.3.25)$$

It is now clear that the average force (3.3.20) is different from the force derived from the average energy (3.3.24). The difference between the two may be referred to as the force due to the entropy change in the process.

3.3.3. The Binding Isotherm

We next introduce ligands into the system. The binding energies to state L and H are U_L and U_H, respectively, and we assume also direct interaction energy U_{12} between the ligands (for simplicity this is independent of the states of the subunits). We define

$$q_L = \exp(-\beta U_L), \qquad q_H = \exp(-\beta U_H), \qquad S = \exp(-\beta U_{12}). \qquad (3.3.26)$$

The grand PF for a single polymer is now

$$\xi = Q(0;0) + [Q(0;1) + Q(1;0)]\lambda + Q(1;1)\lambda^2$$

$$= \sum_{\alpha\beta} Q_\alpha Q_\beta Q_{\alpha\beta} + \left(\sum_{\alpha\beta} Q_\alpha Q_\beta q_\alpha Q_{\alpha\beta} + \sum_{\alpha\beta} Q_\alpha Q_\beta q_\beta Q_{\alpha\beta}\right)\lambda + \sum_{\alpha\beta} Q_\alpha Q_\beta Q_{\alpha\beta} q_\alpha q_\beta S\lambda^2.$$

$$(3.3.27)$$

The sixteen terms in Eq. (3.3.27) correspond to the sixteen states of the system in Fig. 3.9. Each sum in (3.3.27) corresponds to one row in Fig. 3.9.

Another way of writing the PF (3.3.27) is

$$\xi = \sum_{\alpha\beta} Q_\alpha Q_\beta Q_{\alpha\beta}[1 + (q_\alpha + q_\beta)\lambda + q_\alpha q_\beta S\lambda^2]. \qquad (3.3.28)$$

Here each term with specific α, β, e.g., $\alpha = L$, $\beta = L$, corresponds to one state of the polymer, i.e., one column in Fig. 3.9. For instance,

$$\xi_{LL} = Q_L^2 Q_{LL}(1 + 2q_L\lambda + q_L^2 S\lambda^2) \qquad (3.3.29)$$

is the PF for the polymer in state LL; clearly this is the same as the PF of a single-state polymer with two binding sites (section 3.2).

The general form of the binding isotherm is

$$\theta = \frac{\bar{n}}{2} = \frac{\lambda}{2}\frac{\partial \ln \xi}{\partial \lambda} = \frac{\sum\limits_{\alpha\beta} Q_\alpha Q_\beta Q_{\alpha\beta}(q_\alpha\lambda + q_\alpha q_\beta S\lambda^2)}{\xi}. \qquad (3.3.30)$$

Note that even when $S = 1$ the isotherm does not reduce to the Langmuir form. If on the other hand both $S = 1$ and $\eta = 1$, then we find that the PF is

$$\xi = \sum_{\alpha\beta} Q_\alpha Q_\beta \sqrt{Q_{\alpha\alpha}Q_{\beta\beta}}(1 + q_\alpha\lambda)(1 + q_\beta\lambda) = \left[\sum_\alpha Q_\alpha\sqrt{Q_{\alpha\alpha}}\,(1 + q_\alpha\lambda)\right]^2, \quad (3.3.31)$$

which is essentially the same PF as that of the model treated in section 3.2 with $S_L = S_H = 1$. We shall see in the following subsections that the important new features of this model arise when $\eta \neq 1$.

3.3.4. Probabilities

Many probabilities can be defined in this system, of which some are introduced in this subsection. The probability of finding a specific polymer in the conformational state α, β independently of the ligand occupation, is the same as the mole fractions $x_{\alpha\beta}$ of the various species. These are

$$x_{LL} = \frac{Q_L^2 Q_{LL}(1 + 2q_L\lambda + q_L^2 S\lambda^2)}{\xi} \qquad (3.3.32)$$

$$x_{LH} = x_{HL} = \frac{Q_L Q_H Q_{LH}[1 + (q_L + q_H)\lambda + q_L q_H S\lambda^2]}{\xi} \qquad (3.3.33)$$

$$x_{HH} = \frac{Q_H^2 Q_{HH}(1 + 2q_H\lambda + q_H^2 S\lambda^2)}{\xi}. \qquad (3.3.34)$$

Note that here the mole fractions depend on the ligand concentration through λ; in the limit $\lambda \to 0$, the system is empty with respect to the ligands, and these equations reduce to (3.3.3). Clearly the sum $\sum_{\alpha\beta} x_{\alpha\beta} = 1$.

The mole fractions of a single site in states L or H are

$$x_L = x_{LL} + x_{LH} \tag{3.3.35}$$

$$x_H = x_{HH} + x_{LH}; \tag{3.3.36}$$

x_L is the same as the probability of finding a specific polymer with one of its sites being L.

For the sake of the thermodynamic treatment in the next section, it will be necessary to introduce a more detailed notation for the various probability distributions. The state of the entire polymer is characterized by four parameters which we write as, say, $(L, 0; H, 1)$. This is read as follows; the lhs subunit is in the conformational state L and is empty (0), the rhs subunit is in conformational state H and is occupied (1). We use the semicolon between the parameters pertaining to different subunits. When we are interested only in partial characterization we simply eliminate those parameters over which we have summed; for instance, $\Pr(L; H)$ is the probability of finding the lhs subunit in state L and the rhs subunit in state H, independently of the state of occupancy states of the subunit. This quantity is the same as x_{LH} in (3.3.33). On the other hand, $\Pr(1; 1)$ is the probability of finding the lhs subunit and the rhs subunit occupied, independently of the conformational states of the subunits. $\Pr(L, 1;)$ is the probability of finding the lhs subunit in conformational state L and occupied, independently of the state of the rhs subunit. Likewise, $\Pr(; H, 0)$ refers to an empty H conformation of the rhs subunit. This notation is also used for conditional probabilities; for instance, $\Pr(; L, 0/ H, 1;)$ is the conditional probability of finding the rhs subunit empty and in state L, given that the lhs subunit is occupied and in state H. All of these probabilities can be either taken directly from the terms in ξ or constructed by the combination of such terms.

3.3.5. Binding Helmholtz Energies on First and Second Sites

As in subsection 3.2.4, we now define the binding Helmholtz energies on first and second sites. First we bring a ligand G from a fixed position in vacuum onto a specific site, say on the lhs site of an empty polymer. The process is symbolically written as

$$G + (0; 0) \to (1; 0); \tag{3.3.37}$$

compare this with (3.2.33). The corresponding Helmholtz energy change is

$$\Delta A_G^{*(1)} = A(1; 0) - A(0; 0) = -kT \ln \frac{Q(1; 0)}{Q(0; 0)}$$

$$= -kT \ln \left(\frac{\sum_{\alpha\beta} Q_\alpha Q_\beta Q_{\alpha\beta} q_\alpha}{\sum_{\alpha\beta} Q_\alpha Q_\beta Q_{\alpha\beta}} \right)$$

$$= -kT \ln \langle \exp(-\beta B_G) \rangle_{0,0}. \tag{3.3.38}$$

Note that in the definition of $\Delta A_G^{*(1)}$ we used the ratio of the canonical PFs $Q(1;0)$ and $Q(0;0)$. Using the canonical PF of the empty system (in the sense that $\lambda = 0$) given in (3.3.2), we have

$$\Pr(\alpha\,;\beta) = \frac{Q_\alpha Q_\beta Q_{\alpha\beta}}{\displaystyle\sum_{\alpha\beta} Q_\alpha Q_\beta Q_{\alpha\beta}}. \qquad (3.3.39)$$

However, if we take the open system PF (any λ), then from (3.3.27) we have

$$\Pr(\alpha, 0\,;\beta, 0) = \frac{Q_\alpha Q_\beta Q_{\alpha\beta}}{\xi} \qquad (3.3.40)$$

and

$$\Pr(0\,;0) = \sum_{\alpha\beta} \Pr(\alpha, 0\,;\beta, 0) = \frac{\displaystyle\sum_{\alpha\beta} Q_\alpha Q_\beta Q_{\alpha\beta}}{\xi}. \qquad (3.3.41)$$

The conditional probability in the open system is thus

$$\Pr(\alpha\,;\beta/0\,;0) = \frac{\Pr(\alpha, 0\,;\beta, 0)}{\Pr(0\,;0)} = \frac{Q_\alpha Q_\beta Q_{\alpha\beta}}{\displaystyle\sum_{\alpha\beta} Q_\alpha Q_\beta Q_{\alpha\beta}}. \qquad (3.3.42)$$

This result is the same as in (3.3.39), but now we have the explicit condition $(0\,;0)$. This condition did not appear explicitly in (3.3.39), but it was implied by the requirement that $\lambda = 0$, in which case the probability $\Pr(0\,;0)$ in (3.3.41) is unity.

B_G in (3.3.38) is the binding energy of G to say the lhs site (as defined in 3.2.24). In our model, B_G is the same for the lhs and the rhs sites. However, if the two subunits were different, than we should have distinguished between B_G on the rhs and the one on the lhs. Here, it is important to remember that B_G is the binding energy on one of the empty sites. In the next step below we shall also use the same notation for the binding energy on the empty site given that the other site is occupied.

Next we define the binding process to the second site, which is symbolically written

$$G + (1\,;0) \to (1\,;1)\,; \qquad (3.3.43)$$

compare this with (3.2.35). For this the Helmholtz energy change is

$$\Delta A_G^{*(2)} = A(1\,;1) - A(1\,;0) = -kT \ln \frac{Q(1\,;1)}{Q(1\,;0)}$$

$$= -kT \ln\left(\frac{\displaystyle\sum_{\alpha\beta} Q_\alpha Q_\beta Q_{\alpha\beta} q_\alpha q_\beta S}{\displaystyle\sum_{\alpha\beta} Q_\alpha Q_\beta Q_{\alpha\beta} q_\alpha}\right)$$

$$= -kT \ln[S\langle\exp(-\beta B_G)\rangle_{1,0}]$$

$$= -kT \ln\left[\frac{S\langle\exp(-\beta B_{GG})\rangle_{0,0}}{\langle\exp(-\beta B_G)\rangle_{0,0}}\right]. \qquad (3.3.44)$$

In (3.3.44) we have expressed $\Delta A_G^{*(2)}$ in two equivalent forms, first as a conditional average of $\exp(-\beta B_G)$ with the distribution function

$$\Pr(\alpha\,;\beta/1\,;0) = \frac{Q_\alpha Q_\beta Q_{\alpha\beta} q_\alpha}{\sum\limits_{\alpha\beta} Q_\alpha Q_\beta Q_{\alpha\beta} q_\alpha}\,; \tag{3.3.45}$$

i.e., we average over all the conformational states, given that one site is occupied and one site is empty. B_G refers of course to the binding energy on the empty site, here the rhs site, whereas in (3.3.38) it was any site, say the lhs.

The second form on the rhs of (3.3.44) is obtained by expanding the ratio

$$\frac{\sum Q_\alpha Q_\beta Q_{\alpha\beta} q_\alpha q_\beta S}{\sum Q_\alpha Q_\beta Q_{\alpha\beta} q_\alpha} \frac{\sum Q_\alpha Q_\beta Q_{\alpha\beta}}{\sum Q_\alpha Q_\beta Q_{\alpha\beta}} = \frac{\sum \Pr(\alpha\,;\beta/0\,;0)q_\alpha q_\beta S}{\sum \Pr(\alpha\,;\beta/0\,;0)q_\alpha}\,; \tag{3.3.46}$$

i.e., we take the two averages with respect to the conditional probabilities of the empty sites. B_{GG} in (3.3.44) is defined as

$$B_{GG}(\alpha,\beta) = \begin{cases} 2U_L & \text{for } \alpha = L,\ \beta = L \\ U_L + U_H & \text{for } \alpha = L,\ \beta = H \text{ or } \alpha = H,\ \beta = L \\ 2U_H & \text{for } \alpha = H,\ \beta = H. \end{cases} \tag{3.3.47}$$

As in section 3.2, we can define the intrinsic binding constants

$$K_1 = \exp(-\beta\Delta A_G^{*(1)}) \tag{3.3.48}$$

$$K_2 = \exp(-\beta\Delta A_G^{*(2)}) \tag{3.3.49}$$

and rewrite the binding isotherm (3.3.30) as

$$\theta = \frac{K_1\lambda + K_1 K_2\lambda^2}{1 + 2K_1\lambda + K_1 K_2\lambda^2}, \tag{3.3.50}$$

which is formally the same as (3.2.78) with a different meaning assigned to the constants K_1 and K_2 in this model.

3.3.6. Correlation Function and Cooperativity

We define the correlation function between the two ligands on the same polymer as

$$g(1\,;1) = \frac{\Pr(1\,;1)}{[\Pr(\,;1)]^2} = \frac{Q(1\,;1)\lambda^2\xi}{[Q(0\,;1)\lambda + Q(1\,;1)\lambda^2]^2}, \tag{3.3.51}$$

where $\Pr(\,;1)$ is the probability of finding a specific polymer with one site occupied. Note that if the subunits were different we should have $\Pr(\,;1)\Pr(1\,;)$ in the denominator. In our model the values of $\Pr(\,;1)$ and $\Pr(1\,;)$ are the same. In the limit of $\lambda = 0$ (i.e., empty with respect to ligands), both $\Pr(1;1) \to 0$ and $\Pr(;1) \to 0$, but the ratio in (3.3.51) is finite. We thus define

$$g^0(1;1) = \lim_{\lambda \to 0} g(1;1) = \frac{\sum Q_\alpha Q_\beta Q_{\alpha\beta} q_\alpha q_\beta S \sum Q_\alpha Q_\beta Q_{\alpha\beta}}{(\sum Q_\alpha Q_\beta Q_{\alpha\beta} q_\alpha)^2}$$

$$= S \frac{\langle \exp(-\beta B_{GG}) \rangle_{0,0}}{\langle \exp(-\beta B_G) \rangle_{0,0}^2}$$

$$= S \frac{\langle \exp(-\beta B_G) \rangle_{0,1}}{\langle \exp(-\beta B_G) \rangle_{0,0}}. \tag{3.3.52}$$

Note that in the last form on the rhs of (3.3.52) B_G is the binding energy on an empty site. Since the average in the numerator is taken with the condition $(0;1)$, we require that B_G refer to the binding energy on the lhs site.

Using the definitions of $\Delta A_G^{*(1)}$ and $\Delta A_G^{*(2)}$ in (3.3.38) and (3.3.44), respectively, we rewrite (3.3.52) as follows:

$$\exp(-\beta \Delta A_G^{*(2)}) = g^{(0)}(1;1) \exp(-\beta \Delta A_G^{*(1)}), \tag{3.3.53}$$

by analogy with Eq. (3.2.40). The correlation function in (3.3.53) connects the two binding free energies $\Delta A_G^{*(1)}$ and $\Delta A_G^{*(2)}$.

We define the potential of average force between the two ligands (at $\lambda \to 0$) by

$$g^0(1;1) = \exp[-\beta W(1;1)]$$

$$= \exp[-\beta(\Delta A_G^{*(2)} - \Delta A_G^{*(1)})]$$

$$= \exp\{-\beta[A(1;1) - 2A(1;0) + A(0;0)]\}. \tag{3.3.54}$$

Thus $W(1;1)$ is the Helmholtz energy change for the reaction

$$2(0;1) \to (1;1) + (0;0). \tag{3.3.55}$$

We denote, by analogy with (3.2.41), the binding Helmholtz energy of the *pair GG* on the polymer by

$$\exp(-\beta \Delta A_{GG}^*) = \langle \exp(-\beta B_{GG}) \rangle_{0,0}. \tag{3.3.56}$$

We define the indirect correlation function $y(1;1)$ by

$$g^0(1;1) = Sy(1;1). \tag{3.3.57}$$

From (3.3.53) and (3.3.57), we obtain

$$\langle \exp(-\beta B_G) \rangle_{0,1} = y(1;1) \langle \exp(-\beta B_G) \rangle_{0,0}. \tag{3.3.58}$$

Note again that B_G on the rhs of (3.3.58) is the binding energy on, say, the rhs site, but on the lhs of (3.3.58) it must be the binding energy on the lhs site; because of the condition $(0;1)$, only the lhs site is empty.

In (3.3.57) we separated the two contributions to the correlation function. The direct correlation, through $S = \exp(-\beta U_{12})$, and the indirect part, which is due to the response of the system to the binding process. The mechanism of the response is discussed in the following subsections.

We now define *positive* cooperativity with respect to ligand binding in any of the following equivalent forms.

$$W(1;1) < 0 \tag{3.3.59}$$

$$g^0(1;1) - 1 > 0 \tag{3.3.60}$$

$$\Delta A_G^{*(2)} < \Delta A_G^{*(1)} \tag{3.3.61}$$

$$g^0(1;1) = \frac{\Pr(;1/1;)}{\Pr(;1)} > 1. \tag{3.3.62}$$

The last relation should be understood here as the limiting ratio at $\lambda \to 0$. The conditional probability of finding the rhs site occupied given that the lhs site is already occupied is larger than the (unconditional) probability of finding the rhs site occupied.

In contrast to section 3.2.3, where we found that the indirect correlation was always positive, here $y(1;1)$ can be either positive or negative. We now examine some limiting cases:

1. If $q_L = q_H$, then from (3.3.52) and (3.3.57) we have

$$y(1;1) = \frac{q^2 \sum Q_\alpha Q_\beta Q_{\alpha\beta} \sum Q_\alpha Q_\beta Q_{\alpha\beta}}{q^2 (\sum Q_\alpha Q_\beta Q_{\alpha\beta})^2} = 1. \tag{3.3.63}$$

Thus if $U_L = U_H$, independently of the other parameters of the system (E_α and $E_{\alpha\beta}$), there is no indirect correlation. This result means that the indirect correlation is intimately connected with the conformational changes induced by the binding of the ligand. If $U_L = U_H$, there is no effect of G on the equilibrium composition of the L and H forms. This result is essentially the same as in the model treated in section 3.2.3 [see Eq. (3.2.31)].

2. In contrast to the model of section 3.2.3, the indirect correlation depends also on the way the information from one subunit is transmitted through the interactions $E_{\alpha\beta}$ to the other subunits. It is clear that if all $E_{\alpha\beta}$ are equal, then $Q = Q_{LL} = Q_{LH} = Q_{HH}$, and

$$y(1;1) = \frac{Q^2 \sum Q_\alpha Q_\beta q_\alpha q_\beta \sum Q_\alpha Q_\beta}{Q^2 (\sum Q_\alpha Q_\beta q_\alpha)^2} = 1. \tag{3.3.64}$$

But the same result is also obtained under a much weaker condition. It is sufficient that $\eta = Q_{LH}^2 / Q_{LL} Q_{HH}$, defined in (3.3.14), be unity, in which case we can write for any $\alpha\beta$

$$Q_{\alpha\beta} = \sqrt{Q_{\alpha\alpha} Q_{\beta\beta}}, \tag{3.3.65}$$

and hence

$$y(1;1) = \frac{\sum Q_\alpha Q_\beta \sqrt{Q_{\alpha\alpha} Q_{\beta\beta}}\, q_\alpha q_\beta \sum Q_\alpha Q_\beta \sqrt{Q_{\alpha\alpha} Q_{\beta\beta}}}{(\sum Q_\alpha Q_\beta \sqrt{Q_{\alpha\alpha} Q_{\beta\beta}}\, q_\alpha)^2} = 1. \tag{3.3.66}$$

We see that if $\eta = 1$ then, although there exists a conformational change induced by the binding, this effect is not communicated between the two subunits. We have already seen that $\eta = 1$ leads to independence of the subunits [see Eq. (3.3.12)], which means that conformational changes induced by the binding process affects only the subunit on which the ligand binds. This is therefore the same as in the model of section 2.10.

From the above two cases we see that $h = q_H/q_L \neq 1$ and $\eta \neq 1$ are essential for the existence of indirect correlation between the two ligands. The first is responsible for producing conformational change in the subunit, and the second is responsible for transmitting this information from one subunit to the other.

3. In order to produce an indirect correlation, we must have both $h \neq 1$ and $\eta \neq 1$, the simplest case being when $K = Q_H/Q_L = 1$, for which we have

$$y(1;1) = \frac{\sum q_\alpha q_\beta Q_{\alpha\beta} \sum Q_{\alpha\beta}}{\left(\sum Q_{\alpha\beta} q_\alpha\right)^2}. \tag{3.3.67}$$

We shall now see that this can produce either positive or negative correlation. We rewrite (3.3.67) as

$$y(1;1) = 1 + \frac{\sum q_\alpha q_\beta Q_{\alpha\beta} \sum Q_{\alpha\beta} - \left(\sum Q_{\alpha\beta} q_\alpha\right)^2}{\left(\sum Q_{\alpha\beta} q_\alpha\right)^2}. \tag{3.3.68}$$

The numerator in (3.3.68) may be simplified as follows

$$\sum_{\alpha\beta\gamma\delta} (q_\alpha q_\beta Q_{\alpha\beta} Q_{\gamma\delta} - Q_{\alpha\beta} q_\alpha Q_{\gamma\delta} q_\delta) = \sum_{\alpha\beta\gamma\delta} q_\alpha Q_{\alpha\beta} Q_{\gamma\delta} (q_\beta - q_\delta)$$

$$= \sum_{\alpha\gamma} q_\alpha Q_{\alpha H} Q_{\gamma L} (q_H - q_L) + \sum_{\alpha\gamma} q_\alpha Q_{\alpha L} Q_{\gamma H} (q_L - q_H)$$

$$= (q_L - q_H) \sum_{\alpha\gamma} (q_\alpha Q_{\alpha L} Q_{\gamma H} - q_\alpha Q_{\alpha H} Q_{\gamma L})$$

$$= (q_L - q_H)^2 (Q_{LL} Q_{HH} - Q_{LH}^2). \tag{3.3.69}$$

We now use the parameters

$$h = \frac{q_H}{q_L}, \qquad \eta = \frac{Q_{LH}^2}{Q_{LL} Q_{HH}}$$

to rewrite (3.3.68) in the final form

$$y(1;1) = 1 + \frac{(1-h)^2(1-\eta)Q_{HH}Q_{LL}}{[Q_{LL} + Q_{LH}(1+h) + Q_{HH}h]^2}. \tag{3.3.70}$$

From (3.3.70) it is clear that we need both $h \neq 1$ and $\eta \neq 1$ to produce an indirect correlation. Whatever the value of $h \neq 1$, this does not have an effect on the sign of the second term on the rhs of (3.3.70). The parameter that determines the sign of this term is η. If $\eta > 1$, we find that $y < 1$; hence this case produces negative indirect correlation. If $\eta < 1$, we have $y > 1$, corresponding to positive indirect correlation.

The physical reason for this behavior will be clearer from the discussion in section 3.3.8, but qualitatively it is quite simple. If $h = 1$, there cannot be conformational change induced by G, and therefore $y(1;1) = 1$, as in section 3.2.3. Suppose that $U_L < U_H$; then, when G binds, say, to the right subunit it will always shift the equilibrium $L \rightleftarrows H$ toward L. This is the same effect as in sections 2.10 and 3.2.3. The next question is how this shift of the equilibrium of the right subunits is transmitted to the left subunits. The transmission of information depends on the interactions $E_{\alpha\beta}$. If $\eta < 1$, this is equivalent to $2E_{LH} - E_{LL} - E_{HH} > 0$, in which case the conformational change induced by the binding of G on the right subunits will induce conformational changes in the left subunit in the same direction, i.e., in favor of L, and this in turn will enhance the binding of G, i.e., we have positive cooperativity. The opposite effect occurs when $\eta > 1$. Note that in any case h does not affect the sign of the cooperativity, only its magnitude.

The more general case occurs when $Q_L \neq Q_H$. The general result for this case is

$$y(1;1) = \frac{\sum q_\alpha q_\beta Q_\alpha Q_\beta Q_{\alpha\beta} \sum Q_\alpha Q_\beta Q_{\alpha\beta}}{(\sum Q_\alpha Q_\beta Q_{\alpha\beta} q_\alpha)^2}. \tag{3.3.71}$$

If we define $Q'_{\alpha\beta} = Q_\alpha Q_\beta Q_{\alpha\beta}$, then (3.3.71) may be written in the same form as (3.3.70), but replacing $Q_{\alpha\beta}$ by $Q'_{\alpha\beta}$; the result is

$$y(1;1) = 1 + \frac{(1-h)^2(1-\eta)Q'_{HH}Q'_{LL}}{[Q'_{LL} + Q'_{LH}(1+h) + Q'_{HH}h]^2}. \tag{3.3.72}$$

Here again the sign of cooperativity is determined by η. Its value is modified by the quantities $Q_\alpha Q_\beta$ that are now included in (3.3.72). From (3.3.72) it follows that $h = 1$ or $\eta = 1$ is a necessary and sufficient condition for $y(1;1) = 1$.

We have defined the concept of the potential of average force between the two ligands in (3.3.54). In order to justify this term, we need to assume that the ligand–ligand distance can be varied. This may be done in this model in two ways. Either $E_{\alpha\beta}$ could be changed by changing the intersubunit distance, as we have done in section 3.3.2, or the binding energies U_α could change as we change the ligand–subunit distance. The simpler case is the latter. Hence for the present demonstration we assume that $E_{\alpha\beta}$ are fixed, and we vary the distance between the ligand and the subunits.

First we write $W(1;1)$ more explicitly as

$$W(1;1) = W(\mathbf{R}_1, \mathbf{R}_2) = A(\mathbf{R}_1, \mathbf{R}_2) - A(\mathbf{R}_1, 0) - A(0, \mathbf{R}_2) - A(0, 0), \tag{3.3.73}$$

where instead of the notation in (3.3.54) we introduce the locational vector $\mathbf{R}_i$ of the center of the ith ligand. Taking the gradient of $W(\mathbf{R}_1, \mathbf{R}_2)$ with respect to, say, $\mathbf{R}_1$ we obtain

$$-\nabla_1 W(\mathbf{R}_1, \mathbf{R}_2) = -[\nabla_1 A(\mathbf{R}_1, \mathbf{R}_2) - \nabla_1 A(\mathbf{R}_1, 0)]$$

$$= \frac{\sum_{\alpha\beta} Q_\alpha Q_\beta Q_{\alpha\beta} q_\alpha q_\beta S(-\nabla_1 U_{12} - \nabla_1 U_\alpha)}{\sum_{\alpha\beta} Q_\alpha Q_\beta Q_{\alpha\beta} q_\alpha q_\beta S} - \frac{\sum_{\alpha\beta} Q_\alpha Q_\beta Q_{\alpha\beta} q_\alpha (-\nabla_1 U_\alpha)}{\sum_{\alpha\beta} Q_\alpha Q_\beta Q_{\alpha\beta} q_\alpha}$$

$$= \sum_{\alpha\beta} \Pr(\alpha;\beta/1;1)(-\nabla_1 U_{12} - \nabla_1 U_\alpha) - \sum_{\alpha\beta} \Pr(\alpha;\beta/1;0)(-\nabla_1 U_\alpha)$$

$$= -\nabla_1 U_{12} + \sum_{\alpha\beta} [\Pr(\alpha;\beta/1;1) - \Pr(\alpha;\beta/1;0)](-\nabla_1 U_\alpha). \tag{3.3.74}$$

The first term is the direct force exerted on ligand 1 by ligand 2. The second term is an indirect average force. $-\nabla_1 U_\alpha$ is the force exerted by the subunit on ligand 1, averaged over all configurations of the two subunits, once with the distribution $\Pr(\alpha;\beta/1;1)$ and once with $\Pr(\alpha;\beta/1;0)$.

Since $\nabla_1 U_\alpha$ depends only on the index α, we can sum over the index β to obtain

$$\nabla_1 W(\mathbf{R}_1, \mathbf{R}_2) = \nabla_1 U_{12} + \sum_\alpha [\Pr(\alpha;/1;1) - \Pr(\alpha;/1;0)]\nabla_1 U_\alpha. \tag{3.3.75}$$

The meaning of $\Pr(\alpha;/1;0) - \Pr(\alpha;/1;0)$ is further discussed in section 3.3.8. This is the change in the mole fraction of the state α of the lhs subunit caused by adding a ligand at the rhs subunit. It is easy to show that if there are no interactions between the two subunits, or if $\eta = 1$, then the two subunits are effectively independent. Hence

$$\Pr(\alpha;/1; 1) = \Pr(\alpha;/1;) \tag{3.3.76}$$

$$\Pr(\alpha;/1; 0) = \Pr(\alpha;/1;), \tag{3.3.77}$$

and the indirect force in (3.3.75) vanishes. We can conclude that the gradient of $W(1; 1)$ with respect to $\mathbf{R}_1$ consists of two forces: The direct force, due to the direct interaction between the two ligands, and an indirect force. The indirect force is the difference between the conditional average force exerted on the *left*-hand ligand by the *left*-hand subunit when the *right*-hand subunit is empty or occupied.

The concept of the potential average force plays a central role in the theory of liquids and solutions. It consists of two terms, direct and indirect, as in Eq. (3.3.75). However, because of the isotropic nature of the liquid, the average force exerted on one solute, infinitely far from the second solute, is zero. This leads to a slightly simpler form of Eq. (3.3.75) in the case of a solute in a solvent. This is discussed in Chapters 5 and 6.

3.3.7. Energy Change for Binding on First and Second Sites

We define the energy changes for the processes described in (3.3.37) and (3.3.43), i.e., the process of bringing G from a fixed position in the vacuum onto a specific empty site. For the first site this is

$$
\begin{aligned}
\Delta E_G^{*(1)} &= \frac{\partial(\beta \Delta A_G^{*(1)})}{\partial \beta} \\
&= \frac{\sum Q_\alpha Q_\beta Q_{\alpha\beta} q_\alpha (U_\alpha + E_\alpha + E_\beta + E_{\alpha\beta})}{\sum Q_\alpha Q_\beta Q_{\alpha\beta} q_\alpha} - \frac{\sum Q_\alpha Q_\beta Q_{\alpha\beta}(E_\alpha + E_\beta + E_{\alpha\beta})}{\sum Q_\alpha Q_\beta Q_{\alpha\beta}} \\
&= \langle B_G^1 \rangle_{1;0} + \langle E_p^1 + E_p^0 + E_{pp} \rangle_{1;0} - \langle E_p + E_p + E_{pp} \rangle_{0;0}.
\end{aligned}
\tag{3.3.78}
$$

Some care must be exercised in the interpretation of the various averages in (3.3.78). In (3.3.52) we had an average of the form

$$\langle \exp(-\beta B_G) \rangle_{0;1} = \sum_{\alpha\beta} \Pr(\alpha; \beta/0; 1) \exp(-\beta U_\alpha), \tag{3.3.79}$$

and we noted that U_α is the binding energy on the subunit that is empty, i.e., the lhs subunit in this notation. In (3.3.78), we have an average of the form

$$\langle B_G^1 \rangle_{1;0} = \sum_{\alpha\beta} \Pr(\alpha; \beta/1; 0) U_\alpha. \tag{3.3.80}$$

In this case U_α is the binding energy on the subunit that is occupied. This is different from the average

$$\langle B_G^0 \rangle_{1;0} = \sum_{\alpha\beta} \Pr(\alpha; \beta/1; 0) U_\beta; \tag{3.3.81}$$

we used the superscripts 0 and 1 to distinguish between the two averages. Note, however, that $B_G^0(\alpha)$ and $B_G^1(\alpha)$ are the same functions of α as defined in (3.2.24). Similarly we should make the distinction between $\langle E_p^1 \rangle_{1;0}$ and $\langle E_p^0 \rangle_{1;0}$. This is not necessary when the condition is $(0; 0)$ [as in the rhs of (3.3.78)] or when the condition is $(1; 1)$ [as in Eq. (3.3.82), below].

The energy change for the second site is obtained from (3.3.44).

$$\Delta E_G^{*(2)} = \frac{\partial(\beta \Delta A_G^{*(2)})}{\partial \beta}$$

$$= U_{12} + \frac{\sum Q_\alpha Q_\beta Q_{\alpha\beta} q_\alpha q_\beta (U_\alpha + U_\beta + E_\alpha + E_\beta + E_{\alpha\beta})}{\sum Q_\alpha Q_\beta Q_{\alpha\beta} q_\alpha q_\beta}$$

$$- \frac{\sum Q_\alpha Q_\beta Q_{\alpha\beta} q_\alpha (U_\alpha + E_\alpha + E_\beta + E_{\alpha\beta})}{\sum Q_\alpha Q_\beta Q_{\alpha\beta} q_\alpha}$$

$$= U_{12} + \langle B_G \rangle_{1;1} + \langle B_G + E_p + E_p + E_{pp} \rangle_{1;1}$$

$$- \langle B_G^1 + E_p^1 + E_p^0 + E_{pp} \rangle_{1;0}. \tag{3.3.82}$$

The different expressions for $\Delta E_G^{*(1)}$ and $\Delta E_G^{*(2)}$ should be noted carefully. In $\Delta E_G^{*(1)}$, we have the average binding energy of G on the first site with the distribution probability $\Pr(\alpha; \beta/1; 0)$. In addition, we have an energy change in the system caused by the adsorption process. This term will be reinterpreted in section 3.3.8 using a mixture-model approach to the system. The quantity $E_p(\alpha) + E_p(\beta) + E_{pp}(\alpha\beta)$ is the energy of the *empty* polymer in one of the states α, β. The two terms on the rhs of (3.3.78) are two averages of the same quantity, but with different probability distributions. These are $\Pr(\alpha; \beta/1; 0)$ and $\Pr(\alpha; \beta/0; 0)$.

The content of $\Delta E_G^{*(2)}$ in (3.3.82) is similar, except that we have to proceed one step ahead; namely, instead of the transition $(0; 0) \rightarrow (1; 0)$ in (3.3.78) we now have the transition $(1; 0) \rightarrow (1; 1)$. Thus $\Delta E_G^{*(2)}$ contains the interaction U_{12}, the average binding energy of G on the second site with the probability distribution $\Pr(\alpha; \beta/1; 1)$. The last two terms on the rhs of (3.3.82) are again due to the effect induced on the system by the second G. But now, the system already includes one G; therefore, the total energy of the system is $B_G(\alpha) + E_p(\alpha) + E_p(\beta) + E_{pp}(\alpha, \beta)$. This quantity is now averaged with two different probability distributions, $\Pr(\alpha; \beta/1; 0)$ and $\Pr(\alpha; \beta/1; 1)$, corresponding to the system before and after the adsorption of the second ligand.

3.3.8. Induced Conformational Changes in the Two Subunits

In section 3.2.6 we analyzed the conformational changes induced in a single polymer by the binding process. We found there that the ligand always shifts the equilibrium concentrations of L and H toward that state for which the binding energy of the ligand is stronger (more negative). This is intuitively quite clear. The situation is more complex in the present model. Here the shift induced on the *first* subunit on which the ligand binds is in the same direction as in the model of section 3.2.6. However, the induced shift experienced by the *first* subunit induces another shift in the equilibrium concentrations of H and L of the *second* subunit. This effect depends both on the binding energies U_α as well as on the parameter η (which is the parameter responsible for transmitting information between the two subunits). Both of these induced effects affect the thermodynamics of the binding process as well as the extent of cooperativity. We shall now examine these two effects quantitatively.

First we focus on the first subunit, say the left-hand one, and ask what is the change in the mole fraction of the L form upon binding of the first ligand. This effect is similar to the one we have studied in section 2.10. We denote by $d_L^{(1)}$; the difference in the mole fraction of the L form on the subunit that adsorbs the first ligand (here the lhs one). This can be read directly from the corresponding terms in the PF; i.e.,

$$d_{L;}^{(1)} = \text{Pr}(L;/1;0) - \text{Pr}(L;/0;0)$$

$$= \frac{Q_L q_L \sum_\beta Q_\beta Q_{L\beta}}{\sum_{\alpha\beta} Q_\alpha Q_\beta Q_{\alpha\beta} q_\alpha} - \frac{Q_L \sum_\beta Q_\beta Q_{L\beta}}{\sum_{\alpha\beta} Q_\alpha Q_\beta Q_{\alpha\beta}}$$

$$= \frac{Q_L Q_H \sum_{\alpha\beta} Q_\alpha Q_\beta Q_{H\alpha} Q_{L\beta}}{\left(\sum_{\alpha\beta} Q_\alpha Q_\beta Q_{\alpha\beta} q_\alpha\right)\left(\sum_{\alpha\beta} Q_\alpha Q_\beta Q_{\alpha\beta}\right)} (q_L - q_H). \tag{3.3.83}$$

If $U_L < U_H$, then $h = q_H/q_L < 1$ or, equivalently, $q_L - q_H > 0$, in which case $d_{L;}^{(1)} > 0$. This is the expected result. It is similar to the result of section 2.10.

The more difficult question concerns the change in the distribution of the rhs subunit, induced by binding on the lhs subunit, which is defined as

$$d_{;L}^{(1)} = \text{Pr}(;L/1;0) - \text{Pr}(;L/0;0)$$

$$= \frac{Q_L \sum_\alpha Q_\alpha Q_{\alpha L} q_\alpha}{\sum_{\alpha\beta} Q_\alpha Q_\beta Q_{\alpha\beta} q_\alpha} - \frac{Q_L \sum_\alpha Q_\alpha Q_{\alpha L}}{\sum_{\alpha\beta} Q_\alpha Q_\beta Q_{\alpha\beta}}$$

$$= \frac{Q_L^2 Q_H^2 Q_{LL} Q_{HH}}{\left(\sum_{\alpha\beta} Q_\alpha Q_\beta Q_{\alpha\beta} q_\alpha\right)\left(\sum_{\alpha\beta} Q_\alpha Q_\beta Q_{\alpha\beta}\right)} (q_L - q_H)(1 - \eta). \tag{3.3.84}$$

We see again, as in the expression for the indirect correlation, that the sign of this quantity depends on both h and η. A choice of $h < 1$ (i.e., $U_L < U_H$) does not guarantee a positive value of $d_{;L}^{(1)}$. To obtain a positive shift for the second subunit (in the sense of $d_{;L}^{(1)} > 1$) we need also that $\eta < 1$. In this case the binding of the first ligand shifts the equilibrium concentration of both subunits in favor of the L conformation. Similarly, one can define the effect of the second ligand on equilibrium concentrations of the two subunits. These are

$$d_{;L}^{(2)} = \text{Pr}(;L/1;1) - \text{Pr}(;L/1;0) = \frac{Q_L q_L \sum_\alpha Q_\alpha Q_{\alpha L} q_\alpha}{\sum_{\alpha\beta} Q_\alpha Q_\beta Q_{\alpha\beta} q_\alpha q_\beta} - \frac{Q_L \sum_\alpha Q_\alpha Q_\beta Q_{\alpha\beta} q_\alpha}{\sum_{\alpha\beta} Q_\alpha Q_\beta Q_{\alpha\beta} q_\alpha} \tag{3.3.85}$$

$$d_{L;}^{(2)} = \text{Pr}(L;/1;1) - \text{Pr}(L;/1;0) = \frac{Q_L q_L \sum_\beta Q_\beta Q_{L\beta} q_\beta}{\sum_{\alpha\beta} Q_\alpha Q_\beta Q_{\alpha\beta} q_\alpha q_\beta} - \frac{Q_L q_L \sum_\beta Q_\beta Q_{L\beta}}{\sum_{\alpha\beta} Q_\alpha Q_\beta Q_{\alpha\beta} q_\alpha}. \tag{3.3.86}$$

3.3.9. Two Limiting Cases

3.3.9.1. The Concerted Model

The first limiting case was suggested originally by Monod, Wyman, and Changeux (MWC) in 1965.[1] This model requires that the two subunits be either in the L or in the

H state. The conformations of the two subunits change in a concerted way. This is equivalent to the consideration of the first and fourth columns in Fig. 3.9. Mathematically, we can obtain this limiting case by taking $\eta = 0$ in Eq. (3.3.14), which essentially means that the equilibrium concentrations of x_{LH} and x_{HL} are negligible.

The PF of a single polymer may be obtained from (3.3.28) by substituting $Q_{\alpha\beta}^2 = Q_{\alpha\alpha}Q_{\beta\beta}\delta_{\alpha\beta}$. The result is [see also (3.3.29)]

$$\xi_{\text{MWC}} = \sum_a Q_a^2 Q_{aa}(1 + 2q_a\lambda + q_a^2 S\lambda^2) = \xi_{LL} + \xi_{HH}. \qquad (3.3.87)$$

This is essentially the same PF as of the model treated in section 3.2 with the replacement of Q_α [in Eq. (3.2.9)] by $Q_a^2 Q_{aa}$. In essence, the MWC model is equivalent to two state polymers with energies corresponding to $2E_L + E_{LL}$ and $2E_H + E_{HH}$. The fact that we have two subunits does not affect the formalism, except for the redefinitions of the energy levels.

Usually the MWC is applied when there are no direct interactions between the ligands (i.e., $S = 1$) in which case (3.3.87) reduces to

$$\xi_{\text{MWC}} = Q_L^2 Q_{LL}(1 + q_L\lambda)^2 + Q_H^2 Q_{HH}(1 + q_H\lambda)^2. \qquad (3.3.88)$$

If we choose the *L* energy level $2E_L + E_{LL}$ as our zero energy (or, equivalently, define a new PF by $\xi' = \xi / Q_L^2 Q_{LL}$), we may rewrite (3.3.88) in the more familiar form as

$$\xi'_{\text{MWC}} = (1 + K_L C)^2 + \tilde{K}(1 + K_H C)^2, \qquad (3.3.89)$$

where $K_L C = q_L\lambda$, $K_H C = q_H\lambda$, and $\tilde{K}$ is the equilibrium constant for the conversion between the two states, $LL \rightarrow HH$:

$$\tilde{K} = \frac{Q_H^2 Q_{HH}}{Q_L^2 Q_{LL}}. \qquad (3.3.90)$$

From (3.3.72), we see that in the case $\eta = 0$ the indirect correlation function (which is the same as the total correlation function if $S = 1$) is

$$y(1; 1) = 1 + \frac{(1 - h)^2 Q'_{HH} Q'_{LL}}{(Q'_{LL} + Q'_{HH}h)^2} = 1 + \frac{(1 - h)^2 \tilde{K}}{(1 + \tilde{K}h)^2} \geq 1, \qquad (3.3.91)$$

which is essentially the same as (3.2.31) with the replacement of K by $\tilde{K}$. In this case, as in the model of section 3.2 we always have positive (indirect) cooperativity. The molecular reason for this result is quite clear in view of the analysis of the origin of the cooperativity as discussed in subsection 3.3.6. Here a conformational change on one subunit is fully transmitted to the second subunit. In other words, the two subunits respond in a concerted manner, as if they were a single subunit—i.e., as in the model of section 3.2.

3.3.9.2. The Sequential Model

The second extreme case, suggested in 1966 by Koshland, Némethy, and Filmer (KNF), is also known as the sequential model.[2] The mathematical conditions required to obtain this limiting case are quite severe. First, it is assumed that, in the absence of a ligand, one of the conformations is dominant, say the *LL* form. In addition, it is assumed that ligands do not bind to the *L* conformer, but bind very strongly to the *H* conformer.

These assumptions lead to the consideration of the four diagonal states of Fig. 3.9, for which the PF is

$$\xi_{KNF} = Q_L^2 Q_{LL} + 2Q_L Q_H Q_{LH} q_H \lambda + Q_H^2 Q_{HH} q_H^2 \lambda^2. \tag{3.3.92}$$

The empty state is the LL state on the top left corner of Fig. 3.9. The binding of a ligand on any of the subunits will shift its conformation completely from L to H without affecting the conformation of the second subunit. Binding of the two ligands will shift the entire polymer to the state HH. Thus in each binding process there is a complete change of conformation of one subunit; hence the term sequential model.

The mathematical requirements necessary to obtain the KNF model from the general one can be stated as follows. Let $K_{L/H}^0$ be the equilibrium constant for the $H \to L$ conversion of each subunit when it is known to be empty. Likewise let $K_{L/H}^1$ be the equilibrium constant when the subunit is known to be occupied. The conditions of the KNF model can be met by requiring that

$$K_{L/H}^0 \to \infty \tag{3.3.93}$$

and

$$K_{L/H}^1 \to 0. \tag{3.3.94}$$

Figure 3.10 shows the same sixteen states of the polymer in a condensed form. In the diagram of Fig. 3.10 we marked by circles those states (of a single subunit) that violate either condition (3.3.93) or condition (3.3.94); for instance, the state LH in the first row violates the condition (3.3.93) (i.e., an empty subunit cannot be H). On the other hand, the state LH in the third row violates both conditions (i.e., an empty subunit cannot be H and an occupied subunit cannot be L). We see from the diagram that the only states which are consistent with both conditions (3.3.93) and (3.3.94) are the ones along the diagonal, for which we have written the PF (3.3.92).

The conditions (3.3.93) and (3.3.94) can be expressed in terms of the parameters of the model as follows

$$\Pr(L;/0;) = \frac{\Pr(L, 0;)}{\Pr(0;)} = \frac{Q_L \sum_\beta (Q_\beta Q_{L\beta} + Q_\beta Q_{L\beta} q_\beta \lambda)}{\sum_{\alpha\beta} (Q_\alpha Q_\beta Q_{\alpha\beta} + Q_\alpha Q_\beta Q_{\alpha\beta} q_\beta \lambda)}. \tag{3.3.95}$$

$\Pr(L;/0;)$ is the probability of finding the lhs subunit in state L given that the same subunit is empty (because of symmetry, the same expression applies to the rhs subunit).

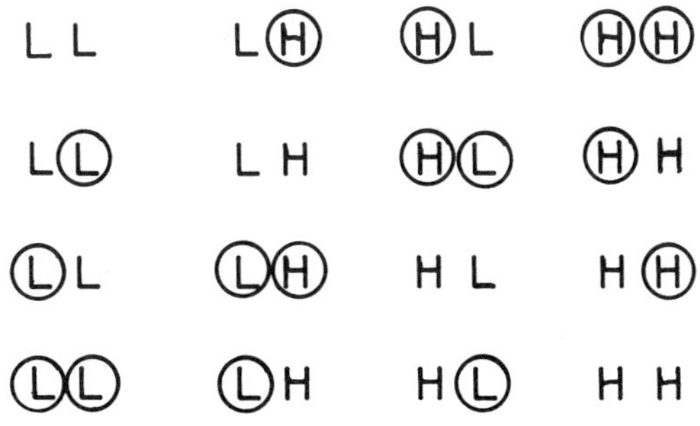

FIGURE 3.10. Same sixteen configurations as in Fig. 3.9. Circled states are inconsistent with the requirements of the sequential model, either Eq. (3.3.93) or (3.3.94).

A similar expression for the H state is

$$\Pr(H;/0;) = \frac{\Pr(H, 0;)}{\Pr(0;)} = \frac{Q_H \sum_{\beta} (Q_{\beta} Q_{H\beta} + Q_{\beta} Q_{H\beta} q_H \lambda)}{\sum_{\alpha\beta} (Q_{\alpha} Q_{\beta} Q_{\alpha\beta} + Q_{\alpha} Q_{\beta} Q_{\alpha\beta} q_{\beta} \lambda)}. \tag{3.3.96}$$

The ratio of (3.3.95) and (3.3.96) is the equilibrium constant (3.3.93), i.e.,

$$K_{L/H}^0 = \frac{\Pr(L;/0;)}{\Pr(H;/0;)} = \frac{Q_L}{Q_H} A(\lambda), \tag{3.3.97}$$

and similarly for the equilibrium constant (3.3.94) we find

$$K_{L/H}^1 = \frac{\Pr(L;/1;)}{\Pr(H;/1;)} = \frac{Q_L q_L}{Q_H q_H} A(S\lambda), \tag{3.3.98}$$

where

$$A(\lambda) = \frac{\sum_{\beta} (Q_{\beta} Q_{L\beta} + Q_{\beta} Q_{L\beta} q_{\beta} \lambda)}{\sum_{\beta} (Q_{\beta} Q_{H\beta} + Q_{\beta} Q_{H\beta} q_{\beta} \lambda)}. \tag{3.3.99}$$

Note that in (3.3.98) we have $A(S\lambda)$ instead of $A(\lambda)$.

One way to meet the requirements (3.3.93) and (3.3.94) is to assume that $Q_{\alpha\beta}$ are all finite quantities. Hence for each λ, $A(\lambda)$ is a finite quantity. Therefore, in order to fulfill (3.3.93), we assume that

$$\frac{Q_L}{Q_H} \to \infty; \tag{3.3.100}$$

i.e., the L state is infinitely more stable than the H state. However, in order to fulfill the condition (3.3.94), we need to assume also that

$$\frac{Q_L}{Q_H} \frac{q_L}{q_H} \to 0, \tag{3.3.101}$$

which means that the ratio q_L/q_H must tend to zero faster than Q_L/Q_H tends to infinity.

The awkwardness of the mathematical requirements can be understood as follows: if L is infinitely more stable than H [see Eq. (3.3.100)], then as we have seen in section 2.10, the system will have an infinite resistance to an induced conformational change. In section 2.10, we have also seen that in order to produce a total conversion from, say, L to H upon binding we need $h = q_H/q_L \to \infty$ for any finite equilibrium constant K. In the sequential model we require that a total conversion $(L \to H)$ is achieved upon binding, in a system which has an infinite resistance to conformational change. This is precisely the reason that we require that $h = (q_H/q_L)$ not only be infinite but a stronger infinity than Q_L/Q_H so that the product in (3.3.101) will be zero. Clearly these are quite severe requirements. We shall discuss this limiting case further in section 3.6.

3.4. THREE IDENTICAL SITES ON A POLYMER HAVING TWO CONFORMATIONAL STATES: TRIPLET CORRELATIONS

This model is an extension of the models treated in sections 2.10 and 3.2. Again we have a polymer that can be in either one of two conformations, L and H. Instead of one site per polymer (section 2.10) or two (section 3.2), we have here three sites per polymer. The generalization is quite straightforward; therefore we shall not dwell on the details of this particular model. Instead we shall elaborate on a new phenomenon that did not occur in the previous models—the triplet potential of average force among the three ligands on the same polymer. The potential of average force, in this particular model, is always negative, and arises from the effect of the ligand on the equilibrium concentrations of the two conformers L and H. We shall also examine the question of pairwise additivity of the potential of average force. This question is easy to analyze in the present model, from which we can learn about the source of the nonadditivity of the potential of average force. We shall return to this question in Chapter 5 in connection with the superposition approximation employed in the theory of liquids.

For the sake of simplicity, we assume that there are only two conformers, L and H. Each has three equivalent binding sites, and the distance between any pair of sites is the same, say R, independently of the conformational state (Fig. 3.11). Furthermore, we assume that the *direct* interaction among the three ligands occupying the same polymer *is* pairwise additive, in the sense that

$$U(1, 2, 3) = U(1, 2) + U(1, 3) + U(2, 3) = 3U(R) = 3U_{12}. \qquad (3.4.1)$$

Although this assumption is not essential, we adopt it to stress that the nonadditivity of the potential of average force is a new phenomenon, independent of whether or not the direct interaction among the three ligands is pairwise additive. Moreover, in order to study the indirect or the polymer-mediated part of the potential of average force, we shall later put $U(R) = 0$ and focus only on the indirect effects.

3.4.1. Binding Thermodynamics

The grand PF of a single polymer is

$$\xi = Q(0) + 3Q(1)\lambda + 3Q(2)\lambda^2 + Q(3)\lambda^3, \qquad (3.4.2)$$

where

$$Q(0) = Q_L + Q_H \qquad (3.4.3)$$

$$Q(1) = Q_L q_L + Q_H q_H \qquad (3.4.4)$$

$$Q(2) = (Q_L q_L^2 + Q_H q_H^2)S \qquad (3.4.5)$$

$$Q(3) = (Q_L q_L^3 + Q_H q_H^3)S^3, \qquad (3.4.6)$$

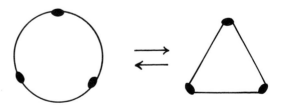

FIGURE 3.11. Two states of a polymer, each having three binding sites.

H   L

and where $Q(k)$ is the canonical partition function of a polymer, k specific sites of which are occupied by ligands. Since the sites are equivalent, we need not specify which of the sites are occupied. In a more general case, where the sites are not equivalent, we need to use a more detailed notation, say $Q(0, 0, 1)$, $Q(0, 1, 0)$, $Q(1, 0, 0)$, and so on, to specify which sites are occupied and which are not.

As before,

$$Q_\alpha = \exp(-\beta E_\alpha), \qquad q_\alpha = \exp(-\beta U_\alpha), \qquad S = \exp(-\beta U_{12}). \qquad (3.4.7)$$

The binding isotherm for this case is

$$\theta = \frac{\bar{n}}{3} = \frac{1}{3} \lambda \frac{\partial \ln \xi}{\partial \lambda} = \frac{Q(1)\lambda + 2Q(2)\lambda^2 + Q(3)\lambda^3}{\xi}. \qquad (3.4.8)$$

If $S = 1$, $Q = Q_L = Q_H$, and $q = q_L = q_H$, then the PF (3.4.2) reduces to

$$\xi = 2Q(1 + 3q\lambda + 3q^2\lambda^2 + q^3\lambda^3) = 2Q(1 + q\lambda)^3, \qquad (3.4.9)$$

and the corresponding isotherm is

$$\theta = \frac{q\lambda}{1 + q\lambda},$$

which is the Langmuir isotherm.

We define the binding Helmholtz energy on the first site as the work required to bring the ligand G from a fixed position in the vacuum to a specific site on an empty polymer; symbolically, the reaction is

$$G + (0, 0, 0) \rightarrow (1, 0, 0), \qquad (3.4.10)$$

and the corresponding Helmholtz energy is

$$\Delta A_G^{*(1)} = -kT \ln \frac{Q_L q_L + Q_H q_H}{Q_L + Q_H} = -kT \ln \langle \exp(-\beta B_G^{(1)}) \rangle_{0,0,0}. \qquad (3.4.11)$$

This is the same as Eq. (3.2.34). Here $B_G^{(1)}$ is the binding energy to the first site. In this particular model, $B_G^{(1)}(\alpha)$, $B_G^{(2)}(\alpha)$, $B_G^{(3)}(\alpha)$ are the same function of α as in (3.2.24). In (3.4.11) it does not matter on which site we place the ligand, but in the following it will be important to keep track of which site is being occupied. The average in (3.4.11) is taken with respect to the distribution x_L^0 and x_H^0 of the empty polymer [which is the same as in Eq. (3.2.34)]. The notation here $(0, 0, 0)$ replaces the notation $(0, 0)$ of section 3.2.

The binding Helmholtz energy on the second site is likewise defined for the process

$$G + (1, 0, 0) \rightarrow (1, 1, 0), \qquad (3.4.12)$$

as

$$\Delta A_G^{*(2)} = -kT \ln \frac{(Q_L q_L^2 + Q_H q_H^2)S}{Q_L q_L + Q_H q_H} = U_{12} - kT \ln \langle \exp(-\beta B_G^{(2)}) \rangle_{1,0,0}. \qquad (3.4.13)$$

Here $\Delta A_G^{*(2)}$ consists of the direct interaction between the two ligands, U_{12}, and an average binding free energy on the second site, given that the first site is occupied. The notation used in (3.4.13) should be noted carefully. In (3.4.11) we had an average of $\exp(-\beta B_G^{(1)})$ with the condition $(0, 0, 0)$, i.e., an empty polymer. Therefore, it does not matter which site is used for placing the first ligand. In (3.4.13) the condition is $(1, 0, 0)$;

i.e., the first site is occupied. Now we have to average over $\exp(-\beta B_G^{(2)})$, and we could equally use $B_G^{(3)}$, but not $B_G^{(1)}$. In other words, once one site is occupied, the second ligand could be placed either on the second or on the third site. These are still equivalent, but not equivalent to the first site.

The Helmholtz energy change for the process $G + (1, 1, 0) \rightarrow (1, 1, 1)$ i.e., binding of the third ligand, is

$$\Delta A_G^{*(3)} = -kT \ln \frac{(Q_L q_L^3 + Q_H q_H^3)S^3}{(Q_L q_L^2 + Q_H q_H^2)S}$$

$$= 2U_{12} - kT \ln \langle \exp(-\beta B_G^{(3)}) \rangle_{1,1,0}. \tag{3.4.14}$$

$\Delta A_G^{*(3)}$ consists of the direct interaction between the third ligand and the two ligands already occupying the polymer, and an average of the quantity $\exp(-\beta B_G^{(3)})$ with the condition $(1, 1, 0)$, i.e., the first and second sites being occupied. Note that in each case $\Delta A_G^{*(k)}$ is defined as the average of $\exp(-\beta B_G^{(k)})$ with a conditional probability distribution of the states of the polymer, given that $k - 1$ sites are already occupied.

We can define the kth intrinsic binding isotherm constant by

$$K_k = \exp(-\beta \Delta A_G^{*(k)}) \tag{3.4.15}$$

with the help of which the binding isotherm (3.4.8) can be written in a more familiar thermodynamic notation as

$$\theta = \frac{K_1 \lambda + 2K_1 K_2 \lambda^2 + K_1 K_2 K_3 \lambda^3}{1 + 3K_1 \lambda + 3K_1 K_2 \lambda^2 + K_1 K_2 K_3 \lambda^3}. \tag{3.4.16}$$

3.4.2. Pair Correlation Functions

As in section 3.2, we can define a multitude of distribution functions in this system. All of these can be either read directly from the terms in equation (3.4.2) or constructed from such terms. For example, the probability of finding a specific polymer empty and in state L is

$$\Pr(L, 0, 0, 0) = \frac{Q_L}{\xi}. \tag{3.4.17}$$

The probability of finding the polymer empty, independently of its conformational state is

$$\Pr(0, 0, 0) = \frac{Q_L + Q_H}{\xi}. \tag{3.4.18}$$

The probability of finding the polymer with exactly one ligand at a specific site, say the first, is

$$\Pr(1, 0, 0) = \frac{(Q_L q_L + Q_H q_H)\lambda}{\xi}. \tag{3.4.19}$$

Note that the probability of finding the polymer with exactly one ligand but at any site is $3\Pr(1, 0, 0) = \Pr(1, 0, 0) + \Pr(0, 1, 0) + \Pr(0, 0, 1)$.

The probability of finding the polymer with one specific site occupied, independently of the occupational state of the other site is

$$Pr(1) = Pr(1, 0, 0) + Pr(1, 1, 0) + Pr(1, 0, 1) + Pr(1, 1, 1).$$

The probability of finding exactly two specific sites of the polymer occupied is

$$Pr(1, 1, 0) \doteq \frac{(Q_L q_L^2 + Q_H q_H^2)S\lambda^2}{\xi}. \tag{3.4.20}$$

The probability of finding any two sites occupied, but the third site empty, is $3Pr(1, 1, 0) = Pr(1, 1, 0) + Pr(1, 0, 1) + Pr(0, 1, 1)$.

The pair correlation function is defined as the limit

$$g^0(1, 1) = \lim_{\lambda \to 0} \frac{Pr(1, 1, 0)}{[Pr(1, 0, 0)]^2} = \frac{Q(2)Q(0)}{[Q(1)]^2}, \tag{3.4.21}$$

and the corresponding potential of average force

$$W(1, 1) = -kT \ln[g^0(1, 1)] = A(1, 1, 0) - 2A(1, 0, 0) + A(0, 0, 0). \tag{3.4.22}$$

Note that in both (3.4.21) and (3.4.22) we require that the third site be empty. We could have used more specific notation, such as $g^0(1, 1, 0)$ or $W(1, 1, 0)$, but these are reserved for triplet correlations as defined in section 3.4.3.

The quantity $W(1, 1)$ as defined in (3.4.22) may be interpreted as the Helmholtz energy change for the reaction, symbolically written as

$$2(1, 0, 0) \to (1, 1, 0) + (0, 0, 0); \tag{3.4.23}$$

i.e., we take two singly occupied polymers and form one empty and one doubly occupied polymer. This is essentially the same reaction (3.2.44), but here we require that the third site be empty.

In analogy with section 3.2, the pair-correlation function $g^0(1, 1)$ may be written as

$$g^0(1, 1) = S \frac{\langle \exp(-\beta B_{GG}) \rangle_{0,0,0}}{\langle \exp(-\beta B_G^{(1)}) \rangle_{0,0,0}^2}$$

$$= S \frac{\langle \exp(-\beta B_G^{(2)}) \rangle_{1,0,0}}{\langle \exp(-\beta B_G^{(1)}) \rangle_{0,0,0}}. \tag{3.4.24}$$

In the first form of the rhs of (3.4.24), the two average quantities are taken with respect to the probability distribution in the empty polymer. More specifically,

$$g^0(1, 1) = \frac{Q(2)Q(0)Q(0)}{Q(0)Q(1)Q(1)} = S \left(\frac{Q_L q_L^2 + Q_H q_H^2}{Q_L + Q_H} \right) \times \left(\frac{Q_L q_L + Q_H q_H}{Q_L + Q_H} \right)^{-2}. \tag{3.4.25}$$

In the second form, on the rhs of (3.4.24), we used a conditional probability distribution in the numerator, i.e.,

$$g^0(1, 1) = \frac{Q(2)}{Q(1)} \frac{Q(0)}{Q(1)}$$

$$= S \left(\frac{Q_L q_L}{Q_L q_L + Q_H q_H} q_L + \frac{Q_H q_H}{Q_L q_L + Q_H q_H} q_H \right) \times \left(\frac{Q_L}{Q_L + Q_H} q_L + \frac{Q_H}{Q_L + Q_H} q_H \right)^{-1}$$

$$\tag{3.4.26}$$

With the help of the pair correlation function we can relate the first and the second binding Helmholtz energies, i.e.,

$$\exp(-\beta \Delta A_G^{*(2)}) = g^0(1, 1) \exp(-\beta \Delta A_G^{*(1)}), \qquad (3.4.27)$$

which has the same formal appearance as similar relations obtained in sections 3.2 and 3.3. The third binding Helmholtz energy defined in (3.4.14) cannot be related to $\Delta A_G^{*(1)}$ or to $\Delta A_G^{*(2)}$ by a relation of the form (3.4.27). In order to find the relation between $\Delta A_G^{*(3)}$ and $\Delta A_G^{*(1)}$ and $\Delta A_G^{*(2)}$ we need to introduce the triplet correlation function. Note that, when $S = 1$, the indirect correlation is the same as the one in section 3.2.

3.4.3. Triplet Correlation Function and Triplet Potential of Average Force

The triplet correlation function is defined by analogy with the definition (3.4.21) by

$$g^0(1, 1, 1) = \lim_{\lambda \to 0} \frac{\Pr(1, 1, 1)}{[\Pr(1, 0, 0)]^3} = \frac{Q(3)Q(0)^2}{Q(1)^3}, \qquad (3.4.28)$$

and the corresponding triplet potential of average force is defined by

$$W(1, 1, 1) = -kT \ln g^0[(1, 1, 1)] = A(1, 1, 1) + 2A(0, 0, 0) - 3A(1, 0, 0). \quad (3.4.29)$$

Thus $W(1, 1, 1)$ is the Helmholtz energy change for the reaction

$$3(1, 0, 0) \to (1, 1, 1) + 2(0, 0, 0). \qquad (3.4.30)$$

The triplet correlation function can be written in various forms. By analogy with (3.4.24), we write

$$g^0(1, 1, 1) = \frac{Q(3)}{Q(0)} \frac{Q(0)^3}{Q(1)^3} = S^3 \frac{\langle \exp(-\beta B_{GGG}) \rangle_{0,0,0}}{\langle \exp(-\beta B_G^{(1)}) \rangle_{0,0,0}^3}, \qquad (3.4.31)$$

where B_{GGG} is the binding energy of the three ligands to an empty polymer. [See the definition of B_{GG} in (3.2.26).]

In a similar fashion one can introduce correlation functions of the form $g^0(1, 1, 0)$, $g^0(1, 0, 0)$, and $g^0(0, 0, 0)$. We shall not need to use these functions here. As in previous sections, we can distinguish between direct and indirect correlation, i.e.,

$$g^0(1, 1, 1) = S^3 y(1, 1, 1), \qquad (3.4.32)$$

and the corresponding indirect part of the potential of average force is defined by $\delta W(1, 1, 1) = -kT \ln y(1, 1, 1) = W(1, 1, 1) - 3U_{12}(R)$.

From the definition (3.4.31) and the explicit expressions for the $Q(k)$ in (3.4.3)–(3.4.6) we can write the general expression for $g^0(1, 1, 1)$ in terms of $K = Q_H/Q_L$, $h = q_H/q_L$, and S:

$$g^0(1, 1, 1) = S^3 \frac{(1 + Kh^3)(1 + K)^2}{(1 + Kh)^3} = S^3 y(1, 1, 1), \qquad (3.4.33)$$

where we have separated the direct correlation S^3 among the three ligands, from the indirect correlation, which is denoted by $y(1, 1, 1)$. The latter can be rewritten as

$$y(1, 1, 1) = 1 + \frac{K(1 - h)^2[(2K + 1)h + K + 2]}{(1 + Kh)^3}. \qquad (3.4.34)$$

From the form of $y(1, 1, 1)$, we can conclude that for any K and h, $y(1, 1, 1) > 1$ or $\delta W(1, 1, 1) < 0$, i.e., the indirect part of the triplet-correlation function is always larger than unity. A plot of the triplet correlation function $y(1, 1, 1)$ as a function of K is shown in Fig. 3.12.

A special case of (3.4.33) occurs when $q_H = q_L$—i.e., $h = 1$—in which case we find

$$g^0(1, 1, 1) = S^3, \qquad y(1, 1, 1) = 1. \tag{3.4.35}$$

Thus for any value of K, if $h = 1$ the indirect correlation function is always unity. This means that there is no indirect correlation or indirect potential of average force if the ligand does not affect the equilibrium concentrations of L and H.

The converse is also true; i.e., if $y(1, 1, 1) = 1$ for any finite value of K, we must have $h = 1$. In other words, $h = 1$ is a necessary and sufficient condition for $y(1, 1, 1) = 1$. To prove that $h = 1$ is necessary, we assume that $y(1, 1, 1) \equiv 1$ for any K; then it follows from (3.4.33) that

$$(1 + Kh^3)(1 + K)^2 = (1 + Kh)^3 \tag{3.4.36}$$

is an identity for any K. Rearranging (3.4.36), we obtain (after division by $K \neq 0$)

$$K(2h^3 - 3h^2 + 1) + h^3 - 3h + 2 = 0. \tag{3.4.37}$$

Since this must hold for any K, we can take the derivative with respect to K to obtain

$$2h^3 - 3h^2 + 1 = 0. \tag{3.4.38}$$

It is easy to see that $h = 1$ is a root of (3.4.38). Factoring $h - 1$ out from (3.4.38), we get

$$(h - 1)(2h^2 - h - 1) = 0. \tag{3.4.39}$$

The three roots of (3.4.38) are

$$h_1 = 1, \qquad h_2 = 1, \qquad h_3 = -\frac{1}{2}.$$

The only physically significant solution ($h > 0$) for our model is $h = 1$, which proves that this is a necessary condition.

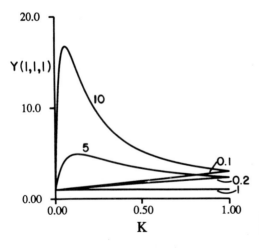

FIGURE 3.12. Triplet correlation function as a function of K for some values of h, Eq. (3.4.34).

Another simple case occurs when $h \neq 1$ but $K = Q_H/Q_L = 1$, in which case (3.4.34) reduces to

$$y(1, 1, 1) = 1 + \frac{3(1 - h)^2(1 + h)}{(1 + h)^2}. \tag{3.4.40}$$

Clearly, in this case,

$$1 \leq y(1, 1, 1) \leq 4. \tag{3.4.41}$$

Thus the minimum value of $y(1, 1, 1)$ occurs when $h = 1$; i.e., when the equilibrium $L \rightleftarrows H$ is not affected by the ligand. If, on the other hand, $h = 0$, the effect of the ligand on the equilibrium $L \rightleftarrows H$ is maximal. The value of $y(1, 1, 1)$ for any K is then, from (3.4.34),

$$y(1, 1, 1) = (1 + K)^2, \tag{3.4.42}$$

and the maximum of this is obtained when $K = 1$, i.e., when the response of the system to the binding of ligands is maximal.

3.4.4. Superposition Approximation: Nonadditivity of the Triplet Potential of Average Force

Triplet and higher correlation functions appear in the theory of liquids (Chapter 5). These are defined as follows: Let $\Pr(\mathbf{R}_1, \mathbf{R}_2, \mathbf{R}_3)$ be the probability density of finding three centers of particles at small elements of volume at $\mathbf{R}_1$, $\mathbf{R}_2$, and $\mathbf{R}_3$. The triplet correlation function is defined by

$$\Pr(\mathbf{R}_1, \mathbf{R}_2, \mathbf{R}_3) = g^{(3)}(\mathbf{R}_1, \mathbf{R}_2, \mathbf{R}_3)\rho^3 \tag{3.4.43}$$

where ρ, the bulk density, is also the probability density of finding one particle at, say, $\mathbf{R}_1$. If the distances between all the pairs of particles are large enough, then

$$\Pr(\mathbf{R}_1, \mathbf{R}_2, \mathbf{R}_3) = \rho^3 \quad \text{or} \quad g^{(3)}(\mathbf{R}_1, \mathbf{R}_2, \mathbf{R}_3) = 1. \tag{3.4.44}$$

At shorter distances, the factorization in (3.4.44) does not apply. There are correlations between each pair and among the three particles. As a better approximation to (3.4.44), one may try a factorization of the form

$$\Pr(\mathbf{R}_1, \mathbf{R}_2, \mathbf{R}_3) = \Pr(\mathbf{R}_1, \mathbf{R}_2) \Pr(\mathbf{R}_1, \mathbf{R}_3) \Pr(\mathbf{R}_2, \mathbf{R}_3) \tag{3.4.45}$$

or, equivalently,

$$g^{(3)}(\mathbf{R}_1, \mathbf{R}_2, \mathbf{R}_3) = g^{(2)}(\mathbf{R}_1, \mathbf{R}_2)g^{(2)}(\mathbf{R}_1, \mathbf{R}_3)g^{(2)}(\mathbf{R}_2, \mathbf{R}_3). \tag{3.4.46}$$

In terms of the potentials of average force, (3.4.46) can be written as

$$W^{(3)}(\mathbf{R}_1, \mathbf{R}_2, \mathbf{R}_3) = W^{(2)}(\mathbf{R}_1, \mathbf{R}_2) + W^{(2)}(\mathbf{R}_1, \mathbf{R}_3) + W^{(2)}(\mathbf{R}_2, \mathbf{R}_3), \tag{3.4.47}$$

which states that the work required to bring the three particles from infinite separation to the final configuration $\mathbf{R}_1$, $\mathbf{R}_2$, $\mathbf{R}_3$ within the liquid (at T, V or T, P constant), is the sum of the pairwise works as written on the rhs of (3.4.47). This approximation is referred to as the superposition approximation for the triplet potential of average force.

We shall now introduce the analogue of the superposition approximation in the model treated in this section. Using the notations introduced in the previous sections, the superposition approximation is written as

$$W(1, 1, 1) = 3W(1, 1). \tag{3.4.48}$$

Note that the notation in (3.4.48) for the arguments differs from the notation in (3.4.47). Here 1 and 0 stand for occupied and empty sites. Since the distances between each pair of sites is the same in this model, we have three times the potential of average force for each pair of ligands. The assumption (3.4.48) can be equivalently formulated in terms of the corresponding correlation functions

$$g^0(1, 1, 1) = [g^0(1, 1)]^3. \tag{3.4.49}$$

In order to examine the approximation (3.4.48) we define the nonadditivity of the triplet potential of average force by

$$W_{n.a} = W(1, 1, 1) - 3W(1, 1). \tag{3.4.50}$$

This can be expressed in various forms as follows [see Eqs. (3.4.21) and (3.4.28)]:

$$\exp(-\beta W_{n.a.}) = \frac{g(1, 1, 1)}{[g(1, 1)]^3} = \frac{Q(3)Q(1)^3}{Q(2)^3 Q(0)}$$

$$= \frac{(Q_L q_L^3 + Q_H q_H^3)(Q_L q_L + Q_H q_H)^3}{(Q_L q_L^2 + Q_H q_H^2)^3 (Q_L + Q_H)}. \tag{3.4.51}$$

In terms of the parameters K and h, Eq. (3.4.51) can be rewritten as

$$\exp(-\beta W_{n.a.}) = \frac{(1 + Kh^3)(1 + Kh)^3}{(1 + Kh^2)^3(1 + K)}. \tag{3.4.52}$$

Clearly, if $h = 1$, then

$$\exp(-\beta W_{n.a.}) = 1. \tag{3.4.53}$$

Recall, however, that if $h = 1$, there is no indirect correlation for both pairs and triplets of ligands; i.e., we already know that

$$g(1, 1, 1) = S^3 \tag{3.4.54}$$

and

$$g(1, 1) = S. \tag{3.4.55}$$

Since we have assumed that the direct interaction is pairwise additive (3.4.33), and since we already know that for $h = 1$ there are no indirect correlations, namely $y(1, 1) = 1$ and $y(1, 1, 1) = 1$, it follows that $W_{n.a.}$ must be zero. The converse of the latter conclusion is also true, i.e., $h = 1$ is also a necessary condition for $W_{n.a.} = 0$.

From the above analysis we conclude that whenever there are indirect correlations, there also exists nonadditivity of the potential of average force. This conclusion is probably true also for the triplet potential of average force in liquids.

A special case of (3.4.52) occurs when $K = 1$. Then

$$\exp(-\beta W_{n.a}) = \frac{1}{2} \frac{(1 + h^3)(1 + h)^3}{(1 + h^2)^3}. \tag{3.4.56}$$

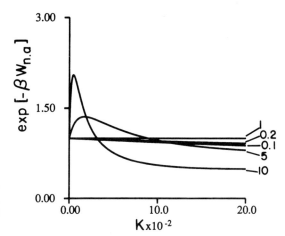

FIGURE 3.13. Nonadditivity of the triplet potential of average force [Eq. (3.4.52)] as a function of K for some values of h.

In this particular case,

$$\tfrac{1}{2} \le \exp(-\beta W_{\text{n.a.}}) \le 1.$$

In the more general case $K \ne 1$, we can rewrite (3.4.52) as

$$\exp(-\beta W_{\text{n.a.}}) = 1 - \frac{K(1-h)^3(1-h^3K^2)}{(1+Kh^2)^3(1+K)}. \tag{3.4.57}$$

We see that for $h = 1$, again, $W_{\text{n.a.}} = 0$. However, for other values of h and K, the nonadditivity can be either positive or negative. Some curves of $\exp(-\beta W_{\text{n.a.}})$ as a function of K are shown in Fig. 3.13.

3.4.5. The Binding Helmholtz Energy on the Third Site

We have defined the binding Helmholtz energy on the first and on the second site in subsection 3.4.1. We also found that the pair correlation function $g^0(1, 1)$ connects the first and the second binding Helmholtz energies by the relation

$$\exp(-\beta \Delta A_G^{*(2)}) = g^0(1, 1) \exp(-\beta \Delta A_G^{*(1)}). \tag{3.4.58}$$

We have defined the binding Helmholtz energy on the third site in (3.4.14). The question we now pose is, how is $\Delta A_G^{*(3)}$ related to $\Delta A_G^{*(2)}$? Is there a relation similar to (3.4.58)?

Using the definition of $\Delta A_G^{*(3)}$ in (3.4.14), we write

$$\exp(-\beta \Delta A_G^{*(3)}) = \frac{Q(3)}{Q(2)} = \frac{Q(3)Q(1)}{Q(2)Q(2)} \frac{Q(2)}{Q(1)} = \frac{Q(3)Q(1)}{Q(2)Q(2)} \exp(-\beta \Delta A_G^{*(2)}). \tag{3.4.59}$$

This is a formal connection between $\Delta A_G^{*(3)}$ and $\Delta A_G^{*(2)}$. We can now identify the factor that connects these two quantities as [see also (3.4.21) and (3.4.51)]

$$\frac{Q(3)Q(1)}{Q(2)^2} = \frac{Q(3)Q(1)^3}{Q(2)^3 Q(0)} \frac{Q(2)Q(0)}{Q(1)^3} = \exp(-\beta W_{\text{n.a.}}) g^0(1, 1). \tag{3.4.60}$$

The required connection between $\Delta A_G^{*(3)}$ and $\Delta A_G^{*(2)}$ is thus

$$\exp(-\beta \Delta A_G^{*(3)}) = \exp(-\beta W_{\text{n.a.}}) g^0(1, 1) \exp(-\beta \Delta A_G^{*(2)}). \tag{3.4.61}$$

The difference between this and the previous connection (3.4.58) is the appearance of the nonadditivity of the potential of average force.

The direct interaction U_{12} can be either negative or positive, giving rise to direct positive or negative cooperativity. We now show that the indirect cooperativity in this model is always positive. To show this, we assume that $U_{12} = 0$. Hence

$$\exp(-\beta \Delta A_G^{*(2)}) = y(1, 1) \exp(-\beta \Delta A_G^{*(1)}). \tag{3.4.62}$$

We have already seen that $y(1, 1) \geq 1$. The qualitative argument was given in section 3.2.4. If $U_L < U_H$, then the first ligand will shift the equilibrium $L \rightleftarrows H$ in favor of L; therefore the binding Helmholtz energy on the second site will be more negative than on the second site. This argument was given in section 3.2 for the two-site polymer. The same argument also holds for the three-site polymer. Therefore we always have

$$\Delta A_G^{*(2)} \leq \Delta A_G^{*(1)}. \tag{3.4.63}$$

The same argument can also be applied to the third ligand; i.e., we show that $\Delta A_G^{*(3)}$ is more negative than $\Delta A_G^{*(2)}$. To prove this formally, we show that the coefficient that connects $\Delta A_G^{*(3)}$ with $\Delta A_G^{*(2)}$ in (3.4.59) is always greater than or equal to unity.

$$\frac{Q(3)Q(1)}{Q(2)^2} = \frac{(1 + Kh^3)(1 + Kh)}{(1 + Kh^2)^2} = 1 + \frac{Kh(1 - h)^2}{(1 + Kh^2)^2} \geq 1. \tag{3.4.64}$$

Therefore we conclude that

$$\Delta A_G^{*(3)} \leq \Delta A_G^{*(2)} \leq \Delta A_G^{*(1)}. \tag{3.4.65}$$

There is an interesting way of casting the relation between $\Delta A_G^{*(3)}$ and $\Delta A_G^{*(2)}$ in a form similar to (3.4.62). We assume that $S = 1$ and that only indirect effects are operative. We note that $y(1, 1)$, viewed as a function of the parameters K and h, is [see Eq. (3.2.31)]

$$y(1, 1) = y(h, K) = 1 + \frac{K(1 - h)^2}{(1 + Kh)^2}. \tag{3.4.66}$$

The connection between $\Delta A_G^{*(3)}$ and $\Delta A_G^{*(2)}$ involves the quantity in (3.4.64), which can be written as

$$y(h, Kh) = 1 + \frac{Kh(1 - h)^2}{(1 + Kh^2)^2}. \tag{3.4.67}$$

We see that relations (3.4.61) and (3.4.62) can be written in terms of the same function y, but evaluated at K and Kh, respectively:

$$\exp(-\beta \Delta A_G^{*(3)}) = y(h, Kh) \exp(-\beta \Delta A_G^{*(2)}), \tag{3.4.68}$$

$$\exp(-\beta \Delta A_G^{*(2)}) = y(h, K) \exp(-\beta \Delta A_G^{*(1)}). \tag{3.4.69}$$

The physical reason is simple. The connection between $\Delta A_G^{*(1)}$ and $\Delta A_G^{*(2)}$ depends on the *initial* equilibrium constant K and on h. When the third ligand binds, the new "initial" equilibrium constant is not K but Kh. Otherwise the same formal connection exists between $\Delta A_G^{*(3)}$ and $\Delta A_G^{*(2)}$.

3.5. THREE SUBUNITS, EACH OF WHICH CAN BE IN ONE OF TWO CONFORMATIONS

This is an extension of the models treated in sections 3.3 and 3.4. It extends the model of section 3.3 by adding one more subunit, and it extends the model of section 3.4 by adding one more degree of freedom, i.e., the transmission of information between the ligands. In section 3.4 we saw that correlation between ligands and the nonadditivity of the triplet potential of average force depends only on the ability of the ligand to induce conformational change in the polymer. In the present model we shall see that the correlation and the extent of nonadditivity depend both on the existence of induced conformational changes and on the ability to transmit this information between the subunits.

To keep the number of molecular parameters of the system as minimal as possible, we assume that the three subunits are identical, and that each can be in one of two states, L and H. The interactions between the subunits are $E_{\alpha\beta}$, with $E_{LH} = E_{HL}$. Each subunit contains only one binding site, and the binding energies of the ligand are U_L and U_H, respectively.

We also assume that the ligands on different subunits are far apart so that direct ligand–ligand interaction can be excluded ($S = 1$ in the notation of section 3.4). Therefore any correlation between ligands is only indirect correlation.

We shall first examine some aspects of the behavior of the empty system and then proceed to the study of the system with ligands.

3.5.1. The Empty System

The partition function of the empty system is

$$\xi(0) = \sum_{\alpha,\beta,\gamma} Q_\alpha Q_\beta Q_\gamma Q_{\alpha\beta} Q_{\beta\gamma} Q_{\gamma\alpha}, \qquad (3.5.1)$$

where Q_α and $Q_{\alpha\beta}$ have the same meaning as in section 3.3. The sum is over all eight possible conformations of the polymer, as in Fig. 3.14. Note that, as in previous sections,

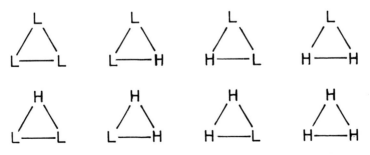

FIGURE 3.14. Eight possible configurations of the empty three-subunit system.

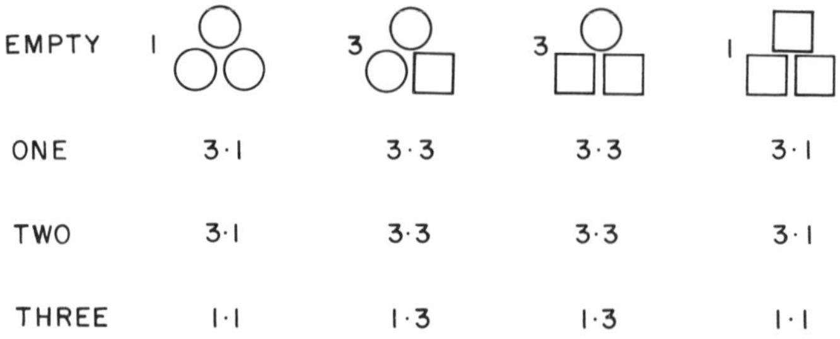

EMPTY		1		3		3		1
ONE		3·1		3·3		3·3		3·1
TWO		3·1		3·3		3·3		3·1
THREE		1·1		1·3		1·3		1·1

FIGURE 3.15. All possible configurations of the polymer and ligands. The configurations of the subunits are shown in the first row only. Subsequent rows correspond to increasing occupation number.

we assume that the polymers are localized. Therefore there are three distinguishable states of the form LLH and three of the form LHH. These numbers are indicated next to each configuration in the first row of Fig. 3.15.

We define the parameter

$$\eta = \frac{Q_{LH}^2}{Q_{LL}Q_{HH}}. \tag{3.5.2}$$

We have seen in section 3.3 that the deviation of η from unity is a measure of the extent of communication between the subunits.

If $\eta = 1$, then the subunits are effectively independent. In that case we write

$$Q_{\alpha\beta} = \sqrt{Q_{\alpha\alpha}Q_{\beta\beta}}, \tag{3.5.3}$$

and (3.5.1) reduces to

$$\xi(0) = \sum_{\alpha\beta\gamma} Q_\alpha Q_\beta Q_\gamma Q_{\alpha\alpha} Q_{\beta\beta} Q_{\gamma\gamma} = \left(\sum_\alpha Q_\alpha Q_{\alpha\alpha}\right)^3. \tag{3.5.4}$$

If $\eta \neq 1$, then we can write

$$Q_{\alpha\beta}^2 = Q_{\alpha\alpha}Q_{\beta\beta}[\eta + \delta_{\alpha\beta}(1 - \eta)], \tag{3.5.5}$$

where $\delta_{\alpha\beta}$ is the Kronecker delta function.

The probability of finding a specific polymer in a specific state, say LLH, is

$$x_{LLH}^0 = \frac{Q_L Q_L Q_H Q_{LH} Q_{LH} Q_{LL}}{\xi(0)}. \tag{3.5.6}$$

Note that by LLH we mean a specific order of the subunits, i.e., only one of the three configurations of this kind in Fig. 3.14.

If $\eta = 1$, then (3.5.6) reduces to

$$x_{LLH}^0 = \frac{Q_L Q_{LL}}{\sum_\alpha Q_\alpha Q_{\alpha\alpha}} \frac{Q_L Q_{LL}}{\sum_\alpha Q_\alpha Q_{\alpha\alpha}} \frac{Q_H Q_{HH}}{\sum_\alpha Q_\alpha Q_{\alpha\alpha}} = x_L^0 x_L^0 x_H^0. \tag{3.5.7}$$

The probability of finding a specific state of the polymer is a product of the probabilities of finding each subunit in the corresponding state. Note, however, that the mole fractions x_L^0 and x_H^0 in (3.5.7) differ from the corresponding mole fractions had the subunits been at infinite separation from each other, i.e., in the absence of interaction. In this limiting case, all $Q_{\alpha\beta} = 1$ and

$$x_L^{0,\infty} = \frac{Q_L}{Q_L + Q_H}, \qquad x_H^{0,\infty} = \frac{Q_H}{Q_L + Q_H}. \tag{3.5.8}$$

We have shown in section 3.3 that $\eta = 1$ was a necessary and a sufficient condition for independence of the subunits. The same holds true in this case. We have already shown that $\eta = 1$ leads to independence. The converse is also true. If we know that $x_{\alpha\beta\gamma} = x_\alpha^0 x_\beta^0 x_\gamma^0$, where x_α^0 are defined by $x_\alpha^0 = \sum_{\beta\gamma} x_{\alpha\beta\gamma}^0$, then by direct substitution in the definitions of these quantities one can prove that $\eta = 1$. The proof requires much more writing than in the case of section 3.3.

We define the potential of average force for the three subunits as the change in the Helmholtz energy for the process of bringing the three subunits from infinite separation from each other to the final configuration, which we denote by d. (In section 3.3.2, d denoted the distance between the two subunits; here d may similarly be used to define the intersubunit distance between each pair of subunits.) Thus

$$\exp[-\beta\Delta A(\infty \to d)] = \frac{\sum\limits_{\alpha\beta\gamma} Q_\alpha Q_\beta Q_\gamma Q_{\alpha\beta} Q_{\beta\gamma} Q_{\gamma\alpha}}{(Q_L + Q_H)^3}. \tag{3.5.9}$$

Using the mole fractions as defined in (3.5.8), we rewrite (3.5.9) as

$$\exp[-\beta\Delta A(\infty \to d)] = \sum_{\alpha\beta\gamma} x_\alpha^{0,\infty} x_\beta^{0,\infty} x_\gamma^{0,\infty} Q_{\alpha\beta} Q_{\beta\gamma} Q_{\gamma\alpha}$$

$$= \sum_{\alpha\beta\gamma} x_{\alpha\beta\gamma}^{0,\infty} \exp[-\beta(E_{\alpha\beta} + E_{\beta\gamma} + E_{\gamma\alpha})], \tag{3.5.10}$$

where $x_{\alpha\beta\gamma}^{0,\infty}$ is the probability distribution of the states when they are empty and separated. This is a generalization of Eq. (3.3.17). As in section 3.3.2, we could also take the gradient of $\Delta A(\infty \to d)$ with respect to the locational vector of one of the subunits and obtain the average force exerted on that subunit.

3.5.2. The System with Ligands

The grand PF of one polymer can be written directly by inspection of the various states of the polymer as depicted in Fig. 3.15. To save some writing, we use the shorthand notation

$$Q_{\alpha\beta\gamma} = Q_\alpha Q_\beta Q_\gamma Q_{\alpha\beta} Q_{\beta\gamma} Q_{\gamma\alpha}. \tag{3.5.11}$$

The PF is thus

$$\xi = \sum_{\alpha\beta\gamma} Q_{\alpha\beta\gamma} + \sum_{\alpha\beta\gamma} Q_{\alpha\beta\gamma}(q_\alpha + q_\beta + q_\gamma)\lambda$$

$$+ \sum_{\alpha\beta\gamma} Q_{\alpha\beta\gamma}(q_\alpha q_\beta + q_\beta q_\gamma + q_\gamma q_\alpha)\lambda^2 + \sum_{\alpha\beta\gamma} Q_{\alpha\beta\gamma} q_\alpha q_\beta q_\gamma \lambda^3. \tag{3.5.12}$$

Each term in (3.5.12) corresponds to one row in Fig. 3.15. The first term is simply $\xi(0)$, for the empty system. The second term corresponds to all of the 24 states with one ligand. There are eight configurations of the subunits, and for each of these we can place a ligand on three different sites. Similarly, we have 24 terms for the polymer with two ligands and eight terms for the fully occupied polymer. Altogether there are 64 terms.

Another way of writing (3.5.12) is

$$\xi = \sum_{\alpha\beta\gamma} Q_{\alpha\beta\gamma}[1 + (q_\alpha + q_\beta + q_\gamma)\lambda + (q_\alpha q_\beta + q_\beta q_\gamma + q_\gamma q_\alpha)\lambda^2 + q_\alpha q_\beta q_\gamma \lambda^3]. \quad (3.5.13)$$

In this representation each term in the sum corresponds to one column in Fig. 3.15.

In the following we shall use the shorthand notation for a polymer in a specific state of occupancy, regardless of the conformational states of the subunit:

$$Q(1;0;0) = \sum_{\alpha\beta\gamma} Q_{\alpha\beta\gamma} q_\alpha \quad (3.5.14)$$

$$Q(1;1;0) = \sum_{\alpha\beta\gamma} Q_{\alpha\beta\gamma} q_\alpha q_\beta. \quad (3.5.15)$$

Note that $(1;0;0)$ refers to the state in which the *first site* is occupied (say the upper subunit in Fig. 3.15), the second subunit empty (say the left-hand subunit) and the third subunit also empty. In general, $Q(1;0;0)$ would be different from $Q(0;1;0)$ and from $Q(0;0;1)$, but in our model the subunits are identical. Therefore we shall use only one of these to represent single occupancy. Similarly, for double occupancy we write $Q(1;1;0)$. The PF in (3.5.12) is now written in a more compact form as

$$\xi = Q(0;0;0) + 3Q(1;0;0)\lambda + 3Q(1;1;0)\lambda^2 + Q(1;1;1)\lambda^3$$

$$= \sum_{i=0}^{3} \binom{3}{i} Q(i)\lambda^i. \quad (3.5.16)$$

It should be noted that $Q(i)$ is the canonical PF for a polymer with i ligands at a *specific* set of i subunits, whereas $\binom{3}{i}Q(i)$ is the canonical PF of a polymer with i ligands. This is different from the notation used in section 2.9 where we denoted by $q(i)$ the PF of a site with i ligands. Thus $q(i)$ of section 2.9 corresponds to $\binom{3}{i}Q(i)$ in this section. The use of, say, $Q(1)$ for any of the three quantities $Q(1;0;0)$, $Q(0;1;0)$, and $Q(0;0;1)$ is possible because of the equivalency of the subunits.

The adsorption isotherm is

$$\theta = \frac{\bar{n}}{3} = \frac{\lambda}{3} \frac{\partial \ln \xi}{\partial \lambda} = \frac{Q(1)\lambda + 2Q(2)\lambda^2 + Q(3)\lambda^3}{\xi}. \quad (3.5.17)$$

We define the binding Helmholtz energy on the first, second, and third sites by analogy with sections 3.3 and 3.4 as

$$\Delta A_G^{*(1)} = -kT \ln \frac{Q(1;0;0)}{Q(0;0;0)} = -kT \ln \left(\frac{\sum_{\alpha\beta\gamma} Q_{\alpha\beta\gamma} q_\alpha}{\sum_{\alpha\beta\gamma} Q_{\alpha\beta\gamma}} \right)$$

$$= -kT \ln \langle \exp(-\beta B_G^{(1)}) \rangle_{0;0;0}. \quad (3.5.18)$$

Note that the average of $\exp(-\beta B_G^{(1)})$ is taken with respect to the probability distribution of the empty system.

$$\Delta A_G^{*(2)} = -kT \ln \frac{Q(1;1;0)}{Q(1;0;0)} = -kT \ln \left(\frac{\sum Q_{\alpha\beta\gamma} q_\alpha q_\beta}{\sum Q_{\alpha\beta\gamma} q_\alpha} \right)$$

$$= -kT \ln \langle \exp(-\beta B_G^{(2)}) \rangle_{1;0;0}, \quad (3.5.19)$$

and

$$\Delta A_G^{*(3)} = -kT \ln \frac{Q(1;1;1)}{Q(1;1;0)} = -kT \ln \left\{ \frac{\sum Q_{\alpha\beta\gamma} q_\alpha q_\beta q_\gamma}{\sum Q_{\alpha\beta\gamma} q_\alpha q_\beta} \right\}$$

$$= -kT \ln \langle \exp(-\beta B_G^{(3)}) \rangle_{1;1;0}. \qquad (3.5.20)$$

The pair correlation function and the corresponding potential of average force are defined by analogy with sections 3.3 and 3.4 as follows:

$$g^0(1, 1) = y(1, 1) = \frac{Q(1;1;0)Q(0;0;0)}{Q(1;0;0)^2}, \qquad (3.5.21)$$

$$W(1, 1) = -kT \ln y(1, 1) = A(1;1;0) - 2A(1;0;0) + A(0;0;0). \qquad (3.5.22)$$

The potential of average force is the Helmholtz energy for the process, which symbolically can be written as

$$2(1;0;0) \rightarrow (1;1;0) + (0;0;0). \qquad (3.5.23)$$

Note that in each case we require a specific configuration of the polymer with respect to its occupancy.

The triplet correlation and the corresponding potential of average force are defined by

$$g^0(1;1;1) = y(1;1;1) = \frac{Q(1;1;1)Q(0;0;0)^2}{Q(1;0;0)^3}, \qquad (3.5.24)$$

$$W(1;1;1) = -kT \ln y(1;1;1) = A(1;1;1) - 3A(1;0;0) + 2A(0;0;0). \qquad (3.5.25)$$

The intrinsic binding constants on the first, second, and third sites are defined by

$$K_i = \exp(-\beta \Delta A_G^{*(i)}). \qquad (3.5.26)$$

In terms of the K_i, the binding isotherm can be written as

$$\theta = \frac{K_1 \lambda + 2K_1 K_2 \lambda^2 + K_1 K_2 K_3 \lambda^3}{1 + 3K_1 \lambda + 3K_1 K_2 \lambda^2 + K_1 K_2 K_3 \lambda^3}, \qquad (3.5.27)$$

which is formally the same as (3.4.16).

The connection between the different $\Delta A_G^{*(i)}$ is, as in section 3.4,

$$\exp(-\beta \Delta A_G^{*(2)}) = y(1;1) \exp(-\beta \Delta A_G^{*(1)}) \qquad (3.5.28)$$

$$\exp(-\beta \Delta A_G^{*(3)}) = \frac{Q(3)Q(1)}{Q(2)^2} \exp(-\beta \Delta A_G^{*(2)})$$

$$= \frac{y(1;1;1)}{y(1;1)^2} \exp(-\beta \Delta A_G^{*(2)}) \qquad (3.5.29)$$

$$= f(1;1;1)y(1;1) \exp(-\beta \Delta A_G^{*(2)}),$$

where

$$f(1;1;1) = \exp(-\beta W_{n.a.}) = \frac{y(1;1;1)}{(y(1;1))^3} \qquad (3.5.30)$$

is a measure of the nonadditivity of the triplet potential of average force.

K_3 and K_2 can also be written in terms of K_1, as

$$K_2 = yK_1 \tag{3.5.31}$$

$$K_3 = fyK_2 = fy^2K_1, \tag{3.5.32}$$

where y is $y(1, 1)$ and $f = f(1; 1; 1)$. The binding isotherm is thus

$$\theta = \frac{K_1\lambda + 2yK_1^2\lambda^2 + fy^3K_1^3\lambda^3}{1 + 3K_1\lambda + 3yK_1^2\lambda^2 + fy^3K_1^3\lambda^3}, \tag{3.5.33}$$

which is formally the same as the binding isotherm of section 3.4 except that here we put $S = 1$, i.e., no direct interactions between the ligands. We have chosen the case $S = 1$ on purpose in order to study the source and the extent of the indirect correlation in this particular model.

Note that if we neglect nonadditivity effects—i.e., we put $f = 1$ in (3.5.33)—the resulting isotherm would be formally identical to an isotherm of a system having three sites with direct interactions between the ligands. In the latter case, y would have represented the direct correlation between two ligands. However, this formal resemblance could be misleading. In the case of systems having only *direct* interactions, the parameters K_1 and y (y represents S in this case) could be chosen independently. On the other hand, if indirect effects exist, then we have seen that nonadditivity always exists. Both K_1 and y depend on the molecular parameters of the system, and therefore they cannot be chosen independently. We shall further discuss this point in section 3.6 in connection with the adsorption isotherm of hemoglobin.

3.5.3. Further Examination of the Correlation Functions and Nonadditivity Effect

As in sections 3.3 and 3.4, we analyze here the pair and the triplet correlation functions in terms of the basic parameters of the system.

$$y(1; 1) = \frac{Q(1; 1; 0)Q(0; 0; 0)}{Q(1; 0; 0)^2} = \frac{\sum Q_{\alpha\beta\gamma}q_\alpha q_\beta \sum Q_{\alpha\beta\gamma}}{[\sum Q_{\alpha\beta\gamma}q_\alpha]^2}. \tag{3.5.34}$$

All the sums are over $\alpha, \beta, \gamma = L, H$ and $Q_{\alpha\beta\gamma} = Q_\alpha Q_\beta Q_\gamma Q_{\alpha\beta}Q_{\beta\gamma}Q_{\gamma\alpha}$.
We define

$$B_{\alpha\beta} = \sum_\gamma Q_{\beta\gamma}Q_{\gamma\alpha}Q_\gamma, \tag{3.5.35}$$

$$Q''_{\alpha\beta} = Q_\alpha Q_\beta Q_{\alpha\beta}B_{\alpha\beta}. \tag{3.5.36}$$

We now introduce the equilibrium constant for the reaction

$$LLX + HHX \rightarrow 2(LHX), \tag{3.5.37}$$

where a polymer in a state LLX means that two of its subunits are in the L conformation and the conformation of the third subunit is unspecified (X can be either L or H). The equilibrium constant for the reaction (3.5.37) in the empty polymer is

$$\eta'' = \frac{\left(\sum_\gamma x^0_{LH\gamma}\right)^2}{\left(\sum_\gamma x^0_{LL\gamma}\right)\left(\sum_\gamma x^0_{HH\gamma}\right)} = \frac{\left(\sum_\gamma Q_{LH\gamma}\right)^2}{\left(\sum_\gamma Q_{LL\gamma}\right)\left(\sum_\gamma Q_{HH\gamma}\right)} = \frac{(Q''_{LH})^2}{Q''_{LL}Q''_{HH}}$$

$$= \frac{Q^2_{LH}}{Q_{LL}Q_{HH}}\frac{B^2_{LH}}{B_{LL}B_{HH}} = \eta\frac{B^2_{LH}}{B_{LL}B_{HH}}. \tag{3.5.38}$$

η is the equilibrium constant for the reaction $LL + HH \rightarrow 2LH$ [see (3.3.13)] in a system consisting of two subunits only. In the presence of a third subunit, the equilibrium constant for the same reaction as written in (3.5.37) is modified by the factor $B^2_{LH}/B_{LL}B_{HH}$.

We now show that $\eta = 1$ if and only if $\eta'' = 1$. First we rewrite (3.3.38) as

$$\eta'' = \eta\frac{(Q_{LL}Q_{LH}Q_L + Q_{HH}Q_{LH}Q_H)^2}{(Q_LQ^2_{LL} + Q^2_{LH}Q_H)(Q_HQ^2_{HH} + Q^2_{LH}Q_L)}$$

$$= \eta^2\frac{(Q_LQ_{LL} + Q_HQ_{HH})^2}{(Q_LQ_{LL} + \eta Q_HQ_{HH})(Q_HQ_{HH} + \eta Q_LQ_{LL})}. \tag{3.5.39}$$

Clearly if $\eta = 1$ we obtain $\eta'' = 1$. Conversely, if $\eta'' = 1$ we obtain from (3.5.39) after rearranging

$$(1 - \eta)[Q_LQ_HQ_{LL}Q_{HH}(1 + \eta) + Q^2_LQ^2_{LL}\eta + Q^2_HQ^2_{HH}\eta] = 0. \tag{3.5.40}$$

Since the expression inside the square bracket is always positive, it follows that $\eta = 1$. With definitions (3.5.35) and (3.5.36), we can cast $y(1; 1)$ in (3.5.34) in the form

$$y(1; 1) = \frac{\sum_{\alpha\beta} Q''_{\alpha\beta}q_\alpha q_\beta \sum_{\alpha\beta} Q''_{\alpha\beta}}{\left(\sum_{\alpha\beta} Q''_{\alpha\beta}q_\alpha\right)^2}, \tag{3.5.41}$$

which is the same as (3.3.71) of section 3.3, except for $Q''_{\alpha\beta}$ replacing $Q'_{\alpha\beta}$. Therefore we can use the same result (3.3.72) for our case, namely,

$$y(1; 1) = 1 + \frac{(1 - h)^2(1 - \eta'')Q''_{HH}Q''_{LL}}{[Q''_{LL} + Q''_{LH}(1 + h) + Q''_{HH}h]^2}. \tag{3.5.42}$$

Thus, exactly as in section 3.3 we can conclude that $h = 1$ or $\eta = 1$ are necessary and sufficient conditions for $y(1; 1) = 1$. In other words, an indirect pair correlation function differs from unity if and only if the ligand produces a conformational change ($h \neq 1$) *and* this information is communicated between the subunits ($\eta \neq 1$).

The triplet correlation function for this model is (note that $S = 1$)

$$y(1; 1; 1) = \frac{(\sum Q_{\alpha\beta\gamma}q_\alpha q_\beta q_\gamma)(\sum Q_{\alpha\beta\gamma})^2}{(\sum Q_{\alpha\beta\gamma}q_\alpha)^3} = \frac{\sum x^0_{\alpha\beta\gamma}q_\alpha q_\beta q_\gamma}{(\sum x^0_{\alpha\beta\gamma}q_\alpha)^3}. \tag{3.5.43}$$

All the sums in (3.5.43) are over all α, β, $\gamma = L$, H, where $x^0_{\alpha\beta\gamma}$ are the mole fractions of the species $\alpha\beta\gamma$ in the empty system (3.5.6). We now show that "either $h = 1$ or $\eta = 1$" is a necessary and sufficient condition for $y(1; 1; 1) = 1$. If $h = 1$, then $q_\alpha = q_\beta = q$ and, because of the normalization $\sum x^0_{\alpha\beta\gamma} = 1$, we obtain $y(1; 1; 1) = 1$. If $\eta = 1$, then we know that $x^0_{\alpha\beta\gamma} = x^0_\alpha x^0_\beta x^0_\gamma$ (3.5.7); therefore, from (3.5.43),

$$y(1; 1; 1) = \frac{\sum x^0_\alpha x^0_\beta x^0_\gamma q_\alpha q_\beta q_\gamma}{(\sum x^0_\alpha x^0_\beta x^0_\gamma q_\alpha)^3} = \frac{(\sum x^0_\alpha q_\alpha)^3}{(\sum x^0_\alpha q_\alpha)^3} = 1. \tag{3.5.44}$$

To show that "either $h = 1$ or $\eta = 1$" is also a necessary condition, we put $y(1; 1; 1) = 1$ in (3.5.43) and obtain

$$\sum_{\alpha\beta\gamma} x^0_{\alpha\beta\gamma} q_\alpha q_\beta q_\gamma - \left(\sum_{\alpha\beta\gamma} x_{\alpha\beta\gamma} q_\alpha\right)^3 = 0. \tag{3.5.45}$$

Define $x^0_\alpha = \sum_{\beta\gamma} x^0_{\alpha\beta\gamma}$ and rewrite (3.5.45) as

$$\sum_{\alpha\beta\gamma} (x^0_{\alpha\beta\gamma} - x^0_\alpha x^0_\beta x^0_\gamma) q_\alpha q_\beta q_\gamma = 0. \tag{3.5.46}$$

Because of the normalization of both $x^0_{\alpha\beta\gamma}$ and $x^0_\alpha x^0_\beta x^0_\gamma$ we can write, for any q,

$$\sum (x^0_{\alpha\beta\gamma} - x^0_\alpha x^0_\beta x^0_\gamma) q^3 = 0. \tag{3.5.47}$$

Combining (3.5.46) and (3.5.47), we have

$$\sum (x^0_{\alpha\beta\gamma} - x^0_\alpha x^0_\beta x^0_\gamma)(q_\alpha q_\beta q_\gamma - q^3) = 0. \tag{3.5.48}$$

If $h \neq 1$, we have to show that $\eta = 1$. Indeed, for $h \neq 1$, all $q_\alpha q_\beta q_\gamma - q^3$ will have a constant sign. For instance, if q is chosen as the smallest of q_L and q_H, then $q_\alpha q_\beta q_\gamma - q^3 > 0$ for any set of α, β, γ. Therefore it follows from (3.5.48) that

$$x^0_{\alpha\beta\gamma} = x^0_\alpha x^0_\beta x^0_\gamma \tag{3.5.49}$$

for each α, β, γ. This means that $x^0_{\alpha\beta\gamma}$ can be factored into products of probabilities of the separate subunits. We have already shown that this is equivalent to $\eta = 1$ (section 3.5.1).

If, on the other hand, $\eta \neq 1$, we have to show that $h = 1$. We prove this by contradiction. Suppose that $h \neq 1$; then it follows that (3.5.49) holds true for any α, β, γ. But from (3.5.49) it follows that $\eta = 1$, in contradiction to our assumption. This concludes the proof that "either $h = 1$ or $\eta = 1$" is a necessary condition for $y(1, 1, 1) = 1$.

Note that in general $x^0_\alpha = \sum_{\beta\gamma} x^0_{\alpha\beta\gamma}$ differs from the quantity x^0_α that appears in (3.5.7). However, in the case $\eta = 1$, the two quantities become identical.

In conclusion, whenever there exists pairwise correlation, there is also triplet correlation as well as nonadditivity effect.

3.5.4. Cooperativity

The concept of cooperativity when applied to systems where only direct interactions between ligands exist, as in the model of section 3.1, is quite simple. If $S > 1$, we can say that there is positive cooperativity. Equivalent ways of saying the same thing are:

1. The binding constant on the second site is larger than on the first site.
2. The probability of finding a polymer doubly occupied given that the polymer is already occupied by one ligand is larger than the probability of finding a polymer with only one ligand.
3. The probability of finding the polymer triply occupied given that it is already doubly occupied is larger than the probability of finding the polymer doubly occupied given that it is already singly occupied.
4. The Helmholtz energy of binding of the third ligand is more negative than that of the second, and that of the second more negative than that of the first.
5. The energy of binding the third ligand is more negative than that of the second, and that of the second more negative than that of the first.
6. The binding isotherm has an S-shape (this is true for $S > 2$).

All these are different manifestations of the direct cooperativity. The situation is more complex when there are also indirect contributions to the correlation between ligands. In such a case one should make a distinction between various *sources* of and different *manifestations* of the cooperativity.

It is common to refer to a system manifesting an S-shaped binding isotherm as a system having *positive* cooperativity. The system is indeed a cooperative system in the sense that ligands behave cooperatively; i.e., there is communication between ligands. However, to say that the system is positively cooperative is sometimes ambiguous. The reason is that different manifestations of cooperativity in the same system can have different signs.

Before giving an example in this particular model it is useful to list possible sources and possible manifestations of cooperativity. Possible sources are:

1. Direct ligand–ligand interaction.
2. Indirect correlation between ligands, communicated by the protein.
3. Indirect correlation between ligands, communicated by the solvent.
4. Indirect correlation between ligands, communicated by both the solvent and the protein.

We have mentioned the solvent because of its crucial importance in an actual system, e.g., hemoglobin in aqueous solution. We shall discuss these effects only in Chapter 8. Here we shall refer only to the first two sources of cooperativity.

Possible different manifestations are

1. $K_1 \lessgtr K_2$ (or, equivalently, $\Delta A G_G^{*(2)} \lessgtr \Delta A_G^{*(1)}$).
2. $K_2 \lessgtr K_3$ (or, equivalently, $\Delta A_G^{*(3)} \lessgtr \Delta A_G^{*(2)}$).
3. $\Delta E_G^{*(2)} \gtrless \Delta E_G^{*(1)}$.
4. $\Delta E_G^{*(3)} \lessgtr \Delta E_G^{*(2)}$.
5. S-shaped binding isotherm.

The sign $\lessgtr$ is used to indicate either larger than, smaller than, or equal to. These different manifestations do not necessarily indicate the same sign of cooperativity. For instance, suppose that we have in this model $S > 1$ (i.e., attractive direct interaction), but if $\eta'' < 1$, then there is a negative indirect correlation (see 3.5.42). If $y(1, 1) < 1$ and $S > 1$, but $y(1, 1)S > 1$, then

$$K_2 = Sy(1, 1)K_1 > K_1, \qquad (3.5.50)$$

which means that there is a positive cooperativity as manifested by the relation between K_1 and K_2. However, for the next binding ligand we have

$$K_3 = Sy(1, 1)fK_2. \tag{3.5.51}$$

If the nonadditivity is such that $f < 1$ in such a way that $Sy(1, 1)f < 1$, we shall find that

$$K_3 < K_2, \tag{3.5.52}$$

which means that there is *negative* cooperativity as manifested by the relation between K_2 and K_3. It is clear that we cannot say that the system as a whole is positively (or negatively) cooperative. The physical reason is that the *system* changes from one event (the first binding) to the second event (the second binding) to the third event. The system can be positively cooperative at one stage and negatively cooperative at another stage.

Clearly, when there are more subunits, there are also more possible manifestations of cooperativity with respect to binding constants (or, equivalently, with respect to $\Delta A_G^{*(i)}$, or to conditional binding probabilities).

Another different manifestation of cooperativity is in the energy change of the binding process. This is sometimes confused with the Helmholtz energy changes for the binding process. Again it is possible that

$$\Delta A_G^{*(2)} < \Delta A_G^{*(1)}, \tag{3.5.53}$$

but

$$\Delta E_G^{*(2)} > \Delta E_G^{*(1)}. \tag{3.5.54}$$

The reason is that $\Delta A_G^{*(1)}$ depends only on an average of the quantity $\langle \exp(-\beta B_G^i) \rangle$, but $\Delta E_G^{*(i)}$ depends on an average of the form $\langle B_G^{(i)} \rangle$ as well as on the contribution from induced conformational change. (In a solvent there is also an induced structural change; see Chapters 5, 7, and 8.) Therefore, in general, the cooperativity defined in terms of the Helmholtz energy of binding does not necessarily have the same sign as cooperativity defined in terms of the energy change of the binding process.

3.6. THE TETRAHEDRAL TETRAMER: A MINIMAL MODEL FOR THE BINDING OF OXYGEN TO HEMOGLOBIN

The mechanism of transporting oxygen from the lungs to other parts of the body has been studied extensively. The main carrier of oxygen is hemoglobin, which is contained in the red blood cells. Myoglobin receives the oxygen in the muscles and reserves it for further use. A great amount of information is now available on this system. The three-dimensional structures of both hemoglobin and myoglobin are known, their oxygen-binding isotherms have been studied under a variety of environmental conditions, and the working mechanism of this system is well understood.

In this section we focus on the central problem of the binding isotherms of hemoglobin and myoglobin. We shall use the simplest theoretical model that mimics the behavior of this system.

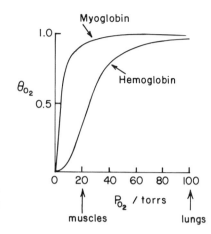

FIGURE 3.16. Schematic binding isotherm of hemoglobin and myoglobin. The partial pressures of oxygen in the
lungs and in the muscles are indicated.

Figure 3.16 shows schematically the oxygen-binding isotherms of hemoglobin and myoglobin. Two important features of these curves are the following: first, the hemoglobin isotherm has the typical S-shaped curve, indicating positive cooperativity. (The adjective "positive" is used here as a qualitative description of the overall form of the curve—see section 3.5.4 for details.) It is clear from this figure that the loading and unloading of oxygen can be accomplished within a relatively narrow range of concentrations, or partial pressures, of oxygen. Roughly at the partial pressure in the muscles (about 20 torr) the fractional occupation of sites is less than $\theta \approx 0.2$. On the other hand, at the partial pressure in the lungs (about 100 torr), the hemoglobin is almost saturated by oxygen ($\theta \approx 1$). We can say that hemoglobin loads ($\theta \approx 1$) and unloads ($\theta \approx 0.2$) in a relatively narrow range of oxygen concentrations, at around some finite value of the oxygen concentration or pressure. The situation is quite different with a typical Langmuir isotherm, as exhibited by myoglobin. Myoglobin can also load and unload at a narrow range of concentrations—but it can do so only around zero concentration or pressure.

The second feature is that the myoglobin curve is everywhere above the hemoglobin curve. In particular, at the low partial pressure of 20 torr, when the hemoglobin occupancy is very low ($\theta \approx 0.2$), the myoglobin curve reaches a near-saturation value of $\theta \approx 1$. This facilitates the transfer of the oxygen from the hemoglobin onto the myoglobin at the partial pressure existing where this transfer takes place.

The minimal model that we shall describe in this section is constructed to mimic these two features. Qualitatively, we have already shown that a Langmuir-type model produces a curve similar to the myoglobin curve. We also know that the allosteric model studied in the preceding section can produce an S-shaped curve similar to the hemoglobin curve. To obtain a quantitative fit between the experimental curve and a theoretical model, it might be necessary not only to adjust the parameters of a specific allosteric model but perhaps also to change the number of subunits in the model. In order to place the myoglobin curve above the hemoglobin curve, it is necessary that the initial slope of the myoglobin be larger than that of hemoglobin. The initial slopes of all the models discussed in this chapter are determined by the first intrinsic binding constant. Clearly one can choose different types of molecules, with different initial binding constants, to mimic the different behavior of hemoglobin and myoglobin. However, if we insist that the binding sites on both hemoglobin and myoglobin be identical (myoglobin is similar to one of the subunits of hemoglobin), then we must choose the model in such a way that the binding constant of oxygen on myoglobin will be much larger than the first binding constant on hemoglobin. This can be achieved with a proper choice of subunit–subunit interactions.

The model studied in the present section consists of a monomeric polymer having one binding site, which represents the myoglobin. The hemoglobin is represented by four identical polymers of the same kind, each having the same binding site as in the monomer. Thus we ignore the small differences between the primary structures of myoglobin and the α and β subunits that constitute the tetrameric hemoglobin molecule. We also ignore other ligands such as hydrogen ions or DPG (2,3-diphosphoglycerate) that are known to affect the binding isotherm of hemoglobin. These additional effects can be taken into account either as part of the solvent effect (see Chapter 8) or treated specifically as allosteric effectors. The latter are discussed in section 3.7.

3.6.1. The Model

Four identical subunits are assumed to form a tetrahedral structure. Each subunit contains one binding site for oxygen. The oxygen–oxygen distances are all equal, but large enough to make any direct interaction negligible. Hence we put $U_{12} = 0$, or $S = 1$, in the following treatment. In reality the hemoglobin operates in aqueous solutions. The solvent might have important consequences for the binding curves. However, in the present treatment the solvent is not taken into account.

Each subunit is presumed to attain one of two conformational states, which we denote by L and H (for low and high energy levels, i.e., $E_L < E_H$). Recall, however, that these are actually Helmholtz energy levels. In the biochemical literature these are often referred to as the tensed T and the relaxed R states.

The notations for the binding energies, intersubunit interactions, correlation functions, and so on are as in the preceding section.

Figure 3.17 shows representations of all the possible states of the hemoglobin molecule. There are sixteen states of the empty polymer. A circle represents a subunit in the L conformation and a square, a subunit in the H conformation. There are four states for which one subunit is H and three are Ls. There are six states for which two subunits are Ls and two are Hs, and so on. Only one of each of these is presented in the first row.

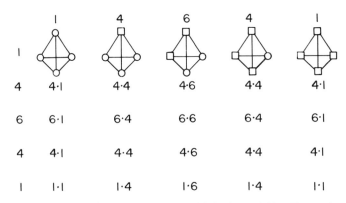

FIGURE 3.17. All possible states of the tetramer model for hemoglobin. The configurations of the subunits are shown in the first row only. Subsequent rows correspond to increasing occupation number.

The numbers next to each configuration are the numbers of ways of selecting $k (0 \leq k \leq 4)$ subunits in the L conformation and $4 - k$ in the H conformation.

The second row shows the number of states of the polymer with one ligand. Again the number of conformational states is denoted below each configuration. These are the same in each row. In addition, there are four possible subunits on which we can place the single ligand. In the third row we have six possible ways of placing two ligands. In the fourth row we have four ways of placing three ligands, and in the fifth row one way of placing four ligands on the four sites. The number of ligands on the polymer is indicated in the left-hand column.

Thus in each row we have all possible configurations of the subunits. In each column we have a fixed configuration of the subunits and all the possible occupational states of the polymer. Altogether we have 256 possible states of the system.

We define

$$Q_{\alpha\beta\gamma\delta} = Q_\alpha Q_\beta Q_\gamma Q_\delta Q_{\alpha\beta} Q_{\beta\gamma} Q_{\gamma\delta} Q_{\delta\alpha} Q_{\alpha\gamma} Q_{\beta\delta} . \tag{3.6.1}$$

This is the canonical PF of an empty polymer with a *specific* set of subunit conformations. For instance, the upper subunit is in state $\alpha (= L, H)$; the left-hand subunit in state $\beta (= L, H)$, the right-hand subunit in the state $\gamma (= L, H)$, and the front subunit (in the tetrahedron of Fig. 3.17) in the state $\delta (= L, H)$. Altogether, there are sixteen terms for the empty polymer. Thus

$$\xi(0) = Q(0) = \sum_{\alpha\beta\gamma\delta} Q_{\alpha\beta\gamma\delta} . \tag{3.6.2}$$

The grand PF of a single polymer is written as

$$\xi = \sum_{i=0}^{4} \binom{4}{i} Q(i) \lambda^i , \tag{3.6.3}$$

where $Q(i)$ is the PF corresponding to one row in Fig. 3.17, having i ligands at i specific sites on the polymer. The sum is over all values of the occupational numbers i. These are written explicitly as

$$Q(1) = \sum Q_{\alpha\beta\gamma\delta} q_\alpha \tag{3.6.4}$$

$$Q(2) = \sum Q_{\alpha\beta\gamma\delta} q_\alpha q_\beta \tag{3.6.5}$$

$$Q(3) = \sum Q_{\alpha\beta\gamma\delta} q_\alpha q_\beta q_\gamma \tag{3.6.6}$$

$$Q(4) = \sum Q_{\alpha\beta\gamma\delta} q_\alpha q_\beta q_\gamma q_\delta . \tag{3.6.7}$$

All summations are as in (3.6.2); i.e., over sixteen possible states of α, β, γ, $\delta = L, H$.

Note that no *direct* interactions are included in these expressions (i.e., $S = 1$ in all possible configurations). Therefore any correlation between ligands must be due to *indirect* effects. We also note that the notation of $Q(i)$ in (3.6.3) differs from the notation used in section 2.9. If we define

$$q(i) = \binom{4}{i} Q(i), \tag{3.6.8}$$

then the grand PF of one polymer becomes

$$\xi = \sum_{i=0}^{4} q(i)\lambda^i, \tag{3.6.9}$$

which is in accordance with the notation of section 2.9. In this section, as in other sections of this chapter, we shall need the quantities $Q(i)$ pertaining to a specific occupation of the ligands. Of course, if we are not interested in the details of the arrangement of ligands on the various sites, the notation used in (3.6.9) is simpler.

3.6.2. Some Special Cases

In order to analyze the properties of this system, it is convenient to rewrite the PF of a single polymer ξ in a different form. We define the quantity

$$\phi(k) = \exp[-\beta E_{pp}(k)], \tag{3.6.10}$$

where $E_{pp}(k)$ is the total interaction energy between the subunits when k of the subunits are in the L conformation and $4 - k$ in the H conformation. The five possible values of $E_{pp}(k)$ are

$$E_{pp}(0) = 6E_{HH}, \qquad \phi(0) = Q_{HH}^6 \tag{3.6.11}$$

$$E_{pp}(1) = 3E_{LH} + 3E_{HH}, \qquad \phi(1) = Q_{LH}^3 Q_{HH}^3 \tag{3.6.12}$$

$$E_{pp}(2) = E_{LL} + E_{HH} + 4E_{LH}, \qquad \phi(2) = Q_{LL} Q_{HH} Q_{LH}^4 \tag{3.6.13}$$

$$E_{pp}(3) = 3E_{LH} + 3E_{LL}, \qquad \phi(3) = Q_{LH}^3 Q_{LL}^3 \tag{3.6.14}$$

$$E_{pp}(4) = 6E_{LL}, \qquad \phi(4) = Q_{LL}^6. \tag{3.6.15}$$

We now write explicitly the coefficients in 3.6.3

$$Q(0) = Q_H^4 \phi(0) + 4Q_H^3 Q_L \phi(1) + 6Q_H^2 Q_L^2 \phi(2) + 4Q_H Q_L^3 \phi(3) + Q_L^4 \phi(4)$$

$$4Q(1) = Q_H^4 \phi(0) 4q_H + 4Q_H^3 Q_L \phi(1)(3q_H + q_L) + 6Q_H^2 Q_L^2 \phi(2)(2q_H + 2q_L)$$
$$+ 4Q_H Q_L^3 \phi(3)(q_H + 3q_L) + Q_L^4 \phi(4) 4q_L$$

$$6Q(2) = Q_H^4 \phi(0) 6q_H^2 + 4Q_H^3 Q_L \phi(1)(3q_L q_H + 3q_H^2) + 6Q_H^2 Q_L^2 \phi(2)(q_L^2 + q_H^2 + 4q_L q_H)$$
$$+ 4Q_H Q_L^3 \phi(3)(3q_L^2 + 3q_L q_H) + Q_L^4 \phi(4) 6q_L^2$$

$$4Q(3) = Q_H^4 \phi(0) 4q_H^3 + 4Q_H^3 Q_L \phi(1)(q_H^3 + 3q_H^2 q_L) + 6Q_H^2 Q_L^2 \phi(2)(2q_H^2 q_L + 2q_L^2 q_H)$$
$$+ 4Q_H Q_L^3 \phi(3)(q_L^3 + 3q_H q_L^2) + Q_L^4 \phi(4) 4q_L^3$$

$$Q(4) = Q_H^4 \phi(0) q_H^4 + 4Q_H^3 Q_L \phi(1) q_H^3 q_L + 6Q_H^2 Q_L^2 \phi(2) q_H^2 q_L^2$$
$$+ 4Q_H Q_L^3 \phi(3) q_H q_L^3 + Q_L^4 \phi(4) q_L^4.$$

We can now sum all these equations, but in order of increasing k; i.e., we sum according to columns with the same value of k. Thus

$$
\begin{aligned}
\xi &= \sum_{i=0}^{4} \binom{4}{i} Q(i) \lambda^i \\
&= Q_H^4 \phi(0)(1 + q_H \lambda)^4 + 4 Q_H^3 Q_L \phi(1)(1 + q_H \lambda)^3 (1 + q_L \lambda) \\
&\quad + 6 Q_H^2 Q_L^2 \phi(2)(1 + q_H \lambda)^2 (1 + q_L \lambda)^2 \\
&\quad + 4 Q_H Q_L^3 \phi(3)(1 + q_H \lambda)(1 + q_L \lambda)^3 + Q_L^4 \phi(4)(1 + q_L \lambda)^4 \\
&= \sum_{k=0}^{4} \binom{4}{k} Q_L^k Q_H^{4-k} \phi(k)(1 + q_L \lambda)^k (1 + q_H \lambda)^{4-k}.
\end{aligned}
\tag{3.6.16}
$$

Thus, instead of summing over the number of ligands i in (3.6.3), we sum over the index k, the number of subunits in the L conformational states. Each term in the latter sum corresponds to a polymer with k subunits in the L and $4 - k$ subunits in the H state, independently of the state of occupancy of the polymer.

Note that if the factor $\phi(k)$ is constant, the sum over k could be performed using the binomial theorem; i.e., if $\phi(k) = c$, where c is a constant independent of k, then (3.6.16) reduces to

$$
\xi = c \sum_{k=0}^{4} \binom{4}{k} Q_L^k Q_H^{4-k} (1 + q_L \lambda)^k (1 + q_H \lambda)^{4-k} = c(Q_L + Q_H + Q_L q_L \lambda + Q_H q_H \lambda)^4.
\tag{3.6.17}
$$

Each factor on the rhs of (3.6.17) is the PF of a single subunit. The factorizing of ξ in this case is the same as if $\phi(k) = c = 1$, i.e., when there are no interactions between the subunits. As we have seen in earlier sections, an effective independence can also be achieved under weaker conditions.

As before, we define the parameter

$$
\eta = \frac{Q_{LH}^2}{Q_{LL} Q_{HH}} = \exp[-\beta(2E_{LH} - E_{LL} - E_{HH})]
\tag{3.6.18}
$$

and the additional two auxiliary parameters

$$
\zeta = \exp[-\beta \tfrac{3}{2}(E_{HH} - E_{LL})]
\tag{3.6.19}
$$

$$
\phi = \exp[-\beta 6 E_{LL}];
\tag{3.6.20}
$$

with the help of these parameters we can rewrite all the $\phi(k)$s as follows:

$$
\begin{aligned}
\phi(0) &= \phi \zeta^4 \\
\phi(1) &= \phi \zeta^3 \eta^{3/2} \\
\phi(2) &= \phi \zeta^2 \eta^2 \\
\phi(3) &= \phi \zeta \eta^{3/2} \\
\phi(4) &= \phi.
\end{aligned}
\tag{3.6.21}
$$

In this presentation, each polymer has one factor ζ for each subunit in the H state and a factor $\eta^{1/2}$ for each LH pair of subunits.

We can now examine some special cases:

3.6.2.1. When $\eta = 1$

If $\eta = 1$, then the PF (3.6.16) may be factored as follows:

$$\xi = \sum_{k=0}^{4} \binom{4}{k} Q_L^k Q_H^{4-k} \phi \zeta^{4-k} (1 + q_L\lambda)^k (1 + q_H\lambda)^{4-k}$$

$$= \phi(Q_L + \zeta Q_H + Q_L q_L\lambda + Q_H q_H\zeta\lambda)^4. \tag{3.6.22}$$

We see that in this case the subunits behave as independent subunits. The effective PF of each subunit is modified by the factor ζ in comparison with (3.6.17). The last result is the same as the one obtained in sections 3.3 and 3.5, namely, $\eta = 1$ makes the subunits effectively independent.

3.6.2.2. The Concerted Model

If $\eta = 0$, the PF (3.6.16) reduces to

$$\xi = \phi[Q_L^4 \zeta^4 (1 + q_L\lambda)^4 + Q_H^4(1 + q_H\lambda)^4]. \tag{3.6.23}$$

In this case the polymer has only two states, either all in the L state or all in the H state. This is the "all or none" or *concerted* model, in which the entire model switches from all L to all H states. The possible states are the first and the fifth columns of Fig. 3.17. In essence this is similar to the model treated in section 3.2, with one polymer having now four binding sites instead of two as in section 3.2.

3.6.2.3. The Sequential Model

As we have seen in section 3.3, it is quite difficult to fulfill the mathematical requirement of the sequential model. In this model we require that in the empty system all the subunits be in the L state, and each occupied subunit is turned from L to H but leaves the empty subunits in the L state.

The extreme conditions required here are physically clear. If the empty polymer is all in L states this means that the L conformation is infinitely more stable than the H state. No other state in the first row of Fig. 3.17 is possible. This requires that $K = Q_H/Q_L$ be zero. From section 2.10, we know that such a system is infinitely resistant to conformational change. In addition, the model requires that each ligand produces a total conformational change $L \rightarrow H$. From section 2.10, we know that for any K, a total conversion $L \rightarrow H$ requires that the binding energy on H be infinitely stronger than on L, i.e., that $h = q_H/q_L = \infty$. But in our case we require that h be not only infinite but a "stronger infinity," since the ligand is expected to produce a *total* conversion on a system *infinitely resistant to conversion*. The precise requirements have already been described in section 3.3. For our particular model, we first put $\lambda = 0$ in (3.6.16) to obtain the PF of the empty system,

$$\xi(0) = \sum_{k} \binom{4}{k} Q_L^k Q_H^{4-k} \phi(k). \tag{3.6.24}$$

In order to have an all-L polymer we must require that

$$K = \frac{Q_H}{Q_L} = 0;\tag{3.6.25}$$

hence

$$\xi(0) = Q_L^4 \phi(0).\tag{3.6.26}$$

On the other hand, in the limit of $\lambda \to \infty$ the system is fully occupied; hence from (3.6.16) we have

$$\xi(\lambda \to \infty) \sim \sum_k \binom{4}{k} Q_L^k Q_H^{4-k} \phi(k) q_L^k q_H^{4-k} \lambda^4.\tag{3.6.27}$$

But now we require that all subunits be in the H state. In general, to obtain an all-H polymer it is sufficient to require that

$$h = \frac{q_H}{q_L} \to \infty.\tag{3.6.28}$$

However, since we already have condition (3.6.25) which favors the all -L polymer, we must require that

$$h \times K \to \infty,\tag{3.6.29}$$

in which case (2.6.27) reduces to one term:

$$\xi(\lambda \to \infty) \to Q_H^4 \phi(4) q_H^4 \lambda^4.\tag{3.6.30}$$

If, in addition, we require that for any finite λ

$$q_H \lambda \gg 1 \quad \text{and} \quad q_L \lambda \ll 1,\tag{3.6.31}$$

then Eq. (3.6.16) reduces to the PF of the sequential model

$$\xi = Q_L^4 \phi(0) + 4 Q_L^3 Q_H \phi(1) q_H \lambda + 6 Q_L^2 Q_H^2 \phi(2)(q_H \lambda)^2$$
$$+ 4 Q_L Q_H^3 \phi(3)(q_H \lambda)^3 + Q_H^4 \phi(4)(q_H \lambda)^4.\tag{3.6.32}$$

Note that conditions (2.6.31) cannot be fulfilled in both limits $\lambda \to 0$ and $\lambda \to \infty$. These two cases were taken care of separately in (2.6.26) and (2.6.30).

3.6.2.4. The A. V. Hill Model

A model introduced by A. V. Hill to describe the binding of oxygen to hemoglobin (Hb) requires that all four (G) subunits be occupied without intermediate states. The reaction is symbolically written

$$\text{Hb} + 4O_2 \rightleftarrows \text{Hb}(O_2)_4.\tag{3.6.33}$$

If the system does not change conformation, then the grand PF for this case is

$$\xi = Q(0)[1 + (\lambda q)^4].\tag{3.6.34}$$

On the other hand, if we know that the oxygenated hemoglobin has a different conformation from that of the deoxygenated hemoglobin, we can write the corresponding PF as

$$\xi = Q_L^4 \phi(0) + Q_H^4 \phi(4) \lambda^4.\tag{3.6.35}$$

This can be formally obtained from (3.6.32) by putting $\eta = 0$ [see also (3.6.21) and (3.6.23)], hence only the first and the last terms in (3.6.32) remain. Clearly this model requires extreme conditions which cannot be met by a physically plausible system.

3.6.3. Binding Thermodynamics and Correlation Functions

As in previous sections, we can define the Helmholtz energy change for binding the first, second, third, and fourth ligand and the corresponding intrinsic binding constants. Pair, triplet, and quadruplet correlation functions can also be defined for this model.

We shall very briefly present the definitions of these quantities. We use a shorthand notation whenever possible. For instance, we shall use $Q(1)$ rather than the more detailed notation $Q(1, 0, 0, 0)$ as we have done in previous sections. Thus the intrinsic binding constants and the corresponding Helmholtz energy changes for binding the ith ligand are

$$K_i = \exp(-\beta \Delta A_G^{*(i)}) = \frac{Q(i)}{Q(i-1)}$$

$$= \langle \exp(-\beta B_G^{(i)}) \rangle_{(i-1)}, \tag{3.6.36}$$

where $\langle \ \rangle_{(i-1)}$ is an average with the probability distribution of states before adding the ith ligand.

With these definitions, we can write the binding isotherm for this model as

$$\theta = \frac{\bar{n}}{4} = \frac{\lambda}{4} \frac{\partial \ln \xi}{\partial \lambda} = \frac{Q(1)\lambda + 3Q(2)\lambda^2 + 3Q(3)\lambda^3 + Q(4)\lambda^4}{Q(0) + 4Q(1)\lambda + 6Q(2)\lambda^2 + 4Q(3)\lambda^3 + Q(4)\lambda^4}$$

$$= \frac{K_1\lambda + 3K_1K_2\lambda^2 + 3K_1K_2K_3\lambda^3 + K_1K_2K_3K_4\lambda^4}{1 + 4K_1\lambda + 6K_1K_2\lambda^2 + 4K_1K_2K_3\lambda^3 + K_1K_2K_3K_4\lambda^4}. \tag{3.6.37}$$

Figure 3.18 shows the binding isotherm corresponding to the tetrahedral model, Eq. (3.6.37). It should be noted that the intrinsic binding constants that appear in (3.6.37) and are defined in (3.6.36) differ from the thermodynamic equilibrium constants that one would get from thermodynamic considerations similar to those of section 2.9.1.

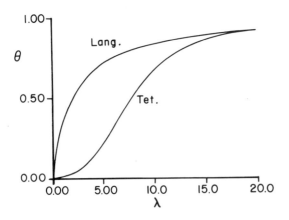

FIGURE 3.18. Binding isotherm for the tetrahedral model with parameters $h = 0.01$, $K = 1$, $Q_{HH}/Q_{LL} = 4$, and $\eta = 0.01$. For comparison the Langmuir isotherm with $K = 1$, $\eta = 1$, $h = 0.01$, and $Q_{HH} = Q_{LL}$ is also drawn.

If we denote by $\bar{K}_i$ the thermodynamic equilibrium constant for binding the ith ligand as in section 2.9.1, we obtain the relation between the two sets of quantities:

$$\bar{K}_i \rho^i = \binom{4}{i} K_i \lambda^i, \tag{3.6.38}$$

where ρ is the density of the gas in an ideal-gas phase.

In order to obtain relations between the various K_i, we introduce the following correlation functions. The correlation functions between i ligands on a polymer where there are exactly i ligands on i specific sites are

$$y(1, 1) = \frac{Q(2)Q(0)}{Q(1)^2} \tag{3.6.39}$$

$$y(1, 1, 1) = \frac{Q(3)Q(0)^2}{Q(1)^3} \tag{3.6.40}$$

$$y(1, 1, 1, 1) = \frac{Q(4)Q(0)^3}{Q(1)^4}. \tag{3.6.41}$$

Note that since we assumed that direct ligand–ligand interaction is negligible, only the indirect correlations are defined above.

As before, the relation between K_1 and K_2 is simply

$$K_2 = \frac{Q(2)}{Q(1)} = \frac{Q(2)Q(0)}{Q(1)^2} \frac{Q(1)}{Q(0)} = y(1, 1)K_1. \tag{3.6.42}$$

The relation between K_3 and K_2 is as in section 3.5:

$$K_3 = \frac{Q(3)}{Q(2)} = \frac{Q(3)Q(1)}{Q(2)^2} \frac{Q(2)}{Q(1)} = \frac{y(1, 1, 1)}{(y(1, 1))^2} K_2 = fy(1, 1)K_2 \tag{3.6.43}$$

where f is related to the nonadditivity of the triplet potential of average force and is defined by

$$f = \frac{y(1, 1, 1)}{(y(1, 1))^3} = \exp[-\beta(W(1, 1, 1) - 3W(1, 1))]. \tag{3.6.44}$$

Finally, the relation between K_4 and K_3 is

$$K_4 = \frac{Q(4)}{Q(3)} = \frac{Q(4)Q(2)}{Q(3)^2} \frac{Q(3)}{Q(2)} = \frac{y(1, 1, 1, 1)y(1, 1)}{y(1, 1, 1)^2} K_3. \tag{3.6.45}$$

Note that if the system obeys a pairwise additivity of both the triplet and the quadruplet potential of average forces, in the sense that

$$y(1, 1, 1) = y(1, 1)^3 \tag{3.6.46}$$

and

$$y(1, 1, 1, 1) = y(1, 1)^6, \tag{3.6.47}$$

then the connection between the K_i reduces to

$$K_2 = yK_1 \tag{3.6.48}$$

$$K_3 = yK_2 = y^2K_1 \tag{3.6.49}$$

$$K_4 = yK_3 = y^3K_1. \tag{3.6.50}$$

Substituting (3.6.48)–(3.6.50) in (3.6.37), we obtain

$$\theta = \frac{K_1\lambda + 3K_1^2 y\lambda^2 + 3K_1^3 y^3\lambda^3 + K_1^4 y^6\lambda^4}{1 + 4K_1\lambda + 6K_1^2 y\lambda^2 + 4K_1^3 y^3\lambda^3 + K_1^4 y^6\lambda^4}. \tag{3.6.51}$$

This is a familiar form of the binding isotherm derived for hemoglobin, based on the assumption of total pairwise additivity of both the triplet and the quadruplet potential of average force, as presented in (3.6.48)–(3.6.50).

If we had complete independence of the subunits, then it would be true that

$$\theta = \frac{K_1\lambda (1 + K_1\lambda)^3}{(1 + K_1\lambda)^4} = \frac{K_1\lambda}{1 + K_1\lambda}, \tag{3.6.52}$$

which is the Langmuir isotherm.

If the interactions are all direct, and if these are pairwise additive, then the corresponding isotherm would be

$$\theta = \frac{K_1\lambda + 3K_1^2 S\lambda^2 + 3K_1^3 S^3\lambda^3 + K_1^4 S^6\lambda^4}{1 + 4K_1\lambda + 6K_1^2 S\lambda^2 + 4K_1^3 S^3\lambda^3 + K_1^4 S^6\lambda^4}. \tag{3.6.53}$$

Here we have a factor S for each ligand–ligand interaction. This is formally the same as (3.6.51) with the replacement of S by y.

This formal resemblance can be misleading. Equation (3.6.53) is the exact isotherm for a system with *direct* interactions only. The two independent parameters of the model are K_1 and S. On the other hand, Eq. (3.6.51) has been derived on the basis of the pairwise additivity (or superposition approximation) assumptions (3.6.46) and (3.6.47). We have already seen that this approximation is unjustified for the indirect correlations. Since we know that in hemoglobin direct interactions are negligible, we have concluded that all correlations are due to indirect interactions, therefore (3.5.51) is incorrect. If we insist on expressing the isotherm in terms of the pair correlation function $y(1, 1)$, we must also include nonadditivity effects [see Eq. (3.6.58) below]. But this is not necessary. A simpler and exact expression can be written in terms of the fundamental parameters of the model. This is essentially Eq. (3.6.37), where the K_i are defined in (3.6.36).

Note also that Eq. (3.6.37) or its thermodynamical equivalent obtained by substituting (3.6.38) is an exact isotherm. However, the constants K_i (or $\bar{K}_i$) are not independent quantities. They are all functions of the molecular parameters of the model.

In order to write the exact binding isotherm for this model in terms of $y(1, 1)$, we define the pairwise nonadditivity of the quadruplet potential of average force in analogy with (3.6.44), as

$$t = \frac{y(1, 1, 1, 1)}{y(1, 1)^6} = \exp\{-\beta[W(1, 1, 1, 1) - 6W(1, 1)]\}. \tag{3.6.54}$$

In terms of f and t the exact relations between the intrinsic binding constants are

$$K_2 = yK_1 \tag{3.6.55}$$

$$K_3 = fyK_2 = fy^2K_1 \tag{3.6.56}$$

$$K_4 = \frac{t}{f^2} yK_3 = \frac{t}{f} y^3K_1. \tag{3.6.57}$$

Compare these with (3.6.48)–(3.6.50). The exact binding isotherm is

$$\theta = \frac{K_1\lambda + 3K_1^2 y\lambda^2 + 3K_1^3 fy^3\lambda^3 + K_1^4 ty^6 K_1}{1 + 4K_1\lambda + 6K_1^2 y\lambda^2 + 3K_1^3 fy^3\lambda^3 + K_1^4 ty^6 K_1}. \tag{3.6.58}$$

Note the difference between (3.6.51) and (3.6.58). A factor f is added for the triplet interaction term, and a factor t for the quadruplet term. These two factors cannot be neglected to obtain (3.6.51). The form (3.6.58) is equivalent to the form (3.6.37) where θ is expressed in terms of the molecular parameters of the model. The latter is to be preferred whenever we need to examine the isotherm numerically.

3.6.4. Comparison with the Square Model

If the four subunits are arranged in a square so that subunit–subunit interactions exist only between nearest neighbors, then most of the equations of this section apply with some reinterpretation. Instead of $Q_{\alpha\beta\gamma\delta}$, defined in (3.6.1) for the tetrahedral model, we now have

$$Q_{\alpha\beta\gamma\delta} = Q_\alpha Q_\beta Q_\gamma Q_\delta Q_{\alpha\beta} Q_{\beta\gamma} Q_{\gamma\delta} Q_{\delta\alpha}. \tag{3.6.59}$$

Note that the two "diagonal" interactions $Q_{\alpha\gamma}$ and $Q_{\beta\delta}$ are missing in (3.6.59). The interpretations of $\phi(k)$ in (3.6.11) to (3.6.15) should also be modified as follows:

$$\phi(0) = Q_{HH}^4 \tag{3.6.60}$$

$$\phi(1) = Q_{LH}^2 Q_{HH}^2 \tag{3.6.61}$$

$$\phi(2) = Q_{LH}^4 \quad \text{or} \quad Q_{LH}^2 Q_{LL} Q_{HH} \tag{3.6.62}$$

$$\phi(3) = Q_{LH}^2 Q_{LL}^2 \tag{3.6.63}$$

$$\phi(4) = Q_{LL}^4. \tag{3.6.64}$$

For $k = 2$, we have two possible configurations; either L or H alternate or two L subunits follow two H subunits. The PF is almost the same as (3.6.16) except for reinterpretation of the $k = 2$ term.

$$\binom{4}{2}\phi(k) = 4Q_{LH}^4 + 2Q_{LH}^2 Q_{LL} Q_{HH}; \tag{3.6.65}$$

i.e., the six equivalent terms in the tetrahedral model are replaced by four terms with all L–H interactions and two terms having one L–L, one H–H, and two L–H interactions. The other $\phi(k)$ to be used in the PF (3.6.16) are taken from (3.6.60)–(3.6.64).

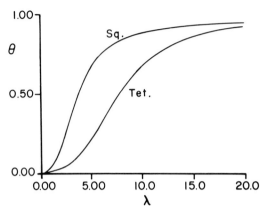

FIGURE 3.19. The binding isotherm of the tetrahedral and square models, with parameters as in Fig. 3.18.

Figure 3.19 shows one example of the binding curve for the square and the tetrahedral model. This particular computation was carried out for $K = 1$, $Q_{HH} = 4Q_{LL}$, $\eta = 0.01$, and $h = 0.01$. Both curves were calculated using Eq. (3.6.37). The tetrahedral and the square model differ in the explicit form of the $Q(i)$, using either (3.6.1) or (3.6.59), respectively. One can loosely say that the tetrahedral model is "more cooperative" in the sense that the binding curve falls below the corresponding curve of the square model.

3.6.5. The Requirement on the First Binding Constants

We noted in the introduction to this section that the binding isotherm of myoglobin is above the curve for hemoglobin. This is equivalent to the statement that the binding constant for myoglobin is larger than the first binding constant for hemoglobin. In the model treated in this section, myoglobin is identical to one of the subunits of hemoglobin. Therefore the two relevant quantities are defined by

$$K(\text{Mb}) = \frac{Q(1)}{Q(0)} = (x_L^{0,\infty} q_L + x_H^{0,\infty} q_H) \tag{3.6.66}$$

and

$$K_1(\text{Hb}) = \frac{Q(1, 0, 0, 0)}{Q(0, 0, 0, 0)} = (x_L^0 q_L + x_H^0 q_H), \tag{3.6.67}$$

where

$$x_L^{0,\infty} = \frac{Q_L}{Q_L + Q_H} \tag{3.6.68}$$

and

$$x_L^0 = \sum_{\beta\gamma\delta} x_{L\beta\gamma\delta}^0 = \frac{\sum\limits_{\beta\gamma\delta} Q_{L\beta\gamma\delta}}{\sum\limits_{\alpha\beta\gamma\delta} Q_{\alpha\beta\gamma\delta}}. \tag{3.6.69}$$

If for concreteness we take $q_L > q_H$, i.e., the conformer L is the one that binds the ligand more strongly than the conformer H, then in order to obtain the required condition

$$K(Mb) > K_1(Hb), \qquad (3.6.70)$$

we need to assume that

$$x_L^{0,\infty} > x_L^0. \qquad (3.6.71)$$

Physically, the last requirement is quite clear. Note that the two averages (3.6.66) and (3.6.67) are formally the same, except for using different distributions of states. In myoglobin, $x_L^{0,\infty}$ and $x_H^{0,\infty}$ are the probabilities of the *empty, isolated* subunits. In hemoglobin, x_L^0 and x_H^0 are the probabilities of the *empty* subunits, but when they are interconnected in the tetramer. In order to meet the requirement (3.6.70), we must assume that when the subunit–subunit interactions are turned on, the equilibrium $L \rightleftarrows H$ is shifted toward H, or, equivalently, that $x_L^{0,\infty} > x_L^0$.

From Figs. 3.18 and 3.19 we see that the binding isotherms for the tetrahedral and the square models fall below the corresponding Langmuir isotherm for the independent monomer (myoglobin). The latter was obtained by choosing the parameters $K = 1$, $Q_{HH} = Q_{LL}$, $\eta = 1$, and $h = 0.01$. The analogy with the experimental curves for hemoglobin and myoglobin in Fig. 16 is clearly demonstrated.

3.7. REGULATORY ENZYMES

One of the most striking facts common to many biosynthetic systems is that enzymes that catalyze the first step in a given synthetic pathway are reversibly inhibited by the end product of the same pathway. This can be presented schematically as follows:

$$A \xrightarrow{E_1} P_1 \xrightarrow{E_2} P_2 \xrightarrow{E_3} P_3 \cdots \longrightarrow B,$$

where E_1 is the enzyme that catalyzes the conversion of the substrate A into the first product P_1. The synthetic pathway produces a series of products P_i. The end product, denoted B, can bind to the first enzyme E_1 and partially inhibits its activity. The qualitative advantage of such a feedback mechanism is quite clear. Suppose we have an unlimited supply of the substrate A, but we are interested in keeping the concentration of the end product B within certain concentration limits, say $\rho_B(min) < \rho_B < \rho_B(max)$. Clearly, the scheme described above can fulfill this requirement. If the concentration of B increases beyond a certain limit, say $\rho_B(max)$, then it binds more effectively to E_1 and thereby inhibits the further production of B. If the concentration ρ_B falls below a certain limit, $\rho_B(min)$, then B desorbs from E_1 and the production of B is reactivated.

Perhaps the best-studied regulatory enzyme of this kind is aspartate transcarbamoylase (ATCase), which catalyzes the first steps in the biosynthetic pathway leading to uridine and cytidine nucleotides. Today, a great amount of information, structural, thermodynamical, and kinetic, is available on this system. Specifically it is known that cytidine triphosphate (CTP) inhibits ATCase. (Other molecules, such as ATP, activate the same enzyme. We shall focus, in this section, on inhibitory effectors only.)

There are many possibilities by which the end product B can deactivate the first enzyme. For instance, B can bind competitively to the same active site on which the substrate A also binds. Or B can bind to a different, allosteric site and mediate its effect on the active site through a conformational change induced in the enzyme (the allosteric effect).

The first possibility, competitive binding, was excluded in the case of CTP binding to ATCase on experimental grounds. It is now known that CTP binds to sites different from the active sites of ATCase. However, the competitive-binding possibility can be excluded on theoretical grounds if we make our requirement of the regulatory system more precise and stringent. In fact, we shall see in the next section that the same argument also excludes a simple allosteric regulatory mechanism. By "simple" we mean one that involves only one regulatory site.

Suppose that we use some measure of the activity of the enzyme as a function of the inhibitor concentration ρ_B. If B binds competitively to the enzyme E_1, then the activity of the enzyme will fall rapidly with ρ_B. A schematic drawing is given in Fig. 3.20a.

Clearly, if we are interested in switching the enzyme on and off by manipulating the concentration of B between $\rho_B = 0$ and ρ'_B, then this mechanism will answer our requirements and a competitive binding of B to the active site will work.

However, if we need to maintain a concentration of B within certain limits around some finite concentration ρ_B^*, say $\rho_B^* - \delta \leq \rho_B \leq \rho_B^* + \delta$, then it is clear that a competitive binding mechanism will not be effective. We shall see that such a mechanism will cause only a small change in the activity of the enzyme (curve a, Fig. 3.20). If, on the other hand, a more sophisticated mechanism is applied, such that a curve of form b in Fig. 3.20 is produced, then clearly the enzyme can be switched on and off within a small interval of concentration of B. We see that the basic trick is the same as the one used in the loading and unloading of oxygen by hemoglobin.

Although the activity of the enzyme is a kinetic property, we can grasp the essence of the regulatory mechanism by the study of a simple equilibrium system. In the following we redefine the concept of activity of the enzyme in such a way that its variation upon the addition of B can be studied by an equilibrium system.

The mechanism, as suggested originally by Monod, Wyman, and Changeux, is the following: Suppose that the enzyme can be in one of two conformations, in equilibrium $L \rightleftarrows H$ (the original nomenclature was T, and R). Suppose that the conformation L is the more active one; for instance, the active site in H is partially or totally blocked for binding the substrate A. Clearly, any factor that shifts the equilibrium concentration toward L will increase the activity of the enzyme. A shift toward H will lead to inhibitory effect. At this point we can leave the kinetic properties of the enzyme and focus on the mechanism of shifting the equilibrium concentrations of L and H by an effector B.

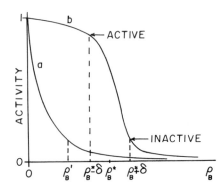

FIGURE 3.20. Activity of an enzyme as a function of the concentration of an inhibitor B. (a) The activity of the enzyme can be switched on and off by varying ρ_B between 0 and ρ'_B. (b) The activity can be switched on and off by varying ρ_B between $\rho_B^* - \delta$ and $\rho_B^* + \delta$.

We shall find in the next section that an inhibitor competing for binding on the active site can regulate the activity of the enzyme but cannot do so at any given narrow range of concentrations. The same conclusion is reached for a "simple" allosteric model. This situation is examined in section 3.7.2. In order to meet the minimal requirements, we need at least two allosteric regulatory sites. More allosteric regulatory sites are needed if the inhibitor is required to operate in a very narrow concentration range around any given concentration ρ_B. This is what nature actually does. Regulatory enzymes are multisubunit proteins. ATCase has six active (catalytic) and six regulatory subunits.

In the beginning of this section, we formulated the problem of regulation of an enzyme by an inhibitor B which is itself an end product of the activity of that enzyme. This does not need to be the general case of an activator or of an inhibitor. Many examples are known in which an effector affects the properties of an enzyme without being a direct or indirect product of that enzyme. Very common effectors include calcium ions, which are known to induce large conformational changes in certain proteins. There is a difference in the purpose of the two types of regulation. In the first, the synthetic pathway needs to produce an end product and maintain its concentration level within a narrow range. In the second case, the activity of an enzyme (or any other function of a protein) is regulated by an external means. It is expected to respond to a narrow change in the concentration of an effector. Thus, although the purposes of the two regulatory mechanisms are different, the mechanism itself could be the same in the two cases.

3.7.1. Regulation by Competitive Binding

This model is essentially the same as the model discussed in section 2.8.6. Let A be the substrate on which the enzyme E acts to produce the product P.

$$A \xrightarrow{E} P. \tag{3.7.1}$$

Let B be an inhibitor that binds to the same active site on which A also binds. The simplest PF of such an enzyme is

$$\xi = Q(0) + Q(A)\lambda_A + Q(B)\lambda_B, \tag{3.7.2}$$

where $Q(A)$ is the PF of the enzyme with the attached ligand A. The binding isotherm for A is

$$\theta_A = \lambda_A \frac{\partial \ln \xi}{\partial \lambda_A} = \frac{Q(A)\lambda_A}{\xi}. \tag{3.7.3}$$

If the enzyme is not affected by the binding (no conformational changes), then we put $Q(A) = q_A Q(0)$ and $Q(B) = q_B Q(0)$, where $q_A = \exp(-\beta U_A)$ and $q_B = \exp(-\beta U_B)$. The binding isotherm (3.7.3) is now written as

$$\theta_A = \frac{q_A \lambda_A}{1 + q_A \lambda_A + q_B \lambda_B}. \tag{3.7.4}$$

Figure 3.21a shows a series of $\theta_A(\lambda_A)$ curves for different values of λ_B (the molecular parameters U_A and U_B are kept constant).

We now choose an arbitrary level of occupancy, say $\theta_A \geq 0.8$, to serve as our reference for the activity of the enzyme. If the occupancy of the enzyme is above 0.8, we shall say that the enzyme is fully active. If it is below, say, 0.2, we shall say that it is not active.

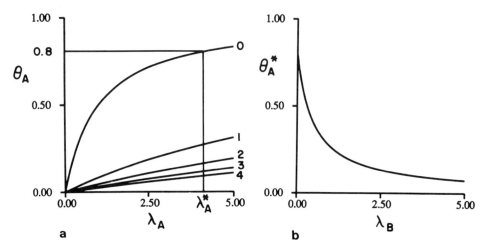

FIGURE 3.21. Binding and regulatory curves for the competitive model. (a) θ_A as a function of λ_A for various values of λ_B (these are indicated next to each curve), based on Eq. (3.7.4) with $q_A = 1$ and $q_B = 10$. (b) The regulatory curve for the same model, with θ_A^* as a function of λ_B, based on Eq. (3.7.7). This function is the drop of the value of θ_A in (a) along the vertical line at the point λ_A^*.

This level of occupancy (i.e., $\theta_A \approx 0.8$) is attained by the enzyme in the absence of inhibitor B when λ_A^* fulfills the equality

$$0.8 = \frac{q_A \lambda_A^*}{1 + q_A \lambda_A^*};$$

(3.7.5)

i.e., when

$$\lambda_A^* = \frac{4}{q_A}.$$

(3.7.6)

We now follow the drop of θ_A from the initial value of $\theta_A = 0.8$ upon adding B along the vertical line at λ_A^* as indicated in Fig. 3.21a. This function is obtained by substituting λ_A^* from (3.7.6) into (3.7.4).

$$\theta_A^*(\lambda_B) = \theta_A^*(\lambda_A = \lambda_A^*, \lambda_B) = \frac{4}{5 + q_B \lambda_B}.$$

(3.7.7)

We shall call the new function $\theta_A^*(\lambda_B)$ the regulatory curve. It gives the drop, caused by adding B from the initial value of $\theta_A = 0.8$ along the vertical line at fixed λ_A^*. Clearly, for any $q_B \neq 0$, the slope of the function is negative and its curvature positive:

$$\frac{\partial \theta_A^*}{\partial \lambda_B} = \frac{-4q_B}{[5 + q_B \lambda_B]^2} \leq 0$$

(3.7.8)

$$\frac{\partial^2 \theta_A^*}{\partial \lambda_B^2} = \frac{8q_B^2}{[5 + q_B \lambda_B]^3} \geq 0.$$

(3.7.9)

This behavior is similar to the Langmuir isotherm, except that it starts at $\theta_A^* = 0.8$ at $\lambda_B = 0$ and it is inverted with respect to the horizontal line $\theta_A^* = 0.8$. If we make the transformation

$$f = 0.8 - \theta^*(\lambda_B) = \frac{0.8 q_B \lambda_B}{5 + q_B \lambda_B}, \tag{3.7.10}$$

we obtain a Langmuir-type function, starting at $f = 0$ for $\lambda_B = 0$ and reaching $f = 0.8$ at $\lambda_B = \infty$. Figure 3.21b shows the regulatory curve $\theta_A^*(\lambda_B)$. Clearly, the enzyme can be switched on and off (in the sense that θ_A^* drops from an initial value of about 0.8 to a final value of about 0.2) by changing λ_B from zero to about $\lambda_B = 4$. However, this Langmuir-type drop of θ_A^* in (3.7.7) cannot serve for regulating the enzyme between two given concentrations of the inhibitor B. We therefore turn next to an allosteric model for regulation.

3.7.2. One Polymer with One Active and One Regulatory Site

We first generalize the model treated in section 3.2. Instead of two equivalent sites, we assume that one site, say the left-hand site in Fig. 3.6, can bind the substrate A and the second site can bind the activator or inhibitor B. This is the simplest model of a heterotropic allosteric effect, i.e., two different ligands communicating through the polymer.

The grand PF for one polymer is

$$\xi = Q(0,0) + Q(A,0)\lambda_A + Q(0,B)\lambda_B + Q(A,B)\lambda_A\lambda_B, \tag{3.7.11}$$

where

$$Q(0,0) = Q_L + Q_H \tag{3.7.12}$$

$$Q(A,0) = Q_L q_{AL} + Q_H q_{AH} \tag{3.7.13}$$

$$Q(0,B) = Q_L q_{BL} + Q_H q_{BH} \tag{3.7.14}$$

$$Q(A,B) = Q_L q_{AL} q_{BL} + Q_H q_{AH} q_{BH}, \tag{3.7.15}$$

where $q_{AL} = \exp(-\beta U_{AL})$, $q_{AH} = \exp(-\beta U_{AH})$, and so on. The four binding energies are U_{AL}, U_{AH}, U_{BL}, and U_{BH}. Note that no direct interaction between A and B is assumed.

The binding isotherm for A is

$$\theta_A = \lambda_A \frac{\partial \ln \xi}{\partial \lambda_A} = \frac{Q(A,0)\lambda_A + Q(A,B)\lambda_A\lambda_B}{\xi}. \tag{3.7.16}$$

A similar isotherm applies to θ_B. We are here interested only in the effect of λ_B on the binding isotherm of A.

Defining the binding constants and the indirect correlation functions by generalization of similar definitions in section 3.2, we get

$$K_A = \frac{Q(A,0)}{Q(0,0)}, \qquad K_B = \frac{Q(0,B)}{Q(0,0)}, \qquad K_{B/A} = \frac{Q(A,B)}{Q(A,0)} \tag{3.7.17}$$

$$y = y(A,B) = \frac{K_{B/A}}{K_B} = \frac{Q(A,B)Q(0,0)}{A(A,0)Q(0,B)}. \tag{3.7.18}$$

In terms of these binding constants, the binding isotherm of A is

$$\theta_A = \frac{K_A\lambda_A + K_{B/A}K_A\lambda_A\lambda_B}{1 + K_A\lambda_A + K_B\lambda_B + K_{B/A}K_A\lambda_A\lambda_B}. \tag{3.7.19}$$

For concreteness, we choose again $\theta_A \geq 0.8$ as the occupancy of the active site for which the enzyme is fully active and $\theta_A \leq 0.2$ as the occupancy for which the enzyme is effectively inactive (we could, of course, choose any two arbitrary values of θ_A).

In the absence of B, i.e., for $\lambda_B = 0$, we achieve full activity at λ_A^*, for which

$$0.8 = \frac{K_A\lambda_A^*}{1 + K_A\lambda_A^*} \tag{3.7.20}$$

or

$$\lambda_A^* = \frac{4}{K_A}. \tag{3.7.21}$$

We now add B and follow the change of θ_A with λ_B along the vertical line at λ_A^*.

Thus we substitute $\lambda_A = \lambda_A^*$ in (3.7.19) and follow the regulatory curve defined by

$$\theta_A^*(\lambda_B) = \theta_A(\lambda_A = \lambda_A^*, \lambda_B) = \frac{4(1 + K_{B/A}\lambda_B)}{1 + 4 + K_B\lambda_B + 4K_{B/A}\lambda_B}. \tag{3.7.22}$$

In terms of the correlation function y, defined in (3.7.18) [compare with (3.2.23)], we have

$$\theta_A^*(\lambda_B) = \frac{4(1 + K_B y\lambda_B)}{1 + 4 + K_B\lambda_B + 4K_B y\lambda_B}. \tag{3.7.23}$$

For $\lambda_B = 0$, we recover $\theta_A^*(\lambda_B = 0) = 0.8$. For $\lambda_B \to \infty$, $\theta_A^*(\lambda_B = \infty)$ approaches a constant limiting value

$$\theta_A^*(\lambda_B = \infty) = \frac{4y}{1 + 4y}. \tag{3.7.24}$$

If $y = 1$, no cooperativity, $\theta_A^*(\lambda_B)$ has a constant value of $\theta^* = 0.8$ for any λ_B. If $y > 1$ (positive cooperativity) then θ_A^* reaches a limiting value larger than the initial value of 0.8. This is the case of an activator which will not concern us here. The case of interest now is when $y < 1$, i.e., negative cooperativity. In this case, addition of B lowers the value of θ_A^* from 0.8 to the final value given by (3.7.24).

It is easy to verify that for $y < 1$ the regulatory curve (3.7.23) has everywhere a negative slope and a positive curvature. This is similar to the Langmuir-type curve we have seen in section 3.7.1. The significance of the condition $y < 1$ in terms of the molecular parameters of the model follows from definition (3.7.18)

$$y(A, B) = \frac{(1 + Kh_Ah_B)(1 + K)}{(1 + Kh_A)(1 + Kh_B)} = 1 + \frac{K(h_B - 1)(h_A - 1)}{(1 + Kh_A)(1 + Kh_B)}, \tag{3.7.25}$$

where $K = Q_H/Q_L$, $h_A = q_{AH}/q_{AL}$, $h_B = q_{BH}/q_{BL}$.

This result should be compared with (3.2.30). There we always had $y(1, 1) \geq 1$. Here also if both $h_B - 1$ and $h_A - 1$ have the same sign, then $y(A, B) \geq 1$. Physically this means that both A and B favor the same conformation, either L or H. In such a case we shall always obtain activation of the enzyme by B. However, if $(h_B - 1)$ has a different sign from that of $(h_A - 1)$, we obtain $y \leq 1$. For instance, if A favors the L form, then B must favor the H form in order to produce negative cooperativity and hence inhibitive regulation effect.

In order to explore the behavior of the binding isotherm (3.7.19) and the regulatory curve (3.7.23), we rewrite these equations in terms of the molecular parameters of the model, i.e.,

$$\theta_A(X_A, X_B) = \frac{\left(\dfrac{1 + Kh_A}{1 + K}\right)X_A + \left(\dfrac{1 + Kh_Ah_B}{1 + K}\right)X_AX_B}{1 + \left(\dfrac{1 + Kh_A}{1 + K}\right)X_A + \left(\dfrac{1 + Kh_B}{1 + K}\right)X_B + \left(\dfrac{1 + Kh_Ah_B}{1 + K}\right)X_AX_B} \quad (3.7.26)$$

and

$$\theta_A^*(X_B) = \frac{4 + 4\left(\dfrac{1 + Kh_Ah_B}{1 + Kh_A}\right)X_B}{5 + \left(\dfrac{1 + Kh_B}{1 + K}\right)X_B + 4\left(\dfrac{1 + Kh_Ah_B}{1 + Kh_A}\right)X_B}, \quad (3.7.27)$$

where $X_A = \lambda_A q_{AL}$ and $X_B = \lambda_B q_{BL}$.

Figure 3.22 shows these two functions for the case of inhibitive regulation, i.e., $(h_A - 1)$ and $(h_B - 1)$ having opposite signs. The results are similar to the case of competitive inhibition in the previous section. Note in particular the steep decrease of θ^* from the chosen initial value of 0.8 to less then 0.2 for $X_B \sim 4$.

This model, in contrast to the model of section 3.7.1, is cooperative (can be either positive or negative), but it is not cooperative enough to produce the required effect; i.e., the enzyme can be switched on and off around $\rho_B = 0$, but not at any arbitrary finite concentration of B. As we shall see in the next section, what is needed is a model in which at least two regulatory sites cooperate between themselves to produce the required

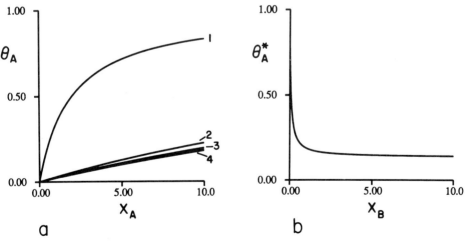

a   b

FIGURE 3.22. Binding and regulatory curves for the model of section 3.7.2. (a) θ_A as a function of X_A for various values of X_B (as indicated next to each curve). Based on Eq. (3.7.26) with parameters $K = 1$, $h_A = 0.01$, and $h_B = 100$. (b) The regulatory curve, Eq. (3.2.27), for the same parameters as in a.

S-shaped analogue of the binding isotherm. It is clear that generalization of the model of section 3.3, where there is exactly one catalytic and one regulatory subunit, will lead to a conclusion similar to the conclusion of this section. The regulatory curve $\theta_A^*(\lambda_B)$ will be essentially the same as that of (3.7.23) with a different interpretation of the coefficient. Therefore in the next section we turn to examining the minimal model that shows an inverted S-shaped regulatory curve.

3.7.3. A Minimal Model for a Regulatory Enzyme

We extend here the model treated in detail in section 3.4. There we studied the homotropic allosteric effect of three identical ligands on a single polymer having two conformations, L and H. Here we use the heterotropic version of the same model. We assume that one site is the catalytic site for A and the two other sites can accommodate the effector B. All the quantities defined in section 3.4 can be used here with the required modifications to account for the different types of ligands. We assume that no direct ligand–ligand interaction exists; hence, all the correlations are due to indirect effects. The binding energies are denoted by U_{AL}, U_{BL}, U_{AH}, and U_{BH}. Other notations are the same as in section 3.4. The grand PF of a single polymer is (see 3.4.2)

$$\xi = Q(0) + Q(A)\lambda_A + 2Q(B)\lambda_B + Q(B, B)\lambda_B^2 + 2Q(A, B)\lambda_A\lambda_B + Q(A, B, B)\lambda_A\lambda_B^2,$$

$$(3.7.28)$$

where

$$Q(0) = Q(0, 0, 0) = Q_L + Q_H$$

$$Q(A) = Q(A, 0, 0) = Q_L q_{AL} + Q_H q_{AH}$$

$$Q(B) = Q(0, B, 0) = Q(0, 0, B) = Q_L q_{BL} + Q_H q_{BH}$$

$$Q(B, B) = Q(0, B, B) = Q_L q_{BL}^2 + Q_H q_{BH}^2$$

$$Q(A, B) = Q(A, B, 0) = Q(A, 0, B) = Q_L q_{AL} q_{BL} + Q_H q_{AH} q_{BH}$$

$$Q(A, B, B) = Q_L q_{AL} q_{BL}^2 + Q_H q_{AH} q_{BH}^2.$$

$$(3.7.29)$$

The binding isotherm of substrate A is

$$\theta_A = \lambda_A \frac{\partial \ln \xi}{\partial \lambda_A} = \frac{Q(A)\lambda_A + 2Q(A, B)\lambda_A\lambda_B + Q(A, B, B)\lambda_A\lambda_B^2}{\xi}$$

$$= \frac{K_A\lambda_A + 2K_A K_B y(A, B)\lambda_A\lambda_B + K_A K_B^2 y(A, B, B)\lambda_A\lambda_B^2}{1 + K_A\lambda_A + 2K_B\lambda_B + K_B^2 y(B, B)\lambda_B^2 + 2K_A K_B y(A, B)\lambda_A\lambda_B + K_A K_B^2 y(A, B, B)\lambda_A\lambda_B^2},$$

$$(3.7.30)$$

where K_A and the K_B are the intrinsic binding constants of A and B on the empty polymer, i.e.,

$$K_A = \frac{Q(A)}{Q(0)}, \qquad K_B = \frac{Q(B)}{Q(0)}, \qquad (3.7.31)$$

and the various correlation functions are defined by analogy with the definitions of section 3.4, i.e.,

$$y(B, B) = \frac{Q(B, B)Q(0)}{[Q(B)]^2}, \qquad y(A, B) = \frac{Q(A, B)Q(0)}{Q(A)Q(B)} \qquad (3.7.32)$$

$$y(A, B, B) = \frac{Q(A, B, B)Q(0)^2}{Q(A)Q(B)^2}. \qquad (3.7.33)$$

A similar binding isotherm applies to B, but we shall not need it here.

Again we choose $\theta_A = 0.8$ as the occupancy level of the active site and solve for λ_A^* at $\lambda_B = 0$:

$$0.8 = \frac{Q(A)\lambda_A^*}{Q(0) + Q(A)\lambda_A^*} \qquad (3.7.34)$$

$$\lambda_A^* = \frac{4Q(0)}{Q(A)} = \frac{4}{K_A}. \qquad (3.7.35)$$

The corresponding regulatory curve is obtained by evaluating $\theta_A(\lambda_A, \lambda_B)$ [Eq. (3.7.30)] at the fixed value of λ_A^*.

$$\theta_A^*(\lambda_B) = Q_A(\lambda_A = \lambda_A^*, \lambda_B)$$

$$= \frac{4 + 8K_B y(A, B)\lambda_B + 4K_B^2 y(A, B, B)\lambda_B^2}{5 + 2K_B\lambda_B + K_B^2 y(B, B)\lambda_B^2 + 8K_B y(A, B)\lambda_B + 4K_B^2 y(A, B, B)\lambda_B^2}. \qquad (3.7.36)$$

If $\lambda_B = 0$, we recover the limiting value of $\theta^* = 0.8$. On the other hand, for $\lambda_B \to \infty$ we have

$$Q_A^*(\lambda_B = \infty) = \frac{4y(A, B, B)}{y(B, B) + 4y(A, B, B)}. \qquad (3.7.37)$$

This should be compared with (3.7.24). As in the model of section 3.7.2, we can obtain both activation and inhibition in this model. Here we are interested in the second possibility only.

We already know that in this model $y(B, B) > 1$ (see section 3.2). In addition, if the initial slope of the regulatory curve is to be negative—i.e., $[\partial\theta_A^*/\partial\lambda_B]_{\lambda_B = 0} < 0$—it follows that $y(A, B) < 1$; i.e., A and B must be negatively cooperative. Furthermore, in order to achieve effective regulation, we should require that θ_A^* at $\lambda_B = \infty$ be below a certain low value, which we consider as the limit of activity of the enzyme. Thus if we set the arbitrary choice of $\theta_A^*(\lambda_B = \infty) \lesssim 0.2$, we obtain the condition that $16y(A, B, B) < y(B, B)$. This means that the B–B correlation must be large enough to overcome sixteen times the correlation between A, B, and B (the number 16 results from the arbitrary choice of 0.2 as our low limit of activity of the enzyme). This simple analysis already demonstrates that in order to achieve an effective inhibitory regulation we need to have a combination of negative cooperativity between A and B and a large positive cooperativity between B and B.

In order to explore the behavior of this system numerically, it is convenient to rewrite the equations in terms of the molecular parameters of the model, namely,

$$K = \frac{Q_H}{Q_L}, \qquad h_A = \frac{q_{AH}}{q_{AL}} \quad \text{and} \quad h_B = \frac{q_{BH}}{q_{BL}}. \qquad (3.7.38)$$

In terms of these we have

$$\theta_A(X_A, X_B) = \frac{(A)X_A + 2(AB)X_AX_B + (ABB)X_AX_B^2}{1 + (A)X_A + 2(B)X_B + (BB)X_B^2 + 2(AB)X_AX_B + (ABB)X_AX_B^2} \tag{3.7.39}$$

and

$$\theta_A^*(X_B) = \frac{4 + 8(AB)X_B + 4(ABB)X_B^2}{5 + 2(B)X_B + (BB)X_B^2 + 8(AB)X_B + 4(ABB)X_B^2}, \tag{3.7.40}$$

where the new symbols are

$$X_A = q_{AL}\lambda_A, \qquad X_B = q_{BL}\lambda_B$$

$$(A) = \frac{1 + Kh_A}{1 + K}, \qquad (B) = \frac{1 + Kh_B}{1 + K}, \qquad (BB) = \frac{1 + Kh_B^2}{1 + K}$$

$$(AB) = \frac{1 + Kh_Ah_B}{1 + Kh_A}, \qquad (ABB) = \frac{1 + Kh_Ah_B^2}{1 + Kh_A}. \tag{3.7.41}$$

Figure 3.23a and b demonstrate the regulatory behavior of this system. To meet the required conditions for this case, we chose $K = 10$, $h_A = 100$, and $h_B = 0.001$. Note that, initially, adding B to the system causes a small change in the binding curve (Fig. 3.23a) the spacing between the curves increases with X_B. This is also shown by the (inverted) S-shaped regulatory curve in Fig. 3.23b.

The regulatory curve in Fig. 3.23b is clearly different from the regulatory curves in Figs. 3.21b and 3.22b. As in the model treated in section 3.4, which was modified in the present section, we cannot expect a sharp transition from active to inactive enzyme, as we would have liked; e.g., see Fig. 3.20. However, it is clear that by generalization of the models treated in sections 3.5 and 3.6 we can obtain sharper transitions within a specific concentration range of the inhibitor B. Actual enzymes normally have more than three

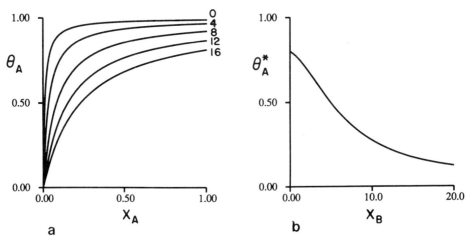

FIGURE 3.23. Binding and regulatory curves for the model of section 3.7.3. (a) θ_A as a function of X_A for various values of X_B (as indicated next to each curve). Based on Eq. (3.7.39) with parameters $K = 10$, $h_A = 100$, $h_B = 0.001$. (b) The regulatory curve [Eq. (3.7.40)] with the same parameters as in a.

subunits. For example, ATCase has six regulatory sites. Such a system can be effective in maintaining the concentration of the end product in quite a narrow range around some required level of concentration.

REFERENCES

1. J. Monod, J. Wyman, and J. P. Changeux, *J. Mol. Biol.*, **12**, 88 (1965).
2. D. Koshland, G. Némethy, and D. Filmer, *Biochemistry*, **5**, 365 (1966).

SUGGESTED READING

A thorough treatment of cooperative phenomena is:

T. L. Hill, *Cooperativity Theory in Biochemistry, Steady State and Equilibrium Systems* (Springer-Verlag, New York, 1985).

4

One-Dimensional Models

4.1. INTRODUCTION

The study of one-dimensional (1-D) systems is of interest for two main reasons. First, there are actual systems that are linear, such as linear polymers, proteins, and nucleic acids. Although all of these occupy three-dimensional space, their main properties are determined by the 1-D sequence of units and bonds. Second, these models are usually easily solvable. Therefore some general properties, theorems, conjectures, approximations, and the like may be tested on a 1-D system. The answers we obtain are sometimes also relevant to three-dimensional systems. Finally, the methods used to solve the 1-D partition functions are elegant and in themselves aesthetically satisfying.

We shall start in the next section with the simplest Ising model. This is a linear sequence of particles, or units, each of which can attain one of two possible states. Interactions between units extend to the nearest neighbor units only.

After studying some of these simplest models—all are formally equivalent—we proceed to introduce the concept of molecular distribution functions. As in Chapter 3, explicit and exact expressions for the molecular distribution functions in terms of molecular parameters can be derived for such systems. In subsequent sections we generalize the Ising model in different directions: Multistate units, triplet interactions, continuous systems, and so on. Finally, we apply these methods to some specific problems such as phase transition, the helix-coil transition, and one aspect of liquid water.

4.2. SIMPLEST ISING MODELS

In this section we study several 1-D systems with nearest-neighbor interactions. Each particle or unit can be in one of two states; up or down for magnetic spins, empty or occupied for the lattice gas, or two conformations L and H for the subunits of a polymer. As we shall soon see, all of these systems are formally equivalent, but we use the different systems to study different aspects of the response of such systems to an external "field." For instance, in the system of magnetic spins we can define the average mole fractions of units in the "up" and "down" states. These mole fractions can be changed by an external magnetic field. Likewise the mole fractions of the subunits of a polymer in the two conformations L and H can be changed by an external pressure (or tension) or by ligand activity.

4.2.1. One-Dimensional Model of Interacting Spins

We consider a chain of N spins, each of which can be in one of two states, up or down. We associate with each spin a variable s_i ($i = 1, 2, \ldots, N$), which is $+1$ if the ith spin is in the state "up" and -1 if its state is "down." The configuration of the entire system is described by the N-dimensional vector $\mathbf{s} = (s_1, s_2, \ldots, s_N)$. For instance a possible configuration of a system of six spins is described by the vector $s = (1, -1, 1, 1, -1, 1)$.

Each spin interacts only with its two nearest neighbors. The interaction between i and $i + 1$ is given by

$$E_{i,i+1} = -\varepsilon s_i s_{i+1}. \tag{4.2.1}$$

We assume here that $\varepsilon > 0$. Thus two spins aligned in the same direction contribute a negative interaction energy $-\varepsilon$. Two spins with opposite direction contribute ε. (The case $\varepsilon > 0$ corresponds to ferromagnetism and the case $\varepsilon < 0$ to antiferromagnetism. These phenomena occur in two- and three-dimensional Ising models but will not concern us here.)

In addition to pairwise interactions (4.2.1), each spin interacts with an external magnetic field H, which is constant throughout the entire system. The energy of the system in any given configuration is thus

$$E(\mathbf{s}) = -\varepsilon \sum_{i=1}^{N} s_i s_{i+1} - H \sum_{i=1}^{N} s_i. \tag{4.2.2}$$

H is chosen in such a way that, for $s_i = +1$, the interaction with the magnetic field is negative; i.e., the state "up" is energetically more favorable in the magnetic field.

The canonical partition function of the system of N spins at temperature T and subjected to an external magnetic field H is

$$Q(N, H, T) = \sum_{\mathbf{s}} \exp[-\beta E(\mathbf{s})], \tag{4.2.3}$$

where the summation is over all states or configurations of the system, i.e., all the possible vectors $\mathbf{s} = (s_1, s_2, \ldots, s_N)$ of the system. Each spin can attain one of the two states $s_i = \pm 1$. We now rewrite the sum (4.2.3) in more detailed form as

$$Q(N, H, T) = \sum_{s_i = \pm 1} \exp[\beta(Hs_1 + \varepsilon s_1 s_2 + Hs_2 + \varepsilon s_2 s_3$$
$$+ \cdots Hs_{N-1} + \varepsilon s_{N-1} s_N + Hs_N)]. \tag{4.2.4}$$

We further assume that N is very large so that edge (surface) effects are negligible. Instead of looking at the linear system of N spins, we close the system into a circle; i.e., the first particle, $i = 1$, interacts with the last particle $i = N$.

The closing of the system requires the addition of the factor $\exp(\beta \varepsilon s_N s_1)$ to the sum (4.2.4). It is easy to see that such an addition has no significant effect on the thermodynamics of the system provided that N is very large.

We notice that each term in the sum (4.2.4) has alternating factors: One factor of $\exp(\beta Hs_i)$ for each *particle*, and one factor of $\exp(\beta \varepsilon s_i s_{i+1})$ for each *bond*, i.e., a pair of consecutive particles. Altogether we have $2N$ factors. It is convenient to rewrite the terms in (4.2.4) as a product of N factors. There are several ways of doing so. One is to assign to each bond the factor

$$\exp(\beta H s_i/2) \exp(\beta \varepsilon s_i s_{i+1}) \exp(\beta H s_{i+1}/2).$$

In this way each term in (4.2.4) consists of N factors, each pertaining to one bond. Alternatively, we can assign to each particle the factor

$$\exp(\beta H s_i) \exp(\beta \varepsilon s_i s_{i+1}).$$

The terms in (4.2.4) then consist of N factors each pertaining to one particle. In most of this chapter the first assignment is used. In section 4.7 we find it more convenient to choose the second assignment.

We thus define the factor corresponding to the $i, i+1$ bond by

$$P_{i,i+1} = \langle s_i | \mathbf{P} | s_{i+1} \rangle = \exp(\beta H s_i/2) \exp(\beta \varepsilon s_i s_{i+1}) \exp(\beta H s_{i+1}/2). \qquad (4.2.5)$$

Since there are two states for each particle $s_i = \pm 1$, there are altogether four possible bonds. The corresponding factors are

$$
\begin{aligned}
\langle +1 | \mathbf{P} | +1 \rangle &= \exp[\beta(\varepsilon + H)] \\
\langle +1 | \mathbf{P} | -1 \rangle &= \exp[\beta(-\varepsilon)] \\
\langle -1 | \mathbf{P} | +1 \rangle &= \exp[\beta(-\varepsilon)] \\
\langle -1 | \mathbf{P} | -1 \rangle &= \exp[\beta(\varepsilon - H)].
\end{aligned}
\qquad (4.2.6)
$$

These define a 2×2 matrix that has the form

$$\mathbf{P} = \begin{pmatrix} \exp[\beta(\varepsilon + H)] & \exp(-\beta\varepsilon) \\ \exp(-\beta\varepsilon) & \exp[\beta(\varepsilon - H)] \end{pmatrix}. \qquad (4.2.7)$$

With the help of the matrix elements P_{ij} we can rewrite (4.2.4) as

$$Q(N, H, T) = \sum_s \exp(\beta H s_1/2) \langle s_1 | \mathbf{P} | s_2 \rangle \cdots \langle s_{N-1} | \mathbf{P} | s_N \rangle \exp[\beta H s_N/2]. \qquad (4.2.8)$$

Summing over all the indices $s_2 s_3 \cdots s_{N-1}$ gives

$$Q(N, H, T) = \sum_{s_1 s_N} \exp(\beta H s_1/2) \langle s_1 | \mathbf{P}^{N-1} | s_N \rangle \exp(\beta H s_N/2). \qquad (4.2.9)$$

If we now add the factor $\exp(\beta \varepsilon s_N s_1)$, closing the cycle, we can simplify (4.2.9) by summing over s_1 and s_N to obtain

$$
\begin{aligned}
Q(N, H, T) &= \sum_{s_1 s_N} \langle s_1 | \mathbf{P}^{N-1} | s_N \rangle \langle s_N | \mathbf{P} | s_1 \rangle \\
&= \sum_{s_1} \langle s_1 | \mathbf{P}^N | s_1 \rangle = \text{Tr } \mathbf{P}^N
\end{aligned}
\qquad (4.2.10)
$$

Thus closing the circle allows us to write the PF of the system as a trace of the matrix $\mathbf{P}^N$. This leads to an elegant solution of the one-dimensional Ising model. The trace of a matrix $\mathbf{A}$ is defined as the sum of all its diagonal elements.

$$\text{Tr } \mathbf{A} = \sum_i \mathbf{A}_{ii}. \qquad (4.2.11)$$

The presentation of the PF as a trace of a 2×2 matrix allows us to solve for the PF in terms of the molecular parameters of the model. To achieve that we use a theorem from matrix algebra that states that if $\mathbf{A}$ is a matrix that diagonalizes the matrix $\mathbf{P}$, it will also diagonalize the matrix $\mathbf{P}^N$.

Let $\mathbf{A}$ be the matrix that diagonalizes $\mathbf{P}$, i.e.,

$$\mathbf{APA}^{-1} = \lambda, \tag{4.2.12}$$

where λ is a diagonalized matrix, i.e.,

$$\lambda = \begin{pmatrix} \lambda_1 & 0 \\ 0 & \lambda_2 \end{pmatrix}, \tag{4.2.13}$$

and $\mathbf{A}^{-1}$ is the inverse of $\mathbf{A}$.

The matrix $\mathbf{A}$ also diagonalizes $\mathbf{P}^N$, i.e.,

$$\mathbf{AP}^N\mathbf{A}^{-1} = \mathbf{APPP} \cdots \mathbf{PA}^{-1}$$

$$= \mathbf{APA}^{-1}\mathbf{APA}^{-1}\mathbf{APA}^{-1} \cdots \mathbf{APA}^{-1} = \lambda^N = \begin{pmatrix} \lambda_1^N & 0 \\ 0 & \lambda_2^N \end{pmatrix}. \tag{4.2.14}$$

where we introduce $\mathbf{A}^{-1}\mathbf{A} = 1$ between each pair of $\mathbf{P}$s.

It is also easy to show that the trace of any product of matrices is invariant to cyclic permutation of matrices, e.g.,

$$\mathrm{Tr}(\mathbf{AB}) = \mathrm{Tr}(\mathbf{BA}). \tag{4.2.15}$$

This follows directly from the definition of a trace:

$$\mathrm{Tr}(\mathbf{AB}) = \sum_i \sum_k A_{ik} B_{ki} = \sum_k \sum_i B_{ki} A_{ik} = \mathrm{Tr}(\mathbf{BA}). \tag{4.2.16}$$

Therefore

$$\mathrm{Tr}(\mathbf{APA}^{-1}) = \mathrm{Tr}(\mathbf{A}^{-1}\mathbf{AP}) = \mathrm{Tr}(\mathbf{P}). \tag{4.2.17}$$

Applying (4.2.17) to the PF (4.2.10), we obtain

$$Q(N, H, T) = \mathrm{Tr}\,\mathbf{P}^N = \mathrm{Tr}(\mathbf{AP}^N\mathbf{A}^{-1}) = \lambda_1^N + \lambda_2^N. \tag{4.2.18}$$

Thus the calculation of the PF reduces to the calculation of the eigenvalues of the 2×2 matrix $\mathbf{P}$. Since we shall be interested only in the limit of large N, we shall need only the larger of the two eigenvalues. Suppose that $\lambda_1 > \lambda_2$. (In the case of $\lambda_1 = \lambda_2$ we have a degeneracy, and this case is discussed in section 4.6.) Then the Helmholtz energy is

$$A = -kT \ln Q = -kT \ln(\lambda_1^N + \lambda_2^N)$$

$$= -kT \ln \lambda_1^N - kT \ln[1 + (\lambda_2/\lambda_1)^N]. \tag{4.2.19}$$

For $N \to \infty$ the second term on the rhs of (4.2.19) can be neglected.

Our task now is to calculate the largest eigenvalue, in terms of the molecular parameters of the system. The eigenvalues of the matrix $\mathbf{P}$ may be obtained from the solution of the secular equation

$$|\mathbf{P} - \lambda \mathbf{I}| = 0, \tag{4.2.20}$$

where $\mathbf{I}$ is the 2×2 unit matrix. More specifically,

$$|\mathbf{P} - \lambda\mathbf{I}| = \begin{vmatrix} \exp[\beta(\varepsilon + H)] - \lambda & \exp[-\beta\varepsilon] \\ \exp[-\beta\varepsilon] & \exp[\beta(\varepsilon - H)] - \lambda \end{vmatrix} = 0 \qquad (4.2.21)$$

or

$$\{\exp[\beta(\varepsilon + H)] - \lambda\}\{\exp[\beta(\varepsilon - H)] - \lambda\} - \exp(-2\beta\varepsilon) = 0.$$

This is a polynomial in λ, the two roots of which are

$$\lambda_{\pm} = \exp(\beta\varepsilon)[\cosh x \pm \sqrt{\sinh^2 x + \exp(-4\beta\varepsilon)}], \qquad (4.2.22)$$

where $x = \beta H$, and we used the identities

$$\sinh x = \frac{e^x - e^{-x}}{2}, \qquad \cosh x = \frac{e^x + e^{-x}}{2}, \qquad \cosh^2 x - \sinh^2 x = 1. \quad (4.2.23)$$

Clearly, the two roots are positive, and $\lambda_+ > \lambda_-$. Thus we arrive at an explicit expression for the Helmholtz energy

$$A = -kT \ln Q = -kT \ln \lambda_+^N; \qquad (4.2.24)$$

i.e., we have the explicit function $A(N, H, T; \varepsilon)$, where ε is the only molecular parameter. (Actually the magnetic moment m_0 is also a molecular parameter, but it has been absorbed into H or, equivalently, chosen as $m_0 = 1$.)

The most important quantity of our system is the magnetization I, defined by

$$I = kT\left(\frac{\partial \ln Q}{\partial H}\right)_{T,N} = kT\left[\frac{\sum\limits_{\mathbf{s}} \exp[-\beta E(\mathbf{s})] \sum\limits_i \beta s_i}{Q}\right] = \sum_{\mathbf{s}} P(\mathbf{s}) \sum_i s_i$$

$$= \left\langle \sum_{i=1}^{N} s_i \right\rangle = N\langle s_1 \rangle. \qquad (4.2.25)$$

The meaning of I is quite simple. For any specific configuration $\mathbf{s}$, we denote by N_+ the number of spins in the "up" state and by N_- the number of spins in the "down" state. Then, for any specific configuration $\mathbf{s}$, we have

$$I(\mathbf{s}) = \sum_{i=1}^{N} s_i = N_+ - N_-. \qquad (4.2.26)$$

Thus I in (4.2.25) is the average of $N_+ - N_-$. (The magnetization is actually this number times the magnetic moment m_0.)

From our solution for the PF (4.2.24), we obtain for the magnetization

$$I(N, H, T) = kTN\frac{\partial \ln \lambda_+}{\partial H} = N\frac{\sinh(\beta H)}{[\sinh^2(\beta H) + \exp(-\beta 4\varepsilon)]^{1/2}}. \qquad (4.2.27)$$

If there are no interactions $\varepsilon = 0$, the spins are independent and

$$I = N\frac{\sinh(\beta H)}{\cosh(\beta H)} = N\tanh(\beta H), \qquad (4.2.28)$$

which is the same as the result obtained in section 2.7.

In the theory of two- and three-dimensional systems of interacting spins, the spontaneous magnetization $I(N, 0, T)$ is an important quantity. This is the magnetization at zero magnetic field. From the result (4.2.27), it is clear that for any $T > 0$, $\sinh(\beta H) = 0$ for $H = 0$; hence $I(N, 0, T) = 0$, i.e., there is no spontaneous magnetization in this model.

4.2.2. Lattice Gas

This is a variation of the simple Ising model. Again we have M units which we refer to as the lattice points. Each lattice point can be either empty or occupied by a particle. The canonical PF of a system of M lattice points with N particles is

$$Q(T, M, N) = \sum_{\mathbf{s}} \exp[-\beta E(\mathbf{s})], \qquad (4.2.29)$$

where again the configuration of the entire system is described by the vector $\mathbf{s} = (s_1 \ldots s_M)$, with $s_i = 0$ and $s_i = 1$ for empty and occupied sites, respectively. Note that here N is the *number* of particles, whereas in section 4.2.1 N was also the number of sites or the length of the Ising model.

Let W_{11} be the interaction energy between two particles on adjacent sites. In this model $W_{01} = W_{00} = 0$; i.e., there is no interaction between a particle and an empty site or between two empty sites. Let N_{11} be the number of pairs of nearest neighbors particles at any given configuration $\mathbf{s}$. Clearly the energy levels of the system are determined by

$$E(\mathbf{s}) = N_{11} W_{11}. \qquad (4.2.30)$$

And the partition function (4.2.29) is

$$Q(T, M, N) = \sum_{\substack{\text{energy} \\ \text{levels}}} \Omega(N_{11}) \exp(-\beta N_{11} W_{11}) = \sum_{\substack{\text{all } \mathbf{s} \text{ with} \\ \text{fixed } N}} \exp(-\beta N_{11} W_{11}). \quad (4.2.31)$$

The first sum on the rhs of (4.2.31) is over all energy levels. Since the energy levels (4.2.30) are determined by the parameter N_{11}, this is the same as a summation over all possible values of N_{11}. $\Omega(N_{11})$ is the number of configurations (states of the system) with a fixed N_{11} or, equivalently, with a fixed energy level $N_{11} W_{11}$. The second sum is over all possible configurations of the system.

In this particular example it is possible to compute $\Omega(N_{11})$ and evaluate the canonical partition function of the system. A more elegant solution is obtained for the grand PF using the same matrix method as was used in previous sections. Thus, we open the system with respect to the particles and write

$$\Xi(T, M, \lambda) = \sum_{N=0}^{M} Q(T, M, N)\lambda^N = \sum_{N=0}^{M} \lambda^N \sum_{\substack{\text{all } \mathbf{s} \text{ with} \\ \text{fixed } N}} \exp(-\beta N_{11} W_{11})$$

$$= \sum_{\text{all } \mathbf{s}} \lambda^N \exp(-\beta N_{11} W_{11}). \qquad (4.2.32)$$

We notice that each term in the last sum on the rhs of (4.2.32) corresponds to one specific vector **s**. For instance, here is how one possible vector **s** would generate its corresponding term:

$$\mathbf{s} = 1 \quad 0 \quad 1 \quad\quad 1 \quad\quad 1 \quad 0 \quad 0. \tag{4.2.33}$$
$$\quad\quad \lambda \quad\quad \lambda \quad q \quad \lambda \quad q \quad \lambda$$

There is one factor λ for each particle and one factor $q = \exp(-\beta W_{11})$ for each particle–particle bond.

The same term (4.2.33) can also be written as

$$
\begin{array}{cccccccc}
1 & 0 & 1 & 1 & 1 & 0 & 0 \\
\lambda^{1/2} & \lambda^{1/2} & \lambda^{1/2} & \lambda^{1/2}q\lambda^{1/2} & \lambda^{1/2}q\lambda^{1/2} & \lambda^{1/2} & 1
\end{array}, \tag{4.2.34}
$$

where now we assign a factor 1 for a pair $(0, 0)$, a factor $\lambda^{1/2}$ for a pair $(0, 1)$ or a pair $(1, 0)$, and a factor λq for a pair $(1, 1)$.

Thus we define the matrix elements

$$\langle 0|\mathbf{P}|0\rangle = 1$$
$$\langle 0|\mathbf{P}|1\rangle = \langle 1|\mathbf{P}|0\rangle = \lambda^{1/2} \tag{4.2.35}$$
$$\langle 1|\mathbf{P}|1\rangle = \lambda q,$$

and the grand PF (4.2.32) is written after closing the cycle (i.e., the site M becomes nearest neighbor to site 1).

$$\Xi(T, M, \lambda) = \sum_{\mathbf{s}} \langle s_1|\mathbf{P}|s_2\rangle\langle s_2|\mathbf{P}|s_3\rangle \cdots \langle s_{M-1}|\mathbf{P}|s_M\rangle\langle s_M|\mathbf{P}|s_1\rangle$$

$$= \operatorname{Tr} \mathbf{P}^M, \tag{4.2.36}$$

where the matrix $\mathbf{P}$ is defined in (4.2.35), or

$$\mathbf{P} = \begin{pmatrix} 1 & \lambda^{1/2} \\ \lambda^{1/2} & \lambda q \end{pmatrix}. \tag{4.2.37}$$

In order to compute the trace of $\mathbf{P}^M$, we need the eigenvalues of the matrix $\mathbf{P}$. The mathematical problem is the same as in section 4.2.1. We denote by γ the eigenvalues of $\mathbf{P}$. (In section 4.2.1 we used λ for the eigenvalues, but here λ stands for the activity of the particles.) Thus we need to solve the secular equation

$$|\mathbf{P} - \gamma\mathbf{I}| = 0, \tag{4.2.38}$$

or

$$\begin{vmatrix} 1 - \gamma & \lambda^{1/2} \\ \lambda^{1/2} & \lambda q - \gamma \end{vmatrix} = 0. \tag{4.2.39}$$

The two solutions of (4.2.39) are

$$\gamma_{\pm} = \frac{(1 + \lambda q) \pm \sqrt{(1 - \lambda q)^2 + 4\lambda}}{2}. \tag{4.2.40}$$

Clearly $\gamma_+ > \gamma_-$. Therefore, for $M \to \infty$ we need only γ_+; hence

$$\Xi(T, M, \lambda) = \gamma_+^M .$$ (4.2.41)

The average number of particles in the system is

$$\bar{N} = \lambda \frac{\partial \ln \Xi}{\partial \lambda} = M \frac{\lambda[qR - q(1 - \lambda q) + 2]}{(1 + \lambda q)R + (1 - \lambda q)^2 + 4\lambda},$$ (4.2.42)

where $R = \sqrt{(1 - \lambda q)^2 + 4\lambda}$.

Equation (4.2.42) can also be written in a more convenient way as

$$\theta = \frac{\bar{N}}{M} = \frac{1 - \gamma_+}{1 + \lambda q - 2\gamma_+}.$$ (4.2.43)

Note that if there are no nearest-neighbor interactions—i.e., $q = 1$—then

$$\theta = \frac{\lambda}{1 + \lambda},$$ (4.2.44)

which is similar to the Langmuir isotherm with zero binding energy.

4.2.3. Lattice Model of a Two-Component Mixture

As in the model of the previous section, we have M sites each of which may be occupied by either an A or a B molecule. We introduce the interaction energies W_{AA}, W_{BB}, and W_{BA} for each type of nearest neighbor. The energy levels of the system are determined by

$$E(\mathbf{s}) = N_{AA}W_{AA} + N_{BB}W_{BB} + N_{AB}W_{AB},$$ (4.2.45)

where $N_{\alpha\beta}$ is the number of nearest-neighbor pairs of the type $\alpha\beta$. If N_A and N_B are the number of A and B molecules, then, since all sites are presumed occupied by either A or B, we must have

$$N_A + N_B = M.$$ (4.2.46)

We also have the two relations

$$2N_A = 2N_{AA} + N_{AB}$$ (4.2.47)

$$2N_B = 2N_{BB} + N_{AB}.$$ (4.2.48)

Each A—A bond contributes two As; each A—B bond contributes one A. If we sum over all N_{AA} bonds and all N_{AB} bonds, we count each A twice; hence relation (4.2.47). A similar argument applies for (4.2.48).

The canonical PF is thus

$$Q(T, M, N_A, N_B) = \sum_{\substack{\text{all states with} \\ N_A + N_B = M}} \exp[-\beta(N_{AA}W_{AA} + N_{BB}W_{BB} + N_{AB}W_{AB})]$$

$$= \exp(-\beta N_A W_{AA}) \exp(-\beta N_B W_{BB}) \sum_{N_{AB}} \Omega(N_{AB})$$

$$\times \exp\left(\frac{-\beta W N_{AB}}{2}\right),$$ (4.2.49)

where the parameter W is defined by

$$W = 2W_{AB} - W_{AA} - W_{BB}. \tag{4.2.50}$$

In (4.2.49) we used relations (4.2.47) and (4.2.48) to rewrite the sum over all states as a sum over all possible values of the parameter N_{AB}. The degeneracy corresponding to the energy level (4.2.45) is denoted by $\Omega(N_{AB})$. As in the previous model, this degeneracy can be calculated for the one-dimensional model. A more elegant treatment is to use the matrix method again. We open the system for both A and B; the corresponding grand PF is

$$\Xi(T, M, \lambda_A, \lambda_B) = \sum_{\substack{N_B = 0 \\ N_A + N_B = M}}^{M} \sum_{N_A = 0}^{M} Q(T, M, N_A, N_B)\lambda_A^{N_A}\lambda_B^{N_B}. \tag{4.2.51}$$

Defining the elements of the matrix P

$$P_{AA} = \langle A|\mathbf{P}|A\rangle = \lambda_A \exp(-\beta W_{AA}) = \lambda_A q_{AA}$$

$$P_{AB} = \langle A|\mathbf{P}|B\rangle = \langle B|\mathbf{P}|A\rangle = \lambda_A^{1/2}\lambda_B^{1/2} \exp(-\beta W_{AB}) = \lambda_A^{1/2}\lambda_B^{1/2}q_{AB} \tag{4.2.52}$$

$$P_{BB} = \langle B|\mathbf{P}|B\rangle = \lambda_B \exp(-\beta W_{BB}) = \lambda_B q_{BB}.$$

We can rewrite the PF (4.2.51) after closing the cycle—i.e., adding the factor $\exp(-\beta W_{s_N s_1})$ to obtain

$$\Xi(T, M, \lambda_A, \lambda_B) = \sum_{\mathbf{s}} \langle s_1|\mathbf{P}|s_2\rangle\langle s_2|\mathbf{P}|s_3\rangle \cdots \langle s_{M-1}|\mathbf{P}|s_M\rangle\langle s_M|\mathbf{P}|s_1\rangle$$

$$= \text{Tr } \mathbf{P}^M = \gamma_+^M, \tag{4.2.53}$$

where γ_+ is the larger of the two eigenvalues of the matrix $\mathbf{P}$. Thus

$$\gamma_+ = \tfrac{1}{2}\{\lambda_A q_{AA} + \lambda_B q_{BB} + [(\lambda_A q_{AA} - \lambda_B q_{BB})^2 + 4\lambda_A\lambda_B q_{AB}^2]^{1/2}\}. \tag{4.2.54}$$

From (4.2.53) and (4.2.54) we can compute all relevant thermodynamic quantities of the system.

A simple case occurs when $W = 0$, i.e., when W_{AB} is the arithmetic average of W_{AA} and W_{BB} or, equivalently,

$$q_{AB}^2 = q_{AA}q_{BB}. \tag{4.2.55}$$

In this case γ_+ reduces to

$$\gamma_+ = \lambda_A q_{AA} + \lambda_B q_{BB}. \tag{4.2.56}$$

This is the symmetrical ideal solution. We shall discuss the thermodynamics of ideal solutions further in Chapter 6. The average number of As (or Bs) in the system is obtained from

$$\bar{N}_A = \lambda_A M \frac{\partial \ln \gamma_+}{\partial \lambda_A}. \tag{4.2.57}$$

The expression for $\bar{N}_A$ is quite cumbersome in the general case. However, it is very simple in the ideal case (4.2.55), for which γ_+ is given by (4.2.56), hence

$$\bar{N}_A = M \frac{\lambda_A q_{AA}}{\lambda_A q_{AA} + \lambda_B q_{BB}}. \tag{4.2.58}$$

The ratio of $\bar{N}_A$ and $\bar{N}_B$ is thus

$$\eta = \frac{\bar{N}_A}{\bar{N}_B} = \frac{\lambda_A q_{AA}}{\lambda_B q_{BB}} = \exp[\beta(\mu_A - \mu_B)] \exp[-\beta(W_{AA} - W_{BB})]. \qquad (4.2.59)$$

This ratio is controlled by the difference in the chemical potentials $\mu_A - \mu_B$ and by the difference $W_{AA} - W_{BB}$. For any fixed values of the molecular parameters $W_{\alpha\beta}$, the ratio η will tend to infinity when $\lambda_A \gg \lambda_B$ and tend to zero when $\lambda_A \ll \lambda_B$.

In the more general (nonideal) case one can perform the derivatives (4.2.57) and, after some algebra, obtain the final results

$$x_A = \frac{\bar{N}_A}{N} = \frac{\lambda_B q_{BB} - \gamma_+}{\lambda_A q_{AA} + \lambda_B q_{BB} - 2\gamma_+} \qquad (4.2.60)$$

$$x_B = \frac{\bar{N}_B}{N} = \frac{\lambda_A q_{AA} - \gamma_+}{\lambda_A q_{AA} + \lambda_B q_{BB} - 2\gamma_+}. \qquad (4.2.61)$$

We shall rederive these relations in section 4.3.1 in connection with the singlet molecular distribution function.

4.2.4. Two-State Equilibrium Modulated by an External Field

As the final example of a system consisting of two-state particles, we consider a linear string of polymers. Each polymer can be in one of two states, long (A) or short (B) with corresponding lengths l_A and l_B, respectively. We denote by Q_A and Q_B the internal PF of the polymer in the A and B states and by W_{AA}, W_{BB}, and W_{AB} the corresponding interaction between nearest pairs. The total length of the system is

$$L = N_A l_A + N_B l_B. \qquad (4.2.62)$$

Note also that we assume that $W_{AB} = W_{BA}$. If the two ends of the polymer are not identical, or similar, then we must distinguish between W_{AB} and W_{BA}.

Since $N = N_A + N_B$ is fixed, then fixing L also fixes N_A, by (4.2.62). Therefore the canonical PF for this system is

$$Q(T, L, N) = Q_A^{N_A} Q_B^{N_B} \sum \exp[-\beta(N_{AA} W_{AA} + N_{BB} W_{BB} + N_{AB} W_{AB})], \qquad (4.2.63)$$

where the sum is over all possible configurations with fixed values of N_A and N_B. In order to remove the condition of fixed N_A or equivalently, of fixed total length L, we transform the variable L into the corresponding one-dimensional pressure, or tension τ. The T, τ, N PF is

$$\Delta(T, \tau, N) = \sum_L Q(T, L, N) \exp(\beta \tau L), \qquad (4.2.64)$$

where the sum is over all possible lengths of the system. If we choose, say, $l_A > l_B$, then the minimum value of L is $N l_B$ and the maximum value is $N l_A$.

The tension τ is defined as positive when the system is stretched. In the three-dimensional system, the pressure is defined as positive when the system is compressed. Therefore we have the factor $\exp(-\beta PV)$ in the T, P, N ensemble, whereas here we have $\exp[\beta \tau L]$.

We now define the four elements of the 2×2 matrix $\mathbf{P}$:

$$P_{AA} = \langle A|\mathbf{P}|A \rangle = Q_A \delta_A \exp(-\beta W_{AA}) = Q_A \delta_A q_{AA}$$

$$P_{AB} = P_{BA} = \langle A|\mathbf{P}|B \rangle = (Q_A \delta_A)^{1/2}(Q_B \delta_B)^{1/2} \exp(-\beta W_{AB}) = (Q_A Q_B \delta_A \delta_B)^{1/2} q_{AB}$$

$$P_{BB} = \langle B|\mathbf{P}|B \rangle = Q_B \delta_B \exp[-\beta W_{BB}] = Q_B \delta_B q_{BB}, \tag{4.2.65}$$

where $\delta_A = \exp(\beta \tau l_A)$, $\delta_B = \exp(\beta \tau l_B)$.

With these definitions we can perform the sum in (4.2.64) simply by writing one specific configuration and summing over all possible configurations. The sum over all L values is the same as the sum over all possible vectors $\mathbf{s}$ of the system. The result after cyclization is

$$\Delta(T, \tau, N) = \sum_{\mathbf{s}} \langle s_1|\mathbf{P}|s_2 \rangle \cdots \langle s_N|\mathbf{P}|s_1 \rangle = \mathrm{Tr}\ \mathbf{P}^N. \tag{4.2.66}$$

The larger of the two eigenvalues of $\mathbf{P}$ can be obtained as before, the result being

$$\gamma_+ = \tfrac{1}{2}\{(P_{AA} + P_{BB}) + [(P_{AA} - P_{BB})^2 + 4P_{AB}^2]^{1/2}\}$$

$$= \tfrac{1}{2}\{(Q_A \delta_A q_{AA} + Q_B \delta_B q_{BB}) + [(Q_A \delta_A q_{AA} - Q_B \delta_B q_{BB})^2$$

$$+ 4 Q_A Q_B \delta_A \delta_B q_{AB}^2]^{1/2}\}. \tag{4.2.67}$$

Note the analogy with the previous models. Here the tension τ plays the role of the activity or the magnetic field in the previous models.

The average length of the system is obtained from

$$\bar{L} = kT \frac{\partial \ln \Xi}{\partial \tau} = NkT \frac{\partial \ln \gamma_+}{\partial \tau}. \tag{4.2.68}$$

The average length per particle can be put in the form

$$\bar{l} = \frac{\bar{L}}{N} = x_A l_A + x_B l_B, \tag{4.2.69}$$

where

$$x_A = \frac{Q_B \delta_B q_{BB} - \gamma_+}{Q_A \delta_A q_{AA} + Q_B \delta_B q_{BB} - 2\gamma_+}, \qquad x_B = 1 - x_A. \tag{4.2.70}$$

The equilibrium constant for the conversion $A \rightleftarrows B$ is

$$\eta = \frac{x_A}{x_B} = \frac{Q_B \delta_B q_{BB} - \gamma_+}{Q_A \delta_A q_{AA} - \gamma_+}. \tag{4.2.71}$$

A simple case occurs when $2W_{AB} = W_{AA} + W_{BB}$ or $q_{AB}^2 = q_{AA}q_{BB}$. (This corresponds to the symmetrical ideal solution of section 4.2.3; see also Chapter 6.) In this case,

$$\gamma_+ = Q_A \delta_A q_{AA} + Q_B \delta_B q_{BB}, \tag{4.2.72}$$

and the equilibrium constant η reduces to

$$\eta = \frac{x_A}{x_B} = \frac{Q_A \delta_A q_{AA}}{Q_B \delta_B q_{BB}} = \frac{Q_A q_{AA}}{Q_B q_{BB}} \exp[\beta \tau (l_A - l_B)]. \tag{4.2.73}$$

If we choose $l_A - l_B > 0$, then increasing the tension favors the longer form, A.

4.3. MOLECULAR DISTRIBUTION FUNCTIONS IN THE ISING MODEL

In this section we introduce a few molecular distribution functions in the one-dimensional Ising model. The general definitions apply to systems for which the units can have any number of states. Specific illustrations will be dealt with for the two-state units treated in section 4.2.

The configuration of the system is described by the vector $\mathbf{s} = (s_1, s_2, \ldots, s_M)$, where s_i may be $+1$ or -1 for the magnetic spins, 0 or 1 in the lattice gas, and A or B in the mixture of two components. Whenever we want to stress that s_i has a specific state we write $s_i = \alpha$ or $s_i = \beta$, etc.

Any one of the partition functions of the Ising model can be written in the general form

$$\Gamma = \sum_{s_1 s_2 \cdots s_M} P_{s_1 s_2} P_{s_2 s_3} \cdots P_{s_M s_1} = \sum_{\mathbf{s}} \prod_{i=1}^{M} P_{s_i s_{i+1}}, \qquad (4.3.1)$$

where Γ can be the canonical PF (section 4.2.2), the grand PF (section 4.2.3), or the T, τ, N PF (section 4.2.4). The summation is over all possible configurations of the system. $P_{s_i s_j}$ are the factors assigned to each of the unit–unit bonds, in the manner discussed in section 4.2. If each unit can be in one of two states, then $P_{s_i s_j}$ are the elements of a 2×2 matrix. If each unit can be in one of three states, then $P_{s_i s_j}$ are the elements of a 3×3 matrix, etc. Examples of 3×3 and higher-order matrices are given in the following sections.

4.3.1. Singlet Distribution Function

According to the general rules of ST, each term in the sum (4.3.1) corresponds to the probability of finding the entire system in some specific configuration. For example, the probability of finding a system of $M = 6$ units in a specific state $\alpha, \alpha, \beta, \alpha, \beta, \beta$ is

$$\Pr(s_1 = \alpha, s_2 = \alpha, s_3 = \beta, s_4 = \alpha, s_5 = \beta, s_6 = \beta) = \frac{P_{\alpha\alpha} P_{\alpha\beta} P_{\beta\alpha} P_{\alpha\beta} P_{\beta\beta} P_{\beta\alpha}}{\Gamma}. \qquad (4.3.2)$$

Note that we always close the cycle, making the Mth unit interact with the first unit.

The probability of finding a *specific* unit, say $i = 1$, in a specific state α, independently of the states of all other units, is

$$\Pr(s_1 = \alpha) = \sum_{s_2 \cdots s_M} \Pr(s_1 = \alpha, s_2, s_3, \ldots, s_M). \qquad (4.3.3)$$

Thus in order to obtain $\Pr(s_1 = \alpha)$, we take the probability distribution of the entire system, fix the state of a specific unit, say $i = 1$, at α, and sum over all possible states of all other units. The resulting probability distribution is called the singlet distribution function.

This quantity may be given a different interpretation, as follows. We rewrite (4.3.3) in a slightly different form:

$$\Pr(s_1 = \alpha) = \sum_{s_1 \cdots s_M} \Pr(s_1, s_2, \ldots, s_M) \delta_{s_1, \alpha}. \qquad (4.3.4)$$

Note that the sum is now over *all* possible states of the system. The Kronecker delta function is defined as

$$\delta_{s_1,\alpha} = \begin{cases} 1 & \text{for } s_1 = \alpha \\ 0 & \text{for } s_1 \neq \alpha. \end{cases} \tag{4.3.5}$$

Clearly the same result is obtained when we choose any other specific unit, say the *i*th unit.

$$\Pr(s_i = \alpha) = \sum_{s_1 \cdots s_M} \Pr(s_1, s_2, \ldots, s_M)\delta_{s_i,\alpha}. \tag{4.3.6}$$

If we sum (4.3.6) over all *i* we obtain *M* times the same value (since all *M* units are equivalent); thus

$$M \Pr(s_1 = \alpha) = \sum_{i=1}^{M} \Pr(s_i = \alpha) = \sum_{s_1 \cdots s_M} \Pr(s_1, \ldots, s_M) \sum_{i=1}^{M} \delta_{s_i,\alpha}$$

$$= \left\langle \sum_{i=1}^{M} \delta_{s_i,\alpha} \right\rangle = \bar{N}(\alpha). \tag{4.3.7}$$

The quantity $\sum_{i=1}^{M} \delta_{s_i,\alpha}$ is referred to as a *counting function*. For each configuration $s_1, \ldots, s_M$ the function $\delta_{s_i,\alpha}$ contributes unity whenever $s_i = \alpha$. Therefore $\sum \delta_{s_i,\alpha}$ is the number of units in the state α in a particular configuration of the system $s_1, \ldots, s_M$. Hence $\langle \sum_i \delta_{s_i,\alpha} \rangle$ is the average number of units in the state α. This is denoted by $\bar{N}(\alpha)$. Combining the two ends of Eq. (4.3.7), we write

$$\Pr(s_1 = \alpha) = \frac{\bar{N}(\alpha)}{M} = x(\alpha), \tag{4.3.8}$$

where $x(\alpha)$ is the mole fraction of units in the state α. Thus $x(\alpha)$ is equal to the probability of finding a *specific* unit, say 1, in the state α. Note that $M \Pr(s_1 = \alpha)$ is *not* the probability of finding *any* unit in state α. In fact, $M \Pr(s_1 = \alpha)$ is not a probability at all.

Our next task is to express $\Pr(s_1 = \alpha)$ in terms of the molecular parameters of the model. Using similar steps as in section 4.2.1, we write (4.3.4) as

$$\Pr(s_1 = \alpha) = \sum_{s_1,\ldots,s_M} \Pr(s_1, \ldots, s_M)\delta_{s_1,\alpha} = \frac{1}{\Gamma}\sum_{s_1} (\mathbf{P}^M)_{s_1,s_1}\delta_{s_1,\alpha}$$

$$= \frac{(\mathbf{P}^M)_{\alpha\alpha}}{\lambda_1^M}. \tag{4.3.9}$$

In (4.3.9) we first sum over $s_2, \ldots, s_M$ to obtain the s_1, s_1 element of the matrix $\mathbf{P}^M$. In contrast to Eq. (4.2.10), where the last summation over s_1 produced the trace of $\mathbf{P}^M$, here, because of the Kronecker delta function, we obtain the $\alpha\alpha$ element of the matrix $\mathbf{P}^M$. We have also replaced the partition function Γ by λ_1^M, where λ_1 is the largest eigenvalue of the matrix $\mathbf{P}$.

Let $|a_i\rangle$ be the eigenvector that corresponds to the eigenvalue λ_i, i.e.,

$$\mathbf{P}|a_i\rangle = \lambda_i|a_i\rangle. \tag{4.3.10}$$

Here $|a_i\rangle$ stands for a column vector, and on the lhs of (4.3.10) we have a product of a matrix and a column vector. Since $\mathbf{P}$ is symmetric, the right and the left eigenvectors that correspond to λ_i are identical.

We use the following identity for the unit matrix

$$\mathbf{I} = \sum_i |a_i\rangle\langle a_i|, \tag{4.3.11}$$

where $\mathbf{I}$ is a unit matrix of the same dimensions as $\mathbf{P}$. $\langle a_i|$ is the transpose of the vector $|a_i\rangle$ (i.e., $\langle a_i|$ is the row vector corresponding to the column vector $|a_i\rangle$). The sum over i is over all eigenvectors of $\mathbf{P}$.

We define the unit vectors corresponding to each of the possible states of a single unit. For instance, if a unit can be in one of three states, A, B, and C, then we assign to each state a unit vector as follows

$$\langle A| = (1, 0, 0), \qquad \langle B| = (0, 1, 0), \qquad \langle C| = (0, 0, 1). \tag{4.3.12}$$

Using these vectors, we can write any matrix element as

$$\mathbf{P}_{AB} = \langle A|\mathbf{P}|B\rangle = (1, 0, 0)\begin{pmatrix} P_{AA} & P_{AB} & P_{AC} \\ P_{BA} & P_{BB} & P_{BC} \\ P_{CA} & P_{CB} & P_{CC} \end{pmatrix}\begin{pmatrix} 0 \\ 1 \\ 0 \end{pmatrix}. \tag{4.3.13}$$

Thus the singlet distribution function in (4.3.9) is written, using the notation (4.3.13) and the identity (4.3.11), as

$$\Pr(s_1 = \alpha) = \frac{1}{\lambda_1^M}\langle \alpha|\mathbf{P}^M|\alpha\rangle$$

$$= \frac{1}{\lambda_1^M}\sum_{i,j}\langle \alpha|a_i\rangle\langle a_i|\mathbf{P}^M|a_j\rangle\langle a_j|\alpha\rangle$$

$$= \frac{1}{\lambda_1^M}\sum_{i,j}\langle \alpha|a_i\rangle\lambda_j^M\delta_{i,j}\langle a_j|\alpha\rangle$$

$$= \frac{1}{\lambda_1^M}\sum_i\langle \alpha|a_i\rangle^2\lambda_i^M$$

$$= \frac{1}{\lambda_1^M}\sum_i a_{i\alpha}^2\lambda_i^M, \tag{4.3.14}$$

where we used the orthogonality condition

$$\langle a_i|a_j\rangle = \delta_{i,j} \tag{4.3.15}$$

and

$$\langle \alpha|a_i\rangle = \langle a_i|\alpha\rangle = a_{i\alpha}, \tag{4.3.16}$$

where $a_{i\alpha}$ is the αth component of the vector $|a_i\rangle$ (or of $\langle a_i|$). Since λ_1 is the largest eigenvalue, we can simplify (4.3.14) at the limit $M \to \infty$:

$$\Pr(s_1 = \alpha) = \sum_i \langle \alpha | a_i \rangle^2 \left(\frac{\lambda_i}{\lambda_1}\right)^M \xrightarrow[M \to \infty]{} \langle \alpha | a_1 \rangle^2 = a_{1\alpha}^2 \qquad (4.3.17)$$

where $a_{1\alpha}$ is the α component of the eigenvector $\langle a_1|$ corresponding to the largest eigenvalue λ_1.

Up to now all the relations apply to a system the elements of which can be in any number of states. In the following, we compute the singlet distribution function for two-state elements only, say α and β, in which case the matrix $\mathbf{P}$ has the form

$$\mathbf{P} = \begin{pmatrix} P_{\alpha\alpha} & P_{\alpha\beta} \\ P_{\beta\alpha} & P_{\beta\beta} \end{pmatrix}. \qquad (4.3.18)$$

The eigenvector belonging to the largest eigenvalue λ_1 is denoted by $|a_1\rangle$, and its components are

$$|a_1\rangle = a_{1\alpha}|\alpha\rangle + a_{1\beta}|\beta\rangle = \begin{pmatrix} a_{1\alpha} \\ a_{1\beta} \end{pmatrix}. \qquad (4.3.19)$$

The normalization condition is

$$a_{1\alpha}^2 + a_{1\beta}^2 = 1, \qquad (4.3.20)$$

which is consistent with the probability interpretation of $a_{1\alpha}^2$ and $a_{1\beta}^2$ in (4.3.17).

The matrix equation (4.3.10) is

$$\begin{pmatrix} P_{\alpha\alpha} & P_{\alpha\beta} \\ P_{\beta\alpha} & P_{\beta\beta} \end{pmatrix} \begin{pmatrix} a_{1\alpha} \\ a_{1\beta} \end{pmatrix} = \lambda_1 \begin{pmatrix} a_{1\alpha} \\ a_{1\beta} \end{pmatrix}. \qquad (4.3.21)$$

Together with the normalization condition (4.3.20), this determines the two unknown components $a_{1\alpha}$ and $a_{1\beta}$. More explicitly, the two equations are

$$P_{\alpha\alpha}a_{1\alpha} + P_{\alpha\beta}a_{1\beta} = \lambda_1 a_{1\alpha} \qquad (4.3.22)$$

$$P_{\beta\alpha}a_{1\alpha} + P_{\beta\beta}a_{1\beta} = \lambda_1 a_{1\beta}. \qquad (4.3.23)$$

These two equations are not independent, since λ_1 has been obtained from the secular equation

$$\begin{vmatrix} P_{\alpha\alpha} - \lambda & P_{\alpha\beta} \\ P_{\beta\alpha} & P_{\beta\beta} - \lambda \end{vmatrix} = 0, \qquad (4.3.24)$$

which is equivalent to

$$(P_{\alpha\alpha} - \lambda)(P_{\beta\beta} - \lambda) - P_{\alpha\beta}P_{\beta\alpha} = 0, \qquad (4.3.25)$$

or

$$\frac{P_{\alpha\alpha} - \lambda}{P_{\alpha\beta}} = \frac{P_{\beta\alpha}}{P_{\beta\beta} - \lambda}. \qquad (4.3.26)$$

This condition makes the two equations (4.3.22) and (4.3.23) identical. Thus in order to solve for $a_{1\alpha}$ and $a_{1\beta}$ we use Eq. (4.3.22) with the normalization condition (4.3.20) to obtain

$$a_{1\alpha} = \frac{-P_{\alpha\beta}a_{1\beta}}{P_{\alpha\alpha} - \lambda} \qquad (4.3.27)$$

$$1 = a_{1\alpha}^2 + a_{1\beta}^2 = a_{1\alpha}^2 + a_{1\alpha}^2 \frac{(P_{\alpha\alpha} - \lambda)^2}{P_{\alpha\beta}^2} \qquad (4.3.28)$$

$$a_{1\alpha}^2 = \frac{P_{\alpha\beta}^2}{P_{\alpha\beta}^2 + (P_{\alpha\alpha} - \lambda_1)^2}. \qquad (4.3.29)$$

We can now use the explicit expression for λ_1, the largest eigenvalue obtained by solving (4.3.24),

$$\lambda_1 = \tfrac{1}{2}\{(P_{\alpha\alpha} + P_{\beta\beta}) + [(P_{\alpha\alpha} - P_{\beta\beta})^2 + 4P_{\alpha\beta}P_{\beta\alpha}]^{1/2}\}. \qquad (4.3.30)$$

Thus by substituting λ_1 from (4.3.30) into (4.3.29) we obtain $a_{1\alpha}^2$ in terms of the molecular parameters of the system. The final result is

$$\Pr(s_1 = \alpha) = a_{1\alpha}^2 = \frac{P_{\alpha\beta}^2}{P_{\alpha\beta}^2 + [\tfrac{1}{2}(P_{\alpha\alpha} - P_{\beta\beta}) - \tfrac{1}{2}R]^2}, \qquad (4.3.31)$$

where

$$R = [(P_{\alpha\alpha} - P_{\beta\beta})^2 + 4P_{\alpha\beta}P_{\beta\alpha})]^{1/2} \qquad (4.3.32)$$

and

$$P(s_1 = \beta) = 1 - a_{1\alpha}^2. \qquad (4.3.33)$$

Another useful form for the singlet distribution function in terms of λ_1 is

$$\Pr(s_1 = \alpha) = \frac{P_{\beta\beta} - \lambda_1}{(P_{\alpha\alpha} - \lambda_1) + (P_{\beta\beta} - \lambda_1)} \qquad (4.3.34)$$

$$\Pr(s_1 = \beta) = 1 - \Pr(s_1 = \alpha). \qquad (4.3.35)$$

These results were used in sections 4.2.3 and 4.2.4. Note, however, that these relatively simple results are valid only for two-state units. The case of three-state units already involves a cubic secular equation, instead of (4.3.24), which is much more difficult to solve for the largest eigenvalue.

4.3.2. Pair Distribution Function

Here we are interested in the following question: What is the joint probability of finding a specific unit i in state α, *and* a second specific unit j ($j \neq i$) in state β? Again, we first derive the general expression for the pair distribution function for the Ising model with any number of states and then we shall particularize to the case of the two-state units.

Since all the units are equivalent, it does not matter which index i we choose. The pair distribution function depends only on the distance $j - i$ between the units. Therefore we choose $i = 1$ and any j, so that the distance is $j - 1$.

The pair distribution function is defined by analogy with 4.3.3 and 4.3.6 as

$$\Pr(s_1 = \alpha, s_j = \beta) = \sum_{s_2, s_3, \ldots, s_{j-1}, s_{j+1}, \ldots, s_M} \Pr(s_1 = \alpha, s_2, s_3, \ldots, s_j = \beta, s_{j+1}, \ldots, s_M)$$

$$= \sum_{s_1, \ldots, s_M} \Pr(s_1, s_2, \ldots, s_M) \delta_{s_1, \alpha} \delta_{s_j, \beta}, \tag{4.3.36}$$

the first sum is over all the indices s_i except s_1 and s_j. The sum is rewritten as a sum over all s_i $(i = 1, \ldots, M)$ after introducing the two delta functions.

Clearly, because of the equivalency of all the units, we could start from any unit i and ask for the pair distribution for the two units, $j - 1$ units apart, i.e.,

$$\Pr(s_i = \alpha, s_{i+j-1} = \beta) = \Pr(s_1 = \alpha, s_j = \beta). \tag{4.3.37}$$

The pair distribution function depends on the *distance* between the units, not on the particular location of the two units.

If we sum over all initial indices i in (4.3.37), we must obtain M identical terms. Therefore by analogy with (4.3.7) we have

$$M \Pr(s_1 = \alpha, s_j = \beta) = \sum_{i=1}^{M} \Pr(s_i = \alpha, s_{i+j-1} = \beta) = \sum_{s_1, \ldots, s_M} \Pr(\mathbf{s}) \sum_{i=1}^{M} \delta_{s_i, \alpha} \delta_{s_{i+j-1}, \beta}$$

$$= \left\langle \sum_{i=1}^{M} \delta_{s_i, \alpha} \delta_{s_{i+j-1}, \beta} \right\rangle. \tag{4.3.38}$$

The sum over i in the last form on the rhs of (4.3.38) is again a *counting function*. We scan all units i. Each unit contributes unity to this sum if it is in state α *and* at the same time has a neighbor at a distance of $j - 1$ units to its right in state β. Therefore the quantity (4.3.38) counts the average number of pairs of units at a distance of $j - 1$ units, such that the left unit is in state α and the right unit in state β. (Of course, left and right are interchangeable; only the distance between the units is important.)

We now follow similar steps to those in the case of the singlet distribution function to obtain

$$\Pr(s_1 = \alpha, s_j = \beta) = \Gamma^{-1} \sum_{s_1, \ldots, s_M} \langle s_1 | \mathbf{P} | s_2 \rangle \langle s_2 | \mathbf{P} | s_3 \rangle \cdots \langle s_M | \mathbf{P} | s_1 \rangle \delta_{s_1, \alpha} \delta_{s_j, \beta}$$

$$= \Gamma^{-1} \sum_{s_1 s_j} \delta_{s_1, \alpha} \delta_{s_j, \beta} \langle s_1 | \mathbf{P}^{j-1} | s_j \rangle \langle s_j | \mathbf{P}^{M-j+1} | s_1 \rangle$$

$$= \lambda_1^{-M} \langle \alpha | \mathbf{P}^{j-1} | \beta \rangle \langle \beta | \mathbf{P}^{M-j+1} | \alpha \rangle. \tag{4.3.39}$$

In the second step on the rhs of (4.3.39), we first sum over all the indices $s_1, \ldots, s_M$ except s_1 and s_j. This produces the elements of the two matrices $\mathbf{P}^{j-1}$ and $\mathbf{P}^{M-j+1}$. The next step is to sum over s_1 and s_j using the basic property of the Kronecker delta function.

Each of the matrix elements in (4.3.39) can be treated as in Eq. (4.3.14) to obtain

$$\langle \alpha | \mathbf{P}^{j-1} | \beta \rangle = \sum_{k,l} \langle \alpha | a_k \rangle \langle a_k | \mathbf{P}^{j-1} | a_l \rangle \langle a_l | \beta \rangle$$

$$= \sum_{k,l} \langle \alpha | a_k \rangle \lambda_l^{j-1} \delta_{k,l} \langle a_l | \beta \rangle$$

$$= \sum_{k} \langle \alpha | a_k \rangle \lambda_k^{j-1} \langle a_k | \beta \rangle, \tag{4.3.40}$$

and similarly for the second factor in (4.3.39):

$$\langle \beta | \mathbf{P}^{M-j+1} | \alpha \rangle = \sum_l \langle \beta | a_l \rangle \lambda_l^{M-j+1} \langle a_l | \alpha \rangle. \tag{4.3.41}$$

(The summation over k and l is over all eigenvalues of the matrix $\mathbf{P}$.)

Note that there is an essential difference between (4.3.40) and (4.3.41). In the first, the exponent j is *finite*. In general we shall be interested in small values of j, $j = 2$ for nearest neighbors or $j = 3$ for next-nearest neighbors. But even if we take a very large j, it is still finite. On the other hand, in (4.3.41) we have M, and in our model we let $M \to \infty$. We shall examine a few particular values of j below. First we write (4.3.39) using (4.3.40) and (4.3.41) as

$$\Pr(s_1 = \alpha, s_j = \beta) = \lambda_1^{-M} \sum_k \langle \alpha | a_k \rangle \lambda_k^{j-1} \langle a_k | \beta \rangle \sum_l \langle \beta | a_l \rangle \lambda_l^{M-j+1} \langle a_l | \alpha \rangle. \tag{4.3.42}$$

We now examine some special cases of (4.3.42) which are of particular interest. We restrict ourselves to the two-state units only.

4.3.2.1. Nearest Neighbors: $j - i = 1$

Here we take $i = 1$ and $j = 2$. Hence the pair distribution function is

$$\Pr(s_1 = \alpha, s_2 = \beta) = \sum_k \langle \alpha | a_k \rangle \lambda_k \langle a_k | \beta \rangle \sum_l \langle \beta | a_l \rangle \langle a_l | \alpha \rangle \frac{\lambda_l^{M-1}}{\lambda_1^M}. \tag{4.3.43}$$

Since λ_1 is the largest eigenvalue, we have, in the limit $M \to \infty$,

$$\lim_{M \to \infty} \sum_l \langle \beta | a_l \rangle \langle a_l | \alpha \rangle \frac{1}{\lambda_1} \left(\frac{\lambda_l}{\lambda_1} \right)^{M-1} = \frac{1}{\lambda_1} \langle \beta | a_1 \rangle \langle a_1 | \alpha \rangle. \tag{4.3.44}$$

Hence

$$\Pr(s_1 = \alpha, s_2 = \beta) = \frac{1}{\lambda_1} \langle \beta | a_1 \rangle \langle a_1 | \alpha \rangle \sum_k \langle \alpha | a_k \rangle \lambda_k \langle a_k | \beta \rangle. \tag{4.3.45}$$

This may be simplified by the following manipulation, using (4.3.11) and (4.3.15):

$$\langle \alpha | \mathbf{P} | \beta \rangle = \sum_{kl} \langle \alpha | a_k \rangle \langle a_k | \mathbf{P} | a_l \rangle \langle a_l | \beta \rangle$$

$$= \sum_{kl} \langle \alpha | a_k \rangle \lambda_l \langle a_k | a_l \rangle \langle a_l | \beta \rangle$$

$$= \sum_{kl} \langle \alpha | a_k \rangle \lambda_l \delta_{k,l} \langle a_l | \beta \rangle$$

$$= \sum_k \langle \alpha | a_k \rangle \lambda_k \langle a_k | \beta \rangle. \tag{4.3.46}$$

Hence from (4.3.45) and (4.3.46) we obtain

$$\Pr(s_1 = \alpha, s_2 = \beta) = \frac{\langle \alpha | \mathbf{P} | \beta \rangle \langle \beta | a_1 \rangle \langle a_1 | \alpha \rangle}{\lambda_1} = \frac{P_{\alpha\beta} a_{1\alpha} a_{1\beta}}{\lambda_1}. \tag{4.3.47}$$

We already have the components of the vector $\langle a_1 |$ in terms of the molecular parameters in (4.3.31). Therefore (4.3.47) can be viewed as an expression of the nearest-neighbors pair distribution function in terms of the molecular parameters of the model. For later treatment it will be useful to use the form (4.3.45). Since the sum over k is only over two eigenvalues, we can write it explicitly as

$$\Pr(s_1 = \alpha, s_2 = \beta)$$

$$= \frac{1}{\lambda_1} \langle \beta | a_1 \rangle \langle a_1 | \alpha \rangle \{ \langle \alpha | a_1 \rangle \lambda_1 \langle a_1 | \beta \rangle + \langle \alpha | a_2 \rangle \lambda_2 \langle a_2 | \beta \rangle \}$$

$$= a_{1\alpha}^2 a_{1\beta}^2 + \frac{\lambda_2}{\lambda_1} a_{1\alpha} a_{1\beta} a_{2\alpha} a_{2\beta} . \tag{4.3.48}$$

One may check that

$$\Pr(s_1 = \alpha) = \sum_{s_2} \Pr(s_1 = \alpha, s_2).$$

With the help of the pair distribution function we can also compute the average size of a block of units in state α. $N \Pr(s_1 = \alpha, s_2 = \beta)$ is the average number of nearest neighbors pairs in different states $\alpha \neq \beta$. This is the same as the average number of transitions from a block of αs to a block of βs, which is also the same as the average number of blocks. Therefore the average size of a block of αs, $\bar{N}(\alpha)$, is given by

$$\bar{N}(\alpha) = \frac{P(s_1 = \alpha)}{P(s_1 = \alpha, s_2 = \beta)} = \frac{a_{1\alpha}^2 \lambda_1}{P_{\alpha\beta} a_{1\alpha} a_{1\beta}} = \frac{\lambda_1}{\lambda_1 - P_{\alpha\alpha}} .$$

4.3.2.2. Next-Nearest Neighbors: $j - i = 2$

From (4.3.42) we have, for the special case $j = 3$,

$$\Pr(s_1 = \alpha, s_3 = \beta) = \frac{1}{\lambda_1^M} \sum_k \langle \alpha | a_k \rangle \lambda_k^2 \langle a_k | \beta \rangle \sum_l \langle \beta | a_l \rangle \lambda_l^{M-2} \langle a_l | \alpha \rangle. \tag{4.3.49}$$

As before, we take the limit

$$\lim_{M \to \infty} \frac{1}{\lambda_1^2} \sum_l \langle \beta | a_l \rangle \left(\frac{\lambda_l}{\lambda_1} \right)^{M-2} \langle a_l | \alpha \rangle = \frac{1}{\lambda_1^2} \langle \beta | a_1 \rangle \langle a_1 | \alpha \rangle, \tag{4.3.50}$$

and following similar steps to (4.3.46) we have

$$\langle \alpha | \mathbf{P}^2 | \beta \rangle = \sum_{kl} \langle \alpha | a_k \rangle \langle a_k | \mathbf{P}^2 | a_l \rangle \langle a_l | \beta \rangle$$

$$= \sum_{kl} \langle \alpha | a_k \rangle \langle a_k | \lambda_l^2 | a_l \rangle \langle a_l | \beta \rangle$$

$$= \sum_{kl} \langle \alpha | a_k \rangle \lambda_l^2 \delta_{k,l} \langle a_l | \beta \rangle$$

$$= \sum_k \langle \alpha | a_k \rangle \lambda_k^2 \langle a_k | \beta \rangle. \tag{4.3.51}$$

Using (4.3.51) in (4.3.49), we obtain

$$\Pr(s_1 = \alpha, s_3 = \beta) = \frac{\langle \alpha | \mathbf{P}^2 | \beta \rangle a_{1\alpha} a_{1\beta}}{\lambda_1^2}, \tag{4.3.52}$$

which is the analogue of (4.3.47).

Another useful form of this distribution function is obtained by writing the sum over k in (4.3.49) explicitly:

$$\Pr(s_1 = \alpha, s_3 = \beta) = \frac{\langle\beta|a_1\rangle\langle a_1|\alpha\rangle\{\langle\alpha|a_1\rangle\lambda_1^2\langle a_1|\beta\rangle + \langle\alpha|a_2\rangle\lambda_2^2\langle a_2|\beta\rangle\}}{\lambda_1^2}$$

$$= a_{1\alpha}^2 a_{1\beta}^2 + \left(\frac{\lambda_2}{\lambda_1}\right)^2 a_{1\alpha}a_{1\beta}a_{2\alpha}a_{2\beta}, \tag{4.3.53}$$

which is the analogue of (4.3.48).

4.3.2.3. Any Finite j

Repeating the same steps as in the previous example, we can obtain the following results for the pair distribution function for two units $j - 1$ units apart.

$$\Pr(s_1 = \alpha, s_j = \beta) = \frac{\langle\alpha|\mathbf{P}^{j-1}|\beta\rangle a_{1\alpha}a_{1\beta}}{\lambda_1^{j-1}}$$

$$= a_{1\alpha}^2 a_{1\beta}^2 + \left(\frac{\lambda_2}{\lambda_1}\right)^{j-1} a_{1\alpha}a_{1\beta}a_{2\alpha}a_{2\beta}, \tag{4.3.54}$$

which is the analogue of (4.3.53) and (4.3.48).

Note that in each case the pair distribution function has the same first term, which can be identified as (see 4.3.17)

$$a_{1\alpha}^2 a_{1\beta}^2 = \Pr(s_1 = \alpha)\,\Pr(s_2 = \beta). \tag{4.3.55}$$

This would have been the pair distribution function for two independent units. The second term on the rhs of (4.3.54) measures the correlation between the two units. Taking as an example the lattice gas of section 4.2.2, we have the explicit solutions (see 4.2.40)

$$\lambda_{1,2} = \frac{1}{2}[(1 + \lambda q) \pm \sqrt{(1 - \lambda q)^2 + 4\lambda}] \tag{4.3.56}$$

$$a_{1\alpha}^2 = \frac{\lambda q - \lambda_1}{1 + \lambda q - 2\lambda_1} \tag{4.3.57}$$

$$a_{1\beta}^2 = 1 - a_{1\alpha}^2. \tag{4.3.58}$$

Note that λ_1 and λ_2 are the eigenvalues of $\mathbf{P}$, and λ is the activity of the lattice gas (α = empty, β = occupied site).

If there are no interactions, then $q = 1$, and

$$\lambda_1 = 1 + \lambda$$

$$\lambda_2 = 0$$

$$a_{1\alpha}^2 = \frac{1}{1 + \lambda} \tag{4.3.59}$$

$$a_{1\beta}^2 = \frac{\lambda}{1 + \lambda}.$$

Clearly in this case there is no correlation between the two units, and (4.3.54) reduces to

$$\Pr(s_1 = \alpha, s_j = \beta) = \frac{\lambda}{(1 + \lambda)^2} = a_{1\alpha}^2 a_{1\beta}^2. \qquad (4.3.60)$$

A second limit is obtained for interacting molecules $q \neq 1$, but at infinite dilution $\lambda \to 0$, in which case

$$\lambda_1 = 1, \qquad \lambda_2 = 0, \qquad a_{1\alpha}^2 = 1, \qquad a_{1\beta}^2 = 0. \qquad (4.3.61)$$

Again we have no correlation between the units, and

$$\frac{\Pr(s_1 = \alpha, s_j = \beta)}{\Pr(s_1 = \alpha)\,\Pr(s_j = \beta)} = 1. \qquad (4.3.62)$$

This may be referred to as the ideal-gas limit.

The third limiting case is obtained for any q and λ, but when j is very large. Since $\lambda_1 > \lambda_2$, then for finite q and λ, the second term on the rhs of (4.3.54) will tend to zero; hence

$$\Pr(s_1 = \alpha, s_j = \beta) = \Pr(s_1 = \alpha)\,\Pr(s_j = \beta); \qquad (4.3.63)$$

i.e., the two units, at large separation, will become uncorrelated. We shall see in Chapter 5 the analogues of all these three limiting cases in connection with the theory of liquids.

4.3.3. Triplet and Higher-Order Distribution Functions

By a straightforward generalization of the arguments given above, we can compute any higher-order distribution function. For instance, for a consecutive triplet of units, we have

$$\Pr(s_1 = \alpha, s_2 = \beta, s_3 = \gamma) = \frac{1}{\lambda_1^M} \sum_{\mathbf{s}} \Pr(s_1, s_2, \ldots, s_M)\,\delta_{s_1,\alpha}\delta_{s_2,\beta}\delta_{s_3,\gamma}$$

$$= \frac{1}{\lambda_1^M} \sum_{s_1,s_2,s_3} \langle s_1|\mathbf{P}|s_2\rangle\langle s_2|\mathbf{P}|s_3\rangle\langle s_3|\mathbf{P}^{M-2}|s_1\rangle\,\delta_{s_1,\alpha}\delta_{s_2,\beta}\delta_{s_3,\gamma}$$

$$= \frac{1}{\lambda_1^M} \langle \alpha|\mathbf{P}|\beta\rangle\langle \beta|\mathbf{P}|\gamma\rangle\langle \gamma|\mathbf{P}^{M-2}|\alpha\rangle, \qquad (4.3.64)$$

and for $M \to \infty$ we have

$$\frac{1}{\lambda_i^M} \langle \gamma|\mathbf{P}^{M-2}|\alpha\rangle = \frac{1}{\lambda_1^M} \sum_{k,l} \langle \gamma|a_k\rangle\langle a_k|\mathbf{P}^{M-2}|a_l\rangle\langle a_l|\alpha\rangle$$

$$= \frac{1}{\lambda_1^M} \sum_{k} \langle \gamma|a_k\rangle\lambda_k^{M-2}\langle a_k|\alpha\rangle$$

$$\xrightarrow[M\to\infty]{} \frac{\langle \gamma|a_1\rangle\langle a_1|\alpha\rangle}{\lambda_1^2}. \qquad (4.3.65)$$

From (4.3.64) and (4.3.65) we obtain

$$\Pr(s_1 = \alpha, s_2 = \beta, s_3 = \gamma) = \frac{1}{\lambda_1^2} P_{\alpha\beta} P_{\beta\gamma} \langle a_1 | \alpha \rangle \langle a_1 | \gamma \rangle. \tag{4.3.66}$$

This should be compared with (4.3.47). This result can be expressed in terms of the singlet and pair distribution functions as follows:

$$\Pr(s_1 = \alpha, s_2 = \beta, s_3 = \gamma) = \frac{P_{\alpha\beta} \langle a_1 | \alpha \rangle \langle a_1 | \beta \rangle}{\lambda_1} \frac{P_{\beta\gamma} \langle a_1 | \beta \rangle \langle a_1 | \gamma \rangle}{\lambda_1} \frac{1}{\langle a_1 | \beta \rangle^2}$$

$$= \frac{\Pr(s_1 = \alpha, s_2 = \beta) \Pr(s_2 = \beta, s_3 = \gamma)}{\Pr(s_2 = \beta)}. \tag{4.3.67}$$

This factoring of the triplet distribution function into pair and single distribution functions is a characteristic feature of the one-dimensional model. In terms of conditional probability, the same result can be stated as follows:

$$\Pr(s_3 = \gamma / s_1 = \alpha, s_2 = \beta) = \frac{\Pr(s_1 = \alpha, s_2 = \beta, s_3 = \gamma)}{\Pr(s_1 = \alpha, s_2 = \beta)}$$

$$= \frac{\Pr(s_2 = \beta, s_3 = \gamma)}{\Pr(s_2 = \beta)}$$

$$= \Pr(s_3 = \gamma / s_2 = \beta). \tag{4.3.68}$$

In other words, the probability of finding unit 3 in state γ, given that unit 2 is in state β *and* unit 1 in state α, is the same as the probability of finding unit 3 in γ given that unit 2 is in β. The information about unit 1 does not affect the conditional probability of the state of unit 3.

This result is a fundamental property of a Markov chain. It can be generalized to any number of units. If we know the states of units $1, 2, \ldots, k$ and we ask for the probability of occurrence of a particular state at unit $k + 1$, all we need to know is the state of the previous unit. We can ignore the "far past" and retain only the "near past," i.e.,

$$\Pr(s_{k+1} = \gamma / s_1 = \alpha, s_2 = \beta, \ldots, s_k = \alpha) = \Pr(s_{k+1} = \gamma / s_k = \alpha). \tag{4.3.69}$$

It should be noted that by "near past" we mean the information available on the unit *closest* to the $k + 1$ unit. For example, if we know the states of units $1, 2, \ldots, k - 2$, then

$$\Pr(s_{k+1} = \gamma / s_1 = \alpha, s_2 = \beta, \ldots, s_{k-2} = \alpha) = \Pr(s_{k+1} = \gamma / s_{k-2} = \alpha); \tag{4.3.70}$$

i.e., we can erase all the information except the latest—here, the state of the $k - 2$ unit.

The result (4.3.67) can be generalized to any number of units, which we write in a shorthand notation as follows:

$$\Pr(1, 2, 3, \ldots, k) = \Pr(k/k - 1) \Pr(k - 1/k - 2) \Pr(k - 2/k - 3) \cdots \Pr(2/1), \tag{4.3.71}$$

where the numbers i stand for the state of unit i. Thus we can conclude that the joint probability of finding k consecutive units in some specific state is determined only by the conditional probabilities of the nearest neighbors pairs.

In Eq. (4.3.68) we are given $s_1 = \alpha$ and $s_2 = \beta$ and we are interested in the probability of $s_3 = \gamma$. But suppose that we are given only $s_1 = \alpha$ and no information on the state of unit 2; what is the conditional probability of finding $s_3 = \gamma$ given only $s_1 = \alpha$?

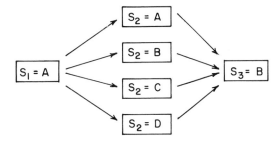

FIGURE 4.1. All possible routes from $s_1 = A$ to $s_3 = B$ through the intermediate unit 2.

To answer this question, we use the following procedure:

$$\Pr(s_3 = \gamma / s_1 = \alpha) = \frac{\Pr(s_3 = \gamma, s_1 = \alpha)}{\Pr(s_1 = \alpha)}$$

$$= \frac{\sum\limits_{s_2} \Pr(s_3 = \gamma, s_2, s_1 = \alpha)}{\Pr(s_1 = \alpha)}$$

$$= \sum_{s_2} \Pr(s_3 = \gamma / s_2) \Pr(s_2 / s_1 = \alpha). \qquad (4.3.72)$$

Thus in order to know the conditional probability of the state of unit 3 given the state of unit 1, we must take all possible routes leading from $s_1 = \alpha$ to $s_3 = \gamma$. For instance, if the units can be in four possible states, A, B, C, D, then given, say, $s_1 = A$ and $s_3 = B$, the sum in (4.3.72) is over all possible states of s_2. Each term in the sum corresponds to one route from unit 1 to unit 3. This is shown schematically in Fig. 4.1.

4.3.4. Correlation Functions

For any j it is useful to define the pair correlation function between the states of two units at a distance of $j - 1$ apart by

$$g(s_1 = \alpha, s_j = \beta) = \frac{\Pr(s_1 = \alpha, s_j = \beta)}{\Pr(s_1 = \alpha) \Pr(s_j = \beta)} = \frac{\langle \alpha | \mathbf{P}^{j-1} | \beta \rangle a_{1\alpha} a_{1\beta}}{\lambda_1^{j-1} a_{1\alpha}^2 a_{1\beta}^2}. \qquad (4.3.73)$$

This quantity measures the correlation between the two units or the extent of dependence between the two units. For $j = 2$, we have

$$g(s_1 = \alpha, s_2 = \beta) = \frac{P_{\alpha\beta}}{\lambda_1 a_{1\alpha} a_{1\beta}}, \qquad (4.3.74)$$

and for very large j we have

$$g(s_1 = \alpha, s_j = \beta) \to 1. \qquad (4.3.75)$$

The pair correlation function can also be defined in terms of the conditional probability of finding the jth unit in state β, given that the first unit, 1, is in state α; i.e.,

$$\Pr(s_j = \beta / s_1 = \alpha) = \frac{\Pr(s_j = \beta, s_1 = \alpha)}{\Pr(s_1 = \alpha)} = g(s_1 = \alpha, s_j = \beta) \Pr(s_j = \beta). \qquad (4.3.76)$$

Thus the correlation function gives the correction to the probability of finding unit j at state β, when it is known that unit 1 is at state α.

The triplet correlation function for three consecutive units is defined by

$$g(s_1 = \alpha, s_2 = \beta, s_3 = \gamma) = \frac{\Pr(s_1 = \alpha, s_2 = \beta, s_3 = \gamma)}{\Pr(s_1 = \alpha) \Pr(s_2 = \beta) \Pr(s_3 = \gamma)}. \qquad (4.3.77)$$

It follows from (4.3.71) that

$$g(s_1 = \alpha, s_2 = \beta, s_3 = \gamma) = g(s_1 = \alpha, s_2 = \beta)g(s_2 = \beta, s_3 = \gamma). \qquad (4.3.78)$$

This result is sometimes referred to as the superposition of the triplet correlation function. However, it is different from the superposition approximation normally used in the theory of liquids. The latter states that (see also section 3.4.4 and Chapter 5):

$$g(s_1 = \alpha, s_2 = \beta, s_3 = \gamma) = g(s_1 = \alpha, s_2 = \beta)g(s_2 = \beta, s_3 = \gamma)g(s_1 = \alpha, s_3 = \gamma). \qquad (4.3.79)$$

Clearly, because of (4.3.78), the superposition approximation as defined in (4.3.79) does not hold. This is because the pair correlation function between two next-nearest neighbors is not unity.

Indeed, from (4.3.72) we find

$$g(s_1 = \alpha, s_3 = \gamma) = \frac{\Pr(s_1 = \alpha, s_3 = \gamma)}{\Pr(s_1 = \alpha) \Pr(s_3 = \gamma)} = \sum_{s_2} g(s_1 = \alpha, s_2) \Pr(s_2)g(s_3 = \gamma, s_2). \qquad (4.3.80)$$

Thus the correlation between units 1 and 3 is the sum over all routes from $s_1 = \alpha$ to $s_3 = \gamma$ as given in (4.3.80), $\Pr(s_2)$ is the singlet distribution function for unit 2. Results (4.3.78) and (4.3.80) may be easily generalized. For instance, the correlation function for four consecutive units, say 1, 2, 3, 4, using a shorthand notation, is

$$g(1, 2, 3, 4) = g(1, 2)g(2, 3)g(3, 4). \qquad (4.3.81)$$

Again this superposition is different from

$$g(1, 2, 3, 4) = g(1, 2)g(2, 3)g(3, 4)g(1, 3)g(1, 4)g(2, 4). \qquad (4.3.82)$$

The pair distribution between two units at distance $j - 1 = 3$ apart is, by generalization of (4.3.80),

$$g(s_1 = \alpha, s_4 = \gamma) = \frac{\Pr(s_1 = \alpha, s_4 = \gamma)}{\Pr(s_1 = \alpha) \Pr(s_4 = \gamma)}$$

$$= \frac{\sum_{s_2 s_3} \Pr(s_1 = \alpha, s_2, s_3, s_4 = \gamma)}{\Pr(s_1 = \alpha) \Pr(s_4 = \gamma)}$$

$$= \sum_{s_2 s_3} g(s_1 = \alpha, s_2, s_3, s_4 = \gamma) \Pr(s_2) \Pr(s_3)$$

$$= \sum_{s_2 s_3} g(s_1 = \alpha, s_2) \Pr(s_2)g(s_2, s_3) \Pr(s_3)g(s_3, s_4 = \gamma). \qquad (4.3.83)$$

4.3.5. Some Examples for the Lattice Gas

We consider the model of section 4.2.2. Each unit can be in one of two states, which we refer to as a "hole" h for an empty unit or w for an occupied state. In this model, the 2×2 matrix is

$$\mathbf{P} = \begin{pmatrix} 1 & \lambda^{1/2} \\ \lambda^{1/2} & \lambda q \end{pmatrix}, \tag{4.3.84}$$

where $q = \exp(-\beta W_{11})$, with W_{11} the interaction between two nearest-neighbor particles. The two eigenvalues are (4.2.40)

$$\gamma_{\pm} = \frac{(1 + \lambda q) \pm \sqrt{(1 - \lambda q)^2 + 4\lambda}}{2}, \tag{4.3.85}$$

where λ is the activity of the molecules.

The singlet molecular distribution for the h and w states is

$$x(h) = \Pr(s_1 = h) = \frac{\lambda}{\lambda + (1 - \gamma_1)^2} \tag{4.3.86}$$

$$x(w) = \Pr(s_1 = w) = 1 - x(h). \tag{4.3.87}$$

In the limit of $\lambda \to 0$, we have

$$x(h) \to 1 \quad \text{and} \quad x(w) \to 0, \tag{4.3.88}$$

as expected. On the other hand, in the limit $\lambda \to \infty$ we find

$$x(h) \to 0 \quad \text{and} \quad x(w) \to 1. \tag{4.3.89}$$

Note the symmetrical roles of h and w in these two limits. We shall see below that this symmetry is more general.

For the nearest-neighbor pair distribution functions we have the following three cases:

$$\Pr(s_1 = h, s_2 = h) = \frac{a_{1h}^2}{\gamma_1} \tag{4.3.90}$$

$$\Pr(s_1 = h, s_2 = w) = \frac{\lambda^{1/2} a_{1h} a_{1w}}{\gamma_1} \tag{4.3.91}$$

$$\Pr(s_1 = w, s_2 = w) = \frac{\lambda q a_{1w}^2}{\gamma_1}, \tag{4.3.92}$$

where

$$a_{1h}^2 = x(h) \quad \text{and} \quad a_{1w}^2 = x(w) = 1 - x(h). \tag{4.3.93}$$

In the limit $\lambda \to 0$, $\gamma_1 \to 1 + \lambda$, hence

$$\Pr(s_1 = h, s_2 = h) \to 1 \tag{4.3.94}$$

$$\Pr(s_1 = h, s_2 = w) \to 0 \tag{4.3.95}$$

$$\Pr(s_1 = w, s_2 = w) \to 0, \tag{4.3.96}$$

which is intuitively clear. But note that the correlation functions are finite in this limit, namely,

$$g(s_1 = h, s_2 = h) \rightarrow 1 \tag{4.3.97}$$

$$g(s_1 = h, s_2 = w) \rightarrow 1 \tag{4.3.98}$$

$$g(s_1 = w, s_2 = w) \rightarrow q. \tag{4.3.99}$$

In particular, the limit of the w—w pair correlation function should be noted. Although the probability distribution (4.3.96) is zero in this limit, the correlation between the two particles is finite. We have seen a similar situation in sections 3.2 and 3.3 and we shall see the analogue of (4.3.99) for liquids in Chapter 5.

Although we shall not introduce the force between the particles—in this model the sites are fixed—it is useful to define the potential of average force by

$$W(s_1, s_2) = -kT \ln g(s_1, s_2). \tag{4.3.100}$$

In terms of the potential of average force the meaning of relations (4.3.97) to (4.3.99) is clear. The work required to bring a pair of holes, or one hole and one particle, to adjacent sites is zero, but for two particles this work is simply the interaction energy W_{11}.

Turning to the $\lambda \rightarrow \infty$ limit, we find

$$\Pr(s_1 = h, s_2 = h) \rightarrow 0 \tag{4.3.101}$$

$$\Pr(s_1 = h, s_2 = w) \rightarrow 0 \tag{4.3.102}$$

$$\Pr(s_1 = w, s_2 = w) \rightarrow 1, \tag{4.3.103}$$

which is as expected intuitively. The corresponding correlation functions are

$$g(s_1 = h, s_2 = h) \rightarrow q \tag{4.3.104}$$

$$g(s_1 = h, s_2 = w) \rightarrow 1 \tag{4.3.105}$$

$$g(s_1 = w, s_2 = w) \rightarrow 1. \tag{4.3.106}$$

Note the symmetry between holes and particles in (4.3.97)–(4.3.99) and (4.3.104)–(4.3.106). The work required to bring two *holes* from infinite separation to nearest neighbors is also W_{11}.

The probability of finding an aggregate, or a cluster of j consecutive particles, is

$$\Pr(s_1 = w \cdots s_j = w) = x(w)^j \prod_{i=1}^{j-1} g(s_i = w, s_{i+1} = w)$$

$$= x(w)(\lambda q/\gamma_1)^{j-1}. \tag{4.3.107}$$

The probability of finding a hole of size j, i.e., a consecutive sequence of j empty sites, is

$$\Pr(s_1 = h \cdots s_j = h) = x(h)^j \prod_{i=1}^{j-1} g(s_i = h, s_{i+1} = h)$$

$$= x(h)\gamma_1^{1-j}. \tag{4.3.108}$$

Clearly, in the $\lambda \to 0$ limit

$$\Pr(s_1 = w \cdots s_j = w) \to 0 \quad \text{and} \quad \Pr(s_1 = h \cdots s_j = h) \to 1, \qquad (4.3.109)$$

and in the $\lambda \to \infty$ limit we have

$$\Pr(s_1 = w \cdots s_j = w) \to 1 \quad \text{and} \quad \Pr(s_1 = h \cdots s_j = h) \to 0. \qquad (4.3.110)$$

This symmetrical behavior is similar to that which we encountered for the special case (4.3.88) and (4.3.89) and similarly for the pair distribution function.

The corresponding correlation functions are

$$g(s_1 = w \cdots s_j = w) = \left[\frac{\lambda q}{\gamma_1 x(w)} \right]^{j-1} \qquad (4.3.111)$$

$$g(s_1 = h \cdots s_j = h) = \left[\frac{1}{\gamma_1 x(h)} \right]^{j-1}. \qquad (4.3.112)$$

Here the limiting results are the following: For $\lambda \to 0$, we have

$$g(s_1 = w \cdots s_j = w) \to q^{j-1} \qquad g(s_1 = h \cdots s_j = h) \to 1, \qquad (4.3.113)$$

and for $\lambda \to \infty$ we find

$$g(s_1 = w \cdots s_j = w) \to 1 \qquad g(s_1 = h \cdots s_j = h) \to q^{j-1}. \qquad (4.3.114)$$

The unexpected result is the last one for the correlation between j consecutive holes. In terms of the potential of average force the result for the limit $\lambda \to 0$ is obvious. The work required to produce an aggregate of j particle is simply $(j-1)W_{11}$. If $W_{11} < 0$, this work is negative. On the other hand, the work required to bring j single-sized holes to form a j-sized hole, or a cavity of size j, is also $(j-1)W_{11}$; i.e., energy is released ($W_{11} < 0$) in the process.

The last results can be rationalized with the help of Fig. 4.2. In the $\lambda \to 0$ limit, we start with exactly two particles in otherwise all empty sites. When this pair is brought to contact, we form one w—w bond; hence the work is W_{11}. On the other hand, in the limit $\lambda \to \infty$, we start with a system of two single-site holes in an otherwise fully occupied system. When this pair of holes is brought to contact, it produces an h—h bond. This process also creates one w—w bond, as can be seen from Fig. 4.2. Therefore when forming a j-sized cavity from j single-sized cavities, we form one w—w bond for each pair of consecutive holes. Hence the process involves the formation of $j-1$ w—w bonds.

The potential of average force between j consecutive particles, at any λ, is defined by

$$W(s_1 = w \cdots s_j = w) = -kT \ln g(s_1 = w \cdots s_j = w). \qquad (4.3.115)$$

FIGURE 4.2. (a) Two particles w are brought from large separation to adjacent sites to form one w—w bond. (b) Two "holes" are brought from large separation to form a h—h "bond."

Likewise, the potential of average force between j consecutive holes is

$$W(s_1 = h \cdots s_j = h) = -kT \ln g(s_1 = h \cdots s_j = h). \tag{4.3.116}$$

Finally, the work required to form a cavity of size j (i.e., j consecutive holes) is

$$W(\text{cavity}) = -kT \ln \Pr(s_1 = h \cdots s_j = h)$$

$$= -kT \ln[x(h)^j g(s_1 = h \cdots s_j = h)]. \tag{4.3.117}$$

4.3.6. An Alternative Way to Obtain the Molecular Distribution Functions from the Partition Function

Consider again the general form of the PF as written in 4.3.1:

$$\Gamma = \sum_s \prod_{i=1}^{M} P_{s_i s_{i+1}}. \tag{4.3.118}$$

We now choose a specific factor $P_{\alpha\beta}$ and take the following derivative

$$\frac{\partial \ln \Gamma}{\partial \ln P_{\alpha\beta}} = \frac{P_{\alpha\beta}}{\Gamma} \sum_s \frac{\partial}{\partial P_{\alpha\beta}} (P_{s_1 s_2} P_{s_2 s_3} \cdots P_{s_M s_1})$$

$$= \frac{P_{\alpha\beta}}{\Gamma} \sum_s M \delta_{s_1,\alpha} \delta_{s_2,\beta} P_{s_2 s_3} \cdots P_{s_M s_1}$$

$$= \frac{M}{\Gamma} \sum_s P_{s_1 s_2} P_{s_2 s_3} \cdots P_{s_M s_1} \delta_{s_1,\alpha} \delta_{s_2,\beta}$$

$$= M \Pr(s_1 = \alpha, s_2 = \beta). \tag{4.3.119}$$

In the first step we differentiate each term in the product $P_{s_1 s_2} \cdots P_{s_M s_1}$ with respect to $P_{\alpha\beta}$. This produces M terms, in each of which one factor has been replaced by the derivative

$$\frac{\partial P_{s_i s_{i+1}}}{\partial P_{\alpha\beta}} = \delta_{s_i,\alpha} \delta_{s_i,\beta}. \tag{4.3.120}$$

Since all of these terms are equivalent, we simply choose one specific term and multiply it by M. This gives the expression which is identical to the definition of the pair distribution function, Eq. (4.3.36).

This procedure is more convenient when we have solved the PF in terms of the matrix elements $P_{\alpha\beta}$, so that the pair distribution function can be computed directly from the final form of the PF. For example, having obtained the PF in the form

$$\Gamma = \lambda_1^M, \tag{4.3.121}$$

where λ_1 is the largest eigenvalue of the corresponding matrix of the system, we may compute

$$\Pr(s_1 = \alpha, s_2 = \beta) = P_{\alpha\beta} \frac{\partial \ln \lambda_1}{\partial P_{\alpha\beta}}. \tag{4.3.122}$$

Once we have the pair distribution function, we can derive the singlet distribution function by summation over all possible values of s_2:

$$\Pr(s_1 = \alpha) = \sum_{s_2} \Pr(s_1 = \alpha, s_2). \tag{4.3.123}$$

4.4. SOME GENERALIZATIONS OF THE ISING MODEL

In this section we extend the treatment of the one-dimensional Ising model in two directions: first, for units having more than two states; second, for units interacting with both nearest neighbors as well as with next-nearest neighbors.

4.4.1. Lattice Gas Model of a Binary Mixture

Consider a linear system of M sites, N_A of which are occupied by molecules of type A and N_B sites are occupied by molecules of type B. If we let $N = N_A = N_B$, then we have a system the PF of which is isomorphous to the simple Ising model studied in section 4.1; i.e., each site can be in either one of two states, A or B, depending on the type of molecule occupying the site. In this section we treat a slightly more general case. Each site can be in one of three states: empty, occupied by A, and occupied by B. Thus we have M sites, N_A sites occupied by A and N_B occupied by B. But $N_A + N_B \leq M$.

The interaction energies for any pair of sites are

$$W_{00} = 0 \qquad W_{A0} = W_{0A} = W_{0B} = W_{B0} = 0$$
$$W_{AA}, W_{BB}, W_{AB} = W_{BA}. \tag{4.4.1}$$

The canonical PF of the system is

$$Q(T, M, N_A, N_B) = \sum_{\text{all states}} \exp(-\beta E_i). \tag{4.4.2}$$

The energy levels are determined by the number of nearest-neighbor pairs of each kind.

$$E = N_{AA}W_{AA} + N_{BB}W_{BB} + N_{AB}W_{AB}. \tag{4.4.3}$$

The sum (4.4.2) is over all possible states consistent with the fixed values of N_A, and N_B.

We now write the grand PF for a system opened with respect to both A and B, i.e.,

$$\Xi(T, M, \lambda_A, \lambda_B) = \sum_{\substack{N_A = 0 \\ N_A + N_B \leq M}}^{M} \sum_{N_B = 0}^{M} Q(T, M, N_A, N_B)\lambda_A^{N_A}\lambda_B^{N_B}$$
$$= \sum_{\mathbf{s}} \langle s_1|\mathbf{P}|s_2\rangle\langle s_2|\mathbf{P}|s_3\rangle \cdots \langle s_M|\mathbf{P}|s_1\rangle. \tag{4.4.4}$$

Thus for each configuration $\mathbf{s}$ of the system, we have a factor λ_A for each site occupied by A, a factor λ_B for a site occupied by B, and a factor $\lambda_0 = 1$ for an empty site. For each pair of neighbor sites we have a factor $\exp(-\beta W_{s_i s_{i+1}})$, where the interaction energies are given in (4.1.1).

Thus the matrix elements are

$$\langle s_i|\mathbf{P}|s_{i+1}\rangle = \lambda_{s_i}^{1/2} \exp(-\beta W_{s_i s_{i+1}})\lambda_{s_{i+1}}^{1/2}. \tag{4.4.5}$$

The sum over all configurations gives the usual trace of the matrix $\mathbf{P}$.

$$\Xi(T, M, \lambda_A, \lambda_B) = \sum_{s_1, s_2, \ldots, s_M} \langle s_1|\mathbf{P}|s_2\rangle\langle s_2|\mathbf{P}|s_3\rangle \cdots \langle s_M|\mathbf{P}|s_1\rangle$$
$$= \sum_{s_1} \langle s_1|\mathbf{P}^M|s_1\rangle = \mathrm{Tr}(\mathbf{P}^M), \tag{4.4.6}$$

where $\mathbf{P}$ is the 3×3 matrix:

$$\mathbf{P} = \begin{pmatrix} 1 & \lambda_A^{1/2} & \lambda_B^{1/2} \\ \lambda_A^{1/2} & \lambda_A q_{AA} & (\lambda_A \lambda_B)^{1/2} q_{AB} \\ \lambda_B^{1/2} & (\lambda_A \lambda_B)^{1/2} q_{AB} & \lambda_B q_{BB} \end{pmatrix}, \tag{4.4.7}$$

where $q_{\alpha\beta} = \exp(-\beta W_{\alpha\beta})$.

In order to solve for the largest eigenvalue we need to find the solution of the secular equation

$$|\mathbf{P} - \lambda \mathbf{I}| = 0, \tag{4.4.8}$$

where $\mathbf{I}$ is a 3×3 unit matrix. The parameter λ should not be confused with the absolute activities λ_A or λ_B.

Equation (4.4.8) is a cubic equation in λ, and normally has three different roots. Since the matrix $\mathbf{P}$ is real and symmetric, the roots are always real and different. The thermodynamic function associated with the grand PF would normally be $\exp(-\beta PV)$, but since the "volume" variable here is M rather than V, it is convenient to define the chemical potential of the sites, μ_M by

$$\exp(-\beta M \mu_M) = \Xi(T, M, \lambda_A, \lambda_B). \tag{4.4.9}$$

Using the definition of Ξ in (4.4.4), we can write also

$$\Xi(T, M, \lambda_A, \lambda_B) = \sum_{N_A} \sum_{N_B} \exp(-\beta A + \beta N_A \mu_A + \beta N_B \mu_B)$$

$$\approx \exp[-\beta(A - N_A \mu_A - N_B \mu_B)], \tag{4.4.10}$$

where in the last equality we take as the average or the most probable values of N_A and N_B. Comparing (4.4.9) and (4.4.10), we obtain

$$A = M\mu_M + N_A \mu_A + N_B \mu_B, \tag{4.4.11}$$

which is consistent with the treatment of the sites as the third species of our system.

4.4.2. Multiple but Degenerate States

In the previous section, we saw that for a system consisting of three-state units we need a 3×3 matrix. It is clear that for an m-state unit, we need an $m \times m$ matrix. This is so because there are $m \times m$ different pairs of units, i.e., $m \times m$ different bonds. In this section we consider one example where the matrix can be reduced considerably even when there are many possible states available for each unit.

Consider the following model as depicted in Fig. 4.3. The particles can assume $m + 1$ orientations with respect to the 1-D line of the system. If the particles are oriented along the line, say with $\phi = 0$, we call their state A. All other states are referred to as B states and denoted by $B_1, B_2, \ldots, B_m$. Altogether there are $m + 1$ states, and in general we need an $(m + 1) \times (m + 1)$ matrix to describe all possible pairs of particles.

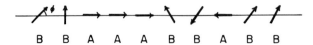

B   B   A   A   A   B   B   A   B   B

FIGURE 4.3. 1-D system of units having different orientations with respect to the 1-D line.

We now assume that the interactions between A and B_i or between different B_is are independent of the index i, i.e.,

$$W_{AB_i} = W_{AB} \qquad \text{for all } i = 1 \cdots m$$

$$W_{B_iB_j} = W_{BB} \qquad \text{for all } i, j = 1 \cdots m. \tag{4.4.12}$$

As an example, suppose that there are two states of B, B_1 and B_2. Then the secular equation for such a system is

$$\begin{vmatrix} P_{AA} - \lambda & P_{AB} & P_{AB} \\ P_{AB} & P_{BB} - \lambda & P_{BB} \\ P_{AB} & P_{BB} & P_{BB} - \lambda \end{vmatrix} = 0. \tag{4.4.13}$$

It is clear that $\lambda = 0$ is one root of this equation. Substituting $\lambda = 0$ leaves two identical rows or columns in this determinant. Therefore we need to solve only a secular equation of a 2×2 matrix. To obtain the required 2×2 matrix, we manipulate the secular equation (4.4.13) as follows;

$$\begin{vmatrix} P_{AA} - \lambda & P_{AB} & P_{AB} \\ P_{AB} & P_{BB} - \lambda & P_{BB} \\ P_{AB} & P_{BB} & P_{BB} - \lambda \end{vmatrix} = \begin{vmatrix} P_{AA} - \lambda & P_{AB} & 0 \\ P_{AB} & P_{BB} - \lambda & \lambda \\ P_{AB} & P_{BB} & -\lambda \end{vmatrix}$$

$$= \lambda \begin{vmatrix} P_{AA} - \lambda & P_{AB} & 0 \\ P_{AB} & P_{BB} - \lambda & 1 \\ P_{AB} & P_{BB} & -1 \end{vmatrix}$$

$$= \lambda \begin{vmatrix} P_{AA} - \lambda & P_{AB} & 0 \\ 2P_{AB} & 2P_{BB} - \lambda & 0 \\ P_{AB} & P_{BB} & -1 \end{vmatrix}$$

$$= -\lambda \begin{vmatrix} P_{AA} - \lambda & P_{AB} \\ 2P_{AB} & 2P_{BB} - \lambda \end{vmatrix} = 0. \tag{4.4.14}$$

we see that apart from the root $\lambda = 0$, we need to solve the secular equation of a 2×2 matrix.

By generalization of the same steps as in (4.4.14), we can reduce the $(m + 1) \times (m + 1)$ secular equation of our system to

$$|\mathbf{P} - \lambda \mathbf{I}| = \lambda^{m-1} \begin{vmatrix} P_{AA} - \lambda & P_{AB} \\ mP_{AB} & mP_{BB} - \lambda \end{vmatrix} = 0. \tag{4.4.15}$$

Thus we have $m - 1$ roots equal to zero and two more obtained by solving the 2×2 secular equation.

The reason for obtaining such a reduction of the matrix is quite simple. Since the interactions between A and B_i and between the B_is are independent of the index i, we can redefine our system as if it consisted of two-state units A and B but take into account the degeneracy of the state B.

Clearly we could have assumed also that A has, say, an n-fold degeneracy, in which case the corresponding secular equation would be

$$\begin{vmatrix} nP_{AA} - \lambda & nP_{AB} \\ mP_{AB} & mP_{BB} - \lambda \end{vmatrix} = \begin{vmatrix} nP_{AA} - \lambda & \sqrt{nm}\,P_{AB} \\ \sqrt{nm}\,P_{AB} & mP_{BB} - \lambda \end{vmatrix} = 0. \qquad (4.4.16)$$

We now generalize this result to the continuous case. With the same model, we let the orientation angle ϕ in Fig. 4.2 change continuously. We define the two states A and B as follows:

$$\text{state } A: \text{ when } \phi \in \Omega_A$$
$$\text{state } B: \text{ when } \phi \in \Omega_B = 2\pi - \Omega_A. \qquad (4.4.17)$$

The physical motivation for this definition is the following: Suppose that the particles can form a bond only when they are oriented in the same direction; i.e., whenever the two particles have an orientational angle of about $\phi = 0$ or about $\phi = \pi$, they form a bond. We call this region Ω_A, or the bonding region. In all other orientations, they cannot form a bond, and for simplicity we assume that the interaction energy is zero.

The configurational PF of the system is thus

$$Z(T, M) = \int \cdots \int d\phi_1 \cdots d\phi_M \exp[-\beta E(\phi_1 \cdots \phi_M)]$$

$$= \sum_s \int_{s_1} \int_{s_2} \cdots \int_{s_M} d\phi_1 \cdots d\phi_M \exp[-\beta E(s_1, s_2, \ldots, s_M)]. \qquad (4.4.18)$$

In the first term on the rhs of (4.4.18) we have the entire range of configurations for all the units. In the second form we first sort out, for each configuration $\phi_1 \cdots \phi_M$, the corresponding states of all the units, say $s_1, s_2, \ldots, s_M$; then we sum over all possible states of all the units. The range of integration indicated under each integral means either the range Ω_A or Ω_B, according to whether s_i is A or B. For instance, for $M = 2$ we have the identity

$$\int_0^{2\pi} \int_0^{2\pi} d\phi_1\, d\phi_2 = \iint_{AA} + \iint_{AB} + \iint_{BA} + \iint_{BB}. \qquad (4.4.19)$$

For $M = 3$, we have eight terms, and for M units we have 2^M terms in the sum on the rhs of (4.4.18).

Since the interaction energies depend only on the states (A or B) of the two units, and not on the specific angles ϕ_i and ϕ_{i+1}, we can perform all the integrations in (4.4.18) to obtain

$$\int_{s_1} \int_{s_2} \cdots \int_{s_M} d\phi_1 \cdots d\phi_M \exp[-\beta E(s_1 \cdots s_M)]$$

$$= \Omega_{s_1} \Omega_{s_2} \cdots \Omega_{s_M} \exp\left[-\beta \sum_{i=1}^M E(s_i, s_{i+1})\right]. \qquad (4.4.20)$$

From now on we can proceed as in the simple two-state Ising model. We define the general matrix elements

$$\langle s_i | \mathbf{P} | s_{i+1} \rangle = (\Omega_{s_i} \Omega_{s_{i+1}})^{1/2} \exp[-\beta E(s_i, s_{i+1})], \qquad (4.4.21)$$

and the PF is written as

$$Z(T, M) = \text{Tr } \mathbf{P}^M, \qquad (4.4.22)$$

where the 2×2 matrix is

$$\mathbf{P} = \begin{pmatrix} \Omega_A P_{AA} & (\Omega_A \Omega_B)^{1/2} P_{AB} \\ (\Omega_A \Omega_B)^{1/2} P_{AB} & \Omega_B P_{BB} \end{pmatrix}. \qquad (4.4.23)$$

The secular equation of this matrix is the generalization of (4.4.16). We see that the discrete degeneracy in (4.4.16) is replaced by the range of degeneracy in (4.4.23). Suppose now that $P_{AA} = q$ and $P_{AB} = P_{BB} = 1$ and we define the ratio $\eta = \Omega_B/\Omega_A$; the corresponding secular equation is then

$$\begin{vmatrix} \Omega_A q - \lambda & (\Omega_A \Omega_B)^{1/2} \\ (\Omega_A \Omega_B)^{1/2} & \Omega_B - \lambda \end{vmatrix} = 0 \qquad (4.4.24)$$

and the largest eigenvalue is

$$\lambda_1 = \frac{\Omega_A}{2}(q + \eta + \sqrt{(q - \eta)^2 + 4\eta}). \qquad (4.4.25)$$

The equilibrium ratio of the two species is (see Eq. 4.3.34)

$$\frac{x_A}{x_B} = \frac{\Omega_B - \lambda_1}{\Omega_A q - \lambda_1}$$

$$= \frac{\eta - q - \sqrt{(q - \eta)^2 + 4\eta}}{q - \eta - \sqrt{(q - \eta)^2 + 4\eta}} \qquad (4.4.26)$$

$$x_A = \frac{1}{2} + \frac{q - \eta}{2\sqrt{(q - \eta)^2 + 4\eta}}. \qquad (4.4.27)$$

For $q \gg \eta$ we have $x_A \to 1$, but for $q \ll \eta$ and $\eta \gg 1$ we have $x_A \to 0$. This is another example of the competition between energy (through q) and entropy (through η). A plot of x_A as a function of the temperature is shown in Fig. 4.4.

4.4.3. Ising Model with Nearest- and Next-Nearest-Neighbor Interactions

We extend the simplest Ising model, where only nearest-neighbor interactions were assumed, to include next-nearest-neighbor interactions.

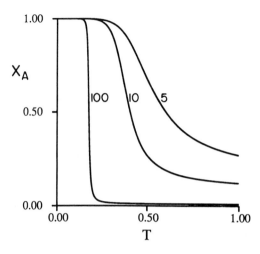

FIGURE 4.4. x_A as a function of temperature (in units of W_{AA}/k) for various values of η, as indicated next to each curve.

Consider again a system with two-states for its units, as depicted in Fig. 4.5. Each unit can have two orientations, A and B. We assume that in addition to pairwise interactions, there are also triplet interactions. We restrict ourselves to one kind of triplet interaction: when three consecutive particles are in state A, there is an additional interaction energy which we denote by W_p. This could occur, for example, when there is a strong polarization effect. Another important case is hydrogen bonding. If two particles in the A orientation form a hydrogen bond with interacting energy W_{AA}, then three successive particles in this orientation would have an interaction energy of $2W_{AA} + W_p$; i.e., the formation of one bond enhances the formation of the second bond ($W_p < 0$). Such a system is often referred to as a cooperative system. Note, however, that even when there are only nearest-neighbor interactions, the system is cooperative in the sense that the conditional probability $\Pr(s_{i+1}/s_i)$ is different from the singlet probability $\Pr(s_{i+1})$. In our case, there is an additional cooperativity in the sense that $\Pr(s_{i+2}/s_{i+1}, s_i)$ is different from $\Pr(s_{i+2}/s_{i+1})$.

The total interaction energy of a system at any particular configuration $\mathbf{s} = s_1 \cdots s_M$ is given by

$$E(\mathbf{s}) = \sum_i E(s_i, s_{i+1}) + \sum_i E(s_{i-1}, s_i, s_{i+1}), \qquad (4.4.28)$$

where now, in addition to the pairwise interactions, we need to consider also triplet interactions, denoted by $E(s_{i-1}, s_i, s_{i+1})$. It is convenient to rewrite $E(\mathbf{s})$ in an equivalent

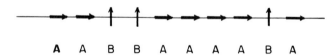

A   A   B   B   A   A   A   A   B   A

FIGURE 4.5. Two-state units A and B with nearest and next-nearest interactions. Three consecutive units in state A have an additional interaction energy W_p.

form as

$$E(\mathbf{s}) = \sum_i \left[\tfrac{1}{2} E(s_i, s_{i+1}) + E(s_{i-1}, s_i, s_{i+1}) + \tfrac{1}{2} E(s_{i-1}, s_i) \right]$$

$$= \sum_i E^*(s_{i-1}, s_i, s_{i+1}). \tag{4.4.29}$$

We assign to the ith unit the energy E^* in such a way that it receives half of the pair interaction from its nearest neighbors, $i-1$ and $i+1$. In addition, we assign the triplet interaction $E(s_{i-1}, s_i, s_{i+1})$ to the particle i. This assignment is arbitrary. The total energy of each configuration $E(\mathbf{s})$ is unaffected by this assignment. We now write the partition function of the system as

$$Q(T, M) = \sum_{\mathbf{s}} \exp[-\beta E(\mathbf{s})]$$

$$= \sum_{\mathbf{s}} \prod_i \exp[-\beta E^*(s_{i-1}, s_i, s_{i+1})]$$

$$= \sum_{\mathbf{s}} \prod_i Q(s_{i-1}, s_i, s_{i+1}), \tag{4.4.30}$$

where we introduce the quantities $Q(s_{i-1}, s_i, s_{i+1})$.

We now wish to rewrite the sum on the rhs of (4.4.30) as the trace of a matrix. To do that we introduce the elements of a 4×4 matrix as follows:

$$P(s_1, s_2, s_2', s_3) = \begin{cases} Q(s_1, s_2, s_3) & \text{if } s_2 = s_2' \\ 0 & \text{if } s_2 \neq s_2', \end{cases} \tag{4.4.31}$$

and conversely for each i,

$$Q(s_{i-1}, s_i, s_{i+1}) = P(s_{i-1}, s_i, s_i, s_{i+1})$$

$$= \sum_{s_i'} P(s_{i-1}, s_i, s_i', s_{i+1}). \tag{4.4.32}$$

These are simple, formal steps. Instead of the eight quantities $Q(s_{i-1}, s_i, s_{i+1})$ for each i, we defined in (4.4.31) sixteen quantities $P(s_{i-1}, s_i, s_i', s_{i+1})$, eight of which are by definition zeros. Thus we have obtained sixteen elements that can be arranged in a 4×4 matrix, which has the form

		3 A	B	A	B
	1 2	A	A	B	B
A A		$Q(A, A, A)$	$Q(A, A, B)$	0	0
A B		0	0	$Q(A, B, A)$	$Q(A, B, B)$
B A		$Q(B, A, A)$	$Q(B, A, B)$	0	0
B B		0	0	$Q(B, B, A)$	$Q(B, B, B)$

$$\mathbf{P} = \qquad\qquad\qquad\qquad\qquad\qquad\qquad\qquad\qquad\qquad\qquad\qquad\qquad\qquad (4.4.33)$$

We see that the matrix contains eight nonzero elements and eight zeros that were introduced by the definition (4.4.31).

The indices of the matrix elements $P(i, j, k, l)$ should be read as follows: i, j from left to right under the column denoted by 1, 2; k, l from the bottom upward in the rows indicated by 2, 3. Thus, $P(A, B, B, A)$ is the element in the second row and third column, and its value is $Q(A, B, A)$ (see 4.4.31). The element $P(A, B, A, A)$ is found in the second row and first column, and its value by definition (4.4.31) is zero.

With definition (4.4.32), we rewrite the partition function as

$$Q(T, M) = \sum_{\mathbf{ss}'} \prod_i P(s_{i-1}, s_i, s_i', s_{i+1}). \tag{4.4.34}$$

Now the sum can be performed by using the rule of matrix multiplication. The rule in this case is

$$\sum_{kl} P(i, j, k, l)P(k, l, m, n) = (\mathbf{P}^2)_{i,j,m,n}. \tag{4.4.35}$$

Thus for $M = 3$, the PF is

$$
\begin{aligned}
Q(T, M = 3) &= \sum_{s_1 s_2 s_3} \sum_{s_1', s_2', s_3'} P(s_1, s_2, s_2', s_3)P(s_2, s_3, s_3', s_1)P(s_3, s_1, s_1', s_2) \\
&= \sum_{s_1 s_2} \sum_{s_2 s_3} P(s_1, s_2, s_2', s_3) \sum_{s_3', s_1'} P(s_2', s_3, s_3', s_1')P(s_3', s_1', s_1, s_2) \\
&= \operatorname{Tr} \mathbf{P}^3
\end{aligned}
\tag{4.4.36}
$$

and, by generalization,

$$Q(T, M) = \operatorname{Tr} \mathbf{P}^M = \sum_{i=1}^{4} \lambda_i^M. \tag{4.4.37}$$

Note that each term in the sum (4.4.36) is zero whenever $s_i \neq s_i'$ for any i. This allows one to exchange primed and unprimed variables without affecting the value of the sum.

In general, we have four different eigenvalues λ_i. A particularly simple case occurs when $W_{AB} = W_{BB} = 0$, but $W_{AA} \neq 0$ and $W_p \neq 0$. For instance, there is hydrogen bonding ($W_{AA} < 0$) and nonadditivity ($W_p < 0$) of the hydrogen bonding, but other interactions are negligible. In this case, the secular equation is

$$
\begin{aligned}
|\mathbf{P} - \lambda \mathbf{I}| &=
\begin{vmatrix}
q_{AA}q_p - \lambda & \sqrt{q_{AA}} & 0 & 0 \\
0 & -\lambda & 1 & 1 \\
\sqrt{q_{AA}} & 1 & -\lambda & 0 \\
0 & 0 & 1 & 1 - \lambda
\end{vmatrix} \\
&= -\lambda
\begin{vmatrix}
q_{AA}q_p - \lambda & \sqrt{q_{AA}} & 0 \\
0 & -\lambda & 1 \\
\sqrt{q_{AA}} & 1 & 1 - \lambda
\end{vmatrix} \\
&= 0,
\end{aligned}
\tag{4.4.38}
$$

where $q_{AA} = \exp(-\beta W_{AA})$ and $q_p = \exp(-\beta W_p)$. Thus, one root is zero and we still have a cubic equation for λ. To explore the behavior of this system we need to solve the cubic equation numerically. We shall return to an equivalent problem in section 4.7. We note here that, in general, if we consider a two-state unit with next-nearest-neighbor interaction, we have eight different configurations for each triplet of consecutive units. This will normally lead to an 8×8 matrix. Of course, for units having more than two states and for interactions that extend beyond the next-nearest neighbors, we shall need to use higher-order matrices.

4.5. ONE-DIMENSIONAL FLUIDS

In this section we proceed from the lattice-type 1-D models to continuous 1-D systems. The latter has the characteristic fluidity of the liquid state.

4.5.1. The Model and Its Solution

Consider a system of N particles in a one-dimensional "box" of length L. The particles are free to move within the limits $(0, L)$ of the system. The location of the center of the ith particle is denoted by X_i, Fig. 4.6.

The total interaction energy of the system is assumed to be the sum of the nearest-neighbor pair interactions, i.e.,

$$U_N(X_1, \ldots, X_N) = \sum_{i=1}^{N-1} U(X_{i+1} - X_i), \tag{4.5.1}$$

where $U(X_{i+1} - X_i)$ is the interaction potential for a pair of consecutive particles i and $i + 1$. We shall always assume that this potential function is infinitely repulsive for $X_{i+1} - X_i \to 0$ and zero for sufficiently long distances.

If we did not assume (4.5.1), i.e., if the potential function U_N is any function symmetrical with respect to interchanging the variables X_i, then the canonical PF of the system is

$$Q(T, L, N) = \frac{1}{N! \Lambda^N} \int_0^L \cdots \int_0^L dX_1 \cdots dX_N \exp(-\beta U_N). \tag{4.5.2}$$

Note that in this case, we have Λ and not Λ^3 for each particle. Here, the integration for each particle ranges from 0 to L. However, in the one-dimensional case, the particles can be put in order in the sense that $X_1 < X_2 < X_3 \cdots X_N$. This allows us to write (4.5.2) in a modified form as follows: For any function $f(X_1 \cdots X_N)$, symmetrical with respect to interchanging the positions of the particles X_i, we can write

$$\int_0^L \cdots \int_0^L dX_1 \cdots dX_N f(X_1 \cdots X_N)$$

$$= N! \int_0^L dX_N \int_0^{X_N} dX_{N-1} \cdots \int_0^{X_2} dX_1 f(X_1 \cdots X_N). \tag{4.5.3}$$

This can be easily verified for small N. For instance, for $N = 2$ the various regions of integrations are indicated in Fig. 4.7.

$$\int_0^L \int_0^L dX_1 \, dX_2 f(X_1, X_2) = \iint_{X_1 > X_2} + \iint_{X_1 < X_2}. \tag{4.5.4}$$

The total region of integration—the square area of Fig. 4.7—is split into two regions: above and below the diagonal line. The integral on the lhs of (4.5.4) is split into the sum

FIGURE 4.6. One-dimensional fluid in a "box" of length L. The location of the center of the ith particle is denoted by X_i.

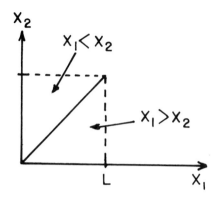

FIGURE 4.7. The two ranges of integration in Eq. (4.5.4).

of two integrals; each corresponds to one of the triangle area of Fig. 4.7. $f(X_1 X_2)$ is presumed to be symmetrical with respect to the two variables $X_1 X_2$. Therefore, for each pair of variables, say $(X_1 = \alpha, X_2 = \beta)$ in one triangle, there is another pair $(X_1 = \beta, X_2 = \alpha)$ for which the integrand has the same value. Therefore, instead of integrating over the entire squared area, it is sufficient to take twice the integral over one of the triangles, i.e.,

$$\int_0^L \int_0^L dX_1\, dX_2\, f(X_1, X_2) = 2\int_{X_1 < X_2} dX_1\, dX_2\, f(X_1, X_2)$$

$$= 2\int_0^L dX_2 \int_0^{X_2} dX_1\, f(X_1, X_2). \qquad (4.5.5)$$

It is easy to visualize the six regions of integration for the case of $N = 3$, and by generalization we apply the identity (4.5.3) for the function $\exp[-\beta U_N(X_1, \ldots, X_N)]$. The canonical partition function is thus

$$Q(T, L, N) = \frac{1}{\Lambda^N}\int_0^L dX_N \int_0^{X_N} dX_{N-1} \cdots \int_0^{X_2} dX_1\, \exp(-\beta U_N). \qquad (4.5.6)$$

This could have been written directly had we imposed an order on the particles. This order can be maintained if the particles cannot cross over one another.

This PF can be transformed to a simpler form by taking the Laplace transform and applying the convolution theorem. In our case, this is equivalent to taking the T, P, N PF. Here P is the "pressure" in the one-dimensional system.

$$\Delta(T, P, N) = C\int_0^\infty Q(T, L, N)\, \exp(-\beta PL)\, dL. \qquad (4.5.7)$$

To simplify the final form of the PF, we add a zeroth particle at the origin X_0 and an $(N+1)$th particle at $X_{N+1} = L$. C is a constant having the dimensions of L^{-1}, so that the entire rhs of (4.5.7) became dimensionless. Since this constant does not affect the properties of the system, we shall simply take $C = 1$. This is equivalent to measuring L in units of C^{-1}.

We now transform to relative coordinates

$$Z_1 = X_1 - 0$$

$$Z_i = X_i - X_{i-1} \tag{4.5.8}$$

$$Z_{N+1} = L - X_N$$

and rewrite the total interaction energy as

$$U(X_0 \cdots X_{N+1}) = \sum_{i=1}^{N+1} U(Z_i). \tag{4.5.9}$$

Since the Jacobian of the transformation of variables (4.5.8) is unity, i.e.,

$$\frac{\partial(X_1 \cdots X_{N+1})}{\partial(Z_1 \cdots Z_{N+1})} = 1,$$

we can rewrite the PF in (4.5.7) as

$$\Delta(T, P, N) = \frac{1}{\Lambda^N} \int_0^\infty dZ_{N+1} \int_0^\infty dZ_N \cdots \int_0^\infty dZ_1$$

$$\times \exp\left[-\beta P \sum_{i=1}^{N+1} Z_i - \beta \sum_{i=1}^{N+1} U(Z_i)\right], \tag{4.5.10}$$

where we have written

$$L = \sum_{i=1}^{N+1} Z_i;$$

note that now the integration over each of the variables extends from zero to infinity.

Since the integrand is a product of $N + 1$ factors, each depends only on one of the variables Z_i; the $(N + 1)$-dimensional integral can be written as a product of $N + 1$ one-dimensional integrals, i.e.,

$$\Delta(T, P, N) = \frac{1}{\Lambda^N} \left\{ \int_0^\infty dR \exp[-\beta PR - \beta U(R)] \right\}^{N+1}. \tag{4.5.11}$$

We define

$$\mathscr{L}(P, T) = \int_0^\infty dR \exp[-\beta PR - \beta U(R)]. \tag{4.5.12}$$

For large systems we can write the PF as

$$\Delta(T, P, N) = \frac{\mathscr{L}(P, T)^N}{\Lambda^N}, \tag{4.5.13}$$

where we ignored one factor $\mathscr{L}(P, T)$. This is the same as closing the one-dimensional system, so that the particle at $X_{N+1} = L$ is identified with the particle at $X_0 = 0$.

4.5.2. Thermodynamics and the Equation of State

The Gibbs energy is obtained immediately from the relation

$$G = -kT \ln \Delta = NkT \ln \Lambda - NkT \ln \mathscr{L}(P, T). \tag{4.5.14}$$

The average length of the system is

$$\bar{L} = \frac{\partial G}{\partial P} = N \frac{\int dR\, R \exp[-\beta PR - \beta U(R)]}{\int dR \exp[-\beta PR - \beta U(R)]}, \qquad (4.5.15)$$

and the chemical potential is

$$\mu = kT \ln \Lambda - kT \ln \mathscr{L}(P, T). \qquad (4.5.16)$$

Results for two simple cases can be obtained immediately. First, if $U(R) = 0$—i.e., no interactions between the particles—then

$$\mathscr{L}(P, T) = (\beta P)^{-1} \qquad (4.5.17)$$

and

$$\mu = kT \ln(\Lambda \beta P) \qquad (4.5.18)$$

$$P\bar{L} = NkT. \qquad (4.5.19)$$

These are the equations of a one-dimensional ideal gas.

The second simple case is for hard rods, i.e., for particles interacting through a pair potential of the form

$$U(R) = \begin{cases} \infty & R \le \sigma \\ 0 & R > \sigma, \end{cases} \qquad (4.5.20)$$

where σ is the length of the particles. For this case we have

$$\mathscr{L}(P, T) = \int_{\sigma}^{\infty} dR \exp(-\beta PR) = \frac{\exp(-\beta P\sigma)}{\beta P} \qquad (4.5.21)$$

$$\mu(P, T) = kT \ln(\Lambda \beta P) + \sigma P \qquad (4.5.22)$$

$$P\bar{L} = NkT + N\sigma P. \qquad (4.5.23)$$

If we denote by $\rho = N/\bar{L}$ the average density in the T, P, N ensemble, we can write the equation of state (4.5.23) as

$$\frac{P}{kT} = \frac{\rho}{1 - \rho\sigma} = \rho[1 + \rho\sigma + (\rho\sigma)^2 + \cdots]. \qquad (4.5.24)$$

This power series converges for $\rho\sigma < 1$. Clearly, when $\rho\sigma \ge 1$, the average length per particle ρ^{-1} is smaller than σ, and this is impossible in this system.

The terms in the square brackets in Eq. (4.5.24) give the successive corrections to the ideal gas pressure due to interactions between pairs, triplets, etc., of particles. This expansion is the virial expansion of the pressure for the one-dimensional system.

We note that the slope of P/kT versus ρ curve is always positive, i.e.,

$$\frac{\partial(P/kT)}{\partial\rho} = \frac{1}{(1 - \rho\sigma)^2} > 0. \qquad (4.5.25)$$

This means that there could be no phase transition in this system. Normally a phase transition is indicated in a P–V diagram, or a P–$\bar{l}$ diagram, where $\bar{l} = \bar{L}/N$ is the length per particle. Equation (4.5.25) is equivalent to the statement that the slope of P as a function of $\bar{l}$ is always negative, i.e.,

$$\frac{\partial(P/kT)}{\partial \bar{l}} = \frac{-\rho^2}{(1 - \rho\sigma)^2} < 0. \qquad (4.5.26)$$

This result is more general and can be obtained for any pair potential $U(R)$. To see this we rewrite (4.5.15) as

$$\bar{l} = \int_0^\infty \Pr(R) R \, dR = \langle R \rangle, \qquad (4.5.27)$$

where $\Pr(R)\, dR$ is the probability of finding two consecutive particles at a distance between R and $R + dR$. Hence from (4.5.15) we obtain

$$\frac{\partial(\bar{l})}{\partial(\beta P)} = -(\langle R^2 \rangle - \langle R \rangle^2) = -\langle (R - \langle R \rangle)^2 \rangle < 0. \qquad (4.5.28)$$

4.5.3. An Alternative Derivation for Hard Rods

For hard particles interacting via a pair potential of the form (4.5.20), the canonical PF (4.5.6) simplifies considerably. Since U_N is either zero or infinity, the integrand $\exp(-\beta U_N)$ is 1, when the configuration $X_1, \ldots, X_N$ is allowable and 0 otherwise. A configuration of the particles is allowable whenever no two particles come closer than the length σ of the rods.

This consideration leads to rewriting the canonical PF as

$$Q(T, L, N) = \frac{1}{\Lambda^N} \int_{(2N-1)\sigma/2}^{L-\sigma/2} dX_N \int \cdots \int_{3\sigma/2}^{X_3 - \sigma} dX_2 \int_{\sigma/2}^{X_2 - \sigma} dX_1. \qquad (4.5.29)$$

The first particle is allowed the region between $\sigma/2$ and $X_2 - \sigma$, the second the region between the first and the third hard rods, etc.

One can now perform the integration step by step. For instance, the first two steps are

$$\int_{\sigma/2}^{X_2 - \sigma} dX_1 = X_2 - \tfrac{3}{2}\sigma \qquad (4.5.30)$$

$$\int_{3\sigma/2}^{X_3 - \sigma} dX_2(X_2 - \tfrac{3}{2}\sigma) = \int_0^{X_3 - (5/2)\sigma} dY_2\, Y_2 = \tfrac{1}{2}(X_3 - \tfrac{5}{2}\sigma)^2. \qquad (4.5.31)$$

In the second step we changed variables, $Y_2 = X_2 - (3/2)\sigma$. Similarly at each step we change variables.

$$Y_i = X_i - (2i - 1)\sigma/2.$$

In this way the integration (4.5.29) can be performed successively and the final result is

$$Q(T, L, N) = \frac{(L - N\sigma)^N}{\Lambda^N N!}. \qquad (4.5.32)$$

The chemical potential is obtained from (4.5.32)

$$\mu(\rho, T) = -kT \frac{\partial \ln Q}{\partial N} = kT \ln \Lambda\rho - kT \ln(1 - \rho\sigma) + \frac{kT\sigma\rho}{1 - \rho\sigma}. \tag{4.5.33}$$

Note that here we have expressed the chemical potential in terms of the variables T and $\rho = N/L$. The same expression may be obtained by substituting the equation of state (4.5.23) into (4.5.22) to reexpress the chemical potential, in the T, P, N ensemble in terms of T and $\rho = N/\bar{L}$, where $\bar{L}$ is the average length of the system in the T, P, N ensemble.

This method can easily be generalized to any number of different components. For two kinds of hard rods of length σ_A and σ_B, the corresponding partition function is

$$Q(T, L, N_A, N_B) = \frac{(L - N_A\sigma_A - N_B\sigma_B)^{N_A + N_B}}{\Lambda_A^{N_A}\Lambda_B^{N_B}N_A!N_B!}. \tag{4.5.34}$$

The chemical potential of, say, the A component is

$$\mu_A(T, \rho_A, \rho_B) = kT \ln \rho_A\Lambda_A - kT \ln(1 - \rho_A\sigma_A - \rho_B\sigma_B) + \frac{kT\sigma_A(\rho_A + \rho_B)}{1 - \rho_A\sigma_A - \rho_B\sigma_B}. \tag{4.5.35}$$

If $\rho_B \to 0$, then we recover Eq. (4.5.33) (pure component A). However, in the limit of $\rho_B \gg \rho_A$ (A very dilute in B), we obtain

$$\mu_A = kT \ln \rho_A\Lambda_A - kT \ln(1 - \rho_B\sigma_B) + \frac{kT\sigma_A\rho_B}{1 - \rho_B\sigma_B}. \tag{4.5.36}$$

The equation of state for the two-component system is obtained directly from (4.5.34):

$$\frac{P}{kT} = \frac{\rho_A + \rho_B}{1 - \rho_A\sigma_A - \rho_B\sigma_B}, \tag{4.5.37}$$

which is the generalization of (4.5.24). Substituting from (4.5.37) into (4.5.35), we may express the chemical potential in the more common variables T, P, x_A as

$$\mu_A(T, P, x_A) = kT \ln x_A\Lambda_A + kT \ln \beta P + \sigma_A P. \tag{4.5.38}$$

Note that for $x_A = 1$ we recover the chemical potential of pure A (Eq. 4.5.22).

For a mixture of two hard rods A and B, we can compute all the solvation thermodynamics of A very diluted in a solvent B. To do that we first write the chemical potential of A in dilute solution (4.5.36) in terms of P, T, i.e.,

$$\mu_A = kT \ln \rho_A\Lambda_A + kT \ln(1 + \beta P\sigma_B) + \sigma_A P, \tag{4.5.39}$$

from which we identify the pseudo chemical potential

$$\mu_A^* = kT \ln(1 + \beta P\sigma_B) + \sigma_A P. \tag{4.5.40}$$

This is the work required to introduce an A solute at a fixed position in a pure B solvent at T, P constants.

The other relevant thermodynamic quantities are

$$S_A^* = -\frac{\partial \mu_A^*}{\partial T} = -k \ln(1 + \beta P\sigma_B) - \frac{P\sigma_B T^{-1}}{1 + \beta P\sigma_B} \tag{4.5.41}$$

$$H_A^* = \mu_A^* + TS_A^* = \sigma_A P + \frac{P\sigma_B}{1 + \beta P\sigma_B} \tag{4.5.42}$$

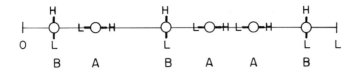

FIGURE 4.8. Two-state "water" molecules in one dimension. Two consecutive molecules in state A can form a "hydrogen bond." In state B they cannot.

$$V_A^* = \frac{\partial \mu_A^*}{\partial P} = \frac{\sigma_B}{1 + \beta P \sigma_B} + \sigma_A \qquad (4.5.43)$$

$$E_A^* = H_A^* - PV_A^* = 0. \qquad (4.5.44)$$

4.5.4. One-Dimensional "Water"

The purpose of this section is to demonstrate, by a very simple model, one of the most outstanding properties of liquid water: A negative thermal expansion coefficient at low temperatures.

Consider first a two state water-like particle, Fig. 4.8. Each molecule has two functional groups: On one side, a hydrogen atom H which can serve as a donor for hydrogen bonding, and on the other side "lone pair electrons" L, through which the molecule can accept a hydrogen bond. The two states of each molecule are indicated by A and B in Fig. 4.8. We assume that the molecules are rigid and can move along the one-dimensional length L. The interaction energy of the pairs A—A, B—B, and B—A are assumed to have an identical component, say, a square well potential, Fig. 4.9.

$$U_{AB}(R) = U_{BA}(R) = U_{BB}(R) = U(R). \qquad (4.5.45)$$

The A—A pair has in addition to $U(R)$, a hydrogen-bond potential function (Fig. 4.9), i.e.,

$$U_{AA}(R) = U(R) + \varepsilon_{HB}G(R), \qquad (4.5.46)$$

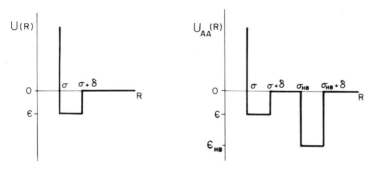

FIGURE 4.9. Square-well potential $U(R)$ (left) and a hydrogen-bond potential $U_{AA}(R)$ (right) defined in Eqs. (4.5.46) and (4.5.47).

where $\varepsilon_{HB} < 0$ is the hydrogen-bond energy and $\mathbf{G}(R)$ is a geometric function of the form

$$\mathbf{G}(R) = \begin{cases} 0 & R \leq \sigma_{HB} = \sigma + \Delta' \\ 1 & \sigma_{HB} \leq R \leq \sigma_{HB} + \delta. \\ 0 & R \geq \sigma_{HB} + \delta \end{cases} \tag{4.5.47}$$

Thus whenever two molecules in the A state approach each other within the range of $\sigma_{HB} \leq R < +\sigma_{HB} + \delta$, they possess an additional interaction energy ε_{HB}.

The canonical PF of system of N molecules in a "volume" of length L and at temperature T is written as

$$Q(T, L, N) = \frac{1}{N! \Lambda^N} Z(T, L, N)$$

$$= \frac{1}{N! \Lambda^N} \sum_{N_A = 0}^{N} \binom{N}{N_A} Z(T, L, N_A, N_B), \tag{4.5.48}$$

where $Z(T, L, N)$ is the configurational PF. In the second line on the rhs of (4.5.48) we sum over all possible values of $N_A(N_B = N - N_A)$, where $Z(T, L, N_A, N_B)$ is the configurational PF of a "frozen-in" system; i.e., there are precisely N_A molecules, say $1, \ldots, N_A$, in state A and the rest $N_A + 1, \ldots, N_A + N_B$ in state B.

$$Z(T, L, N_A, N_B) = \int_0^L \cdots \int_0^L dX^{N_A} dX^{N_B} \exp[-\beta U(N_A, N_B)]. \tag{4.5.49}$$

In order to apply the convolution theorem, as in section 4.5.1, we need to introduce the concept of "ordering of species." The configurational PF can be rewritten by first fixing a specific ordering of the species (S.O.O.S.) and then summing over all possible orderings of species, consistent with the *fixed* value of N_A and N_B. Thus we write

$$Z(T, L, N_A, N_B) = \sum_{\text{all S.O.O.S.}} \int_0^L \cdots \int_0^L$$
$$\underbrace{}_{\text{S.O.O.S.}}$$

$$= \sum_{\text{all S.O.O.S.}} N_A! N_B! \int_0^L dX_N \int_0^{X_N} dX_{N-1} \cdots \int_0^{X_2} dX_1. \tag{4.5.50}$$
$$\underbrace{\phantom{\int_0^L dX_N \int_0^{X_N} dX_{N-1}}}_{\text{S.O.O.S.}}$$

The second step on the rhs of (4.5.50) is the same as in (4.5.3).

A simple example of the identity (4.5.50) for three particles, one A and two Bs is (a and b are the coordinates of A and B, respectively):

$$\int_0^L \int_0^L \int_0^L \exp[-\beta U(a, b_1, b_2)] \, da \, db_1 \, db_2$$

$$= \int_0^L \int_0^L \int_0^L \exp[-\beta U_{AB}(a, b_1) - \beta U_{BB}(b_1, b_2)] \, da \, db_1 \, db_2$$

$$+ \int_0^L \int_0^L \int_0^L \exp[-\beta U_{BA}(b_1, a) - \beta U_{AB}(a, b_2)] \, da \, db_1 \, db_2$$

$$+ \int_0^L \int_0^L \int_0^L \exp[-\beta U_{BB}(b_1, b_2) - \beta U_{BA}(b_2, a)] \, da \, db_1 \, db_2$$

$$= 2 \int_0^L da \int_0^a db_1 \int_0^{b_1} db_2 + 2 \int_0^L db_1 \int_0^{b_1} da \int_0^a db_2$$

$$+ 2 \int_0^L db_1 \int_0^{b_1} db_2 \int_0^{b_2} da. \tag{4.5.51}$$

In the first step we distinguished between all possible orderings of the species. These are ABB, BAB, and BBA. Note that the integrand is different for different orderings of the species. Next, for each fixed ordering of species, we also fixed the ordering of the particles, by limiting the range of integration for each integral. Since now each integrand (not shown in the last step) is symmetric with respect to interchanging particles of the same species, we have to multiply by 2, as in the example of Eq. (4.5.5).

We can now apply the convolution theorem to the integral on the rhs of (4.5.50). We take the T, P, N PF, which is the same as taking the Laplace transform of $Q(T, L, N)$ with respect to the variable L. Following similar steps to those in section 4.5.1, we have

$$\Delta(T, P, N) = \int_0^\infty dL \exp(-\beta PL) Q(T, L, N)$$

$$= \frac{1}{\Lambda^N} \sum_{N_A = 0}^{N} \sum_{\substack{\text{all} \\ \text{S.O.O.S.}}} \mathscr{L}_{s_1 s_2} \mathscr{L}_{s_2 s_3} \cdots \mathscr{L}_{s_N s_1}, \tag{4.5.52}$$

which is a generalization of Eq. (4.5.11). Note, however, that the sum over all N_A and the sum over all specific orderings of species, with a fixed N_A, is the same as the sum over all possible vectors $\mathbf{s} = (s_1, \ldots, s_N)$, where s_i is either A or B.

Hence

$$\Delta(T, P, N) = \frac{1}{\Lambda^N} \sum_{\mathbf{s}} \mathscr{L}_{s_1 s_2} \mathscr{L}_{s_2 s_3} \cdots \mathscr{L}_{s_N s_1}$$

$$= \text{Tr } \mathbf{P}^N, \tag{4.5.53}$$

where the elements of the 2×2 matrix are

$$P_{AA} = \frac{1}{\Lambda} \mathscr{L}_{AA} = \frac{1}{\Lambda} \int_0^\infty dR \exp(-\beta PR) \exp[-\beta U_{AA}(R)] \tag{4.5.54}$$

$$P_{AB} = P_{BA} = P_{BB} = \frac{1}{\Lambda} \int_0^\infty dR \exp(-\beta PR) \exp[-\beta U(R)]. \tag{4.5.55}$$

The eigenvalues of $\mathbf{P}$ are obtained from the equation

$$\begin{vmatrix} P_{AA} - \lambda & P_{AB} \\ P_{AB} & P_{BB} - \lambda \end{vmatrix} = 0, \tag{4.5.56}$$

and the largest eigenvalue is

$$\lambda_1 = \frac{(P_{AA} + P_{BB}) + \sqrt{(P_{AA} - P_{BB})^2 + 4P_{BB}^2}}{2}. \tag{4.5.57}$$

The average length $\bar{L}$ is given by

$$\bar{L} = \frac{\partial G}{\partial P} = -kT\frac{\partial \ln \Delta}{\partial P} = -kTN\frac{\partial \ln \lambda_1}{\partial P}, \tag{4.5.58}$$

and the temperature coefficient of the average length per particle is

$$\alpha = \frac{1}{N}\frac{\partial \bar{L}}{\partial T} = \frac{\partial \bar{l}}{\partial T}. \tag{4.5.59}$$

First we note that for a square-well potential of the form (Fig. 4.9)

$$U(R) = \begin{cases} \infty & R \leq \sigma \\ \varepsilon & \sigma \leq R \leq \sigma + \delta \\ 0 & R > \sigma + \delta \end{cases} \tag{4.5.60}$$

and for $\varepsilon_{HB} = 0$, i.e., no hydrogen bonds, then $P_{AA} = P_{BB}$ and

$$\begin{aligned}\bar{L} &= -kT\frac{\partial \ln \Delta}{\partial P} \\ &= -NkT\frac{\partial \ln \lambda_1}{\partial P} \\ &= N\frac{\int dR\, R \exp[-\beta PR - \beta U(R)]}{\int dR \exp[-\beta PR - \beta U(R)]}. \end{aligned} \tag{4.5.61}$$

This result was already obtained in (4.5.15). In this case we also have [see (4.5.28)]

$$\left(\frac{\partial \bar{L}}{\partial T}\right)_{N,P} = N\frac{\partial \bar{l}}{\partial T} = \frac{-NP}{kT^2}\frac{\partial \bar{l}}{\partial(\beta P)} = \frac{NP}{kT^2}\langle (R - \langle R\rangle)^2\rangle, \tag{4.5.62}$$

which is always positive. Thus we conclude that for the pure B-liquid (or the mixture, but with $\varepsilon_{HB} = 0$), the average length of the system always increases with temperature. This is also what we should expect intuitively. However, we know that liquid water behaves differently at temperatures between 0° and 4°C. To see the molecular reason for this behavior we "turn on" ε_{HB} and examine the change in the thermal expansion of the system. For the specific definitions (4.5.60) and (4.5.47) we can examine the average length as a function of the temperature, numerically. We first rewrite P_{AA} in (4.5.54) as

$$\begin{aligned}P_{AA} &= \frac{1}{\Lambda}\int dR \exp[-\beta PR - \beta U(R) - \beta\varepsilon_{HB}G(R)] \\ &= P_{BB}\langle \exp[-\beta\varepsilon_{HB}G(R)]\rangle, \end{aligned} \tag{4.5.63}$$

where the average $\langle\ \rangle$ is taken with the same probability distribution as in (4.5.27).

Substituting P_{AA} in (4.5.57), we obtain

$$\lambda_1 = P_{BB}\frac{(\langle\ \rangle + 1) + \sqrt{(\langle\ \rangle - 1)^2 + 4}}{2} = P_{BB}P^{HB}, \tag{4.5.64}$$

where $\langle\ \rangle$ is the average quantity appearing in (4.5.63) and P^{HB} is defined in Eq. (4.5.64). If $\varepsilon_{HB} = 0$, then $P^{HB} = 2$.

The average "volume" of the system is thus

$$\bar{L} = -NkT\frac{\partial \ln \lambda_1}{\partial P} = -NkT\frac{\partial \ln P_{BB}}{\partial P} - NkT\frac{\partial \ln P^{HB}}{\partial P}$$

$$= L_{BB} + L^{HB}. \tag{4.5.65}$$

The first term on the rhs of (4.5.65) is the average length of the system, when the hydrogen bonding is switched off, i.e., $\varepsilon_{HB} = 0$. The second is the contribution to the length of the system due to hydrogen bonding.

To demonstrate the behavior of the system numerically, we use the square-well potentials as defined in (4.5.47) and (4.5.60). The results for P_{AA} and P_{BB} are

$$P_{AA} = \frac{\exp[-\beta P\sigma]}{\Lambda\beta P}[1 + (e^{-\beta\varepsilon} - 1)(1 - e^{-\beta P\delta}) + (e^{-\beta\varepsilon_{HB}} - 1)e^{-\beta P\Delta'}(1 - e^{-\beta P\delta})] \tag{4.5.66}$$

$$P_{BB} = \frac{\exp[-\beta P\sigma]}{\Lambda\beta P}[1 + (e^{-\beta\varepsilon} - 1)(1 - e^{-\beta P\delta})].$$

These can be introduced into (4.5.57) to compute λ_1, and therefore all relevant thermodynamic quantities. Figure 4.10 shows the average length as a function of temperature for a specific set of molecular parameters. We see that there exists a region of P and T for which the length per particle decreases with the increase of temperature. This phenomenon disappears at higher pressures.

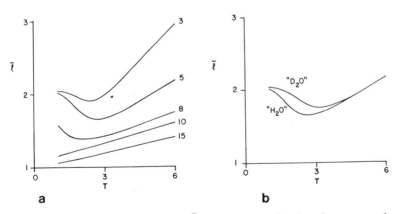

a   **b**

FIGURE 4.10. (a) The average length per particle $\bar{l}$ in units of σ as a function of temperature for various values of the pressure, as indicated next to each curve. The parameters chosen for these illustrations are σ (used as the unit of length), $\Delta = 1$, $\delta = 0.1$, $\varepsilon/k = -1$ K, $\varepsilon_{HB}/k = -10$ K. The pressures are $P\sigma/k = 3, 5, 8, 10,$ and 15 K. (b) Same plot for H_2O and D_2O. All parameters are the same except $\varepsilon_{HB}/k = -12$ K and $P\sigma/k = 5$ K.

Here we pursue the examination of one particularly simple case analytically. We assume that B are hard rods, i.e., $\varepsilon = 0$, and A has a hard-rod diameter σ and, in addition, a hydrogen-bonding capacity at $\sigma_{HB} = \sigma + \Delta'$. We may also assume that $-\beta P\delta$ is small compared to unity; hence for this case we obtain to first order in $\beta P\delta$

$$\lambda_1 = \frac{P_{BB}}{2}[4 + (e^{-\beta\varepsilon_{HB}} - 1)\, e^{-\beta P\Delta'}\beta P\delta]$$

(4.5.67)

$$\bar{l} = -kT\frac{\partial \ln \lambda_1}{\partial P} = -kT\frac{\partial \ln P_{BB}}{\partial P} + \bar{l}^{HB} = \bar{l}_{BB} + \bar{l}^{HB},$$

where $\bar{l}_{BB}$ is the average length of the hard-rod system (i.e., in the absence of hydrogen bonding). The second term is the change in the average length when we turn on the hydrogen bonding.

$$\bar{l}^{HB} = -kT\frac{\partial \ln[4 + (e^{-\beta\varepsilon_{HB}} - 1)\, e^{-\beta P\Delta'}\beta P\delta]}{\partial P}$$

$$\approx \tfrac{1}{4}(e^{-\beta\varepsilon_{HB}} - 1)\, e^{-\beta P\Delta'}(\sigma_{HB} - \bar{l})\beta P\delta.$$

(4.5.68)

The sign of $\bar{l}^{HB}$ is determined by $\sigma_{HB} - \bar{l}$. When the average length per particle is larger than σ_{HB}, then turning on the hydrogen bonding causes an additional attraction to contract the system; hence $\bar{l}^{HB}$ is negative. On the other hand, at higher pressures, when $\bar{l} < \sigma_{HB}$, turning on the hydrogen bonding leads to expansion, i.e., $\bar{l}^{HB}$ is positive.

The temperature dependence of $\bar{l}^{HB}$ is

$$\frac{\partial \bar{l}^{HB}}{\partial T} = \frac{(e^{-\beta\varepsilon_{HB}} - 1)\, e^{-\beta P\Delta}}{4kT^2}[(\sigma_{HB} - \bar{l})(P\Delta' + \varepsilon_{HB}) - P\Delta'].$$

(4.5.69)

Suppose we choose a pressure such as $P\Delta' + \varepsilon_{HB} < 0$. Then we can change the temperature until $\sigma_{HB} - \bar{l} = 0$. This will secure a negative contribution to this term. Furthermore, at any lower temperature $\bar{l}$ will decrease; hence $\sigma_{HB} - \bar{l} > 0$ and the entire expression inside the square brackets becomes negative.

The conclusion reached from this model is relevant to the behavior of real liquid water. One of the most outstanding properties of water is the negative thermal expansion coefficient between $0°$ and $4°C$. No other liquid shows this behavior. We shall see in Chapter 7 that this property results from the peculiar correlation between the average local density and the average binding energy of a water molecule. In this particular model we have built-in this correlation by adding a negative hydrogen-bond energy ($\varepsilon_{HB} < 0$) at a distance larger than the diameter of the particles $\Delta' > 0$. This is of course an artificial model potential. In Chapter 8 we shall see how this feature can be obtained for the potential of average force.

Note also that, at the same conditions, where we get $\partial \bar{l}^{HB}/\partial T < 0$ in Eq. (4.5.69), keeping σ_{HB} constant but increasing the value of $|\varepsilon_{HB}|$, we should expect a more negative value of $\partial \bar{l}^{HB}/\partial T$. This resembles the behavior of D_2O which has a temperature of maximum density higher than that of H_2O (Fig. 4.10b).

4.5.5. One-Dimensional Mixture of Fluids

The canonical PF of a system of c components is

$$Q(T, L, N) = \frac{1}{N! \prod_i \Lambda_i^{N_i}} \int_0^L \cdots \int_0^L \prod_i d\mathbf{X}^{N_i} \exp[-\beta U(\mathbf{N})],$$

(4.5.70)

where we used some shorthand notations: $\mathbf{N} = N_1, \ldots, N_c$, $\mathbf{N}! = \Pi_{i=1}^c N_i!$ and $\Pi_i \, d\mathbf{X}^{N_i} = dX_1 \cdots dX_N$, where $N = \sum_{i=1}^c N_i$. The total energy is assumed to be pairwise additive.

$$U(\mathbf{N}) = \sum_{i=1}^N U_{\alpha_i, \alpha_{i+1}}, \tag{4.5.71}$$

where $U_{\alpha_i, \alpha_{i+1}}$ is the pair potential corresponding to the pair of species α_i and α_{i+1}.

In the one-component system we used the identity (4.5.3) to convert the integral on the lhs to an integral with a specific ordering of the particles, i.e., $0 < X_1 < X_2 < X_3 < \cdots < X_N < L$. In our case, since there are c different species, we have to distinguish between a specific ordering of the species (S.O.O.S.) and a specific ordering of the particles (S.O.O.P.). The configurational integral on the rhs of (4.5.70) can be rewritten by first fixing the order of the species and then summing over all orderings of the species. This is the same procedure we undertook in section 4.5.4 [see in particular Eqs. (4.5.50) and (4.5.51)]. Thus we have

$$\int_0^L \cdots \int_0^L = \sum_{\substack{\text{all orderings} \\ \text{of species} \\ \text{with fixed } N_1, \ldots, N_c}} \underbrace{\int_0^L \cdots \int_0^L}_{\text{S.O.O.S.}}$$

$$= \sum_{\substack{\text{all orderings} \\ \text{of species} \\ \text{with fixed } N_1, \ldots, N_c}} \prod_i N_i! \underbrace{\int_0^L dX_N \cdots \int_0^{X_2} dX_1}_{\text{S.O.O.P.}}, \tag{4.5.72}$$

where in the first step we have fixed a specific ordering of species (S.O.O.S.), and in the second step we have fixed the specific ordering of the particles (S.O.O.P.) and multiplied by the factor $\Pi_i N_i!$. [Note that the integrand in (4.5.70) is symmetric with respect to interchanging particles of the *same* species.]

Applying (4.5.72) to (4.5.70), we have

$$Q(T, L, \mathbf{N}) = \frac{1}{\prod_i \Lambda_i^{N_i}} \sum_{\substack{\text{all orderings} \\ \text{of species} \\ \text{with fixed } N_1, \ldots, N_c}} \underbrace{\int_0^L dX_N \cdots \int_0^{X_2} dX_1}_{\text{S.O.O.P.}} \exp[-\beta U(\mathbf{N})]. \tag{4.5.73}$$

Note that—in contrast to the preceding section, where in (4.5.48) we had an additional sum over N_A—here the number of particles of each species is fixed. In the preceding section we could proceed from Eq. (4.5.52) to (4.5.53) because the sum over all orderings of species *and* the sum over all possible N_A were equivalent to a sum over all configurations s in (4.5.53). Here, we have a sum over all orderings of species in (4.5.73), but $N_1, \ldots, N_c$ are still fixed. Therefore, before taking the Laplace transform, we first open the system with respect to all particles. The corresponding grand PF is

$$\Xi(T, L, \boldsymbol{\lambda}) = \sum_{\mathbf{N}} \prod_i \lambda_i^{N_i} Q(T, L, \mathbf{N}) = \exp(+\beta PL), \tag{4.5.74}$$

where $P(T, L, \boldsymbol{\lambda})$ is the thermodynamic pressure, given as a function of the variables T, L, $\boldsymbol{\lambda}$. We now take the Laplace transform of (4.5.74) with respect to the variable L by introducing the new unspecified variable P^*.

$$\mathscr{L}(\Xi) = \int_0^\infty \exp(-\beta P^* L) \Xi(T, L, \boldsymbol{\lambda}) \, dL. \tag{4.5.75}$$

Applying the convolution theorem, we obtain

$$\mathscr{L}(\Xi) = \int_0^\infty \exp(-\beta P^* L) \sum_N (\lambda_i/\Lambda_i)^{N_i} \sum_{\substack{\text{all orderings} \\ \text{of species} \\ \text{with fixed } N_1,\ldots,N_c}} \underbrace{\int \cdots \int}_{\text{S.O.O.P.}}$$

$$= \sum_{N=0}^\infty \sum_s M_{s_1 s_2} M_{s_2 s_3} \cdots M_{s_N s_1} = \sum_{N=0}^\infty \operatorname{Tr} \mathbf{M}^N, \tag{4.5.76}$$

where the matrix elements are defined for each pair of species $\alpha\beta$ by

$$M_{\alpha\beta}(P^*) = (\lambda_\alpha/\Lambda_\alpha)^{1/2}(\lambda_\beta/\Lambda_\beta)^{1/2} \mathscr{L}_{\alpha\beta}(P^*) \tag{4.5.77}$$

and

$$\mathscr{L}_{\alpha\beta}(P^*) = \int_0^\infty \exp[-\beta P^* R - \beta U_{\alpha\beta}(R)] \, dR. \tag{4.5.78}$$

$\mathscr{L}_{\alpha\beta}$ is the Laplace transform of $\exp[-\beta U_{\alpha\beta}(R)]$. This is the generalization of (4.5.12), but now we choose the arbitrary variable P^* instead of the thermodynamic variable P. The last step in (4.5.76) needs further elaboration. It is easier to visualize for a two-component system, for which we have the identity

$$\sum_{N_A=0}^\infty \sum_{N_B=0}^\infty \sum_{\substack{\text{all orderings} \\ \text{of species} \\ \text{with fixed } N_A, N_B}} = \sum_{N=0}^\infty \sum_{N_A+N_B=N} \sum_{\substack{\text{all orderings} \\ \text{of species} \\ \text{with fixed } N_A, N_B}}$$

$$= \sum_{N=0}^\infty \sum_s . \tag{4.5.79}$$

Note that the sum over "all orderings of species" is restricted to fixed values of N_A and N_B. Once we open the system, we carry the sum over N_A and N_B in two steps. First we fix $N = N_A + N_B$. Now the sum over all N_A and N_B with a fixed N is the same as the sum over all possible vectors $\mathbf{s}$. This is followed by the sum over all N, which is the result (4.5.76).

Assuming that the sum (4.5.76) converges, we rewrite it as

$$\mathscr{L}(\Xi) = \sum_{N=0}^\infty \operatorname{Tr} \mathbf{M}^N = \sum_{N=0}^\infty \sum_{j=1}^c \gamma_j^N = \sum_{j=1}^c [1 - \gamma_j(P^*)]^{-1}, \tag{4.5.80}$$

where γ_j is the jth eigenvalue of the matrix $\mathbf{M}$. We stressed the dependence of γ_j on the variable P^*.

We recall the definition of $\mathscr{L}(\Xi)$ in (4.5.75)

$$\mathscr{L}(\Xi) = \int_0^\infty \exp[-\beta L(P^* - P)] \, dL. \tag{4.5.81}$$

If we choose P^* as the thermodynamic pressure P, then $\mathscr{L}(\Xi)$ is referred to as the generalized PF. It is clear from (4.5.81) that the integral in this case diverges. The reason is that $\Xi(T, L, \lambda)$ is a function of the single extensive variable L. Transforming L into the thermodynamic intensive variable P gives a partition function which is a function of the intensive variables T, P, λ only. However, the Gibbs–Duhem relation states that

$$S \, dT - V \, dP + \sum N_i \, d\mu_i = 0. \tag{4.5.82}$$

Hence the intensive variables T, P, μ or T, P, λ are not independent. For this reason we have denoted by P^* the new variable in the Laplace transform taken in (4.5.75). Suppose that we choose $P^* > P$; then the integral in (4.5.81) converges, and we have

$$\frac{1}{\beta(P^* - P)} = \sum_{j=1}^{c} [1 - \gamma_j(P^*)]^{-1}. \tag{4.5.83}$$

It is clear that, since in the limit $P^* \to P$ the lhs of (4.5.83) diverges, there must be at least one of $\gamma_j(P^*)$ equal to 1. The secular equation of the matrix $\mathbf{M}$ is

$$|\mathbf{M} - \gamma \mathbf{I}| = 0. \tag{4.5.84}$$

Since the elements of $\mathbf{M}$ are functions of T, P, and λ we can use the implicit Eq. (4.5.84) to derive all thermodynamic quantities of interest. We treat a two-component system of A and B for which the secular equation is

$$\begin{vmatrix} M_{AA} - \gamma & M_{AB} \\ M_{AB} & M_{BB} - \gamma \end{vmatrix} = 0, \tag{4.5.85}$$

or equivalently

$$\gamma_\pm = \tfrac{1}{2}[M_{AA} + M_{BB} \pm \sqrt{(M_{AA} - M_{BB})^2 + 4M_{AB}^2}]. \tag{4.5.86}$$

The correct solution can be identified by taking the limit $\lambda_B = 0$, i.e., for pure A we must have

$$\gamma = \tfrac{1}{2}(M_{AA} + M_{AA}) = M_{AA}. \tag{4.5.87}$$

Therefore we take the γ_+ solution of (4.5.87) and equate it to unity to obtain the implicit equation

$$f(T, P, \lambda_A, \lambda_B) = M_{AA} + M_{BB} + \sqrt{(M_{AA} - M_{BB})^2 + 4M_{AB}^2} - 2 = 0. \tag{4.5.88}$$

Taking the total differential of f, we have

$$df = \frac{\partial f}{\partial T} dT + \frac{\partial f}{\partial P} dP + \sum \frac{\partial f}{\partial \lambda_i} d\lambda_i = 0. \tag{4.5.89}$$

Comparing with (4.5.82), we can derive all the thermodynamic quantities of the system from the implicit Eq. (4.5.88).

As an example, the average density of A is

$$\rho_A = \left(\frac{\partial P}{\partial \mu_A}\right)_{T, \mu_B} = \frac{-(\partial f / \partial \mu_A)_{P, T, \mu_B}}{(\partial f / \partial P)_{T, \mu_A, \mu_B}}, \tag{4.5.90}$$

and the entropy of the system is

$$S = \left(\frac{\partial (PV)}{\partial T}\right)_{V, \mu_A, \mu_B} = V\left(\frac{\partial P}{\partial T}\right)_{\mu_A, \mu_B} = -V \frac{(\partial f / \partial T)_{P, \mu_A, \mu_B}}{(\partial f / \partial P)_{T, \mu_A, \mu_B}}. \tag{4.5.91}$$

A particular simple case occurs when

$$\mathscr{L}_{AB}^2 = \mathscr{L}_{AA} \mathscr{L}_{BB}, \tag{4.5.92}$$

or, equivalently,

$$M_{AB}^2 = M_{AA} M_{BB}. \tag{4.5.93}$$

In this case, (4.5.88) reduces to

$$f(T, P, \lambda_A, \lambda_B) = 2(M_{AA} + M_{BB}) - 2 = 0, \tag{4.5.94}$$

the density of A is now

$$\rho_A = \frac{-\beta\lambda_A}{\Lambda_A} \frac{2\mathscr{L}_{AA}}{\partial f/\partial P}, \tag{4.5.95}$$

and the mole fraction of A is

$$x_A = \frac{\rho_A}{\rho_A + \rho_B} = \frac{\lambda_A \mathscr{L}_{AA}}{\Lambda_A}. \tag{4.5.96}$$

This may be written in a more familiar form as

$$\mu_A = kT \ln(\Lambda_A/\mathscr{L}_{AA}) + kT \ln x_A$$
$$= \mu_A^P + kT \ln x_A, \tag{4.5.97}$$

where by virtue of (4.5.16), μ_A^P is the chemical potential of pure A at the same T, P as the mixture. Equation (4.5.97) is the familiar form of the chemical potential of a symmetrical ideal solution. We shall further discuss this kind of ideality in Chapter 6.

4.5.6. Solvation in a One-Dimensional System

The concept of solvation is important in the study of solvent effects on any process taking place in solution. We define *solvation* as the process of transferring a single particle from a fixed point in the vacuum to a fixed point in the solution. The process is carried out at given thermodynamic variables, say T, P constants, T, V constants, etc.

Since our particles are structureless, in the sense that they do not have any internal partition function, it is sufficient to study the pseudochemical potential of the species of which the solvation thermodynamics is under study.

We have already obtained the pseudochemical potentials of a species A in a mixture of hard rods, in section 4.5.3. We generalize here the expression for mixtures of interacting particles.

First we note that the chemical potential for a one-component system (4.5.16) may be written as

$$\mu = kT \ln \rho\Lambda - kT \ln \rho\mathscr{L}(P, T). \tag{4.5.98}$$

Since in the ideal gas limit, we have from (4.5.18) and (4.5.19) that

$$\mu = kT \ln \rho\Lambda, \tag{4.5.99}$$

it follows that the pseudochemical potential for a pure 1-D fluid is

$$\mu^*(P, T) = -kT \ln \rho\mathscr{L}(P, T), \tag{4.5.100}$$

where $\rho = \rho(P, T)$ is the average density in the T, P, N system. Using (4.5.15), we can rewrite (4.5.100) as

$$\mu^*(P, T) = -kT \ln \frac{-\beta\mathscr{L}(P, T)^2}{\mathscr{L}'(P, T)}, \tag{4.5.101}$$

where $\mathscr{L}'(P, T) = \partial\mathscr{L}/\partial P$. We now generalize (4.5.101) for a two-component system, when A is very dilute in B, i.e., for $\lambda_A \ll \lambda_B$. Expanding (4.5.86) to first order in λ_A, we obtain for the density of A, noting that for pure B we have $\lambda_B \mathscr{L}_{BB}/\Lambda_B = 1$,

$$\rho_A = \left(\frac{\partial P}{\partial \mu_A}\right)_T = -\frac{\partial f/\partial \mu_A}{\partial f/\partial P} = -\beta \frac{\lambda_A}{\Lambda_A} \frac{\mathscr{L}_{AB}^2}{\mathscr{L}_{BB}'}, \qquad (4.5.102)$$

where $\mathscr{L}_{BB}'$ is the derivative of $\mathscr{L}_{BB}$ with respect to P. The last equation can be rearranged to obtain

$$\mu_A = kT \ln \rho_A \Lambda_A - kT \ln \frac{-\beta \mathscr{L}_{AB}^2}{\mathscr{L}_{BB}'}, \qquad (4.5.103)$$

from which we identify the pseudochemical potential

$$\mu_A^* = -kT \ln \frac{-\beta \mathscr{L}_{AB}^2}{\mathscr{L}_{BB}'}, \qquad (4.5.104)$$

which is the required generalization of (4.5.101) for a one-component system, as well as of (4.5.40) for mixtures of hard rods. Having $\mu_A^*(P, T)$, we can obtain all other thermodynamic quantities of solvation of A in B.

4.6. PHASE TRANSITION IN A ONE-DIMENSIONAL SYSTEM

We have noted before in this chapter that a phase transition does not occur in a one-dimensional system. This is true for systems consisting of particles the intermolecular interaction of which is of finite range. In this section we study a first-order phase transition in a rather artificial model, i.e., a system with interactions of infinite range. The study of this model is of interest not because of its relevance to any real system, but because it reveals some of the mathematical conditions underlying the thermodynamics of phase transitions. The model used here is a simplified version of a model suggested by T. L. Hill.[1]

4.6.1. The Model and Its Solution

We consider a lattice of M units. Each unit can be in one of two states, denoted by A and B. The units are structureless, i.e., there are no internal degrees of freedom. The nearest-neighbor interaction energies are

$$W_{AA} = \varepsilon_1 < 0 \qquad (4.6.1)$$

$$W_{BB} = 0 \qquad (4.6.2)$$

$$W_{AB} = W_{BA} = \infty. \qquad (4.6.3)$$

The artificiality of the model is in the last assumption. Two adjacent units have zero probability of being in different states. This assumption is equivalent to imposing a uniformity throughout the system; either all are in the A state, or all are in the B state. This is an example of an "all or nothing" model.

We can now introduce an external means that can force transitions between the two states. We choose the chemical potential or the activity of a ligand to produce this change. An equivalent means is an external tension as is discussed in section 4.6.3.

Each unit has a single site that can adsorb one ligand. For simplicity it is assumed that the binding energy of the ligand is U in either the A or on the B states. In addition we have nearest-neighbor interactions between the ligands which are

$$U_{AA} = \varepsilon_2 \tag{4.6.4}$$

$$U_{BB} = 0. \tag{4.6.5}$$

Since the states AB and BA are impossible, we need not describe the ligand–ligand interaction for such a pair.

The qualitative behavior of this system can be easily visualized.

Since $\varepsilon_1 < 0$, the empty system (no ligands) will clearly favor the A state, i.e., all units will be in the A state, or the system will be in the A "phase." If also $\varepsilon_2 < 0$, then we shall still have a one-phase system with interacting ligands. The phase A will be stable for any activity of the ligand. If, on the other hand, we choose $\varepsilon_2 > 0$—i.e., ligand–ligand repulsion—we should expect the following phenomenon. Initially, the system starts in the A phase. As we "turn on" the supply of ligands, they tend to spread on the units, avoiding being at nearest-neighbor units. Upon further increase of the ligand activity, we shall reach a point where ligands will be forced to occupy nearest-neighbor units in spite of the unfavorable energy $\varepsilon_2 > 0$. Further increasing the ligand activity increases this unfavorable ligand–ligand repulsion until we reach a critical level of activity at which this unfavorable repulsion can overcome the energetically favorable A state of the units. This will cause a sharp transition from the A phase to the B phase. Once a B phase becomes more stable, further increasing the ligand activity will produce a simple Langmuir isotherm. This follows from the assumption $U_{BB} = 0$. Of course we could have chosen nonzero W_{BB} and U_{BB}, but to keep the number of parameters minimal we shall work with the parameters as specified above. The essential artificial assumption is (4.6.3). If we force one unit to make a change from A to B, all the other units will be "informed" of this change and will follow suit.

We now turn to the ST of the model that shows this behavior. The canonical PF is

$$Q(T, M, N) = \sum_s \exp[-\beta E(\mathbf{s})], \tag{4.6.6}$$

where the sum is over all configurations with a fixed number of ligands N. Here each unit can be in one of four states: A empty, A occupied, B empty, and B occupied. We shall refer to these states as 1, 2, 3, and 4.

To remove the condition of fixed N, we open the system with respect to the ligands. The grand PF is now

$$\Xi(T, M, \lambda) = \sum_{N=0}^{\infty} Q(T, M, N)\lambda^N$$

$$= \sum_s \lambda^{N(\mathbf{s})} \exp[-\beta E(\mathbf{s})], \tag{4.6.7}$$

where now the sum is over all configurations of the M units. Note that N is not a variable in the second form on the rhs of (4.6.7) and it should be expressed in terms of the variables s_i. This can be done by defining the counting function

$$N(\mathbf{s}) = \sum_{i=1}^{M} (\delta_{s_i,2} + \delta_{s_i,4}),$$ (4.6.8)

where δ is the Kronecker delta function. In the sum above we scan the entire system; a unit either in state 2 (i.e., A-occupied) or in state 4 (i.e., B-occupied) will contribute unity to this sum. Therefore the sum over i counts the number of ligands in the system at any particular configuration $\mathbf{s}$.

We define the matrix elements

$$\langle s_i|\mathbf{P}|s_{i+1}\rangle = (f_{s_i} f_{s_{i+1}})^{1/2} \exp[-\beta E(s_i, s_{i+1})],$$ (4.6.9)

where

$$f_{s_i} = \lambda^{(\delta_{s_i 2} + \delta_{s_i 4})/2}.$$ (4.6.10)

Thus each consecutive pair of units will have an energy parameter $E(s_i, s_{i+1})$ and a factor λ if both units are occupied, and a factor $\lambda^{1/2}$ if only one unit is occupied. The grand PF is thus

$$\Xi(T, M, \lambda) = \sum_{\mathbf{s}} \langle s_1|\mathbf{P}|s_2\rangle\langle s_2|\mathbf{P}|s_3\rangle \cdots \langle s_M|\mathbf{P}|s_1\rangle$$

$$= \mathrm{Tr}\ \mathbf{P}^M.$$ (4.6.11)

We define

$$x = \exp(-\beta\varepsilon_2), \qquad z = \exp(-\beta\varepsilon_1)$$ (4.6.12)

and write the matrix $\mathbf{P}$ explicitly

$$\mathbf{P} = \begin{pmatrix} z & z\lambda^{1/2} & 0 & 0 \\ z\lambda^{1/2} & zx\lambda & 0 & 0 \\ 0 & 0 & 1 & \lambda^{1/2} \\ 0 & 0 & \lambda^{1/2} & \lambda \end{pmatrix}.$$ (4.6.13)

The grand PF is thus

$$\Xi(T, M, \lambda) = \mathrm{Tr}\ \mathbf{P}^M = \sum_{i=1}^{4} \gamma_i^M,$$ (4.6.14)

where γ_i are the eigenvalues of the matrix $\mathbf{P}$. In contrast to previous examples we have retained here all the four eigenvalues. The reason is that we shall view γ_i as a function of the activity λ. The choice of the largest eigenvalue will depend on the value of λ.

The four eigenvalues are obtained from the solution of the secular equation

$$|\mathbf{P} - \gamma\mathbf{I}| = 0.$$ (4.6.15)

Since $\mathbf{P}$ is a 4×4 matrix, we have in general four eigenvalues. In view of the particular choice of the interaction energies, the solution for γ is quite easy. First, we noticed that the determinant in Eq. (4.6.15) can be factorized into a product of two determinants,

$$|\mathbf{P} - \gamma\mathbf{I}| = \begin{vmatrix} z - \gamma & z\lambda^{1/2} \\ z\lambda^{1/2} & zx\lambda - \gamma \end{vmatrix} \times \begin{vmatrix} 1 - \gamma & \lambda^{1/2} \\ \lambda^{1/2} & \lambda - \gamma \end{vmatrix},$$ (4.6.16)

where now we have to solve two 2×2 secular equations. These are

$$(z - \gamma)(zx\lambda - \gamma) - z^2\lambda = 0 \tag{4.6.17}$$

and

$$(1 - \gamma)(\lambda - \gamma) - \lambda = 0. \tag{4.6.18}$$

These two equations provide the four eigenvalues which we group in two pairs:

$$\gamma_{\pm A} = \frac{z}{2}[(1 + \lambda x) \pm \sqrt{(1 - \lambda x)^2 + 4\lambda}] \tag{4.6.19}$$

$$\gamma_{\pm B} = \tfrac{1}{2}[(1 + \lambda) \pm (1 + \lambda)]. \tag{4.6.20}$$

The significance of the two sets of solutions is the following: Suppose the system could be only in phase A, in which case each unit can be either empty or occupied. This is a two-state system for which the secular equation is (4.6.17) and the largest eigenvalue is $\gamma_A = \gamma_{+A}$. The grand PF of the pure A phase is thus

$$\Xi_A(T, M, \lambda) = \gamma_A^M = \exp(-\beta\mu_A M), \tag{4.6.21}$$

where μ_A is the chemical potential of the units of the A phase. Likewise, for the pure B phase, the corresponding secular equation is (4.6.18) and the largest eigenvalue is $\gamma_B = \gamma_{+B} = (1 + \lambda)$. The grand PF in this case is

$$\Xi_B(T, M, \lambda) = \gamma_B^M = (1 + \lambda)^M = \exp(-\beta\mu_B M). \tag{4.6.22}$$

Note that in (4.6.22) we have a simple Langmuir case, whereas in (4.6.21) we have a simple two-state Ising model.

The PF of our system is given in (4.6.14). We see from (4.6.14), (4.6.19), and (4.6.20) that the selection of the largest eigenvalue (among the four γ_i, $i = 1, 2, 3, 4$) depends on the parameters T and λ (as well as on the molecular parameters—but these are considered to be fixed). In the following we assume that $\varepsilon_1 < 0$, $\varepsilon_2 > 0$, and $|\varepsilon_2| > |\varepsilon_1|$. This is the case of interest discussed qualitatively at the beginning of this section.

Consider two limiting cases:

4.6.1.1. λ Very Small

We expand γ_A to linear term in λ and obtain

$$\gamma_A = z(1 + \lambda) \tag{4.6.23}$$

$$\gamma_B = (1 + \lambda). \tag{4.6.24}$$

In our particular choice of parameters, $z > 1$; therefore the largest eigenvalue of $\mathbf{P}$ is γ_A. In view of (4.6.21), the result $\gamma_A > \gamma_B$ is equivalent to the result

$$\mu_A < \mu_B; \tag{4.6.25}$$

i.e., the units in phase A are more stable than those in phase B.

4.6.1.2. λ Very Large

In this limit we take $|\lambda x| \gg 1$ and obtain, from (4.6.19) and (4.6.20),

$$\gamma_A = \lambda xz \tag{4.6.26}$$

$$\gamma_B = \lambda. \tag{4.6.27}$$

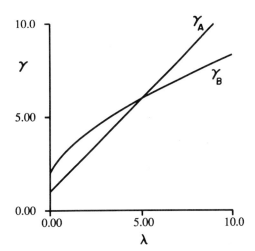

FIGURE 4.11. γ_A and γ_B as functions of λ for $x = 0.1$ and $z = 2$.

In our particular choice of parameters, $z > 1$ and $x < 1$, but $zx < 1$. Therefore, in this limit,

$$\gamma_A < \gamma_B. \tag{4.6.28}$$

In view of (4.6.22), this is equivalent to

$$\mu_A > \mu_B; \tag{4.6.29}$$

i.e., the units in phase B are more stable than those in phase A.

The dependence of γ_A and γ_B on λ is illustrated in Fig. 4.11. Clearly at the point where the two curves intersect, the relative stability of the phases switches from A to B or vice versa. The transition point λ_t occurs when $\gamma_A = \gamma_B$, i.e., when

$$1 + \lambda_t = \frac{z}{2}[(1 + \lambda_t x) + \sqrt{(1 - \lambda_t x)^2 + 4\lambda_t}]. \tag{4.6.30}$$

4.6.2. Thermodynamics

The fundamental connection with thermodynamics is

$$-kT \ln \Xi(T, M, \lambda) = \mu_u M. \tag{4.6.31}$$

Here μ_u may be interpreted as the chemical potential of the units (see section 4.4.1). If γ denotes the largest eigenvalue of $\mathbf{P}$, then

$$-\beta\mu_u = \ln \gamma. \tag{4.6.32}$$

As we have seen, μ_u will switch from μ_A to μ_B as the activity of the ligand λ increases. Figure 4.12 shows the dependence of μ_A and μ_B on λ. The stable phase is indicated by the heavy line, i.e., the curve with the lowest chemical potential. The phase transition occurs at the point of intersection of the μ_A and μ_B curves.

The total differential of $\mu_u M$ is obtained from

$$\mu_u M = A - \mu N \tag{4.6.33}$$

$$d(\mu_u M) = -S\,dT + \mu_u\,dM - N\,d\mu. \tag{4.6.34}$$

The entropy per unit of the system is given by

$$\frac{S}{M} = \frac{-\partial \mu_u}{\partial T} = k \ln \gamma + kT \frac{\partial \ln \gamma}{\partial T}. \tag{4.6.35}$$

The average number of ligands per unit, or the binding isotherm, is given by

$$\theta = \frac{\bar{N}}{M} = \frac{\lambda}{\gamma} \frac{\partial \gamma}{\partial \lambda}. \tag{4.6.36}$$

The energy per particle is

$$\frac{E}{M} = \frac{A}{M} + \frac{TS}{M} = kT\theta \ln \lambda + kT^2 \frac{\partial \ln \gamma}{\partial T}. \tag{4.6.37}$$

Finally, the heat capacity per particle is

$$\frac{C}{M} = \frac{1}{M} \frac{\partial E}{\partial T}. \tag{4.6.38}$$

The binding isotherm is plotted in Fig. 4.13 as a function of λ for one selected values of the molecular parameters of this model. At the transition point λ_t, the system crosses from one isotherm to another.

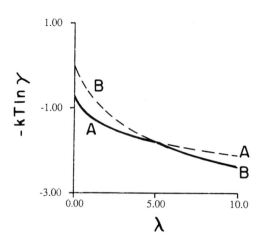

FIGURE 4.12. μ_A and μ_B as functions of λ for $x = 0.1$ and $z = 2$. The bold line indicates the stable phase.

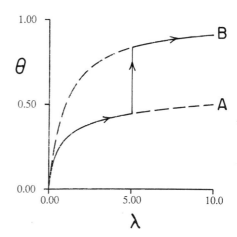

FIGURE 4.13. $\theta_A(\lambda)$ and $\theta_B(\lambda)$ for the two phases. At the transition point λ_t, the system crosses from one isotherm to another.

4.6.3. Phase Transition in the PV Diagram

Now that we have seen how to obtain a phase transition in a one-dimensional system, it is easy to construct a model in which a phase transition takes place in the PV or PL plan. This is the more common phase transition in a one-component system. The simplest model is a modification of the "water" model treated in section 4.5.4. We again assume two orientations of the particles, say horizontal and vertical, but the particles are free to move along the line between 0 and L. The pair potentials are the following, as shown in Fig. 4.14

$$U_{AA}(R) = \begin{cases} \infty & R < \sigma_A \\ \varepsilon_A & \sigma_A \leq R \leq \sigma_A + \delta \\ 0 & R > \sigma_A + \delta \end{cases} \qquad (4.6.39)$$

$$U_{BB}(R) = \begin{cases} \infty & R < \sigma_B \\ \varepsilon_B & \sigma_B \leq R \leq \sigma_B + \delta \\ 0 & R > \sigma_B + \delta \end{cases} \qquad (4.6.40)$$

$$U_{AB}(R) = U_{BA}(R) = \infty. \qquad (4.6.41)$$

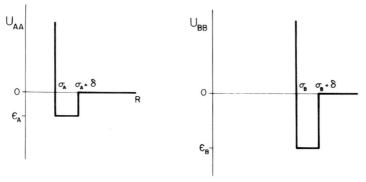

FIGURE 4.14. Pair potentials defined in (4.6.39) and (4.6.40).

The main feature responsible for the occurrence of phase transition is (4.6.41); i.e., the system will be either all in the A phase or all in the B phase.

In this model the treatment of section 4.6.1 applies with one essential difference: Instead of the secular equation (4.6.13), we now have

$$\begin{vmatrix} P_{AA} - \lambda & 0 \\ 0 & P_{BB} - \lambda \end{vmatrix} = 0. \tag{4.6.42}$$

Therefore the eigenvalues can be either

$$\lambda_1 = P_{AA} \tag{4.6.43}$$

or

$$\lambda_2 = P_{BB}, \tag{4.6.44}$$

where

$$P_{AA} = \frac{\exp(-\beta P \sigma_A)}{\beta P} [1 - (e^{-\beta \varepsilon_A} - 1)(e^{-\beta P \delta} - 1)] \tag{4.6.45}$$

$$P_{BB} = \frac{\exp(-\beta P \sigma_B)}{\beta P} [1 - (e^{-\beta \varepsilon_B} - 1)(e^{-\beta P \delta} - 1)]. \tag{4.6.46}$$

Since P_{AA} and P_{BB} are functions of both P and T, the determination of the larger eigenvalue would depend on these variables. From (4.6.45) and (4.6.46), it is clear that if $\sigma_A = \sigma_B$ and, say, $|\varepsilon_A| > |\varepsilon_B|$, then $P_{AA} > P_{BB}$ and the phase A will always be more stable, and no phase transition can occur. Similarly, if $\varepsilon_A = \varepsilon_B$ but, say, $\sigma_A > \sigma_B$, then $P_{AA} < P_{BB}$ and the phase B will be more stable at any P and T. In order to obtain a phase transition we need to choose the two curves of P_{AA} and P_{BB} in such a way that an intersection point occurs at some P, T. At this intersection we obtain the transition from one phase to another. This can occur if we choose different diameters $\sigma_A \neq \sigma_B$ and different interaction energies $\varepsilon_A \neq \varepsilon_B$.

The condition for the existence of phase transition is $\lambda_1 = \lambda_2$ or $P_{AA} = P_{BB}$, i.e.,

$$\exp[-\beta P(\sigma_A - \sigma_B)] = \frac{1 - (e^{-\beta \varepsilon_B} - 1)(e^{-\beta P \delta} - 1)}{1 - (e^{-\beta \varepsilon_A} - 1)(e^{-\beta P \delta} - 1)}. \tag{4.6.47}$$

An approximate estimate of the transition point can be obtained for the case of $\beta P \delta \ll 1$,

$$-\beta P(\sigma_A - \sigma_B) \approx -\beta P \delta (e^{-\beta \varepsilon_A} - e^{-\beta \varepsilon_B}). \tag{4.6.48}$$

Thus if we choose $\sigma_A < \sigma_B$ we must also choose $|\varepsilon_A| < |\varepsilon_B|$ to obtain an intersection point where $\lambda_1 = \lambda_2$. Such a choice is shown in Fig. 4.14.

The qualitative behavior of the system is quite clear. Starting from very low pressure, we can choose the parameters σ_A, σ_B, ε_A, and ε_B such that $P_{AA} < P_{BB}$. Therefore, initially we shall be moving on the $\bar{l} = \bar{l}(P)$ line of the B phase. As we increase the pressure, at constant temperature, $\bar{l}$ cannot be smaller than σ_B; therefore at some point the system as a whole must switch to the A phase—allowing $\bar{l}$ to decrease further. From there on, we shall be moving on the $\bar{l} = \bar{l}(P)$ curve of the A phase. Eventually, for $P \to \infty$, we shall reach the smallest possible length $\bar{l} = \sigma_A$.

4.7. HELIX–COIL TRANSITION IN PROTEINS

4.7.1. The Problem of Protein Denaturation

Proteins are known to undergo a process of denaturation upon heating or changing the solvent condition. By denaturation we refer to the loss of the specific function—say, enzymatic activity—of the protein. It is also known that this process involves the breakdown of the specific three-dimensional structure of the protein, which is essential for its function, i.e., a transition from a well-defined folded form (**F**) into a random or partially random unfolded form (**U**). We denote this process schematically as

$$\mathbf{F} \rightleftarrows \mathbf{U} \qquad (4.7.1)$$

and assume that the transition is reversible.

There are many experimental means by which one can follow the transition from **F** to **U** as a function of temperature, pressure, pH, addition of solutes, etc. For instance, one may follow the absorption spectra of a specific group in the protein as a function of temperature. As the protein unfolds, the environment of that specific group changes and this change will be revealed in the absorption spectra. We can account for the general features of this curve by using a simple, two-state-model approach, due to Schellman.[2]

Let ρ_F and ρ_U be the densities of the **F** and the **U** forms, respectively. At equilibrium we have the equality of the chemical potentials

$$\mu_F = \mu_U, \qquad (4.7.2)$$

or equivalently

$$\left(\frac{\rho_U}{\rho_F}\right)_{eq} = \exp(-\beta\Delta\mu^0), \qquad (4.7.3)$$

where $\Delta\mu^0 = \mu_U^0 - \mu_F^0$ is the change in the standard chemical potential for the reaction (4.7.1). In writing (4.7.3) we have assumed that the protein is very dilute in the solvent. If this is not so, then $\Delta\mu^0$ should be replaced by the difference of pseudochemical potentials $\Delta\mu^* = \mu_U^* - \mu_F^*$.

If ρ is the total density of the protein, $\rho = \rho_F + \rho_U$, then the relative fractions of the protein in the two forms are $x_U = \rho_U/\rho$ and $x_F = \rho_F/\rho$, and (4.7.3) can be rewritten as

$$x_F = [1 + \exp(-\beta\Delta\mu^0)]^{-1}. \qquad (4.7.4)$$

Note that x_F is the fraction of the protein in the F form, not the mole fraction of F in the solution.

We now write

$$\Delta\mu^0 = \Delta H^0 - T\Delta S^0 \qquad (4.7.5)$$

and make the extremely simplifying assumptions that $\Delta\mu^0$ depends linearly on the number of amino acid residues n and that ΔH^0 and ΔS^0 are independent of temperature. Thus we write

$$\Delta\mu^0 = n\Delta\mu_r^0 + C = n(\Delta H_r^0 - T\Delta S_r^0) + C, \qquad (4.7.6)$$

where the subscript r refers to the quantity per amino acid residue.

In terms of fixed values of ΔH_r^0, ΔS_r^0, and C, we can write the explicit dependence of x_F on temperature as

$$x_F = \left[1 + \exp\left(-\frac{n\Delta H_r^0 + C}{kT} + \frac{n\Delta S_r^0}{k}\right)\right]^{-1}.$$ (4.7.7)

The assumptions made above are equivalent to the statement that ΔS_r^0 is related essentially to the ratio of the degeneracies of the two states and ΔH_r^0 is the difference in energies of the two states. For a particular choice of these parameters, we plot x_F as a function of T in Fig. 4.15. Note the similarity with the curve of a two-state model as treated in section 2.4.

The transition temperature T_{tr} can be conveniently defined as the temperature at which $x_F = x_U = \frac{1}{2}$. Hence from (4.7.7) we obtain

$$T_{tr} = \frac{n\Delta H_r^0 + C}{n\Delta S_r^0}.$$ (4.7.8)

The sharpness of the transition can be defined at T_{tr} by

$$\eta = \left(\frac{\partial x_U}{\partial T}\right)_{T=T_{tr}} = \frac{n\Delta H_r^0 - C}{4kT_{tr}^2}.$$ (4.7.9)

Thus for the choice of the parameters of ΔH_r^0, ΔS_r^0, and C in Fig. 4.15, we see that the transition temperature moves to the left and becomes sharper as we increase the number of residues. For $n \to \infty$, $T_{tr} \to \Delta H_r^0 / \Delta S_r^0$ and the transition becomes infinitely sharp, resembling a phase transition.

Although the above theory does not tell us anything about the molecular content of C, ΔH_r^0, and ΔS_r^0, the qualitative rationale of the "melting curves" is quite clear. At very low temperature, $T \to 0$, the energy difference represented by ΔH_r^0 dominates $\Delta \mu_r^0$, and therefore, $x_F \to 1$. At the other extreme limit, $T \to \infty$, the ratio of the degeneracies of the two states represented by $T\Delta S_r^0$ is the dominating term in $\Delta \mu_r^0$; hence in this limit x_F becomes very small. Thus, in a general sense, we have here the same energy–entropy competition as we have seen in the simplest two-state model in section 2.4, as well as in section 4.4. In that model we saw that at $T \to 0$, the mole fraction of the low energy state

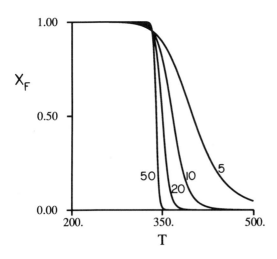

FIGURE 4.15. Typical melting curve of a protein. The mole fraction of the folded form x_F is plotted as a function of temperature. The curves are drawn for various values of n in Eq. (4.7.7), with $C/k = -2000$ K, $\Delta H_r^0/k = 1500$ K, $\Delta S_r^0/k = 6$.

dominates. At $T \to \infty$, the relative fraction of the two species depends on the ratio of the degeneracies of the two states. In the protein case, we have implicitly assumed that the degeneracy of the U form is far greater than the degeneracy of the F form, and that therefore at this limit $x_F \to 0$.

The denaturation of a real protein is extremely complex. It involves the interactions among twenty different amino acid residues as well as interactions between the protein and the solvent. Some of these interactions will be further discussed in Chapter 8. Here we shall focus on one aspect of the process, the so-called helix–coil transition process.

It is known that a large part of the tertiary structure of some proteins consists of packed helical segments. As the temperature increases, these helices unfold and attain a random coil conformation. The transition from the helical to the random coil is similar to the denaturation process. Therefore it is believed that the study of the helix–coil transition is an essential step in the understanding of the denaturation process. It is known, however, that protein denaturation is a far more complex process. It involves the breakdown of an intricate three-dimensional tertiary structure, only part of which is the helix–coil transition. Some proteins do not even contain substantial content of helical structure. Therefore, the study of this particular transition will not give us the answer to the far more complex denaturation process. In the rest of this section we shall focus on the theory of the helix–coil transition only. We shall return to the folding–unfolding process of proteins in Chapter 8.

4.7.2. The Helix–Coil Transition

We describe here a simplified version of the helix–coil $H \to C$ transition theory developed by Lifson and Roig.[3] The main assumptions made in the theory are the following:

1. Instead of a heteropolypeptide we use a homopolypeptide. The units are treated as being identical. Each unit is identified as a single amino acid residue, the boundaries of which are marked by the dotted vertical lines in Fig. 4.16.
2. The state of each unit is determined only by a pair of rotational angles $\Phi_i \psi_i$ about the single bonds, as indicated in Fig. 4.16. No rotation about the partially double bond (the amide bond) is allowed. We also ignore any changes of state that originate from the side chains, denoted by R in Fig. 4.16.
3. The solvent effect is ignored. We note however that the solvent could have a profound effect on the theory. Some aspects of the solvent effects are discussed in Chapter 8.

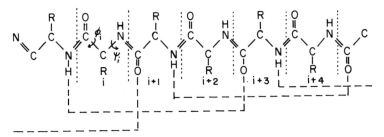

FIGURE 4.16. Schematic illustration of the hydrogen bonds between the NH of the $(i-1)$th residue and the C=O of the $(i+3)$th residue. Dotted lines separate amino-acid residues. Dashed lines indicate hydrogen bonds (only two are shown).

4. The total potential energy of the polymer is assumed to have the form

$$U(\mathbf{\Phi}, \mathbf{\psi}) = \sum_{i=1}^{N} [U^*(\Phi_i, \psi_i) + \varepsilon_{HB}G_i(\Phi_{i-1}, \psi_{i-1}, \Phi_i, \psi_i, \Phi_{i+1}, \psi_{i+1})], \quad (4.7.10)$$

where U^* is the potential of hindered rotation, depending only on the pair of angles $\Phi_i\psi_i$ of one unit; ε_{HB} is a hydrogen-bond energy between a $C = 0$ and an N—H groups, and $\mathbf{G}$ is a geometrical factor, to be defined below.

The entire range of variation of the angles $\Phi_i\psi_i$ is divided into three regions, roughly coinciding with the α-helix, the β-strands, and the inaccessible regions in the Ramachandran map (Fig. 4.17). We thus define three gross states of each unit in terms of the more detailed states defined by $\Phi_i\psi_i$. These are

The helix state (H)	whenever	$(\Phi_i\psi_i) \in \alpha$-helix region
The coiled state (C)	whenever	$(\Phi_i\psi_i) \in \beta$ region
The inaccessible state (I)	whenever	$(\Phi_i\psi_i) \notin \alpha$-helix region and
		$(\Phi_i\psi_i) \notin \beta$ region.

$$(4.7.11)$$

Another way to divide the $\Phi\psi$ plane is: H as above and the entire remaining region as C. Clearly, with these definitions, the microscopic state of the protein described by $(\Phi_1\psi_1 \cdots \Phi_N\psi_N)$ is translated into a coarser description of the polymer in terms of a sequence $HHCHIIHCC$.

We now assume that when a unit is in the I state, its energy $U(I)$ is infinitely positive. Therefore any configuration that includes at least one unit in the I state will have a zero probability; hence, we can ignore all sequence including the I states.

The $\mathbf{G}$ factor in (4.7.10) is now defined as follows

$$G_i(\Phi_{i-1}, \psi_{i-1}, \Phi_i, \psi_i, \Phi_{i+1}, \psi_{i+1}) = \begin{cases} 1 & \text{if } i-1, i \text{ and } i+1 \text{ are in the } H \text{ state} \\ 0 & \text{otherwise.} \end{cases}$$

$$(4.7.12)$$

The qualitative idea underlying definition (4.7.12) is that whenever three successive units are oriented in such a way that they fall in the α-helix region, the $i - 2$ and $i + 2$ units

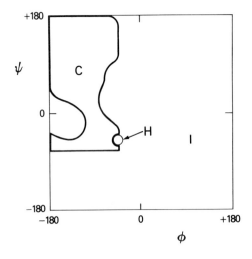

FIGURE 4.17. A schematic Ramachandran map. The H, C, and I regions are indicated.

will assume an orientation in a way that permits them to form a direct hydrogen bond. The energy contributed by this bond is $\varepsilon_{HB} < 0$, and it is arbitrarily assigned to the ith unit. The scheme of hydrogen bonds between the ith and the $(i + 4)$th units is denoted by the broken lines in Fig. 4.16.

Thus the potential function $U(\mathbf{\Phi}, \mathbf{\psi})$ as defined in (4.7.10) includes the energies of the state of each unit and hydrogen-bond energies between the $i - 2$ and $i + 2$ (or i and $i + 4$) units. No other nearest-neighbor or next-nearest-neighbor interactions are presumed in this system. Thanks to the simplifying assumptions made above, we can describe the system by a 4×4 matrix. In Chapter 8 we shall see that inclusion of the solvent would require a substantial modification of the theory.

4.7.3. The Partition Function

The configurational PF of the system of N units at temperature T is

$$Z(N, T) = \int_0^{2\pi} \cdots \int_0^{2\pi} d\Phi_1\, d\psi_1 \cdots d\Phi_N\, d\psi_N \exp[-\beta U(\mathbf{\Phi}, \mathbf{\psi})]. \quad (4.7.13)$$

The first step is to convert from the microscopic description $(\mathbf{\Phi}, \mathbf{\psi})$ to the coarser description in terms of the states H, C and I. For each unit i we can write, as in section 4.4.2,

$$\int_0^{2\pi} \int_0^{2\pi} d\Phi_i\, d\psi_i = \iint_H d\Phi_i\, d\psi_i + \iint_C d\Phi_i\, d\psi_i + \iint_I d\Phi_i\, d\psi_i. \quad (4.7.14)$$

The entire range of integration (the square area of Fig. 4.17) is divided into three nonoverlapping regions. Making this division of the integration region for each i, we can rewrite the PF (4.7.13) as

$$Z(N, T) = \sum_{\mathbf{s}} \iint_{s_1} d\Phi_1\, d\psi_1 \iint_{s_2} d\Phi_2\, d\psi_2 \cdots \iint_{s_N} d\Phi_N\, d\psi_N \exp[-\beta U(\mathbf{\Phi}, \mathbf{\psi})], \quad (4.7.15)$$

where s_i can be H, C, or I. But since the potential energy is infinite for each state I, any term in which an I state appears is zero. Therefore we can ignore these states and view $\mathbf{s}$ as a sequence of Hs and Cs only.

Note that (4.7.15) cannot be factored into a product of integrals. This could have been done if $\mathbf{G} \equiv 0$ in (4.7.10). Nor can we write it as the trace of a 2×2 matrix as we have done in the simple Ising model. The situation is more complicated because of the hydrogen-bonding possibility included in $U(\mathbf{\Phi}, \mathbf{\psi})$.

We next define the following quantities:

$$\iint_C d\Phi_i\, d\psi_i \exp[-\beta U^*(\Phi_i \psi_i)] = u \quad (4.7.16)$$

$$\iint_H d\Phi_i\, d\psi_i \exp[-\beta U^*(\Phi_i, \psi_i) - \beta \varepsilon_{HB} G_i] = \begin{cases} w & \text{if } i-1 \text{ and } i+1 \text{ are } H \\ v & \text{otherwise.} \end{cases} \quad (4.7.17)$$

Thus if the ith unit is in state C, the integrand is $\exp[-\beta U^*(\Phi_i \psi_i)]$ $[\mathbf{G} = 0$, according to (4.7.12)] and the integration extends over the relatively larger region of integration denoted by C. If the ith unit is in state H and both of its immediate neighbors $i - 1$ and

$i + 1$ are also in H, then $\mathbf{G}_i = 1$ and we have assigned a hydrogen-bond energy to this state. The corresponding integral is essentially

$$w = \Omega_H \exp(-\beta \varepsilon_{HB}), \qquad (4.7.18)$$

where Ω_H is the region corresponding to the α-helix area in the Φ, ψ plane. If U^* had been constant in the entire region of the C state, we could also have written

$$u = \Omega_C \exp(-\beta U^*); \qquad (4.7.19)$$

otherwise, U^* in (4.7.19) should be reinterpreted as an average potential energy within the Ω_C region.

More precisely

$$u = \Omega_C \iint_C \frac{d\Phi_i \, d\psi_i}{\Omega_C} \exp[-\beta U^*(\Phi_i, \psi_i)]$$

$$= \Omega_C \langle \exp(-\beta U^*) \rangle, \qquad (4.7.20)$$

where $\langle \exp(-\beta U^*) \rangle$ is an average over all conformations pertaining to the C region. Note that the C region is assumed to be much larger than the H region, i.e.,

$$\Omega_C \gg \Omega_H. \qquad (4.7.21)$$

On the other hand, since the hydrogen-bond energy is much larger than the typical energy U^*, we assume that

$$w \gg u. \qquad (4.7.22)$$

Finally, if i is in state H, but either $i - 1$ or $i + 1$ is not, then $\mathbf{G}_i = 0$, and from (4.7.17) we obtain

$$\iint_H d\Phi_i \, d\psi_i \exp[-\beta U^*(\Phi_i, \psi_i)] = v, \qquad (4.7.23)$$

where v is now interpreted in a similar manner to (4.7.20) as

$$v = \Omega_H \iint_H \frac{d\Phi_i \, d\psi_i}{\Omega_H} \exp[-\beta U^*(\Phi_i, \psi_i)]$$

$$= \Omega_H \langle \exp(-\beta U^*) \rangle. \qquad (4.7.24)$$

Note that the averages in (4.7.20) and (4.7.24) are taken with respect to different probability distributions. Nevertheless, since these two averages are expected to be of similar order of magnitude, the difference between u and v is mainly due to the difference in the regions Ω_C and Ω_H. Hence from (4.7.21) we also assume that

$$u \gg v. \qquad (4.7.25)$$

Thus the inequality (4.7.25) is mainly due to the larger degree of degeneracy of the C state, whereas the inequality (4.7.22) is due to the large energy difference between the two states H and C, in spite of the large difference in the degeneracies of the two states.

With the above definitions of w, u, and v, we can view each term in the sum (4.7.15) as a product of us, vs and ws. For instance, a typical term of (4.7.15) for $N = 5$ would contribute a factor

$$\iint_H d\Phi_1\, d\psi_1 \iint_C d\Phi_2\, d\psi_2 \iint_H d\Phi_3\, d\psi_3 \iint_H d\Phi_4\, d\psi_4 \iint_H d\Phi_5\, d\psi_5 = vuvwv. \quad (4.7.26)$$

Thus each unit in the C state always contributes u; each unit in the H state contributes a factor v if it has a neighbor on either side which is C, and a factor w if both neighbors are in state H. We see that in order to assign the correct factors v, u, and w to each term in (4.7.15), we need to check each triplet of units. Thus, as in section 4.4.3, we define the following quantities:

$$Q(s_{i-1}, s_i, s_{i+1}) = \begin{cases} u & \text{if } s_i = C \\ v & \text{if } s_i = H \text{ and either } s_{i-1} = C \text{ or } s_{i+1} = C \\ w & \text{if } s_i = H \text{ and } s_{i+1} = s_{i-1} = H. \end{cases} \quad (4.7.27)$$

Altogether we have eight elements; $Q(C, C, C) = u$, $Q(H, C, C) = u$, $Q(C, C, H) = u$, $Q(H, C, H) = u$, $Q(H, H, H) = w$, $Q(H, H, C) = v$, $Q(C, H, H) = v$, and $Q(C, H, C) = v$. With these symbols we can rewrite the PF (4.7.15) as

$$Z(N, T) = \sum_s Q(s_1, s_2, s_3) Q(s_2, s_3, s_4) Q(s_3, s_4, s_5) \cdots Q(s_{N-1}, s_N, s_1), \quad (4.7.28)$$

where for simplicity we have closed the cycle, i.e., we identified the $N + 1$ unit with the first unit.

The sum (4.7.28) is still not the trace of a matrix. The situation is exactly the same as in the case treated in section 4.4.3. Therefore we proceed to rewrite the same sum (4.7.28) in an equivalent form which allows one to perform the summation using the rules of matrix multiplication. To do that we introduce the four index quantities defined by

$$P(s_1, s_2, s_2', s_3) = \begin{cases} Q(s_1, s_2, s_3) & \text{if and only if } s_2 = s_2' \\ 0 & \text{if } s_2 \neq s_2', \end{cases} \quad (4.7.29)$$

and conversely

$$Q(s_1, s_2, s_3) = P(s_1, s_2, s_2, s_3) = \sum_{s_2' = H,C} P(s_1, s_2, s_2', s_3). \quad (4.7.30)$$

Note that $P(s_1, s_2, s_2, s_3)$ is the same as $Q(s_1, s_2, s_3)$. We have simply repeated the state index of the central unit. In the last form on the rhs of (4.7.30) we have two terms, $P(s_1, s_2, s_2, s_3)$ and zero. We now write for each i

$$Q(s_{i-1}, s_i, s_{i+1}) = \sum_{s_i' = H,C} P(s_{i-1}, s_i, s_i', s_{i+1}), \quad (4.7.31)$$

and substitute in (4.7.28) to get an expanded sum of the form

$$Z(N, T) = \sum_{s_1 \cdots s_N} \prod_{i=1}^{N} Q(s_{i-1}, s_i, s_{i+1})$$

$$= \sum_{s_1 \cdots s_N} \sum_{s_1' \cdots s_N'} \prod_{i=1}^{N} P(s_{i-1}, s_i, s_i', s_{i+1}). \quad (4.7.32)$$

The sum on the rhs of (4.7.32) is now the trace of a matrix. The elements of this matrix are described by four indices, say $P_{ij,kl}$, and the rule of multiplication is

$$\sum_{kl} P_{ij,kl} P_{kl,nm} = (P^2)_{ij,nm}.$$
(4.7.33)

Applying the multiplication rule in (4.7.33) for the N pairs of indices, we obtain the trace of the matrix $\mathbf{P}$. For instance, for $N = 3$ the sum (4.7.32) is

$$
\begin{aligned}
Z(3, T) &= \sum_{s_1 s_2 s_3} \sum_{s_1', s_2', s_3'} P(s_1, s_2, s_2', s_3) P(s_2, s_3, s_3', s_1) P(s_3, s_1, s_1', s_2) \\
&= \sum_{s_1 s_2} \sum_{s_2' s_3} P(s_1, s_2, s_2', s_3) \sum_{s_3' s_1'} P(s_2', s_3, s_3', s_1') P(s_3', s_1', s_1, s_2) \\
&= \sum_{s_1 s_2} P(s_1, s_2, s_2', s_3) P^2(s_2', s_3, s_1, s_2) = \operatorname{Tr} \mathbf{P}^3.
\end{aligned}
$$
(4.7.34)

Note that each term in the sum (4.7.34) is zero if $s_i = s_i'$ for any i; hence we can exchange primed and unprimed variables without affecting the sum.

By extension to any number N, we can write the PF (4.7.32) as

$$Z(N, T) = \operatorname{Tr} \mathbf{P}^N,$$
(4.7.35)

where the matrix $\mathbf{P}$ is given by

$$
\mathbf{P} =
\begin{array}{cc|cccc}
 & 3 & H & C & H & C \\
1 & 2 & H & H & C & C \\
\hline
H & H & w & v & 0 & 0 \\
H & C & 0 & 0 & u & u \\
C & H & v & v & 0 & 0 \\
C & C & 0 & 0 & u & u \\
\end{array}
$$
(4.7.36)

Note that the eight nonzero elements are those listed in (4.7.27). The additional eight zeros were added by replacing each Q by two Ps in (4.7.30). The reading of the elements of the matrix is as in section 4.4.3. For instance, $P(H, C, C, H) = u$ is in the second row and third column. The matrix elements for which the two central indices are unequal are by definition (4.7.29) zeros. For instance, $P(H, H, C, H) = 0$ is in the first row and in the third column.

Once we have expressed the PF as a trace of a matrix, we can proceed as in previous treatments of the Ising model, by finding the largest eigenvalue of $\mathbf{P}$, etc. The secular equation is

$$|\mathbf{P} - \lambda \mathbf{I}| = 0.$$
(4.7.37)

The secular equation (4.7.37) leads in general to a fourth-degree equation for λ. It is clear from the structure of the matrix $\mathbf{P}$ that one root is always zero; i.e., substituting $\lambda = 0$ in (4.7.37) produces a determinant with two identical columns, so that $\lambda = 0$ is always a solution of (4.7.37). The other three roots are determined by the equation

$$\lambda^3 - \lambda^2(w + u) + \lambda(wu - uv) - uv^2 + uvw = 0.$$
(4.7.38)

To further simplify this equation, we can choose $u = 1$. This amounts to redefining the elements of **P** in units of u. Note however that $u = 1$ means that $U^* = 0$ in (4.7.20) and that Ω_C is chosen as unity. The required equation is now

$$f(\lambda, w, v) = \lambda^3 - \lambda^2(w + 1) + \lambda(w - v) - v^2 + vw = 0. \qquad (4.7.39)$$

Since we have assumed that $v \ll u \ll w$, we shall search for solutions of (4.7.39) near $v = 0$. For $v = 0$, the three roots of (4.7.39) are

$$\lambda_1 = w$$

$$\lambda_2 = 1 \qquad (4.7.40)$$

$$\lambda_3 = 0.$$

For small values of v we expect that the roots will be slightly different from (4.7.40).

Figure 4.18 shows a plot of the function $f(\lambda)$ for $v = 0$ and various values of w. In the general case, $f(\lambda) = 0$ has three different roots. For $v = 0$ and $w = 0$, two of the roots λ_1 and λ_3, coincide, and for $v = 0$ and $w = 1$, the two roots λ_1 and λ_2 coincide.

Figure 4.19 shows the variation of the three roots as a function of w for the three values of $v: v = 0$, $v = 0.1$, and $v = 0.2$. It is easy to see that for very small w, $0 < w \ll 1$, the largest root is $\lambda_2 = 1$, hence the PF is given by

$$Z = \lambda_2^N = 1. \qquad (4.7.41)$$

On the other hand, for very large w, $w \gg 1$, the largest root is $\lambda_1 = w$ and the PF is

$$Z = \lambda_1^N = w^N. \qquad (4.7.42)$$

We now recall that w is essentially the product of Ω_H, measured in units of Ω_C and $\exp(-\beta \varepsilon_{HB})$, measured with respect to the choice of $U^* = 0$. Therefore w is essentially

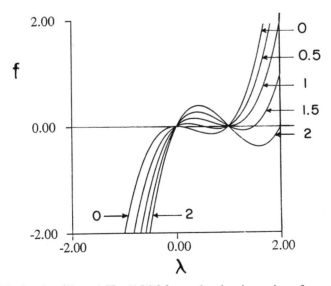

FIGURE 4.18. The function $f(\lambda, w, v)$ [Eq. (4.7.9)] for $v = 0$ and various values of w, as indicated next to each curve.

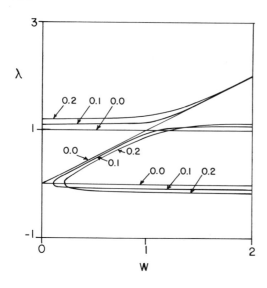

FIGURE 4.19. The variations of the three roots as a function of w for $v = 0$, $v = 0.1$, and $v = 0.2$.

an equilibrium constant for the $C \rightleftarrows H$ transition, i.e.,

$$w = \frac{\Omega_H \exp(-\beta \varepsilon_{HB})}{\Omega_C \exp(-\beta U^*)}. \qquad (4.7.43)$$

Thus for $T \to 0$, the hydrogen bond energy will dominate the equilibrium constant w, and irrespective of the value of the degeneracies Ω_H / Ω_C, the mole fraction of the hydrogen-bonded units is given

$$x(HB) = \frac{\partial \ln \lambda_1}{\partial \ln w} = \frac{\partial \ln w}{\partial \ln w} = 1 \qquad (T \to 0); \qquad (4.7.44)$$

i.e., all units will be hydrogen bonded (by a hydrogen-bonded unit we refer to a unit in the H state when two of its adjacent neighbors are also in the H state).

At the other limit, $T \to \infty$, the energy term becomes unimportant, and w is determined only by the ratio of the degeneracies (irrespective of the hydrogen-bond energy). Since we assume that $\Omega_H \ll \Omega_C$, we have at this limit the PF (4.7.41) and hence

$$x(HB) = \frac{\partial \ln \lambda_2}{\partial \ln w} = 0, \qquad (4.7.45)$$

which means that at $T \to \infty$, the behavior of the system is overwhelmed by the ratio of the degeneracies. As a result all the units will assume the C state.

A plot of $x(HB)$ as a function of w (which is equivalent to a plot against T) is shown in Fig. 4.20 for $v = 0.01$, $v = 0.1$, $v = 0.2$, and $v = 0.3$. Note that at $v = 0$, the transition is infinitely sharp. The mathematical reason is the crossing of the two roots λ_1 and λ_2 at $w = 0$. This is the same phenomenon we have studied in section 4.6.

Note that in Eq. (4.7.39) we have measured w and v with respect to u; i.e., v is redefined as the ratio

$$v = \frac{\Omega_H \langle \exp(-\beta U^*) \rangle}{\Omega_C \langle \exp(-\beta U^*) \rangle} \sim \frac{\Omega_H}{\Omega_C}. \qquad (4.7.46)$$

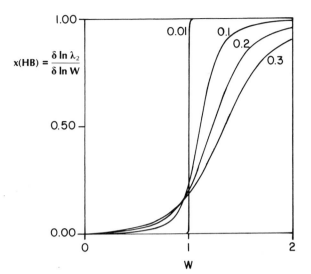

$$x(HB) = \frac{\delta \ln \lambda_2}{\delta \ln W}$$

FIGURE 4.20. x(HB) defined in (4.7.44) as a function of w for $v = 0.01$, $v = 0.1$, $v = 0.2$, and $v = 0.3$.

Thus the significance of $v \to 0$ is that the ratio of the degeneracies $\Omega_H/\Omega_C \to 0$. If we take $v = 0$, then in order to obtain a finite value of w in (4.7.43), we must assume the *HB* energy is infinitely large and negative. These extreme requirements are responsible for obtaining a sharp phase transition like in the case of $v = 0$.

We conclude by pointing out that the model treated in this section strictly pertains to a helix–coil transition in vacuum. The presence of the solvent causes more than rescaling of the various parameters of the model. We shall see in section 8.4 that the same model in a solvent requires a different theory.

REFERENCES

1. T. L. Hill, *Proc. Natl. Acad. Sci. U.S.A.* **57**, 227 (1967).
2. J. A. Schellman, *Compt. Rend. Trav. Lab. Carlsberg, Ser. Chim.* **29**, 230 (1955).
3. S. Lifson and A. Roig, *J. Chem. Phys.* **34**, 1963 (1961).

SUGGESTED READINGS

A variety of one-dimensional systems are discussed in

E. L. Lieb and D. C. Mattis, *Mathematical Physics in One Dimension* (Academic Press, New York, 1966).

Applications of statistical mechanics to polymers are covered in:

T. M. Birshtein and O. B. Ptitsyn, *Conformations of Macromolecules* (Interscience Publishers, New York, 1966).
P. J. Flory, *Statistical Mechanics of Chain Molecules* (Interscience Publishers, New York, 1969).

More specific applications to biopolymers:

D. Poland and A. Scheraga, *Theory of Helix–Coil Transitions in Biopolymers* (Academic Press, New York, 1970).

Theory of Liquids

5.1. INTRODUCTION

In the previous chapters of this book we have dealt mainly with solvable problems. The models could be solved either because they lack interactions (e.g., ideal gases) or because the interactions took place between a small number of particles (Chapter 3) or because the system itself was simple (Chapter 4). From this chapter through the rest of the book we shall be dealing with nonsolvable systems. Liquids, even the simplest ones consisting of hard-sphere particles, pose very difficult theoretical problems. The main difficulty is contained in the configurational partition function, which for simple particles has the form

$$Z = \int \cdots \int \exp[-\beta U(\mathbf{R}^N)]\, d\mathbf{R}^N. \qquad \cdot (5.1.1)$$

Even when we assume pairwise additivity of the total potential energy, this integral cannot be solved for any reasonable pair interaction. Of course the problems become immensely more difficult for nonspherical molecules interacting through nonadditive potentials.

This chapter will not deal with theories of liquids *per se*. Instead we shall present only general relations between thermodynamic quantities and molecular distribution functions. The latter are fundamental concepts which play a central role in the modern theoretical treatment of liquids and solutions. Acquiring familiarity with these concepts should be useful in the study of more complex systems such as aqueous solutions, treated in Chapters 7 and 8. As an exception, a brief outline of the scaled particle theory is presented in section 5.11. This theory, although originally aimed at studying hard-sphere systems, has been used in systems as complex as aqueous protein solutions. The main result that will concern us is the work required to create a cavity in a fluid. This quantity is fundamental in the study of solvation phenomena of simple solutes, as well as very complex ones such as proteins or nucleic acids.

5.2. MOLECULAR DISTRIBUTION FUNCTIONS

The notion of molecular distribution functions (MDF) command a central role in the theory of fluids. Of foremost importance among these are the singlet and the pair distribution functions. The following sections are devoted to describing and surveying the fundamental features of these two functions. We shall also briefly mention the general definitions of higher-order MDFs. These are rarely incorporated into actual applications, since very little is known about their properties.

The pair correlation function, introduced in section 5.2.2, conveys information on the mode of packing of the molecules in the liquid. This information is often referred to as representing a sort of order, or amount of structure that persists in the liquid. More-over, the pair correlation function forms an important bridge between molecular proper-ties and thermodynamic quantities. We start by defining the singlet and pair distribution functions for a system of *rigid*, not necessarily spherical particles.

5.2.1. The Singlet Distribution Function

The results obtained in this section are very simple and intuitively obvious. Neverthe-less, we shall elaborate on the details more than is really necessary. The reason is that, throughout this chaper as well as the rest of the book, some common probabilistic arguments will appear. These are easier to understand in the context of the singlet distri-bution function.

The system under consideration consists of N rigid particles at a given temperature and contained in a volume V. The basic probability density for such a system is (see section 1.5)

$$P(\mathbf{X}^N) = \frac{\exp[-\beta U_N(\mathbf{X}^N)]}{\int \cdots \int d\mathbf{X}^N \exp[-\beta U_N(\mathbf{X}^N)]}. \tag{5.2.1}$$

An average of any function of the configuration $F(\mathbf{X}^N)$ in the T, V, N ensemble is defined by

$$F = \int \cdots \int d\mathbf{X}^N P(\mathbf{X}^N)F(\mathbf{X}^N). \tag{5.2.2}$$

In some cases we use a special symbol such as $\langle F \rangle$ or $\bar{F}$ for an average quantity. However, we refrain from using this notation whenever the meaning of that quantity as an average is evident.

As a very simple example, let us express the average number of particles in a certain region S within the system. (A particle is said to be in the region S whenever its *center* falls within that region.) Let $N(\mathbf{X}^N, S)$ be the number of particles in S at a particular configuration $\mathbf{X}^N$ of the system. One may imagine taking a snapshot of the system at some instant and counting the number of particles that happen to fall within S at that configuration. An illustration is given in Fig. 5.1.

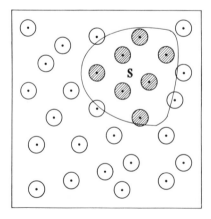

FIGURE 5.1. A region S within the system. A shaded circle indicates a particle whose center falls within S.

The average number of particles in S is, according to (5.2.2),

$$N(S) = \int \cdots \int d\mathbf{X}^N \, P(\mathbf{X}^N) N(\mathbf{X}^N, S). \tag{5.2.3}$$

This relation can be rewritten in a somewhat more complicated form that will turn out to be useful for later applications.

Let us define the characteristic function

$$A_i(\mathbf{R}_i, S) = \begin{cases} 1 & \text{if } \mathbf{R}_i \in S \\ 0 & \text{if } \mathbf{R}_i \notin S \end{cases}. \tag{5.2.4}$$

(The symbol $\in$ means "belongs to" or "contained in.") The quantity $N(\mathbf{X}^N, S)$ can be expressed as

$$N(\mathbf{X}^N, S) = \sum_{i=1}^{N} A_i(\mathbf{R}_i, S), \tag{5.2.5}$$

which simply means that in order to count the number of particles within S, we have to check the location of each particle separately. Each particle whose center falls within S will contribute unity to the sum on the rhs of (5.2.5); hence, the sum counts the number of particles in S. Introducing (5.2.5) into (5.2.3), we obtain

$$\begin{aligned} N(S) &= \int \cdots \int d\mathbf{X}^N \, P(\mathbf{X}^N) \sum_{i=1}^{N} A_i(\mathbf{R}_i, S) \\ &= \sum_{i=1}^{N} \int \cdots \int d\mathbf{X}^N \, P(\mathbf{X}^N) A_i(\mathbf{R}_i, S) \\ &= N \int \cdots \int d\mathbf{X}^N \, P(\mathbf{X}^N) A_1(\mathbf{R}_1, S). \end{aligned} \tag{5.2.6}$$

In (5.2.6) we have used the fact that all the particles are equivalent; hence, the sum over the index i produces N integrals having the same magnitude. We may therefore select one of these integrals, say $i = 1$, and replace the sum by N times one integral.

The mole fraction of particles within S is defined as

$$x(S) = N(S)/N = \int \cdots \int d\mathbf{X}^N \, P(\mathbf{X}^N) A_1(\mathbf{R}_1, S); \tag{5.2.7}$$

$x(S)$ is the average fraction of particles found in S. This quantity may also be assigned a probabilistic meaning that is often useful. To see this, we recall that the function $A_1(\mathbf{R}_1, S)$ employed in (5.2.7) has the effect of reducing the range of integration from V to a restricted range which fulfills the condition of $\mathbf{R}_1$ being located in S. Symbolically, this can be rewritten as

$$\int_V \cdots \int_V d\mathbf{X}^N \, P(\mathbf{X}^N) A_1(\mathbf{R}_1, S) = \int_{R_1 \in S} \cdots \int d\mathbf{X}^N \, P(\mathbf{X}^N) = P_1(S). \tag{5.2.8}$$

We recall that $P(\mathbf{X}^N)$ is the probability density of the occurrence of the "event" $\mathbf{X}^N$. Therefore, integration over all the events $\mathbf{X}^N$ for which the condition $\mathbf{R}_1 \in S$ is fulfilled gives the probability of the occurrence of the condition, i.e., $P_1(S)$ is the probability that a specific particle, say number 1, will be found in S. Of course, we could have chosen in

(5.2.6) any specific particle other than 1. From (5.2.7) and (5.2.8) we arrive at an important relation:

$$x(S) = P_1(S), \qquad (5.2.9)$$

which states that the mole fraction of particles in S equals the probability that a specific particle, say 1, will be found in S.

We now introduce the singlet molecular distribution function, which is obtained from $N(S)$ at the limit of a very small region S. First we note that $A_i(\mathbf{R}_i, S)$ can also be written as

$$A_i(\mathbf{R}_i, S) = \int_S \delta(\mathbf{R}_i - \mathbf{R}') \, d\mathbf{R}', \qquad (5.2.10)$$

where we have exploited the basic property of the Dirac delta function: The integral is unity if $\mathbf{R}_i \in S$ and zero otherwise.

If S is an infinitesimally small region $d\mathbf{R}'$, then we have

$$A_i(\mathbf{R}_i, d\mathbf{R}') = \delta(\mathbf{R}_i - \mathbf{R}') \, d\mathbf{R}'. \qquad (5.2.11)$$

Hence, from (5.2.6) we obtain

$$N(d\mathbf{R}') = d\mathbf{R}' \int \cdots \int d\mathbf{X}^N P(\mathbf{X}^N) \sum_{i=1}^{N} \delta(\mathbf{R}_i - \mathbf{R}'). \qquad (5.2.12)$$

The average local (number) density of particles in the element of volume $d\mathbf{R}'$ is defined by

$$\rho^{(1)}(\mathbf{R}') = N(d\mathbf{R}')/d\mathbf{R}' = \int \cdots \int d\mathbf{X}^N P(\mathbf{X}^N) \sum_{i=1}^{N} \delta(\mathbf{R}_i - \mathbf{R}'), \qquad (5.2.13)$$

which is also referred to as the singlet molecular distribution function.

The meaning of $\rho^{(1)}(\mathbf{R}')$ as local density will prevail in all of our applications. However, in some cases one may also assign to $\rho^{(1)}(\mathbf{R}')$ the meaning of probability density. This must be done with some care, as will be shown below. First, we rewrite (5.2.13) to resemble (5.2.6)

$$\rho^{(1)}(\mathbf{R}') = N \int \cdots \int d\mathbf{X}^N P(\mathbf{X}^N) \delta(\mathbf{R}_1 - \mathbf{R}') = N P^{(1)}(\mathbf{R}'). \qquad (5.2.14)$$

The interpretation of $P^{(1)}(\mathbf{R}') \, d\mathbf{R}'$ follows from the same argument as in the case of $P_1(S)$ in (5.2.8). This is the probability of finding a specific particle, say 1, in $d\mathbf{R}'$ at $\mathbf{R}'$. Hence, $P^{(1)}(\mathbf{R}')$ is often referred to as the specific singlet distribution function.

The next question is: What is the probability of finding *any* particle in $d\mathbf{R}'$? To answer this question, we consider the following events:

Event	Probability of the event
Particle 1 in $d\mathbf{R}'$	$P^{(1)}(\mathbf{R}') \, d\mathbf{R}'$
Particle 2 in $d\mathbf{R}'$	$P^{(1)}(\mathbf{R}') \, d\mathbf{R}'$
$\vdots$	$\vdots$
Particle N in $d\mathbf{R}'$	$P^{(1)}(\mathbf{R}') \, d\mathbf{R}'$

Clearly, by virtue of the equivalence of the particles, we have exactly the same probability for each of the events listed above.

The event "any particle in $d\mathbf{R}'$" means either "particle 1 in $d\mathbf{R}'$" or "particle 2 in $d\mathbf{R}'$", ..., or "particle N in $d\mathbf{R}'$." In probability language, this event is called the *union* of all the events listed above, and is written symbolically as

$$\{\text{any particle in } d\mathbf{R}'\} = \bigcup_{i=1}^{N} \{\text{particle } i \text{ in } d\mathbf{R}'\}. \tag{5.2.15}$$

In general, there exists no simple relation between the probability of a union of events and the probabilities of the individual events. However, if we choose $d\mathbf{R}'$ to be small enough so that no more than a single particle may be found in it at any given time, then all the events listed above become disjointed (i.e., occurrence of one event precludes the possibility of simultaneous occurrence of any other event). In this case, we have the additivity relation for the probability of the union, namely,

$$\Pr\{\text{any particle in } d\mathbf{R}'\} = \sum_{i=1} \Pr\{\text{particle } i \text{ in } d\mathbf{R}'\}$$

$$= \sum_{i=1} P^{(1)}(\mathbf{R}') \, d\mathbf{R}' = NP^{(1)}(\mathbf{R}') \, d\mathbf{R}' = \rho^{(1)}(\mathbf{R}') \, d\mathbf{R}'. \tag{5.2.16}$$

Relation (5.2.16) gives the probabilistic meaning of the quantity $\rho^{(1)}(\mathbf{R}') \, d\mathbf{R}'$, which is contingent upon the choice of a sufficiently small element of volume $d\mathbf{R}'$. The quantity $\rho^{(1)}(\mathbf{R}')$ is referred to as the generic singlet distribution function. Clearly, the generic singlet distribution function is the physically meaningful quantity. We can measure the average number of particles in a given element of volume. We cannot measure the probability of finding a specific particle in a given element of volume.

Some care must be exercised when using the probabilistic meaning of $\rho^{(1)}(\mathbf{R}') \, d\mathbf{R}'$. For instance, the probability of finding a specific particle, say 1, in a region S is obtained from the specific singlet distribution function simply by integration:

$$P_1(S) = \int_S P^{(1)}(\mathbf{R}') \, d\mathbf{R}'. \tag{5.2.17}$$

This interpretation follows from the fact that the events "particle 1 in $d\mathbf{R}'$" and "particle 1 in $d\mathbf{R}''$" are disjoint events (i.e., a specific particle cannot be in two different elements $d\mathbf{R}'$ and $d\mathbf{R}''$ simultaneously). Hence, the probability of the union (particle 1 in S) is obtained as the sum (or integral) of the probabilities of the individual events.

This property is not shared by the generic singlet distribution function, and the integral

$$\int_S \rho^{(1)}(\mathbf{R}') \, d\mathbf{R}' \tag{5.2.18}$$

does not have the meaning of the probability of the event "any particle in S." The reason is that the events "a particle in $d\mathbf{R}'$" and "a particle in $d\mathbf{R}''$" are not disjoint; hence one cannot obtain the probability of their union in a simple fashion. It is for this reason that the meaning of $\rho^{(1)}(\mathbf{R}')$ as a local density at $\mathbf{R}'$ should be preferred. If $\rho^{(1)}(\mathbf{R}') \, d\mathbf{R}'$ is viewed as the average number of particles in $d\mathbf{R}'$, then, clearly, (5.2.18) is the average number of particles in S. The meaning of $\rho^{(1)}(\mathbf{R}') \, d\mathbf{R}'$ as an average number of particles is preserved upon integration; the probabilistic meaning is not.

A particular example of (5.2.18) occurs when S is chosen as the total volume of the system, i.e.,

$$\int_V \rho^{(1)}(\mathbf{R}') \, d\mathbf{R}' = N \int_V P^{(1)}(\mathbf{R}') \, d\mathbf{R}' = N. \tag{5.2.19}$$

The last equality follows from the normalization of $P^{(1)}(\mathbf{R}')$; i.e., the probability of finding particle 1 in any place in V is unity. The normalization condition (5.2.19) can also be obtained directly from (5.2.13).

In a homogeneous fluid, we expect that $\rho^{(1)}(\mathbf{R}')$ will have the same value at any point $\mathbf{R}'$ within the system. (This is true apart from a very small region near the surface of the system, which we always neglect in considering macroscopic systems.) Therefore, we put

$$\rho^{(1)}(\mathbf{R}') = \text{const}, \tag{5.2.20}$$

and from (5.2.19) we obtain

$$\text{const} \times \int_V d\mathbf{R}' = N. \tag{5.2.21}$$

Hence

$$\rho^{(1)}(\mathbf{R}') = N/V = \rho. \tag{5.2.22}$$

The last relation is quite a trivial result for homogeneous systems. It states that the local density at each point $\mathbf{R}'$ is equal to the bulk density ρ. That is, of course, not true in an inhomogeneous system.

In a similar fashion, we may introduce the singlet distribution function for location and orientation, which by analogy to (5.2.14) is defined as

$$\rho^{(1)}(\mathbf{X}') = \int \cdots \int d\mathbf{X}^N \, P(\mathbf{X}^N) \sum_{i=1}^{N} \delta(\mathbf{X}_i - \mathbf{X}')$$

$$= N \int \cdots \int d\mathbf{X}^N \, P(\mathbf{X}^N) \delta(\mathbf{X}_1 - \mathbf{X}')$$

$$= N P^{(1)}(\mathbf{X}'), \tag{5.2.23}$$

where $P^{(1)}(\mathbf{X}')$ is the probability density of finding a specific particle at a given configuration $\mathbf{X}'$.

Again, in a homogeneous and isotropic fluid, we expect that

$$\rho^{(1)}(\mathbf{X}') = \text{const.} \tag{5.2.24}$$

Hence, using the normalization condition

$$\int \rho^{(1)}(\mathbf{X}') \, d\mathbf{X}' = N \int P^{(1)}(\mathbf{X}') \, d\mathbf{X}' = N, \tag{5.2.25}$$

we get

$$\rho^{(1)}(\mathbf{X}') = N/V(8\pi^2) = \rho/8\pi^2. \tag{5.2.26}$$

The connection between $\rho^{(1)}(\mathbf{R}')$ and $\rho^{(1)}(\mathbf{X}')$ is obtained simply by integration over all the orientations

$$\rho^{(1)}(\mathbf{R}') = \int \rho^{(1)}(\mathbf{X}') \, d\mathbf{\Omega}'. \tag{5.2.27}$$

5.2.2. Pair Distribution Function

In this section, we introduce the pair distribution function. We first present its meaning as a probability density and then show how it can be reinterpreted as an average quantity.

Again, the starting point is the basic probability density $P(\mathbf{X}^N)$, see (5.2.1), in the T, V, N ensemble. The specific pair distribution function is defined as the probability density of finding particle 1 at $\mathbf{X}'$ and particle 2 at $\mathbf{X}''$. This can be obtained from $P(\mathbf{X}^N)$ by integrating over all the configurations of the remaining $N - 2$ molecules†

$$P^{(2)}(\mathbf{X}', \mathbf{X}'') = \int \cdots \int d\mathbf{X}_3 \cdots d\mathbf{X}_N P(\mathbf{X}', \mathbf{X}'', \mathbf{X}_3, \ldots, \mathbf{X}_N). \tag{5.2.28}$$

Clearly, $P^{(2)}(\mathbf{X}', \mathbf{X}'') \, d\mathbf{X}' \, d\mathbf{X}''$ is the probability of finding a specific particle, say 1, in $d\mathbf{X}'$ at $\mathbf{X}'$ and another particle, say 2, in $d\mathbf{X}''$ at $\mathbf{X}''$. The same probability applies for any specific pair of two different particles.

Consider the following list of events and their corresponding probabilities.

Event			Probability of the event
particle 1 in $d\mathbf{X}'$	and	particle 2 in $d\mathbf{X}''$	$P^{(2)}(\mathbf{X}', \mathbf{X}'') \, d\mathbf{X}' \, d\mathbf{X}''$
particle 1 in $d\mathbf{X}'$	and	particle 3 in $d\mathbf{X}''$	$P^{(2)}(\mathbf{X}', \mathbf{X}'') \, d\mathbf{X}' \, d\mathbf{X}''$
$\vdots$			
particle 1 in $d\mathbf{X}'$	and	particle N in $d\mathbf{X}''$	$P^{(2)}(\mathbf{X}', \mathbf{X}'') \, d\mathbf{X}' \, d\mathbf{X}''$
particle 2 in $d\mathbf{X}'$	and	particle 1 in $d\mathbf{X}''$	$P^{(2)}(\mathbf{X}', \mathbf{X}'') \, d\mathbf{X}' \, d\mathbf{X}''$
$\vdots$			
particle N in $d\mathbf{X}'$	and	particle $N - 1$ in $d\mathbf{X}''$	$P^{(2)}(\mathbf{X}', \mathbf{X}'') \, d\mathbf{X}' \, d\mathbf{X}''$

$$\tag{5.2.29}$$

The event

$$\{\text{a particle in } d\mathbf{X}' \text{ and another particle in } d\mathbf{X}''\} \tag{5.2.30}$$

is clearly the union of all the $N(N - 1)$ events listed in (5.2.29). However, the probability of the event (5.2.30) is the sum of all the probabilities of the events listed in (5.2.29) *only* if the latter are disjoint. This condition can be realized when the elements of volume $d\mathbf{R}'$ and $d\mathbf{R}''$ (contained in $d\mathbf{X}'$ and $d\mathbf{X}''$, respectively) are small enough so that no more than

---

† We use primed vectors like $\mathbf{X}'$, $\mathbf{X}''$, ... to distinguish them from the vectors X_3, X_4, ... whenever each of the two sets of vectors has a different "status." For instance, in (5.2.28) the primed vectors are fixed in the integrand. Such a distinction is not essential, although it may help to avoid confusion.

one of the events in (5.2.29) may occur at any given time. For this case, we define the generic pair distribution function as

$$\rho^{(2)}(\mathbf{X}', \mathbf{X}'') \, d\mathbf{X}' \, d\mathbf{X}'' = \text{Pr}\{\text{a particle in } d\mathbf{X}' \text{ and a different particle in } d\mathbf{X}''\}$$

$$= \sum_{i \neq j} \text{Pr}\{\text{particle } i \text{ in } d\mathbf{X}' \text{ and particle } j \text{ in } d\mathbf{X}''\}$$

$$= \sum_{i \neq j} P^{(2)}(\mathbf{X}', \mathbf{X}'') \, d\mathbf{X}' \, d\mathbf{X}''$$

$$= N(N - 1) P^{(2)}(\mathbf{X}', \mathbf{X}'') \, d\mathbf{X}' \, d\mathbf{X}''. \tag{5.2.31}$$

The last equality in (5.2.31) follows from the equivalence of all the $N(N - 1)$ pairs of specific and different particles. Using the definition of $P^{(2)}(\mathbf{X}', \mathbf{X}'')$ in (5.2.29), we can transform the definition of $\rho^{(2)}(\mathbf{X}', \mathbf{X}'')$ into an expression which will be interpreted as an average quantity:

$$\rho^{(2)}(\mathbf{X}', \mathbf{X}'') \, d\mathbf{X}' \, d\mathbf{X}''$$

$$= N(N - 1) \, d\mathbf{X}' \, d\mathbf{X}'' \int \cdots \int d\mathbf{X}_3 \cdots d\mathbf{X}_N \, P(\mathbf{X}', \mathbf{X}'', \mathbf{X}_3, \ldots, \mathbf{X}_N)$$

$$= N(N - 1) \, d\mathbf{X}' \, d\mathbf{X}'' \int \cdots \int d\mathbf{X}_1 \cdots d\mathbf{X}_N \, P(\mathbf{X}_1 \cdots \mathbf{X}_N) \delta(\mathbf{X}_1 - \mathbf{X}') \delta(\mathbf{X}_2 - \mathbf{X}'')$$

$$= d\mathbf{X}' \, d\mathbf{X}'' \int \cdots \int d\mathbf{X}^N \, P(\mathbf{X}^N) \sum_{\substack{i=1 \\ i \neq j}}^{N} \sum_{j=1}^{N} \delta(\mathbf{X}_i - \mathbf{X}') \delta(\mathbf{X}_j - \mathbf{X}''). \tag{5.2.32}$$

In the second form on the rhs of (5.2.32), we employ the basic property of the Dirac delta function, so that integration is now extended over all the vectors $\mathbf{X}_1, \ldots, \mathbf{X}_N$. In the third form we have used the equivalence of the N particles, as was done in (5.2.31), to get an average of the quantity

$$d\mathbf{X}' \, d\mathbf{X}'' \sum_{\substack{i=1 \\ i \neq j}}^{N} \sum_{j=1}^{N} \delta(\mathbf{X}_i - \mathbf{X}') \delta(\mathbf{X}_j - \mathbf{X}''), \tag{5.2.33}$$

which can be viewed as a "counting function", i.e., for any specific configuration $\mathbf{X}^N$, this quantity counts the number of *pairs* of particles occupying the elements $d\mathbf{X}'$ and $d\mathbf{X}''$. Hence, the integral (5.2.32) counts the average number of pairs occupying $d\mathbf{X}'$ and $d\mathbf{X}''$. The normalization of $\rho^{(2)}(\mathbf{X}', \mathbf{X}'')$ follows directly from (5.2.32):

$$\int \int d\mathbf{X}' \, d\mathbf{X}'' \, \rho^{(2)}(\mathbf{X}', \mathbf{X}'') = N(N - 1), \tag{5.2.34}$$

which is the exact number of pairs in V. As in the previous section, we note that the meaning of $\rho^{(2)}(\mathbf{X}', \mathbf{X}'')$ as an average quantity is preserved upon integration over any region S. This is not the case, however, when its probabilistic meaning is adopted. For instance, the quantity

$$\int_S \int_S d\mathbf{X}' \, d\mathbf{X}'' \, \rho^{(2)}(\mathbf{X}', \mathbf{X}'') \tag{5.2.35}$$

is the average number of pairs occupying the region S (a factor of $\frac{1}{2}$ should be included if we are interested in different pairs). This quantity is in general not a probability.

It is also useful to introduce the locational (or spatial) pair distribution function, defined by

$$\rho^{(2)}(\mathbf{R}', \mathbf{R}'') = \int\int d\mathbf{\Omega}' \, d\mathbf{\Omega}'' \, \rho^{(2)}(\mathbf{X}', \mathbf{X}''), \qquad (5.2.36)$$

where integration is carried out over the orientations of the two particles. Here $\rho^{(2)}(\mathbf{R}', \mathbf{R}'') \, d\mathbf{R}' \, d\mathbf{R}''$ is the average number of pairs occupying $d\mathbf{R}'$ and $d\mathbf{R}''$ or, alternatively, for infinitesimal elements $d\mathbf{R}'$ and $d\mathbf{R}''$, the probability of finding one particle in $d\mathbf{R}'$ at $\mathbf{R}'$ and a second particle in $d\mathbf{R}''$ at $\mathbf{R}''$. It is sometimes convenient to denote the quantity defined in (5.2.36) by $\bar{\rho}^{(2)}(\mathbf{R}', \mathbf{R}'')$, to distinguish it from the different function $\rho^{(2)}(\mathbf{X}', \mathbf{X}'')$. However, since we specify the arguments of the functions, there should be no reason for confusion as to this notation.

5.2.3. Pair Correlation Function

In most applications, it has been found more useful to employ the pair correlation function, defined below, rather than the pair distribution function itself. Consider the two elements of volume $d\mathbf{X}'$ and $d\mathbf{X}''$ and the intersection of the two events:

$$\{\text{a particle in } d\mathbf{X}'\} \cap \{\text{a particle in } d\mathbf{X}''\}. \qquad (5.2.37)$$

The intersection symbol $\cap$ may be read as "and"; i.e., the combination in (5.2.37) means that the first *and* the second events occur.

Two events are called independent whenever the probability of their intersection is equal to the product of the probabilities of the two events. In general, the two separate events given in (5.2.37) are not independent; the occurrence of one of them may influence the likelihood, or the probability, of occurrence of the other. For instance, if the separation $R = |\mathbf{R}'' - \mathbf{R}'|$ between the two elements is very small (compared to the molecular diameter of the particles), then fulfilling one event strongly reduces the chances of the second.

We now use the following physically plausible contention. In a fluid, if the separation R between two elements is very large, then the two events in (5.2.37) become independent. Therefore, we can write for the probability of their intersection

$$\rho^{(2)}(\mathbf{X}', \mathbf{X}'') \, d\mathbf{X}' \, d\mathbf{X}'' = \Pr\{\text{a particle in } d\mathbf{X}'\} \cap \{\text{a particle in } d\mathbf{X}''\}$$

$$= \Pr\{\text{a particle in } d\mathbf{X}'\} \times \Pr\{\text{a particle in } d\mathbf{X}''\}$$

$$= \rho^{(1)}(\mathbf{X}') \, d\mathbf{X}' \, \rho^{(1)}(\mathbf{X}'') \, d\mathbf{X}'', \qquad \text{for } R \to \infty. \qquad (5.2.38)$$

Or in short,

$$\rho^{(2)}(\mathbf{X}', \mathbf{X}'') \underset{R \to \infty}{=} \rho^{(1)}(\mathbf{X}')\rho^{(1)}(\mathbf{X}'') = (\rho/8\pi^2)^2, \qquad (5.2.39)$$

the last equality holding for a homogeneous and isotropic fluid. If (5.2.39) holds, it is often said that the local densities at $\mathbf{X}'$ and $\mathbf{X}''$ are uncorrelated. (The limit $R \to \infty$ should be taken after the thermodynamic limit, $N \to \infty$, $V \to \infty$, $N/V = \text{const}$, has been taken.)

For any finite distance R, factoring of $\rho^{(2)}(\mathbf{X}', \mathbf{X}'')$ into a product may not be valid. We now introduce the pair correlation function, which measures the extent of deviation from (5.2.39) and is defined by

$$\rho^{(2)}(\mathbf{X}', \mathbf{X}'') = \rho^{(1)}(\mathbf{X}')\rho^{(1)}(\mathbf{X}'')g(\mathbf{X}', \mathbf{X}'')$$
$$= (\rho/8\pi^2)^2 g(\mathbf{X}', \mathbf{X}''). \tag{5.2.40}$$

The second equality holds for a homogeneous and isotropic fluid. A related quantity is the locational pair correlation function, defined in terms of the locational pair distribution function

$$\rho^{(2)}(\mathbf{R}', \mathbf{R}'') = \rho^2 g(\mathbf{R}', \mathbf{R}''). \tag{5.2.41}$$

The relation between $g(\mathbf{R}', \mathbf{R}'')$ and $g(\mathbf{X}', \mathbf{R}'')$ follows from (5.2.36), (5.2.40), and (5.2.41):

$$g(\mathbf{R}', \mathbf{R}'') = \frac{1}{(8\pi^2)^2} \int\int d\mathbf{\Omega}' \, d\mathbf{\Omega}'' \, g(\mathbf{X}', \mathbf{X}''), \tag{5.2.42}$$

which can be viewed as the angle average of $g(\mathbf{X}', \mathbf{X}'')$. Note that this average is taken with the probability distribution $d\mathbf{\Omega}' \, d\mathbf{\Omega}''/(8\pi^2)^2$. This is the probability of finding one particle in orientation $d\mathbf{\Omega}'$ and a second particle in $d\mathbf{\Omega}''$ when they are at infinite separation from each other. At any finite separation, $\mathbf{R}'$ and $\mathbf{R}''$, the probability of finding one particle in $d\mathbf{\Omega}'$ and the second in $d\mathbf{\Omega}''$ given the locations $\mathbf{R}'$ and $\mathbf{R}''$ is

$$\text{Pr}(\mathbf{\Omega}', \mathbf{\Omega}''/\mathbf{R}', \mathbf{R}'') \, d\mathbf{\Omega}' \, d\mathbf{\Omega}'' = \frac{\rho^{(2)}(\mathbf{X}', \mathbf{X}'') \, d\mathbf{\Omega}' \, d\mathbf{\Omega}''}{\int \rho^{(2)}(\mathbf{X}', \mathbf{X}'') \, d\mathbf{\Omega}' \, d\mathbf{\Omega}''} = \frac{g(\mathbf{X}', \mathbf{X}'') \, d\mathbf{\Omega}' \, d\mathbf{\Omega}''}{(8\pi^2)^2 g(\mathbf{R}', \mathbf{R}'')}.$$

It is only for $|\mathbf{R}'' - \mathbf{R}'| \to \infty$ or when g is independent of the orientations that this probability distribution becomes $d\mathbf{\Omega}' \, d\mathbf{\Omega}''/(8\pi^2)^2$.

In this book, we shall be interested only in homogeneous and isotropic fluids. In such a case, there is a redundancy in specifying the full configuration of the pair of particles by 12 coordinates $(\mathbf{X}', \mathbf{X}'')$. It is clear that for any configuration of the pair $\mathbf{X}'$, $\mathbf{X}''$, the correlation $g(\mathbf{X}', \mathbf{X}'')$ is invariant to translation and rotation of the pair as a unit, keeping the relative configuration of one particle toward the other fixed. Therefore, we can reduce to six the number of independent variables necessary for the full description of the pair correlation function. For instance, we may choose the location of one particle at the origin of the coordinate system $\mathbf{R}' = 0$ and fix its orientation, say, at $\Phi' = \theta' = \psi' = 0$. Hence, the pair correlation function is a function only of the six variables $\mathbf{X}'' = \mathbf{R}'', \mathbf{\Omega}''$.

Similarly, the function $g(\mathbf{R}', \mathbf{R}'')$ is a function only of the scalar distance $R = |\mathbf{R}'' - \mathbf{R}'|$. (For instance, $\mathbf{R}'$ may be chosen at the origin $\mathbf{R}' = 0$ and, because of the isotropy of the fluid, the orientation of $\mathbf{R}''$ is of no importance. Therefore, only the separation R is left as the independent variable.) The function $g(R)$, i.e., the pair correlation function expressed explicitly as a function of the distance R, is often referred to as the radial distribution function. This function plays a central role in the theory of fluids.

We now turn to a somewhat different interpretation of the pair correlation function. We define the conditional probability of observing a particle in $d\mathbf{X}''$ at $\mathbf{X}''$, given a particle at $\mathbf{X}'$:

$$\rho(\mathbf{X}''/\mathbf{X}')\, d\mathbf{X}'' = \frac{\rho^{(2)}(\mathbf{X}', \mathbf{X}'')\, d\mathbf{X}'\, d\mathbf{X}''}{\rho^{(1)}(\mathbf{X}')\, d\mathbf{X}'} = \rho^{(1)}(\mathbf{X}'')g(\mathbf{X}', \mathbf{X}'')\, d\mathbf{X}''. \qquad (5.2.43)$$

The last equality follows from the definition of $g(\mathbf{X}', \mathbf{X}'')$ in (5.2.40).

Note that the probability of finding a particle at an exact configuration $\mathbf{X}''$ is zero, which is the reason for taking an infinitesimal element of volume at $\mathbf{X}''$. On the other hand, the conditional probability may be defined for an exact *condition*: "given a particle at $\mathbf{X}'$". This may be seen formally from (5.2.43), where $d\mathbf{X}'$ cancels out once we form the ratio of the two distribution functions. Hence, one can actually take the limit $d\mathbf{X}' \rightarrow 0$ in the definition of the conditional probability.

We recall that the quantity $\rho^{(1)}(\mathbf{X}'')\, d\mathbf{X}''$ is the local density of particles at $\mathbf{X}''$. We now show that the quantity defined in (5.2.43) is the conditional local density at $\mathbf{X}''$, given a particle at $\mathbf{X}'$. In other words, we fix a particle at $\mathbf{X}'$ and view the rest of the $N - 1$ particles as a system subjected to the field of force produced by the first particle. Clearly, the new system is no longer homogeneous, nor isotropic; therefore, the local density may be different at each point of the system. To show this we first define the binding energy B of one particle, say 1 to the rest of the system by

$$U_N(\mathbf{X}_1, \ldots, \mathbf{X}_N) = U_{N-1}(\mathbf{X}_2, \ldots, \mathbf{X}_N) + \sum_{j=2}^{N} U_{1,j}(\mathbf{X}_1, \mathbf{X}_j) = U_{N-1} + B_1. \quad (5.2.44)$$

In (5.2.44), we have split the total potential energy of the system of N particles into two parts: the potential energy of interaction among the $N - 1$ particles and the interaction of one particle, chosen as particle 1, with the $N - 1$ particles. Once we fix the configuration of particle 1 at $\mathbf{X}_1$, the rest of the system can be looked upon as a system in an "external" field defined by B_1.

From the definitions (5.2.1), (5.2.23), (5.2.32), and (5.2.43), we get

$$
\rho(\mathbf{X}''/\mathbf{X}') = \frac{N(N-1) \int \cdots \int d\mathbf{X}^N \exp[-\beta U_N(\mathbf{X}^N)]\delta(\mathbf{X}_1 - \mathbf{X}')\delta(\mathbf{X}_2 - \mathbf{X}'')}{N \int \cdots \int d\mathbf{X}^N \exp[-\beta U_N(\mathbf{X}^N)]\delta(\mathbf{X}_1 - \mathbf{X}')}
$$

$$
= \frac{(N-1) \int \cdots \int d\mathbf{X}_2 \cdots d\mathbf{X}_N \exp[-\beta U_N(\mathbf{X}', \mathbf{X}_2, \ldots, \mathbf{X}_N)]\delta(\mathbf{X}_2 - \mathbf{X}'')}{\int \cdots \int d\mathbf{X}_2 \cdots d\mathbf{X}_N \exp[-\beta U_N(\mathbf{X}', \mathbf{X}_2, \ldots, \mathbf{X}_N)]}
$$

$$
= (N-1) \int \cdots \int d\mathbf{X}_2 \cdots d\mathbf{X}_N \, P^*(\mathbf{X}', \mathbf{X}_2, \ldots, \mathbf{X}_N)\delta(\mathbf{X}_2 - \mathbf{X}''), \quad (5.2.45)
$$

where $P^*(\mathbf{X}', \mathbf{X}_2, \ldots, \mathbf{X}_N)$ is the basic probability density of a system of $N - 1$ particles placed in an "external" field produced by a particle fixed at $\mathbf{X}'$, i.e.,

$$P^*(\mathbf{X}', \mathbf{X}_2, \ldots, \mathbf{X}_N) = \frac{\exp(-\beta U_{N-1} - \beta B_1)}{\int \cdots \int d\mathbf{X}_2 \cdots d\mathbf{X}_N \exp(-\beta U_{N-1} - \beta B_1)}. \quad (5.2.46)$$

This should be compared with the probability density defined in (5.2.1).

We now observe that relation (5.2.45) has the same structure as relation (5.2.23) but with two differences. First, (5.2.45) refers to a system of $N-1$ instead of N particles; second, the system of $N-1$ particles is placed in an external field. Hence, (5.2.45) is simply the local density at $\mathbf{X}''$ of a system of $N-1$ particles placed in the external field B_1. This is an example of a singlet molecular distribution function which is not everywhere constant.

Similarly, for the locational pair correlation function, we have

$$\rho(R''/R') = \rho g(\mathbf{R}', \mathbf{R}''), \quad (5.2.47)$$

which is the conditional average density at $\mathbf{R}''$ given a particle at $\mathbf{R}'$. This interpretation of the pair correlation function will be most useful in the forthcoming applications.

As noted above, the function $g(\mathbf{R}', \mathbf{R}'')$ is a function of the scalar distance $R = |\mathbf{R}'' - \mathbf{R}'|$. (This is true both for spherical particles and for molecular fluids for which an orientational average has been carried out.) Because of the spherical symmetry of the locational pair correlation function, the local density has the same value for any point on the spherical shell of radius R from the center of the fixed particle at $\mathbf{R}'$. It is also convenient to choose as an element of volume a spherical shell of width dR and radius R. The average number of particles in this element of volume is

$$\rho g(\mathbf{R}', \mathbf{R}'') \, d\mathbf{R}'' = \rho g(R) 4\pi R^2 \, dR. \quad (5.2.48)$$

Sometimes, the function $g(R)4\pi R^2$ rather than $g(R)$ is referred to as the radial distribution function. In this book, we use this term only for $g(R)$ itself.

5.3. FEATURES OF THE RADIAL DISTRIBUTION FUNCTION

In this section, we illustrate the general features of the radial distribution function (RDF) $g(R)$ for a system of simple spherical particles. From definitions (5.2.32) and (5.2.40) (applied to spherical particles), we get

$g(\mathbf{R}', \mathbf{R}'')$

$$= \frac{N(N-1)}{\rho^2} \frac{\int \cdots \int d\mathbf{R}_3 \cdots d\mathbf{R}_N \exp[-\beta U_N(\mathbf{R}', \mathbf{R}'', \mathbf{R}_3, \ldots, \mathbf{R}_N)]}{\int \cdots \int d\mathbf{R}_1 \cdots d\mathbf{R}_N \exp[-\beta U_N(\mathbf{R}_1, \ldots, \mathbf{R}_N)]}. \quad (5.3.1)$$

This general relation will be used to extract information on the behavior of $g(R)$ for some simple systems. A more useful expression, which we shall need only for demonstrative purposes, is the density expansion of $g(\mathbf{R}', \mathbf{R}'')$, which reads

$$g(\mathbf{R}', \mathbf{R}'') = \{\exp[-\beta U(\mathbf{R}', \mathbf{R}'')]\}\{1 + B(\mathbf{R}', \mathbf{R}'')\rho + C(\mathbf{R}', \mathbf{R}'')\rho^2 + \cdots\}, \quad (5.3.2)$$

where the coefficients $B(\mathbf{R}', \mathbf{R}'')$, $C(\mathbf{R}', \mathbf{R}'')$, etc., are given in terms of integrals over the so-called Mayer f-function, defined by

$$f(\mathbf{R}', \mathbf{R}'') = \exp[-\beta U(\mathbf{R}', \mathbf{R}'')] - 1. \quad (5.3.3)$$

For instance,

$$B(\mathbf{R}', \mathbf{R}'') = \int_V f(\mathbf{R}', \mathbf{R}_3) f(\mathbf{R}'', \mathbf{R}_3)\, d\mathbf{R}_3. \quad (5.3.4)$$

We now turn to some specific examples.

5.3.1. Ideal Gas

The RDF for an ideal gas can be obtained directly from definition (5.3.1). Putting $U_N \equiv 0$ for all configurations, the integrations become trivial and we get

$$g(\mathbf{R}', \mathbf{R}'') = \frac{N(N-1)}{\rho^2} \frac{\int \cdots \int d\mathbf{R}_3 \cdots d\mathbf{R}_N}{\int \cdots \int d\mathbf{R}_1 \cdots d\mathbf{R}_N} = \frac{N(N-1)V^{N-2}}{\rho^2 V^N} \quad (5.3.5)$$

or

$$g(R) = 1 - (1/N). \quad (5.3.6)$$

As we expect, $g(R)$ is practically unity for any value of R. This is an obvious reflection of the basic property of an ideal gas; i.e., absence of correlation follows from absence of interaction. The term N^{-1} is typical of constant-volume systems. At the thermodynamic limit $N \to \infty$, $V \to \infty$, $N/V = \text{const}$, this term may, for most purposes, be dropped.† Of course, in order to get the correct normalization of $g(R)$, one should use the exact relation (5.3.6), which yields

$$\rho \int_V g(\mathbf{R}', \mathbf{R}'')\, d\mathbf{R}'' = \rho \int_V [1 - (1/N)]\, d\mathbf{R}'' = N - 1, \quad (5.3.7)$$

which is exactly the total number of particles in the system, excluding the one fixed at $\mathbf{R}'$.

---

† In some instances, care must be employed to take the proper limit of infinite systems. For an example, see sections 5.8.4 and 6.7.

5.3.2. Very Dilute Gas

At very low densities, $\rho \to 0$, we may neglect all powers of ρ in the density expansion of $g(R)$, in which case we get, from (5.3.2),

$$g(R) = \exp[-\beta U(R)], \qquad \rho \to 0, \tag{5.3.8}$$

where $U(R)$ is the pair potential operating between two particles. Relation (5.3.8) is essentially the Boltzmann distribution law. Since, at low densities, encounters in which more than two particles are involved are very rare, the pair distribution function is determined solely by the pair potential.

One direct way of obtaining (5.3.8) from definition (5.3.1) (and not through the density expansion) is to consider the case of a system containing only two particles. Letting $N = 2$ in (5.3.1), we get

$$g(R) = \frac{2}{\rho^2} \frac{\exp[-\beta U(R)]}{Z_2}, \tag{5.3.9}$$

where Z_2 is the configurational partition function for the case $N = 2$.

Since we choose $U(R) \to 0$ as $R \to \infty$, we can use (5.3.9) to form the ratio

$$\frac{g(R)}{g(\infty)} = \exp[-\beta U(R)]. \tag{5.3.10}$$

Assuming that at $R \to \infty$, $g(\infty)$ is practically unity, we get from (5.3.10)

$$g(R) = \exp[-\beta U(R)], \tag{5.3.11}$$

which is the same as (5.3.8). Note that (5.3.8) and (5.3.11) have been obtained for two apparently different conditions ($\rho \to 0$ on one hand and $N = 2$ on the other). The identical results for $g(R)$ in the two cases reflects the fact that at very low densities, only interactions between pairs determine the behavior of $g(R)$. In other words, a pair of interacting particles does not "feel" the presence of the other particles. The form of $g(R)$ as $\rho \to 0$ for a system of hard spheres and Lennard–Jones particles is depicted in Fig. 5.2. It is seen that for HS particles as $\rho \to 0$, correlation exists only for $R < \sigma$. For $R > \sigma$, the

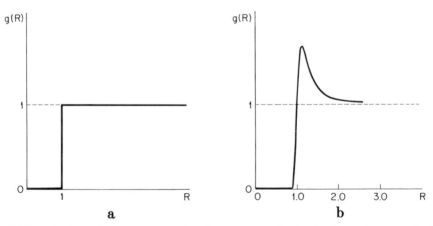

FIGURE 5.2. The form of $g(R)$ at very low densities ($\rho \to 0$). (a) For hard spheres with $\sigma = 1$. (b) For Lennard–Jones particles with parameters $\varepsilon/kT = 0.5$ and $\sigma = 1$.

function $g(R)$ is identically unity. For LJ particles, we observe a single peak in $g(R)$ at the same point for which $U(R)$ has a minimum, namely at $R = 2^{1/6}\sigma$.

5.3.3. Slightly Dense Gas

In the context of this section, a slightly dense gas is a gas properly described by the first-order expansion in density, i.e. up to the linear term in (5.3.2). Before analyzing the content of the coefficient $B(\mathbf{R}', \mathbf{R}'')$ in the expansion of $g(R)$, let us demonstrate its origin by considering a system of three particles. Putting $N = 3$ in definition (5.3.1), we get

$$g(\mathbf{R}', \mathbf{R}'') = \frac{6}{\rho^2} \frac{\int d\mathbf{R}_3 \exp[-\beta U_3(\mathbf{R}', \mathbf{R}'', \mathbf{R}_3)]}{Z_3}, \tag{5.3.12}$$

where Z_3 is the configurational partition function for a system of three particles. Assuming pairwise additivity of the potential energy U_3 and using the definition of the function f in (5.3.3), we can transform (5.3.12) into

$$g(\mathbf{R}', \mathbf{R}'')$$

$$= \frac{6}{\rho^2} \exp[-\beta U(\mathbf{R}', \mathbf{R}'')] \frac{\int d\mathbf{R}_3 [f(\mathbf{R}', \mathbf{R}_3)f(\mathbf{R}_3, \mathbf{R}'') + f(\mathbf{R}', \mathbf{R}_3) + f(\mathbf{R}'', \mathbf{R}_3) + 1]}{Z_3}.$$

$$\tag{5.3.13}$$

Noting again that $U(\mathbf{R}', \mathbf{R}'') = 0$ for $R = |\mathbf{R}'' - \mathbf{R}'| \to \infty$, we form the ratio†

$$\frac{g(R)}{g(\infty)} = \exp[-\beta U(R)] \frac{\int d\mathbf{R}_3 f(\mathbf{R}', \mathbf{R}_3)f(\mathbf{R}_3, \mathbf{R}'') + 2\int d\mathbf{R}_3 f(\mathbf{R}', \mathbf{R}_3) + V}{\lim_{R \to \infty} \left[\int d\mathbf{R}_3 f(\mathbf{R}', \mathbf{R}_3)f(\mathbf{R}_3, \mathbf{R}'') + 2\int d\mathbf{R}_3 f(\mathbf{R}', \mathbf{R}_3) + V \right]}. \tag{5.3.14}$$

Clearly, the two integrals over $f(\mathbf{R}', \mathbf{R}_3)$ and $f(\mathbf{R}'', \mathbf{R}_3)$ are equal and independent of the separation R; i.e.,

$$C \equiv \int_V d\mathbf{R}_3\, f(\mathbf{R}', \mathbf{R}_3) = \int_V d\mathbf{R}_3\, f(\mathbf{R}'', \mathbf{R}_3). \tag{5.3.15}$$

Note also that since $f(R)$ is a short-range function of R, the integral in (5.3.15) does not depend on V.

On the other hand, we have the limiting behavior

$$\lim_{R \to \infty} \int d\mathbf{R}_3\, f(\mathbf{R}', \mathbf{R}_3)f(\mathbf{R}_3, \mathbf{R}'') = 0, \tag{5.3.16}$$

which follows from the fact that the two factors in the integrand contribute to the integral

---

† By $R \to \infty$, we mean here a very large distance compared with the molecular diameter.

only if $\mathbf{R}_3$ is close simultaneously to both $\mathbf{R}'$ and $\mathbf{R}''$, a situation that cannot be attained if $R = |\mathbf{R}'' - \mathbf{R}'| \to \infty$.

Using (5.3.15), (5.3.16), and definition (5.3.4), we now rewrite (5.3.14) as

$$\frac{g(R)}{g(\infty)} = \exp[-\beta U(R)] \frac{B(\mathbf{R}', \mathbf{R}'') + 2C + V}{2C + V}. \qquad (5.3.17)$$

Since C is constant, it may be neglected, as compared with V, in the thermodynamic limit. Also, assuming that $g(\infty)$ is practically unity, we get the final form of $g(R)$ for this case:

$$g(R) = \{\exp[-\beta U(R)]\}[1 + (1/V)B(\mathbf{R}', \mathbf{R}'')], \qquad R = |\mathbf{R}'' - \mathbf{R}'|. \qquad (5.3.18)$$

Note that $1/V$, appearing in (5.3.18), replaces the density ρ in (5.3.2). In fact, the quantity $1/V$ may be interpreted as the density of the free particles (i.e., the particles besides the two fixed at $\mathbf{R}'$ and $\mathbf{R}''$) for the case $N = 3$.

The foregoing derivation of (5.3.18) illustrates the origin of the coefficient $B(\mathbf{R}', \mathbf{R}'')$, which, in principle, results from the simultaneous interaction of three particles [compare this result with (5.3.11)]. This is actually the meaning of the term "slightly dense gas." Whereas in a very dilute gas we take account of interactions between pairs only, here we also consider the effect of interactions among three particles, but not more.

Next we examine the content of the function $B(\mathbf{R}', \mathbf{R}'')$ and its effect on $g(R)$ for a system of hard spheres (HS). In this case, we have

$$f(R) = \begin{cases} -1 & \text{for } R < \sigma \\ 0 & \text{for } R > \sigma \end{cases}. \qquad (5.3.19)$$

Therefore, the only contribution to the integral in (5.3.4) comes from regions in which both $f(\mathbf{R}', \mathbf{R}_3)$ and $f(\mathbf{R}'', \mathbf{R}_3)$ are equal to -1. This may occur for $R < 2\sigma$. The integrand vanishes when either $|\mathbf{R}' - \mathbf{R}_3| > \sigma$ or $|\mathbf{R}'' - \mathbf{R}_3| > \sigma$. Furthermore, for $|\mathbf{R}'' - \mathbf{R}'| < \sigma$, the exponential factor in (5.3.18), $\exp[-\beta U(\mathbf{R}', \mathbf{R}'')]$, vanishes. Thus, the only region of interest is $\sigma \leq R \leq 2\sigma$. Figure 5.3 depicts such a situation. Since the value of the integrand in the region where it is nonzero equals $(-1) \times (-1) = 1$, the integration problem reduces to the geometric problem of computing the volume of the intersection of the two spheres of radius σ. (Note that the *diameter* of the hard spheres is σ.) The solution can be

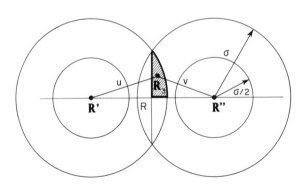

FIGURE 5.3. Two hard spheres of diameter σ at a distance of $\sigma < R < 2\sigma$ from each other. The intersection of the two spheres of radius σ about $\mathbf{R}'$ and $\mathbf{R}''$ is twice the volume of revolution of the shaded area in the figure.

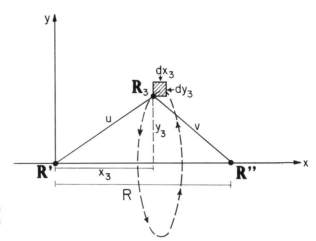

FIGURE 5.4. Bipolar coordinates
used for the integration over the
intersection region of Fig. 5.3.

obtained by transforming to bipolar coordinates. We write

$$u = |\mathbf{R}_3 - \mathbf{R}'|, \qquad v = |\mathbf{R}_3 - \mathbf{R}''|. \tag{5.3.20}$$

Then the element of volume $d\mathbf{R}_3$ can be chosen (noting the axial symmetry of our problem) as a ring of radius y_3 and cross section $dx_3\, dy_3$ (see Fig. 5.4), i.e.,

$$d\mathbf{R}_3 = 2\pi y_3\, dx_3\, dy_3. \tag{5.3.21}$$

The transformation of variables and its Jacobian are

$$u^2 = x_3^2 + y_3^2, \qquad v^2 = y_3^2 + (R - x_3)^2, \qquad \frac{\partial(x_3, y_3)}{\partial(u, v)} = \frac{uv}{y_3 R}. \tag{5.3.22}$$

Hence

$$d\mathbf{R}_3 = (2\pi/R)uv\, du\, dv. \tag{5.3.23}$$

Using this transformation, we rewrite (5.3.4) as

$$B(R) = 2\frac{2\pi}{R} \int_{R/2}^{\sigma} u\, du \int_{R-u}^{u} v\, dv = \frac{4\pi\sigma^3}{3}\left[1 - \frac{3}{4}\frac{R}{\sigma} + \frac{1}{16}\left(\frac{R}{\sigma}\right)^3\right]. \tag{5.3.24}$$

The factor 2 is included in (5.2.24) since we compute only half of the intersection of the two spheres. (This corresponds to the volume of revolution of the shaded area in Fig. 5.3.) Using (5.3.24), we can now write explicitly the form of the radial distribution function for hard spheres at "slightly dense" concentration:

$$g(R) = \begin{cases} 0 & \text{for } R < \sigma \\ 1 + \rho\dfrac{4\pi\sigma^3}{3}\left[1 - \dfrac{3}{4}\dfrac{R}{\sigma} + \dfrac{1}{16}\left(\dfrac{R}{\sigma}\right)^3\right] & \text{for } \sigma < R < 2\sigma. \\ 1 & \text{for } R > 2\sigma \end{cases} \tag{5.3.25}$$

The form of this function is depicted in Fig. 5.5. Further implications of (5.3.25) will be discussed in section 5.4, where we reinterpret the correction term $B(R)$ as an effective "attraction" between two hard spheres.

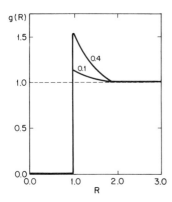

FIGURE 5.5. The form of $g(R)$ for hard-sphere particles ($\sigma = 1$), using the first-order expansion in the density. The two curves correspond to $\rho = 0.1$ and $\rho = 0.4$.

5.3.4. Lennard–Jones Particles at Moderately High Densities

Lennard–Jones (LJ) particles are supposed to resemble closely in behavior real, simple spherical particles such as argon. In this section, we present some further information on the behavior of $g(R)$ and its dependence on density and on temperature. The LJ particles are defined through their pair potential as

$$U_{LJ}(R) = 4\varepsilon\left[\left(\frac{\sigma}{R}\right)^{12} - \left(\frac{\sigma}{R}\right)^{6}\right]. \tag{5.3.26}$$

Figure 5.6 demonstrates the variation of $g(R)$ as we increase the density. The dimensionless densities $\rho\sigma^3$ are recorded next to each curve. One observes that at very low densities, there is a single peak, corresponding to the minimum in the potential function (5.3.26). At successively higher densities, new peaks develop which become more and more pronounced as the density increases. The location of the first peak is essentially unchanged,

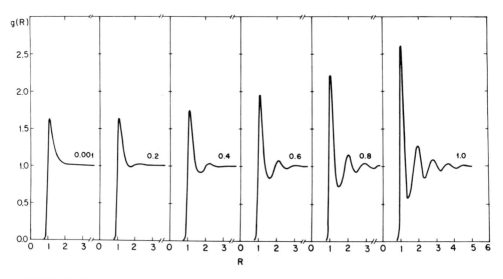

FIGURE 5.6. Dependence of $g(R)$ on the density of the system. The corresponding (number) densities are indicated next to each curve.

even though its height increases steadily. The locations of the new peaks occur roughly at integral multiples of σ, i.e., at $R \sim \sigma$, 2σ, 3σ. . . . This feature reflects the propensity of the spherical molecules to pack, at least locally, in concentric and nearly equidistant spheres about a given molecule. This is a very fundamental property of fluids and deserves further attention.

Consider a random arrangement (configuration) of spherical particles in the fluid. (An illustration in two dimensions is provided in Fig. 5.7.) Now consider a spherical shell of width $d\sigma$ and radius σ, and inquire as to the average number of particles in this element of volume. If the center of the spherical shell has been chosen at random, as on the rhs of the figure, we find that on the average, the number of particles is $\rho 4\pi\sigma^2 \, d\sigma$. On the other hand, if we choose the center of the spherical shell so that it coincides with the center of a particle, then on the average, we find more particles in this element of volume. The drawing on the left illustrates this case for one configuration. One sees that, in this example, there are six particles in the element of volume on the left as compared with two particles on the right. Similarly, we could have drawn spherical shells of width $d\sigma$ at 2σ and again have found excess particles in the element of volume, the origin of which has been chosen at the center of one of the particles. The excess of particles at the distances σ, 2σ, 3σ, etc., from the center of a particle is manifested in the various peaks of the function $g(R)$. Clearly, this effect decays rapidly as the distance from the center increases. We see from Fig. 5.6 that $g(R)$ is almost unity for $R \gtrsim 5\sigma$. This means that correlation between the local densities at two points $\mathbf{R}'$ and $\mathbf{R}''$ extends over a relatively short range.

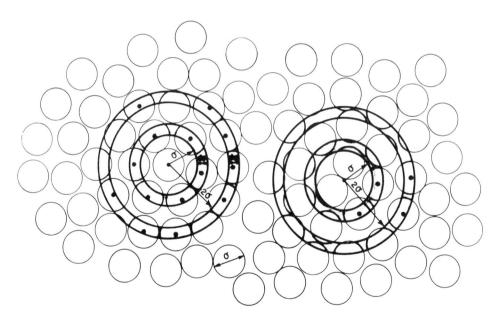

FIGURE 5.7. A random distribution of spheres in two dimensions. Two spherical shells of width $d\sigma$ with radius σ and 2σ are drawn (the diameter of the spheres is σ). On the left, the center of the spherical shell coincides with the center of one particle, whereas on the right, the center of the spherical shell has been chosen at a random point. It is clearly observed that the two shells on the left are filled by centers of particles to a larger extent than the corresponding shells on the right. The average excess of particles in these shells, drawn from the center of a given particle, is manifested by the various peaks of $g(R)$.

At short distances, say $\sigma \lesssim R \lesssim 5\sigma$, in spite of the random distribution of the particles, there is a sort of order revealed by the form of the RDF. This order is often referred to as the local structure of the liquid. The local character of this structure should be stressed, since it contrasts with the long-range order typical of a perfect lattice.

From the definition of $g(R)$, it follows that the average number of particles in a spherical shell of radius R (from a center of a given particle) and width dR is

$$N(dR) = \rho g(R)4\pi R^2 \, dR. \tag{5.3.27}$$

Hence, the average number of particles in a sphere of radius R_M (excluding the particle at the center) is

$$N_{CN}(R_M) = \rho \int_0^{R_M} g(R)4\pi R^2 \, dR. \tag{5.3.28}$$

The quantity $N_{CN}(R_M)$ may be referred to as the coordination number of particles, computed for the particular sphere of radius R_M. A choice of $\sigma \lesssim R_M \lesssim 2\sigma$ will give a coordination number that conforms to the common meaning of this concept.

5.4. POTENTIAL OF AVERAGE FORCE

The concept of potential of average force has been introduced in Chapter 3. Here we introduce the same concept for a pure liquid. The corresponding definitions for mixtures will be introduced in Chapters 6 and 7.

The general definition of the potential of average force between two particles at $\mathbf{X}_1$ and $\mathbf{X}_2$ is

$$W(\mathbf{X}', \mathbf{X}'') = -kT \ln g(\mathbf{X}', \mathbf{X}''), \tag{5.4.1}$$

where $g(\mathbf{X}', \mathbf{X}'')$ is the pair correlation function. In this section we discuss only the potential of average force between spherical particles. We first show that the force operates on a given particle, say at $\mathbf{R}'$, due to the presence of another particle at $\mathbf{R}''$, averaged over all other particles of the system is derived from $W(\mathbf{R}', \mathbf{R}'')$. This will justify the use of the term "potential of average force."

From the definition of $g(\mathbf{R}', \mathbf{R}'')$ in (5.3.1), we have

$$\exp[-\beta W(\mathbf{R}', \mathbf{R}'')] = \frac{N(N-1)}{\rho^2} \frac{\int \cdots \int d\mathbf{R}_3 \cdots d\mathbf{R}_N \exp[-\beta U_N(\mathbf{R}', \mathbf{R}'', \mathbf{R}_3, \ldots, \mathbf{R}_N)]}{Z_N}. \tag{5.4.2}$$

We now take the gradient of $W(\mathbf{R}', \mathbf{R}'')$ with respect to the vector $\mathbf{R}'$, and get

$$-\beta \nabla' W(\mathbf{R}', \mathbf{R}'') = \nabla' \left\{ \ln \int \cdots \int d\mathbf{R}_3 \cdots d\mathbf{R}_N \exp[-\beta U_N(\mathbf{R}', \mathbf{R}'', \mathbf{R}_3, \ldots, \mathbf{R}_N)] \right\}, \tag{5.4.3}$$

where the symbol ∇' stands for the gradient with respect to $\mathbf{R}' = (x', y', z')$, i.e.,

$$\nabla' = \left(\frac{\partial}{\partial x'}, \frac{\partial}{\partial y'}, \frac{\partial}{\partial z'} \right). \tag{5.4.4}$$

We also assume that the total potential energy is pairwise additive; hence, we write

$$U_N(\mathbf{R}', \mathbf{R}'', \mathbf{R}_3, \ldots, \mathbf{R}_N)$$

$$= U_{N-2}(\mathbf{R}_3, \ldots, \mathbf{R}_N) + \sum_{i=3}^{N} [U(\mathbf{R}_i, \mathbf{R}') + U(\mathbf{R}_i, \mathbf{R}'')] + U(\mathbf{R}', \mathbf{R}''). \tag{5.4.5}$$

The gradient of U_N is then

$$\nabla' U_N(\mathbf{R}', \mathbf{R}'', \mathbf{R}_3, \ldots, \mathbf{R}_N) = \sum_{i=3}^{N} \nabla' U(\mathbf{R}_i, \mathbf{R}') + \nabla' U(\mathbf{R}', \mathbf{R}''). \tag{5.4.6}$$

Performing the differentiation in (5.4.3), we get

$$-\nabla' W(\mathbf{R}', \mathbf{R}'') = \frac{\displaystyle \int \cdots \int d\mathbf{R}_3 \cdots d\mathbf{R}_N \exp(-\beta U_N) \left[-\sum_{i=3}^{N} \nabla' U(\mathbf{R}_i, \mathbf{R}') - \nabla' U(\mathbf{R}', \mathbf{R}'') \right]}{\displaystyle \int \cdots \int d\mathbf{R}_3 \cdots d\mathbf{R}_N \exp(-\beta U_N)}.$$

$$\tag{5.4.7}$$

Note that the integration in the numerator of (5.4.7) is over $\mathbf{R}_3 \cdots \mathbf{R}_N$, and the quantity $\nabla'(\mathbf{R}', \mathbf{R}'')$ is independent of these variables. We also introduce the conditional probability density of finding the $N - 2$ particles at a specified configuration, given that two particles are at $\mathbf{R}'$ and $\mathbf{R}''$:

$$P(\mathbf{R}_3, \ldots, \mathbf{R}_N / \mathbf{R}', \mathbf{R}'') = \frac{P(\mathbf{R}', \mathbf{R}'', \mathbf{R}_3, \ldots, \mathbf{R}_N)}{P(\mathbf{R}', \mathbf{R}'')}$$

$$= \frac{\exp[-\beta U_N(\mathbf{R}', \mathbf{R}'', \mathbf{R}_3, \ldots, \mathbf{R}_N)]}{Z_N}$$

$$\times \frac{Z_N}{\displaystyle \int \cdots \int d\mathbf{R}_3 \cdots d\mathbf{R}_N \exp[-\beta U_N(\mathbf{R}', \mathbf{R}'', \mathbf{R}_3, \ldots, \mathbf{R}_N)]}$$

$$= \frac{\exp[-\beta U_N(\mathbf{R}', \mathbf{R}'', \mathbf{R}_3, \ldots, \mathbf{R}_N)]}{\displaystyle \int \cdots \int d\mathbf{R}_3 \cdots d\mathbf{R}_N \exp[-\beta U_N(\mathbf{R}', \mathbf{R}'', \mathbf{R}_3, \ldots, \mathbf{R}_N)]}. \tag{5.4.8}$$

Using (5.4.8), we rewrite (5.4.7) as

$$-\nabla' W(\mathbf{R}', \mathbf{R}'') = -\nabla' U(\mathbf{R}', \mathbf{R}'')$$

$$+ \int \cdots \int d\mathbf{R}_3 \cdots d\mathbf{R}_N\, P(\mathbf{R}_3, \ldots, \mathbf{R}_N / \mathbf{R}', \mathbf{R}'') \sum_{i=3}^{N} [-\nabla' U(\mathbf{R}_i, \mathbf{R}')]$$

$$= -\nabla' U(\mathbf{R}', \mathbf{R}'') + \left\langle -\sum_{i=3}^{N} \nabla' U(\mathbf{R}_i, \mathbf{R}') \right\rangle^{(N-2)}. \tag{5.4.9}$$

In (5.4.9), we expressed $-\nabla' W(\mathbf{R}', \mathbf{R}'')$ as a sum of two terms. The first term is simply the *direct* force exerted on the particle at $\mathbf{R}'$ when the second particle is at $\mathbf{R}''$. The second term is the conditional average force [note that the average has been calculated using the conditional probability density (5.4.8)] exerted on the particle at $\mathbf{R}'$ by all the other particles present in the system. It is an average over all the configurations of the $N-2$ particles (as indicated by the notation $\langle \, \rangle^{(N-2)}$), keeping $\mathbf{R}'$ and $\mathbf{R}''$ fixed. The latter may be referred to as the *indirect* force operating on the particle at $\mathbf{R}'$, which originates from all the other particles excluding the one at $\mathbf{R}''$. The foregoing discussion justifies the designation of $W(\mathbf{R}', \mathbf{R}'')$ as the potential of average force. Its gradient gives the average force, including direct and indirect contributions, operating on the particle at $\mathbf{R}'$.

We can further simplify (5.4.9) by noting that the sum over i produces $N-2$ equal terms, i.e.,

$$\int \cdots \int d\mathbf{R}_3 \cdots d\mathbf{R}_N\, P(\mathbf{R}_3, \ldots, \mathbf{R}_N / \mathbf{R}', \mathbf{R}'') \sum_{i=3}^{N} \nabla' U(\mathbf{R}_i, \mathbf{R}')$$

$$= (N-2) \int \cdots \int d\mathbf{R}_3 \cdots d\mathbf{R}_N\, P(\mathbf{R}_3 \cdots \mathbf{R}_N / \mathbf{R}', \mathbf{R}'') \nabla' U(\mathbf{R}_3, \mathbf{R}')$$

$$= (N-2) \int d\mathbf{R}_3\, \nabla' U(\mathbf{R}_3, \mathbf{R}') \int \cdots \int d\mathbf{R}_4 \cdots d\mathbf{R}_N\, P(\mathbf{R}_3 \cdots \mathbf{R}_N / \mathbf{R}', \mathbf{R}'')$$

$$= (N-2) \int d\mathbf{R}_3\, \nabla' U(\mathbf{R}_3, \mathbf{R}') P(\mathbf{R}_3 / \mathbf{R}', \mathbf{R}'')$$

$$= \int d\mathbf{R} [\nabla' U(\mathbf{R}, \mathbf{R}')] \rho(\mathbf{R} / \mathbf{R}', \mathbf{R}''). \tag{5.4.10}$$

The new quantity introduced in (5.4.10) is the conditional density at a point $\mathbf{R}$, given two particles at $\mathbf{R}'$ and $\mathbf{R}''$. This is a straightforward generalization of the conditional density introduced in Eq. (5.2.45). The total force acting on 1 is thus

$$\mathbf{F}_1 = -\nabla' U(\mathbf{R}', \mathbf{R}'') - \int d\mathbf{R} [\nabla' U(\mathbf{R}, \mathbf{R}')] \rho(\mathbf{R} / \mathbf{R}', \mathbf{R}''). \tag{5.4.11}$$

This form is useful in the study of forces applied to solutes or to groups in proteins in aqueous solutions. The first term is referred to as the direct force and the second term as the solvent-induced forces. (Note that the gradient operates only on $U(\mathbf{R}, \mathbf{R}')$, as we have stressed by the square brackets.)

The form of the function $W(R)$, with $R = |\mathbf{R}'' - \mathbf{R}'|$ for LJ particles, and its density dependence are depicted in Fig. 5.8. At very low densities, the potential of average force is identical to the pair potential; this follows from the negligible effect of all the other particles present in the system. At higher densities, the function $W(R)$ shows successive maxima and minima [corresponding to the minima and maxima of $g(R)$, by virtue of definition (5.4.1)]. The interesting point worth noting is that the indirect force at, say, $R > \sigma$, can be either attractive or repulsive even in the region where the direct force is purely attractive.

There is a fundamental analogy between the concept of the potential force introduced in this section and the corresponding concepts introduced in the allosteric models in Chapter 3. In all of these cases, we average over part of the degrees of freedom of the system and thereby produce new features of the interaction which did not exist in the "bare" (or original) potential function. Here, we have obtained alternating regions of repulsion and attraction at $R \gtrsim \sigma$, whereas in the same region, the original direct potential

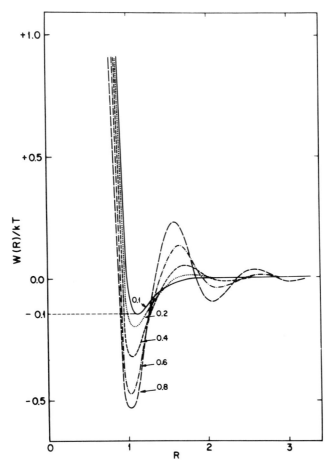

FIGURE 5.8. Schematic illustration of the form of the function $W(R)$ and its density dependence. The number density is indicated next to each curve. Note that for $\rho = 0.1$, the curve of $W(R)/kT$ is almost identical with $U(R)/kT$, with a minimum energy of $W(R)/kT \approx -0.1$. The first minimum becomes deeper and new minima develop as the density increases.

was everywhere attractive. Another feature which results from the averaging process is the possible loss of the pairwise additivity of the interaction. Some examples were discussed in Chapter 3. Further elaboration on this question will be presented in section 5.6.

5.4.1. Hard Spheres and Lennard–Jones Particles

We examine here the origin of the attractive interaction region of $W(R)$ for a system of hard spheres. We recall that the direct interaction $U(R)$ between two hard spheres is either zero, for $R > \sigma$, or infinitely repulsive, for $R \leq \sigma$. We have already seen that at low densities, $g(R)$ is nonunity in the region $\sigma \leq R \leq 2\sigma$. Compare Figs. 5.2 and 5.5. When we convert the function $g(R)$ into $W(R)$, the result is shown in Fig. 5.9. [Note that Figs. 5.5 and 5.9 are both based on Eq. (5.3.25). Therefore the arguments presented below apply only at low enough densities, for which (5.3.25) is a good approximation.]

The analytical form of $W(R)$ is obtained from (5.3.25) and (5.4.1).

$$W(R) = \begin{cases} \infty & \text{for } R < \sigma \\ -kT \ln[1 + \rho B(R)] \approx -kT\rho \dfrac{4\pi\sigma^3}{3}\left[1 - \dfrac{3}{4}\dfrac{R}{\sigma} + \dfrac{1}{16}\left(\dfrac{R}{\sigma}\right)^3\right] & \text{for } \sigma \leq R \leq 2\sigma . \\ 0 & \text{for } R > 2\sigma \end{cases}$$

$$(5.4.12)$$

This function is depicted in Fig. 5.9. Note that the force is attractive everywhere in the range $\sigma \leq R \leq 2\sigma$, i.e.,

$$-\frac{\partial}{\partial R} W(R) = kT\rho \frac{4\pi\sigma^3}{3}\left(-\frac{3}{4\sigma} + \frac{3}{16}\frac{R^2}{\sigma^3}\right) < 0, \qquad \sigma \leq R \leq 2\sigma . \qquad (5.4.13)$$

This is quite a surprising result if we remember that we are dealing with hard-sphere particles. We now examine this phenomenon for a system of three almost hard spheres, two of them fixed at $\mathbf{R}_1$ and $\mathbf{R}_2$ (with $R = |\mathbf{R}_2 - \mathbf{R}_1|$), and a third particle serving as a "solvent." Clearly, this is the simplest solvent we can envisage besides a vacuum. The

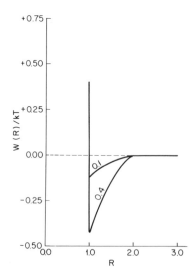

FIGURE 5.9. The forms of $W(R)/kT$ for hard spheres at low densities. The curves are computed for spheres of diameter $\sigma = 1.0$ and two densities, $\rho = 0.1$ and $\rho = 0.4$. Note that for $1 \leq R \leq 2$, we have an attractive potential of average force.

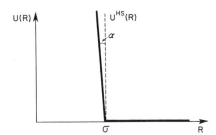

FIGURE 5.10. An almost-hard-sphere potential function. As the angle α tends to zero, one gets the hard-sphere potential $U^{HS}(R)$.

choice of almost hard spheres has been made essentially for convenience, as will become clearer later. The pair potential for our system is depicted in Fig. 5.10. The idea is that for $R > \sigma$, the potential is zero, as it is for hard spheres, whereas for $R < \sigma$, the potential is very strongly repulsive, yet the slope is finite, in contrast to the infinite slope for hard spheres, which is not realistic for any real system.

Figure 5.11 shows the system under consideration. The two fixed particles are denoted by 1 and 2 and are placed at a distance $\sigma \leq R \leq 2\sigma$ (which is the only region of interest for the present discussion). The third particle, which may wander about, is denoted by 3, and a few possible positions of this particle are shown in the figure.

The total force operating on particle 1 is, according to (5.4.9),

$$-\nabla_1 W(\mathbf{R}_1, \mathbf{R}_2) = \int_V d\mathbf{R}_3 \, P(\mathbf{R}_3/\mathbf{R}_1, \mathbf{R}_2)[-\nabla_1 U(\mathbf{R}_1, \mathbf{R}_3)]$$

$$= \int_A + \int_{A'} + \int_B . \tag{5.4.14}$$

Note that for $\sigma < R \leq 2\sigma$, the direct force is zero. In the last form on the rhs of (5.4.14), we have split the integral over V into three integrals over three nonoverlapping regions. The region A includes all points in space for which $|\mathbf{R}_3 - \mathbf{R}_1| < \sigma$ and $|\mathbf{R}_3 - \mathbf{R}_2| < \sigma$, i.e., the region in which particle 3 "touches" both particles 1 and 2. It is obtained by the intersection of the two spheres of radius σ drawn about the centers 1 and 2. Region A' is obtained by reflecting each point in A through the center of particle 1. Region B comprises all the remaining points for which particle 3 interacts with 1 but not with 2.

FIGURE 5.11. Two almost-hard spheres, numbered 1 and 2, at a distance of $\sigma < R < 2\sigma$. A third particle, numbered 3, is shown in various positions: In 3(B) and 3(B'), particle 3 exerts repulsive forces on 1 of equal magnitudes but opposite directions. The probabilities of these two positions are equal since 3 interacts only with 1. On the other hand, in the two positions 3(A) and 3(A'), particle 3 exerts equal forces on 1 in opposite directions, but now the probability of position 3(A) is smaller than that of 3(A').

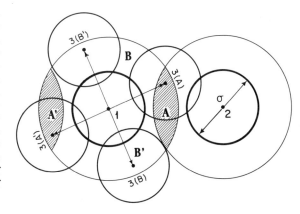

[Other regions in space are of no importance since they do not contribute to the integral (5.4.14).]

Now, for each position of particle 3 in B, there exists a complementary position obtained by reflection through the center of 1. These pairs of positions have the same probability density, since $P(\mathbf{R}_3/\mathbf{R}_1, \mathbf{R}_2)$ depends only on $U(\mathbf{R}_1, \mathbf{R}_3)$ in these regions. Furthermore, the forces exerted on 1 by such a pair of positions are equal in magnitude but opposite in direction. An example of such a pair is shown in Fig. 5.11 and is denoted by $3(B)$ and $3(B')$. It is therefore clear that the contribution to the integral over region B in (5.4.14) is zero.

Turning to regions A and A', we see again that for each point in A, we have a point in A' at which the force exerted by 3 on 1 will have the same magnitude but opposite direction. An example is the pair denoted by $3(A)$ and $3(A')$. But now the probability of finding the particle 3 in A will always be smaller than the probability of finding the particle at the complementary point in A'. The reason is that $P(\mathbf{R}_3/\mathbf{R}_1, \mathbf{R}_2)$ is proportional to $\exp[-\beta U(\mathbf{R}_1, \mathbf{R}_3) - \beta U(\mathbf{R}_2, \mathbf{R}_3)]$ in A, but proportional to $\exp[-\beta U(\mathbf{R}_1, \mathbf{R}_3)]$ in A'. Since the potential is positive in these regions, the probability of finding particle 3 at a point in A is always smaller than the probability of finding it at the complementary point in A'. Hence, the integral over A is larger in absolute magnitude than the integral over A', and the overall direction of the force in (5.4.14) is from $\mathbf{R}_1$ toward $\mathbf{R}_2$; i.e., the indirect force operating between 1 and 2, due to the presence of 3, is attractive.

To summarize, we have started with particles having no direct attractive forces and have arrived at an indirect attractive force, which originates from the averaging over all possible positions of a third particle. The latter, though also exerting only repulsive forces, operates on particle 1 in an asymmetric manner, which leads to a net average attraction between particles 1 and 2.

Of course, we could have chosen the potential function in Fig. 5.10 to be as steep as we wish and hence to approach very closely the potential function for hard spheres. The latter is mathematically inconvenient, however, since we have to deal with the awkward product of infinite forces with zero probabilities under the integral sign in (5.4.14).†

A different and often useful function related to $g(R)$ is defined by

$$y(R) = g(R) \exp[+\beta U(R)]. \qquad (5.4.15)$$

The elimination of the factor $\exp[-\beta U(R)]$ appearing in (5.3.2) is often useful, since the remaining function $y(R)$ becomes everywhere an analytic function of R even for hard spheres. We illustrate this point by considering only the first-order term in the expansion (5.3.2) for hard-sphere particles, where we have

$$y(R) = 1 + B(R)\rho. \qquad (5.4.16)$$

Note that for $R \leq \sigma$, $g(R)$ is zero because of the factor $\exp[-\beta U(R)]$ in (5.3.2) and clearly has a singularity at the point $R = \sigma$. Once we get rid of the singular factor in (5.3.2), we are left with the function $y(R)$, which changes smoothly even in the range $0 \leq R \leq \sigma$. Let us examine the specific example given in (5.4.16).

---

† The arguments presented here are based on the definition of $g(R)$ and hence $W(R)$ in the configurational space. In a real system, where particles possess kinetic energy of translation, one may argue that collisions of particle 3 with 1 are asymmetric because of partial shielding by particle 2. Hence, particle 3 will collide with 1 more often from the left-hand side than from the right-hand side, producing the same net effect of attraction between particles 1 and 2.

The form of $y(R)$ for $R > \sigma$ is identical to the form of $g(R)$ in this region, which has been given in (5.3.25). The only region for which we have to compute $y(R)$ is thus $R \leq \sigma$. Using arguments similar to those in section 5.3 we get, from (5.4.16) and the definition of $B(R)$ in (5.3.4),

$$y(R) = 1 + \frac{4\pi\rho}{R} \int_{R/2}^{\sigma} u \, du \int_{|R-u|}^{u} v \, dv$$

$$= 1 + \rho \frac{4\pi\sigma^3}{3} \left[1 - \frac{3}{4} \frac{R}{\sigma} + \frac{1}{16} \left(\frac{R}{\sigma} \right)^3 \right] \quad \text{for } 0 \leq R \leq \sigma. \quad (5.4.17)$$

The region of integration in (5.4.17) differs from that in (5.3.24) only in the lower limit, $|R - u|$ instead of $R - u$ in (5.3.24). The region of integration is shown in Fig. 5.12 (compare with Fig. 5.3 for $\sigma < R < 2\sigma$), and it consists of twice the volume of revolution of the shaded area in the figure. Note that since the integration over v in (5.4.17) gives $v^2/2$, the sign of $R - u$ in the lower limit is of no importance; hence the integral in (5.4.17) gives the same result as in (5.3.24). Figure 5.13 shows the form of $y(R)$ for HS particles using (5.4.17), and Fig. 5.14 shows $y(R)$ for LJ particles at various densities. Note in particular that the function $y(R)$ is continuous in crossing the point $R = \sigma$ into the physically inaccessible region $R < \sigma$. This feature renders this function of some convenience in the theory of integral equations, where one first solves for $y(R)$ and then constructs $g(R)$ by (5.4.15).

5.4.2. Potential of Average Force and Helmholtz Energy Changes

We have already seen in various places in Chapter 3 that correlation functions are related to the changes in Helmholtz energy. We now derive a similar relation for the pair correlation function of pure liquids.

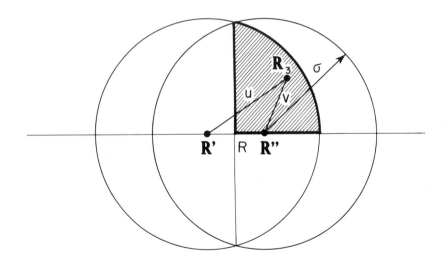

FIGURE 5.12. Region of integration [Eq. (5.4.17)] for two hard spheres of diameter σ at a distance $0 \leq R \leq \sigma$. The integration extends over the volume of revolution of the intersection between the two circles centered at $\mathbf{R}'$ and $\mathbf{R}''$. Half of this intersection is darkened in the figure.

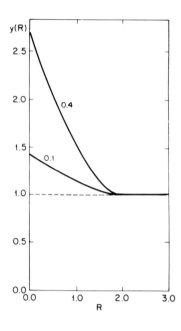

FIGURE 5.13. The form of the function $y(R)$ for hard spheres, using Eq. (5.4.17). Note that the function $y(R)$ changes smoothly as we cross $R = \sigma = 1$ and that it is a monotonically decreasing function of R in the region $0 \leq R \leq \sigma$. The two curves correspond to the same systems as in Fig. 5.5; i.e., for $\sigma = 1.0$, $\sigma = 0.1$, and $\sigma = 0.4$.

Consider a system of N simple spherical particles in a volume V at temperature T. The Helmholtz energy for such a system is

$$\exp[-\beta A(T, V, N)] = (1/N!\Lambda^{3N}) \int \cdots \int d\mathbf{R}^N \exp[-\beta U(\mathbf{R}^N)]. \qquad (5.4.18)$$

Now consider a slightly modified system in which two specific particles, say 1 and 2, have been fixed at the points $\mathbf{R}'$ and $\mathbf{R}''$, respectively. The Helmholtz energy for such a system

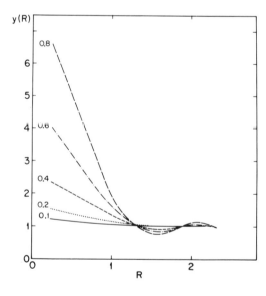

FIGURE 5.14. Schematic variation of $y(R)$ with density for Lennard–Jones particles. (These computations correspond to Lennard–Jones particles in two dimensions with $\sigma = 1.0$ and $\varepsilon/kT = 0.1$. The number density is indicated next to each curve.) Note again the monotonic behavior of $y(R)$ in the region $0 \leq R \leq \sigma$.

is denoted by $A(\mathbf{R'}, \mathbf{R''})$, and we have

$$\exp[-\beta A(\mathbf{R'}, \mathbf{R''})] = \frac{1}{(N-2)! \Lambda^{3(N-2)}}$$

$$\times \int \cdots \int d\mathbf{R}_3 \cdots d\mathbf{R}_N \exp[-\beta U_N(\mathbf{R'}, \mathbf{R''}, \mathbf{R}_3, \ldots, \mathbf{R}_N)]. \quad (5.4.19)$$

Note the differences between (5.4.18) and (5.4.19). Let us denote by $A(R)$, with $R = |\mathbf{R''} - \mathbf{R'}|$, the Helmholtz energy of such a system when the separation between the two particles is R, and form the difference

$$\Delta A(R) = A(R) - A(\infty), \quad (5.4.20)$$

which is the work required to bring the two particles from fixed positions at infinite separation to fixed positions where the separation is R. The process is carried out at constant volume and temperature. From (5.3.1), (5.4.1), (5.4.18), and (5.4.20), we get

$$\exp[-\beta \Delta A(R)] = \frac{\displaystyle\int \cdots \int d\mathbf{R}_3 \cdots d\mathbf{R}_N \exp[-\beta U_N(\mathbf{R'}, \mathbf{R''}, \mathbf{R}_3, \ldots, \mathbf{R}_N)]}{\displaystyle\lim_{R \to \infty} \int \cdots \int d\mathbf{R}_3 \cdots d\mathbf{R}_N \exp[-\beta U_N(\mathbf{R'}, \mathbf{R''}, \mathbf{R}_3, \ldots, \mathbf{R}_N)]}$$

$$= g(R)$$

$$= \exp\{-\beta[W(R) - W(\infty)]\}. \quad (5.4.21)$$

This is an important and useful result. The correlation between two particles at distance R is related to the work required (here at T, V constant) to bring the two particles from infinite separation to a distance R. We recall that in Chapter 3 we obtained similar relationships involving the correlation between ligands and the work required to bring the ligands from different polymers into the same polymer.

Recalling that $g(R)$ is proportional to the probability density of finding the two particles at a distance R, we write

Pr(two particles at R)

$$= C \exp[-\beta(\text{work required to bring the two particles to } R)], \quad (5.4.22)$$

where C is a constant. Thus the probability of finding an event (two particles at R) is related to the work required to create that event. This is a particular example of a much more general relation between the probability of observing an event and the work required to create that event. We shall encounter other examples of these relationships in the following.

In this section we used the T, V, N ensemble to obtain relation (5.4.21). A similar relation can be obtained for any other ensemble. Of particular importance is the analogue of (5.4.21) in the T, P, N ensemble. It has the same form but the events occur in a T, P, N system, and instead of the Helmholtz energy change we need to use the Gibbs energy change.

Because of relation (5.4.21), it is often said that the potential of average force has the character of a free energy. As such, it is both temperature- and pressure-dependent. This is in contrast to the behavior of $U(R)$, which to a good approximation is temperature- and pressure-independent.

Finally we note that the potential of average force is a concept different from the average interaction between two particles. We have already noted the difference between the two concepts in Chapter 3. The same difference applies here.

5.5. A BRIEF SURVEY OF THE METHODS OF EVALUATING g(R)

There exist essentially three sources of information on the form of the function $g(R)$. The first is experimental, based on X-ray or neutron diffraction by liquids. The second is theoretical and usually involves integral equations for $g(R)$ or some related function. The third method may be classified as an intermediate between theory and experiment. It consists of various simulation techniques used for investigating the liquid state.

In this section, we survey the fundamental principles of the various methods. The details are highly technical in character and are not essential for understanding the subject matter of most of this book.

5.5.1. Experimental Methods

The original and most extensively employed method of evaluating $g(R)$ experimentally is the study of X-ray diffraction by liquids. Recently, diffraction of neutrons has also been found useful for this purpose. The principal idea of converting diffraction patterns into pair distribution functions is common to both methods, though they differ in both experimental detail and the scope of information that they provide.

Consider a monochromatic beam of electromagnetic waves with a given wave-number vector $\mathbf{k}_0$ ($\mathbf{k}_0$ is a vector in the direction of propagation of the plane wave with absolute magnitude $|\mathbf{k}_0| = 2\pi/\lambda$, where λ is the wavelength), scattered from a system of N molecules.† The scattered wave-number vector is denoted by $\mathbf{k}$. Two important assumptions are made regarding this process. First, the incident and the scattered beams have the same wavelength, i.e.,

$$|\mathbf{k}_0| = |\mathbf{k}| = 2\pi/\lambda. \qquad (5.5.1)$$

This means that the scattering is elastic of the kind often referred to as "Rayleigh scattering." Second, each ray entering the system is scattered only once. This assumption is essential for obtaining the required relation between the intensity of the scattered beam and the pair distribution function. If multiple scattering occurs, then such a relationship involves higher-order molecular distribution functions.

Let E_j be the complex amplitude of the wave scattered by the jth particle, measured at the point of observation (which is presumed to be far from the system). The total

---

† Actually, the X-rays are scattered by the electrons, and therefore one should start by considering the ensemble of electrons as the fundamental medium for the scattering experiment. However, since all the atoms (or molecules) are considered to be equivalent, one may group together all the electrons belonging to the same atom (or molecule). This grouping procedure is equivalent to a virtual replacement of each atom (or molecule) by a single electron situated at some chosen "center" of the molecule. To compensate for this replacement, one introduces the molecular structure factor (which is the same for all the molecules in the system). In the present treatment, this factor is absorbed in the proportionality constant α' in Eq. (5.5.3).

amplitude measured at the point of observation is

$$E(\mathbf{R}^N) = \sum_{j=1}^{N} E_j = E_0 \sum_{j=1}^{N} \exp(i\phi_j), \qquad (5.5.2)$$

where in E_0 we have included all the factors which have equal values for all the waves scattered, and ϕ_j is the phase of the wave scattered from the jth particle (here, $i = \sqrt{-1}$). We have also explicitly denoted the dependence of E on the configuration $\mathbf{R}^N$.

The intensity of the scattered wave at the point of observation is proportional to the square of the amplitude; hence

$$I(\mathbf{R}^N) = \alpha E(\mathbf{R}^N)E(\mathbf{R}^N)^*$$

$$= \alpha' \sum_{j=1}^{N} \exp(i\phi_j) \sum_{n=1}^{N} \exp(-i\phi_n)$$

$$= \alpha' \left\{ N + \sum_{j \neq n} \exp[i(\phi_j - \phi_n)] \right\}, \qquad (5.5.3)$$

where α and α' are numerical constants. In the last form on the rhs of (5.5.3) we have split the sum into two terms, those with $j = n$ and those with $j \neq n$. The important observation is that $I(\mathbf{R}^N)$ is expressed in (5.5.3) as a sum over terms each of which depends on one pair of particles. According to a general theorem (which will be derived in section 5.8), the average of such an expression over all the configurations of the system can be expressed as an integral involving the pair distribution function. Before taking the average over all the configurations of the system, it is convenient to express the phase difference $\phi_j - \phi_n$ in terms of the positions of the jth and nth particles (the relevant geometry is given in Fig. 5.15).

$$(a + b)2\pi/\lambda = \mathbf{R}_{jn} \cdot (\mathbf{k}_0 - \mathbf{k}) = \phi_j - \phi_n. \qquad (5.5.4)$$

The difference in the wave vectors $\mathbf{k}_0 - \mathbf{k}$ is related to the scattering angle θ through

$$\frac{|\mathbf{S}|}{2|\mathbf{k}|} = \sin(\theta/2), \qquad \mathbf{S} = \mathbf{k}_0 - \mathbf{k}. \qquad (5.5.5)$$

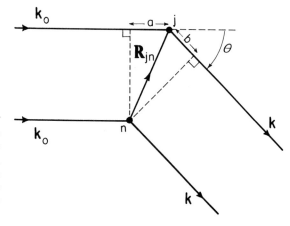

FIGURE 5.15. Schematic illustration of the scattering of an electromagnetic wave by two particles j and n. The wave vectors of the incident and the scattered rays are $\mathbf{k}_0$ and $\mathbf{k}$, respectively. The difference between the optical distance of the two rays scattered from j and n is $a + b$.

Hence, (5.5.3) can be rewritten as

$$I(\mathbf{R}^N, \mathbf{S}) = \alpha' \left[N + \sum_{j \neq n} \exp(i\mathbf{R}_{jn} \cdot \mathbf{S}) \right], \tag{5.5.6}$$

where we have also explicitly introduced the dependence of I on the vector $\mathbf{S}$. Taking the average over all the configurations, say in the T, V, N ensemble, we get

$$I(\mathbf{S}) = \langle I(\mathbf{R}^N, \mathbf{S}) \rangle$$

$$= \alpha' \left[N + \left\langle \sum_{j \neq n} \exp(i\mathbf{R}_{jn} \cdot \mathbf{S}) \right\rangle \right]$$

$$= \alpha' \left[N + \iint d\mathbf{R}_1 \, d\mathbf{R}_2 \, \rho^{(2)}(\mathbf{R}_1, \mathbf{R}_2) \exp(i\mathbf{R}_{12} \cdot \mathbf{S}) \right]. \tag{5.5.7}$$

In the last step of (5.5.7), we have used a theorem that will be proven in more general terms in section 5.8. One may notice that since the pair distribution function depends only on the distance R_{12} and the exponential function depends only on the scalars R_{12}, θ, and λ, one can actually rewrite the integral in (5.5.7) as a one-dimensional integral.

The integral in (5.5.7) is essentially a Fourier transform of the pair distribution function. (The constant α' may also depend on $\mathbf{S}$ through the structure factor included in it.) Hence, one can, in principle, take the inverse transform and express $\rho^{(2)}(R)$ as an integral involving $I(S)$. This process requires some approximations which follow from the incomplete knowledge of the function $I(S)$ over the full range of values of S (or the scattering angles θ).

5.5.2. Theoretical Methods

By theoretical methods, one usually refers to integral equations the solutions of which give $g(R)$ or a function related to $g(R)$. The most successful integral equation for systems consisting of simple particles is the so-called Percus–Yevick[1] (PY) equation. This has been used extensively to compute $g(R)$ for simple liquids such as argon, achieving remarkable agreement with experimental results. The derivation of the PY equation will not be given here. Instead, we discuss only its general structure.

An integral equation for $g(R)$ can be written symbolically as

$$g(R) = F[g(r), U(r); \rho, T], \tag{5.5.8}$$

where F stands for a functional of $g(r)$ and $U(r)$ which is also dependent on thermodynamic parameters, such as ρ, T or P, T, etc.

Once we have chosen a model potential to describe our system of real particles, e.g., Lennard–Jones particles, the functional equation is viewed as an equation involving $g(R)$ as the unknown function. This equation can be solved by either analytical or numerical methods. One version of the PY equation is

$$y(\mathbf{R}_1, \mathbf{R}_2)$$

$$= 1 + \rho \int [y(\mathbf{R}_1, \mathbf{R}_3)f(\mathbf{R}_1, \mathbf{R}_3)][y(\mathbf{R}_3, \mathbf{R}_2)f(\mathbf{R}_3, \mathbf{R}_2) + y(\mathbf{R}_3, \mathbf{R}_2) - 1] \, d\mathbf{R}_3, \tag{5.5.9}$$

where

$$y(\mathbf{R}_1, \mathbf{R}_2) = g(\mathbf{R}_1, \mathbf{R}_2) \exp[+\beta U(\mathbf{R}_1, \mathbf{R}_2)] \qquad (5.5.10)$$

$$f(\mathbf{R}_1, \mathbf{R}_2) = \exp[-\beta U(\mathbf{R}_1, \mathbf{R}_2)] - 1. \qquad (5.5.11)$$

Equation (5.5.9) is an integral equation for y, provided we have chosen a pair potential U. Once we have solved (5.5.9) for y, we can compute g through (5.5.10).

As an illustration of the content of the integral equation, let us substitute on the rhs of (5.5.9) $y \equiv 1$, which is the solution for y at extremely low densities, $\rho \to 0$. (See also the discussion in section 5.3.) This is also a common first step in an iterational procedure for solving Eq. (5.5.9):

$$y(\mathbf{R}_1, \mathbf{R}_2) = 1 + \rho \int f(\mathbf{R}_1, \mathbf{R}_3) f(\mathbf{R}_3, \mathbf{R}_2) \, d\mathbf{R}_3, \qquad (5.5.12)$$

which can be identified with the first-order expansion of y in the density, as discussed in section 5.3. We can continue this process and substitute y from (5.5.12) into the rhs of (5.5.9) to obtain successively higher-order terms in the density expansion of y. This expansion is not exact since it is based on an approximate integral equation. Nevertheless, the solution of (5.5.9) by iteration is considered to give reliable results for simple fluids up to considerably high densities.

5.5.3. Simulation Methods

As noted before, simulation methods may be considered to be intermediate between theory and experiment. In fact, various kinds of simulations have been devised for the purpose of computing $g(R)$ as well as other properties of the liquid. We now make a distinction among three of these methods.

5.5.3.1. Experimental Simulations

Morrell and Hildebrand[2] described a simple and interesting method of computing $g(R)$ for a system of balls of macroscopic size. They used gelatin balls, suspended in cool oil, that were mechanically shaken in their vessel. The coordinates of some of the balls could be measured by taking photographs of the system at various times. From these data, they were able to compute the distribution of distances between the balls, and from this, the radial distribution function. The form of $g(R)$ obtained by this method was remarkably similar to the RDF of argon obtained from the X-ray diffraction pattern. One important conclusion that may be drawn from these experiments is that the mode of packing of the atoms in the liquid follows essentially from geometric considerations of the packing of hard spheres. The fact that atoms have an effective hard-core diameter already dictates the type of packing of the molecules in the liquid. The additional attractive forces operating between real molecules have a relatively small effect on the mode of packing. This conclusion has been confirmed by many authors using considerably different arguments.

5.5.3.2. Monte Carlo Simulations

A closely related "experiment" can be carried out on a computer. Instead of shaking a system of hard balls by mechanical means, one can generate random configurations of particles by computer. The latter method can be used to compute the RDF as well as other thermodynamic quantities of a system.

The specific method devised by Metropolis et al.[3] to compute the properties of liquids is now known as the Monte Carlo (MC) method. In fact, this is a special procedure to compute multidimensional integrals numerically. Consider the computation of any average quantity, say in the T, V, N ensemble:

$$\langle F \rangle = \int \cdots \int d\mathbf{R}^N P(\mathbf{R}^N) F(\mathbf{R}^N), \tag{5.5.13}$$

where $F(\mathbf{R}^N)$ is any function of the configuration $\mathbf{R}^N$. In particular, the pair distribution function (and, hence, the RDF) can be viewed as an average of the form (5.5.13). (See section 5.2.)

The quantity $\langle F \rangle$ can be approximated by a sum over discrete events (configurations)

$$\langle F \rangle \approx \sum_{i=1}^{n} P_i F_i, \tag{5.5.14}$$

where P_i is the probability of the ith configuration and F_i the value of the function $F(\mathbf{R}^N)$ at the ith configuration.

A glance at (5.5.13) reveals that two very severe difficulties arise in any attempt to approximate the average $\langle F \rangle$ by a sum of the form (5.5.14). First, "a configuration" means a specification of $3N$ coordinates (for spherical particles), and, clearly, such a number of coordinates cannot be handled in a computer if N is of the order of 10^{23}. This limitation forces us to choose N of the order of a few hundred. The question of the suitability of such a small sample of molecules to represent a macroscopic system immediately arises. The second difficulty concerns the convergence of the sum in (5.5.14). Suppose we have already chosen N. The question is: How many configurations n should we select in order to ensure the validity of the approximation in (5.5.14)? Again, time limitations impose restrictions on the number n that we can afford in an actual "experiment."

Metropolis et al.[3] suggested two ingenious ways to overcome the two difficulties mentioned above. First, a system with N on the order of a few hundreds possesses surface effects that may dominate the behavior of the sample and therefore may not be suitable to represent a macroscopic property of the liquid. To reduce the surface effects, imposing periodic boundary conditions was suggested. This trick is simpler to visualize in one or two dimensions, where the system is converted into a ring or a torus, respectively. In three dimensions, one may imagine that a particle leaving the surface A enters through an opposite surface A'. If this is done for each of the pairs of opposing surfaces, the system may be viewed as "closed on itself" in the same sense as the two simpler examples mentioned above.

The second idea is both more profound and more interesting. Suppose we have a system of N particles at a relatively high density. The choice of a random configuration for the centers of all particles will almost always lead to the event that at least two particles overlap. Hence, $U(\mathbf{R}^N)$ of that configuration will be large and positive, and therefore $P(\mathbf{R}^N)$ will be very small. Merely picking a sequence of n random configurations would therefore be a very inefficient way of using computer time. Most of the time, we would be handling configurations which are physically almost unattainable. The suggested trick is the following: Instead of choosing configurations randomly, then weighting them with $\exp[-\beta U_N(\mathbf{R}^N)]$, one chooses configurations with probability $\exp[-\beta U_N(\mathbf{R}^N)]$ and weight them evenly.

Let us illustrate the essence of this idea for a one-dimensional integral (see Fig. 5.16)

$$\int_a^b P(x)F(x)\,dx \approx \sum_{i=1}^n P_i F_i. \tag{5.5.15}$$

Suppose that $P(x)$ is almost zero in most of the interval (a, b). In choosing a series of n points at random, we shall be spending most of our time computing the values of the integrand that are nearly zero. A very small fraction of the configurations will hit the important region under the sharp peak.

The modified procedure is to construct a sequence of events (or configurations) in which the probability of the newly chosen event depends on that of the former event. Such a scheme of events is referred to as a Markov chain. Here we have a specific example of such a chain, which we shall very briefly describe.

Consider a given configuration designated by the index i and total potential energy U_i. We now take a new configuration $i + 1$ (say, by randomly moving one particle) and check the difference in the potential energy

$$\Delta U = U_{i+1} - U_i.$$

If $\Delta U < 0$, the new configuration is accepted. If $\Delta U > 0$, we make another check. We select a random number $0 \leq \xi \leq 1$ and make the following decision: If $\exp[-\beta \Delta U] > \xi$, the new configuration is accepted; if $\exp[-\beta \Delta U] < \xi$, the new configuration is rejected. All these possibilities are shown schematically in Fig. 5.17. The new effect of this procedure is to let the choice of the next configuration be biased by its probability. Therefore, in the long run, we will be spending more time working with configurations in the "important" regions of the integrand. Of course, we must have some representatives from the "unimportant" region, but we include them with extreme parsimony.

The Monte Carlo method has been used extensively to investigate the properties of simple fluids such as noble gases as well as more complex liquids such as water. The results of such computations are considered quite reliable even for a system consisting of $N \approx 10^2$ particles. For hard-sphere fluids, where no (real) experimental results can be

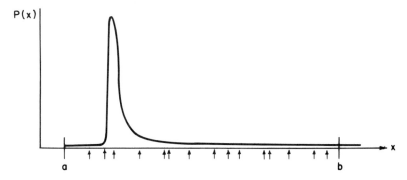

FIGURE 5.16. A function $P(x)$ which has a single sharp peak and is almost zero everywhere else. The arrow indicates random numbers $x_i(a \leq x_i \leq b)$ at which the integrand is evaluated.

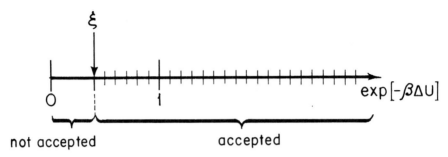

FIGURE 5.17. Schematic illustration of the accepted and unaccepted regions in the Monte Carlo computation. If $\exp(-\beta\Delta U) > 1$, the new configuration is accepted. If $\exp(-\beta\Delta U) < 1$, then one selects a random number $0 \leq \xi \leq 1$ and makes a further test; if $\exp(-\beta\Delta U) > \xi$, the new configuration is accepted; if $\exp(-\beta\Delta U) < \xi$, it is not accepted.

obtained, this kind of (computer) experiment is an important source of information with which theoretical results (say, from integral equations) may be compared.

5.5.3.3. Molecular Dynamics

The molecular dynamics method is conceptually simpler than the Monte Carlo method. Here again, we can compute various averages of the form (5.5.13) and hence the RDF as well. The method consists in a direct solution of the equations of motion of a sample of N ($\approx 10^2$) particles. In principle, the method amounts to computing time averages rather than ensemble averages, and was first employed for simple liquids by Alder and Wainwright.[4] The problem of surface effects is dealt with as in the Monte Carlo method. The sequence of events is now not random, but follows the trajectory which is dictated to the system by the equations of motion. In this respect, this method is of a more general scope, since it permits the computation of equilibrium as well as transport properties of the system.

Consider a given configuration of the system $\mathbf{R}^N$ and a given distribution of momenta at some particular time t. The total force acting on the kth particle is given by the sum of the forces exerted by all the other particles. The equation of motion for the system of N particles is thus

$$\frac{m \, d^2\mathbf{R}_k}{dt^2} = \sum_{j=1, j \neq k}^{N} \mathbf{F}(\mathbf{R}_{kj}) = -\sum_{j=1, j \neq k}^{N} \nabla_k U(\mathbf{R}_{kj}). \tag{5.5.16}$$

Here, m is the mass of each particle. We now let the kth particle move under the influence of the constant force given in (5.5.16) during a short time Δt. Continuation of this process produces a series of configurations over which average quantities may be computed. The essential approximation involves the assumption that the force is constant during Δt. Therefore, it is essential to choose a Δt short enough to validate this approximation. Yet Δt must not be too small, for otherwise the system will tend to equilibrium very sluggishly, and the whole process will require excessive computation time.† There is no absolute

---

† A similar consideration occurs in the Monte Carlo method. To proceed from one configuration to the next, one usually makes a small translation $\Delta\mathbf{R}$ to a given particle selected at random. $\Delta\mathbf{R}$ must be small enough; otherwise the moved particle will, with large probability, enter a position already occupied by another particle, and the new configuration will be rejected. On the other hand, if $\Delta\mathbf{R}$ is too small, the configuration will hardly change from one step to another, and excessive time will be needed to make significant changes of the configuration.

criterion for determining when equilibrium has been reached. One practical test is to start with different initial conditions and see if we end up with the same computed average quantities.

5.6. HIGHER-ORDER MOLECULAR DISTRIBUTION FUNCTIONS

Molecular distribution functions (MDF) of order higher than two rarely appear in actual computations of thermodynamic quantities of liquids. This is not because they are not needed there, but because nearly nothing is known of their analytical properties. The importance of higher-order MDFs is threefold. First, they appear in some relations connecting thermodynamic quantities and molecular properties, such as the relation for the heat capacity or partial molar quantities. Second, some of the common relations between thermodynamic quantities and the pair distribution function hold only by virtue of the assumption of pairwise additivity of the total potential energy. If we give up this assumption and include higher-order potentials, we must use higher-order MDFs (more details are given in section 5.8). Finally, MDFs of higher order do appear in some intermediate steps in various derivations of integral equations. Some knowledge about their properties is therefore a necessity.

Consider again the basic distribution function in the T, V, N ensemble (5.2.1). The *specific* nth-order MDF is defined as the probability density of finding particles 1, 2, ..., n in the configuration $\mathbf{X}^n = \mathbf{X}_1, \mathbf{X}_2, \ldots, \mathbf{X}_n$:

$$P^{(n)}(\mathbf{X}^n) = \int \cdots \int d\mathbf{X}_{n+1} \cdots d\mathbf{X}_N\, P(\mathbf{X}^N). \tag{5.6.1}$$

The corresponding *generic* nth-order MDF is defined by

$$\rho^{(n)}(\mathbf{X}^n) = \frac{N!}{(N-n)!}\, P^{(n)}(\mathbf{X}^n), \tag{5.6.2}$$

which is obtained by generalizing the arguments used in previous sections to define $\rho^{(1)}$ and $\rho^{(2)}$.

Various conditional MDFs can be introduced; for instance, the singlet conditional distribution function for finding a particle at $\mathbf{X}_3$ given two particles at $\mathbf{X}_1$ and $\mathbf{X}_2$ is

$$\rho(\mathbf{X}_3/\mathbf{X}_1, \mathbf{X}_2) = \rho^{(3)}(\mathbf{X}_1, \mathbf{X}_2, \mathbf{X}_3)/\rho^{(2)}(\mathbf{X}_1, \mathbf{X}_2). \tag{5.6.3}$$

This function will appear in the study of solvent-induced interaction in Chapter 7. The quantity defined in (5.6.3) is the average local density of particles at $\mathbf{R}_3$ (with orientation $\mathbf{\Omega}_3$) of a system subjected to an "external" field of force produced by fixing two particles at $\mathbf{X}_1$ and $\mathbf{X}_2$. The arguments for this interpretation are exactly the same as those given in section 5.2.

Higher-order correlation functions and potentials of average forces are defined in analogy with previous definitions for pairs. For instance, for $n = 3$, we define

$$\rho^{(3)}(\mathbf{X}_1, \mathbf{X}_2, \mathbf{X}_3) = \rho^{(1)}(\mathbf{X}_1)\rho^{(1)}(\mathbf{X}_2)\rho^{(1)}(\mathbf{X}_3)g^{(3)}(\mathbf{X}_1, \mathbf{X}_2, \mathbf{X}_3) \tag{5.6.4}$$

and

$$g^{(3)}(\mathbf{X}_1, \mathbf{X}_2, \mathbf{X}_3) = \exp[-\beta W^{(3)}(\mathbf{X}_1, \mathbf{X}_2, \mathbf{X}_3)]. \tag{5.6.5}$$

Using an interpretation similar to that for $W(\mathbf{X}_1, \mathbf{X}_2)$ in section 5.4.2, we call $W^{(3)}(\mathbf{X}_1, \mathbf{X}_2, \mathbf{X}_3)$, the potential of average force for the triplet of particles at $\mathbf{X}_1, \mathbf{X}_2, \mathbf{X}_3$, the average being carried out over all the configurations of the remaining $N - 3$ particles. This function has played an important role in the derivation of integral equations for the pair correlation function. Although we shall not be interested in this topic in the remainder of the book, it is appropriate to point out one analogy between the triplet potential of average force defined in (5.6.5) and the corresponding quantity introduced in Chapter 3. In both cases we start with three interacting particles and assume that the interaction energy is pairwise additive. (This is in general an approximation only. However, we assume the pairwise additivity of the *direct* interaction to stress the new features of the potential of average force.) We now average over all other degrees of freedom of the system. Here, these are the configurations of the remaining $N - 3$ particles. In Chapter 3 we averaged over all possible conformations of the polymers. As in the case of the potential of average force between two particles, the averaging process introduces some new features. Here, we point out the loss of the pairwise additivity of the triplet potential of average force. In Chapter 3 we had an explicit expression for the triplet potential of average force and we found that deviations from pairwise additivity could be quite large. Here we do not have a simple expression for the triplet potential of average force in terms of the molecular parameters of the system. Nevertheless, there is no reason to believe that pairwise additivity of higher order potentials of average force will hold in the liquid state. In the historical development of the theory of liquids, the latter assumption was actually employed, and is known as the Kirkwood superposition approximation. For W^3, one writes

$$W^{(3)}(\mathbf{X}_1, \mathbf{X}_2, \mathbf{X}_3) = W^{(2)}(\mathbf{X}_1, \mathbf{X}_2) + W^{(2)}(\mathbf{X}_1, \mathbf{X}_3) + W^{(2)}(\mathbf{X}_2, \mathbf{X}_3). \qquad (5.6.6)$$

Here, $W^{(2)}$ is the same as W in section 5.4. Recent simulations have shown that (5.6.6) is indeed a poor approximation.

5.7. MOLECULAR DISTRIBUTION FUNCTIONS IN THE GRAND CANONICAL ENSEMBLE

In discussing closed systems, we have stressed the advantage of the generic over the specific MDFs. The latter presupposes the possibility of identification of the individual particles, which of course we cannot achieve in practice. Nevertheless, the specific MDFs were found useful at least as intermediary concepts in the process of formulating the definitions of the generic MDFs. When we turn to open systems, it becomes meaningless to discuss the specific MDFs. Hence, we proceed directly to define the generic MDFs in the grand canonical or T, V, μ ensemble.

We recall that the probability of finding a system in the T, V, μ ensemble with exactly N particles is

$$P(N) = \frac{Q(T, V, N)[\exp(\beta \mu N)]}{\Xi(T, V, \mu)}, \qquad (5.7.1)$$

where $Q(T, V, N)$ and $\Xi(T, V, \mu)$ are the canonical and the grand canonical partition functions, respectively.

The conditional nth-order MDF of finding the configuration $\mathbf{X}^n$, given that the system has N particles, is

$$\rho^{(n)}(\mathbf{X}^n/N) = \frac{N!}{(N-n)!} \frac{\displaystyle\int \cdots \int d\mathbf{X}_{n+1} \cdots d\mathbf{X}_N \exp[-\beta U_N(\mathbf{X}^N)]}{\displaystyle\int \cdots \int d\mathbf{X}^N \exp[-\beta U_N(\mathbf{X}^N)]}. \tag{5.7.2}$$

Clearly, this quantity is defined for $n \le N$ only. The nth-order MDF in the T, V, μ ensemble is defined as the average of (5.7.2) with the weight given in (5.7.1), i.e.,

$$\overline{\rho^{(n)}(\mathbf{X}^n)} = \sum_{N \ge n} P(N) \rho^{(n)}(\mathbf{X}^n/N)$$

$$= \frac{1}{\Xi} \sum_{N \ge n} \frac{N!}{(N-n)!}$$

$$\times \frac{Q(T, V, N)[\exp(\beta\mu N)] \displaystyle\int \cdots \int d\mathbf{X}_{n+1} \cdots d\mathbf{X}_N \exp[-\beta U_N(\mathbf{X}^N)]}{Z_N}. \tag{5.7.3}$$

The bar over $\rho^{(n)}(\mathbf{X}^N)$ denotes average in the T, V, μ ensemble. Recalling that

$$Q(T, V, N) = (q^N/N!)Z_N \tag{5.7.4}$$

and denoting

$$\lambda = \exp(\beta\mu), \tag{5.7.5}$$

we can rewrite (5.7.3) as

$$\overline{\rho^{(n)}(\mathbf{X}^n)} = \frac{1}{\Xi} \sum_{N \ge n} \frac{(\lambda q)^N}{(N-n)!} \int \cdots \int d\mathbf{X}_{n+1} \cdots d\mathbf{X}_N \exp[-\beta U_N(\mathbf{X}^N)]. \tag{5.7.6}$$

The normalization condition for $\overline{\rho^{(n)}(\mathbf{X}^n)}$ is obtained from (5.7.3) by integrating over all the configurations $\mathbf{X}^n$

$$\int \cdots \int d\mathbf{X}^n \, \overline{\rho^{(n)}(\mathbf{X}^n)} = \sum_{N \ge n} P(N) \frac{N!}{(N-n)!} = \left\langle \frac{N!}{(N-n)!} \right\rangle \tag{5.7.7}$$

where here the symbol $\langle \ \rangle$ stands for an average in the T, V, μ ensemble.

Let us work out two simple and important cases. For $n = 1$, we have

$$\int d\mathbf{X}_1 \, \overline{\rho^{(1)}(\mathbf{X}_1)} = \langle N!/(N-1)! \rangle = \langle N \rangle, \tag{5.7.8}$$

which is simply the average number of particles in a system in the T, V, μ ensemble. [Compare with (5.2.19) in the T, V, N ensemble.] Using essentially the same arguments as in section 5.2, we get for a homogeneous and isotropic system

$$\overline{\rho^{(1)}(\mathbf{X})} = \langle N \rangle / 8\pi^2 V = \rho/8\pi^2, \tag{5.7.9}$$

which is the same as in Eq. (5.2.25) but with the replacement of the exact N by the average $\langle N \rangle$.

For $n = 2$, we get from (5.7.7)

$$\iint d\mathbf{X}_1 \, d\mathbf{X}_2 \, \overline{\rho^{(2)}(\mathbf{X}_1, \mathbf{X}_2)} = \langle N!/(N-2)! \rangle = \langle N(N-1) \rangle = \langle N^2 \rangle - \langle N \rangle. \quad (5.7.10)$$

Relations (5.7.8) and (5.7.9) will be applied in the theory of solutions, Chapter 6.

In a similar fashion, one may introduce correlation functions in the T, V, μ ensemble. Of particular importance is the pair correlation function defined by

$$\overline{\rho^{(2)}(\mathbf{X}_1, \mathbf{X}_2)} = \overline{\rho^{(1)}(\mathbf{X}_1)} \, \overline{\rho^{(1)}(\mathbf{X}_2)} \, \overline{g(\mathbf{X}_1, \mathbf{X}_2)}. \quad (5.7.11)$$

One important property of $g(\mathbf{X}_1, \mathbf{X}_2)$, defined in the T, V, μ ensemble, is its limiting behavior at low densities:

$$\overline{g(\mathbf{X}_1, \mathbf{X}_2)} \xrightarrow{\rho \to 0} \exp[-\beta U(\mathbf{X}_1, \mathbf{X}_2)], \quad (5.7.12)$$

which is strictly true without additional terms on the order of $\langle N \rangle^{-1}$.

5.8. MOLECULAR DISTRIBUTION FUNCTIONS AND THERMODYNAMICS

This chapter summarizes the most important relations between thermodynamic quantities and molecular distribution functions for pure liquids. Most of these relations apply to systems obeying the assumption of pairwise additivity for the total potential energy. We shall indicate, however, how to modify the relations when higher-order potentials are to be incorporated in the formal theory. In general, higher-order potentials bring in higher-order MDFs. Since very little is known about the analytical behavior of the latter, such relationships are rarely useful in applications.

Most of the specific derivations carried out in this section apply to systems of simple spherical particles. We shall also point out the appropriate generalizations for nonspherical particles that do not possess internal rotations. For particles with internal rotations, one needs to take the appropriate average over all conformations. This is discussed in Chapter 8.

There is a step common to most of the procedures leading to the relations between thermodynamic quantities and the pair distribution function. Therefore, in the next section we derive a general theorem connecting averages of pairwise quantities and the pair distribution function. In fact, we have already quoted one application of this theorem in section 5.5.1. Further applications will appear in subsequent sections of this chapter.

5.8.1. Average Values of Pairwise Quantities

Consider an average of a general quantity $F(\mathbf{X}^N)$ in the T, V, N ensemble:

$$\langle F \rangle = \int \cdots \int d\mathbf{X}^N \, P(\mathbf{X}^N) F(\mathbf{X}^N), \quad (5.8.1)$$

with

$$P(\mathbf{X}^N) = \frac{\exp[-\beta U_N(\mathbf{X}^N)]}{Z_N}. \quad (5.8.2)$$

A pairwise quantity is a function that is expressible as a sum of terms, each of which depends on the configuration of a pair of particles, namely

$$F(\mathbf{X}^N) = \sum_{i \neq j} f(\mathbf{X}_i, \mathbf{X}_j), \tag{5.8.3}$$

where we have assumed that the sum is over all different pairs. In most of the applications we shall have a factor of $\frac{1}{2}$ in (5.8.3) to account for the fact that this sum counts each pairwise function $f(\mathbf{X}_i, \mathbf{X}_j)$ twice; i.e., $f(\mathbf{X}_1, \mathbf{X}_2)$ appears when $i = 1$ and $j = 2$ and when $i = 2$ and $j = 1$. The sum (5.8.3) is more general, and examples have already appeared in sections 5.2.2 and 5.5.1. In the present treatment, all of the N particles are presumed to be equivalent, so that the function f is the same for each pair of indices. (The extension to mixtures will be discussed at the end of this section.)

Substituting (5.8.3) in (5.8.1), we get

$$\langle F \rangle = \int \cdots \int d\mathbf{X}^N \, P(\mathbf{X}^N) \sum_{i \neq j} f(\mathbf{X}_i, \mathbf{X}_j)$$

$$= \sum_{i \neq j} \int \cdots \int d\mathbf{X}^N \, P(\mathbf{X}^N) f(\mathbf{X}_i, \mathbf{X}_j)$$

$$= N(N-1) \int \cdots \int d\mathbf{X}^N \, P(\mathbf{X}^N) f(\mathbf{X}_1, \mathbf{X}_2)$$

$$= \int d\mathbf{X}_1 \int d\mathbf{X}_2 f(\mathbf{X}_1, \mathbf{X}_2) \left[N(N-1) \int \cdots \int d\mathbf{X}_3 \cdots d\mathbf{X}_N \, P(\mathbf{X}^N) \right]$$

$$= \int d\mathbf{X}_1 \int d\mathbf{X}_2 f(\mathbf{X}_1, \mathbf{X}_2) \rho^{(2)}(\mathbf{X}_1, \mathbf{X}_2). \tag{5.8.4}$$

It is instructive to go through the steps in (5.8.4) carefully since this is a standard procedure in the theory of classical fluids. In the first step we have merely interchanged the signs of summation and integration. In the second step we exploit the fact that all particles are equivalent; thus each term in the sum has the same numerical value, independent of the indices i, j. Hence, we replace the sum over $N(N-1)$ terms by $N(N-1)$ times one integral. In the latter, we have chosen the (arbitrary) indices 1 and 2.

Clearly, due to the equivalence of the particles, we could have chosen any other two indices. The third and fourth steps make use of the definition of the pair distribution function defined in section 5.2.2.

We can rewrite the final result of (5.8.4) as

$$\langle F \rangle = \int d\mathbf{X}' \int d\mathbf{X}'' f(\mathbf{X}', \mathbf{X}'') \rho^{(2)}(\mathbf{X}', \mathbf{X}''), \tag{5.8.5}$$

where we have changed to primed vectors to stress the fact that we do not refer to any specific pair of particles, as might be inferred erroneously from the final form of the rhs of (5.8.4).

A simpler version of (5.8.5) occurs when the particles are spherical so that each configuration $\mathbf{X}$ consists only of the locational vector $\mathbf{R}$. This is the most frequent case in the theory of simple fluids. The corresponding expression for the average in this case is

$$\langle F \rangle = \int d\mathbf{R}' \int d\mathbf{R}'' f(\mathbf{R}', \mathbf{R}'') \rho^{(2)}(\mathbf{R}', \mathbf{R}''). \tag{5.8.6}$$

A common case that often occurs is when the function $f(\mathbf{R}', \mathbf{R}'')$ depends only on the separation between the two points $R = |\mathbf{R}'' - \mathbf{R}'|$. In addition, for homogeneous and isotropic fluids, $\rho^{(2)}(\mathbf{R}', \mathbf{R}'')$ depends only on R. This permits the transformation of (5.8.6) into a one-dimensional integral, which is the most useful form of the result (5.8.6). To do this, we first transform to relative coordinates

$$\bar{\mathbf{R}} = \mathbf{R}', \qquad \mathbf{R} = \mathbf{R}'' - \mathbf{R}'. \tag{5.8.7}$$

Hence

$$\langle F \rangle = \int d\bar{\mathbf{R}} \int d\mathbf{R}\, f(\mathbf{R})\rho^{(2)}(\mathbf{R})$$

$$= V \int d\mathbf{R}\, f(\mathbf{R})\rho^{(2)}(\mathbf{R}). \tag{5.8.8}$$

Next, we transform to polar coordinates:

$$d\mathbf{R} = dx\, dy\, dz = R^2 \sin\theta\, d\theta\, d\phi\, dR. \tag{5.8.9}$$

Since the integrand in the last form of (5.8.8) depends on the scalar R, we can integrate over all the orientations to get the final form

$$\langle F \rangle = V \int_0^\infty f(R)\rho^{(2)}(R)4\pi R^2\, dR$$

$$= \rho^2 V \int_0^\infty f(R)g(R)4\pi R^2\, dR. \tag{5.8.10}$$

It is clear from (5.8.10) that a knowledge of the pairwise function $f(R)$ in (5.8.3) together with the radial distribution function $g(R)$ is sufficient to evaluate the average quantity $\langle F \rangle$. Note that we have taken as infinity the upper limit of the integral in (5.8.10). This is not always permitted. In most practical cases, however, $f(R)$ will be of finite range. Since $g(R)$ tends to unity at distances of a few molecular diameters (excluding the region near the critical point), the upper limit of the integral in (5.8.10) can be taken to be of the order of a few molecular diameters only. Hence, extension to infinity would not affect the value of the integral.

We now briefly mention two straightforward extensions of theorem (5.8.5).

1. For mixtures of, say, two components, a pairwise function is defined as

$$F(\mathbf{X}^{N_A + N_B}) = \sum_{i \neq j} f_{AA}(\mathbf{X}_i, \mathbf{X}_j) + \sum_{i \neq j} f_{BB}(\mathbf{X}_i, \mathbf{X}_j)$$

$$+ \sum_{i=1}^{N_A} \sum_{j=1}^{N_B} f_{AB}(\mathbf{X}_i, \mathbf{X}_j) + \sum_{i=1}^{N_B} \sum_{j=1}^{N_A} f_{BA}(\mathbf{X}_i, \mathbf{X}_j), \tag{5.8.11}$$

where $\mathbf{X}^{N_A + N_B}$ stands for the configuration of the whole system of $N_A + N_B$ particles of species A and B. Here, $f_{\alpha\beta}$ is the pairwise function for the pair of species α and β ($\alpha = A, B$ and $\beta = A, B$). Altogether, we have in (5.8.11) $N_A(N_A - 1) + N_B(N_B - 1) + 2N_A N_B$ terms which correspond to all of the $(N_A + N_B)(N_A + N_B - 1)$ pairs in the system.

Note that in (5.8.11) we have assumed, as in (5.8.3) the summation over $i \neq j$ for pairs of the same species. This is not required for pairs of different species. Using exactly the same procedure as for the one-component system, we get for the average quantity

$$\langle F \rangle = \int d\mathbf{X}' \int d\mathbf{X}'' f_{AA}(\mathbf{X}', \mathbf{X}'') \rho_{AA}^{(2)}(\mathbf{X}', \mathbf{X}'')$$

$$+ \int d\mathbf{X}' \int d\mathbf{X}'' f_{BB}(\mathbf{X}', \mathbf{X}'') \rho_{BB}^{(2)}(\mathbf{X}', \mathbf{X}'')$$

$$+ \int d\mathbf{X}' \int d\mathbf{X}'' f_{AB}(\mathbf{X}', \mathbf{X}'') \rho_{AB}^{(2)}(\mathbf{X}', \mathbf{X}'')$$

$$+ \int d\mathbf{X}' \int d\mathbf{X}'' f_{BA}(\mathbf{X}', \mathbf{X}'') \rho_{BA}^{(2)}(\mathbf{X}', \mathbf{X}''), \qquad (5.8.12)$$

where $\rho_{\alpha\beta}^{(2)}$ are the pair distribution functions for species α and β, which will be introduced in Chapter 6. Note again that if F is, for instance, the total potential energy of the system, we need to take half of the sum (5.8.12).

2. For functions F that depend on pairs as well as on triplets of particles of the form

$$F(\mathbf{X}^N) = \sum_{i \neq j} f(\mathbf{X}_i, \mathbf{X}_j) + \sum_{i \neq j, j \neq k} h(\mathbf{X}_i, \mathbf{X}_j, \mathbf{X}_k). \qquad (5.8.13)$$

The corresponding average is

$$\langle F \rangle = \int d\mathbf{X}' \int d\mathbf{X}'' f(\mathbf{X}', \mathbf{X}'') \rho^{(2)}(\mathbf{X}', \mathbf{X}'')$$

$$+ \int d\mathbf{X}' \int d\mathbf{X}'' \int d\mathbf{X}''' h(\mathbf{X}', \mathbf{X}'', \mathbf{X}''') \rho^{(3)}(\mathbf{X}', \mathbf{X}'', \mathbf{X}'''). \qquad (5.8.14)$$

The arguments leading to (5.8.14) are the same as those for (5.8.4). The new element that enters into (5.8.14) is the triplet distribution function. Similarly, we can write formal relations for average quantities which depend on larger numbers of particles. The result would be integrals involving successively higher-order molecular distribution functions. Unfortunately, even (5.8.14) is rarely useful since we do not have sufficient information on $\rho^{(3)}$. Functions of the form (5.8.13) may occur, for instance, if we include three-body potentials to describe the total potential energy.

5.8.2. Internal Energy

Consider a system in the T, V, N ensemble and assume that the total potential energy of interaction is pairwise additive, namely,

$$U_N(\mathbf{X}^N) = \tfrac{1}{2} \sum_{i \neq j} U(\mathbf{X}_i, \mathbf{X}_j). \qquad (5.8.15)$$

The factor $\tfrac{1}{2}$ is included in (5.8.15) since the sum over $i \neq j$ counts each pair interaction twice.

The partition function for this system is

$$Q(T, V, N) = \frac{q^N}{N!} Z_N = \frac{q^N}{N!} \int \cdots \int d\mathbf{X}^N \exp[-\beta U_N(\mathbf{X}^N)], \qquad (5.8.16)$$

where the momentum partition function is included in q^N.

The internal energy of the system is given by†

$$E = -T^2 \frac{\partial(A/T)}{\partial T}$$

$$= kT^2 \frac{\partial \ln Q}{\partial T} \tag{5.8.17}$$

$$= NkT^2 \frac{\partial \ln q}{\partial T} + kT^2 \frac{\partial \ln Z_N}{\partial T}.$$

The first term on the rhs includes the internal and the kinetic energy of the individual molecules. For instance, for spherical and structureless molecules, we have $q = \Lambda^{-3}$ and, hence,

$$N\varepsilon^K = NkT^2 \left(\frac{\partial \ln q}{\partial T} \right) = \tfrac{3}{2} NkT, \tag{5.8.18}$$

which is the average translational kinetic energy of the molecules.

The second term on the rhs of (5.8.17) is the average energy of interaction among the particles. This can be seen immediately by performing the derivative of the configurational partition function:

$$kT^2 \frac{\partial \ln Z_N}{\partial T} = \frac{\int \cdots \int d\mathbf{X}^N \exp[-\beta U_N(\mathbf{X}^N)] U_N(\mathbf{X}^N)}{Z_N}$$

$$= \int \cdots \int d\mathbf{X}^N P(\mathbf{X}^N) U_N(\mathbf{X}^N). \tag{5.8.19}$$

Hence, the total internal energy is

$$E = N\varepsilon^K + \langle U_N \rangle, \tag{5.8.20}$$

where $N\varepsilon^K$ denotes the first term on the rhs of (5.8.17).

The average potential energy in (5.8.20), with the assumption of pairwise additivity (5.8.15), fulfills the conditions of the previous section; hence, we can immediately apply theorem (5.8.5) to obtain

$$E = N\varepsilon^K + \tfrac{1}{2} \int d\mathbf{X}' \int d\mathbf{X}'' \, U(\mathbf{X}', \mathbf{X}'') \rho^{(2)}(\mathbf{X}', \mathbf{X}''). \tag{5.8.21}$$

For spherical particles, we can transform relation (5.8.21) into a one-dimensional integral. Using the same arguments as were used to derive (5.8.10), we get, from (5.8.21),

$$E = N\varepsilon^K + \tfrac{1}{2} N\rho \int_0^\infty U(R) g(R) 4\pi R^2 \, dR. \tag{5.8.22}$$

---

† Note that E is referred to as the internal energy in the thermodynamic sense. ε^K designates the internal energy of a *single* molecule.

For simple nonpolar particles, $U(R)$ will usually have a range of a few molecular diameters; hence, the main contribution to the integral on the rhs of (5.8.22) comes from the finite region around the origin.

The interpretation of the second term on the rhs of (5.8.22) is as follows. We select a particle and compute its total interaction with the rest of the system. Since the local density of particles at a distance R from the center of the selected particle is $\rho g(R)$, the average number of particles in the spherical element of volume $4\pi R^2\, dR$ is $\rho g(R)4\pi R^2\, dR$. Hence, the average interaction of a given particle with the rest of the system is

$$\int_0^\infty U(R)\rho g(R)4\pi R^2\, dR. \tag{5.8.23}$$

We now repeat the same computation for each particle. Since the N particles are identical, the average interaction of each particle with the medium is the same. However, if we multiply (5.8.23) by N, we will be counting each pair interaction twice. Hence, we must multiply by N and divide by two to obtain the average interaction energy for the whole system:

$$\tfrac{1}{2}N\rho \int_0^\infty U(R)g(R)4\pi R^2\, dR. \tag{5.8.24}$$

Relation (5.8.22) is evidently simpler than the original form of the total energy in (5.8.20). Once we have presumed an analytical form for $U(R)$ and acquired information (from either theoretical or experimental sources) on $g(R)$, we can compute the energy of the system by a one-dimensional integration. The basic simplification achieved in (5.8.21) is due to the assumption of pairwise additivity of the total potential energy. In addition, to get (5.8.22) we have assumed that the particles are spherical. For nonspherical particles, we can still simplify (5.8.21) by transforming to relative coordinates and integrating over the configuration of one particle:

$$E = N\varepsilon^K + \tfrac{1}{2}V8\pi^2 \int d\mathbf{X}\, U(\mathbf{X})\rho^{(2)}(\mathbf{X})$$

$$= N\varepsilon^K + (N\rho/16\pi^2) \int d\mathbf{X}\, U(\mathbf{X})g(\mathbf{X}). \tag{5.8.25}$$

Here, we have a six-dimensional integral instead of a one-dimensional integral as in (5.8.25).

5.8.3. The Pressure Equation

The pressure equation is computed in this section for a system of spherical particles. This choice is made only because of notational convenience. We shall quote the corresponding equation for nonspherical particles at the end of this section.

The pressure is obtained from the Helmholtz energy by

$$P = -\left(\frac{\partial A}{\partial V}\right)_{T,N}, \tag{5.8.26}$$

where

$$A = -kT \ln Q(T, V, N). \tag{5.8.27}$$

Note that the dependence of Q on the volume comes only through the configurational partition function; hence,

$$P = kT \left(\frac{\partial \ln Z_N}{\partial V} \right)_{T,N}. \tag{5.8.28}$$

In order to perform this derivative, we first express Z_N explicitly as a function of V. For macroscopic systems, we assume that the pressure is independent of the geometric form of the system. Hence, for convenience, we choose a cube of edge $V^{1/3}$ so that the configurational partition function is written as

$$Z_N = \int_0^{V^{1/3}} \cdots \int_0^{V^{1/3}} dx_1 \, dy_1 \, dz_1 \cdots dx_N \, dy_N \, dz_N \exp[-\beta U_N(\mathbf{R}^N)]. \tag{5.8.29}$$

We now make the following transformation of variables:

$$x_i' = V^{-1/3} x_i, \qquad y_i' = V^{-1/3} y_i, \qquad z_i' = V^{-1/3} z_i, \tag{5.8.30}$$

so that the limits of the integral in (5.8.29) become independent of V:

$$Z_N = V^N \int_0^1 \cdots \int_0^1 dx_1' \, dy_1' \, dz_1' \cdots dx_N' \, dy_N' \, dz_N' \exp(-\beta U_N). \tag{5.8.31}$$

With the new set of variables, the total potential becomes a function of the volume:

$$U_N = \tfrac{1}{2} \sum_{i \neq j} U(R_{ij}) = \tfrac{1}{2} \sum_{i \neq j} U(V^{1/3} R_{ij}'). \tag{5.8.32}$$

The relation between the distances expressed by the two sets of variables is

$$\begin{aligned} R_{ij} &= [(x_j - x_i)^2 + (y_j - y_i)^2 + (z_j - z_i)^2]^{1/2} \\ &= V^{1/3}[(x_j' - x_i')^2 + (y_j' - y_i')^2 + (z_j' - z_i')^2]^{1/2} \\ &= V^{1/3} R_{ij}'. \end{aligned} \tag{5.8.33}$$

We may now differentiate (5.8.31) with respect to the volume:

$$\begin{aligned} \left(\frac{\partial Z_N}{\partial V} \right)_{T,N} &= N V^{N-1} \int_0^1 \cdots \int_0^1 dx_1' \cdots dz_N' \exp(-\beta U_N) \\ &+ V^N \int_0^1 \cdots \int_0^1 dx_1' \cdots dz_N' [\exp(-\beta U_N)] \left(-\beta \frac{\partial U_N}{\partial V} \right). \end{aligned} \tag{5.8.34}$$

From (5.8.32), we also have

$$\frac{\partial U_N}{\partial V} = \frac{1}{2} \sum_{i \neq j} \frac{\partial U(R_{ij})}{\partial R_{ij}} \frac{\partial R_{ij}}{\partial V}$$

$$= \frac{1}{2} \sum_{i \neq j} \frac{\partial U(R_{ij})}{\partial R_{ij}} \frac{1}{3} V^{-2/3} R'_{ij}$$

$$= \frac{1}{6V} \sum_{i \neq j} \frac{\partial U(R_{ij})}{\partial R_{ij}} R_{ij}. \qquad (5.8.35)$$

Combining (5.8.34) and (5.8.35) and transforming back to the original variables, we obtain

$$\left(\frac{\partial \ln Z_N}{\partial V}\right)_{T,N} = \frac{N}{V} - \frac{\beta}{6V} \int \cdots \int d\mathbf{R}^N \, P(\mathbf{R}^N) \sum_{i \neq j} \frac{\partial U(R_{ij})}{\partial R_{ij}} R_{ij}. \qquad (5.8.36)$$

The second term on the rhs of (5.8.36) is an average of a pairwise quantity. Therefore, we can apply the general theorem (5.8.35), to obtain

$$P = kT \left(\frac{\partial \ln Z_N}{\partial V}\right)_{T,N}$$

$$= kT\rho - \frac{\rho^2}{6} \int_0^\infty R \frac{\partial U(R)}{\partial R} g(R) 4\pi R^2 \, dR. \qquad (5.8.37)$$

This is the final form of the pressure equation for a system of spherical particles obeying the assumption of pairwise additivity for the total potential energy. Note that the first term is the "ideal gas" pressure. The second is due to the effect of the intermolecular forces on the pressure. Note that, in general, $g(R)$ is a function of density, so that this term is not the second-order term in the density expansion of the pressure.

The general relation (5.8.37) is very useful in computing the equation of state of a system once we know the form of the function $g(R)$. Indeed, such computations have been performed to test theoretical methods of evaluating $g(R)$.

In a mixture of c components, the generalization of (5.8.37) is straightforward and leads to

$$P = \sum_{\alpha=1}^{c} kT\rho_\alpha - \frac{1}{6} \sum_{\alpha,\beta=1}^{c} \rho_\alpha \rho_\beta \int_0^\infty \frac{\partial U_{\alpha\beta}(R)}{\partial R} g_{\alpha\beta}(R) 4\pi R^3 \, dR, \qquad (5.8.38)$$

where ρ_α is the density of the α species and $g_{\alpha\beta}(R)$ is the pair correlation function for the pair of species α and β.

For a system of rigid, nonspherical molecules, the derivation of the pressure equation is essentially the same as that for spherical molecules, but with somewhat more notational complication, the result being

$$P = kT\rho - \left(\frac{1}{6V}\right) \int d\mathbf{X}' \int d\mathbf{X}'' [\mathbf{R} \cdot \nabla_{\mathbf{R}} U(\mathbf{X}', \mathbf{X}'')] \rho^{(2)}(\mathbf{X}', \mathbf{X}''), \qquad (5.8.39)$$

where

$$\mathbf{R} = \mathbf{R}'' - \mathbf{R}'. \qquad (5.8.40)$$

5.8.4. The Compressibility Equation

The compressibility relation is one of the simplest and most useful relations between a thermodynamic quantity and the pair correlation function. In this section, we derive this relation and point out some of its outstanding features.

We consider here a system of rigid, nonspherical particles in the T, V, μ ensemble. We stress from the outset that no assumption of additivity of the potential energy is invoked at any stage of the derivation.

We recall the normalization conditions for $\overline{\rho^{(1)}(\mathbf{X}_1)}$ and for $\overline{\rho^{(2)}(\mathbf{X}_1, \mathbf{X}_2)}$ in the T, V, μ ensemble (see section 5.7):

$$\int d\mathbf{X}_1 \, \overline{\rho^{(1)}(\mathbf{X}_1)} = \langle N \rangle \tag{5.8.41}$$

$$\int d\mathbf{X}_1 \, d\mathbf{X}_2 \, \overline{\rho^{(2)}(\mathbf{X}_1, \mathbf{X}_2)} = \langle N^2 \rangle - \langle N \rangle, \tag{5.8.42}$$

where the symbol $\langle \ \rangle$ stands for an average in the T, V, μ ensemble. Squaring (5.8.41) and subtracting it from (5.8.42) yields

$$\int d\mathbf{X}_1 \, d\mathbf{X}_2 [\overline{\rho^{(2)}(\mathbf{X}_1, \mathbf{X}_2)} - \overline{\rho^{(1)}(\mathbf{X}_1)} \, \overline{\rho^{(1)}(\mathbf{X}_2)}] = \langle N^2 \rangle - \langle N \rangle^2 - \langle N \rangle. \tag{5.8.43}$$

Using the relations (homogeneous and isotropic fluid)

$$\overline{\rho^{(1)}(\mathbf{X}_1)} = \frac{\rho}{8\pi^2} \tag{5.8.44}$$

$$\overline{g(\mathbf{X}_1, \mathbf{X}_2)} = \frac{\overline{\rho^{(2)}(\mathbf{X}_1, \mathbf{X}_2)}}{\overline{\rho^{(1)}(\mathbf{X}_1)} \, \overline{\rho^{(1)}(\mathbf{X}_2)}} \tag{5.8.45}$$

$$\overline{g(\mathbf{R}_1, \mathbf{R}_2)} = \frac{1}{(8\pi^2)^2} \int d\mathbf{\Omega}_1, d\mathbf{\Omega}_2 \, \overline{g(\mathbf{X}_1, \mathbf{X}_2)}, \tag{5.8.46}$$

we can rearrange (5.8.43) to obtain

$$\rho^2 \int d\mathbf{R}_1 \, d\mathbf{R}_2 \overline{[g(\mathbf{R}_1, \mathbf{R}_2)} - 1] = \langle N^2 \rangle - \langle N \rangle^2 - \langle N \rangle. \tag{5.8.47}$$

Since $\overline{g(\mathbf{R}_1, \mathbf{R}_2)}$ depends on the scalar separation $R = |\mathbf{R}_2 - \mathbf{R}_1|$, we can rewrite (5.8.47) as

$$1 + \rho \int_V d\mathbf{R} \overline{[g(\mathbf{R})} - 1] = \frac{\langle N^2 \rangle - \langle N \rangle^2}{\langle N \rangle}$$

$$= 1 + \rho \int_0^\infty \overline{[g(R)} - 1]4\pi R^2 \, dR. \tag{5.8.48}$$

Relation (5.8.48) is an important connection between the radial distribution function and fluctuations in the number of particles. In section 1.4, we obtained the relation

$$\langle N^2 \rangle - \langle N \rangle^2 = kTV\rho^2\kappa_T, \tag{5.8.49}$$

where κ_T is the isothermal compressibility of the system. Combining (5.8.49) with (5.8.48), we get the final result

$$\kappa_T = \frac{1}{kT\rho} + \frac{1}{kT}\int_V d\mathbf{R}[\overline{g(R)} - 1] = \frac{1}{kT\rho} + \frac{1}{kT}\int_0^\infty [\overline{g(R)} - 1]4\pi R^2\, dR, \tag{5.8.50}$$

which is known as the compressibility equation. We define the quantity

$$G = \int_0^\infty [\overline{g(R)} - 1]4\pi R^2\, dR. \tag{5.8.51}$$

In terms of G the compressibility equation is written as

$$kT\rho\kappa_T = 1 + \rho G. \tag{5.8.52}$$

Note that the first term on the rhs of (5.8.50) is the compressibility of an ideal gas. That is, for a system obeying the relation $P = \rho kT$, we have

$$\kappa_T = -\frac{1}{V}\left(\frac{\partial V}{\partial P}\right)_{T,N} = \left(\frac{\partial \ln \rho}{\partial P}\right)_T = \frac{1}{kT\rho}. \tag{5.8.53}$$

Hence, the second term on the rhs of (5.8.50) conveys the contribution to the compressibility due to the existence of interaction (and therefore correlation) among the particles.

The compressibility equation has some outstanding features which we note here:

1. We recall that no assumption on the total potential energy has been introduced to obtain (5.8.50). In the previous sections, we found relations between some thermodynamic quantities and the pair correlation functions which were based explicitly on the assumption of pairwise additivity of the total potential energy. We recall, also, that higher-order molecular distribution functions must be introduced if higher-order potentials are not negligible. Relation (5.8.50) does not depend on the additivity assumption; hence, it suffers no modification should high-order potentials be of importance. In this respect, the compressibility equation is far more general than the previously obtained relations (e.g., the energy or the pressure relation).

2. The compressibility equation involves the knowledge of the radial distribution function, even when the system consists of nonspherical particles. We recall that previously obtained relations between, say, the energy or the pressure, and the pair correlation function were dependent on the type of particle under consideration [compare, for instance, relations (5.8.22) and (5.8.25)]. The compressibility of the system depends only on the spatial pair correlation function. If nonspherical particles are considered, it is understood that $\overline{g(R)}$ is obtained by averaging over all orientations (5.8.46).

3. The compressibility equation is a simple integral over $\overline{g(R)}$. It does not require explicit knowledge of $U(R)$ (or higher-order potentials). It is true that $\overline{g(R)}$ is a functional of $U(R)$. However, once we have obtained the former, we can use it directly to compute the compressibility by means of (5.8.50). This is not possible for the computation of, say, the energy.

One of the most important applications of the compressibility equation is to test various theories of computation of $\overline{g(R)}$. We recall that the pressure equation (5.8.37) has been found useful for computing the equation of state of a substance, and hence can be used as a test of the theory that has furnished $\overline{g(R)}$. Similarly, by integrating the compressibility equation, we obtain the equation of state of the system, which may serve as a different test of the theory. Clearly, if we use the exact function $\overline{g(R)}$ in the pressure or in the compressibility equations, we must end up with the same equation of state. However, since we usually have only an approximation for $\overline{g(R)}$, the results of the two equations may be different. Therefore, the discrepancy between the two results obtained with the same $\overline{g(R)}$ using the pressure and the compressibility equations can serve as a sensitive test of the accuracy of the method of computing $g(R)$.

In applying the compressibility equation (5.8.50), care must be exercised to use the pair correlation function $\overline{g(R)}$ as obtained in the grand canonical ensemble, rather than the corresponding function $g(R)$ obtained in a closed system. Although the difference between $\overline{g(R)}$ and $g(R)$ is in a term of the order of N^{-1}—see Eqs. (5.8.59) and (5.8.60) below—this small difference becomes important when integration over the entire volume is performed in the definition of the quantity G (5.8.51).

Let us first demonstrate the source of difficulty by a simple example. Consider an ideal gas in the T, V, N ensemble. In section 5.3 we saw that $g(R)$ in this case has the form

$$g(R) = 1 - \left(\frac{1}{N}\right) \qquad \text{(ideal gas: } T, V, N \text{ ensemble).} \qquad (5.8.54)$$

On the other hand, $\overline{g(R)}$ in the T, V, μ ensemble is

$$\overline{g(R)} = 1 \qquad \text{(ideal gas: } T, V, \mu \text{ ensemble).} \qquad (5.8.55)$$

The difference between the two results (5.8.54) and (5.8.55) arises from the finite number of particles in the T, V, N system. Even if there are no interactions, $(U(\mathbf{R}^N) \equiv 0)$, there is still correlation between the particles. The density at any point in the system is $\rho(\mathbf{R}) = N/V$. The conditional density at $\mathbf{R}$ given a particle at any other point $\mathbf{R}'$ is not $\rho(R) = N/V$ but $(N-1)/V$. Fixing one particle at some point has an effect on the density at any other point merely because the number of particles was reduced from N to $N-1$. Such a correlation does not exist if we open the system, in which case the pair correlation function $\overline{g(R)}$ is unity for an ideal gas.

Clearly, we can always take the infinite-system-size limit of (5.8.54) to obtain

$$\lim_{N \to \infty} g(R) = 1, \qquad (5.8.56)$$

which can be used in the compressibility equation.

Thus, although the difference between $g(R)$ and $\overline{g(R)}$ in (5.8.54) and (5.8.55) is very small, for macroscopic systems $(N \approx 10^{23})$, they produce different results upon integration over a macroscopic volume. The different results obtained by using $g(R)$ and $\overline{g(R)}$ in Eq. (5.8.50) for ideal gas are

$$\kappa_T = \frac{1}{kT\rho} + \frac{1}{kT} \int_V d\mathbf{R} \left(\frac{-1}{N}\right)$$

$$= \frac{1}{kT\rho} - \frac{1}{kT\rho} = 0 \qquad \text{[using } g(R) \text{ from (5.8.54)]} \qquad (5.8.57)$$

$$\kappa_T = \frac{1}{kT\rho} \quad \text{[using } \overline{g(R)} \text{ from (5.8.55)].} \tag{5.8.58}$$

Clearly, only the second result gives the correct compressibility of the ideal gas (5.8.54).

Relations (5.8.54) and (5.8.55) hold for the ideal gas. In the general case, the limiting behavior of $g(R)$ as $R \to \infty$ is

$$g(R) \to 1 - \frac{\rho k T \kappa_T}{N} \quad (T, V, N \text{ ensemble}) \tag{5.8.59}$$

$$g(R) \to 1 \quad (T, V, \mu \text{ ensemble}). \tag{5.8.60}$$

Clearly, (5.8.60) can be obtained from (5.8.59) by taking the infinite-system-size limit ($N \to \infty$). Another way of viewing the origin of the discrepancy between the two results in the T, V, N and the T, V, μ ensembles is in the difference in the normalization conditions for the molecular distribution functions. In particular, in the T, V, N ensemble, we have

$$\langle N^2 \rangle = \langle N \rangle^2 = N^2. \tag{5.8.61}$$

Hence the normalization condition is

$$\int d\mathbf{X}_1 \, d\mathbf{X}_2 [\rho^{(2)}(\mathbf{X}_1, \mathbf{X}_2) - \rho^{(1)}(\mathbf{X}_1)\rho^{(1)}(\mathbf{X}_2)] = -N, \tag{5.8.62}$$

which is equivalent to the normalization condition

$$\rho \int_0^\infty [g(R) - 1] 4\pi R^2 \, dR = -1 \quad (T, V, N \text{ ensemble}). \tag{5.8.63}$$

The last result simply means that the total number of particles around a given particle minus the total number of particles in the system is exactly minus one. This simple calculation does not hold for the open system, where N is not a fixed number.

The corresponding normalization condition in the T, V, μ ensemble is (5.8.47), in which $\langle N^2 \rangle \neq \langle N \rangle^2$. This equation led us to the compressibility equation (5.8.50), which we write again as

$$\rho \int_0^\infty [\overline{g(R)} - 1] 4\pi R^2 \, dR = -1 + kT\rho\kappa_T, \tag{5.8.64}$$

which can be compared with (5.8.63). Clearly the difference is $kT\rho\kappa_T$, which arises from the difference in the limiting behavior of $g(R)$ and $\overline{g(R)}$ in (5.8.59) and (5.8.60).

The reader may wonder why we have dealt only now with the question of the limiting behavior of $g(R)$ as $R \to \infty$. The reason is quite simple. In all of our previous integrals, $g(R)$ appeared with another function in the integrand. For instance, in the equation for the energy, we have an integral of the form

$$\int_0^\infty U(R)g(R) 4\pi R^2 \, dR. \tag{5.8.65}$$

Clearly, since $U(R)$ is presumed to tend to zero, as, say, R^{-6} as $R \to \infty$, it is of no importance whether the limiting behavior of $g(R)$ is given by (5.8.59) or (5.8.60); in both cases the integrand will become practically zero as R becomes large enough so that

$U(R) \approx 0$. The unique feature of the compressibility relation is that only $g(R)$ appears under the sign of integration. Therefore, different results may be anticipated according to the different limiting behavior of $g(R)$ as $R \to \infty$.

In Chapter 6 we shall encounter a generalization of the compressibility equation for mixtures of two or more components. This equation involves quantities $G_{\alpha\beta}$ which are the generalizations of G defined in (5.8.51).

As a corollary of this section, we derive a relation between the density derivative of the chemical potential and an integral involving $g(R)$. Recall the thermodynamic relation (see section 1.4)

$$\left(\frac{\partial \mu}{\partial \rho}\right)_T = \frac{1}{\kappa_T \rho^2}. \tag{5.8.66}$$

Combining (5.8.66) and (5.8.52) yields

$$\left(\frac{\partial \mu}{\partial \rho}\right)_T = \frac{kT}{\rho + \rho^2 G} = kT\left(\frac{1}{\rho} - \frac{G}{1 + \rho G}\right). \tag{5.8.67}$$

Relation (5.8.67) will be generalized in the next chapter to mixtures. Here, we note that by integrating (5.8.67) with respect to the density, we get the chemical potential, i.e.,

$$\mu = \int \frac{kT \, d\rho}{\rho + \rho^2 G} + \text{const.} \tag{5.8.68}$$

Thus, once we have $g(R)$ and its density dependence, we can determine μ from (5.8.68). The constant of integration is evaluated as follows:

We choose a very low density ($\rho_0 \to 0$) in such a way that the chemical potential has the ideal-gas form

$$\mu(\rho_0) = kT \ln(\rho_0 \Lambda^3 q^{-1}) = \mu^{0g} + kT \ln \rho_0. \tag{5.8.69}$$

The chemical potential in (5.8.68) may be obtained by integrating from ρ_0 to the final density ρ, i.e.,

$$\mu(\rho) = \mu(\rho_0) + kT \int_{\rho_0}^{\rho} \left(\frac{1}{\rho'} - \frac{G}{1 + \rho' G}\right) d\rho'$$

$$= \mu^{0g} + kT \ln \rho_0 + kT \int_{\rho_0}^{\rho} \left(\frac{1}{\rho'} - \frac{G}{1 + \rho' G}\right) d\rho'$$

$$= \mu^{0g} + kT \ln \rho - kT \int_{0}^{\rho} \frac{G}{1 + \rho' G} d\rho'. \tag{5.8.70}$$

Note that in the last form on the rhs of (5.8.70) we have replaced the lower limit ρ_0 by $\rho_0 = 0$. This could not have been done when the divergent part $(\rho')^{-1}$ was in the integrand. The last relation will also be generalized for multicomponent systems in the next chapter.

5.9. THE CHEMICAL POTENTIAL

We devote a special section to the chemical potential since this quantity is of central importance in all applications of thermodynamics to chemistry and biochemistry.

5.9.1. The General Expression

The chemical potential is defined, in the T, V, N ensemble, by

$$\mu = \left(\frac{\partial A}{\partial N}\right)_{T,V}. \tag{5.9.1}$$

For reasons that will become clear in the following paragraphs, the chemical potential cannot be expressed as a simple integral involving the pair correlation function.

Consider, for example, the pressure equation that we have derived in section 5.8.3, which we denote by

$$P = P[g(R); \rho, T]. \tag{5.9.2}$$

By this notation, we simply mean that we have expressed the pressure as a function of ρ and T, and also in terms of $g(R)$, which is itself a function of ρ and T.

Since

$$P = -\left(\frac{\partial A}{\partial V}\right)_{T,N} = -\left[\frac{\partial a}{\partial(\rho^{-1})}\right]_T, \tag{5.9.3}$$

where $a = A/N$ and $\rho^{-1} = V/N$, we can integrate (5.9.3) to obtain

$$a = -\int P[g(R); \rho, T]\, d(\rho^{-1}). \tag{5.9.4}$$

We see that in order to express a in terms of $g(R)$, we must know the explicit dependence of $g(R)$ on the density. Thus, if we used the pressure equation in the integrand of (5.9.4), we need a second integration, over the density, to get the Helmholtz energy per particle. The chemical potential follows from the relation

$$\mu = a + Pv \tag{5.9.5}$$

with $v = V/N$.

A second method of computing the chemical potential is to use the energy equation derived in section 5.8.2, which we express as

$$E = E[g(R); \rho, T]. \tag{5.9.6}$$

The relation between the energy per particle and the Helmholtz energy is

$$e = \frac{E}{N} = -T^2 \left\{ \frac{\partial(a/T)}{\partial T} \right\}_\rho, \tag{5.9.7}$$

which can be integrated to obtain

$$\frac{a}{T} = \frac{-1}{N} \int \frac{E[g(R); \rho, T]}{T^2}\, dT. \tag{5.9.8}$$

Again we see that if we use the energy expression [in terms of $g(R)$] in the integrand of (5.9.8), we must also know the dependence of $g(R)$ on the temperature.

The above two illustrations show that in order to obtain a relation between μ and $g(R)$, it is not sufficient to know the function $g(R)$ at a given ρ and T; one needs the more detailed knowledge of $g(R)$ and its dependence on either ρ or T. This difficulty follows from the fact that the chemical potential is not an average of a pairwise quantity, and therefore the general theorem of section 5.8.1 is not applicable here. Let us further elaborate on this point.

The chemical potential can be written as

$$\mu = \left(\frac{\partial A}{\partial N}\right)_{T,V}$$

$$= \lim_{dN \to 0} \frac{A(N + dN) - A(N)}{dN}$$

$$= \frac{A(N + 1) - A(N)}{1}. \tag{5.9.9}$$

All the equalities in (5.9.9) hold for macroscopic systems where the addition of one particle ($dN = 1$) can be considered to be an infinitesimal change in the variable N.

Relation (5.9.9) simply means that in order to compute the chemical potential, it is sufficient to compute the change of the Helmholtz energy upon the addition of one particle. We now use the connection between the Helmholtz energy and the canonical partition function to obtain

$$\exp(-\beta\mu) = \exp\{-\beta[A(T, V, N + 1) - A(T, V, N)]\}$$

$$= \frac{Q(T, V, N + 1)}{Q(T, V, N)}$$

$$= \frac{[q^{N+1}/\Lambda^{3(N+1)}(N+1)!] \int \cdots \int d\mathbf{R}_0 \cdots d\mathbf{R}_N \exp(-\beta U_{N+1})}{(q^N/\Lambda^{3N}N!) \int \cdots \int d\mathbf{R}_1 \cdots d\mathbf{R}_N \exp(-\beta U_N)}. \tag{5.9.10}$$

Note that the added particle has been given the index zero. Using the assumption of pairwise additivity of the total potential, we may split U_{N+1} into two terms

$$U_{N+1}(\mathbf{R}_0, \ldots, \mathbf{R}_N) = U_N(\mathbf{R}_1, \ldots, \mathbf{R}_N) + \sum_{j=1}^{N} U(\mathbf{R}_0, \mathbf{R}_j)$$

$$= U_N(\mathbf{R}_1, \ldots, \mathbf{R}_N) + B(\mathbf{R}_0, \ldots, \mathbf{R}_N), \tag{5.9.11}$$

where we have included all the interactions of the zeroth particle with the rest of the system into the quantity $B(\mathbf{R}_0, \ldots, \mathbf{R}_N)$. Using (5.9.11) and the general expression for the basic probability density in the T, V, N ensemble, we rewrite (5.9.10) as

$$\exp(-\beta\mu)$$

$$= \frac{q}{\Lambda^3(N+1)} \int \cdots \int d\mathbf{R}_0 \, d\mathbf{R}_1 \cdots d\mathbf{R}_N \, P(\mathbf{R}_1, \ldots, \mathbf{R}_N) \exp[-\beta B(\mathbf{R}_0, \ldots, \mathbf{R}_N)].$$

We now transform to relative coordinates

$$\mathbf{R}'_i = \mathbf{R}_i - \mathbf{R}_0, \qquad i = 1, 2, \ldots, N, \tag{5.9.12}$$

and note that $B(\mathbf{R}_0, \ldots, \mathbf{R}_N)$ is actually a function only of the relative coordinates $\mathbf{R}'_1, \ldots, \mathbf{R}'_N$. [For instance, $U(\mathbf{R}_0, \mathbf{R}_j)$ is a function of $\mathbf{R}'_j$ and not of both $\mathbf{R}_0$ and $\mathbf{R}_j$.]

Hence,

$\exp(-\beta\mu)$

$$= \frac{q}{\Lambda^3(N+1)} \int d\mathbf{R}_0 \int \cdots \int d\mathbf{R}'_1 \cdots d\mathbf{R}'_N \, P(\mathbf{R}'_1, \ldots, \mathbf{R}'_N) \exp[-\beta B(\mathbf{R}'_1, \ldots, \mathbf{R}'_N)].$$

$$(5.9.13)$$

Now the integrand is independent of $\mathbf{R}_0$, so that we may integrate over $\mathbf{R}_0$ to obtain the volume. The inner integral is simply the average, in the T, V, N ensemble, of the quantity $\exp(-\beta B)$, i.e.,

$$\exp(-\beta\mu) = \frac{qV}{(N+1)\Lambda^3} \langle \exp(-\beta B) \rangle. \qquad (5.9.14)$$

Putting $\rho = N/V \simeq (N+1)/V$ (macroscopic system), we can rearrange (5.9.14) to obtain the relation

$$\mu = kT \ln(\rho\Lambda^3 q^{-1}) - kT \ln\langle \exp(-\beta B) \rangle. \qquad (5.9.15)$$

This is an important and very useful expression for the chemical potential. We shall soon see some generalizations of this expression (to nonspherical particles, mixture of species, and different ensembles). First we note that the first term on the rhs of (5.9.15) is simply the chemical potential of an ideal gas at the same temperature and density. Clearly if there are no interactions in the system, $B = 0$ and the second term on the rhs is zero. Therefore the second term is the contribution of the interactions among the particles to the chemical potential. The quantity B defined in (5.9.11) is referred to as the total binding energy of the added particle to all the other particles being at some specific configuration $\mathbf{R}_1, \ldots, \mathbf{R}_N$. Note that this binding energy includes both repulsion and attraction depending on the relative position $\mathbf{R}_0$ with respect to all other particles at $\mathbf{R}_1, \ldots, \mathbf{R}_N$. The addition of the new particle at $\mathbf{R}_0$ can also be viewed as the turning on of an "external field" acting on the system of N particles. This external field introduces the factor $\exp(-\beta B)$ in the expression for the chemical potential. More explicitly,

$$\exp[-\beta B(\mathbf{R}_0, \ldots, \mathbf{R}_N)] = \prod_{j=1}^{N} \exp[-\beta U(\mathbf{R}_0, \mathbf{R}_j)]. \qquad (5.9.16)$$

Clearly this is not a pairwise quantity in the sense of (5.8.3), i.e., it is not a *sum*, but a *product* of pairwise functions. This is the inherent reason why we cannot express the chemical potential as a simple integral involving only the pair distribution function.

   We now turn to reinterpret the second term on the rhs of (5.9.15) in three different ways.

5.9.2. Continuous Coupling of the Binding Energy

   In the preceding section we have seen that the chemical potential could be expressed in terms of $g(R)$ provided we also know the dependence of $g(R)$ on either T or ρ. We now derive a third expression due to Kirkwood, which employs the idea of a coupling parameter ξ. The ultimate expression for the chemical potential would be an integral over both R and ξ involving the function $g(R, \xi)$.

We define an auxiliary potential function as follows:

$$U(\xi) = U_N(\mathbf{R}_1, \ldots, \mathbf{R}_N) + \xi \sum_{j=1}^{N} U(\mathbf{R}_0, \mathbf{R}_j), \qquad (5.9.17)$$

which can be compared with (5.9.11). Clearly,

$$U(\xi = 0) = U_N(\mathbf{R}_1, \ldots, \mathbf{R}_N) \qquad (5.9.18)$$

$$U(\xi = 1) = U_{N+1}(\mathbf{R}_0, \ldots, \mathbf{R}_N). \qquad (5.9.19)$$

The idea is that by changing ξ from zero to unity, the function $U(\xi)$ changes continuously from U_N to U_{N+1}. Another way of saying the same thing is that by changing ξ from zero to unity the binding energy of the newly added particle at $\mathbf{R}_0$ is turned on continuously. Note however that within the assumption of pairwise additivity of the total potential energy, the quantity U_N is unaffected by this coupling of the binding energy of the new particle.

Associated with $U(\xi)$ we also define an auxiliary configurational partition function by

$$Z(\xi) = \int \cdots \int d\mathbf{R}_0 \, d\mathbf{R}_1 \cdots d\mathbf{R}_N \exp[-\beta U(\xi)], \qquad (5.9.20)$$

where clearly we have

$$Z(\xi = 0) = \int \cdots \int d\mathbf{R}_0 \, d\mathbf{R}_1 \cdots d\mathbf{R}_N \exp(-\beta U_N) = V Z_N \qquad (5.9.21)$$

and

$$Z(\xi = 1) = Z_{N+1}. \qquad (5.9.22)$$

The expression (5.9.10) for the chemical potential can be rewritten using the above notation as

$$\mu = kT \ln(\rho \Lambda^3 q^{-1}) - kT \ln Z(\xi = 1) + kT \ln Z(\xi = 0) \qquad (5.9.23)$$

or, using the identity

$$kT \ln Z(\xi = 1) - kT \ln Z(\xi = 0) = kT \int_0^1 \frac{\partial \ln Z(\xi)}{\partial \xi} \, d\xi, \qquad (5.9.24)$$

we get

$$\mu = kT \ln(\rho \Lambda^3 q^{-1}) - kT \int_0^1 \frac{\partial \ln Z(\xi)}{\partial \xi} \, d\xi. \qquad (5.9.25)$$

We now differentiate $Z(\xi)$ directly to obtain

$$kT\frac{\partial \ln Z(\xi)}{\partial \xi} = \frac{kT}{Z(\xi)}\int\cdots\int d\mathbf{R}_0\cdots d\mathbf{R}_N \{\exp[-\beta U(\xi)]\}\left[-\beta\sum_{j=1}^{N} U(\mathbf{R}_0,\mathbf{R}_j)\right]$$

$$= -\int\cdots\int d\mathbf{R}_0\cdots d\mathbf{R}_N\, P(\mathbf{R}^{N+1},\xi)\sum_{j=1}^{N} U(\mathbf{R}_0,\mathbf{R}_j)$$

$$= -\sum_{j=1}^{N}\int\cdots\int d\mathbf{R}_0\cdots d\mathbf{R}_N\, P(\mathbf{R}^{N+1},\xi)U(\mathbf{R}_0,\mathbf{R}_j)$$

$$= -N\iint d\mathbf{R}_0\, d\mathbf{R}_1\, U(\mathbf{R}_0,\mathbf{R}_1)\int\cdots\int d\mathbf{R}_2\cdots d\mathbf{R}_N\, P(\mathbf{R}^{N+1},\xi)$$

$$= -\frac{1}{N+1}\iint d\mathbf{R}_0\, d\mathbf{R}_1\, U(\mathbf{R}_0,\mathbf{R}_1)\rho^{(2)}(\mathbf{R}_0,\mathbf{R}_1,\xi)$$

$$= -\rho\int_0^\infty U(R)g(R,\xi)4\pi R^2\, dR. \tag{5.9.26}$$

The formal steps in (5.9.26) are very similar to those in section 5.8 and there is no need to go through them in detail. The only new feature in (5.9.26) is the appearance of the parameter ξ, employed throughout in the distribution functions.

We now combine (5.9.26) with (5.9.25) to obtain the final expression for the chemical potential:

$$\mu = kT\ln(\rho\Lambda^3 q^{-1}) + \rho\int_0^1 d\xi\int_0^\infty U(R)g(R,\xi)4\pi R^2\, dR. \tag{5.9.27}$$

We can also define the *standard chemical potential* in the gas phase by

$$\mu^{0g} = kT\ln(\Lambda^3 q^{-1}) \tag{5.9.28}$$

and

$$kT\ln\gamma^{\text{ideal gas}} = \rho\int_0^1 d\xi\int_0^\infty U(R)g(R,\xi)4\pi R^2\, dR \tag{5.9.29}$$

to rewrite (5.9.27) in the form

$$\mu = \mu^{0g} + kT\ln(\rho\gamma^{\text{ideal gas}}), \tag{5.9.30}$$

where in (5.9.29) we have an explicit expression for the activity coefficient $\gamma^{\text{ideal gas}}$ which measures the extent of deviation of the chemical potential from its ideal-gas form. The quantity $\rho g(R,\xi)$ is the local density of particles around a given particle that is coupled, to the extent of ξ, to the rest of the system. The last sentence means that the total potential energy of the system is given by (5.9.17) with a fixed value of ξ. Note that (5.9.27) is not a simple integral involving $g(R)$. A more detailed knowledge of the function $g(R,\xi)$ is required.

The interpretation of the two terms in (5.9.27) is as follows. Suppose that we have a system of N interacting particles at a given T and ρ. We now add one particle which carries the same momentum and internal partition function as all other particles of the system. This particle initially is uncoupled in the sense of $\xi = 0$ in (5.9.17). The corresponding chemical potential of this particular particle is

$$\mu' = kT \ln(\Lambda^3 q^{-1} V^{-1}). \tag{5.9.31}$$

Note that since we have added only one particle, its density is $\rho' = V^{-1}$. The volume V enters here because the particle can reach any point within the system.

We now turn on the coupling parameter ξ until it reaches the value of unity. The chemical potential of the added particle changes in two ways. First we have the work required to build up the interaction between the particle and the rest of the system. This is the second term on the rhs of (5.9.27). Second, as long as the new particle is distinguishable from all other particles (i.e., $\xi \neq 1$), its density remains fixed, $\rho' = V^{-1}$. At the point $\xi = 1$, it becomes abruptly identical to the other particles. This involves an assimilation Helmholtz energy in the amount (see section 2.3)

$$\Delta A_{\mathrm{ass}} = kT \ln\left(\frac{N}{1}\right). \tag{5.9.32}$$

This, together with the coupling work, converts (5.9.31) into (5.9.27).

5.9.3. The Pseudochemical Potential

The chemical potential is the work (here, at T, V constant) associated with the addition of one particle to a macroscopically large system:

$$\mu = A(T, V, N + 1) - A(T, V, N). \tag{5.9.33}$$

The pseudochemical potential refers to the work associated with the addition of one particle to a fixed position in the system, say at $\mathbf{R}_0$:

$$\mu^* = A(T, V, N + 1; \mathbf{R}_0) - A(T, V, N). \tag{5.9.34}$$

The statistical-mechanical expression for the pseudochemical potential can be expressed, similarly to (5.9.10), as a ratio between two partition functions corresponding to the difference in the Helmholtz energies in (5.9.34), i.e.,

$$\exp(-\beta\mu^*) = \frac{(q^{N+1}/\Lambda^{3N}N!) \displaystyle\int \cdots \int d\mathbf{R}_1 \cdots d\mathbf{R}_N \exp[-\beta U_{N+1}(\mathbf{R}_0, \ldots, \mathbf{R}_N)]}{(q^{N}/\Lambda^{3N}N!) \displaystyle\int \cdots \int d\mathbf{R}_1 \cdots d\mathbf{R}_N \exp[-\beta U_N(\mathbf{R}_1, \ldots, \mathbf{R}_N)]}. \tag{5.9.35}$$

It is instructive to observe the differences between (5.9.10) and (5.9.35). Since the added particle in (5.9.35) is devoid of the translational degree of freedom, it will not bear a momentum partition function. Hence, we have Λ^{3N} instead of $\Lambda^{3(N+1)}$ as in (5.9.10). For the same reason, the integration in the numerator of (5.9.35) is over the N locations $\mathbf{R}_1, \ldots, \mathbf{R}_N$ and not over $\mathbf{R}_0, \ldots, \mathbf{R}_N$ as in (5.9.10). Furthermore, since we have added a particle to a fixed position, it is distinguishable from the other particles; hence, we have $N!$ instead of $(N+1)!$ as in (5.9.10).

Once we have set up the statistical-mechanical expression (5.9.35), the following formal steps are nearly the same as in the previous section. The result is

$$\mu^* = kT \ln q^{-1} - kT \ln\langle \exp(-\beta B)\rangle$$

$$= kT \ln q^{-1} + \rho \int_0^1 d\xi \int_0^\infty U(R)g(R, \xi)4\pi R^2 \, dR, \qquad (5.9.36)$$

which should be compared with (5.9.15) and (5.9.27). Note that we have added the particle to a fixed position $\mathbf{R}_0$; therefore, from the formal point of view, μ^* depends on $\mathbf{R}_0$. However, in a homogeneous fluid, all the points of the system are presumed to be equivalent (except for a small region near the boundaries, which is negligible for our present purposes), and therefore μ^* is effectively independent of $\mathbf{R}_0$.

Combining (5.9.36) with either (5.9.15) or (5.9.27), we obtain the expression for the chemical potential

$$\mu = \mu^* + kT \ln(\rho\Lambda^3). \qquad (5.9.37)$$

This relation has a simple and important interpretation. The work μ required to add a particle to the system is split into two parts. First, we add the particle to a fixed position, say $\mathbf{R}_0$, the corresponding work being μ^*. Next, we remove the constraint imposed by fixing the position of the particle; the corresponding work is the second term on the rhs of (5.9.37). The last quantity was referred to as the liberation Helmholtz energy. Since we are dealing with classical statistics $\rho\Lambda^3 \ll 1$ and therefore the liberation Helmholtz energy is always negative. Note that this term is exactly the same as the corresponding term in an ideal-gas phase (section 2.1). Although this term depends on the dimensionless quantity $\rho\Lambda^3$, there are three conceptually different sources for this Helmholtz energy change. First, there is the gain of the momentum partition function; second, there is the accessibility of the entire volume V; and third, there is the assimilation of the newly added particle when it is released from the fixed position.

5.9.4. Building Up the Density of the System

As a third interpretation of the expression for the chemical potential, we use a relation obtained in section 5.8.4 which we now rewrite as

$$\mu = kT \ln \Lambda^3 q^{-1} + kT \ln \rho - kT \int_0^\rho \frac{G}{1 + \rho'G} \, d\rho'. \qquad (5.9.38)$$

Here, the third term on the rhs of (3.9.38) may be identified with the coupling work; i.e., comparing (5.9.38) with (5.9.15), we have

$$kT \ln\langle \exp(-\beta B)\rangle = kT \int_0^\rho \frac{G}{1 + \rho'G} \, d\rho'. \qquad (5.9.39)$$

The coupling work is interpreted in (5.9.39) as the work required to increase the density from $\rho = 0$ to the final density ρ. A slightly different interpretation is obtained by rewriting (5.9.38) as (see 5.8.70):

$$\mu = (kT \ln \Lambda^3 q^{-1} + kT \ln \rho_0) + kT \int_{\rho_0}^\rho \left(\frac{1}{\rho'} - \frac{G}{1 + \rho'G}\right) d\rho'. \qquad (5.9.40)$$

The expression within the first set of parentheses corresponds to the work required to introduce one particle to an ideal-gas system (ρ_0 very low). The second term is the work involved in changing the density from ρ_0 to the final density. This work is composed of two contributions: first, the change in the assimilation term $kT \ln \rho/\rho_0$ (note that V is constant in the process), and second, the coupling work (5.9.39).

5.9.5. First-Order Expansion of the Coupling Work

We wish to obtain the first-order deviation from the ideal-gas expression for the chemical potential. This may be obtained either from (5.9.27) or from (5.9.38). We know from section 5.3.2 that at the limit of low density we have

$$g(R) = \exp[-\beta U(R)], \tag{5.9.41}$$

and hence

$$g(R, \xi) = \exp[-\beta \xi U(R)]. \tag{5.9.42}$$

Substituting (5.9.42) into (5.9.27), we get an immediate integral over ξ:

$$\int_0^1 d\xi \int_0^\infty U(R) \exp[-\beta \xi U(R)] 4\pi R^2 \, dR = -kT \int_0^\infty \{\exp[-\beta U(R)] - 1\} 4\pi R^2 \, dR. \tag{5.9.43}$$

Using the definition of the second virial coefficient (see section 1.8),

$$B_2(T) = -\tfrac{1}{2} \int_0^\infty \{\exp[-\beta U(R)] - 1\} 4\pi R^2 \, dR, \tag{5.9.44}$$

we can write (5.9.27) for this case as

$$\mu = \mu^{0g} + kT \ln \rho + 2kT B_2(T)\rho. \tag{5.9.45}$$

The last term on the rhs of (5.9.45) is the first-order term in the expansion of the coupling work in the density.

The virial expansion can be recovered from (5.9.45) by using the thermodynamic relation

$$dP = \rho \, d\mu \qquad (T \text{ constant}). \tag{5.9.46}$$

From (5.9.45) we get

$$d\mu = \frac{kT}{\rho} d\rho + 2kT B_2(T) \, d\rho. \tag{5.9.47}$$

Combining (5.9.46) and (5.9.47) yields

$$dP = [kT + 2kT B_2(T)\rho] \, d\rho, \tag{5.9.48}$$

which upon integration between $\rho = 0$ and the final density ρ yields

$$P = kT\rho + kT B_2(T)\rho^2, \tag{5.9.49}$$

which is the leading form of the virial expansion of the pressure.

The same result can be obtained by expanding the third term on the rhs of (5.9.38) to first order in the density, i.e.,

$$-kT \int_0^\rho \frac{G}{1+\rho'G} \, d\rho' = -kT\rho G^0, \qquad (5.9.50)$$

where

$$G^0 = \lim_{\rho \to 0} G. \qquad (5.9.51)$$

From (5.9.50) and (5.9.45) we have

$$G^0 = -2B_2(T) = \int_0^\infty \{\exp[-\beta U(R)] - 1\}4\pi R^2 \, dR. \qquad (5.9.52)$$

5.9.6. Some Generalizations

We now briefly summarize the modifications that must be introduced into the equation for the chemical potential for more complex systems.

1. For systems that do not obey the assumption of pairwise additivity for the potential energy, Eq. (5.9.27) becomes invalid. In a formal way, one can derive an analogous relation involving higher-order molecular distribution functions. This does not seem to be useful at present. However, in many applications for mixtures one can retain the general expression (5.9.15) even when the total potential energy of the solvent does not obey any pairwise additivity assumption. We discuss briefly this case below, and in more detail in Chapter 6.

2. For rigid, nonspherical particles whose potential energy obeys the assumption of pairwise additivity, a relation similar to (5.9.27) holds. However, one must now integrate over the orientation as well as the location of the particle. The generalized relation is

$$\mu = kT \ln(\rho \Lambda^3 q^{-1}) + \int_0^1 d\xi \int d\mathbf{X}'' \, U(\mathbf{X}', \mathbf{X}'')\rho(\mathbf{X}''/\mathbf{X}', \xi). \qquad (5.9.53)$$

Here, q includes the rotational as well as the internal partition function of a single molecule. The quantity $\rho(\mathbf{X}''/\mathbf{X}', \xi)$ is the local density of particles at $\mathbf{X}''$, given a particle at $\mathbf{X}'$ coupled to the extent of ξ. Clearly, the whole integral on the rhs of (5.9.53) does not depend on the choice of $\mathbf{X}'$ (for instance, we can take $\mathbf{R}' = 0$ and $\mathbf{\Omega}' = 0$ and measure $\mathbf{X}''$ relative to this choice).

3. For mixtures of simple liquids there are straightforward generalizations of Eqs. (5.9.15) and (5.9.27). Some of these are discussed in Chapter 6. Here we present one extention of these equations for a very dilute solution of, say, species A in a solvent B.

The binding energy of A to the solvent is defined by

$$B_A = U(\mathbf{R}_A, \mathbf{R}_1 \cdots \mathbf{R}_N) - U(\mathbf{R}_1 \cdots \mathbf{R}_N)$$

$$= \sum_{i=1}^N U(\mathbf{R}_A, \mathbf{R}_i). \qquad (5.9.54)$$

For any specific configuration of the solvent $(\mathbf{R}_1, \ldots, \mathbf{R}_N)$, the binding energy is simply the work required to bring A from a fixed position at infinite distance from the system to the position $\mathbf{R}_A$ within the system. We do not make any assumption on the nature of the total potential energy of the solvent. The only assumption we need is that the solute does not affect the intermolecular interaction among the solvent molecules (say, by polarizing the solvent molecules). In this case we can use the generalizations of (5.9.15) and (5.9.27), which read

$$\mu_A = kT \ln \rho_A \Lambda_A^3 q_A^{-1} - kT \ln \langle \exp[-\beta B_A] \rangle_0$$

$$= kT \ln \rho_A \Lambda_A^3 q_A^{-1} + \rho \int_0^1 d\xi \int_0^\infty U_{AB}(R) g_{AB}(R, \xi) 4\pi R^2 \, dR. \qquad (5.9.55)$$

Note that the average $\langle \ \rangle_0$ is over all the configurations of the solvent molecules, with the probability distribution $P_0(\mathbf{X}^N)$ of the pure solvent. The assumption has been made that the distribution function of solvent configurations is unaffected by the solute. Such an assumption is obviously correct for hard-sphere solutes and a good approximation for simple nonpolar solutes. If the solute A does affect the solvent–solvent interactions (say, by strong polarization), then we can still use the definition of B_A as the difference

$$B_A = U(\mathbf{R}_A, \mathbf{R}_1, \ldots, \mathbf{R}_N) - U(\mathbf{R}_1, \ldots, \mathbf{R}_N), \qquad (5.9.56)$$

but now the second form on the rhs of (5.9.55) does not hold. In addition to the interaction between A and all solvent molecules there is also a term corresponding to the changes in the interactions among the solvent molecules induced by the presence of A.

4. For molecules having internal rotational degrees of freedom (say, polymers), the expression for the chemical potential should be modified (even with pairwise additive interactions) to take into account all possible conformations of the molecules. In particular, the rotational partition function of the molecules (included in q) might be different for different conformations and therefore should be properly averaged. We shall discuss a simple case of such molecules in section 6.15. More complex molecules are treated in Chapter 8.

5.9.7. Other Ensembles

In all the previous sections we have used the definition of the chemical potential in the T, V, N ensemble (Eq. 5.9.1). This was done mainly for convenience. In actual applications, and in particular when comparison with experimental results is required, it is necessary to use the T, P, N ensemble. In that case, the chemical potential is defined by

$$\mu = \left(\frac{\partial G}{\partial N} \right)_{T,P}, \qquad (5.9.57)$$

where G is the Gibbs energy of the system. It is easy to show that the formal split of μ into two terms as in (5.9.15) is maintained. In the T, P, N ensemble, $\rho = N/\langle V \rangle$ where $\langle V \rangle$ is the average volume, and $\langle \ \rangle$ should be interpreted as a T, P, N average. In the T, V, μ ensemble, μ is one of the independent variables used to describe the system. Yet it can also be written in the form (5.9.15), with the reinterpretation of $\rho = \langle N \rangle / V$, where $\langle N \rangle$ is the average in the T, V, μ ensemble (see Appendix E).

5.10. THE WORK REQUIRED TO FORM A CAVITY IN A FLUID

A quantity of considerable interest in the study of fluids is the work required to create a cavity of radius σ at some fixed position $\mathbf{R}_0$. In addition to its importance as a

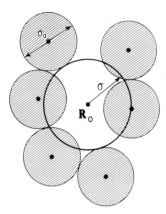

FIGURE 5.18. A cavity of radius σ at $\mathbf{R}_0$. The diameter of the particles is σ_a.

property of the fluid, the work of cavity formation may also be viewed as a first step in the process of solvation. This aspect will be dealt with in Chapters 6 and 7.

5.10.1. Spherical Cavity

A cavity of radius σ at a point $\mathbf{R}_0$ in the fluid is defined as the spherical region of radius σ centered at $\mathbf{R}_0$ from which the centers of all other particles are excluded. Figure 5.18 depicts such a cavity.

It is important to realize that a cavity may seem to be filled in the conventional sense, yet be empty according to the definition given above. Two extreme cases are illustrated in Fig. 5.19. In (a) we have a cavity of radius σ which is completely filled by a molecule of radius $\sigma_a/2$, yet, since no center of a molecule falls in this cavity, it is empty according to our definition. In (b) we have a sphere of radius σ at $\mathbf{R}_0$ which looks empty, but is not a cavity, according to our definition.

The work required to create a cavity of radius σ (keeping T, V, and N constant) is defined by

$$\Delta A_{\mathrm{cav}}(\mathbf{R}_0, \sigma) = A(T, V, N; \mathbf{R}_0, \sigma) - A(T, V, N). \qquad (5.10.1)$$

Here, $A(T, V, N; \mathbf{R}_0, \sigma)$ stands for the Helmholtz energy of a system in the T, V, N ensemble having a cavity of radius σ centered at $\mathbf{R}_0$. Clearly, in a homogeneous fluid, all points in the system are considered to be equivalent (except for a negligible region

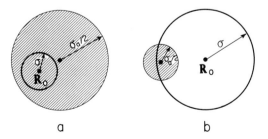

a   b

FIGURE 5.19. (a) A cavity of radius σ "filled" by a particle of diameter σ_a yet empty according to the definition of a cavity. (b) A cavity of radius σ filled according to the definition of a cavity.

near the boundaries of the system), and hence $\Delta A_{cav}(\mathbf{R}_0, \sigma)$ does not depend on $\mathbf{R}_0$. Nevertheless, we shall keep $\mathbf{R}_0$ in our notation to stress the fact that we are concerned with a cavity at a *fixed* position in the system.

From the definition of the concept of a cavity, it follows that all centers of the N particles are excluded from the spherical region of radius σ centered at $\mathbf{R}_0$. Denoting this region by $V(\mathbf{R}_0, \sigma)$, we have from (5.10.1)

$$\exp[-\beta\Delta A_{cav}(\mathbf{R}_0, \sigma)] = \frac{(q^N/\Lambda^{3N}N!) \int \cdots \int_{V - V(\mathbf{R}_0, \sigma)} d\mathbf{R}^N \exp[-\beta U_N(\mathbf{R}^N)]}{(q^N/\Lambda^{3N}N!) \int \cdots \int_V d\mathbf{R}^N \exp[-\beta U_N(\mathbf{R}^N)]}. \quad (5.10.2)$$

Note that the only difference in the two integrals is in the limits of the integration.

The rhs of (5.10.2) can now be interpreted as the probability of observing a cavity of radius σ at $\mathbf{R}_0$. We recall that the probability density of observing any specific configuration $\mathbf{R}^N$ is given by

$$P(\mathbf{R}^N) = \frac{\exp[-\beta U_N(\mathbf{R}^N)]}{\int \cdots \int d\mathbf{R}^N \exp[-\beta U_N(\mathbf{R}^N)]}. \quad (5.10.3)$$

Suppose we choose a region S in our system. The probability of finding the centers of all the particles in S is obtained by summing over all the events (configurations) conforming to the requirement that all $\mathbf{R}_i$ ($i = 1, 2, \ldots, N$) are within S. Hence,

$$P(S) = \int \cdots \int_S d\mathbf{R}^N P(\mathbf{R}^N). \quad (5.10.4)$$

As a particular case, if S coincides with the entire volume of the system, then $P(V) = 1$. Similarly, if all the centers are to be excluded from $V(\mathbf{R}_0, \sigma)$, we have the case in which $S = V - V(\mathbf{R}_0, \sigma)$, and therefore

$$P_{cav}(\mathbf{R}_0, \sigma) = \int \cdots \int_{V - V(\mathbf{R}_0, \sigma)} d\mathbf{R}^N P(\mathbf{R}^N). \quad (5.10.5)$$

From (5.10.2) and (5.10.5), we arrive at the important relation

$$P_{cav}(\mathbf{R}_0, \sigma) = \exp[-\beta\Delta A_{cav}(\mathbf{R}_0, \sigma)]. \quad (5.10.6)$$

On the rhs of (5.10.6), we have the work required to form a cavity of radius σ at $\mathbf{R}_0$, keeping T, V, and N constant. On the lhs, we have the probability of observing such a cavity in a system in the T, V, N ensemble. This is another relation between the probability of finding an event (here a cavity of radius σ) and the work required to create that event. It is quite straightforward to obtain similar relations in other ensembles. For instance, the probability of finding a cavity of radius σ in a T, P, N ensemble is related to the Gibbs energy change for creating such a cavity.

5.10.2. Cavity Formation and the Pseudochemical Potential of a Hard Sphere

We derive here a useful relation between the work required to form a cavity and the pseudochemical potential of a hard-sphere solute. We shall still discuss only a system of spherical particles in the T, V, N ensemble. The results are also valid for more general systems and in other ensembles.

Consider a system of N particles with an effective hard-core diameter σ_a. By "effective" we mean that two molecules at a distance of σ_a from each other feel almost infinite repulsive forces. Certainly there is no unique method of determining the values of σ_a for a real molecule. Nevertheless for, say, Lennard–Jones particles the parameter σ is a reasonable choice for an effective hard-core diameter.

In section 5.9.3, we obtained the pseudochemical potential for a one-component system. We repeat the same process here, but instead of adding the $(N + 1)$th particle, we add a hard-sphere particle of diameter σ_{HS} to a fixed position $\mathbf{R}_0$ in a system of particles having an effective hard-core diameter σ_a. The work associated with this process, keeping T, V, N constant, is given by

$$\exp(-\beta\mu_{HS}^*) = \exp\{-\beta[A(T, V, N; \mathbf{R}_0, \sigma_{HS}) - A(T, V, N)]\}$$

$$= \frac{\int_V \cdots \int dR^N \exp[-\beta U_N(\mathbf{R}^N) - \beta B_{HS}(\mathbf{R}_0)]}{\int_V \cdots \int dR^N \exp[-\beta U_N(\mathbf{R}^N)]}. \tag{5.10.7}$$

Note that since the hard-sphere solute is presumed to have no internal structure, its addition to the system does not introduce an internal partition function. The quantity $B_{HS}(\mathbf{R}_0)$ stands for the total interaction energy between the N molecules in the configuration $\mathbf{R}^N$ and the hard-sphere particle at $\mathbf{R}_0$. More specifically

$$B_{HS}(\mathbf{R}_0) = \sum_{i=1}^{N} U(\mathbf{R}_i, \mathbf{R}_0). \tag{5.10.8}$$

According to our assumption, σ_a is the hard-core diameter of the particles. This statement means that

$$U(\mathbf{R}_i, \mathbf{R}_j) = \infty \quad \text{for } R_{ij} = |\mathbf{R}_j - \mathbf{R}_i| < \sigma_a; i, j = 1, 2, \ldots, N. \tag{5.10.9}$$

For $R_{ij} \geq \sigma_a$, the potential function can have any form. However, the interaction between a particle of the system and the hard-sphere solute is presumed to have the form

$$U(\mathbf{R}_i, \mathbf{R}_0) = \begin{cases} \infty & \text{for } R_{i0} < \sigma = \frac{1}{2}(\sigma_{HS} + \sigma_a), \\ 0 & R_{i0} \geq \sigma = \frac{1}{2}(\sigma_{HS} + \sigma_a), \end{cases} \quad i = 1, 2, \ldots, N. \tag{5.10.10}$$

That is, a particle is not permitted to come closer than σ to the center of the hard-sphere solute. The condition (5.10.10) can be rewritten in an equivalent form as

$$\exp[-\beta U(\mathbf{R}_i, \mathbf{R}_0)] = \begin{cases} 0 & \text{for } R_{i0} < \sigma, \\ 1 & \text{for } R_{i0} \geq \sigma, \end{cases} \quad i = 1, 2, \ldots, N. \tag{5.10.11}$$

Using (5.10.8) and (5.10.11) in (5.10.7), we obtain

$$\exp(-\beta\mu_{HS}^*) = \int \cdots \int_V d\mathbf{R}^N P(\mathbf{R}^N) \exp[-\beta B_{HS}(\mathbf{R}_0)]$$

$$= \int \cdots \int_V d\mathbf{R}^N P(\mathbf{R}^N) \prod_{i=1}^{N} \exp[-\beta U(\mathbf{R}_i, \mathbf{R}_0)]$$

$$= \int \cdots \int_{V - V(\mathbf{R}_0, \sigma)} d\mathbf{R}^N P(\mathbf{R}^N). \tag{5.10.12}$$

The last equality in (5.10.12) follows from the property of the unit step function (5.10.11), which nullifies the integrand whenever $R_{i0} < \sigma$ (for $i = 1, 2, \ldots, N$). Since we have a product of N such factors (each of which operates on one vector $\mathbf{R}_i$) in the integrand of (5.10.12), their effect is to reduce the region of integration, for each $\mathbf{R}_i$, from V to $V - V(\mathbf{R}_0, \sigma)$. Comparing with (5.10.5) and (5.10.12), we arrive at

$$\exp(-\beta\mu_{HS}^*) = P_{cav}(\mathbf{R}_0, \sigma) = \exp[-\beta\Delta A_{cav}(\mathbf{R}_0, \sigma)], \tag{5.10.13}$$

with the condition

$$\sigma = \frac{(\sigma_{HS} + \sigma_a)}{2}. \tag{5.10.14}$$

Thus, the work required to produce a cavity of radius σ at $\mathbf{R}_0$ (which may be chosen at any point in the fluid) in a system of particles with an effective hard-core diameter σ_a is equal to the work required to introduce a hard-sphere particle of diameter σ_{HS} (given by $\sigma_{HS} = 2\sigma - \sigma_a$) into a fixed position $\mathbf{R}_0$.

It is important to realize that the last statement is valid only when we refer to a fixed position for both processes. The equivalence of the two processes is quite clear on intuitive grounds. Creation of a cavity means imposing a restriction on the centers of all particles, keeping them out of a certain region. This constraint is achieved simply by putting a hard-sphere solute at the position $\mathbf{R}_0$, provided that we have properly chosen its diameter by relation (5.10.14). In other words, as far as the solvent (i.e., the particles of the system) is concerned, there is no difference if we create a cavity at $\mathbf{R}_0$ or put a hard-sphere solute there, provided that (5.10.14) is fulfilled.

The above reasoning breaks down if we remove the condition of fixed position for both of the processes. We recall that the work required to add a hard-sphere solute to the system (at T, V, N constant) is equal to the chemical potential of the solute, which can be written as

$$\mu_{HS} = \mu_{HS}^* + kT \ln(\rho_{HS}\Lambda_{HS}^3), \tag{5.10.15}$$

where $\rho_{HS} = 1/V$ is the (number) density of the solute and Λ_{HS}^3 its momentum partition function. We also recall (see section 5.9.3 for more details) that the second term on the rhs of (5.10.15) is the contribution to μ_{HS} associated with the release of the constraint of the fixed position for the solute. The analog of this contribution in the case of a cavity is not evident. In the first place, a "free cavity" has no momentum partition function akin to the quantity Λ_{HS}^3 of a real solute. Second, there is no clear-cut definition of the (number) density of "free cavities of diameter σ." For these reasons, the equivalence of the process of introducing a hard-sphere solute and the creation of a cavity should be restricted only to a fixed position in the liquid.

Having the pseudochemical potential of a hard sphere in any liquid we can always add the liberation Helmholtz energy to obtain the chemical potential of the hard sphere in the same system. The same cannot be done to the quantity ΔA_{cav}. Therefore it is meaningful to speak about the chemical potential of hard spheres in a liquid, but there is no meaning to the analogous concept of the "chemical potential of cavities" in a liquid.

5.10.3. Nonspherical Cavities

In any liquid or mixture of liquids we can introduce a hard particle of any form at some fixed position and orientation. Associated with this process there is a well-defined pseudochemical potential. This hard particle, by its definition, excludes the centers of all other particles from some region which we denote by V^{EX}. Clearly, if the solvent particles are not spherical, or if they have different sizes, then the excluded volume of the hard particle will be different for different orientations of the solvent particles or for particles having different sizes. If we are interested in the work required to create a cavity equivalent to the hard particle, we must take the intersection of all possible V^{EX} with respect to all orientations and all solvent species.

As an example, consider a hard sphere of diameter σ_{HS} placed in a mixture of hard spheres of diameters σ_a and σ_b. The excluded volumes with respect to the two particles are

$$V_a^{\text{EX}} = \frac{4\pi}{3}\left(\frac{\sigma_{\text{HS}} + \sigma_a}{2}\right)^3, \tag{5.10.16}$$

$$V_b^{\text{EX}} = \frac{4\pi}{3}\left(\frac{\sigma_{\text{HS}} + \sigma_b}{2}\right)^3. \tag{5.10.17}$$

The cavity equivalent to placing the hard sphere at some fixed position is the intersection of V_a^{EX} and V_b^{EX}, i.e., $V_a^{\text{EX}} \cap V_b^{\text{EX}}$. In the case of spherical particles this intersection is simply the largest of the two excluded volumes. Similar considerations apply for particles of different shapes and sizes.

In general, the cavity produced by a hard particle α in any solvent is the intersection of all the excluded volumes of α with respect to all solvent particles approaching from all possible orientations. This excluded volume is denoted by V_α^{EX}. The probability of finding such an excluded volume empty is simply obtained by generalization of (5.10.5), namely

$$\Pr(V_\alpha^{\text{EX}}) = \int \cdots \int_{V - V_\alpha^{\text{EX}}} P(\mathbf{X}^N)\, d\mathbf{X}^N. \tag{5.10.18}$$

Since the integrand is always positive, the integral will be larger when the region of integration increases or when the excluded region decreases. If the cavities are spherical, then a larger cavity radius implies a lower probability. In other words, Pr(cavity of radius R) is a decreasing function of R. This follows directly from the definition (5.10.18) and applies to any liquid.

In general, a larger volume of the cavity does not necessarily lead to lower probability. For two excluded volumes for which

$$V_{\alpha_1}^{EX} \supset V_{\alpha_2}^{EX}, \tag{5.10.19}$$

we have

$$\Pr(V_{\alpha_1}^{EX}) < \Pr(V_{\alpha_2}^{EX}). \tag{5.10.20}$$

(5.10.19) means that the region $V_{\alpha_2}^{EX}$ is included in the region $V_{\alpha_1}^{EX}$. Thus for two spheres of radii $R_1 < R_2$,

$$V_{R_1}^{EX} < V_{R_2}^{EX} \tag{5.10.21}$$

and

$$V_{R_1}^{EX} \subset V_{R_2}^{EX}. \tag{5.10.22}$$

Hence

$$\Pr(V_{R_1}^{EX}) > \Pr(V_{R_2}^{EX}). \tag{5.10.23}$$

If the two excluded volumes have the same shape, then the larger one also includes the smaller region, and hence (5.10.20) holds. If the two excluded volumes have different shapes, say a sphere and a rod, then in general we cannot say anything on the relative probabilities, even when we know their relative sizes.

As in the case of spherical cavities, there is a relation between the probability of finding a region V_α^{EX} empty and the corresponding Gibbs or Helmholtz energy change for the creation of such a cavity. In all cases, the cavity referred to must be at some fixed position in the system. It does not matter where we choose that fixed position (if the system is homogeneous), and therefore we sometimes omit the explicit notation $\mathbf{R}_0$.

To see how important it is to fix the position of the cavity, we treat a simple example. We consider a system of N hard spheres of diameter σ in a volume V at temperature T. The density is small enough so that the system's PF is

$$Q(T, V, N) = \frac{V^N}{N!\Lambda^{3N}}. \tag{5.10.24}$$

The partition function of the same system with a cavity V^{EX} at some fixed position $\mathbf{R}_0$ is

$$Q(T, V, N; \mathbf{R}_0) = \frac{1}{N!\Lambda^{3N}} \int \cdots \int_{V - V^{EX}} d\mathbf{R}^N = \frac{(V - V^{EX})^N}{N!\Lambda^{3N}}. \tag{5.10.25}$$

The corresponding Helmholtz energy change is

$$\begin{aligned}
\Delta A &= A(T, V, N; \mathbf{R}_0) - A(T, V, N) \\
&= -kT \ln \frac{Q(T, V, N; \mathbf{R}_0)}{Q(T, V, N)} \\
&= -kT \ln \left(1 - \frac{V^{EX}}{V}\right)^N \\
&= -kT \ln P_{cav}(\mathbf{R}_0). \tag{5.10.26}
\end{aligned}$$

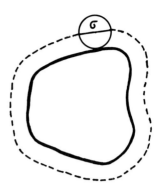

FIGURE 5.20. A cavity formed by a hard nonspherical particle.

In this particular case, ΔA is simply the work required to compress the system from V to $V - V^{\mathrm{EX}}$. The probability of finding a cavity at $\mathbf{R}_0$ is simply

$$P_{\mathrm{cav}}(\mathbf{R}_0) = \left(1 - \frac{V^{\mathrm{EX}}}{V}\right)^N$$

$$\approx 1 - \frac{NV^{\mathrm{EX}}}{V}$$

$$= 1 - \rho V^{\mathrm{EX}}. \tag{5.10.27}$$

The last approximation is valid for any volume V^{EX} provided that it is small compared with the macroscopic volume V; i.e., $V^{\mathrm{EX}}/V \ll 1$.

The above interpretation of P_{cav} becomes invalid if we talk about a cavity (of any form) at *any* point. Clearly the probability of finding a cavity at any point in the system is unity (again we talk about a cavity of microscopic dimensions); there is always some place within V where a small region of volume V^{EX} will be found empty of particles.

As with spherical particles, we can replace the cavity at $\mathbf{R}_0$ by a hard particle of a suitable form (see Fig. 5.20). Such a particle when released has a chemical potential of the form

$$\mu = \mu^* + kT \ln \rho \Lambda^3, \tag{5.10.28}$$

where μ^* is the work required to place the hard particle at some fixed position and orientation. This work will be the same as the work required to form the corresponding cavity [from Eq. (5.10.26)].

5.11. ELEMENTS OF THE SCALED-PARTICLE THEORY (SPT)

In the preceding section we discussd the work required to create a cavity in the liquid. This concept is fundamental in the study of the solvation of solutes in any solvent. The simplest solute is a hard-sphere (HS) particle, and the simple solvent also consists of HS particles. We shall see in section 6.14 that the solvation of any solute in any solvent can always be decomposed into two parts, creating a suitable cavity and then turning on the other parts of the solute–solvent interaction. The scaled-particle theory (SPT) provides an approximate procedure to compute the work required to create a cavity.

Originally, the SPT was devised and used for the study of a hard-sphere (HS) fluid.[5] It was later also found useful and successful for simple fluids such as inert gases in the liquid state. More recently the SPT has also been applied to aqueous solutions. The basic ingredients of the SPT and the nature of the approximation involved are quite simple. We shall present here only a brief outline of the theory, skipping some of the more complicated details.

The starting point of the SPT is the consideration of the work of creating a cavity at some fixed position in the fluid. In a fluid consisting of HS particles of diameter a, a cavity of radius r at $\mathbf{R}_0$ is nothing but a stipulation that no centers of particles may be found in the sphere of radius r centered at $\mathbf{R}_0$. In this sense, creation of a cavity of radius r at $\mathbf{R}_0$ is equivalent to placing at $\mathbf{R}_0$ a HS solute of diameter b such that $r = (a + b)/2$. Note that a HS of zero diameter (or a point HS) produces a cavity of radius $a/2$ in the system. Alternatively, a cavity of radius zero is equivalent to placing a HS of negative diameter $b = -a$. The work required to create such a cavity is equivalent to the work required to introduce a HS solute at $\mathbf{R}_0$. This work is computed by using a continuous process of building up the particle in the solvent. This is the origin of the name "scaled-particle theory."

In a fluid of HS particles, the sole molecular parameter that fully describes the particles is the diameter a.† It is important to bear this fact in mind when the theory is applied to real fluids, in which case one needs at least two molecular parameters to describe the molecules, and more than two parameters for complex molecules such as water. It is a unique feature of the HS fluid that only one molecular parameter is sufficient for its characterization.

The fundamental distribution function in the SPT is $P_0(\mathbf{r})$, the probability that no molecule has its center within the spherical region of radius r centered at some fixed point $\mathbf{R}_0$ in the fluid. Let $P_0(r + dr)$ be the probability that a cavity of radius $r + dr$ is empty. (In all the following, a cavity is always assumed to be centered at some fixed point $\mathbf{R}_0$, but this will not be mentioned explicitly.) This probability may be written as

$$P_0(r + dr) = P_0(r)P_0(dr/r), \tag{5.11.1}$$

where on the rhs of (5.11.1) we have introduced the symbol $P_0(dr/r)$ for the conditional probability of finding the spherical shell of width dr empty, given that the sphere of radius r is empty. The equality (5.11.1) is nothing but the well-known definition of a conditional probability in terms of the joint probability.

We now define an auxiliary function $G(r)$ by the relation

$$4\pi r^2 \rho G(r)\, dr = 1 - P_0(dr/r). \tag{5.11.2}$$

Clearly, since $P_0(dr/r)$ is the conditional probability of finding the spherical shell empty, given that the sphere of radius r is empty, the rhs of (5.11.3) is the conditional probability of finding the center of at least one particle in this spherical shell, given that the sphere of radius r is empty.

Expanding $P_0(r + dr)$ to first order in dr, we get

$$P_0(r + dr) = P_0(r) + \frac{\partial P_0}{\partial r}\, dr + \cdots. \tag{5.11.3}$$

---

† The mass is also a molecular parameter that enters into the theory, but this parameter does not enter in the calculation of the work required to create a cavity.

Hence, from (5.11.1), (5.11.2), and (5.11.3) we obtain

$$\frac{\partial \ln P_0(r)}{\partial r} = -4\pi r^2 \rho G(r). \tag{5.11.4}$$

Thus the function $G(r)$ may be defined either through (5.11.2) or through (5.11.4); the latter may also be rewritten in the integral form

$$\ln P_0(r) - \ln P_0(r = 0) = -\rho \int_0^r 4\pi \lambda^2 G(\lambda)\, d\lambda. \tag{5.11.5}$$

Since the probability of finding a cavity of radius zero is unity (see definition 5.10.5), we get the relation

$$\ln P_0(r) = -\rho \int_0^r 4\pi \lambda^2 G(\lambda)\, d\lambda. \tag{5.11.6}$$

We now recall the relation between $P_0(r)$ and the work $W(r)$ required to form a cavity of radius r—see for instance (5.10.13)—which we write as

$$\begin{aligned}
W(r) &= A(r) - A \\
&= -kT \ln P_0(r) \\
&= kT\rho \int_0^r 4\pi \lambda^2 G(\lambda)\, d\lambda.
\end{aligned} \tag{5.11.7}$$

Since the work required to create a cavity of radius r is the same as the work required to insert a hard sphere of diameter $b = 2r - a$ at $\mathbf{R}_0$, we can write the pseudochemical-potential of the added "solute" in the solvent as

$$\mu_b^* = W(r) = kT\rho \int_0^{(a+b)/2} 4\pi \lambda^2 G(\lambda)\, d\lambda. \tag{5.11.8}$$

In order to get the chemical potential of the solute having diameter b we have to add to (5.11.8) the liberation free energy, namely,

$$\mu_b = \mu_b^* + kT \ln \rho_b \Lambda_b^3. \tag{5.11.9}$$

Note that $\rho_b = 1/V$ is the "solute" density, whereas $\rho = N/V$ is the "solvent" density.

A particular case of (5.11.8) is obtained when we insert a "solute" having a diameter $b = a$, i.e., a solute which is indistinguishable from other particles in the system. In this case we have

$$\mu_a^* = W(r = a) = kT\rho \int_0^a 4\pi \lambda^2 G(\lambda)\, d\lambda. \tag{5.11.10}$$

The integral on the rhs of (5.11.10) describes the work of coupling a new particle to the system using a continuous "charging" parameter λ.

At this stage it is interesting to cite the equation of state for a system of hard spheres of diameter a, namely,

$$\frac{P}{kT} = \rho + \tfrac{2}{3}\pi a^3 \rho^2 G(a); \tag{5.11.11}$$

i.e., the equation of state is determined by the function $G(r)$ at a single point $r = a$. Note that for the chemical potential one needs the entire function $G(\lambda)$, and not just its value at a single point.

The SPT provides an approximate expression for $P_0(r)$ or, equivalently, for $G(\lambda)$. Before presenting this expression we note that an exact expression is available for $P_0(r)$ at very small r. If the diameter of the HS particles is a, then in a sphere of radius $r < a/2$ there can be at most one center of a particle at any given time. Thus for such a small r, the probability of finding the sphere occupied is $4\pi r^3 \rho/3$. Since this sphere may be occupied by at most one center of an HS, the probability of finding it empty is simply

$$P_0(r) = 1 - \frac{4\pi r^3}{3}\rho \qquad \left(\text{for } r \leq \frac{a}{2}\right). \tag{5.11.12}$$

For spheres with a slightly larger radius, namely, for $r \leq a/3^{1/2}$, there can be at most two centers of HSs in it; the corresponding expression for $P_0(r)$ is

$$P_0(r) = 1 - \frac{4\pi r^3}{3}\rho + \frac{\rho^2}{2}\iint_{v(r)} g(\mathbf{R}_1, \mathbf{R}_2)\, d\mathbf{R}_1\, d\mathbf{R}_2, \tag{5.11.13}$$

where $g(\mathbf{R}_1, \mathbf{R}_2)$ is the pair correlation function and the integration is carried out over the region defined by the sphere of radius r. The last equation is valid for a radius smaller than $a/3^{1/2}$. In a formal fashion one can write expressions similar to (5.11.13) for larger cavities, but these involve higher-order molecular distribution functions and therefore are not useful in practical applications.

Clearly the information on $P_0(r)$ in (5.11.12) may be converted to information on $G(r)$. Using relation (5.11.4), we obtain

$$G(r) = \left(1 - \frac{4\pi r^3}{3}\rho\right)^{-1} \qquad \text{for } r \leq \frac{a}{2}, \tag{5.11.14}$$

and for $W(r)$ we have

$$W(r) = -kT\ln\left(1 - \frac{4\pi r^3}{3}\rho\right) \qquad \text{for } r \leq \frac{a}{2}. \tag{5.11.15}$$

We now turn to the other extreme, namely, to very large cavities. In this case the cavity becomes macroscopic and the work required to create it is simply

$$W(r) = Pv(r) \qquad (\text{for } r \to \infty), \tag{5.11.16}$$

where P is the pressure and $v(r)$ is the volume of the cavity.

Another way of obtaining (5.11.16) is to use the basic probability in the grand canonical ensemble. For a very large cavity, we can treat the volume $v(r)$ as the volume of a macroscopic system in the T, v, μ ensemble. The probability of finding the system empty is

$$P_0(r) = \Xi(T, v(r), \mu)^{-1} = \exp\left[\frac{-Pv(r)}{kT}\right], \tag{5.11.17}$$

where Ξ is the grand partition function and the last equality holds for macroscopic systems. Using relation (5.11.7), we obtain

$$W(r) = -kT \ln P_0(r) = Pv(r) \qquad (r \to \infty),$$

which is the same as (5.11.16) obtained from purely thermodynamic considerations.

Equation (5.11.16) is the leading term for a macroscopic volume, i.e., $r \to \infty$. For large cavities one may include a term proportional to $v^{2/3}$ to account for the work required to create the surface area, in which case Eq. (5.11.16) is modified as

$$W(r) = P\tfrac{4}{3}\pi r^3 + 4\pi r^2 \sigma_0, \tag{5.11.18}$$

where σ_0 is the surface tension between the fluid and a hard wall. For a still smaller radius, a correction due to the curvature of the cavity may be introduced into (5.11.18). Also from relation (5.11.4) we may obtain the limit of $G(r)$ at $r \to \infty$:

$$G(r \to \infty) = \frac{P}{kT\rho}. \tag{5.11.19}$$

At this stage we have two exact results for $G(r)$; one for very small r, (5.11.14), and one for very large r, (5.11.19). This information suggests trying to bridge the two ends by a smooth function of r. In fact this was precisely the procedure taken by Reiss, Frisch and Lebowitz.[5] The arguments used by the authors to make a particular choice of such a smooth function are quite lengthy and involved. They assumed that $G(r)$ is a monotonic function of r in the entire range of r. They suggested a trial function of the form

$$G(r) = A + Br^{-1} + Cr^{-2}. \tag{5.11.20}$$

The coefficients A, B, and C were determined by the use of all the available information on the behavior of the function $G(r)$ for a fluid of hard spheres.

The final expression obtained for $G(r)$, after being translated into $W(r)$ [i.e., integration of relation (5.11.7)] is the following:

$$W(r) = K_0 + K_1 r + K_2 r^2 + K_3 r^3, \tag{5.11.21}$$

with the coefficients given by

$$K_0 = kT[-\ln(1-y) + 4.5z^2] - \tfrac{1}{6}\pi Pa^3$$

$$K_1 = -\left(\frac{kT}{a}\right)(6z + 18z^2) + \pi Pa^2$$

$$K_2 = \left(\frac{kT}{a^2}\right)(12z + 18z^2) - 2\pi Pa \tag{5.11.22}$$

$$K_3 = \frac{4\pi P}{3},$$

where a is the diameter of the hard spheres and y and z are defined by

$$y = \frac{\pi\rho a^3}{6}, \qquad z = \frac{y}{(1-y)}. \tag{5.11.23}$$

Thus, in essence, what we have obtained is an approximate expression for the work required to create a cavity of radius r in a fluid of HSs characterized by the diameter a.

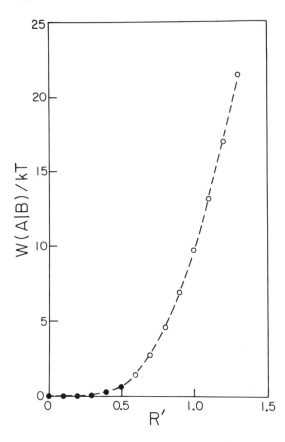

FIGURE 5.21. $W(r')/kT$ as a function of the reduced radius of the cavity.

Figure 5.21 shows $W(r')/kT$ as a function of the reduced radius of the cavity $r' = r/a$, where a is the diameter of the hard spheres. Equation (5.11.15) was used up to $r' = \frac{1}{2}$, and Eq. (5.11.21) for $r' > \frac{1}{2}$. Note that the monotonic increasing function $W(r')$ is a general property of this function, independent of the assumptions of the SPT (see section 5.10).

Using this expression for the particular choice of $r = a$, we obtain an expression for the chemical potential:

$$\mu = kT \ln \rho \Lambda^3 + W(r = a). \qquad (5.11.24)$$

Note that at the point $r = a$, the "solute" becomes identical to the solvent molecules and therefore it is assimilated into the system; i.e., $\rho_b = 1/V$ in (5.11.9) turns into $\rho = N/V$ in (5.11.24).

Relation (5.11.21) is the main result of the SPT that is of interest to us. Strictly, this relation was derived for a fluid consisting of HS particles. We now examine the applicability of this theory to real fluids such as liquid water.

Before doing that, there is one important comment regarding the SPT that should be borne in mind. The computation of the chemical potential (5.11.24) differs from the traditional process in statistical mechanics. Namely, if we work in the T, V, N ensemble, then a purely molecular theory of a fluid should, in principle, provide all the thermodynamic quantities of the system as a function of T, V, and N and of the molecular diameter

a (in the case of HSs). In particular, one should be able to compute both the chemical potential μ and the pressure P in terms of these variables. However, in the above procedure we have expressed the chemical potential μ in terms of T, ρ, and a (ρ replaces V and N, since we are concerned with intensive quantities). In addition we need to use the pressure [in (5.11.21)] as an input quantity. This quantity should, in principle, be an output of the theory in the T, V, N ensemble, i.e., through the relation $P = -\partial A/\partial V$.

The same comment applies if we use the T, P, N ensemble. Here one should be able to compute both the chemical potential (through the derivative $\mu = \partial G/\partial N$) and the average density [through the relation $\rho^{-1} = (\partial \mu/\partial P)_T$]. But in (5.11.21) and (5.11.24) we have expressed μ as a function of T and P, and in addition we need to use ρ as an input, rather than an output, parameter.

In the above sense the SPT is not a purely molecular theory of fluids. This comment should be borne in mind when the theory is applied to complex fluids. In any real liquid, and certainly for water, we need a few molecular parameters to characterize the molecules, say ε and a in a Lennard–Jones fluid, or, in general, a set of molecular parameters $a, b, c, \ldots$. Thus, a proper statistical-mechanical theory of water should provide us with the Gibbs energy as a function of T, P, N and the molecular parameters, $a, b, c, \ldots$, i.e., a function of the form $G(T, P, N; a, b, c, \ldots)$. Instead the SPT makes use of only one molecular parameter, the diameter a. No provision for incorporating other molecular parameters is offered by the theory. This deficiency in the characterization of the molecules is partially compensated for by the use of the measurable density ρ as an input parameter.

Water is currently referred to as being a "structural liquid" (in whichever sense we choose to use this term; for one particular definition, see Chapter 7). One of the main goals in the study of liquid water is the understanding of the role of this peculiar "structure" in the determination of the outstanding properties of this liquid.

Clearly this goal may not be achieved through the use of the SPT. The very fact that we "use" the density of the liquid is equivalent to introducing "structural" information into the theory. Therefore, even if we find that the SPT is successful in predicting some thermodynamic quantities, it cannot be used to explain them on a molecular level. Thus the apparent success of the SPT, even when applied to complex fluids, is not entirely surprising. After all, injecting one macroscopic quantity into the theory is likely to produce other thermodynamic quantities that are at least consistent with the input information. For the computation of entropies, enthalpies, etc., one must also use the temperature dependence of the molecular diameter of the solvent. This quantity is also determined in such a way that the results are consistent with some measurable macroscopic quantity. In this way we further supply the theory with parameters which carry "structural" information on our particular system.

In Chapters 7 and 8 we shall use the general expression (5.11.21) to estimate the work required to create a cavity of radius r in a real solvent. It should be borne in mind that this is an approximate expression even for a hard-sphere fluid. It is certainly a very rough estimate in case of a real fluid. Nevertheless, at present, this is the best method available to compute such quantities.

5.12. PERTURBATION THEORIES OF LIQUIDS

In this section, we present another demonstration of the application of the general theorem of section 5.8.1 to obtain a first-order term in a perturbation expansion of the

Helmholtz energy. Consider a system in the T, V, N ensemble obeying the pairwise additivity assumption for the total potential energy:

$$U(\mathbf{X}^N) = \tfrac{1}{2} \sum_{i \neq j} U(\mathbf{X}_i, \mathbf{X}_j). \tag{5.12.1}$$

Now suppose we can separate the pair potential into two parts,

$$U(\mathbf{X}_i, \mathbf{X}_j) = U^0(\mathbf{X}_i, \mathbf{X}_j) + U^1(\mathbf{X}_i, \mathbf{X}_j). \tag{5.12.2}$$

Hence,

$$U(\mathbf{X}^N) = U^0(\mathbf{X}^N) + U^1(\mathbf{X}^N). \tag{5.12.3}$$

$U^0(\mathbf{X}^N)$ is referred to as the total potential energy of the unperturbed system, whereas $U^1(\mathbf{X}^N)$ is considered to be the perturbation energy. The basic distribution function in the unperturbed system is

$$P^0(\mathbf{X}^N) = \frac{\exp[-\beta U_N^0(\mathbf{X}^N)]}{\displaystyle\int \cdots \int d\mathbf{X}^N \exp[-\beta U_N^0(\mathbf{X}^N)]}, \tag{5.12.4}$$

and the Helmholtz energy of the perturbed and unperturbed systems are given by

$$\exp[-\beta A(T, V, N)] = \frac{q^N}{\Lambda^{3N} N!} \int \cdots \int d\mathbf{X}^N \exp[-\beta U_N(\mathbf{X}^N)] \tag{5.12.5}$$

$$\exp[-\beta A^0(T, V, N)] = \frac{q^N}{\Lambda^{3N} N!} \int \cdots \int d\mathbf{X}^N \exp[-\beta U_N^0(\mathbf{X}^N)]. \tag{5.12.6}$$

Defining the difference

$$A^1(T, V, N) = A(T, V, N) - A^0(T, V, N), \tag{5.12.7}$$

and using relations (5.12.3)–(5.12.6) we get

$$\exp[-\beta A^1(T, V, N)] = \frac{\displaystyle\int \cdots \int d\mathbf{X}^N \exp[-\beta U_N^0(\mathbf{X}^N) - \beta U_N^1(\mathbf{X}^N)]}{\displaystyle\int \cdots \int d\mathbf{X}^N \exp[-\beta U_N^0(\mathbf{X}^N)]}$$

$$= \int \cdots \int d\mathbf{X}^N P^0(\mathbf{X}^N) \exp[-\beta U_N^1(\mathbf{X}^N)]$$

$$= \langle \exp[-\beta U_N^1(\mathbf{X}^N)] \rangle_0. \tag{5.12.8}$$

The symbol $\langle \; \rangle_0$ stands for an average over the unperturbed system (here, in the T, V, N ensemble). Clearly, even when we assume the pairwise additivity (5.12.1), the average in (5.12.8) is not an average of a pairwise quantity.

Up to this point, the formal relation (5.12.8) was general and it applies to any arbitrary division of the pair potential in two parts in (5.12.2). We now choose a more specific division of the pair potential in such a way that $|\beta U^1(\mathbf{X}_i, \mathbf{X}_j)| \ll 1$. The most successful procedure is to use a hard-sphere cutoff at some $R = \sigma$, where σ is the distance below which the interaction becomes steeply repulsive. We can then define the HS part of the interaction by

$$U^{\mathrm{HS}}(R_i) = \begin{cases} \infty & R \leq \sigma \\ 0 & R > \sigma, \end{cases} \qquad (5.12.9)$$

and then define $U^1(\mathbf{X}_i, \mathbf{X}_j)$ simply by the difference

$$U^1(\mathbf{X}_i, \mathbf{X}_j) = U(\mathbf{X}_i, \mathbf{X}_j) - U^{\mathrm{HS}}(R_{ij}). \qquad (5.12.10)$$

If the reference potential U^{HS} has been chosen in such a way that $|\beta U^1| \ll 1$, then one may try to use a first-order approximation for

$$\exp[-\beta U^1(\mathbf{X}_i, \mathbf{X}_j)] = 1 - \beta U^1(\mathbf{X}_i, \mathbf{X}_j) + \cdots. \qquad (5.12.11)$$

And therefore (5.12.8) is also approximated by

$$\exp[-\beta U_N^1(\mathbf{X}^N)] = 1 - \beta U_N^1(\mathbf{X}^N) + \cdots. \qquad (5.12.12)$$

For the present purposes, we assume that $|\beta U_N^1(\mathbf{X}^N)| \ll 1$ for all possible configurations $\mathbf{X}^N$, so that expansion up to the first order as in (5.12.12) is justified. We now have

$$\langle \exp[-\beta U_N^1(\mathbf{X}^N)] \rangle_0 = 1 - \beta \langle U_N^1(\mathbf{X}_N) \rangle_0 + \cdots, \qquad (5.12.13)$$

where the average on the rhs of (5.12.13) is over a pairwise quantity. Hence, using (5.12.1) and the general theorem of section 5.8.1, we get

$$\langle \exp[-\beta U_N^1(\mathbf{X}^N)] \rangle = 1 - \tfrac{1}{2}\beta \int d\mathbf{X}_1 \int d\mathbf{X}_2 \, U^1(\mathbf{X}_1, \mathbf{X}_2)\rho^{0(2)}(\mathbf{X}_1, \mathbf{X}_2), \qquad (5.12.14)$$

where $\rho^{0(2)}(\mathbf{X}_1, \mathbf{X}_2)$ is the pair distribution function for the unperturbed system. Thus, for the Helmholtz energy of the system, we have

$$\begin{aligned} A(T, V, N) &= A^0(T, V, N) + A^1(T, V, N) \\ &= A^0(T, V, N) - kT \ln\langle \exp[-\beta U_N^1(\mathbf{X}^N)] \rangle_0 \\ &= A^0(T, V, N) - kT \ln[1 - \beta \langle U_N^1(\mathbf{X}^N) \rangle_0 + \cdots] \\ &= A^0(T, V, N) + \langle U_N^1(\mathbf{X}^N) \rangle_0 + \cdots \\ &\approx A^0(T, V, N) + \tfrac{1}{2} \int d\mathbf{X}_1 \int d\mathbf{X}_2 \, U^1(\mathbf{X}_1, \mathbf{X}_2)\rho^{0(2)}(\mathbf{X}_1, \mathbf{X}_2). \end{aligned} \qquad (5.12.15)$$

The last relation can be simplified for spherical particles as

$$A(T, V, N) \approx A^0(T, V, N) + \tfrac{1}{2}N\rho \int_0^\infty U^1(R)g^0(R)4\pi R^2 \, dR, \qquad (5.12.16)$$

where $g^0(R)$ is the pair correlation function for the unperturbed system. Relation (5.12.16) is useful whenever we know the Helmholtz energy of the unperturbed system and when the perturbation energy is small compared with kT. It is clear that if we take more terms in the expansion (5.12.12), we end up with integrals involving higher-order molecular distribution functions.

5.13. GENERALIZED MOLECULAR DISTRIBUTION FUNCTIONS[6]

In this section we generalize the concept of molecular distribution to include properties other than the locations and orientations of the particles. We shall mainly focus on the singlet generalized molecular distribution function (MDF), which provides a firm basis for the so-called mixture model approach to liquids. The latter has been used extensively for complex liquids such as water and aqueous solutions.

The general procedure of defining the generalized MDF is the following. We recall the general definition of the nth-order MDF, say in the T, V, N ensemble, which we write in the following two equivalent forms:

$$\rho^{(n)}(\mathbf{S}_1, \ldots, \mathbf{S}_n) = \frac{N!}{(N-n)!} \left(\int \cdots \int d\mathbf{R}_{n+1} \cdots d\mathbf{R}_N \right) P(\mathbf{S}_1, \ldots, \mathbf{S}_n; \mathbf{R}_{n+1}, \ldots, \mathbf{R}_N)$$

$$= \sum_{i_1=1}^{N} \cdots \sum_{\substack{i_n=1 \\ i_1 \neq i_2 \cdots \neq i_n}}^{N} \int \cdots \int d\mathbf{R}^N P(\mathbf{R}^N)[\delta(\mathbf{R}_{i_1} - \mathbf{S}_1) \cdots \delta(\mathbf{R}_{i_n} - \mathbf{S}_n)].$$

$$(5.13.1)$$

Here, $P(\mathbf{R}^N)$ is the basic probability density in the T, V, N ensemble. In the first form on the rhs of (5.13.1) we have made the distinction between fixed variables $\mathbf{S}_1, \ldots, \mathbf{S}_n$ and dummy variables $\mathbf{R}_{n+1}, \ldots, \mathbf{R}_N$ which undergo integration. The second form on the rhs of (5.13.1) is more suitable for pursuing the required generalization. We first recognize that the brackets in the integrand comprise a stipulation on the range of integration, i.e., they serve to extract from the entire configurational space only those configurations (or regions) for which the vector $\mathbf{R}_{i_1}$ attains the value $\mathbf{S}_1, \ldots$ and the vector $\mathbf{R}_{i_n}$ attains the value $\mathbf{S}_n$. Alternatively, we can break up the above statement into the following logical ingredients: We choose a property; here, the property is, "the location of particle i." We then impose a condition on this property, which in this case reads, "the location of particle i is $\mathbf{S}_i$." Next, we combine several of these conditions applied to n particles and arrive at the content of the square brackets in the integral in (5.13.1).

The generalization can be carried out on both the property and the condition imposed on it. The distinction between property and condition is arbitrary and is made for convenience only. In fact, the generalization procedure involves only one concept. This will be demonstrated, along with a few examples, in the next few sections.

5.13.1. The Singlet Generalized Molecular Distribution Function

In this section, we present a special case of the generalization procedure outlined in the previous section. We also establish new notation that will be useful for later applications. Consider the ordinary singlet MDF:

$$N_L^{(1)}(\mathbf{S}_1) \, d\mathbf{S}_1 = d\mathbf{S}_1 \int \cdots \int d\mathbf{R}^N P(\mathbf{R}^N) \sum_{i=1}^{N} \delta(\mathbf{R}_i - \mathbf{S}_1)$$

$$= N \, d\mathbf{S}_1 \int \cdots \int d\mathbf{R}^N P(\mathbf{R}^N)\delta(\mathbf{R}_1 - \mathbf{S}_1). \qquad (5.13.2)$$

Here, $N_L^{(1)}(\mathbf{S}_1)\, d\mathbf{S}_1$ is the average number† of particles occupying the element of volume $d\mathbf{S}_1$. For the present treatment, we limit our discussion to spherical molecules only. As we have already stressed in section 5.2, the quantity defined in (5.13.2) can be assigned two different meanings. The first follows from the first form on the rhs of (5.13.2), which is identified as an average quantity in the T, V, N ensemble. The second form on the rhs of (5.13.2) provides the probability of finding particle 1 in the element of volume $d\mathbf{S}_1$. Clearly, this probability is given by $N_L^{(1)}(\mathbf{S}_1)\, d\mathbf{S}_1/N$.

In order to systematize the procedure of generalization, let us rewrite (5.13.2) in a yet somewhat more complicated way. For each configuration $\mathbf{R}^N$, we define the property of the particle i as

$$L_i(\mathbf{R}^N) = \mathbf{R}_i. \tag{5.13.3}$$

The property of particle i defined in (5.13.3) is simply its location $\mathbf{R}_i$. This is the reason for using the letter L in the definition of the function $L_i(\mathbf{R}^N)$ and as a subscript in (5.13.2).

Next, we define the counting function of the property L by

$$N_L^{(1)}(\mathbf{R}^N, \mathbf{S}_1)\, d\mathbf{S}_1 = \sum_{i=1}^{N} \delta[L_i(\mathbf{R}^N) - \mathbf{S}_1]\, d\mathbf{S}_1, \tag{5.13.4}$$

which is the number of particles whose property L attains a value within $d\mathbf{S}_1$ at $\mathbf{S}_1$, given the specific configuration $\mathbf{R}^N$. The average number (here in the T, V, N ensemble) of such particles is

$$
\begin{aligned}
N_L^{(1)}(\mathbf{S}_1)\, d\mathbf{S}_1 &= \langle N_L^{(1)}(\mathbf{R}^N, \mathbf{S}_1)\rangle\, d\mathbf{S}_1 \\
&= d\mathbf{S}_1 \int \cdots \int d\mathbf{R}^N\, P(\mathbf{R}^N) \sum_{i=1}^{N} \delta[L_i(\mathbf{R}^N) - \mathbf{S}_1],
\end{aligned}
\tag{5.13.5}
$$

which is the same as (5.13.2). For the same property L, we can distinguish among various conditions. For instance, we may count all the particles whose property L has the value within a given region D, instead of $d\mathbf{S}_1$. However, the latter can be obtained by a simple integration (or summation, in discrete cases) over the appropriate region. Therefore, the generalization procedure involves essentially the property of the particles under consideration.

We now illustrate a few examples of properties that may replace L in (5.13.4) and (5.13.5).

5.13.2. Coordination Number

A simple property which has been the subject of many investigations is the coordination number (CN). We recall that the average coordination number can be obtained from the pair distribution function (section 5.3). Here, we are interested in more detailed information on the distribution of CNs.

Let R_C be a fixed number, to serve as the radius of the first coordination shell. If σ is the effective diameter of the particles of the system, a reasonable choice of R_C for our purposes is $\sigma \le R_C \le 1.5\sigma$. The range given above seems to be in conformity with the current concept of the radius of the first coordination sphere around a given particle. In what follows, we assume that R_C has been fixed, and we omit it from the notation.

---

† We use the letter N rather than ρ for density of particles. This is done in order to unify the system of notation for the continuous as well as discrete cases that are treated in this section.

The property to be considered here is the CN of the particle i at a given configuration $\mathbf{R}^N$. This is defined by

$$C_i(\mathbf{R}^N) = \sum_{j=1, j \neq i}^{N} H(|\mathbf{R}_j - \mathbf{R}_i| - R_C), \tag{5.13.6}$$

where $H(x)$ is a unit step function

$$H(x) = \begin{cases} 0 & \text{if } x > 0 \\ 1 & \text{if } x \leq 0. \end{cases} \tag{5.13.7}$$

Each term in (5.13.6) contributes unity whenever $|\mathbf{R}_j - \mathbf{R}_i| < R_C$, i.e., whenever the center of particle j falls within the coordination sphere of particle i. Hence, $C_i(\mathbf{R}^N)$ is the number of particles ($j \neq i$) that falls in the coordination sphere of particle i for a given configuration $\mathbf{R}^N$. Next, we define the counting function for this property by

$$N_C^{(1)}(\mathbf{R}^N, K) = \sum_{i=1}^{N} \delta[C_i(\mathbf{R}^N) - K]. \tag{5.13.8}$$

Here, we have used the Kronecker delta function $\delta(x - K)$ instead of the more common notation $\delta_{x,K}$ for the sake of unity of notation. The meaning of δ as a Dirac or Kronecker delta should be clear from the context. In the sum of (5.13.8) we scan all the particles ($i = 1, 2, \ldots, N$) of the system in a given configuration $\mathbf{R}^N$. Each particle whose CN is exactly K contributes unity to the sum (5.13.8), and zero otherwise. Hence, the sum in (5.13.8) counts all particles whose CN is K for the particular configuration $\mathbf{R}^N$. A schematic illustration of this counting procedure is shown in Fig. 5.22. The average number of

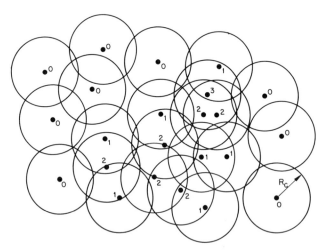

FIGURE 5.22. A two-dimensional example of counting the coordination number of each molecule for a specific configuration of the whole system. A sphere of radius R_C is drawn about the center of each molecule, indicated by the dark circles. The number written near each center is the coordination number corresponding to this particular example. There are altogether 23 particles distributed in the following manner: $N(0) = 9$, $N(1) = 7$, $N(2) = 6$, $N(3) = 1$ (for the configuration shown here).

such particles is

$$N_C^{(1)}(K) = \langle N_C^{(1)}(\mathbf{R}^N, K) \rangle$$

$$= N \int \cdots \int d\mathbf{R}^N \, P(\mathbf{R}^N) \delta[C_1(\mathbf{R}^N) - K]. \qquad (5.13.9)$$

We define the following quantity:

$$x_C(K) = \frac{N_C^{(1)}(K)}{N}$$

$$= \int \cdots \int d\mathbf{R}^N \, P(\mathbf{R}^N) \delta[C_1(\mathbf{R}^N) - K]. \qquad (5.13.10)$$

From the definition of $N_C^{(1)}(K)$ in (5.13.9), it follows that $x_C(K)$ is the mole fraction of particles whose coordination number is equal to K. On the other hand, the second form on the rhs of (5.13.10) provides the probabilistic meaning of $x_C(K)$; i.e., this is the probability that a specific particle, say 1, will be found with CN equal to K.

The quantity $x_C(K)$ can be viewed as a component of a vector

$$\mathbf{x}_C = (x_C(0), x_C(1), \dots), \qquad (5.13.11)$$

which gives the "composition" of the system with respect to the classification according to the CNs; i.e., each component gives the mole fraction of particles with a given CN. The average CN of particles in the system is given by†

$$\langle K \rangle = \sum_{K=0}^{\infty} K x_C(K). \qquad (5.13.12)$$

We also use this example to demonstrate that changes in the condition can be achieved easily. For instance, with the same property (CN), we may ask for the average number of particles whose CN is less than or equal to, say, five. This is obtained from (5.13.9):

$$N_C^{(1)}(K \leq 5) = \sum_{K=0}^{5} N_C^{(1)}(K). \qquad (5.13.13)$$

Therefore, there is no need to treat separately the generalization procedure for the conditions imposed on a given property.

The CN, as defined above, may be viewed as a property conveying the local density around the particles. By a simple transformation, we can make this statement more precise. The local density around particle i for a given configuration $\mathbf{R}^N$ is defined by

$$D_i(\mathbf{R}^N) = C_i(\mathbf{R}^N)(\tfrac{4}{3}\pi R_C^3)^{-1}. \qquad (5.13.14)$$

Hence, the average number of particles having local density equal to η is

$$N_D^{(1)}(\eta) = \left\langle \sum_{i=1}^{N} \delta[D_i(\mathbf{R}^N) - \eta] \right\rangle. \qquad (5.13.15)$$

---

† Note that $\langle K \rangle$, as defined in (5.13.12), coincides with the definition of the average CN given in section 5.3 provided that we choose $\mathbf{R}_C$ of this section to coincide with R_M in section 5.3.

Another quantity conveying a similar meaning will be introduced later, in section 5.13.4.

5.13.3. Binding Energy

The next property is referred to as binding energy (BE) and is defined for particle i and for the configuration $\mathbf{R}^N$ as follows:

$$B_i(\mathbf{R}^N) = U_N(\mathbf{R}_1, \ldots, \mathbf{R}_{i-1}, \mathbf{R}_i, \mathbf{R}_{i+1}, \ldots, \mathbf{R}_N)$$
$$- U_{N-1}(\mathbf{R}_1, \ldots, \mathbf{R}_{i-1}, \mathbf{R}_{i+1}, \ldots, \mathbf{R}_N). \tag{5.13.16}$$

This is the work required to bring a particle from an infinite distance, with respect to the other particles, to the position $\mathbf{R}_i$. For a system of pairwise additive potentials, (5.13.16) reduces to

$$B_i(\mathbf{R}^N) = \sum_{j=1, j \neq i}^{N} U(\mathbf{R}_i, \mathbf{R}_j). \tag{5.13.17}$$

The counting function for this property is

$$N_B^{(1)}(\mathbf{R}^N, v) \, dv = dv \sum_{i=1}^{N} \delta[B_i(\mathbf{R}^N) - v], \tag{5.13.18}$$

which is the number of particles having BE between v and $v + dv$ for the specified configuration $\mathbf{R}^N$. Note that since v is a continuous variable, the δ-function in (5.13.18) is the Dirac delta function. The average number of particles having BE between v and $v + dv$ is

$$N_B^{(1)}(v) \, dv = dv \left\langle \sum_{i=1}^{N} \delta[B_i(\mathbf{R}^N) - v] \right\rangle. \tag{5.13.19}$$

The corresponding mole fraction is

$$x_B(v) \, dv = \frac{N_B^{(1)}(v) \, dv}{N}, \tag{5.13.20}$$

with the normalization condition

$$\int_{-\infty}^{\infty} x_B(v) \, dv = 1. \tag{5.13.21}$$

The function $x_B(v)$ is referred to as the *distribution* of BE. By analogy with the vector (5.13.11), which has discrete components, we often write $\mathbf{x}_B$ for the whole distribution function the components of which are $x_B(v)$.

5.13.4. Volume of the Voronoi Polyhedron

Another local property of interest in the study of liquids is the Voronoi polyhedron (VP), or the Dirichlet region, defined as follows: Consider a specific configuration $\mathbf{R}^N$ and a particular particle i. Let us draw all the segments l_{ij} ($j = 1, \ldots, N, j \neq i$) connecting the centers of particles i and j. Let P_{ij} be the plane perpendicular to and bisecting the line l_{ij}. Each plane P_{ij} divides the entire space into two parts. Denote by V_{ij} the part of space that includes the point $\mathbf{R}_i$. The VP of particle i for the configuration $\mathbf{R}^N$ is defined as the intersection of all the V_{ij} ($j = 1, \ldots, N, j \neq i$):

$$(\text{VP})_i = \bigcap_{j = 1, j \neq i}^{N} V_{ij}(\mathbf{R}_i, \mathbf{R}_j). \tag{5.13.22}$$

A two-dimensional illustration of a construction of a VP is shown in Fig. 5.23. It is clear from the definition that the region $(\text{VP})_i$ includes all the points in space that are nearer to $\mathbf{R}_i$ than to any $\mathbf{R}_j$ ($j \neq i$). Furthermore, each VP contains the center of one and only one particle.

The concept of VP can be used to generate a few local properties;[†] the one we shall be using is the volume of the VP, which we denote by

$$\psi_i(\mathbf{R}^N) = \text{volume of } (\text{VP})_i. \tag{5.13.23}$$

The counting function for this property is

$$N_\psi^{(1)}(\mathbf{R}^N, \phi) \, d\phi = d\phi \sum_{i=1}^{N} \delta[\psi_i(\mathbf{R}^N) - \phi], \tag{5.13.24}$$

and its average is

$$N_\psi^{(1)}(\phi) \, d\phi = d\phi \left\langle \sum_{i=1}^{N} \delta[\psi_i(\mathbf{R}^N) - \phi] \right\rangle. \tag{5.13.25}$$

Clearly, $N_\psi^{(1)}(\phi) \, d\phi$ is the average number of particles whose VP has a volume between ϕ and $\phi + d\phi$.

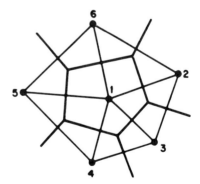

FIGURE 5.23. Construction of the Voronoi polygon of particle 1 in a two-dimensional system of particles.

[†] Note that the form of the VP is also a property which can be considered in the context of this section. Other properties of interest are the number of faces of the VP, the surface area of the VP, etc. The distribution functions defined in this section involve random variables whose values are real numbers. If we choose the form of the VP as a random variable, then its range of variation is the space of geometric figures and not real numbers.

5.13.5. Combination of Properties

One way of generating new properties from previous ones is by combination. For instance, the counting function for BE and the volume of the VP is

$$N_{B,\psi}^{(1)}(\mathbf{R}^N, v, \phi) \, dv \, d\phi = dv \, d\phi \sum_{i=1}^{N} \delta[B_i(\mathbf{R}^N) - v] \delta[\psi_i(\mathbf{R}^N) - \phi], \quad (5.13.26)$$

which counts the number of particles having BE between v and $v + dv$ and the volume of the VP between ϕ and $\phi + d\phi$. The average number of such particles is

$$N_{B,\psi}^{(1)}(v, \phi) \, dv \, d\phi = dv \, d\phi \left\langle \sum_{i=1}^{N} \delta[B_i(\mathbf{R}^N) - v] \delta[\psi_i(\mathbf{R}^N) - \phi] \right\rangle. \quad (5.13.27)$$

Note that although we have combined two properties, we still have a singlet generalized MDF. A related singlet generalized MDF which conveys similar information to the one in (5.13.27) but is simpler for computational purposes, is constructed by the combination of BE and CN, i.e.,

$$N_{B,C}^{(1)}(v, K) \, dv = dv \left\langle \sum_{i=1}^{N} \delta[B_i(\mathbf{R}^N) - v] \delta[C_i(\mathbf{R}^N) - K] \right\rangle. \quad (5.13.28)$$

Note that the first δ on the rhs of (5.13.28) is a Dirac delta function, whereas the second is a Kronecker delta function.

The general procedure of defining generalized MDFs is now clear. We first define a property which is a function definable on the configurational space, say $G_i(\mathbf{R}^N)$, and then introduce its distribution function in the appropriate ensemble.

5.13.6. Some Examples

In this section, we present some examples of singlet generalized MDFs. We confine ourselves to spherical particles in two dimensions. All illustrations given in this section were obtained by a Monte Carlo computation on a two-dimensional system consisting of 36 Lennard–Jones particles for which the pair potential is presumed to have the form

$$U(R) = 4\varepsilon[(\sigma/R)^{12} - (\sigma/R)^6]. \quad (5.13.29)$$

Figure 5.24 shows the distribution of CN, $x_c(K)$, and its density dependence. The Lennard–Jones parameters for these computations are

$$\sigma = 1.0, \qquad \frac{\varepsilon}{kT} = 0.5. \quad (5.13.30)$$

We use σ as our length unit, and hence the densities refer to particles per unit area σ^2.

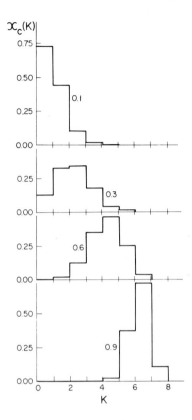

FIGURE 5.24. The singlet functions $x_C(K)$ for spherical particles interacting via a Lennard–Jones potential ($\sigma = 1.0$ and $\varepsilon/kT = 0.5$). The number density is indicated next to each curve.

The coordination radius is chosen as

$$R_C = 1.5\sigma. \tag{5.13.31}$$

These functions behave according to our intuitive expectations. As the density increases, the mole fraction of particles with relatively higher CN increases. Figure 5.25 shows the dependence of $x_B(v)$ on the density. The most pronounced feature of all these curves is the essentially single peak.

Suppose that the intermolecular potential for simple particles can be represented by

$$U(R) = U^{\mathrm{HS}}(R) + U^{\mathrm{SI}}(R), \tag{5.13.32}$$

where the soft interaction (SI) is viewed as a perturbation to the hard-sphere potential function. Clearly, if the perturbation is small, the properties of the system will be dominated by the hard-sphere potential. In particular, we expect that $x_B(v)$ will have essentially a single sharp peak. The average value of v is given by

$$\langle v \rangle = 2 \frac{\langle U_N(\mathbf{R}^N) \rangle}{N}. \tag{5.13.33}$$

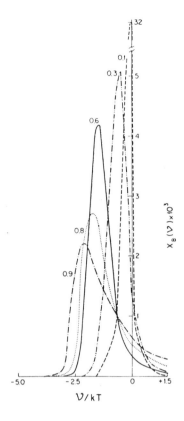

FIGURE 5.25. The singlet functions $x_B(v)$ as a function of density (indicated next to each curve) for spherical particles with parameters $\sigma = 1.0$, $\varepsilon/kT = 0.5$.

This equality can be shown to follow directly from definitions (5.13.19) and (5.13.20); indeed,

$$\langle v \rangle = \int_{-\infty}^{\infty} v x_B(v) \, dv$$

$$= \frac{1}{N} \int_{-\infty}^{\infty} v \, dv \int \cdots \int d\mathbf{R}^N \, P(\mathbf{R}^N) \sum_{i=1}^{N} \delta[B_i(\mathbf{R}^N) - v]$$

$$= \frac{1}{N} \int \cdots \int d\mathbf{R}^N \, P(\mathbf{R}^N) \sum_{i=1}^{N} \int_{-\infty}^{\infty} v \, dv \, \delta[B_i(\mathbf{R}^N) - v]$$

$$= \frac{1}{N} \int \cdots \int d\mathbf{R}^N \, P(\mathbf{R}^N) \sum_{i=1}^{N} B_i(\mathbf{R}^N)$$

$$= \frac{2}{N} \int \cdots \int d\mathbf{R}^N \, P(\mathbf{R}^N) U_N(\mathbf{R}^N)$$

$$= \frac{2\langle U_N(\mathbf{R}^N) \rangle}{N}. \tag{5.13.34}$$

5.13.7. The Mixture Model Approach to Liquids

The general idea of a mixture model (MM) approach is very simple and quite old. In particular, it has been applied extensively in the theory of water and aqueous solutions. In this section, we outline the basic procedure employed in the MM approach.[6]

Consider, for instance, the singlet generalized MDF, based on the coordination number (5.13.10):

$$x_C(K) = \frac{N_C^{(1)}(K)}{N} = \int \cdots \int d\mathbf{R}^N P(\mathbf{R}^N) \delta[C_1(\mathbf{R}^N) - K]. \qquad (5.13.35)$$

This has been interpreted in two ways. First, this is the mole fraction of particles having CN equal to K. Second, $x_C(K)$ is the probability that a specific particle, say 1 will be found with a CN equal to K.

Because of the first interpretation, we may refer to the vector

$$\mathbf{x}_C = (x_C(0), x_C(1), \dots) \qquad (5.13.36)$$

as a quasicomponent distribution function (QCDF). It gives the composition of the system when viewed as a mixture of quasicomponents. The normalization condition for $\mathbf{x}_C$ is

$$\sum_{K=0}^{\infty} x_C(K) = 1. \qquad (5.13.37)$$

Note, however, that since the particles exert strong repulsive forces when brought very close from each other, the occurrence of a large CN for a given particle must be a very improbable event. Therefore the sum over K in (5.13.37) effectively extends from $K = 0$ to, say, $K \approx 12$ (depending of course, on the choice of R_C); i.e., the mole fractions $x_C(K)$, with $K > 12$, will in general be negligible.

A mixture of quasicomponents must be distinguished from a mixture of real components in essentially two respects. First, the quasicomponents do not differ in their chemical composition or in their structure; they are characterized by the nature of their local environment, which, in the above example, is the CN. A more important difference is that a system of quasicomponents cannot be "prepared" in any desired composition; i.e., the components of the vector $\mathbf{x}_C$ cannot be chosen at will. One consequence of this restriction is that quasicomponents have no existence in the pure state.

There exists a certain analogy between a mixture of quasicomponents and a mixture of chemically reacting species. For the sake of simplicity, we recall the case of isomerization reaction, treated in section 2.4. We have stressed there that the distinction between the two species A and B was based on an (arbitrary) classification of all the states of the molecules into two groups. If the classification is such that transitions between the two groups of states is very slow compared with transitions within each group, then it might be possible to isolate the species A and B as "pure" components. This normally involves the introduction of an inhibitor to the reaction $A \rightleftarrows B$. However, whether such an inhibitor actually exists or not, it is of no importance for the formal theory of chemical equilibrium. Therefore, we can use the classification into quasicomponents to view the one-component system as if it were a mixture of species in chemical equilibrium.

In the above example, we elaborated on a classification procedure based on a discrete parameter, the value of the CN. It can be easily generalized to any other *property* used for constructing various singlet generalized MDFs. For instance, the quantity $x_B(v)\, dv$ defined in (5.13.20) is the mole fraction of molecules having BE between v and $v + dv$.

This is an example of a QCDF function based on a continuous parameter, v. Another example is the function $x_\psi(\phi) \, d\phi = N_\psi^{(1)}(\phi) \, d\phi / N$ based on the volume of the VP of the particles.

It should be noted that in all of the above-mentioned examples, the adoption of the MM approach depends solely on the rule of classification of molecules into various groups. In each case, the rule must be unique and exhaustive; i.e., for each configuration, each particle belongs to one and only one group.

In most of the applications introduced in the next chapter, we specialize to the simplest case of MM, which we refer to as the two-structure model. It must be remembered, however, that although we are applying the term "mixture model," we have in fact invoked no modelistic assumptions thus far in our treatment. This will be done in Chapter 7 when we examine various *ad hoc* models for water.

5.13.8. General Relations between Thermodynamics and Quasicomponent Distribution Functions

In this section we shall be working in the T, P, N ensemble, and all the distribution functions are presumed to be defined in this ensemble. We denote by $\mathbf{x}$ either the vector or the function which serves as a QCDF. An appropriate subscript will be used to indicate the property employed in the classification procedure. For instance, using the CN as a property, the components of $\mathbf{x}_C$ are the quantities $x_C(K)$. Similarly, using the BE as a property, the components of $\mathbf{x}_B$ are the quantities $x_B(v)$. When reference is made to a general QCDF, we simply write $\mathbf{x}$ without a subscript. Once a QCDF is given, we can obtain the average number of each quasicomponent directly from the components of the vector $\mathbf{N} = N\mathbf{x}$.

Let E be any extensive thermodynamic quantity expressed as a function of the variables T, P, and N (where N is the total number of molecules in the system). Viewing the same system as a mixture of quasicomponents, we can express E as a function of the new set of variables T, P, and $\mathbf{N}$. For concreteness, consider the QCDF based on the concept of CN. The two possible functions mentioned above are then

$$E(T, P, N) = E(T, P, N_C^{(1)}(0), N_C^{(1)}(1), \ldots). \qquad (5.13.38)$$

For the sake of simplicity, we henceforth use $N(K)$ in place of $N_C^{(0)}(K)$, so that the treatment will be valid for any discrete QCDF. Since E is an extensive quantity, it has the property

$$E(T, P, \alpha N(0), \alpha N(1), \ldots) = \alpha E(T, P, N(0), N(1), \ldots) \qquad (5.13.39)$$

for any real $\alpha \geq 0$; i.e., E is a homogeneous function of order one with respect to the variables $N(0), N(1), \ldots$, keeping T, P constant. For such a function, the Euler theorem states that

$$E(T, P, \mathbf{N}) = \sum_{i=0}^{\infty} \bar{E}_i(T, P, \mathbf{N}) N(i), \qquad (5.13.40)$$

where $\bar{E}(T, P, \mathbf{N})$ is the partial molar (or molecular) quantity defined by

$$\bar{E}(T, P, \mathbf{N}) = \left[\frac{\partial E}{\partial N(i)} \right]_{T, P, N(j)}, \qquad j \neq i. \qquad (5.13.41)$$

In (5.13.40) and (5.13.41) we have stressed the fact that the partial molar quantities depend on the whole vector $\mathbf{N}$.

At this point, we digress to discuss the meaning of the partial derivatives introduced in (5.13.41). We recall that the variables $N(i)$ are *not* independent; therefore, it is impossible to take the derivatives of (5.13.41) experimentally. One cannot, in general, add $dN(i)$ of the i-species while keeping all the $N(j)$, $j \neq i$, constant, a process which can certainly be achieved in a mixture of independent components. However, if we assume that, in principle, E can be expressed in terms of the variables T, P, and $\mathbf{N}$, then $\bar{E}_i$ is the component of the gradient of E along the ith axis. Here, we must assume that in the neighborhood of the equilibrium vector $\mathbf{N}$, there is a sufficiently dense set of vectors (which describe various frozen-in systems) so that the gradient of E exists along each axis.

The generalization of (5.13.40) and (5.13.41) to the case of a continuous QCDF requires the application of the technique of functional differentiation. We introduce the generalized Euler theorem by way of analogy with (5.13.40).

The generalization can be easily visual·zed if we rewrite (5.13.40) in the form

$$E(T, P, \mathbf{N}) = \sum_{i=0}^{\infty} \bar{E}(T, P, \mathbf{N}; i)N(i), \qquad (5.13.42)$$

where we have introduced the (discrete) variable i as one of the arguments of the function $\bar{E}$. If $\mathbf{N}$ is a vector derived from a QCDF based on a continuous variable, say v, then the generalization of (5.13.42) is simply

$$E(T, P, \mathbf{N}) = \int_{-\infty}^{\infty} \bar{E}(T, P, \mathbf{N}; v)N(v)\, dv, \qquad (5.13.43)$$

where $\bar{E}(T, P, \mathbf{N}; v)$ is the functional derivative of $E(T, P, \mathbf{N})$ with respect to $N(v)$, symbolized as

$$\bar{E}(T, P, \mathbf{N}; v) = \frac{\delta E(T, P, \mathbf{N})}{\delta N(v)}. \qquad (5.13.44)$$

By analogy with the discrete case we may assign to $\bar{E}(T, P, N; v)$ the meaning of a partial molar quantity of the appropriate v-species. The functional derivative in (5.13.44) is viewed here as a limiting case of (5.13.41) when the index i refers to a continuous variable.

As an example of (5.13.43), the volume of the system is written as

$$V(T, P, \mathbf{N}_\psi^{(1)}) = \int_0^\infty \phi N_\psi^{(1)}(\phi)\, d\phi. \qquad (5.13.45)$$

Note that this relation is based on the fact that the volumes of the VP of all the particles add up to build the total volume of the system. Here we have an example of an explicit dependence between V and $N_\psi^{(1)}$ that could have been guessed. Therefore, the partial molar volume of the ϕ-species can be obtained by taking the functional derivative of V with respect to $N_\psi^{(1)}(\phi)$:

$$\bar{V}(T, P, \mathbf{N}_\psi^{(1)}; \phi') = \frac{\delta V(T, P, \mathbf{N}_\psi^{(1)})}{\delta N_\psi^{(1)}(\phi')} = \phi'. \qquad (5.13.46)$$

This is a remarkable result. It states that the partial molar volume of the ϕ'-species is exactly equal to the volume of its VP. We note that, in general, the partial molar volume of a species is not related in a simple manner to the actual volume which it contributes to the total volume of the system. We also note that in this particular example, the partial volume $\bar{V}(T, P, \mathbf{N}_\psi^{(1)}; \phi')$ is independent of T, P, $\mathbf{N}_\psi^{(1)}$.

A second example is the average internal energy E, which in the T, P, $\mathbf{N}$ ensemble is given by

$$E(T, P, \mathbf{N}_B^{(1)}) = N\varepsilon^K + \tfrac{1}{2} \int_{-\infty}^{\infty} vN_B^{(1)}(v)\, dv. \tag{5.13.47}$$

Note that, in (5.13.47), E stands for the energy, whereas in previous expressions in this section, we have used E for any extensive thermodynamic quantity.

Since the normalization condition for $\mathbf{N}_B^{(1)}$ is

$$\int_{-\infty}^{\infty} N_B^{(1)}(v)\, dv = N, \tag{5.13.48}$$

we can rewrite (5.13.47) as

$$E(T, P, \mathbf{N}_B^{(1)}) = \int_{-\infty}^{\infty} (\varepsilon^K + \tfrac{1}{2}v)N_B^{(1)}(v)\, dv. \tag{5.13.49}$$

This again is an explicit relation between the energy and the singlet generalized MDF $\mathbf{N}_B^{(1)}$. By direct functional differentiation, we obtain

$$\bar{E}(T, P, \mathbf{N}_B^{(1)}; v') = \frac{\delta E(T, P, \mathbf{N}_B^{(1)})}{\delta N_B^{(1)}(v')} = \varepsilon^K + \tfrac{1}{2}v'. \tag{5.13.50}$$

Thus, the partial molar energy of the v'-species is equal to its average kinetic energy and half of its BE. Here again, the partial molar energy does not depend on composition, although it still depends on T through ε^K.

5.13.9. Reinterpretation of Some Thermodynamic Quantities Using the Mixture Model Approach

In the previous section, we reinterpreted relations (5.13.45) and (5.13.49) as special cases of the generalized Euler theorem, i.e.,

$$V(T, P, \mathbf{N}_\psi^{(1)}) = \int_{0}^{\infty} \phi N_\psi^{(1)}(\phi)\, d\phi, \tag{5.13.51}$$

$$E(T, P, \mathbf{N}_B^{(1)}) = \int_{-\infty}^{\infty} (\varepsilon^K + \tfrac{1}{2}v)N_B^{(1)}(v)\, dv. \tag{5.13.52}$$

Here, by adoption of the MM approach, the quantities ϕ and $(\varepsilon^K + \tfrac{1}{2}v)$ are assigned the meaning of partial molar volume and energy, respectively. We now treat some other thermodynamic quantities which are of importance in the study of aqueous fluids.

Consider first the temperature derivatives of (5.13.51) and (5.13.52):

$$\left(\frac{\partial V}{\partial T}\right)_{P,N} = \int_{0}^{\infty} \phi \frac{\partial N_\psi^{(1)}(\phi)}{\partial T}\, d\phi \tag{5.13.53}$$

$$\left(\frac{\partial E}{\partial T}\right)_{P,N} = NC^K + \frac{1}{2} \int_{-\infty}^{\infty} v \frac{\partial N_B^{(1)}(v)}{\partial T}\, dv. \tag{5.13.54}$$

Since ϕ is independent of temperature, we get in (5.13.53) only the contribution due to the structural changes in the system, i.e., the redistribution of particles among the various species caused by the change in temperature. As a very simple example of the distribution $N_\psi^{(1)}(\phi)$, suppose that the volume of the VP of each particle may have only one of two values, say ϕ_A and ϕ_B, which are independent of temperature. In this case, we have

$$N_\psi^{(1)}(\phi) = N_A\delta(\phi - \phi_A) + N_B\delta(\phi - \phi_B), \tag{5.13.55}$$

and hence

$$\left(\frac{\partial V}{\partial T}\right)_{P,N} = \frac{\partial}{\partial T}\int_0^\infty \phi N_\psi^{(1)}(\phi)\, d\phi$$

$$= \frac{\partial}{\partial T}(N_A\phi_A + N_B\phi_B)$$

$$= (\phi_A - \phi_B)\left(\frac{\partial N_A}{\partial T}\right)_{P,N}, \tag{5.13.56}$$

which means that the temperature dependence of the volume results only from the excitation between the two states A and B. In (5.13.53) we have more complex structural changes taking place among the infinite number of species.

Similarly, the heat capacity† in (5.13.54) receives a contribution from the properties of the single particle (kinetic and internal energies) and a second contribution due to the existence of interactions among the particles. The latter contribution is viewed within the realm of the MM approach as a relaxation term, i.e., a redistribution of particles among the various v species arising from the change in temperature. As a very simple example of a distribution $N_B^{(1)}(v)$, we suppose that the BE may attain only one of two values, say v_1, and v_2, which are independent of temperature. In such a case, relation (5.13.54) reduces to

$$\left(\frac{\partial E}{\partial T}\right)_{P,N} = NC^K + \frac{1}{2}\frac{\partial}{\partial T}\int_{-\infty}^\infty [N_1\delta(v - v_1) + N_2\delta(v - v_2)]\, dv$$

$$= NC^K + \tfrac{1}{2}(v_1 - v_2)\left(\frac{\partial N_1}{\partial T}\right)_{P,N}, \tag{5.13.57}$$

where the second term arises from the thermal excitation from one state to the other.

In a similar fashion, one may consider the pressure derivatives of the volume and the energy in (5.13.51) and (5.13.52) to get

$$\left(\frac{\partial V}{\partial P}\right)_{T,N} = \int_0^\infty \phi \frac{\partial N_\psi^{(1)}(\phi)}{\partial P}\, d\phi, \tag{5.13.58}$$

$$\left(\frac{\partial E}{\partial P}\right)_{T,N} = \frac{1}{2}\int_{-\infty}^\infty v \frac{\partial N_B^{(1)}(v)}{\partial P}\, dv. \tag{5.13.59}$$

---

† If E is expressed as a function of T, V, $\mathbf{N}_B^{(1)}$, we get the heat capacity at constant volume. To get the heat capacity at constant pressure, we must take the enthalpy rather than the energy in (5.13.54). However, for the purpose of this section, we ignore the differences between energy and enthalpy.

In this case, all of the pressure dependences of the volume and the energy are viewed as relaxation terms.

5.13.10. Some Thermodynamic Identities in the Mixture Model Approach

In this section, we consider classification procedures that provide a QCDF that has only two components. For the purpose of this section, as well as for most of the latter applications, we can construct such a two-component system from any of the singlet generalized MDFs that have been previously introduced. For example, using $x_C(K)$, we can regroup the particles with different CN into two classes. First, we select an integer K^* (say $K^* = 5$) and define the following two mole fractions:

$$x_L = \sum_{K=0}^{K^*} x_C(K) \tag{5.13.60}$$

$$x_H = \sum_{K=K^*+1}^{\infty} x_C(K). \tag{5.13.61}$$

Clearly, x_L is the mole fraction of particles having a CN smaller than or equal to K^*. These may be referred to as particles with *low local density*. Similarly, x_H is referred to as the mole fraction of *high-local-density* particles (i.e., particles for which $K > K^*$). In this way, the system is viewed as a mixture of two quasicomponents, L and H. This point of view is called a two-structure model (TSM). In a similar fashion, one can construct TSMs from any other discrete or continuous QCDF. Therefore, the following treatment applies for any TSM, not necessarily the one defined in (5.13.60) and (5.13.61).

Within the realm of TSMs, the following natural question may be asked. Suppose that x_L and x_H are the mole fractions of the two components; how are they expected to respond to variation of temperature or pressure or addition of solute? This question has been the subject of many investigations in connection with the study of aqueous solutions. Here, we shall derive a few identities which will be found useful for later applications.

Given a two-component system of, say, L and H at chemical equilibrium, we have the condition

$$\Delta\mu = \mu_L - \mu_H = 0. \tag{5.13.62}$$

Let $N_L = Nx_L$ and $N_H = Nx_H$ be the average number of L- and H-species respectively, and $N = N_L + N_H$ the total number of particles in the system. Viewing $\Delta\mu$ as a function of the variables T, P, N_L, and N_H, we can write its total differential as

$$0 = d(\Delta\mu)_{eq}$$

$$= \left(\frac{\partial\Delta\mu}{\partial T}\right)_{P,N_L,N_H} dT + \left(\frac{\partial\Delta\mu}{\partial P}\right)_{T,N_L,N_H} dP + \left(\frac{\partial\Delta\mu}{\partial N_L}\right)_{T,P,N_H} dN_L + \left(\frac{\partial\Delta\mu}{\partial N_H}\right)_{T,P,N_L} dN_H. \tag{5.13.63}$$

Clearly, the total differential of $\Delta\mu$ along the equilibrium line (eq) is zero. However, if we freeze in the conversion reaction $L \rightleftarrows H$, then N_L and N_H become virtually independent. Define

$$\Delta S = \bar{S}_L - \bar{S}_H, \quad \Delta H = \bar{H}_L - \bar{H}_H, \quad \Delta V = \bar{V}_L - \bar{V}_H, \quad \mu_{\alpha\beta} = \frac{\partial^2 G}{\partial N_\alpha \partial N_\beta}, \tag{5.13.64}$$

where α and β stand for either L or H. We get from (5.13.63), after some rearrangements,

$$\left(\frac{\partial N_L}{\partial T}\right)_{P,N,\text{eq}} = \frac{\Delta H}{T(\mu_{LL} - 2\mu_{LH} + \mu_{HH})} \tag{5.13.65}$$

$$\left(\frac{\partial N_L}{\partial P}\right)_{T,N,\text{eq}} = -\frac{\Delta V}{(\mu_{LL} - 2\mu_{LH} + \mu_{HH})} \tag{5.13.66}$$

$$\left(\frac{\partial N_L}{\partial N_S}\right)_{T,P,N,\text{eq}} = -\frac{\left[\dfrac{\partial(\Delta\mu)}{\partial N_S}\right]_{T,P,N_L,N_H}}{(\mu_{LL} - 2\mu_{LH} + \mu_{HH})}. \tag{5.13.67}$$

On the lhs of (5.13.65)–(5.13.67), we have a derivative at equilibrium, i.e., the change of N_L with T, P, or N_S along the equilibrium line (N_S is the number of solute molecules added to the system). Note that all the quantities on the rhs of (5.13.65) and (5.13.67) contain partial derivatives pertaining to a system in which the equilibrium has been frozen in.

The quantity $(\mu_{LL} - 2\mu_{LH} + \mu_{HH})$ appearing in the above relations is always positive. We present a direct proof of this contention. Consider a system at equilibrium, with composition N_L and N_H. Now suppose that as a result of a fluctuation (at T, P, N constant), N_L has changed into $N_L + dN_L$ and N_H into $N_H + dN_H$. The condition of stability of the system requires that if we allow the system to relax back to its equilibrium position, the Gibbs energy must decrease, i.e.,

$$G(N_L + dN_L, N_H + dN_H) \geq G(N_L, N_H). \tag{5.13.68}$$

Expanding G to second order in dN_L and dN_H around the equilibrium state N_L, N_H (holding T, P, N fixed), we get

$$G(N_L + dN_L, N_H + dN_H) = G(N_L, N_H) + \left(\frac{\partial G}{\partial N_L}\right)dN_L + \left(\frac{\partial G}{\partial N_H}\right)dN_H$$

$$+ \frac{1}{2}\left(\frac{\partial^2 G}{\partial N_L^2}dN_L^2 + 2\frac{\partial^2 G}{\partial N_L\,\partial N_H}dN_L\,dN_H + \frac{\partial^2 G}{\partial N_H^2}dN_H^2\right) + \cdots. \tag{5.13.69}$$

Using the equilibrium condition (5.13.62), the notation in (5.13.64), the relation $dN_L + dN_H = 0$, and the inequality (5.13.68), we get from (5.13.69)

$$\mu_{LL} - 2\mu_{LH} + \mu_{HH} \geq 0, \tag{5.13.70}$$

which is the condition for a minimum of G.

Finally, we derive some general relations between the pair correlation functions of the various quasicomponents. We begin with the simplest case of a TSM, and denote by L and H the two species and by W any molecule in the system. We denote by $g_{\alpha\beta}(R)$ the r correlation function for the pair of species α and β.† Then, $\rho_\alpha g_{\alpha\beta}(R)$ is the local density of an α molecule at a distance R from a β molecule. Conservation of the total number of W molecules around an L molecule gives

$$\rho_L g_{LL}(R) + \rho_H g_{HL}(R) = \rho_W g_{WL}(R), \tag{5.13.71}$$

where we denote by $g_{WL}(R)$ the pair correlation function between an L molecule and any molecule of the system. Similarly, considering the total density of molecules around an H molecule, we get the equality

$$\rho_H g_{HH}(R) + \rho_L g_{LH}(R) = \rho_W g_{WH}(R). \tag{5.13.72}$$

Multiplying (5.13.71) by ρ_L and (5.13.72) by ρ_H and summing the two equations, we get

$$\rho_L^2 g_{LL}(R) + 2\rho_L \rho_H g_{HL}(R) + \rho_H^2 g_{HH}(R) = \rho_W[\rho_L g_{WL}(R) + \rho_H g_{WH}(R)] = \rho_W^2 g_{WW}(R),$$

$$\tag{5.13.73}$$

where $g_{WW}(R)$ is the ordinary pair correlation function in the system when viewed as a one-component system. Dividing through by ρ_W^2, we get

$$x_L^2 g_{LL}(R) + 2x_L x_H g_{LH}(R) + x_H^2 g_{HH}(R) = g_{WW}(R). \tag{5.13.74}$$

Relations similar to (5.13.74) can be generalized to any number of quasicomponents. In the case of a discrete QCDF, we get the relation

$$\sum_{K'=0}^{\infty} \sum_{K=0}^{\infty} x(K)x(K')g(K, K', R) = g(R), \tag{5.13.75}$$

where $g(R)$ is the ordinary pair correlation function of the system and $g(K, K', R)$ is the pair correlation function between the quasicomponents K and K'. Similarly, for a continuous QCDF, we get

$$\int dv \int dv' \, x(v)x(v')g(v, v', R) = g(R). \tag{5.13.76}$$

REFERENCES

1. J. K. Percus and G. L. Yevick, *Phys. Rev.* **110**, 1 (1958).
2. W. E. Morrell and J. H. Hildebrand, *Science* **80**, 125 (1934).
3. N. A. Metropolis, A. W. Rosenbluth, M. N. Rosenbluth, A. H. Teller, and E. Teller, *J. Chem. Phys.* **21**, 1087 (1953).
4. B. J. Alder and T. E. Weinwright, *J. Chem. Phys.* **27**, 1208 (1957).
5. H. Reiss, H. L. Frisch, and J. L. Lebowitz, *J. Chem. Phys.* **31**, 369 (1959).
6. A. Ben-Naim, *Water and Aqueous Solutions* (Plenum Press, New York, 1974).

---

† Note that by the adoption of the MM approach $g_{\alpha\beta}(R)$ is an ordinary pair correlation function for the two species α, β. However, viewing the same system as a one-component system, $g_{\alpha\beta}(R)$ is considered as a generalized pair correlation function for the two properties α and β, assigned to the two particles.

SUGGESTED READINGS

A thorough and advanced treatment of the theory of the liquid state is contained in

A. Münster, *Statistical Thermodynamics* (Springer-Verlag, Berlin, Vols. 1 and 2, 1969 and 1979).

A review of the scaled-particle theory is found in

H. Reiss, *Adv. Chem. Phys.* **9**, 1 (1966).

Theory of Solutions

6.1. INTRODUCTION

This chapter is concerned with a few aspects of the theory of solutions that are either of fundamental character or useful in the study of aqueous solutions. We begin by generalizing some concepts and relationships from the theory of pure liquids and proceed with aspects that are specific to mixtures and solutions. The terms "mixture" and "solution" are used here almost synonymously. The latter is traditionally used when one component (the solute) is dissolved in the other (the solvent). Perhaps one of the most useful concepts in the theory of solution is the concept of ideal solutions. These were defined originally in terms of experimental observations, such as Raoult's or Henry's laws. We shall develop the theoretical background that led to such ideal behaviors. In section 6.7 we present the Kirkwood–Buff theory of solution—an important tool for the study of simple solutions as well as some aspects of aqueous solutions. The concept of solvation, though traditionally used in the context of extremely dilute solutions, is introduced beginning in section 6.13 and applied to any molecule (not necessarily a solute) in any fluid (not necessarily a solvent). This concept enters whenever we study processes such as chemical equilibrium, adsorption, allosteric effect, and so on, in the liquid state.

6.2. MOLECULAR DISTRIBUTION FUNCTIONS IN MIXTURES: DEFINITIONS

In this section we generalize the concepts of MDF to multicomponent mixtures. As in the case of pure liquids, the fundamental molecular quantities required to determine the MDF are the intermolecular interactions. For pairwise additive systems we need the pair potential function for each pair of species as a function of their relative configurations.

We denote by $U_{AB}(\mathbf{X}', \mathbf{X}'')$ the work required to bring two molecules of species A and B from infinite separation to the final configuration $\mathbf{X}', \mathbf{X}''$. (We adopt the convention that the first vector, $\mathbf{X}'$, describes the configuration of the first species, A; similarly, $\mathbf{X}''$ describes the configuration of the second species, B. This convention will be applied to any other pairwise function as well.)

For spherical molecules, $U_{AB}(\mathbf{R}', \mathbf{R}'')$ is a function of the separation $R = |\mathbf{R}'' - \mathbf{R}'|$ only. Hence, it is clear that

$$U_{AB}(R) = U_{BA}(R). \tag{6.2.1}$$

In principle, the function $U_{AB}(R)$ does not have to bear any resemblance to the corresponding functions $U_{AA}(R)$ and $U_{BB}(R)$ of the pure substances A and B, respectively. However, it is convenient when working with simple molecules—say, Lennard–Jones particles—to adopt the following "combination rules":

$$\sigma_{AB} = \tfrac{1}{2}(\sigma_{AA} + \sigma_{BB}) = \sigma_{BA} \tag{6.2.2}$$

$$\varepsilon_{AB} = (\varepsilon_{AA}\varepsilon_{BB})^{1/2} = \varepsilon_{BA}. \tag{6.2.3}$$

The situation becomes more complex when dealing with nonspherical particles, in which case there is no simple way of relating U_{AB} to U_{AA} and U_{BB}. This is certainly true for aqueous solutions where very little is known about the pair potential for the various species.

A system of two components A and B with composition N_A and N_B, respectively, in a specified configuration $\mathbf{X}^{N_A}$, $\mathbf{X}^{N_B}$ haš a total interaction energy

$$U_{N_A,N_B}(\mathbf{X}^{N_A}, \mathbf{X}^{N_B})$$

$$= \tfrac{1}{2} \sum_{i \neq j} U_{AA}(\mathbf{X}_i, \mathbf{X}_j) + \tfrac{1}{2} \sum_{i \neq j} U_{BB}(\mathbf{X}_i, \mathbf{X}_j) + \sum_{i=1}^{N_A} \sum_{j=1}^{N_B} U_{AB}(\mathbf{X}_i, \mathbf{X}_j). \tag{6.2.4}$$

Here we have assumed pairwise additivity of the total potential energy and adopted the convention that the order of arguments in the parentheses corresponds to the order of species as indicated by the subscript of U. Thus, $\mathbf{X}_j$ in the first sum on the rhs of (6.2.4) is the configuration of the jth molecule ($j = 1, 2, \ldots, N_A$) of species A, whereas $\mathbf{X}_j$ in the last term on the rhs of (4.4) stands for the jth molecule ($j = 1, 2, \ldots, N_B$) of species B.

Some care should be exercised when specifying the configuration of the whole system, which is symbolized in (6.2.4) by $\mathbf{X}^{N_A}$, $\mathbf{X}^{N_B}$. To make this more explicit, we can choose different conventions, such as $(\mathbf{X}_1^A, \mathbf{X}_2^A, \ldots, \mathbf{X}_{N_A}^A, \mathbf{X}_1^B, \mathbf{X}_2^B, \ldots, \mathbf{X}_{N_B}^B)$ $(\mathbf{X}_1, \mathbf{X}_2, \ldots, \mathbf{X}_{N_A}, \mathbf{Y}_1, \mathbf{Y}_2, \ldots, \mathbf{Y}_{N_B})$, or $(\mathbf{X}_1, \mathbf{X}_2, \ldots, \mathbf{X}_{N_A}, \mathbf{X}_{N_A+1}, \mathbf{X}_{N_A+2}, \ldots, \mathbf{X}_{N_A+N_B})$. In the first choice, we use different superscripts for the two species; in the second, we use different symbols for the vectors, i.e., $\mathbf{X}$ for A and $\mathbf{Y}$ for B; and in the last, we use consecutive indices to designate the various species, i.e., $1, 2, \ldots, N_A$ for A and $N_A + 1, N_A + 2, \ldots, N_A + N_B$ for B.

In the following sections, we shall not adhere to any particular choice of convention, but shall adopt the most convenient one for the case under consideration.

The basic probability density in the canonical ensemble is

$$P(\mathbf{X}^{N_A+N_B}) = P(\mathbf{X}^{N_A}, \mathbf{X}^{N_B})$$

$$= \frac{\exp[-\beta U_{N_A,N_B}(\mathbf{X}^{N_A}, \mathbf{X}^{N_B})]}{\displaystyle\int \cdots \int d\mathbf{X}^{N_A} d\mathbf{X}^{N_B} \exp[-\beta U_{N_A,N_B}(\mathbf{X}^{N_A}, \mathbf{X}^{N_B})]}, \tag{6.2.5}$$

where obvious shorthand notation such as $\mathbf{X}^{N_A+N_B}$ and $d\mathbf{X}^{N_B}$ has been used. The singlet distribution function for the A species is defined, in complete analogy with the definition in section 5.2, by

$$\rho_A^{(1)}(\mathbf{X}') = \int \cdots \int d\mathbf{X}^{N_A+N_B} P(\mathbf{X}^{N_A+N_B}) \sum_{i=1}^{N_A} \delta(\mathbf{X}_i^A - \mathbf{X}')$$

$$= N_A \int \cdots \int d\mathbf{X}^{N_A+N_B} P(\mathbf{X}^{N_A+N_B}) \delta(\mathbf{X}_1^A - \mathbf{X}'), \tag{6.2.6}$$

and similarly

$$\rho_B^{(1)}(\mathbf{X}') = N_B \int \cdots \int d\mathbf{X}^{N_A+N_B} P(\mathbf{X}^{N_A+N_B}) \delta(\mathbf{X}_1^B - \mathbf{X}'). \tag{6.2.7}$$

As in the case of a one-component system, $\rho_A^{(1)}(\mathbf{X}')$ is the average density of A molecules in the configuration $\mathbf{X}'$. In a homogeneous and isotropic fluid, we have (see section 5.2 for more details):

$$\rho_A^{(1)}(\mathbf{X}') = \frac{N_A}{V8\pi^2} \qquad (6.2.8)$$

$$\rho_B^{(1)}(\mathbf{X}') = \frac{N_B}{V8\pi^2}. \qquad (6.2.9)$$

The average local density of A molecules at $\mathbf{R}'$ is defined by

$$\rho_A^{(1)}(\mathbf{R}') = \int d\mathbf{\Omega}'\, \rho_A^{(1)}(\mathbf{X}') = \frac{N_A}{V}, \qquad (6.2.10)$$

and a similar definition applies to $\rho_B^{(1)}(\mathbf{R}')$.

In a similar fashion, one defines the pair distribution functions for the four different pairs AA, AB, BA, and BB. For instance,

$$\rho_{AA}^{(2)}(\mathbf{X}', \mathbf{X}'') = \int \cdots \int d\mathbf{X}^{N_A + N_B}\, P(\mathbf{X}^{N_A + N_B}) \sum_{i \neq j} \delta(\mathbf{X}_i^A - \mathbf{X}')\delta(\mathbf{X}_j^A - \mathbf{X}'')$$

$$= N_A(N_A - 1) \int \cdots \int d\mathbf{X}^{N_A + N_B}\, P(\mathbf{X}^{N_A + N_B})$$

$$\times \delta(\mathbf{X}_1^A - \mathbf{X}')\delta(\mathbf{X}_2^A - \mathbf{X}'') \qquad (6.2.11)$$

and, for different species,

$$\rho_{AB}^{(2)}(\mathbf{X}', \mathbf{X}'') = \int \cdots \int d\mathbf{X}^{N_A + N_B}\, P(\mathbf{X}^{N_A + N_B}) \sum_{i=1}^{N_A} \sum_{j=1}^{N_B} \delta(\mathbf{X}_i^A - \mathbf{X}')\delta(\mathbf{X}_j^B - \mathbf{X}'')$$

$$= N_A N_B \int \cdots \int d\mathbf{X}^{N_A + N_B}\, P(\mathbf{X}^{N_A + N_B})\delta(\mathbf{X}_1^A - \mathbf{X}')\delta(\mathbf{X}_1^B - \mathbf{X}''). \qquad (6.2.12)$$

The pair correlation functions $g_{\alpha\beta}(\mathbf{X}', \mathbf{X}'')$, where α and β can be either A or B, are defined by

$$\rho_{\alpha\beta}^{(2)}(\mathbf{X}', \mathbf{X}'') = \rho_\alpha^{(1)}(\mathbf{X}')\rho_\beta^{(1)}(\mathbf{X}'')g_{\alpha\beta}(\mathbf{X}', \mathbf{X}''), \qquad (6.2.13)$$

and the spatial pair correlation functions by

$$g_{\alpha\beta}(\mathbf{R}', \mathbf{R}'') = \frac{1}{(8\pi^2)^2} \int d\mathbf{\Omega}' \int d\mathbf{\Omega}''\, g_{\alpha\beta}(\mathbf{X}', \mathbf{X}''). \qquad (6.2.14)$$

As in a one-component system, the functions $g_{\alpha\beta}(\mathbf{R}', \mathbf{R}'')$ depend only on the scalar distance $R = |\mathbf{R}'' - \mathbf{R}'|$. Hence, for the spatial pair correlation function, we have

$$g_{AB}(R) = g_{BA}(R). \qquad (6.2.15)$$

The conditional distribution functions are defined by

$$\rho_{AB}(\mathbf{X}'/\mathbf{X}'') = \rho_{AB}^{(2)}(\mathbf{X}', \mathbf{X}'')/\rho_B^{(1)}(\mathbf{X}'') = \rho_A^{(1)}(\mathbf{X}')g_{AB}(\mathbf{X}', \mathbf{X}''). \qquad (6.2.16)$$

As in the one-component case, $\rho_{AB}(\mathbf{X}'/\mathbf{X}'')$ may be interpreted as the density of A particles in configuration $\mathbf{X}'$, given a B particle in a fixed configuration $\mathbf{X}''$. In a two-component system, we have *four* conditional distribution functions corresponding to the four pairs of species AA, AB, BA, and BB.

Higher-order molecular distribution functions can easily be defined by a simple generalization of the corresponding definitions in a one-component system.

6.3. MOLECULAR DISTRIBUTION FUNCTIONS IN MIXTURES: PROPERTIES

Most of the properties of the molecular distribution functions discussed in Chapter 5 hold for mixtures as well. In this section, we dwell upon some new features that are specific to multicomponent systems. We shall be mainly interested in the properties of the various pair correlation functions in a mixture of two components with spherical molecules.

Let A and B be two simple spherical molecules interacting through pair potentials which we denote by $U_{AA}(R)$, $U_{AB}(R)$, and $U_{BB}(R)$. For simplicity, we may think of Lennard–Jones particles obeying the following relations:

$$U_{AA}(R) = 4\varepsilon_{AA}\left[\left(\frac{\sigma_{AA}}{R}\right)^{12} - \left(\frac{\sigma_{AA}}{R}\right)^{6}\right] \tag{6.3.1}$$

$$U_{BB}(R) = 4\varepsilon_{BB}\left[\left(\frac{\sigma_{BB}}{R}\right)^{12} - \left(\frac{\sigma_{BB}}{R}\right)^{6}\right] \tag{6.3.2}$$

$$U_{AB}(R) = U_{BA}(R) = 4\varepsilon_{AB}\left[\left(\frac{\sigma_{AB}}{R}\right)^{12} - \left(\frac{\sigma_{AB}}{R}\right)^{6}\right]. \tag{6.3.3}$$

Usually the additional combination rules are assumed, i.e.,

$$\sigma_{AB} = \sigma_{BA} = \tfrac{1}{2}(\sigma_{AA} + \sigma_{BB}) \tag{6.3.4}$$

$$\varepsilon_{AB} = \varepsilon_{BA} = (\varepsilon_{AA}\varepsilon_{BB})^{1/2}. \tag{6.3.5}$$

This is a convenient scheme of potential functions, by the use of which we shall illustrate some of the features that are novel to mixtures. Some features of the various pair correlation functions are similar to those in the one-component system; for instance,

$$g_{AB}(R \le \sigma_{AB}) \approx 0 \tag{6.3.6}$$

$$g_{AB}(R \to \infty) = 1 \tag{6.3.7}$$

$$g_{AB}(R) \xrightarrow{\rho \to 0} \exp[-\beta U_{AB}(R)]. \tag{6.3.8}$$

In the last relation, we require that the *total* density $\rho = \rho_A + \rho_B$ tend to zero to validate the limiting behavior (6.3.8).

Before proceeding to mixtures at high densities, it is instructive to recall the density dependence of $g(R)$ for a one-component system (see section 5.3). We have noticed that the second, third, etc. peaks of $g(R)$ develop as the density increases. The illustrations in section 5.3 were given for Lennard–Jones particles with $\sigma = 1.0$ and increasing (number) density ρ. It is clear, however, that the important parameter determining the form of

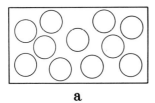 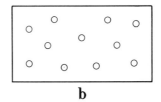

a   b

FIGURE 6.1. Schematic illustration of two systems with the same number density but with different volume density. The system in (a) behaves like a liquid, whereas that in (b) shows gaslike behavior.

$g(R)$ is the dimensionless quantity $\rho\sigma^3$ (assuming for the moment that ε/kT is fixed). This can be illustrated schematically with the help of Fig. 6.1. In the two boxes, we have the same number density, whereas the volume density, defined below, is quite different. Clearly, the behavior of these two systems will differ markedly even when A and B are hard spheres differing only in their diameters. Hence, the form of $g(R)$ will be quite different for these two systems. The reason is that the average separation between the particles in (b) is larger than in (a) when measured in units of σ, although they are almost the same in absolute units.

Now, consider mixtures of A and B (with $\sigma_{AA} \gg \sigma_{BB}$) at different compositions but constant ρ. If we study the dependence of, say, $g_{AB}(R)$ on the mole fraction x_A, we find that at $x_A \approx 1$, $g_{AB}(R)$ behaves as in the case of a high-density fluid, whereas at $x_A \approx 0$, we observe the behavior of the low-density fluid. In order to stress those effects specific to the properties of the mixtures, it is advisable to study the behavior of the pair correlation function when the total "volume density" is constant. The latter is defined as follows. In a one-component system of particles with effective diameter σ, the ratio of the volume occupied by the particles to the total volume of the system is

$$\eta = \frac{N}{V}\frac{4\pi(\sigma/2)^3}{3} = \frac{\rho\pi\sigma^3}{6}. \qquad (6.3.9)$$

The total volume density of a mixture of two components A and B is similarly defined by

$$\eta = \tfrac{1}{6}\pi(\rho_A\sigma_{AA}^3 + \rho_B\sigma_{BB}^3) = \tfrac{1}{6}\pi\rho(x_A\sigma_{AA}^3 + x_B\sigma_{BB}^3). \qquad (6.3.10)$$

In the second equation on the rhs of (6.3.10) we have expressed η in terms of the total (number) density and the mole fractions.

We shall illustrate some of the most salient features of the behavior of the various pair correlation functions in systems of Lennard–Jones particles obeying relations (6.3.1)–(6.3.5) with the parameters

$$\sigma_{AA} = 1.0, \qquad \sigma_{BB} = 1.5$$

$$\frac{\varepsilon_{AA}}{kT} = \frac{\varepsilon_{BB}}{kT} = 0.5, \qquad \eta = 0.45. \qquad (6.3.11)$$

Figure 6.2 shows a set of pair correlation functions for this system at four mole fractions, $x_A = 0.99, 0.8, 0.4,$ and 0.01. Before we discuss some special features of these curves, we recall that the positions of the maxima of $g(R)$ for a one-component system occur roughly at integral multiples of σ. In Fig. 6.2 we have indicated the positions of

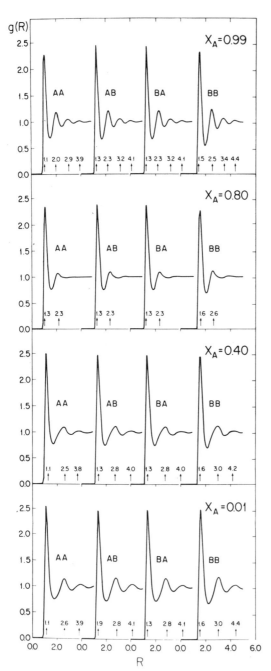

FIGURE 6.2. Pair correlation functions $g_{\alpha\beta}$ for two-component systems of Lennard–Jones particles with the parameters given in (6.13.11). The total volume density η is the same for all the curves. The mole fraction of the component A (with $\sigma_{AA} = 1.0$) is indicated in each row, where all the four $g_{\alpha\beta}(R)$ are shown. The pair of species is indicated next to each curve. The locations of the maxima are shown on the abscissa.

all the maxima of the curves, but in the rest of our discussion, we refer only to the rounded figures which are integral multiples of $\frac{1}{2}\sigma_{AA}$, $\frac{1}{2}\sigma_{BB}$, or combinations of these.

First consider the set of functions $g_{AA}(R)$, $g_{AB}(R)$, $g_{BA}(R)$, and $g_{BB}(R)$ at $x_A = 0.99$. Here, $g_{AA}(R)$ is almost identical with the pair correlation function for pure A. The peaks occur at about σ_{AA}, $2\sigma_{AA}$, $3\sigma_{AA}$, and $4\sigma_{AA}$. Since $\eta = 0.45$ in (6.3.11) corresponds to quite a high density, we have four pronounced peaks. The function $g_{AB}(R)$ has the first peak at σ_{AB}. [The exact value of σ_{AB} is $\frac{1}{2}(\sigma_{AA} + \sigma_{BB}) = 1.25$, but due to errors in the numerical computation and the fact that the minimum of U_{AB} is at $2^{1/6}\sigma_{AB}$, we actually obtain a maximum at about $R = 1.3$.] The second, third, and fourth peaks are determined *not* by multiples of σ_{AB}, but by the addition of σ_{AA}.† That is, the maxima are at $R \approx \sigma_{AB}$, $\sigma_{AB} + \sigma_{AA}$, $\sigma_{AB} + 2\sigma_{AA}$, etc. This is an important feature of a dilute solution of B in A, where the spacing between the maxima is determined by σ_{AA}.

The molecular reason for this is very simple. The spacing between, say, the first and second peaks is determined by the size of the molecule that will most probably fill the space between the two molecules under observation. Because of the prevalence of A molecules in this case, they are the most likely to fill the space between A and B. The situation is depicted schematically in Fig. 6.3. In each row, we show the most likely filling of space between a pair of molecules for the case of $x_A \approx 1$, i.e., for a very dilute solution of B in A. The first row shows the approximate locations of the first three peaks of $g_{AA}(R)$; other rows correspond successively to $g_{AB}(R)$, $g_{BA}(R)$, and $g_{BB}(R)$.

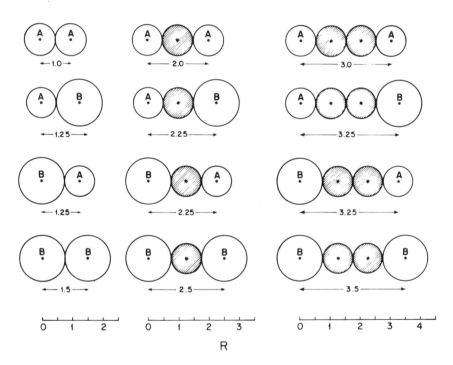

FIGURE 6.3. Configurations corresponding to the first three peaks of $g_{\alpha\beta}(R)$ for a system of B diluted in A (e.g., $x_A = 0.99$) of Fig. 6.2. The two unshaded particles are the ones under observation; i.e., these are the particles for which $g_{\alpha\beta}(R)$ is considered. The shaded particles, which here are invariably of species A, are the ones that fill the spaces between the observed particles. The locations of the expected peaks of $g_{\alpha\beta}(R)$ can be estimated with the help of the scale at the bottom of the figure.

† The second peak of $g_{AB}(R)$ is clearly related to $\sigma_{AB} + \sigma_{AA}$ and the third to $\sigma_{AB} + 2\sigma_{AA}$. If we had chosen $\sigma_{AA} = 1.0$ and $\sigma_{BB} = 2.0$, then we could not have distinguished between $\sigma_{AB} + 2\sigma_{AA}$ and $\sigma_{AB} + \sigma_{BB}$. It is for this reason that we have chosen the values of $\sigma_{AA} = 1.0$ and $\sigma_{BB} = 1.5$, which lead to less ambiguity in the interpretation of the first few peaks.

In the limit of $x_A \to 1$, the component A may be referred to as the solvent and B as the solute. For any pair of species $\alpha\beta$, we can pick up two specific particles (one of species α and the other of species β) and refer to these two particles as a "dimer." From the first row of Fig. 6.3 we see that the most probable configurations of the dimers occur either when the separation is $\sigma_{\alpha\beta}$ or when they are "solvent separated," i.e., when the distances are $R \approx \sigma_{\alpha\beta} + n\sigma_{AA}$, where $n = 1,2,3$.

In Fig. 6.3, the shaded circles correspond to the "solvent" molecules (i.e., the prevalent component A with $\sigma_{AA} = 1.0$), whereas the open circles denote the molecules for which $g(R)$ is observed. The function $g_{BA}(R)$ in the present case must be exactly equal to $g_{AB}(R)$, as is quite evident in Fig. 6.2. Because of the approximate nature of the computations, the curves $g_{AB}(R)$ and $g_{BA}(R)$ may come out a little different; however, theoretically they should be identical.

The second row in Fig. 6.2 corresponds to $x_A = 0.80$. The most remarkable change is the almost complete disappearance of the third and fourth peaks. The second peak is less pronounced than in the case of $x_A = 0.99$. This is an interesting feature, and we return to a further elaboration of it later. We note in connection with this case that the separation between the peaks is still dominated by σ_{AA}, which is in accordance with the expected behavior at composition $x_A = 0.80$. In the third row, we have the curves for $x_A = 0.40$. Here, the separation between the peaks is determined by σ_{BB}. The last row in Fig. 6.2 corresponds to $x_A = 0.01$, i.e., A diluted in B. Clearly, the separation between the peaks is determined by σ_{BB}, since B is now the prevalent component.

Figure 6.4 shows a more detailed composition dependence of $g_{AA}(R)$ in the region $1.2 \le R \le 3.0$. The important point to be noted is the way the location of the second

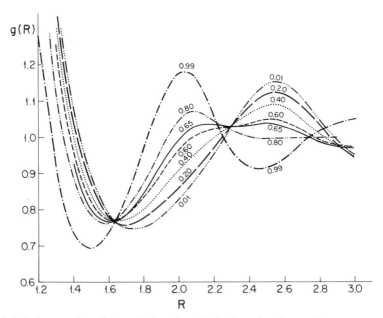

FIGURE 6.4. A close-up view of the variation of $g_{AA}(R)$ in the region $2\sigma_{AA} \lesssim R \lesssim \sigma_{AA} + \sigma_{BB}$ with the composition x_A (as indicated next to each curve). Note that as x_A decreases from 0.99 to 0.01, the second peak at $R \approx 2\sigma_{AA} = 2.0$ gradually disappears. The function is almost flat at about $x_A \approx 0.65$. At lower mole fractions, a new peak evolves at $R \approx \sigma_{AA} + \sigma_{BB} = 2.5$, and becomes more pronounced as $x_A \to 0$.

peak changes from about $\sigma_{AA} + \sigma_{AA}$ at $x_A = 0.99$ (A being the "solvent") to about $\sigma_{AA} + \sigma_{BB}$ at $x_A = 0.01$ (B being the "solvent"). The second peak is maximal for $x_A = 0.99$; it gradually decreases when the composition changes until at about $x_A = 0.65$ the curve is almost flat in the region between $\sigma_{AA} + \sigma_{AA}$ and $\sigma_{AA} + \sigma_{BB}$. When x_A decreases further, a new peak starts to develop at $\sigma_{AA} + \sigma_{BB}$, which reaches its highest value at $x_A = 0.01$.

We stress that the decay of the second peak of $g_{AA}(R)$ as the composition changes is not a result of the decrease in the density of the system. We recall that in a one-component system all the peaks of $g(R)$ except the first one will vanish at $\rho \to 0$. The same is true in the mixture if we let $\rho_A + \rho_B \to 0$. In both cases the disappearance of successive peaks in $g_{\alpha\beta}(R)$ is simply a result of the fact that at $\rho \to 0$ the availability of particles to occupy the space between the selected dimer becomes vanishingly small. The phenomenon we have observed in the mixture at a relatively high volume density ($\eta = 0.45$) is not a result of a lack of particles in the system but a result of the competition between the species to occupy the space between the two selected particles.

To obtain further insight we recall that the location of the second peak is determined principally by the size of the particles that fill the space between the two selected particles. If $x_A = 0.99$, it is most likely that the space will be filled by A molecules. Similarly, for $x_A = 0.01$, it is most probable that B molecules will be filling the space. The strong peak at $2\sigma_{AA}$ in the first case and at $\sigma_{AA} + \sigma_{BB}$ in the second case reflects the high degree of certainty with which the system chooses the species for filling the space between any pair of observed particles. As the mole fraction of A decreases, the B molecules become competitive with A for the "privilege" of filling the space. At about $x_A \approx 0.65$, B is in a state of emulating A. (The fact that this occurs at $x_A \approx 0.65$ and not, say, at $x_A \approx 0.5$, is a result of the difference in σ of the two components. Since B is "larger" than A, its prevalence as volume occupant is effective at $x_B \approx 0.35 < 0.5$.) The fading of the second peak manifests the inability of the system to make a decision as to which kind of particle should be filling the space between the two selected particles. We shall see in the next section an equivalent interpretation in terms of the force acting between the two particles.

6.4. POTENTIAL OF AVERAGE FORCE IN MIXTURES

In section 5.4 we defined the potential of average force between two tagged particles in a one-component system. This definition can be extended to any pair of species; for example, for the AA pair, the potential of average force is defined by

$$g_{AA}(R) = \exp[-\beta W_{AA}(R)]. \tag{6.4.1}$$

Similar definitions apply to other pairs of species. Repeating exactly the same procedure as in section 5.4, we can show that the gradient of $W_{AA}(R)$ is related to the average force operating between the two tagged particles. The generalization of the expression (5.4.11) is quite straightforward. The force acting on the first A particle at $\mathbf{R}'$, given a second A particle at $\mathbf{R}''$, can be written as (see 5.4.11)

$$\mathbf{F}_1 = -\nabla' U_{AA}(\mathbf{R}', \mathbf{R}'') - \int d\mathbf{R}_A \, \nabla' U_{AA}(\mathbf{R}_A, \mathbf{R}') \rho(\mathbf{R}_A/\mathbf{R}', \mathbf{R}'')$$

$$- \int d\mathbf{R}_B \, \nabla' U_{BA}(\mathbf{R}_B, \mathbf{R}') \rho(\mathbf{R}_B/\mathbf{R}', \mathbf{R}''). \tag{6.4.2}$$

The first term on the rhs of (6.4.2) is simply the direct force exerted on the first A at $\mathbf{R}'$, by the second A at $\mathbf{R}''$. The average force exerted by the solvent now has two terms, instead of one in (5.4.11). The quantity $-\nabla'U_{AA}(\mathbf{R}_A, \mathbf{R}')$ is the force exerted by any A particle (other than the selected two As) located at $\mathbf{R}_A$ on the particle at the fixed position $\mathbf{R}'$ and $\rho(\mathbf{R}_A/\mathbf{R}', \mathbf{R}'')$ is the conditional density of A particles at $\mathbf{R}_A$, given two As at $\mathbf{R}'$ and $\mathbf{R}''$. Integration over all locations of $\mathbf{R}_A$ gives the average force exerted by the A component on the A particle at $\mathbf{R}'$. Similarly the third term on the rhs of (6.4.2) is the average force exerted by the B component on the A particle at $\mathbf{R}'$. The combination of the two last terms can be referred to as the "solvent" induced force. (The term "solvent" is used here for all the particles in the system except the two selected tagged particles.)

Two extreme cases of (6.4.2) are the following: If $\rho_B \to 0$, then $\rho(\mathbf{R}_B/\mathbf{R}', \mathbf{R}'') \to 0$ also and the third term on the rhs of (6.4.2) vanishes. This is the case of a pure A. The "solvent" in this case will be all the A particles other than the two tagged particles at $\mathbf{R}'$ and $\mathbf{R}''$.

The second extreme case occurs when $\rho_A \to 0$. Note, however, that we still have two As at fixed positions $\mathbf{R}', \mathbf{R}''$, but otherwise the solvent (here in the conventional sense) is pure B. We have the case of an extremely dilute solution of A in pure B. Note also that at the limit $\rho_A \to 0$, both the pair and the singlet distribution functions of A tend to zero, i.e.,

$$\rho_{AA}^{(2)}(\mathbf{R}', \mathbf{R}'') \to 0 \tag{6.4.3}$$

$$\rho_A^{(1)}(\mathbf{R}') \to 0. \tag{6.4.4}$$

However, the pair correlation function as well as the potential of average force are finite at this limit. This is similar to the situation discussed in Chapter 3 and in Appendix I. We can think of $W_{AA}(R)$ in the limit $\rho_A \to 0$ as the work required to bring two As from infinite separation to the distance R in a pure solvent B at constant T and V (or T, P depending on the ensemble we use).

As in the case of pure liquids, the solvent-induced force can be attractive or repulsive even in regions where the direct force is negligible. An attractive force corresponds to a positive slope of $W(R)$, or, equivalently, to a negative slope of $g(R)$. Thus the region to the right of each peak of $g_{\alpha\beta}(R)$ in Fig. 6.2 is the attractive region, and to the left is the repulsive region. These regions change when the composition of the system changes. Specifically, for $x_A \to 1$ we have the second peak of $g_{AA}(R)$ at about $\sigma_{AA} + \sigma_{AA} \sim 2$. On the other hand, for $x_A \to 0$, the second peak of $g_{AA}(R)$ is at $\sigma_{AA} + \sigma_{AB} \sim 2.5$ (see Fig. 6.4). Clearly, there are regions that are attractive for $x_A \to 1$ (say $2 \lesssim R \lesssim 2.5$) but become repulsive at $x_A \to 0$. Therefore when we change the composition of the system continuously, there are regions in which the two terms on the rhs of (6.4.2) produce forces in different directions. The result is a net diminishing of the overall solvent-induced force between the two tagged A-particles. This corresponds to the flattening of $g(R)$ or of $W(R)$ that we have observed in Fig. 6.4 at $x_A \sim 0.65$.

Another useful way of examining the behavior of, say, $g_{AA}(R)$ in a mixture of A and B is to look at the first-order expansion of $g_{AA}(R)$ in ρ_A and ρ_B. The generalization of (5.3.2) or (5.3.18) for two-component systems is

$$g_{AA}(R) = \exp[-\beta U_{AA}(R)]\left[1 + \rho_A \int f_{AA}(\mathbf{R}', \mathbf{R}_A)f_{AA}(\mathbf{R}_A, \mathbf{R}'') \, d\mathbf{R}_A\right.$$

$$\left. + \rho_B \int f_{AB}(\mathbf{R}', \mathbf{R}_B)f_{BA}(\mathbf{R}_B, \mathbf{R}'') \, d\mathbf{R}_B + \cdots\right], \tag{6.4.5}$$

where $f_{\alpha\beta}$ is defined by extension of (5.3.3) as

$$f_{\alpha\beta}(\mathbf{R}', \mathbf{R}'') = \exp[-\beta U_{\alpha\beta}(\mathbf{R}', \mathbf{R}'')] - 1. \tag{6.4.6}$$

In (6.4.5) we denoted by $\mathbf{R}'$ and $\mathbf{R}''$ the locations of the two tagged particles and $R = |\mathbf{R}'' - \mathbf{R}'|$. Following similar reasoning given in section 5.3.3, we expect that the first integral will contribute an attractive region (even when As and Bs are hard spheres) at $\sigma_{AA} \leq R \leq 2\sigma_{AA}$, whereas the second integral will have the attractive region at $\sigma_{AB} \leq R \leq 2\sigma_{AB}$.

6.5. MIXTURES OF VERY SIMILAR COMPONENTS

In this section, we consider a system of two components in the T, P, N_A, N_B ensemble. Similar arguments and results apply to multicomponent systems. We have chosen the T, P, N_A, N_B ensemble because the isothermal–isobaric systems are the most common ones in actual experiments.

By *very similar* components we mean, in the present context, that the potential energy of interaction among a group of n molecules in a configuration $\mathbf{X}^n$ is independent of the species we assign to each configuration $\mathbf{X}_i$. For example, the pair potential $U_{AA}(\mathbf{X}', \mathbf{X}'')$ is nearly the same as the pair potential $U_{AB}(\mathbf{X}', \mathbf{X}'')$ or $U_{BB}(\mathbf{X}', \mathbf{X}'')$, provided that the configuration of the pair is the same in each case. Clearly, we do not expect that this property will be fulfilled exactly for any pair of different real molecules. However, for molecules differing in, say, isotopic constitution, it may hold to a good approximation.

The chemical potential of A is defined by

$$\mu_A = \left(\frac{\partial G}{\partial N_A}\right)_{T,P,N_B} = G(T, P, N_A + 1, N_B) - G(T, P, N_A, N_B), \tag{6.5.1}$$

where G is the Gibbs energy and the last equality is valid by virtue of the same reasoning as given in section 5.9.

The connection between the chemical potential and statistical mechanics follows directly from the definition of the chemical potential,

$$\exp(-\beta\mu_A) = \frac{\Delta(T, P, N_A + 1, N_B)}{\Delta(T, P, N_A, N_B)}$$

$$= \frac{q_A \int dV \int d\mathbf{X}^{N_A+1} d\mathbf{X}^{N_B} \exp[-\beta PV - \beta U_{N_A+1,N_B}(\mathbf{X}^{N_A+1}, \mathbf{X}^{N_B})]}{\Lambda_A^3 (N_A + 1) \int dV \int d\mathbf{X}^{N_A} d\mathbf{X}^{N_B} \exp[-\beta PV - \beta U_{N_A,N_B}(\mathbf{X}^{N_A}, \mathbf{X}^{N_B})]}, \tag{6.5.2}$$

where Λ_A^3 and q_A are the momentum and the internal partition function of an A molecule, respectively. An obvious shorthand notation has been used for the total potential energy of the system. The configuration $\mathbf{X}^{N_A}$, $\mathbf{X}^{N_B}$ denotes the total configuration of N_A molecules of type A and N_B molecules of type B.

Next, consider a system of N particles of type A only. The chemical potential for such a system (with the same P and T as before) is

$$\exp(-\beta\mu_A^p) = \frac{q_A \int dV \int d\mathbf{X}^{N+1} \exp[-\beta PV - \beta U_{N+1}(\mathbf{X}^{N+1})]}{\Lambda_A^3(N+1) \int dV \int d\mathbf{X}^N \exp[-\beta PV - \beta U_N(\mathbf{X}^N)]}, \quad (6.5.3)$$

where we have denoted by μ_A^p the chemical potential of pure A at the same P and T as above.

Now let us choose N in (6.5.3) equal to $N_A + N_B$ in (6.5.2). The assumption of *very similar* implies, according to its definition, the two equalities

$$U_{N+1}(\mathbf{X}^{N+1}) = U_{N_A+1,N_B}(\mathbf{X}^{N_A+1}, \mathbf{X}^{N_B}) \quad (6.5.4)$$

$$U_N(\mathbf{X}^N) = U_{N_A,N_B}(\mathbf{X}^{N_A}, \mathbf{X}^{N_B}). \quad (6.5.5)$$

Using (6.5.4) and (6.5.5) in (6.5.2) and (6.5.3), we get for the ratio of the latter pair of equations

$$\exp(-\beta\mu_A + \beta\mu_A^p) \approx \frac{(N+1)}{(N_A+1)}. \quad (6.5.6)$$

Rearranging (6.5.6) and noting that for macroscopic systems

$$x_A = \frac{N_A}{N} \approx \frac{(N_A+1)}{(N+1)},$$

we get

$$\mu_A(T, P, x_A) = \mu_A^p(T, P) + kT \ln x_A. \quad (6.5.7)$$

Here we have expressed the chemical potential in terms of the intensive parameters T, P, and x_A. A system for which relation of the form (6.5.7) is obeyed by each component is called a symmetrical ideal solution.

Relation (6.5.7) is important since it gives an explicit dependence of the chemical potential on the composition, the fruitfulness of which was recognized long ago. This relation has been obtained at the expense of the strong assumption that the two components are very similar. We know from experiment that a relation such as (6.5.7) holds also under much weaker conditions.

We shall see in section 6.8 that relations of the form (6.5.7) could be obtained under much weaker assumption on the "similarity" of the two components. In fact, relation (6.5.7) could not have been so useful had it been restricted to the extreme case of very similar components, such as two isotopes.

An alternative derivation of (6.5.7) which uses essentially the same assumption is the following. We write the general expression for the chemical potential of A in pure A denoted by μ_A^p as

$$\mu_A^p = \mu_A^* + kT \ln \rho_A \Lambda_A^3 = W(A|A) + kT \ln \rho_A^p \Lambda_A^3 q_A^{-1}. \quad (6.5.8)$$

This is the same as Eq. (5.9.15) or (5.9.27). We use the notation $W(A|A)$ to designate the coupling work of A against an environment which is pure A, ρ_A^P being the density of pure A. A straightforward generalization of (6.5.8) for a two-component mixture is

$$\mu_A = W(A|A + B) + kT \ln \rho_A \Lambda_A^3 q_A^{-1}, \qquad (6.5.9)$$

where $W(A|A + B)$ is the coupling work of A against an environment of A and B at the same P and T. Suppose now that we replace every B in this environment by A. If the two components are very similar, the particle being coupled would not be able to notice the change in its environment; therefore, $W(A|A + B)$ must be the same as $W(A|A)$ in (6.5.8). Therefore, substituting $W(A|A)$ from (6.5.8) into (6.5.9), we obtain

$$\mu_A = \mu_A^P - kT \ln \rho_A^P \Lambda_A^3 q_A^{-1} + kT \ln \rho_A \Lambda_A^3 q_A^{-1}$$
$$= \mu_A^P + kT \ln x_A, \qquad (6.5.10)$$

where

$$x_A = \frac{\rho_A}{\rho_A^P} = \frac{\rho_A}{\rho_A + \rho_B} \qquad (6.5.11)$$

This is the same result as (6.5.7). Although we have used the assumption of "very similarity" of A and B, it is clear that the requirements in the second derivation are somewhat weaker. We need only that $W(A|A + B)$ be the same for any replacement of A and B in the environment of A. This condition is weaker since it involves only an average quantity and not the bare pair potential. We shall make this statement more precise in section 6.8.

The symmetrical ideal behavior is equivalent to Raoult's law. Suppose that the mixture of A and B is in equilibrium with an ideal-gas phase. Let P_A be the partial pressure of A. The chemical potential of A in the gas phase is

$$\mu_A^g = kT \ln\left(\frac{\Lambda_A^3 q_A^{-1} P_A}{kT}\right). \qquad (6.5.12)$$

From the equilibrium condition $\mu_A^g = \mu_A(T, P, x_A)$, we obtain from (6.5.7) and (6.5.12)

$$P_A = P_A^0 x_A. \qquad (6.5.13)$$

The proportionality constant P_A^0 can be identified as the vapor pressure of pure A at the same temperature and total pressure P.

6.6. VERY DILUTE SOLUTION OF A IN B

In the preceding section we derived the characteristic expression for the chemical potential in a symmetrical ideal solution. Here, we derive another ideal behavior, which is obtained whenever one component (the solute) is very dilute in the second component (the solvent).

We start from the general expression (6.5.9) and take the limit $\rho_A \to 0$. Of course, if we take this limit we obtain pure B, and the chemical potential of A is not defined.

However, if we start from pure B and add one A, we obtain the following expression for the chemical potential

$$\mu_A = W(A|B) + kT \ln \rho_A \Lambda_A^3 q_A^{-1}. \tag{6.6.1}$$

Note that the coupling work of A is against pure B. We shall later in this chapter see that this expression is valid not only in the extreme condition of $N_A = 1$, but also in a solution for which $N_A \ll N_B$ in such a way that correlations between A particles can be neglected.

Equation (6.6.1) is written in thermodynamic notation as

$$\mu_A = \mu_A^{0\rho} + kT \ln \rho_A. \tag{6.6.2}$$

This limit is known as the dilute-ideal limit. It is equivalent to the region in which Henry's law is obeyed.

In order to obtain Henry's law in the familiar form, we transform variables from ρ_A to x_A. Since the solution is very dilute, $\rho_A \ll \rho_B$, and

$$\rho_A = x_A(\rho_A + \rho_B) \approx x_A \rho_B^P, \tag{6.6.3}$$

where ρ_B^P is the density of pure B. Substituting (6.6.3) into (6.6.2), we obtain

$$\mu_A = [W(A|B) + kT \ln \rho_B^P \Lambda_A^3 q_A^{-1}] + kT \ln x_A$$
$$= \mu_A^{0x} + kT \ln x_A. \tag{6.6.4}$$

If the mixture is at equilibrium with an ideal-gas mixture, then we have the equality $\mu_A^g = \mu_A$. Hence, from (6.5.12) and (6.6.4) we obtain

$$P_A = K_H x_A, \tag{6.6.5}$$

where K_H is the Henry constant. Note that, in contrast to (6.5.13), we cannot identify K_H with the vapor pressure of pure A. Equation (6.6.5) is valid only for very small value of $x_A \ll 1$.

The striking similarity between the expressions (6.5.7) and (6.6.4) is a common source of confusion. Both have the form

$$\mu_A = C + kT \ln x_A. \tag{6.6.6}$$

However, the constant C is the chemical potential of *pure* A in (6.5.7). In (6.6.4) C is simply a constant defined by the square brackets. The first includes the coupling work of A against A and the density of pure A. The second includes the coupling work of A against B and the density of pure B.

6.7. THE KIRKWOOD–BUFF THEORY OF SOLUTIONS

The Kirkwood–Buff[1] (KB) theory of solutions provides new relations between thermodynamic quantities and molecular distribution functions. Moreover, these relations are very general and indeed enjoy all of the advantages that we listed in connection with the compressibility equation (section 5.8). Because of its importance, we shall recapitulate the main features of these relations that make them powerful:

1. The theory is valid for any kind of particle, not necessarily spherical.
2. Only the *spatial* pair correlation functions appear in the relations, even when the particles are not spherical.
3. No assumption of additivity of the total potential energy is invoked at any stage of the derivation, hence its more universal validity over theories that explicitly depend on the assumption of pairwise additivity.

As a theory of solutions it is conveniently applied to the entire range of compositions. This feature makes it more useful than the McMillan–Mayer theory of solutions developed in section 6.11.

Because of these features, the KB theory has become a powerful tool in the study of particularly complex fluids such as water and aqueous solutions. We demonstrate such applications in the following chapter. Here, we derive the formal relations. The arguments we use throughout are basically generalizations of those used in deriving the compressibility equation in section 5.8.

6.7.1. General Derivation

Consider a grand canonical ensemble characterized by the variables T, V, and $\boldsymbol{\mu}$, where $\boldsymbol{\mu} = (\mu_1, \mu_2, \ldots, \mu_c)$ stands for the vector comprising the chemical potentials of all the c components of the system. The normalization conditions for the singlet and the pair distribution functions follow directly from their definitions. Here, we use the indices α and β to denote the species, i.e., $\alpha, \beta = 1, 2, \ldots, c$. Hence, the two normalization conditions are

$$\int \overline{\rho_\alpha^{(1)}(\mathbf{X}')}\, d\mathbf{X}' = \langle N_\alpha \rangle \tag{6.7.1}$$

$$\int \overline{\rho_{\alpha\beta}^{(2)}(\mathbf{X}', \mathbf{X}'')}\, d\mathbf{X}'\, d\mathbf{X}'' = \begin{cases} \langle N_\alpha N_\beta \rangle & \text{if } \alpha \neq \beta \\ \langle N_\alpha(N_\alpha - 1) \rangle & \text{if } \alpha = \beta \end{cases}$$

$$= \langle N_\alpha N_\beta \rangle - \langle N_\alpha \rangle \delta_{\alpha\beta}, \tag{6.7.2}$$

where the symbol $\langle \ \rangle$ stands for an average in the grand canonical ensemble. In (6.7.2), we must make a distinction between two cases: $\alpha \neq \beta$ and $\alpha = \beta$. The two cases can be combined into a single equation by using the Kronecker delta function $\delta_{\alpha\beta}$. For homogeneous and isotropic fluids, we also have the relations

$$\overline{\rho_\alpha^{(1)}(\mathbf{X}')} = \frac{\rho_\alpha}{8\pi^2} \tag{6.7.3}$$

$$\overline{\rho_{\alpha\beta}^{(2)}(\mathbf{X}', \mathbf{X}'')} = \frac{\rho_\alpha \rho_\beta \overline{g_{\alpha\beta}(\mathbf{X}', \mathbf{X}'')}}{(8\pi^2)^2}, \tag{6.7.4}$$

where ρ_α is the average number density of molecules of species α, i.e., $\rho_\alpha = \langle N_\alpha \rangle / V$. We also recall the definition of the spatial pair correlation function

$$\overline{g_{\alpha\beta}(\mathbf{R}', \mathbf{R}'')} = (8\pi^2)^{-2} \int d\boldsymbol{\Omega}'\, d\boldsymbol{\Omega}''\, \overline{g_{\alpha\beta}(\mathbf{X}', \mathbf{X}'')},$$

which, as usual, is a function of the scalar distance $R = |\mathbf{R}'' - \mathbf{R}'|$.

From (6.7.1) and (6.7.2) we obtain

$$\int \overline{\rho_{\alpha\beta}^{(2)}(\mathbf{X}', \mathbf{X}'')} \, d\mathbf{X}' \, d\mathbf{X}'' - \int \overline{\rho_\alpha^{(1)}(\mathbf{X}')} \, d\mathbf{X}' \int \overline{\rho_\beta^{(1)}(\mathbf{X}'')} \, d\mathbf{X}''$$

$$= \int [\overline{\rho_{\alpha\beta}^{(2)}(\mathbf{X}', \mathbf{X}'')} - \overline{\rho_\alpha^{(1)}(\mathbf{X}')\rho_\beta^{(1)}(\mathbf{X}'')}] \, d\mathbf{X}' \, d\mathbf{X}''$$

$$= \langle N_\alpha N_\beta \rangle - \langle N_\alpha \rangle \delta_{\alpha\beta} - \langle N_\alpha \rangle \langle N_\beta \rangle. \tag{6.7.6}$$

Using relations (6.7.3)–(6.7.5), we can simplify (6.7.6) to

$$\rho_\alpha \rho_\beta \int [\overline{g_{\alpha\beta}(\mathbf{R}', \mathbf{R}'')} - 1] \, d\mathbf{R}' \, d\mathbf{R}'' = \langle N_\alpha N_\beta \rangle - \langle N_\alpha \rangle \delta_{\alpha\beta} - \langle N_\alpha \rangle \langle N_\beta \rangle. \tag{6.7.7}$$

We now define the quantity

$$\overline{G_{\alpha\beta}} = \int_0^\infty [\overline{g_{\alpha\beta}(R)} - 1] 4\pi R^2 \, dR. \tag{6.7.8}$$

Combining (6.7.7) and (6.7.8) we get

$$\bar{G}_{\alpha\beta} = V \left(\frac{\langle N_\alpha N_\beta \rangle - \langle N_\alpha \rangle \langle N_\beta \rangle}{\langle N_\alpha \rangle \langle N_\beta \rangle} - \frac{\delta_{\alpha\beta}}{\langle N_\alpha \rangle} \right). \tag{6.7.9}$$

This is a connection between the cross fluctuations in the number of particles of various species and an integral involving only the spatial pair correlation functions for the corresponding pair of species α and β.

Before we derive the connection between the $\bar{G}_{\alpha\beta}$s and thermodynamics, it should be stressed that all the distribution functions used in this section are defined in the open system. This has been indicated by a bar over the various distribution functions. If we were in a closed system, the normalization conditions (6.7.1) would have changed into

$$\int \rho_\alpha^{(1)}(\mathbf{X}') \, d\mathbf{X}' = N_\alpha \tag{6.7.10}$$

$$\int \rho_{\alpha\beta}^{(2)}(\mathbf{X}', \mathbf{X}'') \, d\mathbf{X}' \, d\mathbf{X}'' = N_\alpha N_\beta - N_\alpha \delta_{\alpha\beta}, \tag{6.7.11}$$

and instead of relation (6.7.9), we have for the (unbarred) $G_{\alpha\beta}$ the result

$$G_{\alpha\beta}^{(\text{closed})} = \frac{V}{N_\alpha} \delta_{\alpha\beta}. \tag{6.7.12}$$

The reason for this fundamentally different behavior of $G_{\alpha\beta}$ in the closed and opened systems is the same as in the one-component system discussed in section 5.8.

Relation (6.7.12) can be rewritten as

$$\rho_A G_{AA}^{(\text{closed})} = -1 \tag{6.7.13}$$

$$\rho_A G_{AB}^{(\text{closed})} = 0. \tag{6.7.14}$$

Thus, in a closed system, placing an A at a fixed position changes the average number of A particles in the entire surroundings of A by exactly -1. Placing an A at a fixed position does not change the total number of Bs in its entire surroundings. This is a direct consequence of the closure of the system with respect to the number of particles. We shall come back to the interpretation of these quantities in section 6.8. Since we shall use only the $G_{\alpha\beta}$ defined in the open system, we remove the bar over $\bar{G}_{\alpha\beta}$ in all the following uses of the KB theory. Whenever reference is made to a closed system, we shall use the notation of (6.7.13) and (6.7.14).

Next, we establish a connection between the fluctuations in the number of molecules and thermodynamic quantities. We start with the grand canonical partition function for a c-component system:

$$\Xi(T, V, \boldsymbol{\mu}) = \sum_{\mathbf{N}} Q(T, V, \mathbf{N}) \exp(\beta \boldsymbol{\mu} \cdot \mathbf{N}), \tag{6.7.15}$$

where $\mathbf{N} = N_1, N_2, \ldots, N_c$ and the summation is over each of the N_i from zero to infinity. The exponential function includes the scalar product

$$\boldsymbol{\mu} \cdot \mathbf{N} = \sum_{i=1}^{c} \mu_i N_i. \tag{6.7.16}$$

The average number of, say, α molecules in this ensemble is

$$\langle N_\alpha \rangle = \Xi^{-1} \sum_{\mathbf{N}} N_\alpha Q(T, V, \mathbf{N}) \exp(\beta \boldsymbol{\mu} \cdot \mathbf{N})^{\cdot}$$

$$= kT \left[\frac{\partial \ln \Xi(T, V, \boldsymbol{\mu})}{\partial \mu_\alpha} \right]_{T,V,\mu_\alpha'}, \tag{6.7.17}$$

where $\boldsymbol{\mu}_\alpha'$ stands for the set $\mu_1, \mu_2, \ldots, \mu_c$ excluding μ_α.

Differentiating (6.7.17) with respect to μ_β, we get

$$kT \left(\frac{\partial \langle N_\alpha \rangle}{\partial \mu_\beta} \right)_{T,V,\mu_\beta'} = \Xi^{-1} \sum_{\mathbf{N}} N_\alpha N_\beta Q(T, V, \mathbf{N}) \exp(\beta \boldsymbol{\mu} \cdot \mathbf{N}) - \langle N_\alpha \rangle \langle N_\beta \rangle$$

$$= \langle N_\alpha N_\beta \rangle - \langle N_\alpha \rangle \langle N_\beta \rangle. \tag{6.7.18}$$

By symmetry of the arguments with respect to interchanging the indices α and β, we have

$$kT \left(\frac{\partial \langle N_\alpha \rangle}{\partial \mu_\beta} \right)_{T,V,\mu_\beta'} = kT \left(\frac{\partial \langle N_\beta \rangle}{\partial \mu_\alpha} \right)_{T,V,\mu_\alpha'}$$

$$= \langle N_\alpha N_\beta \rangle - \langle N_\alpha \rangle \langle N_\beta \rangle. \tag{6.7.19}$$

Combining relations (6.7.19) with (6.7.9), we get

$$B_{\alpha\beta} \equiv \frac{kT}{V} \left(\frac{\partial \langle N_\alpha \rangle}{\partial \mu_\beta} \right)_{T,V,\mu_\beta'} = kT \left(\frac{\partial \rho_\alpha}{\partial \mu_\beta} \right)_{T,\mu_\beta'}$$

$$= \rho_\alpha \rho_\beta G_{\alpha\beta} + \rho_\alpha \delta_{\alpha\beta}. \tag{6.7.20}$$

Note that $G_{\alpha\beta} = G_{\beta\alpha}$, by virtue of the symmetry with respect to interchanging the α and β indices in either (6.7.9) or (6.7.20).

Relation (6.7.20) is already a connection between thermodynamics and molecular distribution functions. However, since the derivatives in (6.7.20) are taken at constant chemical potentials, these relations are of importance mainly in osmotic systems. Here we are interested in derivatives at constant temperature and pressure. To obtain these require some simple manipulations in partial derivatives. We define the elements of the matrix $\mathbf{A}$ by

$$A_{\alpha\beta} \equiv \frac{V}{kT}\left(\frac{\partial \mu_\alpha}{\partial \langle N_\beta \rangle}\right)_{T,V,\mathbf{N}_\beta'} = \frac{1}{kT}\left(\frac{\partial \mu_\alpha}{\partial \rho_\beta}\right)_{T,\boldsymbol{\rho}_\beta'}, \qquad (6.7.21)$$

where again we use $\mathbf{N}_\beta'$ and $\boldsymbol{\rho}_\beta'$ to denote vectors from which we have excluded the components N_β and ρ_β, respectively. Using the chain rule of differentiation, we get the identities

$$\delta_{\alpha\gamma} = \left(\frac{\partial \mu_\alpha}{\partial \mu_\gamma}\right)_{T,\mu_\gamma'} = \sum_{\beta=1}^{c}\left(\frac{\partial \mu_\alpha}{\partial \rho_\beta}\right)_{T,\boldsymbol{\rho}_\beta'}\left(\frac{\partial \rho_\beta}{\partial \mu_\gamma}\right)_{T,\boldsymbol{\mu}_\gamma'}$$

$$= \sum_{\beta=1}^{c} A_{\alpha\beta}B_{\beta\gamma}, \qquad (6.7.22)$$

where the elements $B_{\alpha\beta}$ are defined in (6.7.20). Equation (6.7.22) can be written in matrix notation as

$$\mathbf{A}\cdot\mathbf{B} = \mathbf{I}, \qquad (6.7.23)$$

with $\mathbf{I}$ the unit matrix. From (6.7.23), we can solve for $\mathbf{A}$ if we know $\mathbf{B}$. Taking the inverse of the matrix $\mathbf{B}$, we get for the elements of $\mathbf{A}$ the relation

$$A_{\alpha\beta} = \frac{B^{\alpha\beta}}{|\mathbf{B}|}, \qquad (6.7.24)$$

where $B^{\alpha\beta}$ stands for the cofactor of the element $B_{\alpha\beta}$ in the determinant $|\mathbf{B}|$. [The cofactor of $B_{\alpha\beta}$ is obtained by eliminating the row and the column containing $B_{\alpha\beta}$ in the determinant $|\mathbf{B}|$ and multiplying† the result by $(-1)^{\alpha+\beta}$.] The existence of the inverse of the matrix $\mathbf{B}$ is equivalent to a condition of stability of the system. Since the $B_{\alpha\beta}$ are already expressible in terms of the $G_{\alpha\beta}$ through (6.7.20), relation (6.7.24) also connects $A_{\alpha\beta}$ with the molecular quantities $G_{\alpha\beta}$.

Next, we transform the volume, as an independent variable, into the pressure. This can be achieved via the thermodynamic identity‡

$$\left(\frac{\partial \mu_\alpha}{\partial N_\beta}\right)_{T,V,\mathbf{N}_\beta'} = \left(\frac{\partial \mu_\alpha}{\partial N_\beta}\right)_{T,P,\mathbf{N}_\beta'} + \left(\frac{\partial \mu_\alpha}{\partial P}\right)_{T,\mathbf{N}}\left(\frac{\partial P}{\partial N_\beta}\right)_{T,V,\mathbf{N}_\beta'}. \qquad (6.7.25)$$

Using the identity

$$\left(\frac{\partial P}{\partial N_\beta}\right)_{T,V,\mathbf{N}_\beta'}\left(\frac{\partial N_\beta}{\partial V}\right)_{T,P,\mathbf{N}_\beta'}\left(\frac{\partial V}{\partial P}\right)_{T,\mathbf{N}} = -1 \qquad (6.7.26)$$

---

† Here α and β must take numerical values; otherwise, $(-1)^{\alpha+\beta}$ is meaningless. In the following applications, we shall take α and β to stand for, say, components A and B, respectively. In this case, we may assign the number 1, say, to A, and the number 2 to B.

‡ From here on, for convenience of notation, we use N_α for $\langle N_\alpha \rangle$.

and the definition of the partial molar volume

$$\bar{V}_\alpha = \left(\frac{\partial V}{\partial N_\alpha}\right)_{T,P,N_\alpha'} = \left(\frac{\partial \mu_\alpha}{\partial P}\right)_{T,N}, \tag{6.7.27}$$

we get from (6.7.25) the relation

$$\mu_{\alpha\beta} \equiv \left(\frac{\partial \mu_\alpha}{\partial N_\beta}\right)_{T,P,N_\beta}$$

$$= \left(\frac{\partial \mu_\alpha}{\partial N_\beta}\right)_{T,V,N_\beta} - \frac{\bar{V}_\alpha \bar{V}_\beta}{V\kappa_T} \tag{6.7.28}$$

where κ_T is the isothermal compressibility of the system

$$\kappa_T = -\frac{1}{V}\left(\frac{\partial V}{\partial P}\right)_{T,N}. \tag{6.7.29}$$

We now have all the necessary relations to express the thermodynamic quantities $\mu_{\alpha\beta}$, $\bar{V}_\alpha$, and κ_T in terms of the $G_{\alpha\beta}$. The general solution is quite involved. Therefore, we specialize to the case of two components, A and B.

6.7.2. Two-Component Systems

Let us collect all the relations for this special case. From (6.7.28) and (6.7.24),

$$\mu_{\alpha\beta} = \frac{kT}{V}\frac{B^{\alpha\beta}}{|\mathbf{B}|} - \frac{\bar{V}_\alpha \bar{V}_\beta}{V\kappa_T}, \qquad \alpha = A,B; \qquad \beta = A,B. \tag{6.7.30}$$

We also have the two Gibbs–Duhem relations

$$\rho_A \mu_{AA} + \rho_B \mu_{AB} = 0 \tag{6.7.31}$$

$$\rho_A \mu_{AB} + \rho_B \mu_{BB} = 0 \tag{6.7.32}$$

and the identity

$$\rho_A \bar{V}_A + \rho_B \bar{V}_B = 1. \tag{6.7.33}$$

Relations (6.7.30)–(6.7.33) comprise seven equations (the first comprises four equations), from which we can solve for the seven thermodynamic quantities $\mu_{\alpha\beta}$ ($\alpha = A, B$; $\beta = A, B$), $\bar{V}_\alpha$ ($\alpha = A, B$), and κ_T.

Before solving these equations, we write the explicit form of the determinant

$$|\mathbf{B}| = \begin{vmatrix} \rho_A + \rho_A^2 G_{AA} & \rho_A\rho_B G_{AB} \\ \rho_A\rho_B G_{AB} & \rho_B + \rho_B^2 G_{BB} \end{vmatrix}$$

$$= \rho_A\rho_B[1 + \rho_A G_{AA} + \rho_B G_{BB} + \rho_A\rho_B(G_{AA}G_{BB} - G_{AB}^2)] \tag{6.7.34}$$

and the various cofactors

$$B^{AA} = \rho_B + \rho_B^2 G_{BB}, \qquad B^{AB} = B^{BA} = -\rho_A\rho_B G_{AB}, \qquad B^{BB} = \rho_A + \rho_A^2 G_{AA}. \tag{6.7.35}$$

Note the different meanings assigned to B in (6.7.35).

It will be convenient to write†

$$\eta = \rho_A + \rho_B + \rho_A \rho_B (G_{AA} + G_{BB} - 2G_{AB}) \tag{6.7.36}$$

$$\zeta = 1 + \rho_A G_{AA} + \rho_B G_{BB} + \rho_A \rho_B (G_{AA} G_{BB} - G_{AB}^2). \tag{6.7.37}$$

By straightforward algebra, we can solve Eqs. (4.60)–(4.63) and express all the thermodynamic quantities $\mu_{\alpha\beta}$, $\bar{V}_\alpha$, and κ_T, in terms of the $G_{\alpha\beta}$:

$$\kappa_T = \frac{\zeta}{kT\eta} \tag{6.7.38}$$

$$\bar{V}_A = \frac{1 + \rho_B(G_{BB} - G_{AB})}{\eta} \tag{6.7.39}$$

$$\bar{V}_B = \frac{1 + \rho_A(G_{AA} - G_{AB})}{\eta} \tag{6.7.40}$$

$$\mu_{AA} = \frac{\rho_B kT}{\rho_A V\eta}, \qquad \mu_{BB} = \frac{\rho_A kT}{\rho_B V\eta}, \qquad \mu_{AB} = \mu_{BA} = -\frac{kT}{V\eta}. \tag{6.7.41}$$

This completes the process of expressing the thermodynamic quantities in terms of the molecular quantities. Let us examine a few limiting cases. In the limit $\rho_B \to 0$, we have

$$\lim_{\rho_B \to 0} \eta = \rho_A, \qquad \lim_{\rho_B \to 0} \zeta = 1 + \rho_A G_{AA}^0. \tag{6.7.42}$$

Hence, (6.7.38) reduces to

$$\lim_{\rho_B \to 0} \kappa_T = \frac{1 + \rho_A G_{AA}^0}{kT\rho_A}, \tag{6.7.43}$$

which is just the compressibility equation for a one-component system (G_{AA}^0 being the limiting value of G_{AA} as $\rho_B \to 0$). Similarly, from (6.7.39) and (6.7.40) we get

$$\lim_{\rho_B \to 0} \bar{V}_A = \frac{1}{\rho_A}, \qquad \lim_{\rho_B \to 0} \bar{V}_B = \frac{1 + \rho_A^0(G_{AA}^0 - G_{AB}^0)}{\rho_A^0}. \tag{6.7.44}$$

Thus, for A, we simply get the molar (strictly molecular) volume of pure A, whereas for B, we get the partial molar volume at infinite dilution.

We now derive some relations which will prove useful in later applications of the theory. All of the following relations are obtainable by the application of simple identities

† From the stability conditions of the system, it can be proven that $\eta > 0$ and $\zeta > 0$ always. The first follows from the stability condition applied to the chemical potential. We must have $\mu_{AB} < 0$; hence, $\eta > 0$. Furthermore, since $\kappa_T > 0$, it follows from (6.7.38) that $\zeta > 0$ also.

between partial derivatives, such as

$$\rho_A \left(\frac{\partial \mu_A}{\partial \rho_B} \right)_{T,P} + \rho_B \left(\frac{\partial \mu_B}{\partial \rho_B} \right)_{T,P} = 0 \tag{6.7.45}$$

$$\left(\frac{\partial \mu_A}{\partial \rho_B} \right)_{T,\mu_B} \left(\frac{\partial \rho_B}{\partial \mu_B} \right)_{T,\mu_A} \left(\frac{\partial \mu_B}{\partial \mu_A} \right)_{T,\rho_B} = -1 \tag{6.7.46}$$

$$\left(\frac{\partial \mu_B}{\partial \rho_B} \right)_{T,P} = \left(\frac{\partial \mu_B}{\partial \rho_B} \right)_{T,\mu_A} + \left(\frac{\partial \mu_B}{\partial \mu_A} \right)_{T,\rho_B} \left(\frac{\partial \mu_A}{\partial \rho_B} \right)_{T,P}. \tag{6.7.47}$$

From (6.7.45)–(6.7.47), we eliminate the required derivative at constant P and T, namely

$$\left(\frac{\partial \mu_B}{\partial \rho_B} \right)_{T,P} = \frac{\rho_A (\partial \mu_B / \partial \rho_B)_{T,\mu_A} (\partial \mu_A / \partial \rho_B)_{T,\mu_B}}{\rho_A (\partial \mu_A / \partial \rho_B)_{T,\mu_B} - \rho_B (\partial \mu_B / \partial \rho_B)_{T,\mu_A}}. \tag{6.7.48}$$

On the rhs of (6.7.48), we have only quantities that are expressible in terms of the $G_{\alpha\beta}$ through (6.7.20). Carrying out this substitution yields

$$\left(\frac{\partial \mu_B}{\partial \rho_B} \right)_{T,P} = \frac{kT}{\rho_B (1 + \rho_B G_{BB} - \rho_B G_{AB})} = kT \left(\frac{1}{\rho_B} - \frac{G_{BB} - G_{AB}}{1 + \rho_B G_{BB} - \rho_B G_{AB}} \right). \tag{6.7.49}$$

The second form on the rhs of (6.7.49) will turn out to be particularly useful for the study of very dilute solutions of B in A.

Using (6.7.45) we also get

$$\left(\frac{\partial \mu_A}{\partial \rho_B} \right)_{T,P} = -\frac{\rho_B}{\rho_A} \left(\frac{\partial \mu_B}{\partial \rho_B} \right)_{T,P} = \frac{-kT}{\rho_A (1 + \rho_B G_{BB} - \rho_B G_{AB})}. \tag{6.7.50}$$

Similarly, if we interchange the roles of A and B, we obtain

$$\left(\frac{\partial \mu_A}{\partial \rho_A} \right)_{T,P} = \frac{kT}{\rho_A (1 + \rho_A G_{AA} - \rho_A G_{AB})} \tag{6.7.51}$$

$$\left(\frac{\partial \mu_B}{\partial \rho_A} \right)_{T,P} = \frac{-kT}{\rho_B (1 + \rho_A G_{AA} - \rho_A G_{AB})}. \tag{6.7.52}$$

Note that

$$\left(\frac{\partial \mu_B}{\partial \rho_A} \right)_{T,P} \neq \left(\frac{\partial \mu_A}{\partial \rho_B} \right)_{T,P}. \tag{6.7.53}$$

The relation between these two derivatives can be obtained by taking the ratio of (6.7.50) and (6.7.52):

$$\left(\frac{\partial \mu_A}{\partial \rho_B}\right)_{T,P} = \left(\frac{\partial \mu_B}{\partial \rho_A}\right)_{T,P} \frac{\rho_B(1 + \rho_A G_{AA} - \rho_A G_{AB})}{\rho_A(1 + \rho_B G_{BB} - \rho_B G_{AB})}$$

$$= \left(\frac{\partial \mu_B}{\partial \rho_A}\right)_{T,P} \frac{\rho_B}{\rho_A} \frac{\bar{V}_B}{\bar{V}_A} \tag{6.7.54}$$

Another useful relation is

$$\left(\frac{\partial \rho_A}{\partial \rho_B}\right)_{T,P} = \frac{(\partial \rho_A/\partial \mu_A)_{T,P}}{(\partial \rho_B/\partial \mu_A)_{T,P}}$$

$$= -\frac{1 + \rho_A G_{AA} - \rho_A G_{AB}}{1 + \rho_B G_{BB} - \rho_B G_{AB}}$$

$$= -\frac{\bar{V}_B}{\bar{V}_A}. \tag{6.7.55}$$

Finally, we obtain the derivative of the chemical potential with respect to the mole fraction via

$$\left(\frac{\partial \mu_A}{\partial x_A}\right)_{T,P} = \left(\frac{\partial \mu_A}{\partial \rho_A}\right)_{T,P}\left(\frac{\partial \rho_A}{\partial x_A}\right)_{T,P} = \left(\frac{\partial \mu_A}{\partial \rho_A}\right)_{T,P} (\rho_A + \rho_B)^2 \bar{V}_B. \tag{6.7.56}$$

In the last form on the rhs of (6.7.56), we have used the identity

$$\left(\frac{\partial x_A}{\partial \rho_A}\right)_{T,P} = \frac{1}{(\rho_A + \rho_B)^2 \bar{V}_B} = \frac{1}{\rho^2 \bar{V}_B}. \tag{6.7.57}$$

From (6.7.56), (6.7.40), and (6.7.51), we get

$$\left(\frac{\partial \mu_A}{\partial x_A}\right)_{T,P} = \frac{kT\rho^2}{\rho_A \eta} = kT\left(\frac{1}{x_A} - \frac{\rho_B \Delta_{AB}}{1 + \rho_B x_A \Delta_{AB}}\right), \tag{6.7.58}$$

where we used the notation

$$\Delta_{AB} = G_{AA} + G_{BB} - 2G_{AB}. \tag{6.7.59}$$

Relation (6.7.58) will be useful for the study of various concepts of ideality carried out in the next sections.

6.7.3. Inversion of the Kirkwood–Buff Theory[2]

The Kirkwood–Buff theory of solutions was originally formulated to obtain thermodynamic quantities from molecular distribution functions. This formulation is useful whenever distribution functions are available either by analytical calculations or from computer simulations. The inversion procedure of the same theory reverses the role of the thermodynamic and molecular quantities—i.e., it allows the evaluation of integrals over the pair correlation functions from thermodynamic quantities. These integrals G_{ij}, referred to as the Kirkwood–Buff integrals, were found useful in characterizing mixtures

on a molecular level. They are also used in the theory of preferential solvation to characterize the local environment of a molecule in a mixture of a two-component solvent (see
section 6.13).

The main result of the KB theory can be symbolically written as:

$$G_{ij} \to \{\bar{V}_i, \kappa_T, \partial\mu_i/\partial\rho_j\}. \tag{6.7.60}$$

Having information on G_{ij}s one can compute the thermodynamic quantities. However, since the quantities G_{ij} are not available from experiment, the theory could have
been used only in rare cases where G_{ij} were obtained from theoretical work. In principle,
having an approximate theory for computing the various pair correlation functions $g_{ij}(R)$,
it is possible to evaluate the integrals G_{ij} and then compute the thermodynamic quantities
through the KB theory. Comparison between the thermodynamic quantities thus
obtained and the corresponding experimental data could serve as a test of the theory that
provides the pair correlation functions.

The inversion procedure may be symbolically written as

$$\{\bar{V}_i, \kappa_T, \partial\mu_i/\partial\rho_j\} \to G_{ij}. \tag{6.7.61}$$

In this form the thermodynamic quantities are used as input to compute the molecular quantities G_{ij}. Since it is relatively easier to measure the required thermodynamic
quantities, the inversion procedure provides a new tool to investigate the characteristics
of the local environments of each species in a multicomponent system.

A brief outline of the inversion procedure is given below. We consider a two component system, say of water (W) and ethanol (E). The isothermal compressibility κ_T, the
partial molar volumes $\bar{V}_E$, $\bar{V}_W$, and the derivatives of the chemical potentials are given
in Eqs. (6.7.38)–(6.7.41). These are

$$\kappa_T = \frac{\zeta}{kT\eta} \tag{6.7.62}$$

$$\bar{V}_E = \frac{1 + \rho_W(G_{WW} - G_{WE})}{\eta} \tag{6.7.63}$$

$$\bar{V}_W = \frac{1 + \rho_E(G_{EE} - G_{WE})}{\eta} \tag{6.7.64}$$

$$\mu_{EE} = \frac{\rho_W kT}{\rho_E \eta V}, \qquad \mu_{WW} = \frac{\rho_E kT}{\rho_W \eta V}, \qquad \mu_{EW} = \mu_{WE} = -\frac{kT}{\eta V}, \tag{6.7.65}$$

where $\mu_{\alpha\beta} = (\partial\mu_\alpha/\partial N_\beta)_{T,P,N}$. Here we have written six equations, but only three of them
are independent, since we have the following thermodynamic identities:

$$\rho_W \mu_{WW} + \rho_E \mu_{WE} = 0 \tag{6.7.66}$$

$$\rho_W \mu_{WE} + \rho_E \mu_{EE} = 0 \tag{6.7.67}$$

$$\rho_E \bar{V}_E + \rho_W \bar{V}_W = 1. \tag{6.7.68}$$

Thus in essence we have three independent equations relating the thermodynamic
quantities to the three quantities G_{WW}, G_{EE}, and $G_{EW} = G_{WE}$ (for any given temperature
and composition of the system). An inversion procedure is possible in which the three
G_{ij}s may be computed from the above-mentioned thermodynamic quantities.

First we assume that the vapor above the mixture at room temperature may be considered to be an ideal-gas mixture. Thus, for the chemical potential of the water we write:

$$\mu_W^l = \mu_W^g = \mu_W^{0g} + kT \ln P_W. \tag{6.7.69}$$

Using the equation for μ_{WW} in (6.7.65) and the thermodynamic identity

$$\left(\frac{\partial \mu_W}{\partial N_W}\right)_{T,P,N_E} = \left(\frac{\partial \mu_W}{\partial x_W}\right)_{T,P} \left(\frac{\partial x_W}{\partial N_W}\right)_{T,P,N_E}$$

$$= \left(\frac{\partial \mu_W}{\partial x_W}\right)_{T,P} \frac{N_E}{(N_E + N_W)^2}, \tag{6.7.70}$$

we obtain

$$\left(\frac{\partial \mu_W}{\partial x_W}\right)_{T,P} = \frac{kT\rho^2}{\rho_W \eta}, \tag{6.7.71}$$

where $\rho = (N_W + N_E)/V$ is the total number density of molecules in the system. Introducing the assumption of ideality for the gas phase, we obtain from (6.7.69) and (6.7.71)

$$x_W \frac{\partial \ln P_W}{\partial x_W} = \frac{\rho}{\eta}, \tag{6.7.72}$$

from which we can compute η if we have the partial vapor pressure of one component as a function of the composition.

Note that we can use information on the vapor pressure of either component, since the corresponding derivatives are related by the identity:

$$x_W \frac{\partial \ln P_W}{\partial x_W} = x_E \frac{\partial \ln P_E}{\partial x_E}. \tag{6.7.73}$$

Having computed η from (6.7.72), we use relation (6.7.62) to compute ζ from the experimental values of the isothermal compressibilities.

Next we use the following identity, which can be easily obtained from (6.7.63) and (6.7.64)

$$\bar{V}_W \bar{V}_E = \frac{\zeta - \eta G_{WE}}{\eta^2} \tag{6.7.74}$$

and which permits the evaluation of G_{WE} in terms of η, ζ, and the product of the two partial molar volumes of E and W. Finally, from (6.7.63) and (6.7.64) we may get G_{WW} and G_{EE}. The above procedure for two-component mixtures may be extended to any number of components.

6.8. SYMMETRIC IDEAL SOLUTIONS: NECESSARY AND SUFFICIENT CONDITIONS

In section 6.5 we showed that in a mixture of "very similar" components, the chemical potential of each component has the form

$$\mu_\alpha = \mu_\alpha^p + kT \ln x_\alpha, \tag{6.8.1}$$

where μ_α^p is the chemical potential of pure† species α (at the same temperature and pressure as the solution under observation) and x_α is its mole fraction. In this section, we confine ourselves to the treatment of mixtures of two components A and B characterized by the thermodynamic variables T, P, N_A, N_B.

6.8.1. Necessary and Sufficient Conditions

A symmetric ideal (SI) solution is defined as a solution for which the chemical potential of each component obeys relation (6.8.1) in the entire range of compositions, keeping T, P constant, i.e.,

$$\mu_\alpha(T, P, x_\alpha) = \mu_\alpha^p(T, P) + kT \ln x_\alpha, \qquad 0 \le x_\alpha \le 1; \qquad T, P \text{ constant.} \quad (6.8.2)$$

For most practical applications, it is also useful to require that (6.8.2) hold also in a small neighborhood of T and P so that we may differentiate (6.8.2) with respect to these variables. There is no need to require the validity of (6.8.2) for all T, P, since no real mixture is expected to fulfill such an exaggerated requirement.

Since (6.8.2) holds for any x_A, we can differentiate it to obtain, say, for component A,

$$\left(\frac{\partial \mu_A}{\partial x_A}\right)_{T,P} = \frac{kT}{x_A}, \qquad 0 \le x_A \le 1; \qquad T, P \text{ constant.} \quad (6.8.3)$$

If (6.8.3) holds, then by integration we get

$$\mu_A = kT \ln x_A + c(T, P), \qquad 0 \le x_A \le 1.$$

The constant of integration c is identified by substituting $x_A = 1$. Hence, (6.8.2) and (6.8.3) are equivalent definitions of SI solutions.

We now recall the result (6.7.58) of the preceding section:

$$\left(\frac{\partial \mu_A}{\partial x_A}\right)_{T,P} = kT\left(\frac{1}{x_A} - \frac{x_B \rho \Delta_{AB}}{1 + \rho x_A x_B \Delta_{AB}}\right), \quad (6.8.4)$$

where $\rho = \rho_A + \rho_B$. A comparison of (6.8.3) with (6.8.4) shows that at any finite density‡ ρ, a necessary and a sufficient condition for a SI solution (in a binary system at T, P constant) is

$$\Delta_{AB} = 0 \qquad \text{for } 0 \le x_A \le 1. \quad (6.8.5)$$

The condition (6.8.5) is clearly a sufficient condition for SI solutions, since by substitution in (6.8.4), we get (6.8.3). Conversely, in order that (6.8.3) and (6.8.4) be equivalent, we must have

$$x_B \rho \Delta_{AB} = 0 \qquad \text{for } 0 \le x_A \le 1. \quad (6.8.6)$$

Since ρ is presumed to be nonzero, (6.8.6) implies (6.8.5) and hence (6.8.5) is also a necessary condition for SI solutions.

---

† The superscript p on μ_α^p stands for the *pure* species. We reserve the symbol μ_α^0 to denote the standard chemical potential.

‡ The case of very low densities, $\rho \to 0$, will be discussed separately in section 6.10. Here, we are interested in solutions at liquid densities. (Note, however, that here ρ is not an independent variable; it is determined by T, P, x_A.) Also, we shall assume throughout that all the $G_{\alpha\beta}$ are finite quantities.

The condition (6.8.5) is very general for SI solutions. It should be recognized that this condition does not depend on any modelistic assumption for the solution. For instance, within the lattice models of solutions we find a *sufficient* condition for SI solutions of the form

$$W = W_{AA} + W_{BB} - 2W_{AB} = 0, \qquad (6.8.7)$$

where $W_{\alpha\beta}$ are the interaction energies between the species α and β situated on adjacent lattice points.

We now define the concept of "similarity" between two components A and B whenever they fulfill condition (6.8.5). We shall soon see that the concept of "similarity" defined here implies a far less stringent requirement on the two components than does the concept of "very similar" defined in section 6.5.

The condition "very similar" was defined by the requirement

$$U_{AA} = U_{AB} = U_{BA} = U_{BB}. \qquad (6.8.8)$$

We have seen in section 6.5 that (6.8.8) is a sufficient condition for an SI solution. Let us examine the following series of conditions:

$$
\begin{aligned}
(a) \quad & U_{AA} = U_{AB} = U_{BA} = U_{BB} \\
(b) \quad & g_{AA} = g_{AB} = g_{BA} = g_{BB} \\
(c) \quad & G_{AA} = G_{AB} = G_{BB} \\
(d) \quad & G_{AA} + G_{BB} - 2G_{AB} = 0
\end{aligned}
\qquad \text{(for } 0 \le x_A \le 1\text{).} \qquad (6.8.9)
$$

Each of the conditions in (6.8.9) follows from its predecessor. Symbolically, we can write

$$(a) \Rightarrow (b) \Rightarrow (c) \Rightarrow (d). \qquad (6.8.10)$$

The first relation, $(a) \Rightarrow (b)$, follows directly from the formal definition of the pair correlation function. The second relation, $(b) \Rightarrow (c)$, follows from the definition of $G_{\alpha\beta}$ [see Eq. (6.7.8)], and the third relation $(c) \Rightarrow (d)$ is obvious.

Since we have shown that condition (d) is a sufficient condition for SI solutions, any condition that precedes (d) will also be a sufficient condition. In particular, (a) is a sufficient condition for SI solutions (a conclusion that was derived directly in section 6.5). In general, the arrows in (6.8.10) may not be reversed. For instance, condition (c) implies an equality of the integrals, which is a far weaker requirement than equality of the integrands $g_{\alpha\beta}$ in (b). It is also obvious that (d) is much weaker than (c); i.e., the $G_{\alpha\beta}$ may be quite different and yet fulfill (d).

We now elaborate on the meaning of condition (6.8.5) for "similarity" between two components on a molecular level. First, we note that the concept of "very similar" defined in section 6.5 is independent of temperature or pressure. This is not the case, however, for the concept of "similarity"; Δ_{AB} can be equal to zero at one P, T but different from zero at other P or T.

We recall the definition of $G_{\alpha\beta}$ in Eq. (6.7.8). Suppose we pick up an A molecule and observe the local density in spherical shells around this molecule. The local density of, say, B molecules at a distance R is $\rho_B g_{BA}(R)$; hence, the average number of B particles in a spherical shell of width dR at distance R from an A particle is $\rho_B g_{BA}(R)4\pi R^2 \, dR$. On the other hand, $\rho_B 4\pi R^2 \, dR$ is the average number of B particles in the same spherical shell, the origin of which has been chosen at random. Therefore, the quantity $\rho_B[g_{BA}(R) - 1]4\pi R^2 \, dR$ measures the excess (or deficiency) in the number of B particles in a spherical shell of volume $4\pi R^2 \, dR$ around an A molecule, relative to the number that would have been measured there using the bulk density ρ_B. Hence, the quantity $\rho_B G_{BA}$ is the average excess of B particles around A. Similarly, $\rho_A G_{AB}$ is the average excess of A particles around B. Thus, G_{AB} is the average excess of A (or B) particles around B (or A) per unit density of A (or B). Therefore, G_{AB} is a measure of the *affinity* of A toward B (and vice versa). A similar meaning is ascribed to G_{AA} and G_{BB}.

Note that the above meaning ascribed to $G_{\alpha\beta}$ is valid only when this quantity has been defined in the open system. In a closed system, if we place an A at the origin of our coordinate system, the total deficiency of As in the entire volume is exactly -1. The total deficiency of Bs in the entire volume is exactly zero. These statements are expressed in Eqs. (6.7.13) and (6.7.14). In both cases the correlation due to intermolecular interactions between particles extends to a distance of a few molecular diameters (except in the region near the critical point). Denote the correlation distance by R_c, then we can write the local change in the number of As around an A at the origin by

$$\Delta N_{AA} = \rho_A \int_0^{R_c} [g_{AA}(R) - 1]4\pi R^2 \, dR. \tag{6.8.11}$$

In the open system, $g_{AA}(R)$ is practically unity at $R > R_c$, therefore, the local change ΔN_{AA} can be equated to the global change $\rho_A G_{AA}$ in the entire volume. On the other hand, if ΔN_{AA} is defined in the closed system, it still has the meaning of the local change in the number of As in the sphere of radius R_c. It is also true that this is the change due to molecular interactions in the system. However, this meaning cannot be retained if we extend the upper limit of the integral from R_c to infinity. The reason is the same as in the one-component system, discussed in section 5.8.4. In a closed system there is a long-range correlation of the order of $1 - N^{-1}$ due to the finite number of particles. Therefore, if we extend the limit of integration from R_c to infinity, we add to the integral a finite quantity, the result of which is that ρG_{AA} would not be a measure of the *local* change in the number of As around an A.

We now return to the meaning of condition (6.8.5) for the SI solutions. $\Delta_{AB} = 0$ is equivalent to the statement that the affinity of A toward B is the arithmetic average of the affinities of A toward A and B toward B. This is true for all compositions $0 \le x_A \le 1$ at a given T, P.

We end this section by considering the phenomenological characterization of SI solutions by their partial molar entropies and enthalpies.

From definition (6.8.1), we get by differentiation

$$\bar{S}_\alpha = -\left(\frac{\partial \mu_\alpha}{\partial T}\right)_P = S_\alpha^p - k \ln x_\alpha \tag{6.8.12}$$

$$\bar{H} = \mu_\alpha + T\bar{S}_\alpha = H_\alpha^p \tag{6.8.13}$$

$$\bar{V}_\alpha = \left(\frac{\partial \mu_\alpha}{\partial P}\right)_T = V_\alpha^p, \tag{6.8.14}$$

where S_α^p, H_α^p, and V_α^p are the molar quantities of pure α. It is easily verified that relations (6.8.12) and (6.8.13), if obeyed throughout the entire range of compositions, form an equivalent definition of SI solutions.

An alternative, although equivalent, way of describing ideal solutions phenomenologically is by introducing excess thermodynamic functions, defined by

$$G^E = G - G^{\text{ideal}} = G - \left[\sum_{\alpha=1}^{c} N_\alpha(\mu_\alpha^p + kT \ln x_\alpha) \right] \tag{6.8.15}$$

$$S^E = S - S^{\text{ideal}} = S - \left[\sum_{\alpha=1}^{c} N_\alpha(S_\alpha^p - kT \ln x_\alpha) \right] \tag{6.8.16}$$

$$V^E = V - V^{\text{ideal}} = V - \sum_{\alpha=1}^{c} N_\alpha V_\alpha^p \tag{6.8.17}$$

$$H^E = H - H^{\text{ideal}} = H - \sum_{\alpha=1}^{c} N_\alpha H_\alpha^p. \tag{6.8.18}$$

The ideal solutions are characterized by zero excess thermodynamic functions. In thermodynamics, ideal solutions are often characterized by their excess entropy and enthalpy. Note that to obtain these, we need to assume the differentiability of (6.8.2) with respect to temperature.

6.8.2. Small Deviations from Symmetric Ideal (SI) Solutions

Any solution of two components can be viewed as a SI solution, with a correction term which takes into account the extent of dissimilarity between the two components. More precisely, we can integrate relation (6.8.4) to obtain

$$\mu_A(T, P, x_A) = \mu_A^p(T, P) + kT \ln x_A + kT \int_0^{x_B} \frac{\rho x_B' \Delta_{AB}}{1 + \rho x_A' x_B' \Delta_{AB}} dx_B', \tag{6.8.19}$$

where we put $dx_B' = -dx_A'$, and we note that in general ρ and Δ_{AB} are composition dependent. The constant of integration has been obtained by putting $x_B = 0$; hence, $\mu_A^p(T, P)$ is identified as the chemical potential of pure A at the given T and P. We can define an activity coefficient by

$$kT \ln \gamma_A^S = kT \int_0^{x_B} \frac{\rho x_B' \Delta_{AB}}{1 + \rho x_B' x_A' \Delta_{AB}} dx_B' \tag{6.8.20}$$

and rewrite (6.8.19) as

$$\mu_A(T, P, x_A) = \mu_A^p(T, P) + kT \ln(x_A \gamma_A^S). \tag{6.8.21}$$

The superscript S on γ_A^S indicates that this activity coefficient is for symmetric ideal behavior and should be distinguished from other activity coefficients introduced later which have different meanings.

In the general case, (6.8.20) is not expected to be useful, since we know nothing about the analytical dependence of ρ or Δ_{AB} on composition. We therefore confine ourselves to a discussion of first-order deviations from SI solutions.

We recall that SI solutions are characterized by the condition $\Delta_{AB} = 0$. It is therefore natural to measure deviations from "similarity" by the parameter Δ_{AB}. By first-order deviations from SI solutions, we mean the cases in which

$$\rho x_A x_B \Delta_{AB} \ll 1 \qquad \text{for } 0 \leq x_A \leq 1. \tag{6.8.22}$$

Note again that we exclude here the case of ideal-gas mixtures attainable as $\rho \to 0$ (see also section 6.10). Also, since condition (6.8.22) is required to hold throughout the entire range of composition, it is, in fact, a condition on Δ_{AB} only.

In practical cases, we have two components which are quite similar (in the usual sense) but not "similar" (in the sense of the definition of the preceding section), and therefore we expect (6.8.22) to be a valid approximation. In fact, first-order theories of regular and athermal solutions are special cases of (6.8.22), although their phenomenological characterization is more general.

A special and simple case occurs when $\rho \Delta_{AB}$ is independent of composition and hence the integral in (6.8.19) can be evaluated immediately:

$$\mu_A(T, P, x_A) = \mu_A^p(T, P) + kT \ln x_A + \tfrac{1}{2} kT \rho x_B^2 \Delta_{AB}$$
$$= \mu_A^p(T, P) + kT \ln(x_A \gamma_A^S), \tag{6.8.23}$$

where γ_A^S includes only the first-order deviations from SI solutions.

The limiting behavior of γ_A^S in (6.8.23) is

$$\lim_{\Delta_{AB} \to 0} \gamma_A^S = 1, \qquad T, P, x_A \text{ constant.} \tag{6.8.24}$$

6.9. DILUTE IDEAL SOLUTIONS

In this section, we discuss a different class of ideal solutions which have been of central importance in the study of solution thermodynamics. We shall refer to a dilute ideal (DI) solution whenever one of the components is very dilute in the solvent. The term "very dilute" depends on the system under consideration, and we shall define it more precisely in what follows. The solvent may have a single component or be a mixture of several components. Here, however, we confine ourselves to two-component systems. The solute, say A, is the component diluted in the solvent B.

The very fact that we make a distinction between the solute A and the solvent B means that the system is treated unsymmetrically with respect to A and B. We shall be mainly concerned with the behavior of the solute, and only briefly mention the relevant relations for the solvent.

6.9.1. Limiting Behavior of the Chemical Potential

The characterization of a DI solution can be carried out along different but equivalent routes. Here, we have chosen the Kirkwood–Buff theory to provide the basic relations from which we derive the limiting behavior of DI solutions. The appropriate relations needed are (6.7.20) (6.7.51) and (6.7.24) which, when specialized to a two-component system, can be rewritten as

$$\left(\frac{\partial \mu_A}{\partial \rho_A}\right)_{T,\mu_B} = \frac{kT}{\rho_A^2 G_{AA} + \rho_A} = kT\left(\frac{1}{\rho_A} - \frac{G_{AA}}{1 + \rho_A G_{AA}}\right) \tag{6.9.1}$$

$$\left(\frac{\partial \mu_A}{\partial \rho_A}\right)_{T,P} = kT\left(\frac{1}{\rho_A} - \frac{G_{AA} - G_{AB}}{1 + \rho_A G_{AA} - \rho_A G_{AB}}\right) \tag{6.9.2}$$

$$\left(\frac{\partial \mu_A}{\partial \rho_A}\right)_{T,\rho_B} = kT\left(\frac{1}{\rho_A} - \frac{G_{AA} + \rho_B(G_{AA}G_{BB} - G_{AB}^2)}{1 + \rho_A G_{AA} + \rho_B G_{BB} + \rho_A\rho_B(G_{AA}G_{BB} - G_{AB}^2)}\right). \tag{6.9.3}$$

Since we are interested in the limiting behavior $\rho_A \to 0$ we have separated the singular part, ρ_A^{-1}, as a first term on the rhs.

Note that the response of the chemical potential to variations in the density ρ_A is different for each set of thermodynamic variables. The three derivatives in (6.9.1)–(6.9.3) correspond to three different processes. The first corresponds to a process in which the chemical potential of the solvent is kept constant (the temperature being constant in all three cases) and therefore is useful in the study of osmotic experiments. This is the simplest expression of the three and it should be noted that if we simply drop the condition of μ_B constant, we get the appropriate derivative for the pure A component system. This is not an accidental result; in fact, this is the case where strong resemblance exists between the behavior of the solute A in a solvent B under constant μ_B and a system A in a vacuum which replaces the solvent. We return to this analogy in section 6.11.

The second derivative, (6.9.2), is the most important one from the practical point of view, since it is concerned with a system under constant pressure. The third relation, (6.9.3), is concerned with a system under constant volume, which is rarely useful in practice.

A common feature of all the derivatives (6.9.1)–(6.9.3) is the ρ_A^{-1} divergence as $\rho_A \to 0$ (which is the reason for the convenient form in which we have written them; note also that we always assume here that all the $G_{\alpha\beta}$ are finite quantities).

For sufficiently low solute density, $\rho_A \to 0$, the first term on the rhs of each of Eqs. (6.9.1)–(6.9.3) is the dominant one; hence, we get the limiting form of these equations:

$$\left(\frac{\partial \mu_A}{\partial \rho_A}\right)_{T,\mu_B} = \left(\frac{\partial \mu_A}{\partial \rho_A}\right)_{T,P} = \left(\frac{\partial \mu_A}{\partial \rho_A}\right)_{T,\rho_B} = \frac{kT}{\rho_A}, \qquad \rho_A \to 0, \tag{6.9.4}$$

which, upon integration, yields

$$\mu_A(T, \mu_B, \rho_A) = \mu_A^0(T, \mu_B) + kT \ln \rho_A$$

$$\mu_A(T, P, \rho_A) = \mu_A^0(T, P) + kT \ln \rho_A \qquad (\rho_A \to 0) \tag{6.9.5}$$

$$\mu_A(T, \rho_B, \rho_A) = \mu_A^0(T, \rho_B) + kT \ln \rho_A.$$

A few comments regarding Eqs. (6.9.5) are now in order.

1. The precise condition ρ_A must satisfy if (6.9.4) is to hold clearly depends on the independent variables we have chosen to describe the system. For instance, if $\rho_A G_{AA} \ll 1$ in (6.9.1), then the term $G_{AA}/(1 + \rho_A G_{AA})$ is negligible compared with ρ_A^{-1}, and we may assume the validity of (6.9.4). The corresponding requirement for (6.9.2) is that $\rho_A(G_{AA} - G_{AB}) \ll 1$, which is clearly different from the previous condition, if only because

the latter depends on G_{AA} as well as on G_{AB}. Similarly, the precise condition under which the limiting behavior of (6.9.3) is obtained involves all three $G_{\alpha\beta}$. We can define a DI solution, for each case, whenever ρ_A is sufficiently small that the limiting behavior of either (6.9.4) or (6.9.5) is valid.

2. Once the limiting behavior (6.9.5) has been attained, we see that all three equations have the same formal form, i.e., a constant of integration, independent of ρ_A, and a term of the form $kT \ln \rho_A$. This is quite a remarkable observation, which holds only in this limiting case. This uniformity of the chemical potential already disappears in the first-order deviation from a DI solution, a topic discussed in the next subsection.

3. The quantities μ_A^0 that appear in (6.9.5) are constants of integration, and as such depend on the thermodynamic variables, which are different in each case. We may refer to these quantities as the "standard chemical potentials" of A in the corresponding set of thermodynamic variables. It is important to realize that these quantities, in contrast to the μ_A^p of the previous section, are *not* chemical potentials of A in any real system. Therefore, we refer to μ_A^0 merely as a constant of integration.

4. Instead of starting with relations (6.9.1)–(6.9.3), we could have started from relation (6.7.58) of the Kirkwood–Buff theory, namely

$$\left(\frac{\partial \mu_A}{\partial x_A}\right)_{T,P} = kT\left(\frac{1}{x_A} - \frac{\rho_B \Delta_{AB}}{1 + \rho_B x_A \Delta_{AB}}\right). \tag{6.9.6}$$

A limiting behavior of (6.9.6) is obtained for $x_A \to 0$, in which case we have

$$\left(\frac{\partial \mu_A}{\partial x_A}\right)_{T,P} = \frac{kT}{x_A}, \qquad x_A \to 0 \tag{6.9.7}$$

which, upon integration in the region for which (6.9.7) is valid, yields

$$\mu_A(T, P, x_A) = \mu_A^{0x}(T, P) + kT \ln x_A, \qquad x_A \to 0. \tag{6.9.8}$$

Again, $\mu_A^{0x}(T, P)$ is merely a constant of integration. It is different from $\mu_A^0(T, P)$ in (6.9.5), and therefore it is wise to use a different superscript to stress this difference. The exact relation between μ_A^0 and μ_A^{0x} can be obtained by noting that $x_A = \rho_A/\rho$, where ρ is the total density of the solution. Hence, from (6.9.8) we get

$$\mu_A = \mu_A^{0x}(T, P) + kT \ln \rho_A - kT \ln \rho$$
$$= \mu_A^0(T, P) + kT \ln \rho_A. \tag{6.9.9}$$

Hence,

$$\mu_A^0(T, P) = \mu_A^{0x}(T, P) - kT \ln \rho_B^p, \tag{6.9.10}$$

where we put ρ_B^p in place of ρ, as is permissible since as $\rho_A \to 0$, $\rho \to \rho_B^p$.

Instead of ρ_A or x_A as a concentration variable, we can use the molality of A, which is related to x_A (for dilute solutions) by

$$m_A = \frac{1000 x_A}{M_B}, \tag{6.9.11}$$

with M_B the molecular weight of B. From (6.9.11) and (6.9.8), we get

$$\mu_A = \mu_A^{0x}(T, P) + kT \ln\left(\frac{M_B m_A}{1000}\right)$$

$$= \left[\mu_A^{0x}(T, P) + kT \ln\left(\frac{M_B}{1000}\right)\right] + kT \ln m_A$$

$$= \mu_A^{0m}(T, P) + kT \ln m_A, \tag{6.9.12}$$

where we have introduced a new standard chemical potential μ_A^{0m}, which is different from both μ_A^0 and μ_A^{0x}.

We now discuss briefly the behavior of the solvent in a DI solution of a two-component system. The simplest way of obtaining the chemical potential of the solvent B is to apply the Gibbs–Duhem relation, which, in combination with (6.9.7), yields

$$x_B\left(\frac{\partial \mu_B}{\partial x_A}\right)_{T,P} + x_A\left(\frac{\partial \mu_A}{\partial x_A}\right)_{T,P} = -x_B\left(\frac{\partial \mu_B}{\partial x_B}\right)_{T,P} + kT = 0, \qquad x_A \to 0. \tag{6:9.13}$$

Hence, upon integration in the region for which (6.9.13) is valid, we get

$$\mu_B(T, P, x_B) = C(T, P) + kT \ln x_B, \qquad x_A \to 0. \tag{6.9.14}$$

Since the condition $x_A \to 0$ is the same as the condition $x_B \to 1$, we can substitute $x_B = 1$ in (6.9.14) to identify the constant of integration as the chemical potential of pure B at the given T and P, i.e.,

$$\mu_B(T, P, x_B) = \mu_B^p(T, P) + kT \ln x_B, \qquad x_B \to 1. \tag{6.9.15}$$

Note that (6.9.15) has the same form as, say, (6.8.1), except for the restriction $x_B \to 1$ in the former.

6.9.2. Small Deviations from Dilute Ideal Solutions

In this section, we discuss first-order deviations from DI solutions. In fact, these nonideal cases are of foremost importance in practical applications. There exist formal statistical-mechanical expressions for the higher-order deviations of DI behavior; however, their practical value is questionable since they usually involve higher-order molecular distribution functions. As in the previous section, we derive all the necessary relations from the Kirkwood–Buff theory.

Consider again the general relations (6.9.1)–(6.9.3) of the preceding subsection. In each of these, we can expand the nonsingular term on the rhs in power series of ρ_A. The leading term in each case is

$$\left(\frac{\partial \mu_A}{\partial \rho_A}\right)_{T,\mu_B} = kT\left(\frac{1}{\rho_A} - G_{AA}^0 + \cdots\right) \tag{6.9.16}$$

$$\left(\frac{\partial \mu_A}{\partial \rho_A}\right)_{T,P} = kT\left[\frac{1}{\rho_A} - (G_{AA}^0 - G_{AB}^0) + \cdots\right] \tag{6.9.17}$$

$$\left(\frac{\partial \mu_A}{\partial \rho_A}\right)_{T,\rho_B} = kT\left\{\frac{1}{\rho_A} - \frac{G_{AA}^0 + \rho_B^0[G_{AA}^0 G_{BB}^0 - (G_{AB}^0)^2]}{1 + \rho_B^0 G_{BB}^0} + \cdots\right\}. \tag{6.9.18}$$

The superscript zero in (6.9.16)–(6.9.18) stands for the limiting value of the corresponding quantity as $\rho_A \to 0$. Note that the limit $\rho_A \to 0$ is taken under different conditions in each case; i.e., T and μ_B are constants in the first, T and P are constants in the second, etc.

These relations can be integrated, in the region of ρ_A for which the first-order expansion is valid, to obtain

$$\mu_A(T, \mu_B, \rho_A) = \mu_A^0(T, \mu_B) + kT \ln \rho_A - kT G_{AA}^0 \rho_A + \cdots \qquad (6.9.19)$$

$$\mu_A(T, P, \rho_A) = \mu_A^0(T, P) + kT \ln \rho_A - kT(G_{AA}^0 - G_{AB}^0)\rho_A + \cdots \qquad (6.9.20)$$

$$\mu_A(T, \rho_B, \rho_A) = \mu_A^0(T, \rho_B) + kT \ln \rho_A - kT\left[G_{AA}^0 - \frac{\rho_B^0(G_{AB}^0)^2}{1 + \rho_B^0 G_{BB}^0}\right]\rho_A + \cdots. \qquad (6.9.21)$$

A comparison of these relations with (6.9.5) clearly shows that the uniformity shown in (6.9.5) breaks down once we consider deviations from DI behavior. The first-order terms in (6.9.19)–(6.9.20) depend on the thermodynamic variables we choose to describe our system. We can now introduce activity coefficients to account for the first-order deviations from the DI behavior. These are defined by

$$kT \ln \gamma_A^D(T, \mu_B, \rho_A) = -kT G_{AA}^0 \rho_A \qquad (6.9.22)$$

$$kT \ln \gamma_A^D(T, P, \rho_A) = -kT(G_{AA}^0 - G_{AB}^0)\rho_A \qquad (6.9.23)$$

$$kT \ln \gamma_A^D(T, \rho_B, \rho_A) = -kT\left(G_{AA}^0 - \frac{\rho_B^0(G_{AB}^0)^2}{1 + \rho_B^0 G_{BB}^0}\right)\rho_A, \qquad (6.9.24)$$

so that Eqs. (6.9.19)–(6.9.21) can be written as

$$\mu_A(T, \mu_B, \rho_A) = \mu_A^0(T, \mu_B) + kT \ln[\rho_A \gamma_A^D(T, \mu_B, \rho_A)] \qquad (6.9.25)$$

$$\mu_A(T, P, \rho_A) = \mu_A^0(T, P) + kT \ln[\rho_A \gamma_A^D(T, P, \rho_A)] \qquad (6.9.26)$$

$$\mu_A(T, \rho_B, \rho_A) = \mu_A^0(T, \rho_B) + kT \ln[\rho_A \gamma_A^D(T, \rho_B, \rho_A)]. \qquad (6.9.27)$$

We note first that the activity coefficients in (6.9.22)–(6.9.24) differ fundamentally from the activity coefficient introduced in Eq. (6.8.20). To stress this difference, we have used the superscript D to denote deviations from DI behavior and the superscript S to denote deviations from SI behavior. Furthermore, each of the activity coefficients defined in (6.9.22)–(6.9.24) depends on the thermodynamic variables, say T and μ_B or T and P. This has been stressed in the notation. In practical applications, however, one usually knows which variables have been chosen, in which case one can drop the arguments in the notation for γ_A^D.

The limiting behavior of the activity coefficients defined in (6.9.22)–(6.9.24) is, for example,

$$\lim_{\rho_A \to 0} \gamma_A^D(T, P, \rho_A) = 1 \qquad T, P \text{ constant} \qquad (6.9.28)$$

Suppose we choose a system defined by T, V, N_A, N_B, where $N_A \ll N_B$ in such a way that the system is DI. The densities are $\rho_A = N_A/V$ and $\rho_B = N_B/V$. Now, suppose we take a T, P, N_A, N_B system but choose P in such a way that the average volume $\langle V \rangle$ is equal to the exact volume of the previous system. Likewise, we choose a T, μ_B, V, N_A system in which the average number of solvent molecules $\langle N_B \rangle$ is equal to N_B in the first system. All these three systems will be equivalent from the thermodynamic point of view. In particular, the standard chemical potentials in the three equations (6.9.5) are equivalent.

We now add some additional As to these three systems, keeping T, ρ_B constant in the first, T, P constant in the second, and T, μ_B constant in the third. As long as the ideal limiting behavior is maintained, Eqs. (6.9.5) are equivalent. Once deviations from DI behavior appear, then the extent of deviation will be different in each of these systems—they become unequivalent.

Consider next the content of the first-order contribution to the activity coefficients in (6.9.22)–(6.9.24). Note that all of these include the quantity G_{AA}^0. Recall that G_{AA}^0 is a measure of the solute-solute affinity. In the limit of DI, the quantity G_{AA}^0 is still finite, but its effect on the activity coefficient vanishes at the limit $\rho_A \to 0$. It is quite clear on qualitative grounds that the standard chemical potential is determined by the solvent–solvent and solvent–solute affinities (this will be shown more explicitly in the next section). Thus, the effect of solute–solute affinity becomes operative only when we increase the solute concentration so that the solute molecules "see" each other, which is the reason for the appearance of G_{AA}^0 in (6.9.22)–(6.9.23). In addition to G_{AA}^0, relation (6.9.23) also includes G_{AB}^0 and relation (6.9.24) also includes G_{BB}^0.

The quantity G_{AA} is often referred to as representing the solute–solute interaction. In this book, we reserve the term "interaction" for the direct intermolecular interaction operating between two particles. Thus, two hard-sphere solutes of diameter σ do not *interact* with each other at a distance $R > \sigma$, yet the solute–solute affinity conveyed by G_{AA} may be different from zero. Furthermore, the use of the term solute–solute interaction for G_{AA} may lead to some misinterpretations. Consider for instance an imperfect gas, where deviations from the ideal gas law are noticeable. Here, one can imagine a process of "switching off" all the interactions between the particles. Such a process will immediately turn our system into an ideal gas. Care must be employed when extending the analogy to solutions. Suppose we have a dilute solution of A in B in which deviations from DI behavior are significant. Here, if we "switch off" the interactions between the solute particles, we do *not* obtain a DI solution. It is difficult to point out the precise quantity that should be "switched off" in this case. As an example, in (6.9.22), we should "switch off" the correlation between the solutes, i.e., put $g_{AA}(R) - 1 \equiv 0$ in order to produce a zero solute–solute affinity, and hence DI solutions. The situation is clearly more complex with relations (6.9.23) and (6.9.24). Therefore, care must be exercised in identifying DI solutions as arising from the absence of solute–solute interactions.

Another common misinterpretation of experimental results is the following. Suppose we measure deviations from a DI solution in a T, P, N_A, N_B system. The corresponding activity coefficient is given by (6.9.23); the same quantity is often referred to as the excess chemical potential of the solute. One then expands the activity coefficient (or the excess chemical potential) to first order in ρ_A and interprets the first coefficient as a measure of the extent of "solute–solute interaction." Clearly, such an interpretation is valid for an osmotic system provided we understand "interaction" in the sense of affinity, as pointed out above. However, in the T, P, N_A, N_B system, the first-order coefficient depends on the difference $G_{AA}^0 - G_{AB}^0$. It is in principle possible that G_{AA}^0 be, say, positive, whereas the first-order coefficient in (6.9.23) can be positive, negative, or zero. This clearly invalidates the interpretation of the first-order coefficient in (6.9.23) in terms of solute–solute correlation. Similar expansions are common for the excess enthalpies and entropies where the significance of the first-order coefficient is not known explicitly.

From the practical point of view, the most important set of thermodynamic variables is, of course, T, P, ρ_A, employed in (6.9.23). However, relation (6.9.22) is also useful, and has enjoyed considerable attention in osmotic experiments, where μ_B is kept constant. We have already noted, in section 6.9, that this set of variables provides relations that bear a remarkable analogy to the virial expansion of various quantities of real gasses. We demonstrate this point by extracting the first-order expansion of the osmotic pressure π in the solute density ρ_A. This can be obtained by the use of the thermodynamic relation

$$\left(\frac{\partial \pi}{\partial \rho_A}\right)_{T,\mu_B} = \rho_A \left(\frac{\partial \mu_A}{\partial \rho_A}\right)_{T,\mu_B}. \tag{6.9.29}$$

Using either (6.9.16) or the more general expression (6.9.1) in (6.9.29), we get

$$\left(\frac{\partial \pi}{\partial \rho_A}\right)_{T,\mu_B} = \frac{kT}{\rho_A G_{AA} + 1} \xrightarrow{\rho_A \to 0} kT(1 - G_{AA}^0 \rho_A + \cdots). \tag{6.9.30}$$

Upon integration, we get

$$\frac{\pi}{kT} = \rho_A - \tfrac{1}{2} G_{AA}^0 \rho_A^2 + \cdots, \tag{6.9.31}$$

which is better known in the form

$$\frac{\pi}{kT} = \rho_A + B_2^* \rho_A^2 + \cdots, \tag{6.9.32}$$

where B_2^* is the analog of the second virial coefficient in the density expansion of the pressure (see section 1.9)

$$\frac{P}{kT} = \rho + B_2 \rho^2 + \cdots. \tag{6.9.33}$$

We now turn to a brief treatment of the chemical potential of the solvent B for a system deviating slightly from the DI behavior. The simplest way of doing this is to use relation (6.7.58), from the Kirkwood–Buff theory, which, when written for the B component and expanded to first order in x_A, yields

$$\left(\frac{\partial \mu_B}{\partial x_B}\right)_{T,P} = kT\left(\frac{1}{x_B} - \rho^0 \Delta_{AB}^0 x_A + \cdots\right), \qquad x_A \to 0, \tag{6.9.34}$$

where ρ^0 and Δ_{AB}^0 are the limiting values of ρ and Δ_{AB} as $x_A \to 0$. Integrating (6.9.34) yields

$$\mu_B(T, P, x_B) = \mu_B^p(T, P) + kT \ln x_B + \int_0^{x_A} \rho^0 \Delta_{AB}^0 x_A' \, dx_A', \qquad x_A \to 0. \tag{6.9.35}$$

Here, $\rho^0 \Delta_{AB}^0$ is independent of composition; hence we can integrate to obtain

$$\mu_B(T, P, x_A) = \mu_B^p(T, P) + kT \ln x_B + \tfrac{1}{2} \rho_B^0 \Delta_{AB}^0 x_A^2, \qquad x_A \to 0. \tag{6.9.36}$$

Note that in the limit $\rho_A \to 0$, we can replace ρ^0 by ρ_B^0, the density of the pure solvent B at the given T, P. Note, however, the difference between (6.9.36) and relation (6.8.23), which looks very similar. Here, the expansion is valid for small x_A, but otherwise the value of $\rho_B^0 \Delta_{AB}^0$ is unrestricted. In (6.8.23), on the other hand, we have a first order expansion in Δ_{AB}, which is required to hold for all compositions $0 \le x_A \le 1$.

6.10. A COMPLETELY SOLVABLE EXAMPLE

This section is devoted to illustrating the three fundamentally different types of ideal mixtures. The first and simplest class is that of the ideal-gas (IG) mixtures, which, as in the case of an ideal gas, are characterized by the complete absence (or neglect) of all intermolecular forces. This class is of least importance in the study of solution chemistry.

The second class, referred to as symmetric ideal (SI) solutions, occurs whenever the various components are "similar" to each other. There are no restrictions on the magnitude of the intermolecular forces or on the densities. The third class, dilute ideal (DI) solutions, consists of those solutions for which at least one component is very diluted in the remaining solvent, which may be a one-component or a multicomponent system. Again, there is no restriction on the strength of the intermolecular forces, the total density, or the degree of similarity of the various components.

The occurrence of all three types of ideality can be demonstrated by the use of Eq. (6.7.58) from the Kirkwood–Buff theory of solutions. We limit the discussion to a two-component system:

$$\left(\frac{\partial \mu_A}{\partial x_A}\right)_{T,P} = kT\left(\frac{1}{x_A} - \frac{\rho x_B \Delta_{AB}}{1 + \rho x_A x_B \Delta_{AB}}\right). \tag{6.10.1}$$

We distinguish between the following three limiting cases of this equation:

1. At the limit $\rho \to 0$, the particles are, on the average, very far from each other; hence, the intermolecular forces have a negligible effect on the properties of the mixture. This case is referred to as an ideal-gas mixture. As noted in Chapter 1, the theoretical definition of an ideal gas is different and requires the complete absence of intermolecular interactions, a situation that is never realized in real systems. Clearly, letting $\rho \to 0$ in (6.10.1), we get the typical ideal behavior

$$\left(\frac{\partial \mu_A}{\partial x_A}\right)_{T,P} = \frac{kT}{x_A}. \tag{6.10.2}$$

2. For a real system with any type of intermolecular force at a finite density ρ, the second term on the rhs of (6.10.1) may vanish if the two components are "similar" in the sense of section 6.8; i.e., if $\Delta_{AB} = 0$ for the entire range of compositions, we get the SI solutions. (Note the essential difference between this and the previous case.)
3. For a real system of two components that are not similar but in which $x_A \to 0$, we get from (6.10.1) the DI solutions.

Any arbitrary mixture of two components can be viewed as deviating from one of the ideal reference cases. This can be written symbolically as

$$\begin{aligned} \mu_A &= c_1 + kT \ln(x_A \gamma_A^I) \\ &= c_2 + kT \ln(x_A \gamma_A^S) \\ &= c_3 + kT \ln(x_A \gamma_A^D), \end{aligned} \tag{6.10.3}$$

where the constants c_i are independent of x_A. Here, γ_A^I, γ_A^S, and γ_A^D are activity coefficients that incorporate the correction due to nonideality† (I for IG, S for SI, and D for DI).

We now consider a particular example of a system which, on the one hand, is not trivial, since interactions between particles are taken into account, yet is sufficiently simple that all three activity coefficients can be written in an explicit form.

We choose a two-component system for which the pressure (or the total density) is sufficiently low that the pair correlation function for each pair of species has the form

$$g_{\alpha\beta}(R) = \exp[-\beta U_{\alpha\beta}(R)], \tag{6.10.4}$$

where, for simplicity, we assume that all the pair potentials are spherically symmetrical. The general expression for the chemical potential of, say, A in this system is obtained by a simple extension of the one-component expression given in section 5.9 (for the purpose of this section, we assume that the internal partition function is unity):

$$\mu_A = kT \ln(\rho_A \Lambda_A^3) + \rho_A \int_0^1 d\xi \int_0^\infty U_{AA}(R) g_{AA}(R, \xi) 4\pi R^2 \, dR$$

$$+ \rho_B \int_0^1 d\xi \int_0^\infty U_{AB}(R) g_{AB}(R, \xi) 4\pi R^2 \, dR$$

$$= kT \ln(\rho_A \Lambda_A^3) + 2kT B_{AA} \rho_A + 2kT B_{AB} \rho_B, \tag{6.10.5}$$

where we have used (6.10.4) with $\xi U_{\alpha\beta}(R)$ replacing $U_{\alpha\beta}(R)$ so that integration over ξ becomes immediate. Also, we have used the more familiar notation

$$B_{\alpha\beta} = -\tfrac{1}{2} \int_0^\infty \{\exp[-\beta U_{\alpha\beta}(R)] - 1\} 4\pi R^2 \, dR. \tag{6.10.6}$$

It is possible to analyze Eq. (6.10.5) for the various kinds of ideality, but for the purpose of this section, we prefer to transform it first so that μ_A is expressed as a function of T, P, and x_A. To do this, we use the analog of the virial expansion for mixtures, which reads (section 1.9)

$$\frac{P}{kT} = (\rho_A + \rho_B) + [x_A^2 B_{AA} + 2 x_A x_B B_{AB} + x_B^2 B_{BB}](\rho_A + \rho_B)^2 + \cdots. \tag{6.10.7}$$

The factor in the square brackets can be viewed as the second virial coefficient for the mixture of two components. We now invert this relation by assuming an expansion of the form

$$\rho = \rho_A + \rho_B = \left(\frac{P}{kT}\right) + CP^2 + \cdots. \tag{6.10.8}$$

---

† For the purpose of demonstration, we have chosen T, P, x_A as the thermodynamic variables. A parallel treatment can be carried out for any other set of thermodynamic variables.

This is substituted on the rhs of (6.10.7), and on equating coefficients of equal powers of P, we get

$$\rho = \rho_A + \rho_B = \left(\frac{P}{kT}\right) - (x_A^2 B_{AA} + 2x_A x_B B_{AB} + x_B^2 B_{BB})\left(\frac{P}{kT}\right)^2 + \cdots. \quad (6.10.9)$$

Substituting $\rho_A = x_A \rho$ and $\rho_B = x_B \rho$ in (6.10.5) and using the expansion (6.10.9) for ρ, we get the final form of the chemical potential:

$$\mu_A(T, P, x_A) = kT \ln(x_A \Lambda_A^3) + kT \ln\left(\frac{P}{kT}\right)$$

$$+ kT \ln\left[1 - (x_A^2 B_{AA} + 2x_A x_B B_{AB} + x_B^2 B_{BB})\frac{P}{kT}\right]$$

$$+ (2kTB_{AA}x_A + 2kTB_{AB}x_B)\left[\left(\frac{P}{kT}\right) - (x_A^2 B_{AA} + 2x_A x_B B_{AB} + x_B^2 B_{BB})\left(\frac{P}{kT}\right)^2\right]$$

$$= kT \ln(x_A \Lambda_A^3) + kT \ln\left(\frac{P}{kT}\right) + PB_{AA} - Px_B^2(B_{AA} + B_{BB} - 2B_{AB}), \quad (6.10.10)$$

where, in the last form of (6.10.10), we have retained only first-order terms in the pressure (except for the logarithmic term). We now view expression (6.10.10) in various ways, according to the choice of the reference ideal state.

6.10.1. Ideal-Gas Mixture as a Reference System

In $P \to 0$ (or if no interactions exist, so that all $B_{\alpha\beta} = 0$), we get from (6.10.10)

$$\mu_A^{IG} = kT \ln(x_A \Lambda_A^3) + kT \ln\left(\frac{P}{kT}\right)$$

$$= \mu_A^{0g}(T, P) + kT \ln x_A. \quad (6.10.11)$$

Here, $\mu_A^{0g}(T, P)$ is defined as the IG standard chemical potential (note its dependence on both T and P).

Comparing (6.10.10) with (6.10.11), we see that the correction due to deviations from the IG mixture can be included in the activity coefficient

$$kT \ln \gamma_A^I = PB_{AA} - Px_B^2(B_{AA} + B_{BB} - 2B_{AB}), \quad (6.10.12)$$

and hence

$$\mu_A(T, P, x_A) = \mu_A^{0g}(T, P) + kT \ln(x_A \gamma_A^I), \quad (6.10.13)$$

which is a particular case of the first form on the rhs of (6.10.3). Here γ_A^I measures deviations from ideal gas behavior due to all interactions between the molecules.

6.10.2. Symmetric Ideal Solution as a Reference System

Suppose that the two components A and B are "similar" in the sense of section 6.8, which, in this case, means [see Eq. (6.85)]

$$B_{AA} + B_{BB} - 2B_{AB} = 0 \tag{6.10.14}$$

This is the SI solution. Substituting in (6.10.10), we obtain

$$\mu_A^{SI} = kT \ln(x_A \Lambda_A^3) + kT \ln\left(\frac{P}{kT}\right) + PB_{AA}$$

$$= \mu_A^P(T, P) + kT \ln x_A. \tag{6.10.15}$$

Note that μ_A^P is the chemical potential of pure A at this particular T and P, and therefore it includes the effect of A–A interactions through B_{AA}. If, on the other hand, the system is not SI, then we define the activity coefficient as

$$kT \ln \gamma_A^S = -Px_B^2(B_{AA} + B_{BB} - 2B_{AB}), \tag{6.10.16}$$

and (6.10.10) is now rewritten as

$$\mu_A(T, P, x_A) = \mu_A^P + kT \ln(x_A \gamma_A^S), \tag{6.10.17}$$

where γ_A^S is a measure of the deviations due to the dissimilarity between the components. Equation (6.10.17) is a particular case of the second form on the rhs of (6.10.3).

6.10.3. Dilute Ideal Solution as a Reference System

Here A is diluted in B, i.e., $x_A \to 0$, or $x_B \to 1$. In this case, we have a DI solution. Equation (6.10.10) reduces to

$$\mu_A^{DI} = kT \ln(x_A \Lambda_A^3) + kT \ln\left(\frac{P}{kT}\right) + P(2B_{AB} - B_{BB})$$

$$= \mu_A^{0x} + kT \ln x_A, \tag{6.10.18}$$

where μ_A^{0x} is defined in (6.10.18) for this particular example. Since dilute solutions are preferably expressed in terms of ρ_A rather than x_A, we can transform (6.10.18) by substituting $x_A = \rho_A/\rho$ in (6.10.18) and using (6.10.9) to obtain

$$\mu_A^{DI} = kT \ln(\rho_A \Lambda_A^3) + 2PB_{AB} = \mu_A^{0\rho} + kT \ln \rho_A. \tag{6.10.19}$$

Note that μ_A^{0x} and $\mu_A^{0\rho}$ do not include the term B_{AA} which reflects the extent of solute–solute interactions. The word "interaction" is appropriate in the present context since, in the present limiting case, we know that $g_{\alpha\beta}(R)$ depends only on the direct interaction between the pair of species α and β, as we have assumed in (6.10.4).

The activity coefficients defined for the two representations (6.10.18) and (6.10.19) are obtained from comparison with (6.10.10) as

$$kT \ln \gamma_A^{D,x} = 2x_A P(B_{AA} + B_{BB} - 2B_{AB}) \tag{6.10.20}$$

$$kT \ln \gamma_A^{D,\rho} = 2kT\rho_A(B_{AA} - B_{AB}). \tag{6.10.21}$$

Note that in (6.10.20) and (6.10.21), we retained only the first-order terms in x_A and in ρ_A, respectively. Using these activity coefficients, the chemical potential in (6.10.10) can be rewritten in two alternative forms:

$$\mu_A(T, P, x_A) = \mu_A^{0x} + kT \ln(x_A \gamma_A^{D,x}) \qquad (6.10.22)$$

$$\mu_A(T, P, \rho_A) = \mu_A^{0\rho} + kT \ln(\rho_A \gamma_A^{D,\rho}) \qquad (6.10.23)$$

The notation $\gamma_A^{D,x}$ and $\gamma_A^{D,\rho}$ has been introduced to distinguish between the two cases.

6.11. THE MCMILLAN–MAYER THEORY OF SOLUTIONS

The McMillan–Mayer[3] (MM) theory is essentially a formal generalization of the theory of real gases. We recall that in the theory of real gases we have an expansion of the pressure in power series in the density of the form

$$\frac{P}{kT} = \rho + B_2(T)\rho^2 + B_3(T)\rho^3 + \cdots, \qquad (6.11.1)$$

where B_j is the jth virial coefficient (see section 1.8). Likewise, the MM theory leads to an expansion of the osmotic pressure in power series in the solute density. For a two-component system of A diluted in B, the analogue of the virial expansion is

$$\frac{\pi}{kT} = \rho_A + B_2^*(T, \lambda_B)\rho_A^2 + B_3^*(T, \lambda_B)\rho_A^3 + \cdots, \qquad (6.11.2)$$

where B_j^* is called the jth virial coefficient of the osmotic pressure. Note that these virial coefficients depend on both the temperature and the solvent activity λ_B [or the solvent chemical potential $\lambda_B = \exp(\beta\mu_B)$].

In the case of real gasses, the terms in the expansion (6.11.1) correspond to successive corrections to the ideal-gas behavior, due to interactions among pairs, triplets, quadruplets, etc., of particles. One of the most remarkable results of the statistical-mechanical theory of real gases is that the coefficients B_j depend on the properties of a system containing exactly j particles. For instance, $B_2(T)$ can be computed from a system of two particles in the system.

In a similar fashion, the coefficients B_j^* can be expressed in terms of the properties of j solute particles in a pure solvent B at a given activity λ_B. If $\lambda_B \to 0$, we recover the expansion (6.11.1) from (6.11.2), i.e., the vacuum filling the space between the particles in (6.11.1) is a special case of a solvent in (6.11.2). It is in this sense that (6.11.2) is a generalization of (6.11.1).

Inherent in the MM theory is the distinction between the solute and the solvent. (Either or both may consist of one or more components. Here we treat only the case of one solute A in one solvent B.) Furthermore, the theory is useful for quite low solute densities. The most useful case is the expansion up to ρ_A^2, i.e., the first-order deviation from the dilute ideal behavior. Higher-order corrections to the ideal-gas equations are sometimes useful if we know the interaction energy among j particles. The situation is less satisfactory for the higher order corrections to the dilute ideal behavior. As we shall see, B_2^* is expressible as an integral over the pair correlation function for two solutes in a pure solvent. B_3^* requires the knowledge of the triplet correlation function for three solutes in pure solvent. Since we know almost nothing on the triplet (and higher) correlation functions, the expansion (6.11.2) is useful in actual application up to the second-order term in the solute density.

6.11.1. Derivation

We now derive a general result which expresses the GPF of a two-component system in a form which is strikingly analogous to the expression of the GPF of a one-component system. Consider a two-component system of a solute A and a solvent B at a given temperature and activities λ_A and λ_B, respectively. The GPF of such a system is defined by

$$
\begin{aligned}
\Xi(T, V, \lambda_B, \lambda_A) &= \sum_{N_A \geq 0} \sum_{N_B \geq 0} Q(T, V, N_B, N_A)\lambda_A^{N_A}\lambda_B^{N_B} \\
&= \sum_{N_A \geq 0} \sum_{N_B \geq 0} \frac{z_A^{N_A} z_B^{N_B}}{N_A! N_B!} \int \exp[-\beta U(N_B, N_A)] \, d\mathbf{X}^{N_B} \, d\mathbf{X}^{N_A} \\
&= \sum_{N_A \geq 0} \frac{z_A^{N_A}}{N_A!} \int d\mathbf{X}^{N_A} \left\{ \sum_{N_B \geq 0} \frac{z_B^{N_B}}{N_B!} \int \exp[-\beta U(N_B, N_A)] \, d\mathbf{X}^{N_B} \right\}, \quad (6.11.3)
\end{aligned}
$$

where we defined

$$
z_\alpha = \frac{\lambda_\alpha q_\alpha}{\Lambda_\alpha^3 (8\pi^2)} \tag{6.11.4}
$$

and used the shorthand notation $U(N_B, N_A)$ for the total interaction energy among the $N_A + N_B$ particles in the system.

At the limit of the dilute ideal solution, we already know the relation between λ_A and ρ_A, i.e.,

$$
\mu_A = W(A|B) + kT \ln \rho_A \Lambda_A^3 q_A^{-1}, \tag{6.11.5}
$$

or equivalently

$$
\lambda_A = \exp(\beta \mu_A) = \rho_A \Lambda_A^3 q_A^{-1} \exp[\beta W(A|B)]. \tag{6.11.6}
$$

We have the limiting behavior of z_A from (6.11.4) and (6.11.5)

$$
\gamma_A^0 = \lim_{\rho_A \to 0} \frac{z_A 8\pi^2}{\rho_A} = \exp[\beta W(A|B)]. \tag{6.11.7}
$$

Thus γ_A^0, defined in (6.11.7), is related to the coupling work of A against pure B.

We now recall the definition of the molecular distribution function of n_A solute particles in an open system (see section 5.7)

$$
\begin{aligned}
\rho^{(n_A)}(\mathbf{X}^{n_A}) &= \Xi^{-1} \sum_{N_A \geq n_A} \sum_{N_B \geq 0} \frac{z_A^{N_A} z_B^{N_B}}{(N_A - n_A)! N_B!} \int \exp[-\beta U(N_B, N_A)] \, d\mathbf{X}^{N_A - n_A} \, d\mathbf{X}^{N_B} \\
&= z_A^{n_A} \Xi^{-1} \sum_{N_A' \geq 0} \sum_{N_B \geq 0} \frac{z_A^{N_A'} z_B^{N_B}}{N_A'! N_B!} \int \exp[-\beta U(N_B, N_A' + n_A)] \, d\mathbf{X}^{N_A'} \, d\mathbf{X}^{N_B}, \quad (6.11.8)
\end{aligned}
$$

where $\Xi = \Xi(T, V, \lambda_B, \lambda_A)$. In the second form on the rhs of (6.11.8), we have changed variables from $N_A \geq n_A$ to $N'_A = N_A - n_A \geq 0$. $\rho^{(n_A)}(\mathbf{X}^{n_A})$ is the probability density of finding n_A particles at a configuration $\mathbf{X}^{n_A} = \mathbf{X}_1 \cdots \mathbf{X}_{n_A}$ in an open system characterized by the variables $T, V, \lambda_B, \lambda_A$. The n_A-particle correlation function is defined by

$$g^{(n_A)}(\mathbf{X}^{n_A}) = \frac{\rho^{(n_A)}(\mathbf{X}^{n_A})}{[\rho_A^{(1)}(\mathbf{X})]^{n_A}}$$

$$= \left(\frac{z_A 8\pi^2}{\rho_A}\right)^{n_A} \Xi^{-1} \sum_{N_A \geq 0} \sum_{N_B \geq 0} \frac{z_A^{N_A} z_B^{N_B}}{N_A! N_B!}$$

$$\times \int \exp[-\beta U(N_B, N_A + n_A)]\, d\mathbf{X}^{N_A}\, d\mathbf{X}^{N_B}, \qquad (6.11.9)$$

where ρ_A is the average density of A (in the open system), and we have replaced N'_A by N_A; otherwise, the sum in (6.11.8) is the same as the sum in (6.11.9).

We now take the limit $\rho_A \to 0$ (or $z_A \to 0$; the two are proportional to each other at low density). In this limit, $z_A 8\pi^2/\rho_A \to \gamma_A^0$, and all the terms in the sum (6.11.9) are zeros except those for which $N_A = 0$. Therefore, denoting

$$g^{(n_A)}(\mathbf{X}^{n_A}, z_A = 0) = \lim_{\rho_A \to 0} g^{(n_A)}(\mathbf{X}^{n_A}, z_A), \qquad (6.11.10)$$

we obtain from (6.11.9) the expression

$$g^{(n_A)}(\mathbf{X}^{n_A}, z_A = 0) = (\gamma_A^0)^{n_A} \Xi(T, V, \lambda_B)^{-1} \left\{ \sum_{N_B \geq 0} \frac{z_B^{N_B}}{N_B!} \int \exp[-\beta U(N_B, n_A)]\, d\mathbf{X}^{N_B} \right\}.$$
$$(6.11.11)$$

It should be realized that in a pure solvent ($\rho_A = 0$, or $z_A = 0$) there are no solute particles and therefore one cannot define the correlation function among solute particles. The quantity defined in (6.11.10) is the correlation function among n_A solute particles when the density of all the remaining solute particles goes to zero. In other words, $g^{(n_A)}(\mathbf{X}^{n_A}, z_A = 0)$ is the correlation function for exactly n_A solute particles at configuration $\mathbf{X}^{n_A}$ in a pure solvent B.

We now observe that the expression in curly brackets in (6.11.11) is the same as in (6.11.3) except for the replacement of N_A by n_A. Hence we rewrite (6.11.3) using (6.11.11) as

$$\Xi(T, V, \lambda_B, \lambda_A) = \sum_{N_A \geq 0} \frac{(z_A/\gamma_A^0)^{N_A}}{N_A!} \Xi(T, V, \lambda_B) \int g^{(N_A)}(\mathbf{X}^{N_A}; z_A = 0)\, d\mathbf{X}^{N_A}. \quad (6.11.12)$$

We now define the N_A-particle potential of average force in a pure solvent B by

$$W(\mathbf{X}^{N_A}; z_A = 0) = -kT \ln g^{(N_A)}(\mathbf{X}^{N_A}; z_A = 0). \qquad (6.11.13)$$

We also recall the fundamental relation between the pressure of a system and the GPF

$$P(T, V, \lambda_B, \lambda_A)V = -kT \ln \Xi(T, V, \lambda_B, \lambda_A) \qquad (6.11.14)$$

$$P(T, V, \lambda_B)V = -kT \ln \Xi(T, V, \lambda_B). \qquad (6.11.15)$$

In (6.11.14), $P(T, V, \lambda_B, \lambda_A)$ is the pressure of a system characterized by the variables $T, V, \lambda_B, \lambda_A$, whereas $P(T, V, \lambda_B)$ in (6.11.15) is the corresponding pressure of the pure solvent B at T, V, λ_B. The difference between these two pressures is the osmotic pressure, thus

$$\pi V = [P(T, V, \lambda_B, \lambda_A) - P(T, V, \lambda_B)]V = -kT \ln\left[\frac{\Xi(T, V, \lambda_B, \lambda_A)}{\Xi(T, V, \lambda_B)}\right]. \quad (6.11.16)$$

Defining

$$\zeta_A = \frac{z_A}{\gamma_A^0}, \quad (6.11.17)$$

we can rewrite (6.11.12) as

$$\exp(\beta \pi V) = \sum_{N_A \geq 0} \frac{\zeta_A^{N_A}}{N_A!} \int \exp[-\beta W(\mathbf{X}^{N_A}; z_A = 0)] \, d\mathbf{X}^{N_A}. \quad (6.11.18)$$

This result should be compared with the corresponding expression of the GPF of a *one-component* system

$$\exp(\beta P V) = \sum_{N \geq 0} \frac{z^N}{N!} \int \exp[-\beta U(\mathbf{X}^N)] \, d\mathbf{X}^N. \quad (6.11.19)$$

Here the pressure of a one-component open system is related to integrals over the potentials of N particles $U(\mathbf{X}^N)$, in the same manner as the osmotic pressure is related to integrals over the potentials of average force of N_A solute particles, $W(\mathbf{X}^{N_A}; z_A = 0)$ in a pure solvent B. This is the main result of the MM theory.

This formal analogy assures us that if we expand the osmotic pressure in power series in the solute density ρ_A, the analogy between the virial expansion of the pressure and the virial expansion of the osmotic pressure will be maintained. For instance, B_2 is given by an integral over the pair potential $U(\mathbf{X}_1, \mathbf{X}_2)$; likewise, B_2^* will be given by the same form of an integral, but one in which the pair potential $U(\mathbf{X}_1, \mathbf{X}_2)$ is replaced by the potential of average force $W_{AA}(\mathbf{X}_1, \mathbf{X}_2; z_A = 0)$ for two solute particles in a pure solvent.

In the virial expansion of the pressure, each coefficient B_j depends on the properties of a system of exactly j molecules. Likewise, in the virial expansion of the osmotic pressure, each coefficient B_j^* depends on the properties of j solutes in a pure solvent. It is true that in the expressions B_j^*, the solvent does not feature explicitly. This apparent simplification is quite deceiving, however. It should be realized that the potential of average force is dependent on the solvent properties. Therefore, in any actual calculation one must evaluate the relevant potential of average force, taking into account the properties of the solute as well as of the solvent.

As we shall soon see, the virial expansion of the osmotic pressure, although formally exact, is not very useful beyond the first-order correction to the DI limiting case. Higher-order correction terms involve higher-order potential of average forces about the analytical form of which very little is known.

The first-order term involves the pair potential of average force. This may be sometimes approximated by viewing the solvent as a continuum. An example of such an approximation is the Debye–Hückel theory of ionic solutions.

6.11.2. The Virial Expansion of the Osmotic Pressure

The osmotic pressure is defined experimentally as the excess pressure required to be exerted on a solution to maintain its equilibrium with respect to the flow of solvent from the pure solvent into the solution.

Consider two systems characterized by the variables T, P, N_B, N_A, and T, P, N_B, respectively. If we bring the two systems into contact through a partition permeable to B only, then the solvent B will flow from the pure phase into the solution, Fig. 6.5.

This follows from the condition of stability; i.e., we always have

$$\mu_B(T, P, N_B) > \mu_B(T, P, N_B, N_A), \qquad (6.11.20)$$

and therefore B will flow from the pure phase into the solution. Equation (6.11.20) is equivalent to the condition

$$\left(\frac{\partial \mu_B}{\partial N_A}\right)_{T,P,N_B} \leq 0. \qquad (6.11.21)$$

This is one of the stability conditions.

At constant P, T we cannot maintain the equilibrium in the system of Fig. 6.5(a). The solvent will flow into the solution indefinitely. To stop the flow we can increase the pressure on the solution. Since[†]

$$\left(\frac{\partial \mu_B}{\partial P}\right)_{T,N_A,N_B} = \bar{V}_B > 0, \qquad (6.11.22)$$

increasing the pressure increases the chemical potential of B. The osmotic pressure is defined as the excess pressure required to be exerted on the solution to compensate exactly

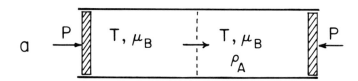

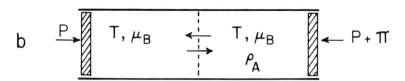

FIGURE 6.5. (a) Two compartments at T, μ_B (left) and $T, \mu_B \rho_A$ (right). Solvent will flow from left to right. (b) Osmotic experiment. Additional pressure on the solution prevents the flow of solvent.

---

† Note that in general the partial molar volume can also be negative. Here we consider dilute solutions for which $\bar{V}_B \approx V_B^P > 0$.

for the decrease of the chemical potential of B due to the presence of the solute, Fig. 6.5b, i.e:

$$\mu_B(T, P + \pi, N_B, N_A) = \mu_B(T, P, N_B). \tag{6.11.23}$$

The relation between π and statistical mechanics is obtained for a system such as the one in Fig. 6.5b, when the two compartments at equilibrium are characterized by the variables $T, V, \lambda_B, \lambda_A$ and T, V, λ_B, respectively. The corresponding relation is (6.11.16), which we rewrite as

$$\exp(\beta \pi V) = \frac{\Xi(T, V, \lambda_B, \lambda_A)}{\Xi(T, V, \lambda_B)}$$

$$= \frac{\sum\limits_{N_A \geq 0} \sum\limits_{N_B \geq 0} Q(T, V, N_B, N_A) \lambda_B^{N_B} \lambda_A^{N_A}}{\sum\limits_{N_B \geq 0} Q(T, V, N_B) \lambda_B^{N_B}}$$

$$= 1 + \sum\limits_{N_A \geq 1} Q_{N_A}^* \lambda_A^{N_A}, \tag{6.11.24}$$

where $Q_{N_A}^*$ is defined by

$$Q_{N_A}^* = Q^*(T, V, \lambda_B, N_A)$$

$$= \frac{\sum\limits_{N_B \geq 0} Q(T, V, N_B, N_A) \lambda_B^{N_B}}{\sum\limits_{N_B \geq 0} Q(T, V, N_B) \lambda_B^{N_B}}. \tag{6.11.25}$$

Equation (6.11.24) is formally the same as the expression of the pressure in section 1.8, except for $Q_{N_A}^*$ replacing Q_N and λ_A replacing λ. Therefore if we repeat the same steps as in section 1.9 we should arrive at formally the same expansion as in section 1.8. We briefly repeat the derivation to obtain the expansion of the osmotic pressure up to the first-order deviation from the dilute ideal limit.

Expanding $\beta \pi V$ from (6.11.24) to the second order in λ_A, we obtain

$$\beta \pi V = Q_1^* \lambda_A + [Q_2^* - \tfrac{1}{2} Q_1^{*2}] \lambda_A^2 + \cdots. \tag{6.11.26}$$

The average density of A is obtained from

$$\rho_A = \frac{\bar{N}_A}{V}$$

$$= \frac{\lambda_A}{V} \frac{\partial \ln \Xi(T, V, \lambda_B, \lambda_A)}{\partial \lambda_A}$$

$$= \frac{\lambda_A}{V} \frac{\partial (\beta \pi V)}{\partial \lambda_A}$$

$$= \frac{Q_1^*}{V} \lambda_A + \frac{2}{V} (Q_2^* - \tfrac{1}{2} Q_1^{*2}) \lambda_A^2 + \cdots. \tag{6.11.27}$$

As in section 1.8, we invert the power series (6.11.27) to obtain an expansion of λ_A in power series of ρ_A:

$$\lambda_A = \frac{V}{Q_1^*}\rho_A - 2(Q_2^* - \tfrac{1}{2}Q_1^{*2})\frac{V^2}{Q_1^{*3}}\rho_A^2 + \cdots, \tag{6.11.28}$$

which is substituted in (6.11.26) to obtain the required expansion

$$\beta\pi = \rho_A - [Q_2^* - \tfrac{1}{2}Q_1^{*2}]\frac{V}{Q_1^{*2}}\rho_A^2 + \cdots$$

$$= \rho_A + B_2^*\rho_A^2 + \cdots. \tag{6.11.29}$$

A more explicit expression for B_2^* can be obtained by rewriting Q_2^* and Q_1^*, which from (6.11.24) and (6.11.18) can be written as

$$Q_1^* = \frac{q_A}{\Lambda_A^3 \gamma_A^0 (8\pi^2)}\int d\mathbf{X}_1 \tag{6.11.30}$$

$$Q_2^* = \frac{1}{2}\left(\frac{q_A}{\Lambda_A^3 \gamma_A^0 8\pi^2}\right)^2 \int \exp[-\beta W_{AA}^{(2)}(\mathbf{X}_1, \mathbf{X}_2)]\, d\mathbf{X}_1\, d\mathbf{X}_2. \tag{6.11.31}$$

Hence

$$B_2^* = -\frac{V}{Q_1^{*2}}[Q_2^* - \tfrac{1}{2}Q_1^{*2}]$$

$$= \frac{-1}{2(8\pi^2)^2 V}\int \{\exp[-\beta W_{AA}^{(2)}(\mathbf{X}_1, \mathbf{X}_2)] - 1\}\, d\mathbf{X}_1\, d\mathbf{X}_2$$

$$= -\tfrac{1}{2}\int_0^\infty \{\exp[-\beta W_{AA}^{(2)}(R)] - 1\}4\pi R^2\, dR, \tag{6.11.32}$$

where $W_{AA}^{(2)}(R)$ is the potential of average force between two solutes A at a distance R, in pure solvent at a given T and μ_B. As expected, the coefficient B_2^* is related to $W_{AA}^{(2)}(R)$ in the same manner as B_2 is related to $U(R)$ for real gases. Alternatively, if we let $\lambda_B \to 0$ in (6.11.32), $W_{AA}^{(2)}(R) \to U_{AA}(R)$ and $B_2^* \to B_2$.

Another way of interpreting B_2^* is the following: We recall the normalization conditions of $\rho_{AA}^{(2)}$ and $\rho_A^{(1)}$ in the open system (section 6.7)

$$\int \overline{\rho_{AA}^{(2)}(\mathbf{X}_1, \mathbf{X}_2)}\, d\mathbf{X}_1\, d\mathbf{X}_2 = \langle N_A^2 \rangle - \langle N_A \rangle \tag{6.11.33}$$

$$\int \overline{\rho_A^{(1)}(\mathbf{X}_1)}\, d\mathbf{X}_1 = \langle N_A \rangle. \tag{6.11.34}$$

Hence

$$\int [\overline{\rho_{AA}^{(2)}(\mathbf{X}_1, \mathbf{X}_2)} - \overline{\rho_A^{(1)}(\mathbf{X}_1)\rho_A^{(1)}(\mathbf{X}_2)}]\, d\mathbf{X}_1\, d\mathbf{X}_2 = \langle N_A^2 \rangle - \langle N_A \rangle^2 - \langle N_A \rangle. \tag{6.11.35}$$

We have already obtained the expansion of $\langle N_A \rangle$ to second order in λ_A in (6.11.27):

$$\langle N_A \rangle = Q_1^* \lambda_A + 2(Q_2^* - \tfrac{1}{2}Q_1^{*2})\lambda_A^2 + \cdots . \qquad (6.11.36)$$

We use the following identity

$$\lambda_A \frac{\partial}{\partial \lambda_A}\left(\lambda_A \frac{\partial \ln \Xi}{\partial \lambda_A}\right) = \lambda_A \frac{\partial}{\partial \lambda_A}(\langle N_A \rangle)$$

$$= \langle N_A^2 \rangle - \langle N_A \rangle^2 \qquad (6.11.37)$$

to obtain an expansion of

$$\langle N_A^2 \rangle - \langle N_A \rangle^2 = \lambda_A \frac{\partial}{\partial \lambda_A} \langle N_A \rangle$$

$$= Q_1^* \lambda_A + 4(Q_2^* - \tfrac{1}{2}Q_1^{*2})\lambda_A^2 + \cdots . \qquad (6.11.38)$$

Hence using (6.11.36) and (6.11.38) in (6.11.35) we obtain

$$\int \left[\overline{\rho_{AA}^{(2)}(\mathbf{X}_1, \mathbf{X}_2)} - \overline{\rho_A^{(1)}(\mathbf{X}_1)}\,\overline{\rho_A^{(1)}(\mathbf{X}_2)}\right] d\mathbf{X}_1 \, d\mathbf{X}_2 = 2(Q_2^* - \tfrac{1}{2}Q_1^{*2})\lambda_A^2 + \cdots$$

$$= 2(Q_2^* - \tfrac{1}{2}Q_1^{*2})\left(\frac{V}{Q_1^*}\right)^2 \rho_A^2 + \cdots$$

$$= -2V\rho_A^2 B_2^* + \cdots . \qquad (6.11.39)$$

Using the notation of section 6.7, we arrive at the relation

$$B_2^* = -\tfrac{1}{2}G_{AA}^0, \qquad (6.11.40)$$

where G_{AA}^0 is the limit of G_{AA} when $\rho_A \to 0$.

Finally, we note that had we kept the second-order correction to the ideal dilute limiting behavior of the osmotic pressure, we would have obtained

$$\beta \pi = \rho_A + B_2^* \rho_A^2 + B_3^* \rho_A^3 + \cdots , \qquad (6.11.41)$$

where B_3^* has the same form as B_3 in section 1.8, but for using $W^{(3)}$ and $W^{(2)}$ in place of the potential functions U_3 and U. However, in contrast to the case of real gases, where the expression for B_3 could be simplified if the assumption of pairwise additivity of U_3 is introduced [Eq. (1.8.24)], here, no such simplification can be achieved without further assumptions. The reason is that $W^{(3)}$ can always be written as

$$W^{(3)}(\mathbf{X}_1, \mathbf{X}_2, \mathbf{X}_3) = U_3(\mathbf{X}_1, \mathbf{X}_2, \mathbf{X}_3) + \delta W(\mathbf{X}_1, \mathbf{X}_2, \mathbf{X}_3), \qquad (6.11.42)$$

where δW is the indirect work (in the appropriate ensemble) required to bring the three particles from infinite separation to the final configuration $\mathbf{X}_1, \mathbf{X}_2, \mathbf{X}_3$. Even for systems for which U_3 is pairwise additive, the quantity $W^{(3)}$ in general is not pairwise additive. This is the reason why the usefulness of the expansion (6.11.41) is limited to the second virial term only.

6.12. ELECTROLYTE SOLUTIONS

Aqueous electrolyte solutions have been studied from the very beginning of physical chemistry. These solutions have been investigated extensively by both experimental and theoretical methods. They are also the main media in which most biochemical processes occur. From the thermodynamic point of view, electrolyte solutions have two distinct features that separate them from other solutions and therefore deserve special considera- tion. First, the dissociation into ions leads to deviations from the ideal dilute behavior. This may be easily taken care of by recounting the number of free solute particles in the solution. Second, after we have accounted for the dissociation into ions, the deviations from the ideal dilute behavior occur at relatively lower solute densities, and their depend- ence on the solute density is anomalous compared with nonionic solutes. These deviations will be the main subject matter of this section.

6.12.1. Dissociation into Ions

Nowadays, it is a well-established fact that electrolyte solutes dissociate into ions. We shall deal only with the so-called strong electrolytes, which are presumed to be completely dissociated, at least in the region of concentrations in which we shall be interested in this section, i.e., below, say, 0.1 molar.

To demonstrate the effect of the dissociation on the thermodynamic behavior, con- sider a solute A, very diluted in a solvent B. In the limit of very low solute density, the system obeys Henry's law in the form

$$P_A = K_A x_A, \tag{6.12.1}$$

where P_A is the partial pressure of A, x_A its mole fraction in the solution, and K_A is the Henry's constant. The theoretical significance of K_A may be obtained from the equilibrium condition

$$\mu_A^l = \mu_A^g \tag{6.12.2}$$

or

$$W(A \mid B) + kT \ln \rho_B^p \Lambda_A^3 + kT \ln x_A = kT \ln\left(\frac{P_A \Lambda_A^3}{kT}\right), \tag{6.12.3}$$

from which we identify

$$K_A = kT\rho_B^p \exp[\beta W(A \mid B)]; \tag{6.12.4}$$

i.e., K_A is proportional to the pure solvent density ρ_B^p and is related to the coupling work of A against pure B.

Experimentally, K_A can be determined simply from the limiting slope of the partial pressure P_A as a function of x_A, i.e.,

$$K_A = \lim_{x_A \to 0}\left(\frac{P_A}{x_A}\right). \tag{6.12.5}$$

If A is an electrolyte, say HCl, then a plot of P_A as a function of x_A would not lead to the typical linear dependence (6.12.1) at the limit of $x_A \to 0$. Instead, we find that at this limit

$$P_A = K_A x_A^2. \tag{6.12.6}$$

The molecular interpretation of K_A may be obtained by considering again the equilibrium condition (6.12.2), but now we write the chemical potential of the solute A, say HCl, as

$$\mu^l_{HCl} = \mu_{H^+} + \mu_{Cl^-}. \qquad (6.12.7)$$

Hence the analogue of (6.12.3) is

$$W(H^+, Cl^- | B) + kT \ln \rho_{H^+} \rho_{Cl^-} \Lambda^3_{H^+} \Lambda^3_{Cl^-} = kT \ln \left(\frac{P_A \Lambda^3_A q_A^{-1}}{kT} \right), \qquad (6.12.8)$$

where we assumed that HCl is completely dissociated in the liquid phase, but is in a molecular form in the gaseous phase. q_A is the vibrational–rotational PF of HCl. Since $\rho_{H^+} = \rho_{Cl^-} = \rho^p_B x_A$, we can identify K_A from (6.12.6) and (6.12.8) as

$$K_A = \frac{kT (\rho^p_B)^2 \Lambda^3_{H^+} \Lambda^3_{Cl^-} q_A \exp[\beta W(H^+, Cl^- | B)]}{\Lambda^3_A} \qquad (6.12.9)$$

$W(H^+, Cl^- | B)$ is the coupling work of H^+ and Cl^- at infinite separation from each other in a pure solvent B. This may be determined from experimental data (see section 6.13).

The characteristic behavior (6.12.6) of a 1:1 electrolyte merely reflects the fact that at infinite dilution, each solute A produces two free particles, here H^+ and Cl^-, in the solution. Similar effects can be observed with respect to any other colligative property of the solution.

For a general salt of the type $C_{\nu^+} A_{\nu^-}$ that dissociates into

$$C_{\nu^+} A_{\nu^-} \rightarrow \nu_+ C^{z+} + \nu_- A^{z-}, \qquad (6.12.10)$$

where z_+ and z_- are the valences of the cation and the anion, respectively, i.e., the charge of C is $z_+ e$ and of A is $z_- e$, where $-e$ is the charge of the electron. The total number of free species in the solution is thus

$$\nu = \nu_+ + \nu_-. \qquad (6.12.11)$$

Hence the corresponding Henry's law (presuming that this salt has a measurable vapor pressure) is

$$P_{salt} = K x^\nu_{salt}. \qquad (6.12.12)$$

We see that the hypothesis that electrolytes dissociate into ions, put forward by Arrhenius almost a hundred years ago, can account for the deviations from the ideal behavior of extremely dilute solutions.

6.12.2. Deviations from Ideality Due to Long-Range Interactions

For nonionic solutes at densities higher than the limiting dilute ideal case, we may introduce activity coefficients into Eq. (6.12.1) to account for deviations from Henry's law. Thus we write

$$P_A = K_A x_A \gamma_A, \qquad (6.12.13)$$

where γ_A is the activity coefficient (see section 6.9 for more details on the notation of the various activity coefficients). We have also seen that γ_A may be expanded in power series in the solute density of the form

$$kT \ln \gamma_A = a_1 \rho_A + a_2 \rho_A^2 + a_3 \rho_A^3 + \cdots, \tag{6.12.14}$$

where the coefficients a_i are related to the virial coefficients of the osmotic pressure. The physical significance of these coefficients is that as we increase the solute concentration, we can account for the deviations from ideality by successive correction terms that depends on two particles, three particles, etc. It turns out that such systematic corrections cannot be applied to ionic solutions. The reason is that the ions have a long-range coulombic interaction. These long range interactions cause deviations from ideality at relatively lower concentrations, compared with nonionic solutes, and the limiting corrections to the ideal behavior depends on the collective interactions among many ions.

To appreciate the nature of the problem, suppose we take a system containing exactly two particles in a volume V. The pressure of such a system is given by

$$\beta P = \rho + B_2(T, R_M)\rho^2, \tag{6.12.15}$$

where

$$B_2(T, R_M) = -\tfrac{1}{2} \int_0^{R_M} \{\exp[-\beta U(R)] - 1\} 4\pi R^2 \, dR \tag{6.12.16}$$

and $\rho = 2/V$. Here, for simplicity we took a spherical volume of radius R_M. Equation (6.12.15) will be exact whether or not the two particles are charged. Since the volume is finite, the quantity $B_2(T, R_M)$ is finite even when $U(R)$ is a long-range potential. Also, if we let $R_M \to \infty$ but keep the same two particles in the system, we obtain

$$\lim_{R_M \to \infty} B_2(T, R_M) \left(\frac{2}{V}\right)^2 = 0; \tag{6.12.17}$$

i.e., at the limit of $R_M \to \infty$ we obtain the ideal gas behavior $\beta P = \rho$, again, whether or not the particles are charged. The anomalous behavior of charged particles occurs when we attempt to take the thermodynamic limit of (6.12.15), i.e., if we let $R_M \to \infty$ and $N \to \infty$ keeping the density fixed at the same value, $\rho = 2/V$. In this limit, the behavior of Eq. (6.12.15) is different for charged or uncharged particles. For uncharged particles, since ρ is fixed, and since $U(R)$ is of relatively short range, the passage to the limit $R_M \to \infty$ will lead to a constant value of

$$B_2(T) = \lim_{R_M \to \infty} B_2(T, R_M) \tag{6.12.18}$$

and therefore expression (6.12.15) will correctly account for the first-order deviation from the ideal gas behavior. What happens when the particles are charged? At the thermodynamic limit, ρ is fixed, but if $U(R)$ is a coulombic potential of the form

$$U(R) = \frac{\alpha}{R}, \tag{6.12.19}$$

then in the limit (6.12.18), $B_2(T)$ will diverge to infinity. To see this more clearly, we can expand the exponent in (6.12.16) using the potential (6.12.19) to obtain

$$B_2(T, R_M) = -\tfrac{1}{2} \int_0^{R_M} \left[\left(-\beta\frac{\alpha}{R} \right) + \frac{1}{2}\left(\beta\frac{\alpha}{R} \right)^2 + \frac{1}{6}\left(\beta\frac{\alpha}{R} \right)^3 + \cdots \right] 4\pi R^2\, dR$$

$$= a_1 R_M^2 + a_2 R_M + a_3 \ln R_M + a_4 R_M^{-1} + \cdots . \qquad (6.12.20)$$

Clearly, if we let $R_M \to \infty$, this expansion diverges to infinity. This is the reason why the virial expansion (6.12.15) cannot describe the behavior of a macroscopic system of charged particles. Note that the integral diverges also for potential of the form α/R^2 or α/R^3 but converges for α/R^4. The typical van der Waals interactions behave as α/R^6. It should be noted that the integrands in (6.12.20) also diverge at the lower limit of the integral. This is not a real difficulty since we do not expect the coulombic interaction (6.12.19) to hold for $R \to 0$ (strictly point charges). In reality, the particles always have a finite volume and therefore for sufficiently small R, $U(R)$ actually diverges to $+\infty$ and $\exp[-\beta U(R)] \to 1$ so that the lower limit of the integral does not cause any difficulty for real particles.

6.12.3. The Debye–Hückel Theory

We consider a system of N solvent molecules and any number of electrolyte solutes. We denote by N_i the number of ions of species i having charge $z_i e$ ($-e$ being the charge of an electron).

Consider for simplicity the chemical potential of a $1:1$ salt, which we denote by CA (for cation and anion),

$$\mu_{CA} = W(C|l) + W(A|l) + kT \ln \rho_C \rho_A \Lambda_C^3 \Lambda_A^3, \qquad (6.12.21)$$

where $W(\alpha|l)$ is the coupling work of the species α (C or A) to the entire liquid l. We assumed also that the ions do not have internal structure so that $q_C = q_A = 1$.

If the solution is extremely dilute with respect to CA, then the coupling work $W(C|l)$ and $W(A|l)$ in (6.12.21) can be replaced by the corresponding coupling work against the pure solvent w which we denote by $W(C|w)$ and $W(A|w)$, respectively. For any finite concentration of the salt we may rewrite (6.12.21) as

$$\mu_{CA} = W(C|w) + W(A|w) + kT \ln \rho_C \rho_A \Lambda_C^3 \Lambda_A^3 \gamma_C \gamma_A, \qquad (6.12.22)$$

where

$$kT \ln \gamma_A = W(A|l) - W(A|w) \qquad (6.12.23)$$

$$kT \ln \gamma_C = W(C|l) - W(A|w), \qquad (6.12.24)$$

i.e., the activity coefficients γ_A and γ_C correct for the deviations from the ideal dilute limit. Note that in general the extent of deviation from ideality will depend on the independent variables we use to describe the system (section 6.9). Here we did not specify which variables have been chosen.

We are now interested in the first-order deviation from the ideal dilute limit. We assume that as we increase the salt concentration slightly beyond the ideal dilute limit, the long-range coulomb interaction between the ions causes the initial deviations from ideality; i.e., at these concentrations the system would have been dilute ideal if the solute were not electrolytes. Thus, our main goal is to calculate the deviations due solely to the interionic interactions. We shall now discuss the chemical potential of each ion separately. This is done for convenience only. Eventually we must combine these chemical potentials to obtain the chemical potential of the entire salt. Thus we write for the 1:1 salt

$$\mu_{CA} = \mu_C + \mu_A,$$
(6.12.25)

where

$$\mu_a = W(a \mid w) + kT \ln \rho_a \Lambda_a^3 \gamma_a.$$
(6.12.26)

μ_a formally corresponds to the work of adding one ion of species a, keeping all other N_i constant. This process includes the electrical work of charging the system, which depends on the form of the volume of the system. However, whatever the charging work is, it will be canceled out eventually when we form the combination (6.12.25).

We estimate $kT \ln \gamma_a$ (a being either C or A). We recall the general expression for the coupling work of a particle of species a

$$W(a \mid l) = -kT \ln \langle \exp(-\beta B_a) \rangle_l,$$
(6.12.27)

where B_a is the total binding energy of the particle of species a, at some fixed position, say $\mathbf{R}_a$, to the entire solvent molecules, and the average is over all configurations of all the other solute and solvent particles in the liquid except for one a, which is held at a fixed position. We write

$$B_a = \sum_{i=1}^{N} U(\mathbf{R}_a, \mathbf{X}_i) + \sum_{j=1} U(\mathbf{R}_a, \mathbf{R}_j) = B_a^w + B_a^{el},$$
(6.12.28)

where the first sum includes the interactions between the a particle and all solvent molecules, and the second term includes all the interactions with the other ions in the solution. Since we are interested in very dilute solutions, the latter essentially consists of the electrostatic interactions between the a particle and all other ions. We now rewrite the average quantity in (6.12.27) as (for more details see section 6.14)

$$-kT \ln \langle \exp[-\beta(B_a^w + B_a^{el})] \rangle_l = -kT \ln \langle \exp(-\beta B_a^w) \rangle_l \langle \exp(-\beta B_a^{el}) \rangle_{l, B_a^w}$$
$$= W(a \mid w) + W(a \mid el),$$
(6.12.29)

where the first average is over all particles in the solution except the specific ion of type a. If the solution is very dilute, then this term is the same as the coupling work of a to pure solvent w. The second term includes the conditional average of the quantity $\exp(-\beta B_a^{el})$ over all configurations of the particles in the solution, given that the solute–solvent interactions have already been switched on. Thus we identify the activity coefficient of the species a with

$$kT \ln \gamma_a = W(a \mid el).$$
(6.12.30)

Note that we have implicitly assumed that the activity coefficient is due to the electrostatic interactions only. Other parts of the solute–solute interactions are presumed to be unimportant at very low concentrations.

The expression for $W(a \mid el)$ can be rewritten as follows. We introduce a charging (or coupling) parameter λ in Eq. (6.12.28) defined by

$$B_a(\lambda) = B_a^w + \lambda B_a^{el},$$
(6.12.31)

where for $\lambda = 0$, the interaction between α and all the other ions is switched off. For $\lambda = 1$, we recover (6.12.28). Thus for any value of λ we can write

$$W(\alpha, \lambda | el) = -kT \ln \langle \exp(-\beta\lambda B_\alpha^{el}) \rangle_{l, B_\alpha^w} \qquad (6.12.32)$$

$$\frac{\partial W}{\partial \lambda} = \frac{\int B_\alpha^{el} \exp(-\beta U_N - \beta B_\alpha^w - \beta\lambda B_\alpha^{el}) \, d\mathbf{X}^l}{\int \exp(-\beta U_N - \beta B_\alpha^w - \beta\lambda B_\alpha^{el}) \, d\mathbf{X}^l} = \langle B_\alpha^{el} \rangle_\lambda, \qquad (6.12.33)$$

or

$$W(\alpha | el) = \int_0^\lambda \langle B_\alpha^{el} \rangle_\lambda \, d\lambda. \qquad (6.12.34)$$

$\langle B_\alpha^{el} \rangle_\lambda$ is the conditional average of the quantity B_α^{el}, defined in (6.12.28), over all configurations of the particles in the system, except α, given that the solute–solvent interaction of α is already fully turned on, and that the interaction between α and all ions is coupled to the extent λ. It should be noted that this process does not correspond to a physical process. Charging the ion α would lead to changes in the interaction of α with both the solvent and the solute molecules. Here, we are interested only in the latter part of the charging work.[†]

The Debye–Hückel theory is an approximate way of computing $W(\alpha | el)$, by turning on λ from zero to unity. The process is carried out at constant temperature, volume, and all number of particles.

6.12.4. The Poisson–Boltzmann Equation

Let an ion of type α be placed at the center of our coordinate system. Let $\rho_i(\mathbf{R}/\alpha)$ be the conditional density of ions of type i at $\mathbf{R}$, given an ion of type α at the center (Fig. 6.6). If ρ_i is the bulk density of ions of type i, then we have the relation

$$\rho_i(\mathbf{R}/\alpha) = \rho_i g_{i\alpha}(\mathbf{R}) = \rho_i \exp[-\beta W_{i\alpha}(\mathbf{R})], \qquad (6.12.35)$$

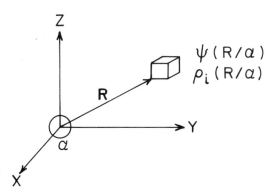

FIGURE 6.6. The density $\rho_i(\mathbf{R}/\alpha)$ measured relative to an ion of type α at the origin.

[†] Another way of computing $W(\alpha | el)$ is to charge once against the entire solution l, and once against pure water w, and then compute the difference as in (6.12.23) and (6.12.24).

where g_{ia} is the pair correlation function and W_{ia} is the potential of average force between the species i and α.

The essential approximation introduced in the Debye–Hückel theory is the statement that

$$W_{ia}(\mathbf{R}) = z_i e \psi(\mathbf{R}/\alpha), \qquad (6.12.36)$$

where $z_i e$ is the charge on the ion of type i and $\psi(\mathbf{R}/\alpha)$ is the average electrical potential at $\mathbf{R}$ given α at the origin. The average is taken over all configurations of the particles in the system except the one placed at the origin. Note that $\psi(\mathbf{R}/\alpha)$ includes two contributions; one of which is due to the point charge of α at the origin and the second of which is due to the distribution of ions around α, the so-called ionic atmosphere. The latter is designated by $\psi^{\text{atm}}(\mathbf{R}/\alpha)$, and we write

$$\psi(\mathbf{R}/\alpha) = \frac{z_a e}{\varepsilon R} + \psi^{\text{atm}}(\mathbf{R}/\alpha). \qquad (6.12.37)$$

In calculating the activity coefficient of the species α, we shall need only the electric potential due to the ionic atmosphere, which we denoted by $\psi^{\text{atm}}(\mathbf{R}/\alpha)$. First we need to solve for the total potential function $\psi(\mathbf{R}/\alpha)$. To evaluate $\psi(\mathbf{R}/\alpha)$, we use the Poisson equation from electrostatics, which reads

$$\nabla^2 \psi(\mathbf{R}/\alpha) = \frac{-4\pi \rho^c(\mathbf{R}/\alpha)}{\varepsilon}, \qquad (6.12.38)$$

where ∇^2 is the Laplace operator $(\partial^2/\partial x^2 + \partial^2/\partial y^2 + \partial^2/\partial z^2)$, $\rho^c(\mathbf{R}/\alpha)$ is the average charge density at $\mathbf{R}$ given α at the center, and ε is the macroscopic dielectric constant of the solvent. The average charge density is computed from the relation

$$\rho^c(\mathbf{R}/\alpha) = z_a e \delta(\mathbf{R}) + \sum_i (z_i e) \rho_i(\mathbf{R}/\alpha). \qquad (6.12.39)$$

The first term on the rhs of (6.12.39) is the contribution to the charge density due to the point charge at the origin, where $\delta(\mathbf{R})$ is the Dirac delta function. The second term sums over all ionic species i. Since we are interested in large distances $|\mathbf{R}|$, we can ignore the contribution from the first term on the rhs of (6.12.39). Substituting (6.12.35) and (6.12.36) into (6.12.39) and then substituting (6.12.39) into (6.12.38), we obtain

$$\nabla^2 \psi(\mathbf{R}/\alpha) = \frac{-4\pi}{\varepsilon} \sum_i (z_i e) \rho_i \exp[-\beta z_i e \psi(\mathbf{R}/\alpha)]. \qquad (6.12.40)$$

This is the Poisson–Boltzmann equation.

At this stage it is appropriate to reflect on the nature of the approximation introduced into (6.12.36).

The potential of average force is defined as the work (in the appropriate ensemble) required to bring particle i from infinity to a location $\mathbf{R}$ relative to α at the origin of our coordinate system. If the distance $|\mathbf{R}|$ is short, then $W_{ia}(\mathbf{R})$ includes contributions due to the electrostatic interactions as well as indirect interactions due to the solvent. The latter are present even when the particles i and α are uncharged (see section 6.4). In making the approximation (6.12.36), we have to consider the following three points:

1. $z_i e \psi(\mathbf{R}/\alpha)$ is the work required to bring the charge $z_i e$ from infinity to $\mathbf{R}$ given a fixed electrical potential $\psi(\mathbf{R}/\alpha)$. In general, the potential of average force may be written as

$$W_{i\alpha}(\mathbf{R}) = U_{i\alpha}(\mathbf{R}) + \delta W_{i\alpha}(\mathbf{R})$$
$$= U_{i\alpha}^*(\mathbf{R}) + U_{i\alpha}^{el}(\mathbf{R}) + \delta W_{i\alpha}^*(\mathbf{R}) + \delta W_{i\alpha}^{el}(\mathbf{R}), \qquad (6.12.41)$$

where U is the direct and δW is the indirect contribution to the potential of average force. We have further split each of these into the short-range (denoted by asterisk) and the long range (denoted by el) contributions and identify the electrostatic work with the long-range effects only, i.e.,

$$(z_i e)\psi(\mathbf{R}/\alpha) = U_{i\alpha}^{el}(\mathbf{R}) + \delta W_{i\alpha}^{el}(\mathbf{R}). \qquad (6.12.42)$$

Thus we have ignored the short-range contributions U^* and δW^* to $W_{i\alpha}$. Clearly, the short-range contributions would not be important when $|\mathbf{R}|$ is very large.

2. As we bring $(z_i e)$ to $\mathbf{R}$, the electrostatic work would have been $(z_i e)\psi(\mathbf{R}/\alpha)$ had $\psi(\mathbf{R}/\alpha)$ been the electrical potential at $\mathbf{R}$ *after* we have brought $(z_i e)$ to $\mathbf{R}$. However, in the Debye–Hückel theory, we compute $\psi(\mathbf{R}/\alpha)$ through Eq. (6.12.40) *before* we bring $(z_i e)$ to $\mathbf{R}$. For a point charge at the origin, $\psi(\mathbf{R}/\alpha)$ will be a spherically symmetrical function. This is because the distribution of the ionic atmosphere around the origin will be spherically symmetrical. However, if we bring a finite charge $(z_i e)$ to a short distance $|\mathbf{R}|$, then the distribution of ions in the ionic atmosphere will be changed; it will be affected by both ions i and α. The final distribution of ions will have cylindrical symmetry about the line connecting the centers of i and α. Clearly, only for very large distances $|\mathbf{R}|$ can we expect that the effect of the ion i on the electric potential $\psi(\mathbf{R}/\alpha)$ will be negligible (Fig. 6.7).

3. $\psi(\mathbf{R}/\alpha)$ is computed through the Poisson–Boltzmann equation (6.12.40), which in turn depends on the validity of the Poisson equation, (6.12.38). The Poisson equation is essentially equivalent to the coulombic interaction between the charge at the origin and the charge density at $\mathbf{R}$. For short distances $|\mathbf{R}|$, the Poisson equation could be applied, but one must take an effective dielectric constant which depends on the distance, i.e., $\varepsilon(\mathbf{R})$. It is only for large distances that we can assume that $\varepsilon(\mathbf{R}) \to \varepsilon$, where ε is the macroscopic dielectric constant of the solvent.

To summarize, if we were to make an estimate of $W_{i\alpha}(\mathbf{R})$ at short distances $|\mathbf{R}|$, the three comments made above should invalidate the approximation (6.12.36). However, since the theory will be applicable to very low densities, the average interionic distances are quite large (compared with the molecular diameters of our particles, see below). Therefore, in this limit the approximation (6.12.36) becomes valid; this is the essential reason for the success of the Debye–Hückel theory.

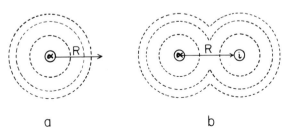

FIGURE 6.7. (a) Distribution of ions around α fixed at the origin. (b) Distribution of ions around α and i at fixed locations.

a   b

The Poisson–Boltzmann equation is a difficult nonlinear differential equation. Even if we solve it exactly, it will lead to some inconsistencies in the solution, resulting from the approximations introduced above. Fortunately, since we are interested only in the limit of low-solute densities, where the average interionic distances are large, we can further assume that

$$|\beta z_i e \psi(\mathbf{R}/a)| \ll 1, \tag{6.12.43}$$

which allows the expansion of the exponent in (6.12.41) to the first order

$$\exp[-\beta z_i e \psi(\mathbf{R}/a)] = 1 - \beta z_i e \psi(\mathbf{R}/a). \tag{6.12.44}$$

Introducing (6.12.44) into (6.12.40), we obtain

$$\nabla^2 \psi(\mathbf{R}/a) = \frac{-4\pi}{\varepsilon} \left[\sum_i (z_i e)\rho_i + \sum_i - \beta z_i^2 e^2 \rho_i \psi(\mathbf{R}/a) \right]. \tag{6.12.45}$$

The first sum on the rhs of (6.12.45) is zero, due to the total electrical neutrality of the system, i.e.,

$$\sum_i (z_i e)\rho_i = 0. \tag{6.12.46}$$

The equation that is left is referred to as the linearized Poisson–Boltzmann equation:

$$\nabla^2 \psi(\mathbf{R}/a) = \left(\frac{4\pi e^2}{\varepsilon k T} \sum_i z_i^2 \rho_i \right) \psi(\mathbf{R}/a)$$
$$= \kappa^2 \psi(\mathbf{R}/a), \tag{6.12.47}$$

where we have introduced the positive parameter

$$\kappa = \left(\frac{4\pi e^2}{\varepsilon k T} \sum_i z_i^2 \rho_i \right)^{1/2}. \tag{6.12.48}$$

κ has the dimension of reciprocal length. This is the reason for referring to κ as the Debye radius. As we shall soon see, this may be interpreted as the effective radius of the ionic atmosphere.

The most general solution of Eq. (6.12.47) is

$$\psi(R/a) = \frac{A}{R} \exp(-\kappa R) + \frac{B}{R} \exp(\kappa R), \tag{6.12.49}$$

where we now view $\psi(R/a)$ as a function of the scalar distance $R = |\mathbf{R}|$. Since we require that $\psi(R/a) = 0$ at $R \to \infty$, we must have $B = 0$. To determine the constant A, we assume that all ions are rigid spheres of diameter σ (this is called the restrictive primitive model of ionic solution). Hence, the condition of electrical neutrality of the entire system requires that

$$\int_\sigma^\infty \rho^c(R/a) 4\pi R^2 \, dR + z_a e = 0, \tag{6.12.50}$$

which simply means that the sum of the charge at the origin and the total charge beyond $R \geq \sigma$ must be zero.

From (6.12.39), (6.12.40), (6.12.43), and (6.12.44), we have

$$\rho^c(R/\alpha) = \sum_i (z_i e)\rho_i(R/\alpha) = \sum_i z_i e[1 - \beta z_i e\rho_i \psi(R/\alpha)]$$

$$= -\sum_i \beta z_i^2 e^2 \rho_i \psi(R/\alpha). \qquad (6.12.51)$$

Inserting this result into (6.12.50) we have, noting (6.12.49) (with $B = 0$),

$$-\sum_i \beta z_i^2 e^2 \rho_i \int_\sigma^\infty \psi(R/\alpha) 4\pi R^2 \, dR = -\sum_i \beta z_i^2 e^2 \rho_i \int_\sigma^\infty A \exp[-\kappa R] 4\pi R^2 \, dR = -z_\alpha e,$$

$$(6.12.52)$$

since

$$\int_\sigma^\infty \exp[-\kappa R] R \, dR = \frac{e^{\kappa\sigma}(1 + \kappa\sigma)}{\kappa^2}. \qquad (6.12.53)$$

We can eliminate A from (6.12.52)

$$A = \frac{z_\alpha e}{\varepsilon} \frac{\exp(\kappa\sigma)}{1 + \kappa\sigma}. \qquad (6.12.54)$$

Hence the solution for $\psi(R/\alpha)$ is

$$\psi(R/\alpha) = \frac{z_\alpha e}{\varepsilon R} \frac{\exp[-\kappa(R - \sigma)]}{1 + \kappa\sigma}. \qquad (6.12.55)$$

This is the total potential at a distance R from an ion of species α.

For calculating the activity coefficient from Eqs. (6.12.30) and (6.12.34), we shall need the value of $\psi^{atm}(R/\alpha)$ defined in (6.12.37) at the point $R = \sigma$, i.e.,

$$\psi^{atm}(R = \sigma/\alpha) = \psi(R = \sigma/\alpha) - \frac{z_\alpha e}{\varepsilon\sigma} = \frac{-z_\alpha e\kappa}{\varepsilon(1 + \kappa\sigma)}. \qquad (6.12.56)$$

Also, for $R \gg \sigma$ and $\kappa\sigma \ll 1$, we can take the approximate form of $\psi(R/\alpha)$ from (6.12.55):

$$\psi(R/\alpha) = \frac{z_\alpha e}{\varepsilon R} \exp(-\kappa R). \qquad (6.12.57)$$

This should be compared with the potential at R due to the ion at the origin, which is simply $z_\alpha e/\varepsilon R$. The factor $\exp(-\kappa R)$ is called the screening effect due to the ionic atmosphere. Using (6.12.57) in (6.12.55), we have for the potential of average force

$$W_{i\alpha}(R) = \frac{z_i z_\alpha e^2}{\varepsilon R} \exp(-\kappa R), \qquad (6.12.58)$$

which is the coulombic interaction $z_i z_\alpha e^2/\varepsilon R$ between the two ions times the screening effect due to the ionic atmosphere. Note that the long-range potential of average force between i and α due to the solvent only is

$$W_{i\alpha}^0(R) = \frac{z_i z_\alpha e^2}{\varepsilon R}. \qquad (6.12.59)$$

In (6.12.58) we have the potential of average force resulting from both the solvent and the solutes (through κ). The contribution of the solute–solute interactions is thus

$$W_{ia}(R) - W_{ia}^0(R) = \frac{z_i z_a e^2}{\varepsilon R} [\exp(-\kappa R) - 1]. \tag{6.12.60}$$

6.12.5. Calculation of the Activity Coefficient

In Eqs. (6.12.30) and (6.12.34), we expressed the activity coefficient of the ion of type α as

$$kT \ln \gamma_\alpha = W(\alpha \,|\, el) = \int_0^1 \langle B_\alpha^{el} \rangle_\lambda \, d\lambda, \tag{6.12.61}$$

where $\langle B_\alpha^{el} \rangle_\lambda$ is the conditional average of the ion–ion binding energy of an ion of species α, given that the solute–solvent interaction of α has already been fully coupled to the system [see Eq. (6.12.33)]. We now "charge" the α particle using the parameter λ, but we have to exclude the work required to charge the ion itself and the work required to couple the ion to the solvent. What we need in (6.12.61) is the work required to charge α against the ionic atmosphere only. Equation (6.12.56) is the electric potential due to the ionic atmosphere at $R = \sigma$. Since no ion can penetrate into the sphere of radius $R = \sigma$, $\psi^{\text{atm}}(R = \sigma)$ is also the ionic atmosphere potential at any $R \leq \sigma$. Hence we put

$$\langle B_\alpha^{el} \rangle_\lambda = \lambda z_\alpha e \psi^{\text{atm}}(R = \sigma) = \frac{-\lambda z_\alpha^2 e^2 \kappa}{\varepsilon(1 + \kappa\sigma)} \tag{6.12.62}$$

and integrate (6.12.61) to obtain

$$kT \ln \gamma_\alpha = \frac{-z_\alpha^2 e^2 \kappa}{\varepsilon(1 + \kappa\sigma)} \int_0^1 \lambda \, d\lambda = \frac{-z_\alpha^2 e^2 \kappa}{2\varepsilon(1 + \kappa\sigma)}. \tag{6.12.63}$$

This is the main result of the Debye–Hückel theory. Recalling the definition of κ in Eq. (6.12.48), the activity coefficient may be rewritten in a more useful form as

$$\ln \gamma_\alpha = -\frac{(2\pi)^{-1} e^3 z_\alpha^2 I^{1/2}}{(\varepsilon kT)^{3/2}(1 + \kappa\sigma)}, \tag{6.12.64}$$

where we defined the ionic strength

$$I = \tfrac{1}{2} \sum_i \rho_i z_i^2 \tag{6.12.65}$$

for a 1:1 electrolyte, $\rho_+ = \rho_- = \rho$; hence

$$I = \tfrac{1}{2}(\rho_+ + \rho_-) = \rho, \tag{6.12.66}$$

where ρ is the density of the salt in the solution. In most actual applications (see the numerical example below), $\kappa\sigma \ll 1$ and the activity coefficient is given by

$$\ln \gamma_\alpha = -A z_\alpha^2 I^{1/2}, \tag{6.12.67}$$

where A includes constants independent of the ionic species. Thus at the limit of low concentrations, the activity coefficient is proportional to the square root of the ionic strength, which for 1:1 electrolyte is simply the square root of the solute density.

For a simple electrolyte CA, the activity coefficient of the salt is defined by

$$\gamma_{\pm}^2 = \gamma_C \gamma_A, \tag{6.12.68}$$

which measures the deviation from the ideal behavior of the entire solute CA [see Eq. (6.12.61)]. From (6.12.67) and (6.12.29) we have

$$\ln \gamma_{\pm}^2 = -A(z_+^2 + z_-^2)I^{1/2} \tag{6.12.69}$$

for a $1:1$ electrolyte, $|z_+| = |z_-|$, and therefore

$$\ln \gamma_{\pm} = -A|z_+ z_-|I^{1/2}. \tag{6.12.70}$$

For a general electrolyte (6.12.10)

$$C_{\nu^+}A_{\nu^-} = \nu_+ C^{z_+} + \nu_- C^{z_-}; \tag{6.12.71}$$

the chemical potential of the salt is decomposed into the ionic components

$$\mu_{\text{salt}} = \nu_+ \mu_+ + \nu_- \mu_-$$
$$= \nu_+ W(+|w) + \nu_- W(-|w) + kT \ln[(\rho_+ \Lambda_+^3 \gamma_+)^{\nu_+}(\rho_- \Lambda_-^3 \nu_-)^{\nu_-}]. \tag{6.12.72}$$

The activity coefficient of the salt is defined by

$$\gamma_{\text{salt}}^{\nu} = \gamma_+^{\nu_+} \gamma_-^{\nu_-}. \tag{6.12.73}$$

Using (6.12.67) for γ_+ and γ_-, we obtain the activity coefficient of the general salt (6.12.71) as

$$\ln \gamma_{\text{salt}} = -A|z_+ z_-|I^{1/2}, \tag{6.12.74}$$

which is the same as (6.12.70), but here $|z_+| \neq |z_-|$.

6.12.6. The Concept of the Ionic Atmosphere

In Eq. (6.12.51), we had an expression for the charge density as a function of the distance from the ion α at the origin. Using the solution $\psi(R/\alpha)$ from (6.12.57), we write

$$\rho^C(R/\alpha) = -\sum_i \beta z_i^2 e^2 \rho_i \frac{z_\alpha e}{\varepsilon R} \exp(-\kappa R). \tag{6.12.75}$$

The average charge in a spherical shell of volume $4\pi R^2\, dR$ is

$$\rho^C(R/a)4\pi R^2\, dR = \frac{-\kappa^2 e z_a}{4\pi} R\exp(-\kappa R)\, dR. \tag{6.12.76}$$

The function $f(R) = R\exp(-\kappa R)$ has a maximum at (see Fig. 6.8)

$$\frac{d}{dR}(R\exp(-\kappa R)) = 0, \tag{6.12.77}$$

$$R_{\max} = \kappa^{-1}. \tag{6.12.78}$$

This justifies the interpretation of κ^{-1} as the effective radius of the ionic atmosphere.

Another interpretation of κ^{-1} may be given as follows. For $\kappa\sigma \ll 1$, we may expand the exponent to the first order to obtain

$$\psi(R/a) = \frac{z_a e}{\varepsilon R}(1 - \kappa R) = \frac{z_a e}{\varepsilon R} - \frac{z_a e}{\varepsilon \kappa^{-1}}. \tag{6.12.79}$$

In this presentation, the total electrical potential at R is the sum of two contributions—the direct potential due to a at the center and an effective screening potential, having the opposite sign to that of the first term—which arise from the ionic atmosphere. The latter may be viewed as the constant potential within a sphere of radius κ^{-1} due to a net charge of $-z_a e$ on its surface. This is another interpretation of the effective radius of the ionic atmosphere.

As a numerical example, we consider a $1:1$ electrolyte at $25°C$ in water for which $\varepsilon = 78.3$,

$$e^2 = (1.602 \times 10^{-19})^2(\text{coulomb})^2 = 2.307 \times 10^{-18}\ \text{J Å/molecule}$$

$$k = 1.3807 \times 10^{-23}\ \text{J K}^{-1}, \qquad T = 298\ \text{K}.$$

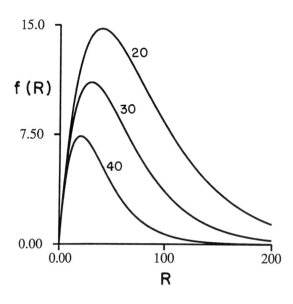

FIGURE 6.8. The function $f(R) = R\exp(-\kappa R)$ for three values of κ. Note that the maximum at $R = 1/\kappa$ moves to the right for smaller κ.

For this case,

$$\frac{\kappa^2}{\rho} = \frac{180 \text{ Å}}{\text{particle}}. \tag{6.12.80}$$

Thus for $\rho = 0.01 \text{ mol/L} = 6.02 \times 10^{-6} \text{ particles/Å}^3$,

$$\kappa^{-1} = 30.4 \text{ Å}. \tag{6.12.81}$$

If the molecular diameter of an ion is taken as 3 Å, then

$$\kappa\sigma = \frac{3}{30} = \frac{1}{10}, \tag{6.12.82}$$

which justifies the neglect of $\kappa\sigma$ with respect to unity. If we take a much lower solute density of about $\rho = 10^{-4} \text{ mol/L}$, we obtain $\kappa^{-1} \sim 300 \text{ Å}$, which is very large compared with a typical molecular diameter of a few angstroms.

6.12.7. Excess Thermodynamic Quantities

The excess Gibbs energy is defined by

$$\Delta G^{el} = G(T, V, \mathbf{N}) - G^{\text{ideal}}(T, V, N)$$
$$= \sum_i N_i kT \ln \gamma_i. \tag{6.12.83}$$

Note that the independent variables chosen here are T, V, $\mathbf{N}$; ΔG^{el} is due to the charging process against the ionic atmosphere potential only, at constant volume [see Eq. (6.12.61)]. When the charging work is taken at constant V, κ is also constant. However, if the charging process is carried out at P constant, the volume V changes and therefore κ would be a function of λ (see section 6.12.8).

Using the result (6.12.63) for the activity coefficient, we obtain

$$\Delta G^{el} = -\sum_i \frac{N_i z_i^2 e\kappa}{2\varepsilon} = \frac{-kTV\kappa^3}{8\pi}. \tag{6.12.84}$$

We now write the total Gibbs energy of the system as

$$G(T, V, \mathbf{N}) = N_w \mu_w + \sum_i N_i \mu_i$$
$$= N_w \mu_w + \sum_i N_i W(i|w) + \sum kTN_i \ln \rho_i \Lambda_i^3 + \Delta G^{el}$$
$$= G^0 + \sum_i kTN_i \ln \rho_i \Lambda_i^3 - \frac{kTV\kappa^3}{8\pi}, \tag{6.12.85}$$

where the sum is over all ionic species in the system.

We now use the thermodynamic relation

$$dG = -S \, dT + V \, dP. \tag{6.12.86}$$

Viewing G as a function of V, T we have

$$\left(\frac{\partial P}{\partial V}\right)_{T,\mathbf{N}} = \frac{1}{V}\left(\frac{\partial G}{\partial V}\right)_{T,\mathbf{N}}, \tag{6.12.87}$$

differentiating (6.12.85) with respect to V (note that κ is a function of V through the densities; see (6.12.48), and then integrating (6.12.87), we obtain

$$P = \sum_i kT\rho_i - \frac{\kappa^3}{24\pi} + C. \qquad (6.12.88)$$

There are two limiting cases of interest. First, if all $\rho_i = 0$ and $\rho_w \to 0$, then we approach the ideal-gas limit; therefore at this limit $P \to 0$ and thus $C = 0$. Since at this limit the dielectric constant is unity, we define

$$\kappa^0 = \kappa(\varepsilon = 1), \qquad (6.12.89)$$

and therefore the total pressure of the gas (or plasma, i.e., ionic particles with no solvent) is

$$P^{\text{gas}} = \sum_i kT \ln \rho_i - \frac{kT(\kappa^0)^3}{24\pi}. \qquad (6.12.90)$$

The other limit of interest occurs when all $\rho_i \to 0$, but the density of the solvent is $\rho_w \to \rho_w^P$, in which case the pressure in (6.12.88) must tend to the pressure of the pure solvent, hence

$$\lim_{\text{all } \rho_i \to 0} P = C = P^{\text{solvent}}. \qquad (6.12.91)$$

Using (6.12.91) in (6.12.88), we identify the osmotic pressure of the solution as the difference

$$\pi = P - P^{\text{solvent}} = \sum_i kT\rho_i - \frac{kT\kappa^3}{24\pi}. \qquad (6.12.92)$$

Note that here κ is $\kappa(\varepsilon)$ with the value of ε of the pure solvent. The first term on the rhs of (6.12.92) is the osmotic pressure of an ideal solution. The correction term is the change in the osmotic pressure due to the ionic interactions only.

The Helmholtz energy change associated with the charging process is obtained as follows: For the charging process at constant volume we have

$$\Delta A^{el} = \Delta G^{el} - V\Delta P^{el}, \qquad (6.12.93)$$

where the changes ΔA^{el}, ΔG^{el}, and ΔP^{el} all refer to the process at constant volume. From (6.12.84) and (6.12.92), we obtain

$$\Delta A^{el} = -\frac{kTV\kappa^3}{8\pi} + V\frac{kT\kappa^3}{24\pi} = \frac{-kTV\kappa^3}{12\pi}. \qquad (6.12.94)$$

In the next section we shall derive the same expression (6.12.94) by a different charging process, known as the Debye charging process.

6.12.8. The Debye Charging Process

In section 6.12.5 we computed the activity coefficient by charging one ion α in a system in which all the other ions were fully charged. Here we compute the work of charging all ions simultaneously. We start with the system at T, V, $\mathbf{N}$ with the charging parameter λ and compute the work required to charge all the ions against the ionic atmosphere, i.e.,

$$\Delta A^{el} = A(\lambda = 1) - A(\lambda = 0)$$

$$= -\int_0^1 \sum_i \psi^{atm}(R = \sigma, \lambda/\alpha) N_i z_i e \, d\lambda, \qquad (6.12.95)$$

where $\psi^{atm}(R = \sigma, \lambda/\alpha)$ is the electric potential due to the ionic atmosphere, as in (6.12.56), but now all the ions are charged to the extent λ. From (6.12.56) (neglecting $\kappa\sigma$) we have

$$\psi^{atm}(R = \sigma, \lambda/\alpha) = \frac{-\lambda z_\alpha e \kappa(\lambda)}{\varepsilon}. \qquad (6.12.96)$$

Note that in contrast to the charging process of section 6.12.5 where only the electric charge of α was changed, here all the charges of all the ions are changed simultaneously; hence, κ changes during the charging process. Since κ is proportional to the sum of the charges of all the particles, it follows from the definition (6.12.48) that

$$\kappa(\lambda) = \lambda\kappa(1) = \lambda\kappa. \qquad (6.12.97)$$

Hence

$$\Delta A^{el} = -\sum_i \frac{N_i z_i^2 e^2 \kappa}{\varepsilon} \int_0^1 \lambda^2 \, d\lambda = -\sum_i \frac{N_i z_i^2 e^2 \kappa}{3\varepsilon}, \qquad (6.12.98)$$

and using the definition of κ in (6.12.48) we obtain

$$\Delta A^{el} = -kT \frac{V\kappa^3}{12\pi}, \qquad (6.12.99)$$

in agreement with (6.12.94).

6.13. SOLVATION THERMODYNAMICS

6.13.1. Definition of the Solvation Process

The concept of solvation is of central importance in the study of processes in the liquid state. In Chapters 2, 3, and 4 we discussed various processes taking place in the vacuum or in an ideal-gas phase. Examples are the adsorption and allosteric phenomena, chemical equilibrium, and the helix–coil transition. However, in most systems of interest in chemistry and biochemistry, a solvent is always present. The presence of the solvent leads to modifications in the theory. The modification might be relatively minor or very thorough depending on the specific process and the extent of the coupling between the solvent and the particles involved in the process. We shall discuss some specific solvent effects in Chapter 8, especially for solvents which are themselves quite complex—i.e., aqueous solutions. In this section we lay the groundwork for studying solvent effects in general. We shall introduce the definition of the thermodynamic quantities associated with the solvation process and methods of obtaining these quantities, either from theory or from experiment, and present a few examples.

Our starting point is the general expression for the chemical potential of a molecule s in any liquid l, which we write as

$$\mu_s = \mu_s^* + kT \ln \rho_s \Lambda_s^3$$
$$= W(s|l) + kT \ln q_s^{-1} + kT \ln \rho_s \Lambda_s^3. \qquad (6.13.1)$$

In writing (6.13.1), we already commit ourselves to a semiclassical system; i.e., the translational degrees of freedom are treated as classical, while the internal degrees of freedom may be treated either classically or quantum mechanically. The internal PF, q_s, includes vibrational, rotational, and electronic degrees of freedom. In this section we shall not discuss molecules having internal rotations, e.g., butane, or polymers. We shall always assume that internal degrees of freedom are not affected by the solvent. This is clearly an approximation. It is expected to be a good approximation for simple solvents at relatively low pressure. However, we shall use this approximation even for aqueous solutions, where water molecules are known to affect the internal degrees of freedom either of the water molecules themselves or of solutes with which they strongly interact, say through hydrogen bonding.

From now on we shall assume that q_s is the same PF of a single molecule s, whether in an ideal gas or in the liquid. Our focus in the study of solvation will be on the coupling work $W(s|l)$. We have seen in Chapters 5 and 6 that this quantity conveys the effect of molecular interactions between the particles on the chemical potential. In the limit of very low densities, or when interactions are negligible, this term vanishes and the chemical potential (6.13.1) reduces to that of a molecule of species s in an ideal gas. In section 5.9.3 we presented a convenient interpretation of the pseudochemical potential μ_s^* as the work involved in placing s at a fixed location in the liquid. Within the classical treatment of our systems the process of fixing the location of a specific particle is meaningful. We shall adopt this interpretation of μ_s^* throughout the book.

We define the solvation process of a molecule s in a fluid l as the process of transfering the molecule s from a fixed position in an ideal gas phase g into a fixed position in the fluid or liquid phase l. The process is carried out at constant temperature T and pressure P. Also, the composition of the system is unchanged.

When such a process is carried out, we shall say that the molecule s is being solvated by the liquid phase l. If s is a simple spherical molecule, it is sufficient to require that the center of the molecule be fixed. On the other hand, if s is a more complex molecule, such as n-alkane or a protein, we require that the center of mass of the molecule be at a fixed position. We note, also, that in complex molecules the geometrical location of the center of mass might change upon changing the conformation of the molecule. In such cases we need to distinguish between the process of solvation of the molecule at a particular conformation and an average solvation process over all possible conformations of the molecule.

One could also define the solvation process as above, but at constant volume rather than constant pressure. The definition given above is the one which may be related more directly to experimental quantities. However, for some theoretical considerations it might be more convenient to treat the constant-volume solvation process. The relation between the two is discussed in Appendix E.

In defining a particular solvation process of a molecule s, we must specify the temperature, the pressure, and the composition of the liquid phase. There is no restriction whatsoever on the concentration of s in the system. This may be very dilute s in l, in which case the term solute in its conventional sense might apply for s. It may be a concentrated solution of s in l, or even a pure liquid s. Clearly, in the latter cases the conventional sense of the term solute becomes inappropriate. However, what remains unchanged is the conceptual meaning of the term solvation as a measure of the interaction between s and its entire surroundings l.

As we shall see in section (6.13.2), at the limit of very dilute solutions some of the thermodynamic quantities of solvation, as defined above, coincide with the conventional quantities of solvation as defined in thermodynamics. This is not the case, however, for higher concentrations of s in l. While conventional thermodynamics cannot be applied to these systems, the theoretical definition along with the pertinent thermodynamic quantities, can be applied without any restrictions on the concentration of s. In this sense the new definition generalizes the concept of solvation beyond its traditional limits.

It will be useful to introduce at this point the concept of a *solvaton*. The solvaton s is that particular molecule s the solvation of which is studied. This term is introduced to stress the distinction between the molecule serving as our "test particle" and other molecules of the same species that might be in the surroundings of the solvaton.

Having defined the process of solvation, we proceed to introduce the corresponding thermodynamic quantities. We shall henceforth talk about solvation entropy, solvation energy, solvation volume, and so on, meaning the change in the corresponding thermodynamic quantity associated with the solvation process as defined above.

First, and of foremost importance, is the Gibbs energy of solvation of s in l. This is defined as

$$\Delta G_s^* = \mu_s^{*l} - \mu_s^{*ig}, \tag{6.13.2}$$

where μ_s^{*l} and μ_s^{*ig} are the pseudochemical potential (PCP) of s in the liquid and in an ideal-gas (ig) phase, respectively. Clearly, ΔG_s^* is the Gibbs energy change for transferring s from a fixed position in an ideal-gas phase into a fixed position in the liquid phase l.

Since in the ideal-gas phase we have

$$\mu_s^{*ig} = -kT \ln q_s, \tag{6.13.3}$$

it follows that ΔG_s^*, as defined above, includes all the effects due to the interaction between s and its entire environment.

A relatively simple situation arises when the solvaton has no internal degrees of freedom or when these are effectively unaffected by the surroundings. In these cases we may write

$$\mu_s^{*l} = W(s|l) - kT \ln q_s. \tag{6.13.4}$$

Hence the solvation Gibbs energy reduces to

$$\Delta G_s^* = W(s|l) \tag{6.13.5}$$

which is simply the coupling work of s to l.

The statistical mechanical expression for $W(s|l)$ is:

$$W(s|l) = -kT \ln \left\langle \exp\left(-\frac{B_s}{kT}\right) \right\rangle, \tag{6.13.6}$$

where B_s is the total binding energy of s to the entire system l at some specific configuration and $\langle \ \rangle$ signifies an ensemble average over all possible configurations of all the molecules in the system, excluding the solvaton s, which is the molecule s under observation. Expression (6.13.6) may be interpreted as the average Gibbs energy of interaction of s with l. This should be clearly distinguished from the average interaction energy between s and l.

At this point we introduce a notation for two limiting cases of the quantity ΔG_s^*, namely,

$$\Delta G_s^{*0} = \lim_{\rho_s \to 0} \Delta G_s^* \qquad (6.13.7)$$

and

$$\Delta G_s^{*p} = \lim_{\rho_s \to \rho_s^p} \Delta G_s^* . \qquad (6.13.8)$$

The first quantity is the low-density limit, i.e., the limit when s is very dilute in l. This is the region where conventional thermodynamics is applicable.

The second limit occurs in the case of pure s, i.e., when the solvaton s is being solvated only by s molecules. This case cannot be treated by purely thermodynamic methods. Here, ρ_s^p denotes the number density of pure s at the specified temperature and pressure.

Having defined the Gibbs energy of solvation in Eq. (6.13.2), we can define all the other thermodynamic quantities of solvation by using standard thermodynamic relationships. The most important quantities are the first derivatives of the Gibbs energy, i.e.,

$$\Delta S_s^* = - \left(\frac{\partial \Delta G_s^*}{\partial T} \right)_P \qquad (6.13.9)$$

$$\Delta H_s^* = \Delta G_s^* + T \Delta S_s^* \qquad (6.13.10)$$

$$\Delta V_s^* = \left(\frac{\partial \Delta G_s^*}{\partial P} \right)_T . \qquad (6.13.11)$$

One may also add the Helmholtz energy of solvation ($\Delta A_s^* = \Delta G_s^* - P\Delta V_s^*$) and the internal energy of solvation ($\Delta E_s^* = \Delta H_s^* - P\Delta V_s^*$). However, for most of the specific systems discussed in this book, the term $P\Delta V_s^*$ is usually negligible with respect to ΔH_s^* or ΔG_s^*. Therefore, the distinction between ΔG_s^* and ΔA_s^* or between ΔH_s^* and ΔE_s^* is usually insignificant.

Finally, we note that all the thermodynamic quantities of solvation as defined above pertain to exactly the same process, the solvation process as defined in this section. We stress, however, that the process of solvation is not experimentally feasible; i.e., we cannot carry out this process in the laboratory. For this reason it cannot be handled within the realm of classical thermodynamics. It is accessible only through statistical-mechanical considerations. Fortunately, as we shall see in the following section, statistical mechanics does provide a simple connection between solvation quantities and experimentally measurable quantities.

6.13.2. Calculation of the Thermodynamic Quantities of Solvation from Experimental Data

Having defined the solvation process in the preceding section, we now turn to the question of evaluating the pertinent thermodynamic quantities from experimental data. We discuss in this section the case of a solvaton s which does not undergo dissociation in any of the phases g and l. The more complicated case of dissociable solvatons (such as ionic solutes) will be discussed separately in section 6.16.

Consider two phases α and β in which s molecules are distributed. We do not impose any restrictions on the concentration of s in the two phases. The only assumption being made is the applicability of classical statistical mechanics. At equilibrium, assuming that the two phases are at the same temperature and pressure, we have the following equation for the chemical potential of s in the two phases:

$$\mu_s^\alpha = \mu_s^\beta . \tag{6.13.12}$$

In the traditional thermodynamic treatment one usually imposes the restriction of a very dilute solution of s in the two phases. However, here we shall use the general expression (6.13.1) for the chemical potential of s in the two phases. Applying Eq. (6.13.1) to Eq. (6.13.12), we obtain

$$\mu_s^{*\alpha} + kT \ln \rho_s^\alpha = \mu_s^{*\beta} + kT \ln \rho_s^\beta \tag{6.13.13}$$

or, after rearrangement,

$$\Delta G_s^{*\beta} - \Delta G_s^{*\alpha} = (\mu_s^{*\beta} - \mu_s^{*ig}) - (\mu_s^{*\alpha} - \mu_s^{*ig})$$

$$= \mu_s^{*\beta} - \mu_s^{*\alpha} = kT \ln \left(\frac{\rho_s^\alpha}{\rho_s^\beta} \right)_{eq} . \tag{6.13.14}$$

Here ρ_s^α and ρ_s^β are the number densities of s in the two phases, at equilibrium. Relation (6.13.14) provides a very simple way of computing the difference in the solvation Gibbs energies of s in the two phases α and β from the measurement of the two densities ρ_s^α and ρ_s^β at equilibrium. The most useful particular example of (6.13.14) occurs when one of the phases, say α, is an ideal gas. In such a case $\Delta G_s^{*\alpha} = 0$ and relation (6.13.14) reduces to

$$\Delta G_s^{*\beta} = kT \ln \left(\frac{\rho_s^{ig}}{\rho_s^\beta} \right)_{eq} . \tag{6.13.15}$$

We note, however, that there is no restriction on the density of s in phase β, but ρ_s^{ig} must be low enough to ensure that this phase (α) is an ideal gas. A specific example is a liquid–vapor equilibrium in a one-component system. If the vapor pressure is low enough, we may safely assume that the vapor behaves as an ideal gas. In such a case we rewrite Eq. (6.13.15) as

$$\Delta G_s^{*P} = kT \ln \left(\frac{\rho_s^{ig}}{\rho_s^P} \right)_{eq} , \tag{6.13.16}$$

where ΔG_s^{*P} is the solvation Gibbs energy of s in its own pure liquid.

Another extreme case is the very dilute solution of s in phase β, say argon in water, for which we have the limiting form of Eq. (6.13.15) that reads

$$\Delta G_s^{*0} = kT \ln \left(\frac{\rho_s^{ig}}{\rho_s^\beta} \right)_{eq} . \tag{6.13.17}$$

This relation is identical in *form* to an equation derived from thermodynamics. However, we stress that *conceptually* the two relations differ from each other, and further elaboration on this point is given below.

Having obtained the Gibbs energy of solvation through one of the relations cited above, it is a straightforward matter to calculate other thermodynamic quantities of solvation using the standard relations. For instance, the solvation entropy might be calculated from the temperature dependence of the $\Delta G_s^{*\beta}$, i.e.,

$$\Delta S_s^{*\beta} = -\left\{\frac{\partial}{\partial T}\left[kT\ln\left(\frac{\rho_s^{ig}}{\rho_s^{\beta}}\right)_{eq}\right]\right\}_p. \tag{6.13.18}$$

We note that the differentiation in this equation is carried out at constant pressure. One must distinguish between this derivative and the derivative along the liquid–vapor equilibrium line. The relation between the two quantities is discussed in section 6.13.3.

We now turn to a brief comparison between the solvation quantities as defined above on the one hand, and the conventional standard thermodynamics of solutions on the other. Quite a few conventional quantities have been employed in the literature. We shall discuss in this section only one of these, the most frequently used.

Let C_s be any units used to measure the concentration of s in the system. The most common units are the molarity ρ_s, the molality m_s, and the mole fraction x_s.

In the conventional thermodynamic approach, we make use of the fact that Henry's law is obeyed at the limit of a very dilute solution of s. In this concentration range the chemical potential (CP) of the species s is given by

$$\mu_s = \mu_s^{0c} + kT\ln C_s, \tag{6.13.19}$$

where μ_s^{0c} is referred to as the standard chemical potential of s, based on the concentration scale C. This is formally defined as the limit

$$\mu_s^{0c} = \lim_{C_s \to 0}(\mu_s - kT\ln C_s). \tag{6.13.20}$$

Let α and β be two phases in which the concentrations of s are C_s^α and C_s^β, respectively. If the limiting expression (6.13.20) applies for both phases, we may define the Gibbs energy change for the process of transferring one s from α to β as

$$\Delta G\left[\begin{matrix}\alpha \to \beta \\ C_s^\alpha, C_s^\beta\end{matrix}\right] = \mu_s^\beta - \mu_s^\alpha = \mu_s^{0c\beta} - \mu_s^{0c\alpha} + kT\ln\left(\frac{C_s^\beta}{C_s^\alpha}\right). \tag{6.13.21}$$

In all that follows we assume that the temperature and pressure are the same in the two phases. Hence, these will be omitted from our notation. On the left-hand side of relation (6.13.21), we do specify the concentrations of s in the two phases. These can be chosen freely as long as they are within the range of validity of Eq. (6.13.19).

Next, we make a choice of a standard process. In principle one can choose any specific values of C_s^α and C_s^β to characterize this standard process. A simple and very common choice is $C_s^\alpha = C_s^\beta$, for which relation (6.13.21) reduces to

$$\Delta G\left[\begin{matrix}\alpha \to \beta \\ C_s^\alpha = C_s^\beta\end{matrix}\right] = \mu_s^{0c\beta} - \mu_s^{0c\alpha}. \tag{6.13.22}$$

This quantity is measurable provided we can measure the ratio C_s^β/C_s^α at equilibrium between the two phases α and β and provided both C_s^α and C_s^β at equilibrium are small enough to ensure ideal behavior. If these conditions are met, then in applying relation

(6.13.21) to the case of equilibrium between the two phases, where $\mu_s^\alpha = \mu_s^\beta$, we obtain

$$\mu_s^{0c\beta} - \mu_s^{0c\alpha} = -kT \ln\left(\frac{C_s^\beta}{C_s^\alpha}\right)_{eq}. \tag{6.13.23}$$

Thus Eq. (6.13.23) provides the means of measuring the standard Gibbs energy of transfer of s from α to β at $C_s^\alpha = C_s^\beta$.

We now define one special case, most commonly employed in the literature, to which we refer as the *x*-process. This is defined as the process of transferring an s molecule from an ideal gas at 1 atm pressure to a hypothetical dilute ideal solution in which the mole fraction of s is unity. (The temperature T and pressure 1 atm are the same in the two phases.)

The relation between the standard Gibbs energy of the *x*-process and the solvation Gibbs energy is obtained from the general expression (6.13.1), i.e.,

$$\Delta G_s(\text{x-process}) = \left[\mu_s^{*l} - \mu_s^{*g} + kT \ln\left(\frac{\rho_s^l}{\rho_s^g}\right)\right]_{\substack{P_s = 1 \\ x_s = 1}}, \tag{6.13.24}$$

where we must substitute $P_s = 1$ atm and $x_s = 1$ in Eq. (6.13.24). To achieve that, we transform variables as follows: Since the *x*-process applies for ideal gases, we put

$$\rho_s^g = \frac{P_s}{kT}, \tag{6.13.25}$$

where P_s is the partial pressure of s in the gaseous phase. Furthermore, the *x*-process applies to dilute ideal solutions; hence we may put

$$x_s^l = \frac{\rho_s^l}{\sum \rho_i^l} \approx \frac{\rho_s^l}{\rho_B^l}, \tag{6.13.26}$$

where ρ_B^l is the number density of the solvent B.

With these transformations of the variables we rewrite Eq. (6.13.24) in the form

$$\Delta G_s(\text{x-process}) = \left[\mu_s^{*l} - \mu_s^{*g} + kT \ln\left(\frac{x_s^l \rho_B^l kT}{P_s}\right)\right]_{\substack{P_s = 1 \\ x_s = 1}}$$

$$= \mu_s^{*l} - \mu_s^{*g} + kT \ln(kT\rho_B^l)$$

$$= \Delta G_s^* + kT \ln(kT\rho_B^l). \tag{6.13.27}$$

Note that since we put $P_s = 1$ atm, $kT\rho_B^l$ must also be expressed in units of atmospheres.

Equation (6.13.27) is the required connection between the Gibbs energy of the *x*-process and the solvation Gibbs energy. Again, we note that this equality holds only for the ideal-gas and ideal solutions. To obtain the corresponding relations for the other thermodynamic quantities associated with the processes defined above, we start from the general equation, Eq. (6.13.1), for the chemical potential. The partial molar (or molecular) entropy of s is obtained from

$$\bar{S}_s = -\frac{\partial \mu_s}{\partial T}$$

$$= -\frac{\partial \mu_s^*}{\partial T} - k \ln \rho_s \Lambda_s^3 - kT \frac{\partial \ln(\rho_s \Lambda_s^3)}{\partial T}. \tag{6.13.28}$$

Note that the differentiation is performed at constant pressure and composition of the system.

We denote by S_s^* the entropy change that corresponds to the process of adding one s molecule to a fixed position in the system. On performing the differentiation with respect to temperature in Eq. (6.13.28), we obtain

$$\bar{S}_s = S_s^* - k \ln \rho_s \Lambda_s^3 + kT\alpha_p + \tfrac{3}{2}k, \tag{6.13.29}$$

where $\alpha_p = V^{-1} \partial V / \partial T$ is the thermal expansion coefficient of the system at constant pressure.

Applying Eq. (6.13.29) for an ideal-gas phase and for an ideal dilute solution, we may derive the entropy changes associated with the standard processes as defined above. For the solvation process, we simply have

$$\Delta S_s^* = S_s^{*l} - S_s^{*g}$$

$$= -\partial \Delta G_s^* / \partial T. \tag{6.13.30}$$

For the x-process we have

$$\Delta S_s(x\text{-process}) = [\bar{S}_s^l - \bar{S}_s^g]_{\substack{x_s = 1 \\ P_s = 1}}$$

$$= S_s^{*l} - S_s^{*g} - k \ln(kT\rho_B^l) + kT\alpha_p^l - k$$

$$= \Delta S_s^* - k \ln(kT\rho_B^l) + kT\alpha_p^l - k. \tag{6.13.31}$$

A glance at the expression for $\Delta S_s(x\text{-process})$ shows that it contains the solvent density ρ_B^l under the logarithm sign. Thus for a series of solvents with decreasing densities ΔS_s (x-process) will diverge to infinity, clearly an undesirable feature for a quantity that is presumed to measure the solvation entropy of a molecule s. On the other hand, ΔS_s^* tends to zero as the solvent density decreases to zero. To obtain the enthalpies of the various processes, we simply form the combination $\bar{H}_s = \mu_s + T\bar{S}_s$ and apply it to the processes of interest. The results are

$$\Delta H_s^* = H_s^{*l} - H_s^{*g} = \Delta G_s^* + T\Delta S_s^* \tag{6.13.32}$$

$$\Delta H_s(x\text{-process}) = \Delta H_s^* + kT^2\alpha_p^l - kT. \tag{6.13.33}$$

Taking the derivative of the general expression for the chemical potential with respect to pressure, we obtain the partial molar (or molecular) volume of s, i.e.,

$$\bar{V}_s = \left(\frac{\partial \mu_s}{\partial P}\right)_T$$

$$= \left(\frac{\partial \mu_s^*}{\partial P}\right)_T + kT\left(\frac{\partial \ln \rho_s}{\partial P}\right)_T$$

$$= V_s^* + kT\kappa_T^l, \tag{6.13.34}$$

where V_s^* is the volume change due to the addition of one s molecule at a fixed position in the system and κ_T^l is the isothermal compressibility of the phase l,

$$\kappa_T^l = -\frac{1}{V}\left(\frac{\partial V}{\partial P}\right)_T \tag{6.13.35}$$

which, for an ideal-gas phase, reduces to

$$\kappa_T^{ig} = \frac{1}{P}. \tag{6.13.36}$$

The volume changes for the two processes of interest are

$$\Delta V_s^* = \partial \Delta G_s^* / \partial P = V_s^{*l} - V_s^{*g} \tag{6.13.37}$$

$$\Delta V_s(x\text{-process}) = \Delta V_s^* + kT(\kappa_T^l - 1/\text{atm}). \tag{6.13.38}$$

From the quantities derived above, one may construct the internal energy of solvation ($\Delta E_s^* = \Delta H_s^* - P\Delta V_s^*$) and the Helmholtz energy of solvation ($\Delta A_s^* = \Delta G_s^* - P\Delta V_s^*$) and compare these with the corresponding conventional quantities. As noted above, the difference between ΔE_s^* and ΔH_s^* and that between ΔA_s^* and ΔG_s^* are usually very small and may be neglected for most systems of interest discussed in this book.

We now summarize the essential differences between the statistical-mechanical approach to the study of solvation and the conventional thermodynamic approach based on various standard processes. First and foremost is the simple fact that the solvation process as defined in this section is the most direct means of probing the interaction of a solvaton with its environment. The pertinent thermodynamic quantities of solvation tend to zero when the solvent density goes to zero (i.e., when there is no interaction between the solvaton and its environment). This is not the case for the conventional thermodynamic quantities, some of which even diverge to plus or minus infinity in this limit. Second, the solvation process is meaningful in the entire range of concentration of s from very dilute s to pure liquid s. The thermodynamic quantities are restricted to the extreme limit of very dilute solutions. Thus in thermodynamics we may speak of the solvation of say, ethanol, in very dilute solution in water, or water in very dilute solution in ethanol. No such restriction on the concentration exists in the study of solvation as defined above.

Finally, we note the relation between the solvation Gibbs energy and the Henry's law constant. The Henry's law constant in its most common form is defined as

$$K_H = \lim_{x_s^l \to 0} \left(\frac{P_s}{x_s^l}\right), \tag{6.13.39}$$

where P_s is the partial pressure, x_s^l is the mole fraction of s in the system, and the limit takes x_s^l into the range where Henry's law becomes valid.

The general expression for the solvation Gibbs energy in this case is

$$\Delta G_s^{*0} = kT \ln \left(\frac{\rho_s^{ig}}{\rho_s^l}\right)_{eq}. \tag{6.13.40}$$

Assuming that we have a sufficiently dilute solution of s in l such that Henry's law in the form $P_s = K_H x_s^l$ is obeyed, we can transform Eq. (6.13.40) into

$$\Delta G_s^{*0} = kT \ln \left(\frac{P_s}{kT x_s^l \rho_B^l}\right)$$

$$= kT \ln \left(\frac{K_H}{kT\rho_B^l}\right). \tag{6.13.41}$$

This is the required connection between the values of K_H and the solvation Gibbs energy. (Here we have assumed, for simplicity, that the solvent consists of one component with a number density ρ'_B. If the solvent contains several components, excluding s, then ρ'_B should be replaced by the sum of the number densities of all the other components.) We note also that from relations (6.13.27) and (6.13.41) we also have

$$\Delta G_s(x\text{-process}) = kT \ln K_H. \tag{6.13.42}$$

Thus, information on K_H is essentially equivalent to information on the Gibbs energy change for the x-process.

6.13.3. Solvation of Inert-Gas Molecules

The simplest nontrivial solvation phenomenon is the case of a hard-sphere (HS) particle in a fluid of HS particles. This case was discussed in connection with the scaled-particle theory in section 5.11. Here we note that the solvation Gibbs energy of an HS solvaton in an HS solvent is always positive, and it increases monotonically as a function of the size of the HS solvaton or, equivalently, the radius of the corresponding cavity (see section 5.11).

The solvation of an HS solvaton in a real fluid is important in the study of the solvation thermodynamic of any real solvaton. The solvation process of a real particle may be performed in two steps: first, the creation of a suitable cavity, then turning on the other parts of the interactions between the solvaton and the (real) solvent. In order to define the size of the cavity we need to assign an effective hard-core diameter to the solvent molecules. For simple spherical molecules, such as the noble gases, a natural choice of the effective diameter might be the van der Waals or Lennard–Jones diameter of the molecules. For more complex molecules there is no universal way of defining an effective HS diameter to be assigned to the solvent molecules.

In spite of this ambiguous concept of the effective HS diameter, a great amount of work has been carried out using the methods of the SPT to calculate the solvation thermodynamics of an HS in various real fluids, ranging from simple noble gases to water and even to ionic melts.

Basically, the SPT is used in these cases to compute the cavity work as a first step in the solvation process. The second step consists of turning on the soft part of the interaction between the solvaton and its entire environment. The formal split of the solvation process into these two parts is discussed in section 6.14.

The simplest real molecules for which the solvation quantities can be measured are the inert gases. Two kinds of experimental data that we can use to evaluate solvation quantities are available for these systems. First are the vapor–liquid densities of the two phases along the coexistence equilibrium line. Second are PVT data on inert-gas liquids. Both are used in the following. Let ρ_s^g and ρ_s^l be the densities of the inert gas s in the gas and liquid phases, respectively. Using our general expression (6.13.1) for the chemical potential we have

$$\Delta\Delta G_s^* = \mu_s^{*l} - \mu_s^{*g}$$
$$= \Delta G_s^{*l} - \Delta G_s^{*g} = W(s|l) - W(s|g)$$
$$= kT \ln \left(\frac{\rho_s^g}{\rho_s^l}\right)_{\text{eq}}. \tag{6.13.43}$$

Thus, knowledge of the density ratio is sufficient for obtaining the difference in the solvation Gibbs energies of s in the two phases (or, equivalently, the Gibbs energy change for the transfer of s from a fixed position in g to a fixed position in l, at temperature T and corresponding equilibrium pressure P). Since the inert gas molecules are presumed to possess essentially only translational degrees of freedom, $\Delta\Delta G_s^*$ is also equal to the difference in the coupling work of s in the two phases.

In Fig. 6.9 we have plotted $\Delta\Delta G_s^*$ as a function of $T - T_{tr}$, where T_{tr} is the triple-point temperature for each of the gases neon, argon, krypton, and xenon.

We note that as $T \to T_{tr}$, the density of s in the gaseous phase becomes very low. In this region one can assume that the gaseous phase is an ideal gas, i.e., that $\Delta G_s^{*g} = W(s|g) \approx 0$ and hence $\Delta\Delta G_s^*$ is simply the solvation Gibbs energy of s in the liquid phase (near $T = T_{tr}$ and at the corresponding low pressure). Thus we always have

$$\Delta G_s^{*l} = \lim_{T \to T_{tr}} \Delta\Delta G_s^*. \tag{6.13.44}$$

Another limiting case is the critical point. As we approach the critical temperature T_{cr}, the densities of s in the two phases approach each other; hence

$$\lim_{T \to T_{cr}} \Delta\Delta G_s^* = 0. \tag{6.13.45}$$

The tendency is clearly discernible for all the gases in Fig. 6.9.

Another set of data that we can use is the density of the liquid phase and the total pressure of the system. In order to convert these data into solvation thermodynamics, we require the densities of the gas in the gaseous phase. We therefore assumed that for

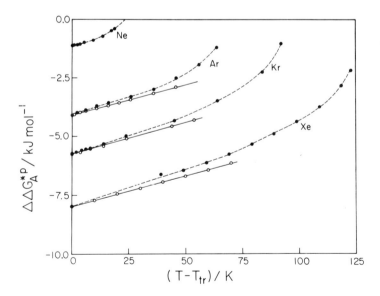

FIGURE 6.9. Values of $\Delta\Delta G_A^*$ in kJ mol^{-1} as a function of $(T - T_{tr})/K$ for the inert gases in their own liquids. The triple points for neon, argon, krypton, and xenon were taken as 24.55, 83.78, 115.95, and 161.3 K, respectively. The full points correspond to computations based on Eq. (6.13.43), and the open circles to computations based on Eq. (6.13.46).

pressures up to 1 atm the gaseous phase is ideal. Hence we compute the solvation Gibbs energy directly from the available data, i.e.,

$$\Delta G_s^{*l} = kT \ln\left(\frac{P}{kT\rho_s^l}\right)_{eq}. \qquad (6.13.46)$$

In Fig. 6.9 we have added the open circles, which were computed by Eq. (6.13.46). As expected, the two sets of data converge at the limit $T \to T_{tr}$, where the gaseous phase becomes ideal. Clearly, as $T - T_{tr}$ increases, $\Delta\Delta G_s^*$ is no longer equal to ΔG_s^{*l} and Eq. (6.13.46) is no longer valid; hence, the two sets of curves diverge.

In order to compute the entropy and enthalpy of solvation, we need the temperature derivative of $\Delta\Delta G_s^*$ at constant pressure. The data available, though, are along the equilibrium line. The two derivatives are related to each other by

$$\left(\frac{d\Delta\,\Delta G_s^*}{dT}\right)_{eq} = \left(\frac{\partial\Delta\Delta G_s^*}{\partial T}\right)_P + \left(\frac{\partial\Delta\Delta G_s^*}{\partial P}\right)_T \left(\frac{dP}{dT}\right)_{eq}. \qquad (6.13.47)$$

Both of the straight derivatives (taken along the equilibrium line) may be computed from the available experimental data, provided we can assume ideal-gas behavior in the gaseous phase. Hence from now on we assume that $\Delta\Delta G_s^*$ is essentially ΔG_s^*, so Eq. (6.13.47) reduces to

$$\left(\frac{d\Delta\,G_s^*}{dT}\right)_{eq} = -\Delta S_s^* + \Delta V_s^*\left(\frac{dP}{dT}\right)_{eq}, \qquad (6.13.48)$$

where ΔS_s^* and ΔV_s^* are the entropy and volume of solvation of s in the pure liquids, taken at the limit of low pressure (or, equivalently, using the experimental data near $T \sim T_{tr}$).

Clearly, Eq. (6.13.48) cannot be solved for both ΔS_s^* and ΔV_s^*. Fortunately, however, ΔV_s^* may be computed independently from the molar volume of s (V_s^l) and the compressibility κ_T^l of the pure liquid s using the relation

$$\Delta V_s^* = V_s^l - kT\kappa_T^l. \qquad (6.13.49)$$

In Table 6.1 we present the thermodynamic quantities of solvation of some inert gases, as well as methane, near the triple point for each gas. The figures in brackets are approximate values obtained by either neglecting $kT\kappa_T^l$ in Eq. (6.13.49) or neglecting $\Delta V_s^*(dP/dT)$ in Eq. (6.13.48).

TABLE 6.1
Values of T_{tr}, ΔG_s^*, $T\Delta S_s^*$, ΔH_s^*, and ΔV_s^* for Neon, Argon, Krypton, Xenon, and Methane near Their Corresponding Triple Points T_{tr}

	T_{tr} (K)	ΔG_s^* (kJ mol^{-1})	$T\Delta S_s^*$ (kJ mol^{-1})	ΔH_s^* (kJ mol^{-1})	ΔV_s^* (cm^3 mol^{-1})
Neon	24.55	−1.144	[−0.44]	[−1.58]	[16.18]
Argon	83.80	−4.097	−2.202	−6.299	26.86
Krypton	115.95	−5.733	[−2.66]	[−8.53]	[34.22]
Xenon	161.3	−7.991	[−3.94]	[−11.89]	[44.29]
Methane	90.68	−5.655	−2.502	−8.157	34.47

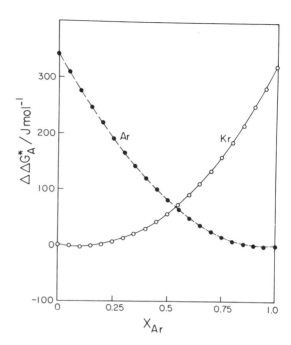

FIGURE 6.10. Values $\Delta\Delta G_A$ in J mol^{-1} for argon and krypton as a function of the mole fraction of argon at $T = 115.77$ K.

A glance at Table 6.1 shows that all values of ΔG_s^* are negative and their absolute magnitude increases with the mass of the molecules. This is in contrast to what we have seen for the solvation Gibbs energy in the HS systems. Clearly, the negative sign is due to the attractive part of the intermolecular potential. Also, it is evident from this table that $T\Delta S_s^*$ values are relatively small compared with ΔH_s^*. This means that the energetic factor, rather than the entropic factor, is dominating ΔG_s^*, i.e., the sign of ΔG_s^* is determined by ΔH_s^*, which is almost identical with ΔE_s^*. The $P\Delta V_s^*$ term is usually negligible for most of the systems discussed in this book (see Appendix E).

Having computed the thermodynamic quantities of solvation of inert gases in their own liquids, we now follow the variation in the solvation quantities when we add a second component to the system. Figure 6.10 shows the variation of $\Delta\Delta G_A^*$ for both argon and krypton in the argon–krypton mixtures at $T = 115.77$ K. These quantities are defined by

$$\Delta\Delta G_A^* = \Delta G_A^*(\text{in mixture}) - \Delta G_A^{*P}. \tag{6.13.50}$$

In section 6.16 we present some more data on the solvation of ionic solutes. In Chapter 7 we also discuss the solvation of simple solutes as well as water in liquid water.

6.14. CONDITIONAL SOLVATION AND THE PAIR CORRELATION FUNCTION

Consider again a solvent l at any given temperature, volume, and composition. Let s be a simple spherical molecule. The solvation Helmholtz energy of s is given by

$$\exp[-\beta\Delta A_s^*(\mathbf{R}_1)] = \langle\exp[-\beta B_s(\mathbf{R}_1)]\rangle_0. \tag{6.14.1}$$

Note that we have explicitly introduced the location $\mathbf{R}_1$ at which we have placed the solvaton s. This is in general not necessary since all points in the solvent are equivalent. In this section, however, we shall produce inhomogeneity by introducing two particles at $\mathbf{R}_1$ and $\mathbf{R}_2$; therefore, the recording of their locations is important. Also, in this section we treat the very dilute solution of s in a solvent. Theoretically, we can think of having just one s in a pure solvent. The symbol $\langle\ \rangle_0$ stands for an average over all the configurations of the solvent molecules in the T, V, N ensemble, i.e.,

$$\langle \exp[-\beta B_s(\mathbf{R}_1)]\rangle_0 = \int \cdots \int d\mathbf{X}^N P_0(\mathbf{X}^N) \exp[-\beta B_s(\mathbf{R}_1, \mathbf{X}^N)], \qquad (6.14.2)$$

where $P_0(\mathbf{X}^N)$ is the probability density of finding a configuration $\mathbf{X}^N$, i.e.,

$$P_0(\mathbf{X}^N) = \frac{\exp[-\beta U(\mathbf{X}^N)]}{\int \cdots \int d\mathbf{X}^N \exp[-\beta U(\mathbf{X}^N)]}. \qquad (6.14.3)$$

Next, suppose we have one solute s at $\mathbf{R}_1$ and we introduce a second s at a different location $\mathbf{R}_2$. The corresponding work is obtained by taking the ratio of the two partition functions

$$\exp[-\beta \Delta A_s^*(\mathbf{R}_2/\mathbf{R}_1)]$$

$$= \frac{Q(T, V, N; \mathbf{R}_1, \mathbf{R}_2)}{Q(T, V, N; \mathbf{R}_1)}$$

$$= \frac{\int \cdots \int d\mathbf{X}^N \exp[-\beta U(\mathbf{X}^N) - \beta B_s(\mathbf{R}_1) - \beta B_s(\mathbf{R}_2) - \beta U(\mathbf{R}_1, \mathbf{R}_2)]}{\int \cdots \int d\mathbf{X}^N \exp[-\beta U(\mathbf{X}^N) - \beta B_s(\mathbf{R}_1)]}$$

$$= \exp[-\beta U(\mathbf{R}_1, \mathbf{R}_2)]\langle \exp[-\beta B_s(\mathbf{R}_2)]\rangle_{\mathbf{R}_1}, \qquad (6.14.4)$$

where $U(\mathbf{R}_1, \mathbf{R}_2)$ is the direct interaction between the two solutes at $\mathbf{R}_1$, $\mathbf{R}_2$, and the symbol $\langle\ \rangle_{\mathbf{R}_1}$ stands for a conditional average, i.e., an average over all configurations of the solvent molecules, given one solute at $\mathbf{R}_1$. The conditional density is now

$$P(\mathbf{X}^N/\mathbf{R}_1) = \frac{\exp[-\beta U(\mathbf{X}^N) - \beta B_s(\mathbf{R}_1)]}{\int \cdots \int d\mathbf{X}^N \exp[-\beta U(\mathbf{X}^N) - \beta B_s(\mathbf{R}_1)]}. \qquad (6.14.5)$$

Note that $P(\mathbf{X}^N/\mathbf{R}_1)$ may be viewed as the probability density in a presence of an "external" field produced by s at $\mathbf{R}_1$. If the "external" field is zero, we recover (6.14.3). Equation (6.14.4) can be rewritten in another equivalent form as

$$\exp[-\beta \Delta A_s^*(\mathbf{R}_2/\mathbf{R}_1)] = \exp[-\beta U(\mathbf{R}_1, \mathbf{R}_2)] \frac{\langle \exp[-\beta B_{ss}(\mathbf{R}_1, \mathbf{R}_2)]\rangle_0}{\langle \exp[-\beta B_s(\mathbf{R}_1)]\rangle_0}, \qquad (6.14.6)$$

where $B_{ss}(\mathbf{R}_1, \mathbf{R}_2) = B_s(\mathbf{R}_1) + B_s(\mathbf{R}_2)$. Equation (6.14.16) can be rearranged into

$$\exp[-\beta \Delta A_s^*(\mathbf{R}_2/\mathbf{R}_1)] = g(\mathbf{R}_1, \mathbf{R}_2) \exp[-\beta \Delta A_s^*(\mathbf{R}_1)]. \qquad (6.14.7)$$

Thus the correlation function connects between the solvation Helmholtz energy on the first "site" and on the second "site." This is formally the same relation as we encountered several times in Chapter 3. Equation (6.14.7) can also be read as

$$W(\mathbf{R}_1, \mathbf{R}_2) = \Delta A_s^*(\mathbf{R}_2/\mathbf{R}_1) - \Delta A_s^*(\mathbf{R}_1). \tag{6.14.8}$$

Thus the potential of average force is the same as the solvation Helmholtz energy at $\mathbf{R}_2$ given a particle at $\mathbf{R}_1$, minus the solvation Helmholtz energy at $\mathbf{R}_1$.

In the above examples, the conditional solvation Helmholtz energy includes the direct interaction between the two solute particles, as well as the effect of the solvent. In some applications it will be found useful to exclude the direct interaction.

This occurs whenever we want to estimate the contributions to the solvation Helmholtz energy of each part of a combined solute. In our definitions of both $\Delta A_s^*(\mathbf{R}_1)$ and $\Delta A_s^*(\mathbf{R}_2/\mathbf{R}_1)$, we transferred one solute s from a fixed position in an ideal gas into the liquid. Now suppose that we are given a pair of solutes at a distance $R = |\mathbf{R}_2 - \mathbf{R}_1|$ in an ideal gas. This pair of solutes can be viewed as a single molecule, as in section 6.14.2 below, or a protein molecule, as in Chapter 8. We now wish to know the contribution of each particle, 1 and 2, to the Helmholtz energy of solvation of the pair. The latter is

$$\exp[-\beta A_{ss}^*(\mathbf{R}_1, \mathbf{R}_2)] = \langle \exp[-\beta B_{ss}(\mathbf{R}_1, \mathbf{R}_2)] \rangle_0. \tag{6.14.9}$$

Now, instead of transferring the pair as a single entity, we first transfer one particle. The solvation Helmholtz energy change is almost the same as in (6.14.1), but we must also add the energy required to break the s–s bond in the vacuum. In the second step we transfer the second particle; the corresponding work is now exactly as in (6.14.6), where we now gain the s–s bond energy in the liquid. Therefore in the entire process, the direct interaction between the two solutes cancels out. This is schematically shown in Fig. 6.11.

Instead of Eq. (6.14.7), we now write the corresponding relation excluding the direct interaction; namely, we define

$$y(\mathbf{R}_1, \mathbf{R}_2) = g(\mathbf{R}_1, \mathbf{R}_2) \exp[\beta U(\mathbf{R}_1, \mathbf{R}_2)] \tag{6.14.10}$$

and rewrite (6.14.7) as

$$\langle \exp[-\beta B_s(\mathbf{R}_2)] \rangle_{\mathbf{R}_1} = y(\mathbf{R}_1, \mathbf{R}_2) \langle \exp[-\beta B_s(\mathbf{R}_1)] \rangle_0. \tag{6.14.11}$$

This is the analogue of relation (3.3.58) of Chapter 3.

Note that the average on the rhs of (6.14.11) is simply as in (6.14.1); i.e., this is the same as the solvation Helmholtz energy of one s particle in a pure solvent. On the lhs of

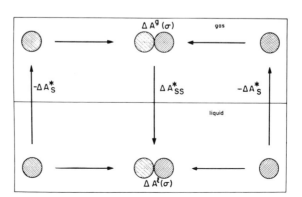

FIGURE 6.11. Cyclic process involving the quantities ΔA_{ss}^* and ΔA_s^* in Eqs. (6.14.7) and (6.14.9).

(6.14.11) we have the conditional solvation Helmholtz energy as in (6.14.4), but excluding the direct interaction between the two solutes. Note that the solute at $\mathbf{R}_1$ still affects the distribution function (6.14.5), and therefore the average on the lhs of (6.14.11) is different from the average on the rhs of (6.14.11). We shall later refer to this conditional average as a conditional Helmholtz energy of solvation, meaning the quantity $\Delta A_{ss}^*(\mathbf{R}_2, \mathbf{R}_1) - U(\mathbf{R}_2, \mathbf{R}_1)$. We shall do so whenever the direct interaction will not enter into the final results.

Another useful form of (6.14.11) is

$$
y(\mathbf{R}_1, \mathbf{R}_2) = \frac{\langle \exp[-\beta B_s(\mathbf{R}_2)] \rangle_{\mathbf{R}_1}}{\langle \exp[-\beta B_s(\mathbf{R}_1)] \rangle_0}
$$

$$
= \frac{\langle \exp[-\beta B_{ss}(\mathbf{R}_1, \mathbf{R}_2)] \rangle_0}{\langle \exp[-\beta B_s(\mathbf{R}_1)] \rangle_0^2}. \tag{6.14.12}
$$

Note that both averages on the rhs of (6.14.12) are taken with the distribution function of the pure solvent (6.14.3). If $|\mathbf{R}_1 - \mathbf{R}_2| \to \infty$, the two solutes are uncorrelated and the average in the numerator of (6.14.12) can be factored into a product of two averages

$$
\langle \exp[-\beta B_{ss}(\mathbf{R}_1, \mathbf{R}_2)] \rangle_0 = \langle \exp[-\beta B_s(\mathbf{R}_1)] \rangle_0^2 \tag{6.14.13}
$$

and

$$
y(\mathbf{R}_1, \mathbf{R}_2) = 1, \qquad |\mathbf{R}_1 - \mathbf{R}_2| \to \infty. \tag{6.14.14}
$$

6.14.1. Conditional Solvation Helmholtz Energy: Hard and Soft Parts

One of the earliest attempts to interpret the anomalous properties of aqueous solutions was based on splitting the solvation process into two parts: a cavity formation and introduction of the solute into the cavity. In the present section we shall discuss the statistical-mechanical basis for such a split of the solvation process.

In essence, the molecular basis for splitting the solvation into two (or more) steps stems from the recognition of the two (or more) parts of the solute–solvent intermolecular potential function, the so-called hard and soft parts of the interactions.

For simplicity let us assume that the solute–solvent pair potential is a function of the distance R only, and that this function may be written as

$$
U_{sb}(R) = U_{sb}^H(R) + U_{sb}^S(R), \tag{6.14.15}
$$

where the "hard" part of the potential is defined through a choice of effective hard-core diameter for the solute σ_s and solvent σ_b, respectively, namely

$$
U_{sb}^H(R) = \begin{cases} \infty & \text{for } R \le \sigma_{sb} \\ 0 & \text{for } R > \sigma_{sb} \end{cases}, \tag{6.14.16}
$$

where $\sigma_{sb} = (\sigma_s + \sigma_b)/2$. One can always find a small enough value of σ_{sb} such that, for $R \le \sigma_{sb}$, the potential function rises so steeply as to justify an approximation of the form (6.14.16). The "soft" part, U_{sb}^S, is next defined through Eq. (6.14.15).

We now write the solvation Helmholtz energy of s as

$$
\Delta A_s^* = -kT \ln \langle \exp(-\beta B_s) \rangle_0
$$

$$
= -kT \ln \langle \exp(-\beta B_s^H - \beta B_s^S) \rangle_0, \tag{6.14.17}
$$

where B_s^H and B_s^S are the "hard" and "soft" parts of the binding energies of s to all the solvent molecules.

The average quantity in relation (6.14.17) can be viewed as an average of a product of two functions. If these were independent (in the sense of the independence of two random variables), one could rewrite the average as a product of two averages, i.e.,

$$\langle \exp(-\beta B_s^H) \exp(-\beta B_s^S) \rangle_0 = \langle \exp(-\beta B_s^H) \rangle_0 \langle \exp(-\beta B_s^S) \rangle_0, \quad (6.14.18)$$

and ΔA_s^* could be split into the corresponding factors

$$\Delta A_s^* = \Delta A_s^{*H} + \Delta A_s^{*S}. \quad (6.14.19)$$

However, the factorization in Eq. (6.14.18) is in general invalid, and therefore a split of ΔA_s^* in terms of a hard and a soft part in the form of (6.14.19) is not justified. Instead, we rewrite the average in (6.14.17) as

$$\langle \exp(-\beta B_s^H) \exp(-\beta B_s^S) \rangle_0$$

$$= \frac{\int d\mathbf{X}^N \exp(-\beta U_N - \beta B_s^H) \exp(-\beta B_s^S)}{\int d\mathbf{X}^N \exp(-\beta U_N)}$$

$$= \frac{\int d\mathbf{X}^N \exp(-\beta U_N - \beta B_s^H)}{\int d\mathbf{X}^N \exp(-\beta U_N)} \frac{\int d\mathbf{X}^N \exp(-\beta U_N - \beta B_s^H) \exp(-\beta B_s^S)}{\int d\mathbf{X}^N \exp(-\beta U_N - \beta B_s^H)}$$

$$= \langle \exp(-\beta B_s^H) \rangle_0 \langle \exp(-\beta B_s^S) \rangle_H, \quad (6.14.20)$$

where the second average on the rhs of Eq. (6.14.20) is a conditional average, using the probability distribution function

$$P(\mathbf{X}^N / \mathbf{X}_s^H) = \frac{P(\mathbf{X}^N, \mathbf{X}_s^H)}{P(\mathbf{X}_s^H)} = \frac{\exp(-\beta U_N - \beta B_s^H)}{\int d\mathbf{X}^N \exp(-\beta U_N - \beta B_s^H)}, \quad (6.14.21)$$

with $P(\mathbf{X}^N / \mathbf{X}_s^H)$ the probability density of finding a configuration $\mathbf{X}^N$ of the N particles given a hard particle at $\mathbf{X}_s^H$. Thus, with the aid of Eq. (6.14.21), the solvation Helmholtz energy may be split as

$$\Delta A_s^* = \Delta A_s^{*H} + \Delta A_s^{*S/H}, \quad (6.14.22)$$

where ΔA_s^{*H} is the Helmholtz energy of solvation of the hard part of the potential, and $\Delta A_s^{*S/H}$ is the conditional Helmholtz energy of solvation of the soft part of the potential. It is the work required to couple, or to switch on, the soft part of the solute–solvent interaction, given that the hard part of the potential has already been solvated. Clearly, the procedure outlined above may be generalized to include various contributions to the pair potential. An important generalization of Eq. (6.14.15) might be the inclusion of electrostatic interactions, e.g., charge–dipole interaction between an ionic solute and the dipole moment of a solvent molecule. The generalization of Eq. (6.14.15) might look like this:

$$U_{sb}(R) = U_{sb}^{H}(R) + U_{sb}^{S}(R) + U_{sb}^{el}(R).$$ (6.14.23)

Correspondingly, the solvation Helmholtz energy will be written, in a generalization of expression (6.14.22), as follows:

$$\Delta A_s^* = \Delta A_s^{*H} + \Delta A_s^{*S/H} + \Delta A_s^{*el/S,H},$$ (6.14.24)

where $\Delta A_s^{*el/S,H}$ is the conditional Helmholtz energy of solvation of the electrostatic part of the interaction, given that a solvaton with the soft and hard parts (excluding the electrostatic part) has already been placed at a fixed position in the solvent.

We recall that the solvation Helmholtz energy of a hard particle in a solvent is related to the probability of finding a cavity of suitable size in the liquid (see section 5.10). From (6.14.17) and (6.14.20) we have

$$\exp(-\beta \Delta A_s^*) = \Pr(\text{cavity})\langle \exp(-\beta B_s^S) \rangle_H.$$ (6.14.25)

This equation is the analogue of Eq. (2.8.59). There, the probability of finding a site empty (cavity on the surface) was $1 - \theta$, which has the same role as $\Pr(\text{cavity})$ in (6.14.25). As we noted in Chapter 2, the factorization into two terms corresponding to the hard and soft works is a result of the recognition of the two parts of the interaction potential. The same is clearly true in the present case.

Equation (6.14.25) is useful in actual estimation of the solvation Helmholtz energy. The cavity work is usually estimated by the SPT (section 5.11). If the soft interaction is small, i.e., if $|\beta B_s^S| \ll 1$, then we may estimate

$$\langle \exp(-\beta B_s^S) \rangle_H \approx 1 - \beta \langle B_s^S \rangle_H,$$ (6.14.26)

where $\langle B_s^S \rangle_H$ is the conditional average binding energy of the soft part of the interaction between the solute s and the solvent.

6.14.2. Conditional Solvation Helmholtz Energy: Group Additivity

Researchers in solvation thermodynamics have long attempted to assign group contributions of various parts of a solute molecule to the thermodynamic quantities of solvation. In this section we examine the molecular basis of such a group-additivity approach. As we shall see, the problem is essentially the same as that treated in the previous section; i.e., it originates from a split of the solute–solvent intermolecular potential function into two or more parts.

Let us start with a solute of the form X—Y, where X and Y are two groups, say CH_3 and CH_3 in ethane or CH_3 and OH in methanol. We assume that the solute—solvent interaction may be split into two parts as follows:

$$U(X—Y, i) = U(X, i) + U(Y, i),$$ (6.14.27)

where $U(X, i)$ and $U(Y, i)$ are the interaction energies between the groups X and Y and the ith solvent molecule, respectively.

As in the preceding section, where we had split the interaction energy into a hard and a soft part, we also have here an element of ambiguity as to the exact manner in which this split may be achieved. However, assuming that an equation of the form

(6.14.27) may be applied, we can proceed to split the total binding energy of the solute X—Y into two parts, simply by summing the interactions between the solute and all solvent molecules at some particular configuration, say $\mathbf{X}^N$; thus,

$$B_{X-X} = \sum_{i=1}^{N} U(X, i) + \sum_{i=1}^{N} U(Y, i)$$

$$= B_X + B_Y. \qquad (6.14.28)$$

The solvation Helmholtz energy of the solute is now written as

$$\Delta A^*_{X-Y} = -kT \ln\langle \exp(-\beta B_{X-Y}) \rangle_0$$

$$= -kT \ln\langle \exp(-\beta B_X) \exp(-\beta B_Y) \rangle_0. \qquad (6.14.29)$$

We see that, as in the preceding section, we have again an average of a product of two functions. This, in general, may not be factorized into a product of two average quantities. If this could have been done, then relation (6.14.29) could have been written as a sum of two terms, i.e.,

$$\Delta A^*_{X-Y} = -kT \ln\langle \exp(-\beta B_X) \rangle_0 - kT \ln\langle \exp(-\beta B_Y) \rangle_0$$

$$= \Delta A^*_X + \Delta A^*_Y. \qquad (6.14.30)$$

Such a group additivity, though very frequently assumed to hold for ΔA^*_{X-Y} (as well as for other thermodynamic quantities), has clearly no molecular basis, even if we assume that a split of the potential function in expression (6.14.27) is exact.

The reason is quite simple: since the two groups X and Y are very close to each other, their solvation atmospheres must be correlated, and therefore the average in (6.14.29) cannot be factored into a product of two averages. In order to achieve some form of additivity similar to relation (6.14.30), we rewrite Eq. (6.14.29) as follows:

$$\Delta A^*_{X-Y} = -kT \ln\langle \exp(-\beta B_X - \beta B_Y) \rangle_0$$

$$= -kT \ln\left[\frac{\int d\mathbf{X}^N \exp(-\beta U_N) \exp(-\beta B_X) \exp(-\beta B_Y)}{\int d\mathbf{X}^N \exp(-\beta U_N) \exp(-\beta B_X)} \right.$$

$$\left. \times \frac{\int d\mathbf{X}^N \exp(-\beta U_N) \exp(-\beta B_X)}{\int d\mathbf{X}^N \exp(-\beta U_N)} \right]$$

$$= -kT \ln\left[\int d\mathbf{X}^N P(\mathbf{X}^N/\mathbf{X}_X) \exp(-\beta B_Y) \int d\mathbf{X}^N P(\mathbf{X}^N) \exp(-\beta B_X) \right]$$

$$= \Delta A^*_{Y/X} + \Delta A^*_X. \qquad (6.14.31)$$

The significance of Eq. (6.14.31) is the following: The solvation Helmholtz energy of the solute X—Y is written [exactly, presuming the validity of relation (6.14.28)] as a sum of two contributions. First, the solvation Helmholtz energy of the group X, and second, the conditional solvation Helmholtz energy of the group Y at an adjacent point near the group X given that the group X is already at a fixed point in the liquid.

Clearly, since the second group, Y, is brought to a location very near X, one cannot ignore the effect of X on the (conditional) solvation of Y. It is only in the very extreme cases when X and Y are very far apart that we could assume that their solvation Helmholtz energies will be strictly additive, as in Eq. (6.14.30).

Because of symmetry we could, of course, replace Eq. (6.14.31) by the equivalent relation

$$\Delta A^*_{X-Y} = \Delta A^*_{X/Y} + \Delta A^*_Y. \tag{6.14.32}$$

The procedure outlined above may evidently be generalized to include large molecules with many groups. We shall make use of such a procedure for proteins in Chapter 8.

Note that since ΔA^*_{X-Y} applies to the entire molecule X—Y, the conditional solvation Helmholtz energy $\Delta A^*_{X/Y}$ does not include the direct bond energy between X and Y. In this sense this conditional Helmholtz energy differs from the one defined in (6.14.4).

Finally, in this section we used the T, V, N ensemble to obtain various Helmholtz energies of solvation. Rewriting the same relations in the T, P, N ensemble will provide the corresponding solvation Gibbs energies. The latter is more useful in the treatment of actual experimental data.

6.15. THE SOLVATION HELMHOLTZ ENERGY OF A MOLECULE HAVING INTERNAL ROTATIONAL DEGREES OF FREEDOM

In most of our discussion of solvation phenomena, we assumed that the internal partition function is not affected by the solvent; i.e., q_s in Eq. (6.13.1) was assumed to be the same in the gas or in the liquid state. This assumption is approximately correct for the internal partition function of simple molecules in a simple solvent. There is one important exception where we must take into account the solvent effects even in simple solvents: the case in which the molecule can have different conformations, each with a different rotational PF. Clearly, for large polymers, the rotational PF of the extended polymer differs significantly from the rotational PF of a compact conformer of the same molecule. Since these two conformations have different binding energies to the solvent, the relative weights given to each conformation will be different in the gas or in the liquid state. We shall demonstrate this effect for a small molecule such as butane (Fig. 6.12), and then generalize to larger polymers.

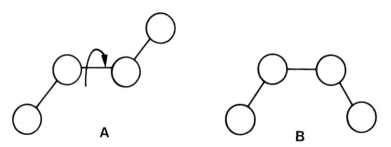

FIGURE 6.12. Two conformations of butane, treated as two isomers A and B.

Let s be a molecule with internal rotational degrees of freedom. We assume that the vibrational, electronic, and nuclear partition functions are separable and independent of the configuration of the molecules in the system. We define the pseudochemical potential (PCP) of a molecule with a fixed conformation $\mathbf{P}_s$ as the change in the Helmholtz energy for the process of introducing s into the system l (at fixed T, V), in such a way that its center of mass is at a fixed position $\mathbf{R}_s$. If we release the constraint on the fixed position of the center of mass, we can define the chemical potential of the $\mathbf{P}_s$ conformer in the gas and liquid phases as follows:

$$\mu_s^g(\mathbf{P}_s) = \mu_s^{*g}(\mathbf{P}_s) + kT \ln \rho_s^g \Lambda_s^3 \qquad (6.15.1)$$

$$\mu_s^l(\mathbf{P}_s) = \mu_s^{*l}(\mathbf{P}_s) + kT \ln \rho_s^l \Lambda_s^3. \qquad (6.15.2)$$

Note that the rotational partition function of the entire molecule as well as the internal partition functions of s are included in the PCP. In classical systems, the momentum partition function Λ_s^3 is independent of the environment, whether it is a gas or a liquid phase.

The solvation Helmholtz energy of the $\mathbf{P}_s$ conformer is defined as

$$\Delta\mu_s^*(\mathbf{P}_s) = \mu_s^{*l}(\mathbf{P}_s) - \mu_s^{*g}(\mathbf{P}_s)$$

$$= -kT \ln\langle \exp[-\beta B_s(\mathbf{P}_s)]\rangle_0; \qquad (6.15.3)$$

i.e., this is the Helmholtz energy of transferring an s molecule, frozen in its $\mathbf{P}_s$ conformation, from a fixed position in g into a fixed position in l. If we also assume that all vibrational, electronic, and nuclear degrees of freedom are not affected by this transfer from g to l, we can write the second equality in Eq. (6.15.3).

Next, we wish to find the relation between $\Delta\mu_s^*(\mathbf{P}_s)$ and the experimental Helmholtz energy of solvation of the molecule s, irrespective of its conformation. We carry out the derivation in two steps, and for convenience we use the T, V, N ensemble. Suppose first that s can attain only two conformations A and B, say the *cis* and *trans* conformations of a given molecule at equilibrium. The PCP is the change in the Helmholtz energy for placing an A molecule at a fixed position l. The corresponding statistical-mechanical expression is

$$\exp(-\beta\mu_A^{*l}) = \frac{q_A \int d\mathbf{X}^N \, d\mathbf{\Omega}_A \exp[-\beta U_N(\mathbf{X}^N) - \beta B_A(\mathbf{X}^N) - \beta U^*(A)]}{(8\pi^2) \int d\mathbf{X}^N \exp[-\beta U_N(\mathbf{X}^N)]}, \qquad (6.15.4)$$

where $B_A(\mathbf{X}^N)$ is the binding energy of A to the rest of the system of N molecules at configuration $\mathbf{X}^N$. (Note that N is the sum of all molecules in the system including any s molecules, but excluding only the solvaton, which is an A molecule.) $U^*(A)$ denotes the intramolecular potential or the internal rotation potential function of s at the state A. $U^*(A)$ can be defined with respect to a state where all the nuclei of the molecules are at infinite separation from each other. We also assume that q_A has the form

$$q_A = q_v q_e q_{r,A}, \qquad (6.15.5)$$

where q_v and q_e are presumed to be independent of the conformation as well as of the solvent. $q_{r,A}$ is the rotational partition function of the entire molecule in state A. Clearly, since the two conformations have different moments of inertia, they will also have different rotational PFs.

Since integration over Ω_A produces $8\pi^2$, Eq. (6.15.4) may be rewritten as

$$\exp(-\beta\mu_A^{*l}) = q_A \exp[-\beta U^*(A)]\langle\exp(-\beta B_A)\rangle_0, \qquad (6.15.6)$$

where the average is over all configurations of the N molecules in the system. Likewise, for the gaseous phase we have

$$\exp(-\beta\mu_A^{*g}) = q_A \exp[-\beta U^*(A)]. \qquad (6.15.7)$$

Hence, the solvation Helmholtz energy of A is obtained from Eqs. (6.15.6) and (6.15.7) in the form

$$\exp(-\beta\Delta\mu_A^{*l}) = \langle\exp(-\beta B_A)\rangle_0, \qquad (6.15.8)$$

and a similar expression holds true for B.

To obtain the connection between $\Delta\mu_A^{*l}$, $\Delta\mu_B^{*l}$, and $\Delta\mu_s^{*l}$, we start with the equilibrium condition

$$\mu_s^l = \mu_A^l = \mu_B^l, \qquad (6.15.9)$$

or, equivalently,

$$\mu_s^{*l} + kT \ln \rho_s^l \Lambda_s^3 = \mu_A^{*l} + kT \ln \rho_A^l \Lambda_s^3 = \mu_B^{*l} + kT \ln \rho_B^l \Lambda_s^3, \qquad (6.15.10)$$

where $\rho_s^l = \rho_A^l + \rho_B^l$; and ρ_A^l and ρ_B^l are the densities of A and B at equilibrium. Equation (6.15.10) may be rearranged to yield

$$\exp(-\beta\mu_s^{*l}) = \frac{\rho_s^l}{\rho_A^l}\exp(-\beta\mu_A^{*l}), \qquad (6.15.11)$$

and

$$\exp(-\beta\mu_s^{*l}) = \frac{\rho_s^l}{\rho_B^l}\exp(-\beta\mu_B^{*l}). \qquad (6.15.12)$$

On multiplying Eq. (6.15.11) by x_A^l and Eq. (6.15.12) by x_B^l and adding the resulting two equations (where $x_A^l = \rho_A^l/\rho_s^l$ and $x_B^l = 1 - x_A^l$), we arrive at

$$\exp(-\beta\mu_s^{*l}) = \exp(-\beta\mu_A^{*l}) + \exp(-\beta\mu_B^{*l}). \qquad (6.15.13)$$

This equation is equivalent to the statement that the partition function of a system with one additional s particle at a fixed position is the sum of the partition function of the same system with one A particle at a fixed position and the partition function of the same system with one B particle at a fixed position.

We now write the corresponding expression for the ideal-gas phase, namely,

$$\exp(-\beta\mu_s^{*g}) = \exp(-\beta\mu_A^{*g}) + \exp(-\beta\mu_B^{*g}). \qquad (6.15.14)$$

Taking the ratio between expressions (6.15.13) and (6.15.14) with the aid of Eqs. (6.15.6)–(6.15.8), we obtain

$$\exp(-\beta\Delta\mu_s^{*l}) = \frac{q_A \exp[-\beta U^*(A)]\exp(-\beta\Delta\mu_A^{*l}) + q_B \exp[-\beta U^*(B)]\exp(-\beta\Delta\mu_B^{*l})}{q_A \exp[-\beta U^*(A)] + q_B \exp[-\beta U^*(B)]},$$

$$(6.15.15)$$

or, equivalently,

$$\exp(-\beta\Delta\mu_s^{*l}) = y_A^g \exp(-\beta\Delta\mu_A^{*l}) + y_B^g \exp(-\beta\Delta\mu_B^{*l}), \qquad (6.15.16)$$

where y_A^g and y_B^g are the equilibrium mole fractions of A and B in the gaseous phase. These are defined by

$$y_A^g = \frac{q_A \exp[-\beta U^*(A)]}{q_A \exp[-\beta U^*(A)] + q_B \exp[-\beta U^*(B)]} \qquad (6.15.17)$$

and

$$y_B^g = 1 - y_A^g.$$

Equation (6.15.16) is the required relation between the solvation Helmholtz energy of the solute s and the solvation Helmholtz energies of the two conformers A and B. Note that if q_v and q_e are the same for the two conformations, they will cancel in (6.5.17) and in (6.15.15). What remains is only the rotational PF of the two conformers. Generalization to the case with n discrete conformations gives

$$\exp(-\beta\Delta\mu_s^{*l}) = \frac{\sum_{i=1}^{n} q_i \exp[-\beta U^*(i)]\langle\exp(-\beta B_i)\rangle_0}{\sum_i q_i \exp[-\beta U^*(i)]}$$

$$= \sum_{i=1}^{n} y_i^g \exp(-\beta\Delta\mu_i^{*l}). \qquad (6.15.18)$$

The generalization to the continuous case is

$$\exp(-\beta\Delta\mu_s^{*l}) = \frac{\int d\mathbf{P}_s\, q(\mathbf{P}_s)\exp[-\beta U^*(\mathbf{P}_s)]\langle\exp[-\beta B(\mathbf{P}_s)]\rangle_0}{\int d\mathbf{P}_s\, q(\mathbf{P}_s)\exp[-\beta U^*(\mathbf{P}_s)]}$$

$$= \int d\mathbf{P}_s\, y^g(\mathbf{P}_s)\exp[-\beta\Delta\mu^{*l}(\mathbf{P}_s)]$$

$$= \langle\langle\exp[-\beta B(\mathbf{P}_s)]\rangle_0\rangle, \qquad (6.15.19)$$

where $q(\mathbf{P}_s)$ denotes the rotational, vibrational, etc., partition function of a single s molecule at a specific conformation $\mathbf{P}_s$, $y^g(\mathbf{P}_s)\, d\mathbf{P}_s$ is the mole fraction of s molecules at conformations between $\mathbf{P}_s$ and $\mathbf{P}_s + d\mathbf{P}_s$, and in the final expression on the rhs of relation (6.15.19) we have rewritten the integral as a double average quantity: one over all configurations of the N molecules (excluding the solvaton) and the second over all conformations of the s molecule with distribution function $y^g(\mathbf{P}_s)$.

The chemical potential of s in an ideal gas phase can now be written as

$$\mu_s^g = \mu_s^{*g} + kT \ln \rho_s \Lambda_s^3$$

$$= -kT \ln\left\{\int d\mathbf{P}_s\, q(\mathbf{P}_s)\exp[-\beta U^*(\mathbf{P}_s)]\right\} + kT \ln \rho_s \Lambda_s^3$$

$$= -kT \ln q_{int}^g + kT \ln \rho_s \Lambda_s^3, \qquad (6.15.20)$$

where the term in the curly brackets can be interpreted as the internal partition function of a single s in the gas phase. This is denoted by q_{int}^g.

In the liquid state we have the corresponding equation for the chemical potential of s,

$$\mu_s^l = \mu_s^{*l} + kT \ln \rho_s \Lambda_s^3$$

$$= -kT \ln \left\{ \int d\mathbf{P}_s \, q(\mathbf{P}_s) \exp[-\beta U^*(\mathbf{P}_s)] \langle \exp[-\beta B(\mathbf{P}_s)] \rangle_0 \right\} + kT \ln \rho_s \Lambda_s^3. \quad (6.15.21)$$

Here we cannot separate the internal PF from the coupling work. The reason is that each conformation has a different binding energy to the solvent. In a formal way we can use the definition of q_{int}^g from (6.15.20) to rewrite (6.15.21) as

$$\mu_s^l = -kT \ln q_{\text{int}}^g + \Delta \mu_s^{*l} + kT \ln \rho_s \Lambda_s^3. \quad (6.15.22)$$

Here the first term on the rhs is the same as in (6.15.20) but $\Delta \mu_s^{*l}$ includes both the coupling work of all the conformations as well as the effect of the solvent on the internal degrees of freedom on the molecule.

6.16. SOLVATION THERMODYNAMICS OF COMPLETELY DISSOCIABLE SOLUTES

In section 6.13 we defined the process of solvation as the process of transferring a single molecule from a fixed position in an ideal-gas phase to a fixed position in the liquid. For solutes which do not dissociate into fragments, the solvation Helmholtz or Gibbs energy is related to experimental quantities by the equation (see 6.13.15)

$$\Delta G_s^{*l} = kT \ln(\rho_s^{i,g}/\rho_s^l)_{\text{eq}}. \quad (6.16.1)$$

We now wish to generalize this relation to solutes that dissociate in the liquid. The most important solutes of this kind are electrolytes.

We consider a molecule which is only in a state of a dimer D in the gaseous phase. When introduced into the liquid, it completely dissociates into two fragments A and B. Of foremost importance is the case of ionic solutes—e.g., D may be HCl—then A and B are H^+ and Cl^-, respectively. For simplicity we assume that A and B do not have any internal degrees of freedom. The generalization to the case of multi-ionic solutes and polynuclear ions is quite straightforward.

The relevant solvation process has been defined in section 6.13. Since the particles are presumed not to possess any internal degrees of freedom, the Gibbs energy of solvation of the pair of ions is simply the coupling work of A and B to the liquid phase l; thus

$$\Delta G_{AB}^* = W(A, B|l). \quad (6.16.2)$$

Also, for simplicity, we shall use below the T, V, N ensemble. Hence we shall derive an expression for ΔA_{AB}^*. However, it is easy to show that ΔA_{AB}^* at T, V constant is the same quantity as ΔG_{AB}^* at T, P constant (see Appendix E).

Consider now a system of N_w solvent molecules and N_D solute molecules contained in a volume V at temperature T. By "solute" we mean those molecules D for which we wish to evaluate the solvation thermodynamic quantities. The "solvent," which is usually water, may be any liquid or any mixture of liquids and could contain any number of other solutes besides D. In the most general case, N_w will be the total number of molecules in the system except those that are counted as "solutes" in N_D. However, for notational simplicity we shall treat only two-component systems, w and D.

We shall need the following partition functions. First, we need the partition function of the system of N_w solvent molecules and N_D solutes at given V, T without any further restrictions:

$$Q(T, V, N_w, N_D) = \frac{q_w^{N_w} q_A^{N_A} q_B^{N_B} \int d\mathbf{X}^{N_w} d\mathbf{X}^{N_A} d\mathbf{X}^{N_B} \exp[-\beta U(N_w, N_A, N_B)]}{\Lambda_w^{3N_w} \Lambda_A^{3N_A} \Lambda_B^{3N_B} N_A! N_B! N_w!}, \quad (6.16.3)$$

where we have assumed that the solute molecules are completely dissociated into A and B in the liquid; $U(N_w, N_A, N_B)$ is the total potential energy of interaction among the N_w, N_A, and N_B molecules at a specific configuration (with $N_D = N_A = N_B$). If the solvent contains more than one component, then q_w and Λ_w should be interpreted as products of the corresponding quantities for all the components present in the solvent.

We next add one dimer D, or equivalently one A and one B, to the liquid in two ways: first, without any further restrictions, and second, with the constraint that A and B be placed at two fixed positions $\mathbf{R}_A$ and $\mathbf{R}_B$, respectively. The corresponding partition functions are

$$Q(T, V, N_w, N_D + 1)$$

$$= \frac{q_w^{N_w} q_A^{N_A + 1} q_B^{N_B + 1} \int d\mathbf{R}_A \, d\mathbf{R}_B \int d\mathbf{X}^{N_w + N_A + N_B} \exp[-\beta U(N_w, N_D + 1)]}{\Lambda_w^{3N_w} \Lambda_A^{3(N_A + 1)} \Lambda_B^{3(N_B + 1)} N_w! (N_A + 1)! (N_B + 1)!} \quad (6.16.4)$$

and

$$Q(T, V, N_w, N_D + 1; \mathbf{R}_A, \mathbf{R}_B)$$

$$= \frac{q_w^{N_w} q_A^{N_A + 1} q_B^{N_B + 1} \int d\mathbf{X}^{N_w + N_A + N_B} \exp[-\beta U(N_w, N_D + 1)]}{\Lambda_w^{3N_w} \Lambda_A^{3N_A} \Lambda_B^{3N_B} N_w! N_A! N_B!}. \quad (6.16.5)$$

Note the difference in the various factors in Eqs. (6.16.4) and (6.16.5). In Eq. (6.16.4), we have two more integrations over $\mathbf{R}_A$ and $\mathbf{R}_B$, and also one more Λ_A^3 and Λ_B^3. In addition, we have $(N_A + 1)!$ and $(N_B + 1)!$ in Eq. (6.16.4), but only $N_A!$ and $N_B!$ in (6.16.5). All these differences arise from the constraint we have imposed on the locations $\mathbf{R}_A$ and $\mathbf{R}_B$ in Eq. (6.16.5). The total potential energies of interaction in the two systems are related by the equation

$$U(N_w, N_D + 1) = U(N_w, N_D) + U_{AB}(\mathbf{R}_A, \mathbf{R}_B) + B_D(\mathbf{R}_A, \mathbf{R}_B), \quad (6.16.6)$$

where the configuration of the system, except for $\mathbf{R}_A$ and $\mathbf{R}_B$, is denoted symbolically by (N_w, N_D) and $(N_w, N_D + 1)$. The quantity $U_{AB}(\mathbf{R}_A, \mathbf{R}_B)$ is the direct interaction potential between A and B, being at $\mathbf{R}_A$ and $\mathbf{R}_B$, respectively; $B_D(\mathbf{R}_A, \mathbf{R}_B)$ is the "binding energy," i.e., the total interaction energy between the pair A and B at $(\mathbf{R}_A, \mathbf{R}_B)$ and all the other particles in the system at some specific configuration (the latter is omitted from the notation for simplicity).

We are now ready to derive the expression for the chemical potential of the solute D in the liquid l. This is obtained from Eqs. (6.16.3) and (6.16.4), i.e.,

$$\mu_D^l = -kT \ln \left[\frac{Q(T, V, N_w, N_D + 1)}{Q(T, V, N_w, N_D)} \right]$$

$$= -kT \ln \left\{ \left[q_A q_B \int d\mathbf{R}_A \, d\mathbf{R}_B \exp[-\beta U_{AB}(\mathbf{R}_A, \mathbf{R}_B)] \right. \right.$$

$$\left. \times \int d\mathbf{X}^{N_w + N_A + N_B} \exp[-\beta U(N_w, N_D) - \beta B_D(\mathbf{R}_A, \mathbf{R}_B)] \right]$$

$$\left. \times \left[\Lambda_A^3 \Lambda_B^3 (N_A + 1)(N_B + 1) \int d\mathbf{X}^{N_w + N_A + N_B} \exp[-\beta U(N_w, N_D)] \right]^{-1} \right\}$$

$$= -kT \ln \left\{ \frac{q_A q_B}{\Lambda_A^3 \Lambda_B^3 (N_A + 1)(N_B + 1)} \right.$$

$$\left. \times \int d\mathbf{R}_A \, d\mathbf{R}_B \exp[-\beta U_{AB}(\mathbf{R}_A, \mathbf{R}_B)] \langle \exp(-\beta B_D) \rangle_* \right\}, \qquad (6.16.7)$$

where the symbol $\langle \ \rangle_*$ indicates an average over all configurations of the $N_w + N_A + N_B$ particles, excluding only the solvaton AB at $\mathbf{R}_A$ and $\mathbf{R}_B$.

Equation (6.16.7) may be transformed into a simpler and more convenient form as follows. The pair distribution function for the species A and B in the liquid is defined by

$$\rho_{AB}^{(2)}(\mathbf{R}_A, \mathbf{R}_B) = (N_A + 1)(N_B + 1) \frac{\int d\mathbf{X}^{N_w + N_A + N_B} \exp[-\beta U(N_w, N_D + 1)]}{\int d\mathbf{X}^{N_w + (N_A + 1) + (N_B + 1)} \exp[-\beta U(N_w, N_D + 1)]}.$$

$$(6.16.8)$$

If the pair correlation function at infinite separation is denoted by $\rho_{AB}^{(2)}(\infty)$, then we can write the following ratio:

$$\frac{\rho_{AB}^{(2)}(\mathbf{R}_A, \mathbf{R}_B)}{\rho_{AB}^{(2)}(\infty)} = \frac{\int d\mathbf{X}^{N_w + N_A + N_B} \exp[-\beta U(N_w, N_D) - \beta B_D(\mathbf{R}_A, \mathbf{R}_B) - \beta U_{AB}(\mathbf{R}_A, \mathbf{R}_B)]}{\int d\mathbf{X}^{N_w + N_A + N_B} \exp[-\beta U(N_w, N_D) - \beta B_D(\infty) - \beta U_{AB}(\infty)]},$$

$$(6.16.9)$$

where we have used Eq. (6.16.6) and also introduced the notations $B_D(\infty)$ and $U_{AB}(\infty)$ for the binding energy and interaction energy at infinite separation. Using now the same probability distribution as was used in relation (6.16.7), we may rewrite Eq. (6.16.9) in the form

$$\frac{\rho_{AB}^{(2)}(\mathbf{R}_A, \mathbf{R}_B)}{\rho_{AB}^{(2)}(\infty)} = \frac{\exp[-\beta U_{AB}(\mathbf{R}_A, \mathbf{R}_B)] \langle \exp[-\beta B_D(\mathbf{R}_A, \mathbf{R}_D)] \rangle_*}{\exp[-\beta U_{AB}(\infty)] \langle \exp[-\beta B_D(\infty)] \rangle_*}. \qquad (6.16.10)$$

Equation (6.16.10) is now introduced into relation (6.16.7) to yield

$$\mu'_D = -kT \ln \left(\frac{q_A q_B \int d\mathbf{R}_A \, d\mathbf{R}_B \, \rho^{(2)}_{AB}(\mathbf{R}_A, \mathbf{R}_B) \exp[-\beta U_{AB}(\infty)] \langle \exp[-\beta B_D(\infty)] \rangle_*}{\Lambda_A^3 \Lambda_B^3 (N_A + 1)(N_B + 1) \rho^{(2)}_{AB}(\infty)} \right).$$

(6.16.11)

Since we define our zero potential energy at $R_{AB} = \infty$, we have $U_{AB}(\infty) = 0$. Also, we have $\rho^{(2)}_{AB}(\infty) = \rho_A \rho_B$. The normalization condition for $\rho^{(2)}_{AB}$,

$$\int d\mathbf{R}_A \, d\mathbf{R}_B \, \rho^{(2)}_{AB}(\mathbf{R}_A, \mathbf{R}_B) = (N_A + 1)(N_B + 1)$$

(6.16.12)

is now employed to obtain the final form:

$$\mu'_D = -kT \ln \langle \exp[-\beta B_D(\infty)] \rangle_* + kT \ln \left(\frac{\rho_A \rho_B \Lambda_A^3 \Lambda_B^3}{q_A q_B} \right).$$

(6.16.13)

Note that in relation (6.16.7) we wrote μ'_D as an average over all possible positions of A and B; the simplification achieved in Eq. (6.16.13) was rendered possible because there is equilibrium among all pairs of A and B at any specific configuration $\mathbf{R}_A, \mathbf{R}_B$. This fact allows us to choose one configuration $R_{AB} = \infty$ and use it to obtain the simplified form of the chemical potential μ'_D.

We can now also identify the liberation Helmholtz energy of the pair A and B. For any specific configuration this may be obtained from the ratio of the partition functions (6.16.4) and (6.16.5), i.e.,

$$\Delta A(\text{Lib}) = -kT \ln \frac{Q(T, V, N_w, N_D + 1)}{Q(T, V, N_w, N_D + 1; \mathbf{R}_A, \mathbf{R}_B)}$$

$$= -kT \ln \left(\frac{\int d\mathbf{R}_A \, d\mathbf{R}_B \int d\mathbf{X}^{N_w + N_A + N_B} \exp[-\beta U(N_w, N_D + 1)]}{\Lambda_A^3 \Lambda_B^3 (N_A + 1)(N_B + 1) \int d\mathbf{X}^{N_w + N_A + N_B} \exp[-\beta U(N_w, N_D + 1)]} \right)$$

$$= kT \ln[\Lambda_A^3 \Lambda_B^3 \rho^{(2)}(\mathbf{R}_A, \mathbf{R}_B)],$$

(6.16.14)

which is a generalization of the expression for the liberation Helmholtz energy of one particle in section (2.2). Here, the liberation Helmholtz energy depends on the particular configuration of A and B from which these particles are being released. For our application we need only the liberation energy at infinite separation, namely,

$$\Delta A(\text{Lib}, \infty) = kT \ln(\Lambda_A^3 \Lambda_B^3 \rho_A \rho_B).$$

(6.16.15)

Using relation (6.16.15) in Eq. (6.16.13), we may identify the solvation Helmholtz energy of the pair A and B, i.e.,

$$\mu^*_D = \Delta A^*_{AB} - kT \ln q_A q_B = -kT \ln \langle \exp[-\beta B_D(\infty)] \rangle_* - kT \ln q_A q_B.$$

(6.16.16)

This is a generalization of Eq. (6.13.4).

The chemical potential of D in the gaseous phase (where it is assumed that it exists only in dimeric form) is simply

$$\mu_D^g = kT \ln \rho_D^g \Lambda_D^3 q_D^{-1}, \tag{6.16.17}$$

where ρ_D^g is the number density and Λ_D^3 the momentum partition function of D; q_D includes the rotational, vibrational, and electronic partition functions of D. For all practical purposes, we may assume that the dimer D at room temperature is in its electronic and vibrational ground states. We also assume that the rotation may be treated classically; hence we write

$$q_D = q_{\text{rot}} \exp(-\tfrac{1}{2}\beta h\nu) \exp(-\beta \varepsilon_{\text{el}})$$

$$= q_{\text{rot}} \exp(\beta D_0), \tag{6.16.18}$$

where $\varepsilon_{\text{el}} = U_{AB}(\sigma) - U_{AB}(\infty)$ is the energy required to bring A and B from infinite separation to the equilibrium distance $R_{AB} = \sigma$. The experimental dissociation energy, as measured relative to the vibrational zero-point energy, is

$$D_0 = -\tfrac{1}{2}h\nu - \varepsilon_{\text{el}} = -\tfrac{1}{2}h\nu - [U_{AB}(\sigma) - U_{AB}(\infty)], \tag{6.16.19}$$

where $\tfrac{1}{2}h\nu$ is the zero-point energy for the vibration of D. We note that in the case of an ionic solution we can either separate the two *ions* from σ to ∞ or first separate the two neutral *atoms* and then add the ionization energy and electron affinity of the cation and anion, respectively. The latter procedure is adopted in the numerical calculations reported in Table 6.2. Using the equilibrium condition for D in the two phases, and putting $q_A = q_B = 1$, we have

$$0 = \mu_D^l - \mu_D^g = [-kT \ln(\rho_D^g \Lambda_D^3 q_{\text{rot}}^{-1}) - \tfrac{1}{2}h\nu] + [U_{AB}(\infty) - U_{AB}(\sigma)]$$

$$- kT \ln\langle \exp[-\beta B_D(\infty)]\rangle_* + [kT \ln \rho_A \rho_B \Lambda_A^3 \Lambda_B^3]. \tag{6.16.20}$$

From Eq. (6.16.20) we can eliminate the required Helmholtz energy of solvation of the pair AB to obtain the final relation that has been used in the computations reported in Table 6.2, namely,

$$\Delta A_{AB}^* = -kT \ln\langle \exp[-\beta B_D(\infty)]\rangle_*$$

$$= -D_0 + kT \ln \left(\frac{\Lambda_D^3}{\Lambda_A^3 \Lambda_B^3 q_{\text{rot}}} \right) \left(\frac{\rho_D^g}{\rho_A^l \rho_B^l} \right)_{\text{eq}}. \tag{6.16.21}$$

We see that in this case we need not only the number densities of D and of A and B at equilibrium, but also the molecular quantities Λ_A^3, Λ_B^3, Λ_D^3, q_{rot}, and D_0.

6.17. PREFERENTIAL SOLVATION

The problem of preferential solvation (PS) arises in almost any physical chemical study of solutes in mixed solvents. The study could be thermodynamic, spectroscopic, or kinetic, where the behavior of the solute (e.g., chemical shift, diffusion, reactivity, etc.,) depends on the composition of the solvent. In order to understand how the solvent composition affects the solute behavior, we need to know the composition that the solute "sees," i.e., the composition in its immediate vicinity. This is, in general, different from the bulk composition of the mixed solvent.

TABLE 6.2

Gibbs Energy of Solvation of HCl, HBr, and HI in Aqueous Solutions of Various Concentrations at 25°C [a]

x_D^l	ρ_D^l (mol dm^{-3})	ρ_D^g (mol dm^{-3})	ΔG_{AB}^* (kJ mol^{-1})
HCl			
0.083	4.556	2.895×10^{-6}	-1419.5
0.098	5.367	7.935×10^{-6}	-1417.78
0.112	6.122	1.878×10^{-5}	-1416.3
0.126	6.872	4.600×10^{-5}	-1414.65
0.139	7.562	1.030×10^{-4}	-1413.12
0.153	8.304	2.229×10^{-4}	-1411.67
0.165	8.924	4.600×10^{-4}	-1410.24
0.178	9.616	9.059×10^{-4}	-1408.93
0.190	10.210	1.694×10^{-3}	-1407.67
0.201	10.772	3.024×10^{-3}	-1406.50
0.213	11.36	5.306×10^{-3}	-1405.37
HBr			
0.105	5.556	9.698×10^{-8}	-1388.09
0.120	6.318	2.754×10^{-7}	-1386.14
0.134	7.022	5.535×10^{-7}	-1384.94
0.147	7.659	2.245×10^{-6}	-1381.90
0.151	7.850	2.951×10^{-6}	-1381.34
HI			
0.083	4.248	1.608×10^{-5}	-1336.31
0.153	7.2	1.776×10^{-4}	-1332.98

[a] x_D^l is the mole fraction of the acid in the solution; ρ_D^l and ρ_D^g are the molar concentrations of the acid in the liquid and gas phases.

The simplest approach to the problem of PS is to follow some property of a solute in a mixed solvent. For example, if δ_A is the NMR chemical shift (or other spectroscopic quantity) characteristic of the solute S in a pure solvent A, and δ_B the corresponding quantity for pure solvent B, then one might relate the observed chemical shift of S in a mixture of A and B, $\delta_{A,B}$, to δ_A and δ_B by the equation

$$\delta_{A,B} = x_A(\text{local})\delta_A + [1 - x_A(\text{local})]\delta_B, \qquad (6.17.1)$$

where $x_A(\text{local})$ defined in Eq. (6.17.1) is a measure of the local composition of the solution near the solute. This may or may not be different from the bulk composition x_A of the solvent mixture, x_A being the mole fraction of the component A in the mixture.

Although Eq. (6.17.1) can serve as an operational definition of $x_A(\text{local})$, it does not really tell us what *is* the local composition in the vicinity of the solute S. The reason is that there is no theoretical support for the assumption that $\delta_{A,B}$ is an average of δ_A and δ_B as implied in Eq. (6.17.1). Therefore, the approximation involved in using Eq. (6.17.1) will, in general, give different results for different properties of S in mixtures of A and B.

What we need is an unambiguous definition of, and a method of measuring, the local composition of the solvent, which is independent of a specific measurable property of S.

In this section we first define the PS in terms of the Kirkwood–Buff integrals for a three-component system: a solute S and a two-component *solvent*, say of A and B. Then we generalize the definition of PS for a two-component system, of A and B only.

6.17.1. Formulation of the Problem for a Three-Component System

Consider a mixture of two components, N_A molecules of A and N_B molecules of B, at some temperature T and pressure P. In such a system, the composition measured by the mole fraction $x_A = N_A/(N_A + N_B)$, will be the same at any point $\mathbf{R}_0$ within the system.

We shall refer to x_A as the bulk composition of the system. Next, consider a very dilute solution of a solute S in our two-component mixture. (In principle, we could have discussed also high concentrations of S in the system. However, this will require consideration of three components, S, A, and B in the surroundings of S. For simplicity, we treat only the case of very dilute solutions, so that S "sees" practically only As and Bs.)

The question we now ask is the following: What is the composition of the liquid in the immediate vicinity of the solute S? Clearly, since the affinity of S to A might be different from its affinity to B, we should expect that the composition near the solute S will differ from the bulk composition x_A.

The main question that will concern us in this section is how to define the local region in the vicinity of S in which the composition is expected to be affected by the presence of S. Once we have defined the region, we shall provide a method of measuring the composition in this region.

Consider first a simple spherical solute, say argon, in a two-component solvent, say water and ethanol. Let $d\mathbf{R}' = dx'\, dy'\, dz'$ be an element of volume at a distance R' from the center of S. The average number densities of A and B in this element of volume will be

$$\rho_A(R') = \rho_A(\text{bulk})g_{AS}(R') \tag{6.17.2}$$

$$\rho_B(R') = \rho_B(\text{bulk})g_{BS}(R'), \tag{6.17.3}$$

where $\rho_A(\text{bulk})$ and $\rho_B(\text{bulk})$ are the bulk densities of A and B, respectively, and $g_{AS}(R')$ and $g_{BS}(R')$ are the radial distribution functions for the pair of species A, S and B, S, respectively.

Clearly, if we had full information about these two radial distribution functions, we could have defined the local composition at any distance R' from the center of S by

$$x_A(R') = \frac{\rho_A(R')}{\rho_A(R') + \rho_B(R')}. \tag{6.17.4}$$

Unfortunately, there is no experimental data on the separate radial distribution functions in two or more component systems. Even if we had such information, it would have been too detailed to be useful for our purposes. Instead, we are interested in the overall composition in the local neighborhood of the solute, which roughly coincides with the region in which $g(R)$ is significantly different from unity. Fortunately, this information may be obtained from thermodynamic quantities through the Kirkwood–Buff integral (section 6.7).

$$G_{\alpha\beta} = \int_0^\infty [g_{\alpha\beta}(R) - 1]4\pi R^2\, dR, \tag{6.17.5}$$

where $g_{\alpha\beta}(R)$ is the radial distribution function for the pair of species α and β. The integration is extended from zero to infinity. However, in most practical cases, $g_{\alpha\beta}(R)$ differs from unity only at distances of the order of magnitude of a few molecular diameters. Therefore, the main contribution to the integral comes from the region in which $g_{\alpha\beta}$ differs considerably from unity. This region can be conveniently referred to as the *correlation region*.

Let R_c be the largest correlation distance for any pair of species in the system. We define the correlation volume as

$$V_{\text{cor}} = \frac{4\pi R_c^3}{3}. \tag{6.17.6}$$

Since all the pair correlation functions are practically equal to unity at $R \geq R_c$, we may write the average number of A particles in the correlation volume around S as

$$\bar{N}_{A,S}(R_c) = \rho_A \int_0^{R_c} g_{AS}(R)4\pi R^2 \, dR$$

$$= \rho_A \int_0^{\infty} [g_{AS}(R) - 1]4\pi R^2 \, dR + \rho_A \int_0^{R_c} 4\pi R^2 \, dR$$

$$= \rho_A G_{AS} + \rho_A V_{\text{cor}}. \tag{6.17.7}$$

Equation (6.17.7) simply means that the average number of As in the correlation volume is the sum of the average number of As in the same region, before placing S at its center, plus the change in the number of As in the same region caused by placing S in the center.

Using a similar definition for $\bar{N}_{B,S}(R_c)$, we write

$$\bar{N}_{B,S}(R_c) = \rho_B G_{BS} + \rho_B V_{\text{cor}}. \tag{6.17.8}$$

We define the local composition in the correlation region around S as

$$x_{A,S}(\text{local}) = \frac{\bar{N}_{A,S}(R_c)}{\bar{N}_{A,S}(R_c) + \bar{N}_{B,S}(R_c)}$$

$$= \frac{x_A G_{AS} + x_A V_{\text{cor}}}{x_A G_{AS} + x_B G_{BS} + V_{\text{cor}}}, \tag{6.17.9}$$

where x_A is the bulk composition of the system.

The local composition $x_{A,S}(\text{local})$ can now be compared with x_A to determine the preferential solvation of S. If $x_{A,S}(\text{local}) > x_A$, we say that S is preferentially solvated by A, and if $x_{A,S}(\text{local}) < x_A$, S is preferentially solvated by B.

Thus we define the preferential solvation of S with respect to A simply by

$$\delta_{A,S} = x_{A,S}(\text{local}) - x_A = \frac{x_A x_B (G_{AS} - G_{BS})}{x_A G_{AS} + x_B G_{BS} + V_{\text{cor}}}. \tag{6.17.10}$$

Clearly, the sign and extent of preferential solvation might depend on the composition of the solvent. Figure 6.13 depicts a few possible cases where there are positive, negative, or mixed signs of preferential solvation according to whether $x_{A,S}(\text{local})$ is above or below the diagonal line.

Note that

$$\delta_{A,S} \to 0 \quad \begin{cases} \text{if } x_A \to 0 \\ \text{or } x_B \to 0 \\ \text{or } G_{AS} - G_{BS} \to 0. \end{cases} \tag{6.17.11}$$

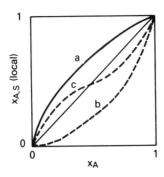

FIGURE 6.13. Schematic dependence of the local composition $s_{A,S}$ (local) as a function of the bulk composition x_A. The diagonal line corresponds to the case when there is no preferential solvation of S. Curves a and b correspond to positive and negative preferential solvation. In curve c, the preferential solvation changes sign as x_A changes.

The only quantity left ambiguous in (6.17.10) is V_{cor}. Clearly, if we choose a very large correlation volume, we obtain

$$\delta_{A,S} \to 0 \qquad \text{for } V_{\text{cor}} \to \infty. \tag{6.17.12}$$

On the other hand, for too small V_{cor}, the approximate equality of $g_{AS}(R) \approx 1$ for $R \geq R_C$ will not hold. This suggests that we consider the first-order term in the expansion of $\delta_{A,S}$ in (6.17.10) in power series about $\varepsilon \equiv V_{\text{cor}}^{-1}$, thus,

$$\delta_{A,S} = 0 + \varepsilon x_A x_B (G_{AS} - G_{BS}) + \cdots . \tag{6.17.13}$$

We define the limiting linear preferential solvation as

$$\delta_{A,S}^0 = \frac{\partial \delta_{A,S}}{\partial \varepsilon}\bigg|_{\varepsilon = 0} = x_A x_B (G_{AS} - G_{BS}). \tag{6.17.14}$$

Since $x_A x_B > 0$, the sign of $\delta_{A,S}^0$ is determined by the sign of $G_{AS} - G_{BS}$, and this is, of course, independent of the correlation volume. Thus we have defined in (6.17.14) a quantity that unambiguously measures the preferential solvation of S with respect to a two-component solvent. Since G_{AS} measures the affinity of S toward A, the difference $G_{AS} - G_{BS}$ measures the difference between the affinities of S toward A and B. We next turn to the question of measurability of the quantity $G_{AS} - G_{BS}$.

6.17.2. Relation between Preferential Solvation and Measurable Quantities

We assume that S is highly diluted in a solvent mixture of A and B at composition x_A. Let l denote the liquid mixture of A and B. In this system, the chemical potential of the solute S is

$$\mu_S^l(T, P, x_A, \rho_S) = \mu_S^{*l}(T, P, x_A) + kT \ln \rho_S \Lambda_S^3. \tag{6.17.15}$$

We now take the derivative of (6.17.14) with respect to N_A:

$$\left(\frac{\partial \mu_S^l}{\partial N_A}\right)_{T,P,N_B,N_S} = \left(\frac{\partial \mu_S^{*l}}{\partial N_A}\right)_{T,P,N_B,N_S} - \frac{kT \bar{V}_A}{V},$$

where $\bar{V}_A$ is the partial molar volume of A in the mixture. This derivative as well as $\bar{V}_A$ may be expressed in terms of Kirkwood–Buff integrals. The algebra is quite lengthy; therefore, we present the final result here (some details are discussed in Appendix J):

$$\lim_{\rho_S \to 0} \left(\frac{\partial \mu_S^{*l}}{\partial x_A} \right)_{T,P} = \frac{kT(\rho_A + \rho_B)^2}{\eta} (G_{BS} - G_{AS}), \qquad (6.17.16)$$

where η is defined in section 6.7.

Let μ_S^{*g} be the pseudochemical potential of S in an ideal gas phase. Clearly, μ_S^{*g} is independent of x_A. Therefore, we may rewrite (6.17.16) as

$$\lim_{\rho_S \to 0} \left(\frac{\partial \Delta G_S^*}{\partial x_A} \right)_{P,T} = \frac{kT(\rho_A + \rho_B)^2}{\eta} (G_{BS} - G_{AS}), \qquad (6.17.17)$$

where $\Delta G_S^* = \mu_S^{*l} - \mu_S^{*g}$ is the solvation Gibbs energy of S in our system. Thus, by measuring the slope of the solvation Gibbs energy as a function of x_A, we can extract the required difference $G_{AS} - G_{BS}$.

Note that η is a measurable quantity through the inversion of the Kirkwood–Buff theory. Since $\eta > 0$, the entire quantity $kT(\rho_A + \rho_B)^2/\eta$ is always positive. Therefore, the sign of the derivative is the same as the sign of $G_{BS} - G_{AS}$.

Figure 6.14 shows the solvation Gibbs energy of methane in mixtures of water and ethanol as a function of the mole fraction of ethanol throughout the entire range of

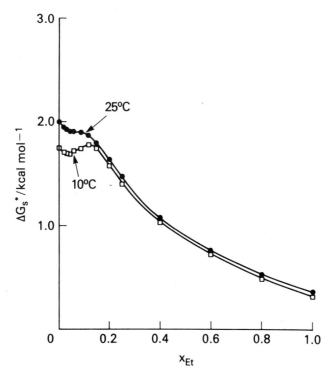

FIGURE 6.14. Solvation Gibbs energy of methane in mixtures of water and ethanol, as a function of the mole fraction of ethanol, at two temperatures.

composition. Regions in which the slope is positive correspond to $G_{BS} - G_{AS} > 0$, which in this example means that methane is preferentially solvated by water. Note that this occurs only in a very small region, say $0.1 \lesssim x_{Et} \lesssim 0.15$. In most of the composition range, methane is preferentially solvated by ethanol.

Relation (6.17.17) is useful for a solute for which the solvation Gibbs energy can be determined. If we are interested in proteins as solutes, then (6.17.17) is impractical. However, we can still measure the solvation free energy of S in l relative to, say, pure A (e.g., solvation of protein in a solution relative to the solvation in pure water). The relevant relation is

$$\Delta\Delta G_S^* = \Delta G_S^{*l} - \Delta G_S^{*A} = kT \ln\left(\frac{\rho_S^A}{\rho_S^l}\right)_{eq}. \tag{6.17.18}$$

Thus by measuring the density of S in l and in pure A at equilibrium with respect to the pure, say, solid S, we can determine $\Delta\Delta G_S^*$ from (6.17.18). Since ΔG_S^{*A} is independent of x_A, we can apply (6.17.18) in (6.17.17) to obtain

$$\lim_{\rho_S \to 0} \frac{\partial(\Delta\Delta G_S^*)}{\partial x_A} = \frac{kT(\rho_A + \rho_B)^2}{\eta} (G_{BS} - G_{AS}), \tag{6.17.19}$$

which is the useful relation for complex solutes such as proteins. Note that if A and B form a symmetrical ideal solution, i.e., when

$$G_{AA} + G_{BB} - 2G_{AB} = 0 \tag{6.17.20}$$

throughout the entire range of composition, Eq. (6.17.17) as well as (6.17.19) reduces to

$$\lim_{\rho_S \to 0} \frac{\partial\Delta G_S^*}{\partial x_A} = kT(\rho_A + \rho_B)(G_{BS} - G_{AS}). \tag{6.17.21}$$

Thus, although A and B are "similar" in the sense of (6.17.20), they can still have different affinities toward a third component S. It is only in the case when $G_{BS} = G_{AS}$ that there is no preferential solvation, and the solvation Gibbs energy of S becomes independent of composition.

Up to this point, we were interested in the difference $G_{BS} - G_{AS}$ only. However, the Kirkwood–Buff theory allows us to express both G_{BS} and G_{AS} in terms of measurable quantities. Again, the algebra involved is quite lengthy. We therefore present the final result only. First we express the partial molar volume of S in the limit of very dilute solution in terms of the Kirkwood–Buff integrals. This relation is

$$\bar{V}_S^0 = \lim_{\rho_S \to 0} \bar{V}_S = kT\kappa_T - \rho_A \bar{V}_A G_{SA} - \rho_B \bar{V}_B G_{SB}. \tag{6.17.22}$$

In Eqs. (6.17.17) and (6.17.22), all the quantities κ_T, η, ρ_A, ρ_B, $\bar{V}_A$, $\bar{V}_B$, $\bar{V}_S^0$, and $\partial\Delta G_S^*/\partial x_A$ are experimentally determinable. Hence, these two equations may be used to eliminate the required quantities G_{BS} and G_{AS}. Thus, defining

$$a = \left(\frac{\partial\Delta G_S^*}{\partial x_A}\right)_{P,T} \qquad b = \frac{kT(\rho_A + \rho_B)^2}{\eta}, \quad \text{and} \quad c = kT\kappa_T, \tag{6.17.23}$$

we may solve for G_{AS} and G_{BS}. The results are

$$G_{BS} = c - \bar{V}_S^0 + \frac{a}{b}\rho_A\bar{V}_A \qquad\qquad (6.17.24)$$

$$G_{AS} = c - \bar{V}_S^0 + \frac{a}{b}\rho_B\bar{V}_B, \qquad\qquad (6.17.25)$$

which are the required quantities.

6.17.3. Preferential Solvation in a Two-Component System

In the previous subsection we discussed the theory of preferential solvation of a solute S in a two-component system. In the traditional concept of solvation thermodynamics, only very dilute solutions could be treated. Therefore, the minimum number of components required for such a study is three: a solute and two solvents.

However, the question of PS can also be asked in a *two*-component system, say of A and B. At any composition x_A, we may focus on one A molecule and ask what is the PS of A with respect to the two components, A and B, in its immediate vicinity. Likewise, we may focus on one B molecule and ask the same, but independent, question of the PS of B with respect to the two components A and B. In this sense, the treatment of the two-component system is a "generalization" of the corresponding three-component system.

Since our mixture may have an arbitrary composition, the last two questions concerning the PS of A and B could not be dealt with within the traditional concept of solvation. Fortunately, the theoretical treatment can be easily extended to the entire range of compositions. In a two-component system, we define the average number of A and B molecules around an A molecule by [note Eq. (6.17.7)]

$$\bar{N}_{A,A} = \rho_A \int_0^{R_C} g_{A,A}(R)4\pi R^2 \, dR \qquad\qquad (6.17.26)$$

$$\bar{N}_{B,A} = \rho_B \int_0^{R_C} g_{B,A}(R)4\pi R^2 \, dR, \qquad\qquad (6.17.27)$$

where ρ_A and ρ_B are the number densities of A and B, respectively, and $g_{\alpha\beta}$ is the spatial pair correlation function for the pair of species α and β. In the following treatment we focus on a single A molecule which we refer to as an A-solvaton. A similar treatment applies to a B-solvaton.

For any radius R_c, we define the local mole fraction of A-molecules around an A-solvaton by

$$x_{A,A}(R_c) = \frac{\bar{N}_{A,A}(R_c)}{\bar{N}_{A,A}(R_c) + \bar{N}_{B,A}(R_c)}. \qquad\qquad (6.17.28)$$

Next we define the PS of an A-solvaton with respect to A-molecules simply by the deviation of the local from the bulk composition, i.e.,

$$\delta_{A,A} = x_{A,A}(R_c) - x_A. \qquad\qquad (6.17.29)$$

As in section 6.17.2, we rewrite $\bar{N}_{A,A}$ as

$$\bar{N}_{A,A} = \rho_A \int_0^{R_C} g_{AA}(R)4\pi R^2 \, dR$$

$$= \rho_A \int_0^{\infty} [g_{AA}(R) - 1]4\pi R^2 \, dR + \rho_A V_{\text{cor}}$$

$$= \rho_A \int_0^{\infty} [g_{AA}(R) - 1]4\pi R^2 \, dR + \rho_A V_{\text{cor}}$$

$$= \rho_A G_{AA} + \rho_A V_{\text{cor}}. \tag{6.17.30}$$

Hence, for the PS of an A-solvaton we have

$$\delta_{A,A} = \frac{\bar{N}_{AA}}{\bar{N}_{A,A} + \bar{N}_{B,A}} - x_A$$

$$= \frac{x_A x_B (G_{AA} - G_{BA})}{x_A G_{AA} + x_B G_{AB} + V_{\text{cor}}}. \tag{6.17.31}$$

Similarly, for the PS of a B-solvaton in the same system, we have

$$\delta_{A,B} = \frac{\bar{N}_{A,B}}{\bar{N}_{A,B} + \bar{N}_{BB}} - x_A$$

$$= \frac{x_A x_B (G_{AB} - G_{BB})}{x_A G_{AB} + x_B G_{BB} + V_{\text{cor}}}. \tag{6.17.32}$$

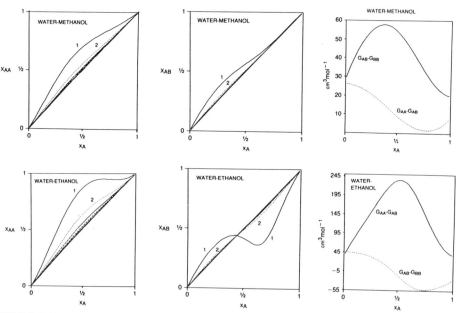

FIGURE 6.15. Preferential solvation of water–methanol and water–ethanol as a function of the mole fraction of water (component A).

Note that $G_{AB} = G_{BA}$, and that $\delta_{A,A}$ and $\delta_{A,B}$ are in general independent quantities but $\delta_{A,A} = -\delta_{B,A}$ and $\delta_{B,B} = -\delta_{A,B}$, which follows from definitions (6.17.31) and (6.17.32). Since all the $G_{\alpha\beta}$ are computable from thermodynamic quantities, using the inversion of the Kirkwood–Buff theory of solution, one can compute $\delta_{A,A}$ and $\delta_{A,B}$ for any choice of V_{cor}.

As in section (6.17.2) we define the linear coefficient of the PS as

$$\delta_{A,A}^0 = x_A x_B (G_{AA} - G_{AB}) \tag{6.17.33}$$

$$\delta_{A,B}^0 = x_A x_B (G_{AB} - G_{BB}). \tag{6.17.34}$$

Thus, besides the product $x_A x_B$, the difference $G_{AA} - G_{AB}$ characterizes the linear coefficients of the PS of A, and likewise $G_{AB} - G_{BB}$ characterizes the linear coefficient of the PS of B. By using these quantities as our measures of the PS, we have eliminated the need to make a choice of R_c. In fact, these quantities include the effect of the solvaton on the entire range for which the correlation between the various molecules extends.

Figure 6.15 shows two examples of PS in the systems of water–methanol and water–ethanol. In the first case, both the water and the ethanol are preferentially solvated by water. In the second case, water is preferentially solvated by water throughout the entire range of compositions. On the other hand, the preferential solvation of ethanol switches from water to ethanol as the concentration of water increases. The various curves correspond to different choices of correlation radius. The curve numbered 1 corresponds roughly to the first coordination spheres, etc. Usually, after 3–5 coordination spheres, the deviation of the PS curve from the diagonal line is negligible. This roughly corresponds to the limit of the correlation distance, which is of the order of a few molecular diameters.

REFERENCES

1. J. G. Kirkwood and F. P. Buff, *J. Chem. Phys.* **19**, 774 (1951).
2. A. Ben-Naim, *J. Chem. Phys.* **67**, 4884 (1977).
3. W. G. McMillan and J. E. Mayer, *J. Chem. Phys.* **13**, 276 (1945).

SUGGESTED READINGS

Older approaches to the theory of mixtures are

E. A. Guggenheim, *Mixtures* (Oxford University Press, Oxford, England, 1952).
I. Prigogin, *The Molecular Theory of Solutions* (North-Holland, Amsterdam, 1968).

More modern reviews of the theory of mixtures are

J. S. Rowlinson and F. L. Swinton, *Liquids and Liquid Mixtures*, 3rd ed. (Butterworth, London, 1982).
E. Matteoli and G. A. Mansoori (eds), *Fluctuation Theory of Mixtures* (Taylor and Francis, New York, 1990).

More specialized treatments of solvation phenomena include

Y. Marcus, *Ion Solvation* (John Wiley, New York, 1985).
A. Ben-Naim, *Solvation Thermodynamics* (Plenum Press, New York, 1987).

Water and Aqueous Solutions

7.1. INTRODUCTION

Water is the most abundant liquid on earth and the one of most importance for biological systems. From the theoretical point of view, it poses many difficult challenges to the physical chemist. These difficulties are above and beyond the usual difficulties in the theory of simple fluids. Nevertheless, because of its unique properties and importance, water has been studied more than any other liquid. In this chapter we present some aspects of the theory of liquid water and simple aqueous solutions. More complex systems such as proteins in water will be further discussed in the next chapter.

Since at present we cannot expect to have a complete theory of water and aqueous solutions, we must satisfy ourselves with partial theories aimed at specific aspects of these systems. Therefore, in choosing the molecular model for the system, we need to decide on which properties of the system we are going to focus our theory. There exists a hierarchy of levels from which we can choose to start. Some possible choices are:

1. Start from a collection of electrons and nuclei.
2. Start from atomic hydrogen and oxygen or the corresponding ions H^+ and O^{-2}.
3. Start with a model for water molecules as our fundamental particles, with or without rigid geometry or internal degrees of freedom such as stretching and bending of bond lengths and angles.
4. Start with a mixture of more complex aggregates of water molecules, such as monomers, dimers, trimers, etc.

In principle, each of these can be used to formulate an exact theory of water. The choice of the particular "level" depends on the questions we want to ask about the system. If we are interested only in explaining some macroscopic thermodynamic properties of pure water, we might be satisfied with the choice of a relatively simple mixture model. If we want to compute the pair correlation function, then a rigid model for water molecules may be used. If we are also interested in the dielectric properties of pure water or the solvation of ions in water, we need to assign an electric dipole moment, or perhaps a quadrupole moment, to our rigid water molecule. If we want to allow for dissociation into ions, then clearly a rigid model for water molecules will not be appropriate, and we need to consider a lower level of treatment such as a collection of H^+ and O^{-2}. Finally, if we are interested in the chemical reactivity of water molecules, we must start from the more elementary description of the system in terms of electrons and various nuclei and solve the Schrödinger equation for all the molecules involved in the chemical reaction.

The theory of liquid water and aqueous solutions has been developed mainly along two routes, both of which have some merits and involve serious approximations. The older approach, recognizing from the outset the immense difficulties involved in treating a system consisting of strongly interacting molecules, has focused on aggregates of water molecules, or clusters. All the strong interactions (hydrogen bonding), are included within the clusters. In this way our one-component system is converted into a mixture of clusters, the interaction between which is relatively weaker than the typical hydrogen-bond inter-action. All bonding within the cluster is therefore viewed as part of the internal degrees of freedom of a single cluster. We shall describe one representative of this approach within the framework of the mixture model theory of liquids in section 7.9.

The more modern approach is to start with single water molecules, usually having rigid geometry, assume some kind of effective pair interaction, and work out the statistical mechanics of a collection of such particles. Here we outline some of the basic assumptions that are involved in most theories, without which it is doubtful that any progress could have been achieved in this field.

The most fundamental starting point for any theory of water should be the quantum-mechanical partition function for the whole system. However, since at present we cannot solve the Schrödinger equation for such a complex system, we resort to the partially classical partition function, written as

$$Q(T, V, N) = \frac{q^N}{(8\pi^2)^N \Lambda^{3N} N!} \int \cdots \int d\mathbf{X}^N \exp[-\beta U_N(\mathbf{X}^N)], \qquad (7.1.1)$$

where we have separated the internal partition function q from the configurational parti-tion function. The former can be treated by classical or quantum mechanical methods. In the following discussion, we assume the validity of the partition function in this form.

The quantity q can be further factored into rotational, vibrational, and electronic partition functions. We will not need explicit knowledge of this quantity. It should be recognized, however, that a molecule experiencing a strong interaction with its neighbor-ing molecules may be perturbed to a significant extent. The very writing of (7.1.1) implies that we disregard these effects. This is the meaning of the separation of q^N from the configurational partition function.

The next problem is concerned with the content of the quantity $U_N(\mathbf{X}^N)$. It is well known that major progress in the theory of simple fluids could not have been achieved without the assumption of pairwise additivity for the total potential energy, written as

$$U_N(\mathbf{X}^N) = \sum_{i<j} U(\mathbf{X}_i, \mathbf{X}_j). \qquad (7.1.2)$$

It is widely believed that such an assumption is a good approximation for simple nonpolar fluids such as liquid argon. In a formal manner, one may write the total potential energy $U_N(\mathbf{X}^N)$ as a sum of contributions due to pairs, triplets, quadruplets, etc., of particles.

$$U_N(\mathbf{X}^N) = \sum_{i<j} U(\mathbf{X}_i, \mathbf{X}_j) + \sum_{i<j<k} U^{(3)}(\mathbf{X}_i, \mathbf{X}_j, \mathbf{X}_k) + \cdots + U^{(N)}(\mathbf{X}_1, \ldots, \mathbf{X}_N), \qquad (7.1.3)$$

where $U^{(n)}(\mathbf{X}_1, \mathbf{X}_2, \ldots, \mathbf{X}_n)$ is referred to as the nth-order potential.

Consider as a simple example a system of three particles. The total potential energy of the system at a given configuration can be written as

$$U_3(\mathbf{X}_1, \mathbf{X}_2, \mathbf{X}_3) = U(\mathbf{X}_1, \mathbf{X}_2) + U(\mathbf{X}_1, \mathbf{X}_3) + U(\mathbf{X}_2, \mathbf{X}_3) + U^{(3)}(\mathbf{X}_1, \mathbf{X}_2, \mathbf{X}_3). \qquad (7.1.4)$$

Recent quantum-mechanical computations for a triplet of water molecules have shown that deviations from the additivity assumption for such a system are quite appreciable.

Similarly, it is also expected that higher-order potentials may contribute to a significant extent to the total potential energy. The main reason for the nonadditivity of the potential energy of interaction is the cooperativity of the hydrogen bonding. A water molecule engaged in a hydrogen bond (HB) with a second water molecule also polarizes the second molecule in such a way as to affect its capability of forming a HB with a third molecule. In other words, the energy involved in the formation of two HBs between a triplet of water molecules at some specific configuration can be larger than the energy involved in the formation of two separate HBs. In a formal way one can incorporate all nonadditivity effects by introducing the idea of an *effective* pair potential, i.e., one assumes that the total energy can be written as

$$U_N(\mathbf{X}^N) = \sum_{i<j} U_{\text{eff}}(\mathbf{X}_i, \mathbf{X}_j). \tag{7.1.5}$$

Since U_{eff} includes contributions from higher-order potentials, it will, in general, depend on T and ρ. Such pair potentials have been constructed essentially from two sources: either from quantum-mechanical calculations of the energy of a pair of water molecules at some selected configurations or by assuming some model potential function, the parameters of which are determined by fitting the results to some experimental data. This is currently an active and changing field of research. We shall describe very briefly the essential features of such a potential function in section 7.4.

Once we have chosen an effective pair potential, we have, in fact, committed ourselves to a model of particles. That is, we define the particles of our system in terms of their (effective) pair potential. This is exactly the same procedure that one employs when studying Lennard–Jones or hard-sphere particles.

The next difficulty is technical in essence. It involves the actual computation of average quantities by one of the numerical methods of statistical mechanics. Use of an angle-dependent pair potential function introduces only a minor modification in the formal appearance of the theory but vastly increases the amount of computational time required to accomplish a project which would normally take a relatively short time in the angle-independent case.

These difficulties become more severe when we proceed to study aqueous solutions. Here again, various potential functions must be used to describe the interaction between pairs of molecules of different species. Because of these difficulties, it is no wonder that scientists have searched for other routes to study liquid water and aqueous solutions. Some of these routes are based on the mixture-model approach to these systems and will be described later in this chapter.

7.2. SURVEY OF SOME PROPERTIES OF PURE WATER

The volume of information available on the properties of water is very large. We present here only a few experimental facts that we shall need in this and in the next chapter. The geometry of a water molecule is depicted in Fig. 7.1. The nuclei of the oxygen and two hydrogen atoms form an isosceles triangle. The equilibrium O—H bond length is 0.957 Å and the H—O—H angle is 104.52°. Deviations from the equilibrium values cited above for water at room temperature are very small. Hence, we normally assume that a single water molecule has a fixed and rigid geometry and that this geometry is maintained in the liquid and solid states.

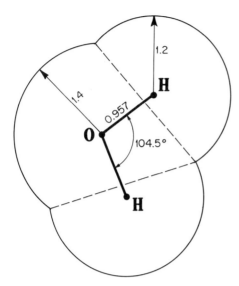

FIGURE 7.1. Geometry of a single water molecule. The O—H distance and the H—O—H angle are indicated, as are the van der Waals radii of the hydrogen and the oxygen.

The center of the oxygen nucleus is usually chosen as the center of the molecule (which is slightly different from the center of mass of the molecule). Figure 7.1 depicts the water molecule according to the van der Waals radii assigned to the oxygen (1.4 Å) atoms. It is sometimes convenient to view a water molecule as a sphere of radius 1.41 Å. (This is about half of the average distance between two closest water molecules, as manifested by the first peak of the radial distribution function of water; see below.)

An important characteristic feature of the charge distribution in a water molecule follows from the hybridization of the $2s$ and the $2p$ orbitals of the oxygen atom. As a result of this hybridization, two lobes of charges are created by the unshared electron pairs (or lone-pair electrons), which are symmetrically located above and below the molecular plane. These two directions, together with the two O—H directions, give rise to the tetrahedral geometry which is one of the most prominent structural features of the water molecule and is probably responsible for the particular mode of packing of the molecules in the liquid and solid states.

In the solid state, at least nine forms of ice are known. Ordinary ice, referred to as hexagonal ice, denoted as I_h, has a structure isomorphous to the wurtzite form of ZnS. The most important feature of the structure of ice that is relevant to our study of water is the local tetrahedral geometry around each oxygen atom. That is, each oxygen atom is surrounded by four oxygen atoms situated at the vertices of a regular tetrahedron, at a distance of 2.76 Å from the central oxygen. This basic unit of geometry is show in Fig. 7.2.

It is convenient to introduce four unit vectors originating from the center of the oxygen atom and pointing toward the four corners of the regular tetrahedron. Let $\mathbf{h}_{ik}(k = 1, 2)$ be the two unit vectors belonging to the ith molecule and pointing approximately along the two O—H directions. The remaining two vectors $\mathbf{l}_{ik}(k = 1, 2)$ act along the directions of the lone pairs of electrons. Using the terminology of hydrogen-bond formation, we identify the $\mathbf{h}_{ik}$ as the directions along which molecule i forms a hydrogen bond as a donor molecule, whereas $\mathbf{l}_{ik}$ is the direction along which the same molecule participates as an acceptor for hydrogen-bonding. We shall refer to these directions as the four "arms" of a water molecule.

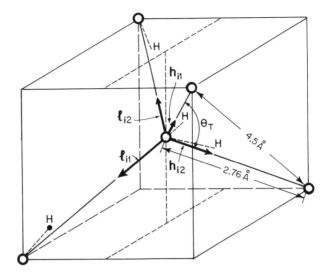

FIGURE 7.2. Basic geometry around a water molecule in ordinary ice. The four unit vectors $\mathbf{h}_{i1}$, $\mathbf{h}_{i2}$, $\mathbf{l}_{i1}$, and $\mathbf{l}_{i2}$ point to the four vertices of a regular tetrahedron.

Note that the H—O—H angle of 104.54° is slightly smaller than the characteristic tetrahedral angle of 109.46° (see Fig. 7.2).

A variety of interesting properties are manifested by the high-pressure polymorphs of ice. For the purpose of the study of liquid water, it is useful to remember that a large number of structures can be formed around a water molecule in the solid state. In particular, we draw attention to the fact that both open- and close-packed structures are possible. There is no doubt that the open structure of ice, I_h, is maintained because of strong directional forces (hydrogen bonds) operating along the directions of the four unit vectors, as depicted in Fig. 7.2. The high-pressure polymorphs of ice are characterized by relatively higher densities. That is, each water molecule experiences a higher local density than in ice I_h. In spite of the fact that the number of nearest neighbors is larger in these structures, their internal energies are higher than those of ice I_h, which indicates that the average binding energy of a water molecule in an open structure may be stronger than the binding energy of the same molecule in a more closely packed structure. We will see later that the relation between local density and binding energy is an important aspect of the mode of packing of water molecules in the liquid state.

The most unusual property of water is the temperature dependence of the density between 0 and 4°C. The fact that ice contracts upon melting, though a rare phenomenon, is not a unique property of water. The outstanding property of water is the continual *increase* in density upon increase of temperature between 0 and 4°C. Relevant experimental data are presented in Fig. 7.3. Note that the relative decrease in volume is quite small compared with the molar volume at these temperatures. This anomalous phenomenon of water disappears at high pressures, and the temperature dependence of the volume becomes normal.

The coefficient of isothermal compressibility $\kappa_T = [\partial(\ln \rho)/\partial P]_T$ decreases between 0 and 4°C (at 1 atm), whereas the compressibility of most liquids increases monotonically as temperature increases. The experimental values of κ_T for water are shown in Fig. 7.4.

The value of the heat capacity of water (both C_V and C_P) is much higher than the value expected from the sum of the contributions due to the various degrees of freedom of a water molecule (Fig. 7.5). Neglecting the contribution from vibrational excitation,

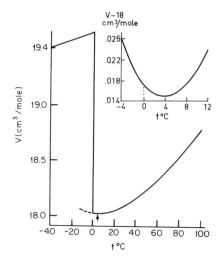

FIGURE 7.3. The temperature dependence of the molar volume of water.

one expects that a mole of water will contribute $3R/2$ due to translational degrees of freedom and $3R/2$ due to the rotational degrees of freedom ($R = 1.98$ cal/mol°K is the gas constant). Thus, the estimated heat capacity of water due to its kinetic degrees of freedom is about $3R$, i.e., 5.96 cal/mol°K, which is indeed quite close to the value of the heat capacity of water vapor. The actual value of C_V of water in the liquid state is about 18 cal/mol°K, which is quite large compared with the corresponding values of C_V for either the solid or the gaseous phases. It is also worthwhile noting that the C_P of liquid water passes through a minimum at about 35°C, whereas C_V decreases monotonically from about 18.1 at 0°C to about 16.2 cal/mol°K at 100°C.

Other well-known outstanding properties of water are its high melting and boiling temperatures (compared, for instance, with the isoelectronic sequence of hydride molecules such as NH_3, HF, and CH_4), the high value of the heat and entropy of vaporization, and the high value of the dielectric constant.

All of these properties reflect the role of strong and highly directional forces, which dictate a unique mode of packing of water molecules in the liquid state. In the following sections, we elaborate on the various theoretical tools that can be employed to illuminate the origin of these properties from the molecular point of view.

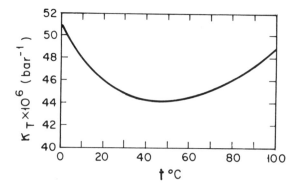

FIGURE 7.4. The isothermal compressibility of water as a function of temperature.

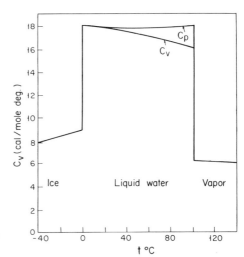

FIGURE 7.5. The heat capacity of water as a function of temperature.

7.3. THE RADIAL DISTRIBUTION FUNCTION OF WATER

The most important experimental information on the mode of packing of water molecules in the liquid state is contained in the radial distribution function, which is obtained by processing X-ray or neutron scattering data. The full orientation–dependent pair correlation function cannot be obtained from such an experiment. Instead, only information on the spatial pair correlation function is obtainable. We recall the definition of this function,

$$g(\mathbf{R}_1, \mathbf{R}_2) = \frac{1}{(8\pi^2)^2} \iint d\mathbf{\Omega}_1 \, d\mathbf{\Omega}_2 \, g(\mathbf{R}_1, \mathbf{\Omega}_1, \mathbf{R}_2, \mathbf{\Omega}_2), \qquad (7.3.1)$$

where the function $g(\mathbf{R}_1, \mathbf{R}_2)$ is a function of the scalar distance $R = |\mathbf{R}_2 - \mathbf{R}_1|$.

Water, as a heteroatomic liquid, produces a diffraction pattern that reflects the combined effects of O—O, O—H, and H—H correlations. Thus, in principle, we have three distinct atom–atom pair correlation functions: $g_{OO}(R)$, $g_{OH}(R)$, and $g_{HH}(R)$. Experimental data cannot, at present, be resolved to obtain these three functions separately. Therefore, the only information obtained is a weighted average of these three functions. In all our future reference to the experimental radial distribution function, we shall always refer to $g_{OO}(R)$ or simply to $g(R)$. In some simulation calculations, more detailed information on $g_{OO}(R)$, $g_{OH}(R)$, and $g_{HH}(R)$ have been obtained.

Figure 7.6 shows the radial distribution function of water and argon as a function of the reduced distance $R^* = R/\sigma$, where σ has been taken as 2.82 Å for water and 3.4 Å for argon. [These "effective" diameters are roughly the locations of the first maximum of the function $g(R)$ for water and argon, respectively.] Note that the location of the first peak of $g(R)$ of water at 2.82 Å is close to 2.76 Å, the value of the nearest-neighbor distance of O—O in ice I_h. This indicates a high probability of obtaining nearest neighbors at the correct position, as in ice.

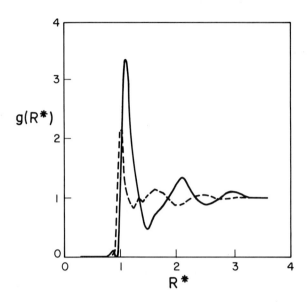

FIGURE 7.6. Radial distribution function $g(R^*)$ for water (dashed line) at 4°C and 1 atm and for argon (solid line) at 84.25°K and 0.71 atm, as a function of the reduced distance $R^* = R/\sigma$.

The most important differences between the two curves in Fig. 7.6 are the following:

1. The first coordination number, defined by

$$n_{CN} = \rho \int_0^{R_M} g(R)4\pi R^2 \, dR, \qquad (7.3.2)$$

where R_M may be taken conveniently as the location of the first minimum (following the first maximum at $R^* \approx 1$), measures the average number of centers of molecules in a sphere of radius R_M drawn about the center of a given molecule (the latter is not included in the counting). The value of n_{CN} for water at 4°C is about 4.4, whereas the value for argon (at 84.25°K and 0.71 bar) is about 10. The value of 4.4 for the coordination number is strongly reminiscent of the exact coordination number, 4, in the ice I_h structure. This is one indication that a significant degree of the structure is preserved upon the melting of ice.

2. The average coordination number as defined above slightly increases with an increase in temperature. This is in contrast to the normal behavior of argon, where the average coordination number decreases with the increase of temperature, as in Fig. 7.7. We shall see that this fact indicates an important feature of the correlation between local density and local binding energy (see section 7.9).

3. The location of the second peak of $g(R)$ of argon is at about $R^* = 2$. In Chapter 5, we showed that this phenomenon reflects the tendency of the spherical molecules to pack in roughly concentric and equidistant spheres around a given particle. In contrast to the spherical case, the spatial correlation between water molecules is mainly determined by the strong directional forces acting between the molecules. This is reflected in the location of the second peak of $g(R)$, which for water is $R^* = 1.6$ or $R \approx 4.5$ Å. This distance coincides almost exactly with the distance of the second-nearest neighbors in the ice I_h lattice. Indeed, a simple computation (see Fig. 7.2) shows that the second neighbors in ice are found to be at a distance of

$$2 \times 2.76 \times \sin\left(\frac{\theta_T}{2}\right) = 4.5, \qquad (7.3.3)$$

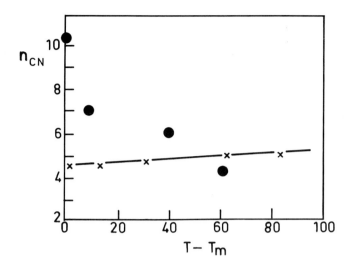

FIGURE 7.7. The average coordination number of argon (●) and of water (×) as a function of temperature, above the melting temperature (T_m).

where θ_T is the tetrahedral angle ($\theta_T = 109.46°$). The difference in the nature of the mode of packing of molecules in water and that of spherical particles is shown schematically in Fig. 7.8. It is thus seen that the strong directional forces (hydrogen bonds) dictate a particular pattern of packing of the water molecules.

These three features of the radial distribution function lead to the following conclusion: The basic geometry around a single molecule in water is, to a large extent, similar to that of ice. This is to say that, on the average, each molecule has a coordination number of about four, and, furthermore, there is a high probability that triplets of molecules will be found with nearly the same geometry as triplets of molecules at successive lattice points in ice. This conclusion pertains only to the *local* environment of a water molecule; no information whatsoever is furnished by $g(R)$ on the structure of the extended layer of molecules. In other words, if one were to sit at the center of a water molecule and observe the local geometry in a sphere of radius, say 5 Å, one would see most of the time a picture very similar to the one seen from an ice molecule, with frequent distortions caused by thermal agitation typical of the liquid state. The distortions become so large at distances greater than, say, 5 Å that one can hardly recognize a characteristic pattern of ice beyond that distance.

The temperature dependence of $g(R)$ of water is shown in Fig. 7.9. There is a gradual shift in the location of the first peak from 2.84 Å at 4°C to about 2.94 Å at 100°C. A more characteristic feature is the rapid decay of the second peak, which is almost unrecognizable at 100°C.

7.4. EFFECTIVE PAIR POTENTIAL FOR WATER

In this section we present some of the characteristics of an effective pair potential that can be used in a molecular theory of liquid water. As is the case for any liquid, neither theory nor experiment provide us with an analytical form of the entire pair

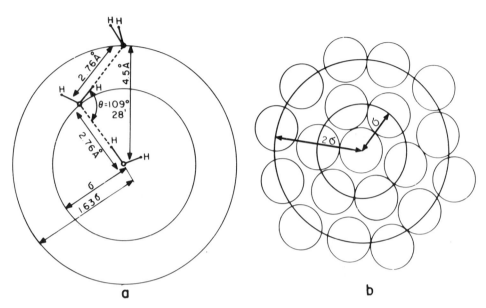

FIGURE 7.8. Schematic description of the distribution of nearest and second-nearest neighbors (a) in water and (b) in a simple fluid. The tetrahedral orientation of the hydrogen bond induces a radial distribution of nearest and second-nearest neighbors at σ and 1.63σ, respectively, $\sigma = 2.76$ Å being the O—O distance in ice I_h. The almost equidistant and concentric nature of the packing of particles in a simple fluid produces the first and second peaks of $g(R)$ at σ and 2σ.

potential. Therefore, the general procedure is to assume an analytical form of such a potential function and then determine the parameters of this function by using either theoretical or experimental information. For instance, for simple spherical atoms, an analytical function of the Lennard–Jones type is believed to present the overall distance dependence of the pair potential. The parameters ε and σ can then be determined by fitting the curve either to some experimental data—e.g., the second virial coefficient—or to some theoretical calculations of the interaction energies, when such a calculation is feasible with reasonable accuracy.

The same procedure is, in principle, employed in the case of water. Only the degree of complexity is far larger than in the case of simple fluids. Again, essentially two sources of information can be used. One is to compute the interaction energy of a pair of water molecules at some few hundreds of configurations and then fit these results to an analytical function. The second is to guess an analytical form and then determine the parameters of this function (often referred to as a model function) that best fit to some experimental quantities, e.g., second virial coefficient, dipole moment, spectroscopic data, etc.

Because of the rather complex nature of such a six-dimensional function, it is clear that many model functions can be chosen to give results which are in reasonable agreement with experiments. Indeed there are a few potential functions by the use of which some of the outstanding properties of liquid water can be successfully reproduced. We shall not describe these here. Instead, we shall describe only the essential features that we expect from such a function; then we present a qualitative general form of such a potential to which we shall refer as the primitive model of water.

Let us consider first the various ingredients that are expected to contribute to the interaction energy between two water molecules.

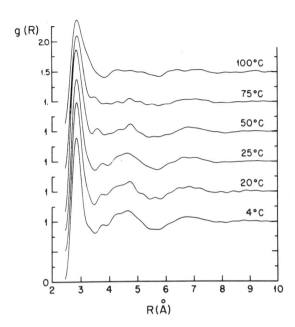

FIGURE 7.9. The temperature dependence of $g(R)$ of water.

1. At very short distances, say $R < 2.4$ Å, the two molecules exert strong repulsive forces on each other, thereby preventing excessive interpenetration. A reasonable description of the potential energy in this region can be achieved with a Lennard–Jones function of the form

$$U_{\mathrm{LJ}}(R) = 4\varepsilon \left[\left(\frac{\sigma}{R}\right)^{12} - \left(\frac{\sigma}{R}\right)^{6} \right]. \tag{7.4.1}$$

Since water and neon are isoelectronic molecules, it is appropriate to take for ε and σ in (7.4.1) the corresponding values for neon, namely,

$$\varepsilon = 5.01 \times 10^{-15}\,\mathrm{erg} = 7.21 \times 10^{-2}\,\mathrm{kcal/mol}, \qquad \sigma = 2.82\,\text{Å}. \tag{7.4.2}$$

2. At a large distance, say a few molecular diameters, the interaction between the two molecules can be described as resulting from a few ideal electric multipoles. The most important term is obviously the dipole–dipole interaction, namely,

$$U_{\mathrm{DD}}(\mathbf{X}_1, \mathbf{X}_2) = R_{12}^{-3}[\boldsymbol{\mu}_1 \cdot \boldsymbol{\mu}_2 - 3(\boldsymbol{\mu}_1 \cdot \mathbf{u}_{12})(\boldsymbol{\mu}_2 \cdot \mathbf{u}_{12})], \tag{7.4.3}$$

where $\boldsymbol{\mu}_i$ is the dipole moment vector of the ith particle and $\mathbf{u}_{12}$ is a unit vector along the direction $\mathbf{R}_2 - \mathbf{R}_1$. The dipole moment for a single water molecule is $1.84D = 1.84 \times 10^{-18}$ esu cm.

In principle, one can add to (7.4.3) interactions between higher multipoles and thus arrive at a more precise description of the interaction between the two molecules.

3. The intermediate range of distances, say 2.4 Å $\lesssim R \lesssim 4$ Å, is the most difficult one to describe analytically. We know that two water molecules can engage in hydrogen bonding. The HB energy is stronger than the typical interaction energy between simple, nonpolar molecules, yet it is much weaker than typical energies of chemical bonds. It is also highly orientation-dependent; hence, any function that is supposed to describe the interaction energy in this range must be a function of at least six coordinates, consisting

of the separation and the relative orientation of the two molecules. We use the term HB to refer to the potential function operative in this intermediate range of distances.

A rough estimate of the order of magnitude of the HB energy can be obtained from the heat of sublimation of ice, which is about 11.65 kcal/mol. Since one mole of ice contains two "moles" of hydrogen bonds, then, assuming that all the interaction energy comes from the nearest-neighbor HBs, we arrive at the estimate of 5.82 kcal/mol or about $10kT$ at room temperature.

Recognizing the fact that the hydrogen bond occurs only along four directions pointing to the vertices of a regular tetrahedron, we can use the four unit vectors introduced in section 7.2 (see Fig. 7.2) to construct a HB potential function. This we denote by $U_{HB}(\mathbf{X}_1, \mathbf{X}_2)$, so that the full effective pair potential is a superposition of three terms.

$$U(\mathbf{X}_1, \mathbf{X}_2) = U_{LJ}(R_{12}) + U_{DD}(\mathbf{X}_1, \mathbf{X}_2) + U_{HB}(\mathbf{X}_1, \mathbf{X}_2). \qquad (7.4.4)$$

Figure 7.10 depicts two waterlike molecules in a favorable orientation for the formation of a hydrogen bond. This means that the O—O distance is about 2.76 Å; that one molecule, serving as a donor, has its O—H direction at about the direction of $\mathbf{R}_{12} = \mathbf{R}_2 - \mathbf{R}_1$, and that the second molecule, serving as an acceptor, has the lone-pair direction along $-\mathbf{R}_{12}$.

Mathematically we define the *primitive pair potential* for water as follows:

$$U(\mathbf{X}_1, \mathbf{X}_2) = U_{LJ}(R_{12}) + \varepsilon_{HB}G(\mathbf{X}_1, \mathbf{X}_2), \qquad (7.4.5)$$

where ε_{HB} is the hydrogen bond energy ($\varepsilon_{HB} \approx -6$ kcal/mol ≈ 25 kJ/mol) and $G(\mathbf{X}_1, \mathbf{X}_2)$ is essentially a geometric function defined as

$$\mathbf{G}(\mathbf{X}_1, \mathbf{X}_2) = \begin{cases} 1 & \text{whenever molecules 1 and 2} \\ & \text{are in favorable configuration to form a HB} \\ 0 & \text{otherwise} \end{cases} \qquad (7.4.6)$$

In order to formulate an analytical form for the function $\mathbf{G}(\mathbf{X}_1, \mathbf{X}_2)$, it is convenient to think of a water molecule as having four "arms", i.e., four selected directions along which they can form HBs. These arms are along the four unit vectors introduced in Fig.

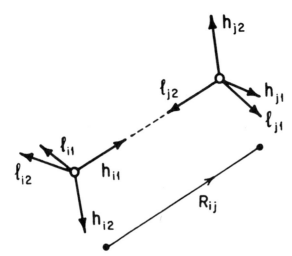

FIGURE 7.10. Two waterlike molecules in a favorable configuration for the formation of a HB.

7.2. We can also distinguish between a donor arm $\mathbf{h}_{i1}$ or $\mathbf{h}_{i2}$ and an acceptor arm $\mathbf{l}_{i1}$ or $\mathbf{l}_{i2}$ of the ith water molecule.

In terms of these vectors, the function $\mathbf{G}(\mathbf{X}_1, \mathbf{X}_2)$ is assumed to have the following general form:

$$\mathbf{G}(\mathbf{X}_i, \mathbf{X}_j) = G_{\sigma'}(R_{ij} - R_{\mathrm{H}}) \left\{ \sum_{\alpha, \beta = 1}^{2} G_\sigma[(\mathbf{h}_{i\alpha} \cdot \mathbf{u}_{ij}) - 1] G_\sigma[(\mathbf{l}_{j\beta} \cdot \mathbf{u}_{ij}) + 1] \right.$$

$$\left. + \sum_{\alpha, \beta = 1}^{2} G_\sigma[(\mathbf{l}_{i\alpha} \cdot \mathbf{u}_{ij}) - 1] G_\sigma[(\mathbf{h}_{j\beta} \cdot \mathbf{u}_{ij}) + 1] \right\}, \tag{7.4.7}$$

where the function $G(x)$ may be chosen either as step function of the form

$$G_\sigma(x) = \begin{cases} 1 & \text{for } |x| < \sigma \\ 0 & \text{for } |x| \geq \sigma \end{cases} \tag{7.4.8}$$

or as an unnormalized Gaussian function defined by

$$G_\sigma(x) = \exp(-x^2/2\sigma^2). \tag{7.4.9}$$

The above function, though cumbersome in appearance, is quite simple in content. Consider first the function $G_{\sigma'}(R_{ij} - R_{\mathrm{H}})$, where R_{H} is the intermolecular distance at which we expect a hydrogen bond to be formed. A reasonable choice is $R_{\mathrm{H}} = 2.76$ Å. This function attains its maximal value of unity. It drops to zero [either abruptly (7.4.8) or continuously (7.4.9)] for $|R_{ij} - R_{\mathrm{H}}| > \sigma'$. Next, we stipulate the relative orientation of the pair of molecules. The factor $G_\sigma[(\mathbf{h}_{i\alpha} \cdot \mathbf{u}_{ij}) - 1]$ attains its maximum value whenever the unit vector $\mathbf{h}_{i\alpha}$ (i.e., the donor arm $\mathbf{h}_{i\alpha}$, $\alpha = 1, 2$, of the ith molecule) is in the direction of the unit vector $\mathbf{u}_{ij} = \mathbf{R}_{ij}/R_{ij}$. Similarly, $G_\sigma[(\mathbf{l}_{j\beta} \cdot \mathbf{u}_{ij}) + 1]$ attains its maximum value whenever the direction of the acceptor arm $\mathbf{l}_{j\beta}$ is in the direction $-\mathbf{u}_{ij}$. Thus, the product of these three functions attains a value close to unity only if, simultaneously, R_{ij} is about R_{H}, the direction of $\mathbf{h}_{i\alpha}$ is about that of $\mathbf{u}_{ij}$, and the direction of $\mathbf{l}_{j\beta}$ is about that of $-\mathbf{u}_{ij}$. Such a configuration is said to be "favorable" for HB formation. Clearly, if all of the above three conditions are fulfilled, then the interaction energy is about $\varepsilon_{\mathrm{HB}}$. The sum of the various terms in the curly brackets of (7.4.7) arises from the total of eight possible favorable directions for HB formation (four when i is a donor and four when i is an acceptor). The variances σ and σ' are considered as parameters that, in principle, can be determined once detailed knowledge of the variation of the HB energy with distance and orientation is available. Note that of the eight terms in the curly brackets, only one may be appreciably different from zero at any given configuration $\mathbf{X}_i, \mathbf{X}_j$. This can always be achieved by choosing a sufficiently small value for σ.

A choice of the square well function (7.4.8) leads to an on–off definition of a HB; i.e., two water molecules are either HB'd or not. This simplified view is sometimes convenient in applications. The continuous definition of the HB potential using (7.4.9) provides a gradual change from a HB'd to a non-HB'd pair of water molecules. This is clearly a more realistic view that can be described as stretching and bending of the hydrogen bond.

Note also that in Eq. (7.4.7) we accounted only for configurations that are favorable for HBing. In a more realistic function one might add another eight terms to (7.4.7) to account for the electrostatic repulsion whenever two donor or two acceptor arms approach each other. In some of our order-of-magnitude estimates presented in this and in the next chapter we shall not need any detailed description of the potential function. We shall use only the qualitative description given to Eq. (7.4.5).

7.5. SECOND VIRIAL COEFFICIENTS OF WATER

The virial coefficients of water are defined as the coefficients in the density expansion of the pressure:

$$\frac{P}{kT} = \rho + B_2(T)\rho^2 + B_3(T)\rho^3 + \cdots . \tag{7.5.1}$$

The second virial coefficient is given by (section 1.8)

$$B_2(T) = -\frac{1}{16\pi^2} \int d\mathbf{X}_2 \{\exp[-\beta U(\mathbf{X}_1, \mathbf{X}_2)] - 1\}. \tag{7.5.2}$$

The second virial coefficient is a sixfold integral involving the pair potential for two water molecules. The numerical evaluation of such integrals is more difficult than that of the corresponding one-dimensional integral for a system of spherical particles. Furthermore, the integral in (7.5.2), even when evaluated exactly, cannot be expected to be a stringent test of the potential function. Indeed, there can be an infinite number of functions $U(\mathbf{X}_1, \mathbf{X}_2)$ which will lead to the same virial coefficient $B_2(T)$.

Figure 7.11 shows $B_2(T)$ as a function of T. The curve is based on an empirical equation of the form

$$B_2(T)/\text{cm}^3 \, \text{g}^{-1} = 2.062 - (2.9017 \times 10^3/T) \exp[1.7095 \times 10^5/T^2]. \tag{7.5.3}$$

Note that at the limit of very high temperatures

$$B_2(T \to \infty) \approx 37.1 \, \text{cm}^3/\text{mol}. \tag{7.5.4}$$

For some approximate estimates we do not need the explicit form of the pair potential as presented in the previous section. Instead, we use directly the information contained in $B_2(T)$, which effectively conveys the major contribution of the pair potential.

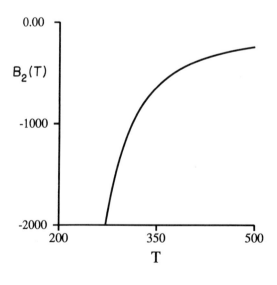

FIGURE 7.11. The form of $B_2(T)$ for water, based on Eq. (7.5.3).

Using the primitive pair potential in the form

$$U(\mathbf{X}_1, \mathbf{X}_2) = U_{LJ}(R_{12}) + \varepsilon_{HB}G(\mathbf{X}_1, \mathbf{X}_2) \tag{7.5.5}$$

in (7.5.2) we can split the integral into two contributions

$$B_2(T) = -\frac{1}{16\pi^2}\int d\mathbf{X}_2\{\exp[-\beta U(\mathbf{X}_1, \mathbf{X}_2)] - 1]\}$$

$$\approx -\frac{1}{16\pi^2}\left\{(-1)\int d\mathbf{X}_2 + [\exp(-\beta\varepsilon_{HB}) - 1]\int\cdots\int d\mathbf{X}_2\right\}. \tag{7.5.6}$$

$$\text{HB region}$$

The first term corresponds to the region where $G(\mathbf{X}_1, \mathbf{X}_2) = 0$ and $U_{LJ}(R_{12}) \to \infty$. In this region, the integrand is (-1) and if we choose $\sigma_w \approx 2.8$ Å as the effective diameter of the water molecule, we have

$$\frac{1}{16\pi^2}\int d\mathbf{X}_2 = \frac{1}{2}\int_0^{\sigma_w} 4\pi R^2\, dR = \frac{4\pi\sigma_w^3}{6} \approx 28 \text{ cm}^3/\text{mol}. \tag{7.5.7}$$

This is roughly half of the excluded volume between two spheres of diameter 2.8 Å. Note that if we choose a diameter of about 3 Å, we obtain a value closer to the estimated value of $B_2(T = \infty)$ made in (7.5.4). The second part of the integral in (7.5.5) is approximated as follows. We neglect the LJ energy compared with ε_{HB} and assume that the HB interaction is realized only from a small region $\Delta\mathbf{X}$ (Appendix A).

Since we have eight equivalent regions from which molecule 2 can form a HB with molecule 1 (see section 7.4), we define the HB region per bond $\Delta\mathbf{X}$ as

$$8\Delta\mathbf{X} = \int\cdots\int d\mathbf{X}_2; \tag{7.5.8}$$

$$\text{HB region}$$

with (7.5.7) and (7.5.8) we can rewrite (7.5.6) as

$$B_2(T) = \frac{4\pi\sigma_w^3}{6} - \frac{8}{2}\{\exp(-\beta\varepsilon_{HB}) - 1\}\frac{\Delta\mathbf{X}}{8\pi^2}. \tag{7.5.9}$$

In some calculations, it is not necessary to specify the range of distances and orientations from which a HB can be formed. Instead, one can use the HB region $\Delta\mathbf{X}$ that includes the entire region (both location and orientation) from which one arm of molecule 1 can form a HB with one arm of molecule 2. The value of $\Delta\mathbf{X}$ may be estimated from the experimental value of the second virial coefficient. For instance, at $T = 298$ K, the value of B_2 is estimated from (7.5.3)†

$$B_2(T = 298 \text{ K}) = -1165 \text{ cm}^3/\text{mol}. \tag{7.5.10}$$

† Note that this estimate is very uncertain. The analytical function (7.5.3) represents the experimental values of B_2 in the range 323–733 K. Around room temperature $B_2(T)$ is a very steep function of T.

Using the value of 28 cm^3/mol from (7.5.7) or of $B_2(T = \infty)$ from (7.5.4) in (7.5.9), we obtain

$$\frac{\Delta \mathbf{X}}{8\pi^2} [\exp(-\beta \varepsilon_{\text{HB}}) - 1] = \frac{28 + 1165}{4} \sim 298 \text{ cm}^3/\text{mol}. \qquad (7.5.11)$$

Thus for any choice of ε_{HB} we can estimate $\Delta \mathbf{X}$. Two estimates of $\Delta \mathbf{X}$ are

$$\frac{\Delta \mathbf{X}}{8\pi^2} = 5.1 \times 10^{-3} \text{ cm}^3/\text{mol for } \varepsilon_{\text{HB}} = -6.5 \text{ kcal/mol} \qquad (7.5.12)$$

$$\frac{\Delta \mathbf{X}}{8\pi^2} = 9.4 \times 10^{-4} \text{ cm}^3/\text{mol for } \varepsilon_{\text{HB}} = -7.5 \text{ kcal/mol}. \qquad (7.5.13)$$

Clearly, the stronger we choose the HB energy, the smaller is the HB region necessary to fit the experimental value of the second virial coefficient. See also Appendix A.

7.6. THE STRUCTURE OF WATER AND THE MIXTURE MODEL APPROACH TO THE THEORY OF WATER

The concept of the structure of water (SOW) is ubiquitous in the literature on water and aqueous solutions. There have been many definitions of this concept. We present here one possible definition which is based on the form of the pair potential as discussed in section 7.4 and can be computed by any simulation technique.

Using the definition of the pair potential in the form

$$U(\mathbf{X}_1, \mathbf{X}_2) = U_{\text{LJ}}(R_{12}) + \varepsilon_{\text{HB}} G(\mathbf{X}_1, \mathbf{X}_2), \qquad (7.6.1)$$

we define the following function:

$$\psi_i(\mathbf{X}^N) = \sum_{\substack{j=1 \\ j \neq i}}^{N} G(\mathbf{X}_i, \mathbf{X}_j). \qquad (7.6.2)$$

For each configuration $\mathbf{X}^N$ of the N water molecule, we choose one particular molecule, say, the ith one. Since $G(\mathbf{X}_i, \mathbf{X}_j)$ is unity whenever molecule j is in a favorable configuration to form an HB with i, the sum on the rhs of (7.6.2) counts all the water molecules that are HB'd to i at a specific configuration of the entire system. Since we have chosen σ in Eq. (7.4.7) to be small enough so that two water molecules can form at most one HB between them, the value of $\psi(\mathbf{X}^N)$ can roughly change between zero and four. [In the case of the discrete definition of HB (7.4.8), G can be either zero or one, hence, ψ_i can take only the discrete values of 0, 1, 2, 3, 4. For the continuous definition (7.4.9), G changes continuously from zero to one; hence, ψ_i can also take on any value between 0 and 4.]

We now define the average value of $\psi(\mathbf{X}^N)$ in, say, the T, V, N ensemble as

$$\langle \psi \rangle = \int d\mathbf{X}^N P(\mathbf{X}^N) \psi_1(\mathbf{X}^N). \qquad (7.6.3)$$

This is the average number of HBs formed by a specific water molecule (say, number 1), in a system of pure water at T, V, N.

The quantity $\langle \psi \rangle$ is related to the total average number of HBs in the system, which we denote by $\langle N_{HB} \rangle$. This is defined as

$$\langle N_{HB} \rangle = \tfrac{1}{2} \int d\mathbf{X}^N \, P(\mathbf{X}^N) \sum_{\substack{i=1 \\ i \neq j}}^{N} \sum_{j=1}^{N} \mathbf{G}(\mathbf{X}_i, \mathbf{X}_j)$$

$$= \tfrac{1}{2} \int d\mathbf{X}_1 , d\mathbf{X}_2 \, \rho^{(2)}(\mathbf{X}_1, \mathbf{X}_2)\mathbf{G}(\mathbf{X}_1, \mathbf{X}_2). \qquad (7.6.4)$$

Since the sum over all i and j (with $i \neq j$) counts the number of all different pairs of molecules, the integral is the average number of pairs of molecules the configuration of which is within the favorable range to form a HB. The factor $\tfrac{1}{2}$ is included because the sum in the integrand counts each pair twice.

The second form on the rhs of (7.6.4) converts the average into an integral over the pair distribution function. This last form suggests that $\mathbf{G}$ may be viewed as a characteristic function defining the HB; i.e., $\mathbf{G}$ is unity or zero according to whether $(\mathbf{X}_i, \mathbf{X}_j)$ fall in the HB region or not. The relation between $\langle N_{HB} \rangle$ and $\langle \psi \rangle$ is obtained from (7.6.4) and (7.6.2); i.e.,

$$\langle N_{HB} \rangle = \tfrac{1}{2} \int d\mathbf{X}^N \, P(\mathbf{X}^N) \sum_{i=1}^{N} \sum_{\substack{j=i \\ j \neq 1}}^{N} \mathbf{G}(\mathbf{X}_i, \mathbf{X}_j)$$

$$= \tfrac{1}{2} N \int d\mathbf{X}^N \, P(\mathbf{X}^N)\psi_1(\mathbf{X}^N)$$

$$= \tfrac{1}{2} N \langle \psi \rangle. \qquad (7.6.5)$$

Thus either $\langle \psi \rangle$ or $\langle N_{HB} \rangle$ can serve as a definition of the SOW. This definition conforms with our intuitive expectation of such a concept. If $\langle \psi \rangle \approx 4$, each molecule is on the average HB'd to four other molecules. This is a highly structured state. On the other hand, $\langle \psi \rangle \approx 0$ corresponds to the lowest-structured state.

We now use the definition of $\psi(\mathbf{X}^N)$ in (7.6.2) to construct an exact mixture model approach to liquid water. (This is exact within the definition of the primitive pair potential introduced in section 7.4.) In section 5.13 we showed that any quasicomponent distribution function can serve as the means for constructing a mixture model for any liquid. Specifically, for water, we construct the following mixture model. First we define the counting function

$$N_n(\mathbf{X}^N) = \sum_{i=1}^{N} \delta[\psi_i(\mathbf{X}^N) - n], \qquad (7.6.6)$$

where δ can be either a Kronecker delta or a Dirac delta function depending on whether we choose the discrete or the continuous definition of the HB [i.e., Eqs. (7.4.8), (7.4.9)]. $N_n(\mathbf{X}^N)$ is the number of molecules which form n HBs when the system is at a specific configuration $\mathbf{X}^N$. The average number of such molecules is

$$\langle N_n \rangle = \int d\mathbf{X}^N \, P(\mathbf{X}^N) \sum_{i=1}^{N} \delta[\psi_i(\mathbf{X}^N) - n]$$

$$= N \int d\mathbf{X}^N \, P(\mathbf{X}^N)\delta[\psi_1(\mathbf{X}^N) - n]. \qquad (7.6.7)$$

From (7.6.7) we can define the mole fraction x_n corresponding to $\langle N_n \rangle$, i.e.,

$$x_n = \frac{\langle N_n \rangle}{N}. \tag{7.6.8}$$

The average number of HBs in the system is thus

$$\frac{N}{2} \sum_{n=0}^{4} n x_n = \frac{N}{2} \int d\mathbf{X}^N \, P(\mathbf{X}^N) \sum_{n=0}^{4} n \delta[\psi_1(\mathbf{X}^N) - n]$$

$$= \frac{N}{2} \int d\mathbf{X}^N \, P(\mathbf{X}^N) \psi_1(\mathbf{X}^N)$$

$$= \langle N_{HB} \rangle, \tag{7.6.9}$$

which is the same as $\langle N_{HB} \rangle$ defined in (7.6.5). We shall discuss in section 7.8 a method of estimating x_n and hence $\langle N_{HB} \rangle$. It is clear from the definitions of these quantities that they can all be computed by any simulation technique applied to liquid water.

We now turn to a relation between the solvation Gibbs energy of water and the quantity $\langle \psi \rangle$ or, equivalently, $\langle N_{HB} \rangle$ in pure water. This estimate is based on the assumption that H_2O and D_2O may each be represented by a molecular model with a pair potential of the form (7.4.5). The two liquids are assumed to have the same pair potential except for a difference in the HB energy ε_{HB}, which we denote by ε_D and ε_H for D_2O and H_2O respectively.

Starting with the expression for the solvation Gibbs energy of water in pure water (see sections 6.13 and 7.7 below),

$$\Delta G_w^* = -kT \ln \langle \exp(-\beta B_w) \rangle_0, \tag{7.6.10}$$

we differentiate with respect to ε_{HB} (at constant P, T, N) to obtain, after some elementary algebra,

$$\left(\frac{\partial \Delta G_w^*}{\partial \varepsilon_{HB}} \right) = \langle N_{HB} \rangle_w - \langle N_{HB} \rangle_0. \tag{7.6.11}$$

Note that the two averages on the rhs of (7.6.11) are over all the configurations of N water molecules. In $\langle N_{HB} \rangle_w$, we have one more water molecule, the solvaton, fixed at some location. Clearly, the average number of HBs in the system is not changed by releasing the solvaton, so that the difference on the rhs of (7.6.11) is simply the difference in the average number of HBs in a system $(T, P, N+1)$ and in (T, P, N). Therefore, using relation (7.6.5), we obtain

$$\frac{\partial \Delta G_w^*}{\partial \varepsilon_{HB}} = \frac{N+1}{2} \langle \psi \rangle_0 - \frac{N}{2} \langle \psi \rangle_0$$

$$= \tfrac{1}{2} \langle \psi \rangle_0. \tag{7.6.12}$$

Thus the derivative of ΔG_w^* with respect to the parameter ε_{HB} is a measure of the structure of water.

We can now use (7.6.12) to estimate the structure of water. The assumption is made that $(\varepsilon_D - \varepsilon_H)$ is small enough so that we can expand ΔG_w^* to first order in $\varepsilon_D - \varepsilon_H$ to obtain

$$\Delta G_{D_2O}^* - \Delta G_{H_2O}^* \approx \tfrac{1}{2}\langle\psi\rangle_0(\varepsilon_D - \varepsilon_H). \qquad (7.6.13)$$

This relation may be used either to estimate the extent of structure $\langle\psi\rangle_0$ from the data on the solvation Gibbs energies and on $\varepsilon_D - \varepsilon_H$, or to estimate the molecular quantity $\varepsilon_D - \varepsilon_H$ if we know the average structure.

Another useful result may be obtained by differentiating $\langle N_{HB}\rangle_0$ with respect to ε_{HB}. The result is

$$\left(\frac{\partial\langle N_{HB}\rangle_0}{\partial(-\varepsilon_{HB})}\right)_{T,P,N} = \beta[\langle(\langle N_{HB}\rangle_0 - N_{HB})^2\rangle_0] \geq 0, \qquad (7.6.14)$$

which means that the stronger the HB energy, the larger the structure of the solvent. This is consistent with current belief that D_2O is a more structured liquid than H_2O.

As an example of the application of (7.6.13), we use the solvation Gibbs energies of H_2O and D_2O in their own liquids, at 25°C (see section 7.7)

$$\Delta G_{D_2O}^* - \Delta G_{H_2O}^* = -26.82 + 26.47 = 0.35 \text{ kJ mol}^{-1},$$

and taking $\varepsilon_D - \varepsilon_H \approx 0.93$ kJ mol^{-1} (based on the difference in the enthalpies of sublimation of light and heavy ice) we obtain

$$\langle\psi\rangle \approx 0.75. \qquad (7.6.15)$$

This is too low a value for the average number of HBs formed by a specific water molecule. It is not clear whether this is a result of the simplifying assumptions made in the model or due to the inadequacy of the first order expansion used in Eq. (7.6.13). We shall see in the next section that $\langle\psi\rangle$ is closer to 4, indicating a high degree of structure.

The important and intuitively obvious result is (7.6.14), which states the average structure of the system is a monotonically increasing function of the parameter $|\varepsilon_{HB}|$.

7.7. SOLVATION OF WATER IN PURE WATER

The experimental values of the solvation Gibbs energies of water in water are obtainable from the densities of the liquid and the vapor pressure of water at various temperatures. It is assumed that the vapor at equilibrium with liquid water at most temperatures of interest may be viewed as ideal gas, in which we have (see section 6.13)

$$\Delta G_w^{*p} = kT \ln\left(\frac{\rho_w^g}{\rho_w^l}\right)_{eq}$$

$$= kT \ln\left(\frac{P_w}{kT\rho_w^l}\right)_{eq}, \qquad (7.7.1)$$

where P_w is the vapor pressure of water at the given temperature. Table 7.1 presents some selected values of the solvation thermodynamic quantities of H_2O and D_2O as a function of temperature along the vapor–liquid equilibrium line. We note that the large negative

TABLE 7.1
Solvation Thermodynamical Quantities of H_2O and D_2O as a Function of Temperature[a]

Temperature (C°)		ΔG_w^* (kJ mol^{-1})	ΔS_w^* (J K^{-1} mol^{-1})	ΔH_w^* (kJ mol^{-1})	$\Delta C_{p,w}^*$ (J K^{-1} mol^{-1})
0	H_2O	−27.792	−54.56	−42.695	—
5	H_2O	−27.521	−53.92	−42.52	35.4
	D_2O	−27.94	−57.8	−44.018	43.3
10	H_2O	−27.253	−53.28	−42.338	36.6
	D_2O	−27.653	−57.01	−43.797	44.7
20	H_2O	−26.727	−51.97	−41.962	38.0
	D_2O	−27.091	−55.43	−43.339	46.2
25	H_2O	−26.468	−51.33	−41.771	38.4
	D_2O	−26.816	−54.64	−43.107	46.4
30	H_2O	−26.213	−50.68	−41.578	38.6
	D_2O	−26.545	−53.87	−42.875	46.5
40	H_2O	−25.713	−49.43	−41.193	38.5
	D_2O	−26.014	−52.37	−42.413	46.0

[a] The pressure corresponds to the vapor pressure at equilibrium with the liquid at each temperature.

values of ΔG_w^* of water are primarily determined by the large enthalpy of solvation, and this in turn reflects the strong binding energy of a water molecule to its environment.

We shall now estimate the conditional solvation Gibbs or Helmholtz energy per arm of a water molecule. This quantity will be useful in various applications in the theory of aqueous solutions. Since $P\Delta V_w^*$, where ΔV_w^* is the solvation volume, is usually negligible compared with ΔG_w^*, we shall actually compute the solvation Helmholtz energy and assume that $\Delta G_w^* \approx \Delta A_w^*$ (Appendix F).

We start from the primitive pair potential (Eq. 7.4.4), which we write as

$$U(\mathbf{X}_i, \mathbf{X}_j) = U_{LJ}(R_{ij}) + \varepsilon_{HB}\mathbf{G}(\mathbf{X}_i, \mathbf{X}_j)$$

$$= U_{LJ} + U^{HB}, \tag{7.7.2}$$

where U_{LJ} is the Lennard–Jones part, defined in 7.4.1.

In the second form on the rhs of (7.7.2), we denoted by U^{HB} the HB pair interaction between two water molecules. The solvation Helmholtz energy of water in pure water is now written as (see section 6.14)

$$\Delta A_w^* = -kT \ln\langle \exp(-\beta B_w)\rangle_0$$

$$= -kT \ln\langle \exp[-\beta(B_{LJ} + B^{HB})]\rangle_0$$

$$= -kT \ln\langle \exp(-\beta B_{LJ})\rangle_0 - kT \ln\langle \exp(-\beta B^{HB})\rangle_{LJ}. \tag{7.7.3}$$

The total binding energy of the water-solvaton at $\mathbf{X}_0$ is written as

$$B_w(\mathbf{X}_0) = \sum_{j=1}^{N} U(\mathbf{X}_0, \mathbf{X}_j)$$

$$= B_{LJ} + B^{HB}$$

$$= B_{LJ} + \sum_{l=1}^{4} B_l^{HB}, \tag{7.7.4}$$

where B_{LJ} is the total Lennard–Jones binding energy of the solvaton to all the other molecules in the system and B^{HB} is the corresponding binding energy due to the turning on the HBing part of the pair potential, i.e., turning on all four arms of the solvaton. The latter is further written as a sum over the binding energy of the four arms of the solvaton.

For simple fluids, we used an expression similar to (7.7.3), where the first term was the solvation Helmholtz energy of the hard sphere and the second term was the conditional solvation of the soft interaction, given that the hard interaction has been coupled to the system (section 6.14). Here, we substitute (7.7.4) into (7.7.3) to rewrite ΔA_w^* as

$$\Delta A_w^* = \Delta A_{LJ}^* - kT \ln \left\langle \exp\left(-\beta \sum_{l=1}^{4} B_l^{HB}\right) \right\rangle_{LJ}. \tag{7.7.5}$$

The assumption is now made that the conditional average on the rhs of (7.7.5) may be factorized into a product of four averages each of which pertains to the binding energy of one specific arm, i.e.,

$$\left\langle \exp\left(-\beta \sum_{l=1}^{4} B_l^{HB}\right) \right\rangle_{LJ} = \prod_{l=1}^{4} \langle \exp(-\beta B_l^{HB}) \rangle_{LJ}. \tag{7.7.6}$$

Thus effectively we have assumed that the conditional solvation of the four arms of the solvaton are independent. We shall discuss this assumption further in Appendix B.

Substituting (7.7.6) into (7.7.5), we rewrite the solvation Helmholtz energy of a water molecule as

$$\Delta A_w^* = \Delta A_{Ne}^* + \sum_{l=1}^{4} \Delta A^{*HB}(l/LJ) \tag{7.7.7}$$

where we have identified the LJ part of the solvation Helmholtz energy ΔA_{LJ}^* with the solvation Helmholtz energy of a neon atom in water. Note that in the definition of ΔA_{LJ}^*, only the LJ-binding energy B_{LJ} of the solvaton enters [see the first term on the rhs of (7.7.3)]. However, the average is taken with the probability distribution of the configurations of all the N water molecules. Therefore, we can take for ΔA_{LJ}^* the solvation Helmholtz (or Gibbs) energy of neon in pure water.

The second term on the rhs of (7.7.7) is the conditional solvation Helmholtz energy for the process of turning on of the four arms of the solvaton. Strictly, we have to distinguish between the two donor arms and the two acceptor arms (section 7.4). However, for all our purposes, we can disregard this distinction and assume that this term is simply four times the conditional solvation Helmholtz energy of one arm. Hence we write

$$\Delta A_w^* = \Delta A_{Ne}^* + 4\Delta A^{*HB}(a/LJ). \tag{7.7.8}$$

The conditional Helmholtz energy of solvation of one arm may be determined either by using experimental values of ΔA_w^* and ΔA_{Ne}^* or by theoretical calculation.

The experimental value of the conditional solvation Helmholtz (or Gibbs) energy per arm of H_2O is at 25°C

$$\Delta A^{*HB}(a/LJ) = \frac{\Delta A_w^* - \Delta A_{Ne}^*}{4}$$

$$= \frac{-26.5 - 11.2}{4}$$

$$= -9.42 \text{ kJ/mol}$$

$$= -2.25 \text{ kcal/mol}. \tag{7.7.9}$$

The corresponding value for D_2O is

$$\Delta A^{*HB}(a/LJ) = \frac{-26.8 - 11}{4}$$

$$= -9.45 \text{ kJ/mol}. \tag{7.7.10}$$

It is of interest to note that had we used the same calculation for H_2S and H_2Se as solvatons in water, the corresponding values are much smaller. Thus for H_2S we have

$$\Delta A^{*HB}(a/LJ) = \frac{\Delta A^*_{H_2S} - \Delta A^*_{Ar}}{4}$$

$$= \frac{-2.3 - 8.4}{4}$$

$$= -2.7 \text{ kJ/mol}$$

$$= -0.64 \text{ kcal/mol}, \tag{7.7.11}$$

and for H_2Se we find

$$\Delta A^{*HB}(a/LJ) = \frac{\Delta A_{H_2Se} - \Delta A^*_{Kr}}{4}$$

$$= \frac{-1.8 - 6.9}{4}$$

$$= -2.17 \text{ kJ/mol}$$

$$= -0.52 \text{ kcal/mol}. \tag{7.7.12}$$

We now proceed to a theoretical estimate of $\Delta A^{*HB}(a/LJ)$ for water in water. We start from the definition of this quantity in (7.7.6) and (7.7.8).

$$\exp[-\beta\Delta A^{*HB}(a/LJ)] = \langle\exp(-\beta B_a^{HB})\rangle_{LJ}$$

$$= \int d\mathbf{X}^N \, P(\mathbf{X}^N/LJ) \exp(-\beta B_a^{HB}). \tag{7.7.13}$$

Here, B_a^{HB} is the HBing binding energy of one arm (a) to the entire system of N water molecules being at some specific configuration $\mathbf{X}^N$. The average in (7.7.13) is taken with respect to the conditional distribution $P(\mathbf{X}^N/LJ)$, i.e., the distribution of the configuration $\mathbf{X}^N$, given an LJ particle at some fixed position.

The range of integration over all the configurations of the N water molecules may be divided into two regions as follows:

$$\int d\mathbf{X}^N \, P(\mathbf{X}^N/LJ) \exp(-\beta B_a^{HB}) = \int_{\substack{\text{one } w \\ \text{HB'd to } a}} + \int_{\substack{\text{no } w \\ \text{HB'd to } a}}. \tag{7.7.14}$$

The first integral is over all configurational space for which at least one water molecule (w) is HB'd to the arm (a). Since at most one molecule can form a HB to a single arm at any time, this is the same as the region from which exactly one water molecule is HB'd to the arm. Furthermore, since there are N equivalent water molecules, we can rewrite this term as

$$\int_{\substack{\text{one } w \\ \text{HB'd to } a}} = \sum_{i=1}^{N} \int_{\substack{\text{ith } w \\ \text{HB'd to } a}} = N \int_{\substack{\text{molecule 1} \\ \text{HB'd to } a}} d\mathbf{X}^N P(\mathbf{X}^N/\text{LJ}) \exp[-\beta U^{\text{HB}}(a, \mathbf{X}_1)], \quad (7.7.15)$$

where in the last form on the rhs of (7.7.14) we have chosen one water molecule, say number 1, and took N times the integral over all configurational space for which molecule 1 is HB'd to a. Furthermore, since the arm a can form only one HB at a time, the binding energy B_a^{HB} is simply $U^{\text{HB}}(a, \mathbf{X}_1)$, i.e., the HB part of the potential between the arm a and the specific water molecule at $\mathbf{X}_1$.

The second integral on the rhs of (7.7.13) can now be written as

$$\int_{\substack{\text{no } w \\ \text{HB'd to } a}} d\mathbf{X}^N P(\mathbf{X}^N/\text{LJ}) = 1 - \sum_{i=1}^{N} \int_{\substack{\text{ith } w \\ \text{HB'd to } a}} d\mathbf{X}^N P(\mathbf{X}^N/\text{LJ}). \quad (7.7.16)$$

Since this integral is over all configurational space for which no water molecule is HB'd to the arm a, B_a^{HB} is zero in the entire range of configurations. What remains on the lhs of (7.7.16) is an integral over the probability distribution function $P(\mathbf{X}^N/\text{LJ})$. This integral is simply the probability of finding a configuration of all the water molecules such that no water molecule can form a HB with the solvaton, given that the LJ solvaton is at some fixed position. It should be noted that the HB part of the potential was defined in terms of the geometrical function $\mathbf{G}(\mathbf{X}_i, \mathbf{X}_j)$ in section 7.5. When we refer to water molecules forming a HB with the solvaton, we mean that a water molecule is in a favorable configuration such that it can form a HB with one arm of the solvaton, even when that solvaton is coupled only with respect to the LJ part of the potential function.

Since the lhs of (7.7.16) is a probability of an event, it can be written as one minus the probability of the complementary event, i.e., that at least one water molecule is HB'd to a. The latter is further written as a sum of probabilities; i.e., either water numbered 1 is HB'd, water numbered 2 is HB'd, etc. Combining (7.7.14) and (7.7.15) with (7.7.16), we can rewrite (7.7.13) as

$$\exp[-\beta \Delta A^{*\text{HB}}(a/\text{LJ})] = 1 + N \int_{\substack{\text{molecule 1} \\ \text{HB'd to } a}} d\mathbf{X}^N \{\exp[-\beta U^{\text{HB}}(a, \mathbf{X}_1)] - 1\} P(\mathbf{X}^N/\text{LJ})$$

$$= 1 + \int d\mathbf{X}_1 \{\exp[-\beta U^{\text{HB}}(a, \mathbf{X}_1)] - 1\} \rho_w(\mathbf{X}_1/\text{LJ}). \quad (7.7.17)$$

In the last form on the rhs of (7.7.17), we used the definition of the conditional density of water molecules at $\mathbf{X}_1$, given a LJ particle at the center of our coordinate system. Note also that since U^{HB} is zero except for a very small configurational space, the integrand is zero beyond this region, therefore we have removed the restriction on the range of the integration.

The conditional density can be written as

$$\rho_w(\mathbf{X}_1/\text{LJ}) = \frac{1}{8\pi^2} \rho_w g_{w,\text{LJ}}(\mathbf{X}_1, \text{LJ}). \quad (7.7.18)$$

The average correlation function of water around a LJ particle at a distance of about 2.8 Å is estimated by simulation to be about 2. Assuming now that the HB function is discrete (see section 7.4), we can rewrite the integral in (7.7.17) as

$$\exp[-\beta \Delta A^{*HB}(a/LJ)] = 1 + [\exp(-\beta \varepsilon_{HB}) - 1]\frac{2(2\Delta X)\rho_w}{8\pi^2}. \tag{7.7.19}$$

Here, ρ_w is the number density of water and ΔX is the configurational range for the formation of one HB between two water molecules (see section 7.5). Since one arm can form two different HBs with one water molecule, we have the factor of 2 in (7.7.19).

One can use a specific model potential function to estimate the rhs of (7.7.19). Fortunately, since the combination $[\exp(-\beta \varepsilon_{HB}) - 1]\Delta X/8\pi^2$ has already been estimated from the second virial coefficient, we need not know the values of each factor. Using the value of $298 \, \text{cm}^3/\text{mol}$ given in (7.5.11), we estimate for $T = 298 \, \text{K}$

$$\Delta A^{*HB}(a/LJ) = -kT\ln(1 + 298 \times 4 \times 5.55 \times 10^{-2})$$

$$= -10.4 \, \text{kJ/mol}$$

$$= -2.5 \, \text{kJ/mol}, \tag{7.7.20}$$

where $\rho_w = 5.55 \times 10^{-2} \, \text{mol/cm}^3$ is the number density of water at 25°C. This result is in good agreement with the experimental value given in (7.7.9). Note that in this calculation we used the value of $g_{w,LJ} \approx 2$, which is probably too large. With a smaller value we could have improved somewhat the agreement between the theoretical value (7.7.20) and the experimental value (7.7.9).

7.8. DISTRIBUTION OF SPECIES OF WATER MOLECULES

In section 5.13 we have seen that any quasicomponent distribution function can be used as a basis for constructing an exact mixture model for any liquid. We now develop a mixture model approach which is particularly suitable for liquid water. A variety of approximate versions of such mixture model approaches have been used in the development of theories of water and aqueous solutions.

The classification of water molecules according to the number of HBs in which they participate has been given in section 7.6. Here we shall be interested in the distribution of these species. Let ρ_w be the total number density of water molecules and ρ_n the number density of water molecules participating in n HBs. The mole fraction of such species is $x_n = \rho_n/\rho_w$ and was defined in section 7.6. The chemical potential of the nth species is written as

$$\mu_n = \mu_n^* + kT\ln \rho_n\Lambda^3, \tag{7.8.1}$$

where μ_n^* is the pseudochemical potential of the nth species. Since our particles have the same internal degrees of freedom, both the momentum and the internal PF of a single molecule are common to all species.

At equilibrium, we have the condition

$$\mu_w = \mu_0 = \mu_1 = \mu_2 = \mu_3 = \mu_4, \tag{7.8.2}$$

where μ_w is the chemical potential of the bulk water. Specifically, for the non-HB'd species $(n = 0)$, we write

$$\mu_0 = \mu_0^* + kT \ln \rho_0 \Lambda^3$$
$$= W(0|w) + kT \ln \rho_0 \Lambda^3 q_w^{-1}$$
$$= \Delta A_{Ne}^* + kT \ln \rho_0 \Lambda^3 q_w^{-1}, \tag{7.8.3}$$

where $W(0|w)$ is the coupling work of the $n = 0$ species, which we identify as the solvation Helmholtz energy of a LJ or a neon particle.

The pseudochemical potential of the water molecules is written as

$$\exp(-\beta\mu_w^*) = \frac{Q(T, V, N; \mathbf{R}_0)}{Q(T, V, N)} = \frac{\sum_{n=0}^{4} \binom{4}{n} Q(T, V, N; n)}{Q(T, V, N)}, \tag{7.8.4}$$

where $Q(T, V, N)$ is the PF of N water molecules at T, V. $Q(T, V, N; \mathbf{R}_0)$ is the PF of the same system with an additional water molecule—the solvaton—placed at some fixed position $\mathbf{R}_0$. This PF can now be written as a sum over all possible states of the solvaton. $Q(T, V, N; n)$ is the PF of the system for which the solvaton at $\mathbf{R}_0$ (not indicated in the notation) has exactly n *specific* arms engaged in HBs with the "solvent." Defining the pseudochemical potential of nth species by

$$\exp(-\beta\mu_n^*) = \binom{4}{n} \frac{Q(T, V, N; n)}{Q(T, V, N)}, \tag{7.8.5}$$

we can rewrite (7.8.4) as

$$\exp(-\beta\mu_w^*) = \sum_{n=0}^{4} \exp(-\beta\mu_n^*). \tag{7.8.6}$$

Equation (7.8.6) may be viewed as a generalization of relation (6.15.13). There we had two isomers having different internal properties; here the five "isomers" have the same internal properties but differ in their coupling work, i.e., the number of HBs in which they participate.

The ratio of the two PFs in (7.8.5) is related to the solvation Helmholtz energy of a solvaton with n *specific* arms engaged in HBs, i.e.

$$\frac{Q(T, V, N; n)}{Q(T, V, N)} = q_w \exp[-\beta\Delta A^*(n \text{ specific arms HB'd})]. \tag{7.8.7}$$

We now use the approximation of section 7.7, that the arms are independent; hence, as in Eq. (7.7.8), we write

$$\Delta A^*(n \text{ specific arms HB'd}) = \Delta A_{Ne}^* + n\Delta A^{*\text{HB}}(1, \text{HB/LJ}), \tag{7.8.8}$$

where $\Delta A^{*\text{HB}}(1, \text{HB/LJ})$ is the conditional solvation Helmholtz energy of one specific arm known to be HB'd.

Combining (7.8.8), (7.8.7), and (7.8.4) we have

$$\exp(-\beta\mu_w^*) = \sum_{n=0}^{4} \binom{4}{n} q_w \exp[-\beta\Delta A_{Ne}^* - \beta n \Delta A^{*HB}(1, HB/LJ)]. \quad (7.8.9)$$

From the equilibrium condition (7.8.2) it follows that

$$\mu_w^* - \mu_n^* = kT \ln\left(\frac{\rho_n}{\rho_w}\right)$$

$$= kT \ln x_n, \quad (7.8.10)$$

where x_n is the average mole fraction of water molecules engaged in exactly nHBs. Defining

$$Y = \exp[-\beta\Delta A^{*HB}(1, HB/LJ)], \quad (7.8.11)$$

we obtain from (7.8.5), (7.8.9), and (7.8.10) the final result

$$x_n = \frac{\binom{4}{n} Y^n}{(1+Y)^4}. \quad (7.8.12)$$

The average number of HBs formed by a selected water molecule is

$$\langle \psi \rangle = \sum_{n=0}^{4} n x_n = \frac{4y}{1+y}. \quad (7.8.13)$$

We recall the experimental value of $\Delta A^{*HB}(a/LJ)$, which is the conditional solvation Helmholtz energy per arm of a water molecule. The relation between this quantity and Y is obtained from (7.8.9), i.e.

$$\exp[-\beta 4\Delta A^{*HB}(a/LJ)] = \sum_{n=0}^{4} \binom{4}{n} \exp[-\beta n \Delta A^{*HB}(1, HB/LJ)]$$

$$= (1+Y)^4. \quad (7.8.14)$$

Thus from the experimental value of $\Delta A^{*HB}(a/LJ)$ in (7.7.9), we can compute the distribution of the various species, as well as the average quantity $\langle \psi \rangle$. These values are shown for H_2O and D_2O at 25°C and 1 atm in Table 7.2.

TABLE 7.2
Values of the Mole Fractions of the Various
Species of Water Molecule and the Average
Structure Computed from Eqs (7.8.12) and
(7.8.13) for H_2O and D_2O at 25°C

	H_2O	D_2O
x_0	2.49×10^{-7}	2.36×10^{-7}
x_1	4.37×10^{-5}	4.18×10^{-5}
x_2	2.87×10^{-3}	2.79×10^{-3}
x_3	8.36×10^{-2}	8.24×10^{-2}
x_4	0.913	0.915
$\langle \psi \rangle$	3.909	3.913

In this approximation we find that most of the water molecules are in the $n = 4$ state. The value of $\langle \psi \rangle$ is probably too high. It should be noted that within the primitive model for water, the distribution of species depends only on the solvation Helmholtz energy of the arms and not on the solvation Helmholtz energy of the LJ, or the neon part.

7.9. APPLICATION OF THE MIXTURE MODEL APPROACH

There is a long history of the theory of water and aqueous solutions based on various mixture model (MM) approaches. One of the earliest documented explanations of some anomalous properties of water is due to Röntgen, who in 1892 proposed to view liquid water as consisting of two kinds of molecules, one of which he referred to as "ice-molecules." The general idea of explaining the properties of water by viewing it as a mixture of species probably originated much earlier.

Beginning in the early 1960s, efficient simulation techniques developed for the study of simple liquids were also applied to aqueous fluids. This approach requires us to start from a model for water molecules defined in terms of their pair potential. (This method is sometimes referred to as *ab initio*, or a continuous, theory of water.)

In this section we present an example of the application of a two-structure model, based on the exact MM approach to the theory of liquids (section 5.13). Then we extract a particular MM which can be viewed as an approximation of the general exact MM approach. The latter, because of its simplicity and solvability, is useful in the study of some thermodynamic aspects of both pure water and aqueous solutions of simple solutes.

7.9.1. Construction of an Exact Two-Structure Model

The simplest version of the MM approach and the one that has been most often applied is the two-structure model (TSM). We start with some examples of exact TSMs that are derivable from any one of the quasicomponent distribution functions discussed in section 5.13. Consider, for instance, the distribution based on the concept of coordination number (CN). We have denoted by $x_c(K)$ the mole fraction of molecules having a CN equal to K. The vector $\mathbf{x}_c = (x_c(0), x_c(1), \ldots)$ describes the *composition* of the system when viewed as a mixture of quasicomponents, distinguishable according to their CN. Instead, we may distinguish between two groups of molecules: those whose CN is smaller than or equal to some number, say K^*, and those whose CN is larger than K^*. The two corresponding mole fractions are

$$x_L = \sum_{K=0}^{K^*} x_C(K), \qquad x_H = \sum_{K=K^*+1}^{\infty} x_C(K), \qquad (7.9.1)$$

where x_L may be referred to as the mole fraction of molecules with low (L) local density and x_H as that of molecules with relatively high (H) local density. The new vector composed of two components (x_L, x_H) is also a quasicomponent distribution function and gives the composition of the system when viewed as a mixture of two components, which we may designate as L and H molecules. Starting with the same vector $x_C = (x_C(0), x_C(1), \ldots)$, we may, of course, derive many other TSMs differing from the one in (7.9.1). A possibility which may be useful for liquid water is

$$x_A = x_C(4), \qquad x_B = 1 - x_A, \tag{7.9.2}$$

where we distinguish between molecules with CN equal to four in one group (A) and all other molecules in the second group (B).

Instead of the CN we can use the number of HBs for our classification into species. The five species defined in sections 7.6 and 7.8 can also be used to construct an exact TSM. For instance, we can consider the four HB'd species as one species of the TSM and refer to all the other species (with $n = 0, 1, 2, 3$) collectively as a second species. An approximate version of this TSM is worked out in subsection 7.9.3.

As a second example, consider the quasicomponent distribution function based on the concept of binding energy (BE). We recall that the vector (or the function) $\mathbf{x}_B$ gives the composition of the system when viewed as a mixture of molecules differing in their BE. Thus, $x_B(v)\, dv$ is the mole fraction of molecules with BE between v and $v + dv$. A possible TSM constructed from this function is, dropping subscript B in $x_B(v)$,

$$x_A = \int_{-\infty}^{v^*} x(v)\, dv \qquad x_B = 1 - x_A, \tag{7.9.3}$$

where x_A is the mole fraction of molecules whose BE is below a certain value v^* and x_B is the mole fraction whose BE is larger than v^*. Again, we have a new quasicomponent distribution function composed of two components (x_A, x_B).

The above examples illustrate the general procedure by which we construct a TSM from any quasicomponent distribution function. From now on, we assume that we have made a classification into two components, L and H, without referring to a specific example. The arguments we use will be independent of any specific classification procedure. We will see that in order for such a TSM to be useful in interpreting the properties of water, we must assume that each component in itself behaves "normally" (in the sense discussed below). The anomalous properties of water are then interpreted in terms of structural changes that take place in the liquid.

Let N_L and N_H be the equilibrium numbers of L and H molecules, respectively, and N_W be the total number of water molecules in the system, $N_W = N_L + N_H$. Viewing the system as a mixture of two components, we write the volume of the system, using the Euler theorem, as

$$V = N_L \bar{V}_L + N_H \bar{V}_H, \tag{7.9.4}$$

where $\bar{V}_L$ and $\bar{V}_H$ are the partial molar (or molecular) volumes of L and H, respectively. The total differential of $V(T, P, N_L, N_H)$ is

$$dV = \left(\frac{\partial V}{\partial T}\right)_{P, N_L, N_H} dT + \left(\frac{\partial V}{\partial P}\right)_{T, N_L, N_N} dP + \left(\frac{\partial V}{\partial N_L}\right)_{T, P, N_H} dN_L + \left(\frac{\partial V}{\partial N_H}\right)_{T, P, N_L} dN_H. \tag{7.9.5}$$

The temperature dependence of the volume, along the equilibrium line with respect to the reaction $L \rightleftarrows H$ (keeping P and N_W constant), is

$$\left(\frac{\partial V}{\partial T}\right)_{P, N_W, \mathrm{eq}} = \left(\frac{\partial V}{\partial T}\right)_{P, N_L, N_H} + (\bar{V}_L - \bar{V}_H)\left(\frac{\partial N_L}{\partial T}\right)_{P, N_W, \mathrm{eq}}. \tag{7.9.6}$$

The first term on the rhs of (7.9.6) gives the temperature dependence of the volume of the frozen-in system and, in view of (7.9.4), can be written as

$$\left(\frac{\partial V}{\partial T}\right)_{P,N_L,N_H} = N_L\left(\frac{\partial \bar{V}_L}{\partial T}\right)_{P,N_L,N_H} + N_H\left(\frac{\partial \bar{V}_H}{\partial T}\right)_{P,N_L,N_H}. \qquad (7.9.7)$$

This quantity may be referred to as the "frozen-in" part of the total temperature dependence of the volume. It must be stressed that the splitting of $(\partial V/\partial T)$ in (7.9.6) into two terms depends strongly on the classification procedure employed to define the species L and H. Therefore, any reference to a frozen part has meaning only for the particular classification that has been employed.

The second term on the rhs of (7.9.6) may be referred to as the relaxation part of the total temperature dependence of the volume. It is quite clear that this term contains the contribution to the temperature dependence of the volume that is associated with the "structural changes" in the system.

In the context of this section, the term "structural changes" refers only to the interconversion between the two species L and H arising from the change in temperature. We denote by

$$\alpha^f = \frac{1}{V}\left(\frac{\partial V}{\partial T}\right)_{P,N_L,N_H} \qquad (7.9.8)$$

the "frozen-in" part of the coefficient of thermal expansion. Using the identities of Chapter 5 we can rewrite the coefficient of thermal expansion as

$$\alpha = \frac{1}{V}\left(\frac{\partial V}{\partial T}\right)_{P,N_W,eq} = \alpha^f + \frac{(\bar{V}_L - \bar{V}_H)(\bar{H}_L - \bar{H}_H)x_L x_H \eta}{kT^2}, \qquad (7.9.9)$$

where

$$\eta = \rho_L + \rho_H + \rho_L \rho_H (G_{LL} + G_{HH} - 2G_{LH}). \qquad (7.9.10)$$

In order to use (7.9.9) for interpreting the anomalous negative temperature dependence of the volume of water between 0 and 4°C, we must choose a classification procedure so that (1) the "frozen-in" term α^f behaves normally, i.e., each of the derivatives in (7.9.7) is positive; such an assumption is invoked either explicitly or implicitly in the various *ad hoc* TSMs for water; (2) the relaxation term in (7.9.9) must be negative and large enough (in absolute magnitude) to overcompensate for the positive value of the frozen part.

Regarding the first condition, there is no general rule that guarantees that α^f is normal. In the simplified *ad hoc* TSMs, this assumption seems to be quite natural. For instance, if one species is icelike and the second one is, say, monomeric, then it is natural to assume that each of the partial molar volumes has a normal temperature dependence.

The analysis of the second condition leads to somewhat more interesting insights as to the origin of negative temperature dependence of the volume. Consider first the sign of the relaxation term in (7.9.9). Clearly, in order that this term be negative, we must have (note that $\eta > 0$)

$$(\bar{V}_L - \bar{V}_H)(\bar{H}_L - \bar{H}_H) < 0. \qquad (7.9.11)$$

This means that $\bar{V}_L - \bar{V}_H$ and $\bar{H}_L - \bar{H}_H$ must have opposite signs. Next, in order to get a large negative value for the relaxation term in (7.9.9), two conditions must be fulfilled: (1) The two components L and H must not be very similar, since otherwise the differences $\bar{V}_L - \bar{V}_H$ and $\bar{H}_L - \bar{H}_H$ will be very small. (2) Neither of the mole fractions x_L or x_H may be too small, since otherwise the product $x_L x_H$ will be small, and we will get a small relaxation term in (7.9.9).

Let us further examine the above two conditions. Suppose we define L and H in such a way that L has a low local density and H has a relatively high local density. Then, we expect that $\bar{V}_L - \bar{V}_H > 0$. Using the Kirkwood–Buff expressions for $\bar{V}_L$ and $\bar{V}_H$ (section 6.7) and the identities (5.13.71) and (5.13.72), we can express the difference in the partial molar volumes in terms of molecular distribution functions, namely

$$\bar{V}_L - \bar{V}_H = \frac{\rho_H G_{HH} - \rho_H G_{LH} - \rho_L G_{LL} + \rho_L G_{HL}}{\eta}$$

$$= \frac{\rho_W}{\eta}(G_{WH} - G_{WL}). \tag{7.9.12}$$

Thus, the requirement that $\bar{V}_L - \bar{V}_H$ be positive is equivalent to the statement that the overall excess of water molecules around H is larger than the overall excess of water molecules around L. This is precisely the meaning of the requirement that H is a molecule with a relatively higher local density than L.

It is clear that fulfillment of one condition, say $\bar{V}_L - \bar{V}_H > 0$, can be achieved by proper definition of the two components. The question that arises is: Under what circumstances is the second condition, $\bar{H}_L - \bar{H}_H < 0$, also fulfilled?

In order to understand the characteristic nature of the packing of water molecules in the liquid state, consider first a simple fluid such as argon. Suppose that we define two components L and H having a low and high local density, respectively. For instance, we use (7.9.1) with $K^* = 6$. In this case, we have $\bar{V}_L - \bar{V}_H > 0$ by virtue of the definition of the two components. However, since an argon molecule, having more than six neighbors, is likely to have a lower partial molar enthalpy, we also expect that $\bar{H}_L - \bar{H}_H > 0$ (i.e., the conversion of a molecule from H to L is accompanied by an increase in enthalpy). Such a TSM would not work for our purposes, since (7.9.11) is not fulfilled. A different choice that may give the correct signs is obtained with, say, $K^* = 13$. Clearly, in this case, we also have $\bar{V}_L - \bar{V}_H > 0$. (The H component has a very high local density.) In addition, because of this particular choice of H, it is clear that the transformation $H \to L$ is likely to involve a decrease in enthalpy (since most of the neighbors must exert strong repulsive forces on the H-molecule, so that its transfer to an L releases energy). Hence, $\bar{H}_L - \bar{H}_H$ is likely to be negative. This choice seems to satisfy our conditions (7.9.11) for a negative relaxation term. However, with the particular choice of $K^* = 13$, it is very likely that $x_H \approx 0$ (a CN higher than 12 must be a very rare event). Hence, the product $x_L x_H$ is very small and the whole relaxation term in (7.9.10) will be negligibly small, though having a correct sign.

The situation in water is different since we know that the strong directional forces (hydrogen bonds) are responsible for maintaining low local density. Hence, it seems possible to define two components L and H in such a way that

$$\bar{V}_L - \bar{V}_H > 0, \qquad \bar{H}_L - \bar{H}_H < 0 \tag{7.9.13}$$

and, at the same time, none of the mole fractions x_L and x_H is too small. If this is indeed possible, we will have a large, negative relaxation term in (7.9.10) which will dominate the temperature dependence of the volume. Of course, in order to make a more quantitative statement about the magnitudes of $\bar{V}_L - \bar{V}_H$, $\bar{H}_L - \bar{H}_H$, and x_L, we must specialize to a particular model.

Two pieces of experimental evidence are in conformity with the requirements in (7.9.13). One is the relation between the molar volume and energy of the high-pressure polymorphs of ice pointed out in section 7.2. There we noted that structures having a relatively large coordination number have also internal energies less negative than the corresponding internal energy of ordinary hexagonal ice.

Second, we pointed out the slight increase in the average coordination number of water as we increase the temperature (see Fig. 7.7), i.e.,

$$\frac{d}{dT} [\sum Kx_C(K)] > 0, \tag{7.9.14}$$

which means that the distribution $x_C(K)$ is shifted toward higher values of K as we increase the temperature. On the other hand, since the heat capacity is always positive, addition of heat to any system at equilibrium will shift any distribution of species toward those that have higher energy (see section 2.4.5). In other words, the structural change in the system will always be in that direction so that part of the heat is absorbed by that change. Therefore, using the distribution $x_C(K)$ used in (7.9.14), we can conclude that shift towards higher coordination number is correlated with the shift towards less negative binding energy.

On a molecular level, the condition (7.9.13) can be formulated in terms of the quasicomponent distribution function $x_{B,C}(v, K)$. For a normal liquid, we expect that the average binding energy of species having a fixed coordination number will be a decreasing function of K. On the other hand, in water we expect that this function will increase with K, at least for some range of P and T. This has indeed been demonstrated for two-dimensional water-like particles, and recently also by Monte Carlo simulation on water.

Thus the unique temperature dependence of the volume of water is due to the unique packing of water molecules in such a way that an open local structure is connected with large (negative) binding energy. This unique packing has been built in the model of section 4.5.4 to show a negative thermal expansion coefficient in a one-dimensional system. We shall also see that this unique property is also responsible for one outstanding property of aqueous solutions of nonpolar solutes (section 7.12).

Consider next the heat capacity (at constant pressure) which, in the TSM formalism, can be obtained as follows. The total enthalpy of the system is

$$H = N_L \bar{H}_L + N_H \bar{H}_H \tag{7.9.15}$$

and the heat capacity is given by

$$C_P = \left(\frac{\partial H}{\partial T} \right)_{P, N_W, \text{eq}}$$

$$= \left(N_L \frac{\partial \bar{H}_L}{\partial T} + N_H \frac{\partial \bar{H}_H}{\partial T} \right) + (\bar{H}_L - \bar{H}_H) \left(\frac{\partial N_L}{\partial T} \right)_{P, N_W, \text{eq}}, \tag{7.9.16}$$

where, again, we have split the heat capacity into two terms. The first is the heat capacity of the frozen-in system and the second is the corresponding relaxation term. We stress again that the split into these two terms depends on the choice of the particular classification procedure used to define L and H.

Using the identity (5.13.65), we can rewrite the relaxation term in (7.9.4) as

$$(\bar{H}_L - \bar{H}_H)\left(\frac{\partial N_L}{\partial T}\right)_{P,N_W,eq} = \frac{(\bar{H}_L - \bar{H}_H)^2}{T(\mu_{LL} - 2\mu_{LH} + \mu_{HH})}$$

$$= \frac{(\bar{H}_L - \bar{H}_H)^2 x_L x_H \eta V}{kT^2}. \qquad (7.9.17)$$

Clearly, the relaxation term in (7.9.17) is always positive, independent of the definition of L and H. Of course, the magnitude of the relaxation term depends on the particular choice of the classification procedure. In order to explain the high value of the heat capacity of liquid water, one assumes that the "frozen-in" term, i.e., the first term on the rhs of (7.9.16), has a normal value. The excess heat capacity is then attributed to the relaxation term. In order for the latter to be large, we must have two components which differ appreciably in their partial molar enthalpy [otherwise, $(\bar{H}_L - \bar{H}_H)^2$ cannot be large] and none of the mole fractions x_L and x_H can be too small.

It should be noted that the requirements for a large value of the relaxation term of the heat capacity are somewhat weaker than those needed for a large relaxation term of the thermal expansion coefficient. It is likely that the peculiar relation between $\bar{V}_L - \bar{V}_H$ and $\bar{H}_L - \bar{H}_H$, supplemented by the requirement that x_L and x_H not be too small, is one of the most strikingly unique features of liquid water. On the other hand, the requirement that $(\bar{H}_L - \bar{H}_H)^2$ be large is certainly a result of the existence of strong hydrogen bonds in water. Such a property is not unique to water, and indeed liquid ammonia, for example, also has a relatively high heat capacity, which probably has the same origin as that for liquid water.

As a final example, we present the expression for the isothermal compressibility in the TSM:

$$\kappa_T = -\frac{1}{V}\left(\frac{\partial V}{\partial P}\right)_{T,N_W,eq}$$

$$= -\frac{1}{V}\left[N_L\frac{\partial \bar{V}_L}{\partial P} + N_H\frac{\partial \bar{V}_H}{\partial P}\right] - \frac{1}{V}(\bar{V}_L - \bar{V}_H)\left(\frac{\partial N_L}{\partial P}\right)_{T,N_W,eq}. \qquad (7.9.18)$$

The relaxation term can be rewritten, using identity (5.13.66) as

$$-\frac{1}{V}(\bar{V}_L - \bar{V}_H)\left(\frac{\partial N_L}{\partial P}\right)_{T,N_W,eq} = -\frac{(\bar{V}_L - \bar{V}_H)^2}{V(\mu_{LL} - 2\mu_{LH} + \mu_{HH})}$$

$$= -\frac{(\bar{V}_L - \bar{V}_H)^2 x_L x_H \eta}{kT}. \qquad (7.9.19)$$

We note that this term is always negative, independent of the particular definitions of L and H.

One general comment applicable to all the above discussions is concerned with the nature of the mixture of L and H. The question is whether the mixture may be assumed to be ideal and, if so, in what sense. First, suppose that we have defined L and H in such a way that one component is very dilute in the other. This can be easily achieved. For instance, if we take $K^* = 12$ in (7.9.1), then it is likely that $x_H \approx 0$. Hence, such a solution will be dilute ideal and we get

$$\mu_{LL} - 2\mu_{LH} + \mu_{HH} = \frac{kT}{N_W x_L x_H}. \qquad (7.9.20)$$

However, the fact that one component is very dilute in the other implies that the product $x_L x_H$ is very small. Hence, all the relaxation terms in (7.9.9), (7.9.17), and (7.9.19) are very small, and the whole treatment is rendered useless.

A second possibility is to define L and H as similar, so that the solution is symmetrically ideal, i.e.,

$$G_{LL} + G_{HH} - 2G_{LH} = 0. \qquad (7.9.21)$$

Hence, $\eta = \rho_L + \rho_H = \rho_W$ [in (7.9.9), (7.9.17), and (7.9.19)]. The latter form of the relaxation term has been used by many authors.

Some care must be exercised in applying the symmetric ideal assumption to the TSM of water. We saw that in order to get large relaxation terms in (7.9.9), (7.9.17), and (7.9.19) we must assume that the two components L and H are markedly different. This, in general, will contradict the assumption of similarity in the sense of (7.9.21). We recall that condition (7.9.21) is essentially a condition on the similarity of the local environments of L and H. In a real mixture of two components A and B, similarity of the molecules A and B implies similarity in their local environments as well. In the MM approach for water, we start with two components which are identical in their chemical composition. (Both L and H are water molecules!) Therefore, it is very tempting to assume that these two components also form a symmetric ideal solution. However, since by their definitions we require the two components to be different in their local environments, the validity of (7.9.21) cannot be guaranteed.

7.9.2. A Prototype of an Interstitial Lattice Model for Water

Lattice models for liquids are rarely used nowadays. The same is true of lattice models for water. Nevertheless, the model presented in this section is of interest for three reasons: First, it presents a prototype of an interstitial model having features in common with many models proposed for water and used successfully to explain some of the outstanding properties of water and aqueous solutions. Second, this model demonstrates some general aspects of the mixture model approach to the theory of water, for which explicit expressions for all the thermodynamic quantities in terms of molecular properties may be obtained. Finally, the detailed study of this model has a didactic virtue, being an example of a simple and solvable model.

The interstitial model has a serious drawback, however, which is similar to the shortcomings of applying a lattice model to a fluid in general. It is therefore important to make a clear-cut distinction between results that pertain strictly to the model and results that have more general validity. In the following we shall describe the model and briefly outline some of the results. The details are left to the reader.

The system consists of N_W water molecules at a specified temperature T and pressure P. A total of N_L molecules participate in the formation of a regular lattice with a well-defined structure, which for our purposes need not be specified. (It may be an ice I_h structure, as in the Samoilov model, or a clathrate type of structure, as in the Pauling model.) The lattice is presumed to contain empty spaces, or holes. We let N_0 be the number of holes per lattice molecule; i.e., $N_0 N_L$ is the total number of holes formed by N_L lattice molecules. We assume that the system is macroscopically large, so that surface effects are negligible; hence, N_0 is considered to be independent of N_L. In the case of ice I_h, we have $N_0 = \frac{1}{2}$. Each hole is surrounded by 12 molecules, but each molecule participates in six holes.

The remaining $N_H = N_W - N_L$ molecules are assumed to occupy the holes and may be referred to as interstitial molecules. A schematic illustration of such a model is shown in Fig. 7.12.

To keep the complexity of the model at a minimum, we require the following simplifying assumptions: (1) All the holes have the same structure. (2) A hole can accommodate at most one water molecule in such a way as not to distort the lattice structure to a significant extent. (3) The interstitial molecules do not "see" each other; i.e., there is no direct interaction between interstitial molecules in adjacent holes. Hence, occupancy of a certain hole does not affect the chances of an adjacent hole being empty or filled. (4) The lattice molecules are assumed to hold the equilibrium lattice points, and vibrational excitation is negligible. (5) An interstitial molecule is assumed to be situated in a fixed position in the hole.

Clearly, the condition $N_H \leq N_0 N_L$ must be satisfied. Alternatively, the mole fraction of interstitial molecules is restricted to vary between the limiting values

$$0 \leq x_H = \frac{N_H}{N_W} \leq \frac{N_0}{(1 + N_0)}. \tag{7.9.22}$$

For instance, in the model depicted in Fig. 7.12 $N_0 = \frac{1}{2}$; i.e., each lattice particle belongs to three holes, but each hole is built up of six particles, so that each particle contributes half a hole. Thus the mole fraction of interstitial molecules cannot exceed $\frac{1}{3}$.

The total energy of the system E_T is composed only of the interaction energies among the molecules

$$E_T = N_L E_L + N_H E_H, \tag{7.9.23}$$

where E_L is the lattice energy per lattice molecule and E_H is the interaction energy of an interstitial molecule with its surroundings.

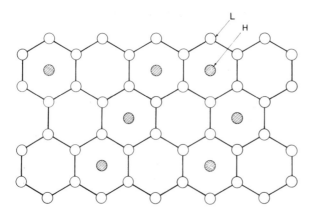

FIGURE 7.12. Schematic illustration of an interstitial model for water in two dimensions.

Thus, the canonical partition function for this model is

$$Q(T, V, N_W) = \sum_{\text{configurations}} \exp[-\beta(N_L E_L + N_H E_H)]$$

$$= \sum_{E_T} \Omega(E_T, V, N_W) \exp[-\beta(N_L E_L + N_H E_H)], \qquad (7.9.24)$$

where, in the second form on the rhs, we have collected all the terms in the partition function corresponding to the same value of the total energy E_T, which is Ω-fold degenerate. Clearly, the total energy depends only on the number N_L (E_L and E_H are assumed fixed by the model, and $N_H = N_W - N_L$). Hence, to compute the degeneracy of a given energy E_T, it is sufficient to compute the number of ways we can arrange N_H interstitial molecules in $N_0 N_L$ holes. This number is

$$\Omega(E_T, V, N_W) = \frac{(N_0 N_L)!}{N_H!(N_0 N_L - N_H)!}. \qquad (7.9.25)$$

Furthermore, in this particular model, the total volume V can be written as

$$V = N_L V_L, \qquad (7.9.26)$$

where V_L is the volume of the system per lattice molecule. (Interstitial molecules do not contribute to the volume.) The condition (7.9.26) considerably simplifies the evaluation of the partition function in (7.9.21). If we determine the total volume V, then from (7.9.26) the number N_L is determined as well. This, in turn, determines the total energy through (7.9.23). Therefore, the partition function (7.9.24) reduces to a single term, namely

$$Q(T, V, N_W) = \frac{(N_0 N_L)!}{N_H!(N_0 N_L - N_H)!} \exp[-\beta(N_L E_L + N_H E_H)]. \qquad (7.9.27)$$

Next, we write the T, P, N_W partition function. This is more convenient for later applications in which various thermodynamic quantities are evaluated:

$$\Delta(T, P, N_W) = \sum_{V = V_{\min}}^{V_{\max}} Q(T, V, N_W) \exp(-\beta P V)$$

$$= \sum_{N_L = N_{L_{\min}}}^{N_W} \frac{(N_0 N_L)!}{N_H!(N_0 N_L - N_H)!} \exp[-\beta(N_L E_L + N_H E_H + P N_L V_L)], \qquad (7.9.28)$$

where, in the second form on the rhs, we have transformed the summation over all possible volumes of the system to a sum over N_L, which determines the volume through (7.9.26). The summation is carried out from the minimum value of $N_{L_{\min}} = N_W/(1 + N_0)$ [see (7.9.22)] to the maximum value of N_W. This completes our description of the model.

Having the PF of the system, we can derive all the required thermodynamic quantities by standard relations. As in any mixture model approach (section 5.13), we start with a one-component system. The independent variables are T, P and N_W. We then choose to refer to a lattice water molecule as L and to an interstitial molecule as an H molecule. Once we have made this classification, we may adopt the point of view that our system is a mixture of two components, L and H molecules.

Let us rewrite (7.9.28) in the form

$$\Delta(T, P, N_W) = \sum_{N_L} \Delta(T, P, N_W; N_L), \tag{7.9.29}$$

where each term $\Delta(T, P, N_W; N_L)$ is the partition function of a system in which the conversion reaction $L \rightleftarrows H$ has been frozen-in.

Using a connection between this partition function and the Gibbs energy, we can rewrite (7.9.29) as

$$\exp[-\beta G(T, P, N_W)] = \sum_{N_L} \exp[-\beta G(T, P, N_W; N_L)]. \tag{7.9.30}$$

We denote by N_L^* (and $N_H^* = N_W - N_L^*$) the value of N_L for which $G(T, P, N_W; N_L)$ is minimum. Taking this minimum, we obtain

$$\frac{(N_0 N_L^*)^{N_0} N_H^*}{(N_0 N_L^* - N_H^*)^{N_0 + 1}} = \exp[\beta(E_L - E_H + PV_L)]$$

$$\equiv K(T, P). \tag{7.9.31}$$

The conventional mole fractions of the L and H species are

$$x_L = \frac{N_L}{N_W}, \qquad x_H = 1 - x_L. \tag{7.9.32}$$

However, the equilibrium condition (7.9.31) can be written more conveniently in terms of the "mole fractions" of empty and occupied holes, which are defined, respectively, as

$$y_0 = \frac{(N_0 N_L - N_H)}{N_0 N_L}, \qquad y_1 = \frac{N_H}{N_0 N_L}. \tag{7.9.33}$$

With these variables, the equilibrium condition (7.9.31) is

$$\frac{y_1^*}{(y_0^*)^{N_0 + 1}} = K(T, P), \tag{7.9.34}$$

which has a simple interpretation as an equilibrium constant for the reaction

$$(N_0 + 1)[\text{empty holes}] \rightleftarrows [\text{occupied holes}]. \tag{7.9.35}$$

The stoichiometry of this reaction is understood as follows: In order to create one mole of occupied holes, we must cancel one mole of empty holes. In addition, in order to fill these holes, we need one mole of molecules, which come from the lattice; hence, N_0 empty holes must also be destroyed. Altogether, we need $N_0 + 1$ moles of empty holes to be converted to one mole of occupied holes.

Once we have determined N_L^*, we can replace the sum in (7.9.30) by a single term

$$\exp[-\beta G(T, P, N_W)] = \sum_{N_L} \exp[-\beta G(T, P, N_W; N_L)]$$

$$= \exp[-\beta G(T, P, N_W; N_L^*)]. \tag{7.9.36}$$

In the general case, it is difficult to solve (7.9.31) explicitly. As a simple case, we take $N_0 = 1$ and rewrite (7.9.31) in terms of the mole fraction $x_L = N_L/N_W$:

$$\frac{x_L^*(1 - x_L^*)}{(2x_L^* - 1)^2} = K. \tag{7.9.37}$$

The solution of (7.9.37) is

$$x_L^* = \frac{1}{2} \pm \frac{1}{2}\left(\frac{1}{4K + 1}\right)^{1/2}. \tag{7.9.38}$$

There are two possible solutions, only one of which is physically acceptable. In order to determine the correct sign, we can check one limiting case. For instance, for given T and P, if $E_L \ll E_H$, then $K \to 0$. In this case, x_L^* must tend to unity. From (7.9.38), we have

$$x_L^* \xrightarrow{K \to 0} \frac{(1 \pm 1)}{2}. \tag{7.9.39}$$

Hence the plus sign is the acceptable one in (7.9.38). The chemical potentials of the components L and H are

$$\mu_L = \left(\frac{\partial G}{\partial N_L}\right)_{T,P,N_H}$$

$$= E_L + PV_L - kT[N_0 \ln(N_0 N_L) - N_0 \ln(N_0 N_L - N_H)]$$

$$= E_L + PV_L - kT N_0 \ln \frac{N_0 x_L}{N_0 x_L - x_H} \tag{7.9.40}$$

$$\mu_H = \left(\frac{\partial G}{\partial N_H}\right)_{T,P,N_L}$$

$$= E_H - kT[\ln(N_0 N_L - N_H) - \ln N_H]$$

$$= E_H + kT \ln \frac{x_H}{N_0 x_L - x_H}. \tag{7.9.41}$$

Some points are worth our attention in (7.9.40) and (7.9.41):

1. The mixture of L and H is not ideal in any of the senses discussed in Chapter 6. In particular, it is not symmetric ideal. Note that L and H are both water molecules. However, by their very definition, they differ markedly in their local environment; hence, it is unlikely that they obey the condition for symmetric ideal solutions.

2. The chemical potentials μ_L and μ_H are definable only in the frozen-in system, since we require N_H and N_L to be kept constant in (7.9.40) and (7.9.41), respectively. In the special case $N_L = N_L^*$ and $N_H = N_H^*$, we get [note (7.9.31)]

$$\mu_L = \mu_H. \tag{7.9.42}$$

The total entropy of the system is

$$-S = \left(\frac{\partial G}{\partial T}\right)_{P,N_W}$$

$$= \left(\frac{\partial G}{\partial T}\right)_{P,N_L,N_H} + \left[\left(\frac{\partial G}{\partial N_L}\right)_{T,P,N_H} - \left(\frac{\partial G}{\partial N_H}\right)_{T,P,N_L}\right]\left(\frac{\partial N_L}{\partial T}\right)_{P,N_W}$$

$$= \left(\frac{\partial G}{\partial T}\right)_{P,N_L,N_H}. \tag{7.9.43}$$

The total volume is

$$V = \left(\frac{\partial G}{\partial P}\right)_{T,N_W} = \left(\frac{\partial G}{\partial P}\right)_{T,N_L,N_H} = N_L V_L, \tag{7.9.44}$$

which is an obvious result for this model.

The enthalpy and the internal energy are similarly obtained:

$$H = G + TS = N_L E_L + N_H E_H + P N_L V_L \tag{7.9.45}$$

$$E = H - PV = N_L E_L + N_H E_H. \tag{7.9.46}$$

The temperature dependence of the volume is

$$\left(\frac{\partial V}{\partial T}\right)_{P,N_W,\text{eq}} = \left(\frac{\partial V}{\partial T}\right)_{P,N_L,N_H} + \left[\left(\frac{\partial V}{\partial N_L}\right)_{T,P,N_H} - \left(\frac{\partial V}{\partial N_H}\right)_{T,P,N_L}\right]\left(\frac{\partial N_L}{\partial T}\right)_{P,N_W,\text{eq}}. \tag{7.9.47}$$

In our particular model we have the result

$$\left(\frac{\partial V}{\partial T}\right)_{P,N_W,\text{eq}} = \frac{V_L(E_L - E_H + PV_L)(N_0 N_L^* - N_H^*)x_L^* x_H^*}{kT^2 N_0}. \tag{7.9.48}$$

Since $V_L > 0$, $N_0 N_L \geq N_H$, and $x_L x_H \geq 0$, the sign of (7.9.48) is determined by the quantity $E_L - E_H + PV_L$. Now, in most interstitial models, $E_L - E_H$ is of the order of about -10^3 cal/mole. At 1 atm and with V_L on the order of 18 cm³/mole, we have $PV_L \approx 18/41 = 0.44$ cal/mole. Thus, the PV_L term is negligible compared to $E_L - E_H$ and, therefore, the sign of (7.9.48) is negative. This is the essence of the "explanation" of the temperature dependence of the volume of water. If we assume that V_L is temperature dependent, so that instead of (7.9.48) we have

$$\left(\frac{\partial V}{\partial T}\right)_{P,N_W,\text{eq}} = N_L \left(\frac{\partial V_L}{\partial T}\right)_{P,N_L,N_H} + (\bar{V}_L - \bar{V}_H)\left(\frac{\partial N_L}{\partial T}\right)_{P,N_W,\text{eq}}, \tag{7.9.49}$$

and if we assume that the first term on the rhs of (7.9.49) is positive (i.e., the lattice expands with increasing temperature), then we have competition between two terms of opposite sign, which happen to cancel out at about 4°C. The heat capacity at constant pressure is given by

$$C_P = \left(\frac{\partial H}{\partial T}\right)_{P,N_W,\text{eq}} = \left(\frac{\partial H}{\partial T}\right)_{P,N_L,N_H} + (\bar{H}_L - \bar{H}_H)\left(\frac{\partial N_L}{\partial T}\right)_{P,N_W,\text{eq}}. \tag{7.9.50}$$

This is the general expression for the heat capacity in the mixture model approach.

In our model, the first term on the rhs of (7.9.50) is zero and for the second term we have

$$C_P = (E_L - E_H + PV_L)\left(\frac{\partial N_L}{\partial T}\right)_{P,N_W,eq}$$

$$= \frac{(E_L - E_H + PV_L)^2(N_0N_L^* - N_H^*)x_L^*x_H^*}{kT^2N_0}. \qquad (7.9.51)$$

In this particular model, the heat capacity is due to thermal excitation from the H to the L state.

As a final example of quantities in this group, we compute the isothermal compressibility for this model:

$$\kappa_T = -\frac{1}{V}\left(\frac{\partial V}{\partial P}\right)_{T,N_W,eq}$$

$$= -\frac{1}{V}\left[\left(\frac{\partial V}{\partial P}\right)_{T,N_L,N_H} + (\bar{V}_L - \bar{V}_H)\left(\frac{\partial N_L}{\partial P}\right)_{T,N_W,eq}\right]$$

$$= \frac{V_L^2(N_0N_L^* - N_H^*)x_L^*x_H^*}{kTVN_0}, \qquad (7.9.52)$$

which is always positive.

We conclude this section with a general comment on interstitial models. The study of such models is useful and quite rewarding in that it gives us insight into the possible mechanism by which water reveals its anomalous behavior. One should not overstress, however, the significance of the numerical results obtained from the model as an indication of the extent of the reality of the model. It is possible, by a judicious choice of the molecular parameters, to obtain thermodynamic results which are in agreement with experimental values measured for real water. Such agreement can be achieved by quite different models. The important point is the qualitative explanation that the model is capable of offering of the various properties of water. We shall return to the same model in section 7.12 to explain some aspects of aqueous solutions of simple solutes.

7.10. AQUEOUS SOLUTIONS OF SIMPLE SOLUTES: PROPERTIES

The study of very dilute solutions of simple solutes is of interest for various reasons. First, these solutions reveal some anomalous properties in comparison with nonaqueous solutions and have therefore presented an attractive challenge to scientists. Second, the study of dilute aqueous solutions viewed as systems deviating slightly from pure water provides some further information helpful in the study of water itself. Finally, understanding the thermodynamics of these systems on a molecular level is a first step in the study of biochemical systems.

One cannot expect the theoretical development of the study of dilute aqueous solutions to be better than that for pure water. Nevertheless, because of their interest and importance, there has been significant progress in the understanding of these relatively simple systems. Part of the outcome of this research has also been applied to the study of more complex aqueous solutions of biochemical molecules. This will be discussed in Chapter 8.

7.10.1. Survey of Some Properties of Simple Aqueous Solutions

In this section we survey some of the unusual properties of aqueous solutions of simple nonpolar solutes such as argon, methane, and the like. The solubility of such solutes, as measured by the Ostwald absorption coefficient, is markedly smaller in water than in a typical organic liquid. (By typical or normal organic liquids, we mean alkanes, alkanols, benzene and its simple derivatives, and so on.)

The Ostwald absorption coefficient is defined as the ratio of the number densities of a solute s in the liquid l and the gaseous g phases at equilibrium, i.e.,

$$\gamma_s = \left(\frac{\rho_s^l}{\rho_s^g}\right)_{eq}. \tag{7.10.1}$$

The solvation Gibbs energy of s in l is directly related to γ_s through

$$\Delta G_s^* = kT \ln\left(\frac{\rho_s^g}{\rho_s^l}\right)_{eq} = -kT \ln \gamma_s. \tag{7.10.2}$$

Thus the statement of a low solubility, in terms of γ_s, is equivalent to a relatively large and positive value of the solvation Gibbs energy, say of argon in water, as compared with other organic solvents. In Fig. 7.13, we plot the solvation Gibbs energies of argon and xenon in a series of alkanols as a function of n, the number of carbon atoms in the n-alkanol molecule (with $n = 0$ corresponding to liquid water). We see that the values of ΔG_s^* change very little as a function of n. If we attempt to extrapolate the value of ΔG_s^* for water by taking the limit $n \rightarrow 0$, we find that these values (indicated by arrows in Fig. 7.13) are much lower than the experimental values. Thus a distinctly abrupt change in ΔG_s^* is observed when we pass from the series of n-alkanols to liquid water. In other words, water, though formally belonging to this homologous series with $n = 0$,

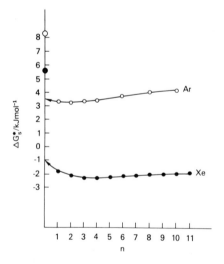

FIGURE 7.13. Solvation Gibbs energies of argon and xenon in n-alkanols as a function of n, the number of carbon atoms in the n-alkanol. The corresponding values for water are indicated at $n = 0$. All values correspond to 25°C and 1 atm.

has unusual properties and cannot be viewed as one of the members of the homologue series of n-alkanols.

The second striking difference between aqueous and nonaqueous solutions is the temperature dependence of the solvation Gibbs energy. This is manifested in the relatively large entropy of solvation of the inert gases in water as compared with other liquids. In Fig. 7.14 we plotted values of ΔH_s^* and $T\Delta S_s^*$ for xenon in a series of n-alkanols. Again, we see that there is a very small variation in these values as we change n. (The variation between the extreme values is believed to lie within the range of experimental error.) The corresponding values of ΔH_s^* and $T\Delta S_s^*$ in water are markedly lower than the values that one would have extrapolated for the case of $n = 0$.

It is noteworthy that although both ΔH_s^* and ΔS_s^* are derived from the temperature dependence of ΔG_s^*, they reflect distinctly different aspects of the solvation process. In thermodynamic terms, we have

$$\Delta S_s^* = -\left(\frac{\partial \Delta G_s^*}{\partial T}\right)_P \tag{7.10.3}$$

$$\Delta H_s^* = \Delta G_s^* + T\Delta S_s^*$$

$$= kT^2 \left(\frac{\partial \ln \gamma_s}{\partial T}\right)_P. \tag{7.10.4}$$

Hence ΔS_s^* measures directly the temperature coefficient of the solvation Gibbs energy, while ΔH_s^* is a measure of the temperature coefficient of γ_s or the concentration ratio of s between the two phases.

The third unusual property of aqueous solutions of inert gas molecules is the relatively large partial molar heat capacity of the solute in water. This is equivalent to a large heat capacity of solvation of the inert gases in water.

If we define the heat capacity of solvation as

$$\Delta C_{p,s}^* = \frac{\partial \Delta H_s^*}{\partial T}, \tag{7.10.5}$$

we find that for inert gas molecules in typical organic liquids, $\Delta C_{p,s}^*$ are, within the limits of experimental error, nearly zero. On the other hand, at about room temperature

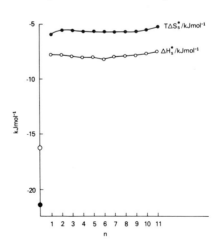

FIGURE 7.14. Values of $T\Delta S_s^*$ in kJ mol^{-1} and ΔH_s^* in kJ mol^{-1} for xenon in water ($n = 0$) and in a series of n-alkanols (n being the number of carbon atoms of the n-alkanols). All values pertain to 20°C and 1 atm.

TABLE 7.3
Solvation Volume $\Delta V_s^*/(\mathrm{cm^3\,mol^{-1}})$ for Methane
and Ethane at 25°C at Infinite Dilution in Some
Solvents

Solvent	Methane	Ethane
Water	36.17	50.07
Carbon tetrachloride	49.06	63.36
n-Hexane	56.03	65.33
Benzene	54.61	70.61

ΔH_s^* has a clear-cut positive slope as a function of temperature, which gives rise to values of $\Delta C_{p,s}^*$ in the range of 150 to 250 J K^{-1} mol^{-1} for the inert gas molecules in water at room temperature.

The volume of solvation is defined by

$$\Delta V_s^* = \left(\frac{\partial \Delta G_s^*}{\partial P}\right)_T. \tag{7.10.6}$$

Table 7.3 shows some values of ΔV_s^* for methane and ethane in some liquids at 25°C. These values are somewhat smaller in water as compared with the other liquids for which the relevant data are available. The differences are, however, not very large, as we have noted for the other quantities of solvation.

The solvation Gibbs energy of small nonpolar molecules in D_2O is smaller than in H_2O. Table 7.4 presents some relevant data. These findings are somewhat unexpected. We shall discuss this topic in section 7.13.

Figure 7.15 presents some data on the solvation Gibbs energy of various hydrocarbon molecules in water. For each homologous series, ΔG_s^* is linear in n, the number of carbon atoms, except for the smallest member of the series. In section 7.14 we shall discuss this behavior in terms of the hydrophobic interaction between simple nonpolar

TABLE 7.4
Values of ΔG_s^*, ΔS_s^* and ΔH_s^* for Neon, Argon, Krypton, and Xenon in H_2O and D_2O at 25°C

		ΔG_s^* (kJ mol^{-1})	ΔS_s^* (J K^{-1} mol^{-1})	ΔH_s^* (kJ mol^{-1})
Neon				
	H_2O	11.19	−45.4	−2.3
	D_2O	10.99	−43.6	−2.0
Argon				
	H_2O	8.40	−65.1	−11.0
	D_2O	8.15	−68.5	−12.26
Krypton				
	H_2O	6.94	−70.5	−14.1
	D_2O	6.99	−68.4	−13.4
Xenon				
	H_2O	5.62	−81.9	−18.8
	D_2O	5.64	−70.9	−15.5

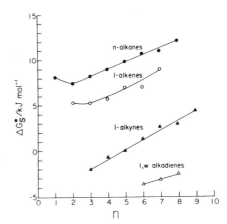

FIGURE 7.15. Solvation Gibbs energies of some hydrocarbons in water as a function of n, the number of carbon atoms in the alkyl chains. All values pertain to $T = 298.15$ K.

solutes. Denoting by ΔG_n^* the Gibbs energy of solvation of a molecule with chain length n, we may define

$$\delta G(n + CH_3 \rightarrow n + 1) = \Delta G_{n+1}^* - \Delta G_n^* - \Delta G_{CH_3}^*, \qquad (7.10.7)$$

where δG is the indirect work required to bring a methyl group from infinite separation from a hydrocarbon of size n to form a hydrocarbon of size $n + 1$. Since all of the quantities on the rhs of Eq. (7.10.7) pertain to the same solvent, the approximate constancy of the slopes of the curves in Fig. 7.15 is essentially a reflection of the fact that δG is constant, almost independent of n and of the functional group of a particular homologous series.

In Fig. 7.16 we present some data on the effect of branching and cyclization on ΔG_s^* of a given alkane. We note that most branched alkanes fall above the curve of the n-alkanes (with the same n), while the cycloalkanes fall below this curve. (A few of the branched alkanes are also below the curve, but in these cases their values are very close to the values of the normal alkanes. In these cases, we judge that the difference between the ΔG_s^* values is quite within the limits of the experimental error.)

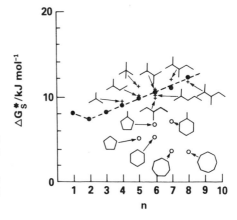

FIGURE 7.16. Solvation Gibbs energies of alkanes in water as a function of n, the number of carbon atoms in the alkyl chains. All values pertain to $T = 298.15$ K. Branched alkanes are represented by plus signs (+), cyclic by open circles (O), and normal alkanes by full circles (●).

We shall see in section 7.14 that the indirect part of the work required to make a cyclic from a normal alkane is approximately given by

$$\delta G(\text{cyclization}) = \Delta G^*_{\text{cyclic}} - \Delta G^*_{\text{normal}}. \tag{7.10.8}$$

Currently, there exists no theory of aqueous solutions that offers a satisfactory explanation of the outstanding properties of these systems. There are, however, some partial interpretations of some of the properties. Here we shall mention, only briefly, two approaches: both of them view the solvation process as comprising two steps. The first, and older one, splits the process of solvation (or rather the process of solution in the original treatment) into a cavity formation and a soft interaction between the solute and solvent. The second is based on the recognition that liquid water has some degree of structure and that this structure might be affected by the process of solvation. Hence one splits the process of solvation into two steps: first the solute is introduced into the solvent to which the structure has been frozen in, and then the structure is allowed to relax to its final equilibrium value in the presence of the solute.

Both approaches contain an element of ambiguity. In the first, it is the size of the cavity that is presumed to accommodate the solute. For real solutes and solvents, there is no clear-cut way of defining the size of such a cavity. One usually assigns an effective hard sphere diameter to the molecules involved and then defines the radius of the appropriate cavity as the arithmetic mean of the solute and solvent diameters. Once the choice of the cavity radius has been made, one can proceed in a rigorous and formal way and split all the thermodynamic quantities of solvation into two contributions. This approach is identical to the one described in section 6.14. The second approach requires the definition of the structure of water. We shall describe some aspects of this approach in sections 7.11–7.13.

7.10.2. Hydrophobicity and Conditional Hydrophobicity

The terms hydrophobic (HϕO) and hydrophilic (HϕI) solutes have been used in the literature with a variety of meanings. We shall introduce here a simple definition in terms of the partition coefficients of a given solute in two solvents. Qualitatively, having two immiscible solvents, water w and some reference solvent α, we add a solute s and ask how it is distributed between the two liquids. The simplest measure of this distribution is the ratio of the number densities of s in the two phases, at equilibrium. This is related to the difference in the solvation Gibbs energies of s in the two phases; thus

$$\eta_s = \left(\frac{\rho_s^w}{\rho_s^\alpha}\right)_{\text{eq}} = \exp[-\beta(\Delta G_s^{*w} - \Delta G_s^{*\alpha})], \tag{7.10.9}$$

where ΔG_s^{*w} and $\Delta G_s^{*\alpha}$ are the solvation Gibbs energies in w and α respectively. If we choose the two phases α and w to have the same volumes, then the ratio on the lhs of (7.10.9) gives the ratio of the number of solute molecules in the two phases.

Clearly the larger the ratio η_s, the larger is the relative affinity of s for the two phases. Since we are interested in the relative affinities of various solutes toward water, we can choose a constant reference solvent α and measure the ratio η_s for various solutes. The simplest reference solvent is an ideal gas, for which $\Delta G_s^{*\alpha} = 0$ and η_s in (7.10.9) reduces to the Ostwald coefficient (7.10.1)

$$\gamma_s = \left(\frac{\rho_s^w}{\rho_s^{ig}}\right)_{eq}$$

$$= \exp(-\beta \Delta G_s^{*w}). \qquad (7.10.10)$$

Having various solutes we can construct a hydrophobicity (or hydrophilicity) scale according to the value of γ_s or, equivalently, according to the value of ΔG_s^*.

Within this definition, argon is more hydrophobic (or less hydrophilic) than krypton at, say, 25°C. Thus the more positive ΔG_s^*, the more hydrophobic is the solute. The more negative ΔG_s^*, the more hydrophilic is the solute. One can conveniently choose $\Delta G_s^* = 0$ as the dividing line between HϕO and HϕI solutes, but this is not necessary. What is important is the relative hydrophobicity, or the difference between the solvation Gibbs energies of two solutes in the *same* solvent. In the above definition, the same solvent is water. Clearly, the concept can be extended to any solvent or mixture of solvents.

One important extension of the concept of hydrophobicity is useful in the study of solvent effects on biochemical processes. This is the relative hydrophobicity of groups such as methyl, ethyl, hydroxyl, etc., attached to a polymer. Formally we can define the same quantity ΔG_s^* as before, but now the solvent is not the pure water, but the molecules of solvent next to the polymer. This extension of the concept of the hydrophobicity of groups or radicals leads naturally to the concept of the conditional solvation Gibbs energy. Thus instead of ΔG_s^{*w}, the work required to transfer s from a fixed position in the gaseous phase to a fixed position in water, we now need to transfer the group s to a fixed position adjacent to the polymer. Clearly, because of the effect of the polymer on its immediate environment, the conditional solvation Gibbs energy will be different from the solvation Gibbs energy. This is similar to the difference in Gibbs energies in any two solvents. Therefore a hydrophobicity scale based on the solvation Gibbs energy might be very different from a hydrophobicity scale based on conditional solvation Gibbs energy.

For groups or radicals such as methyl, ethyl, etc., we cannot measure the partition coefficient analogue to γ_s. We shall discuss an approximate measure of the conditional solvation Gibbs energy in Chapter 8 and in Appendices G and H.

7.11. FORMAL STATISTICAL MECHANICAL EXPRESSIONS FOR THE SOLVATION QUANTITIES IN WATER

In this section we derive some general statistical mechanical expressions for the thermodynamic quantities of solvation, which are independent of any modelistic assumptions. We shall later examine special cases for these relations for either a molecular model for water (in terms of a model pair potential) or for a specific mixture model view of liquid water.

We start with a very dilute system of a solute s in water. By very dilute we mean that the ideal dilute limiting behavior is attained. Formally, it is sufficient to treat a system containing just one solute s in a solvent containing N water molecules. For convenience we use the T, V, N ensemble, but similar expressions can be obtained in the T, P, N ensemble (where the Gibbs energy and the enthalpy of solvation replace the

Helmholtz energy and the energy of solvation discussed here within the T, V, N ensemble). In section 6.13 we derived the expression for the Helmholtz energy of solvation of s in the T, V, N ensemble, which reads

$$\Delta A_s^* = -kT\ln\langle\exp(-\beta B_s)\rangle_0, \tag{7.11.1}$$

where

$$\langle\exp(-\beta B_s)\rangle_0 = \frac{\int d\mathbf{X}^N \exp[-\beta U_N(\mathbf{X}^N) - \beta B_s(\mathbf{R}_s, \mathbf{X}^N)]}{\int d\mathbf{X}^N \exp[-\beta U_N(\mathbf{X}^N)]}$$

$$= \int d\mathbf{X}^N P(\mathbf{X}^N)\exp[-\beta B_s(\mathbf{R}_s, \mathbf{X}^N)]. \tag{7.11.2}$$

We have denoted by $B_s(\mathbf{R}_s, \mathbf{X}^N)$ the total binding energy of s to all the N water molecules at a specific configuration $\mathbf{X}^N$. The solute s is presumed to be at some fixed position $\mathbf{R}_s$. We note also that the average $\langle\ \rangle_0$ is taken with the probability distribution of the N water molecules of pure liquid water, i.e., in the absence of a solute s at $\mathbf{R}_s$. The subscript zero serves to distinguish this average from the conditional average introduced below.

The solvation entropy of s is obtained by taking the derivative of Eq. (7.11.1) with respect to T, i.e.,

$$\Delta S_s^* = -\left(\frac{\partial\Delta A_s^*}{\partial T}\right)_{V,N}$$

$$= k\ln\langle\exp(-\beta B_s)\rangle_0 + \frac{1}{T}(\langle B_s\rangle_s + \langle U_N\rangle_s - \langle U_N\rangle_0), \tag{7.11.3}$$

where the symbol $\langle\ \rangle_s$ signifies a conditional average over all configurations of the solvent molecules, given a solute s at a fixed position $\mathbf{R}_s$; that is, we use the conditional distribution defined by

$$P(\mathbf{X}^N/\mathbf{R}_s) = \frac{\exp[-\beta U_N(\mathbf{X}^N) - \beta B_s(\mathbf{R}_s, \mathbf{X}^N)]}{\int d\mathbf{X}^N \exp[-\beta U_N(\mathbf{X}^N) - \beta B_s(\mathbf{R}_s, \mathbf{X}^N)]}. \tag{7.11.4}$$

The solvation energy is obtained from Eqs. (7.11.1) and (7.11.4) in the form

$$\Delta E_s^* = \Delta A_s^* + T\Delta S_s^* = \langle B_s\rangle_s + \langle U_N\rangle_s - \langle U_N\rangle_0. \tag{7.11.5}$$

As noted before (see also Appendix F) for all the systems of interest in this book, there is not much difference between the values of ΔG_s^* and ΔA_s^* or between ΔH_s^* and ΔE_s^*.

The above expressions for ΔA_s^*, ΔS_s^*, and ΔE_s^* are very general and apply to any solute in any solvent. Note the similarity with the corresponding expressions for the various binding models in section 2.10.

The solvation energy as presented in Eq. (7.11.5) has a very simple interpretation. It consists of an average binding energy of s to the system and a change in the average total interaction energy among all the N water molecules, induced by the solvation process. For any solvent, this change in the total potential energy may be reinterpreted as a "structural change" induced by the solute s on the solvent. Here, the term structural changes refers to the redistribution of the solvent molecules into various quasicomponents. Specifically, using the binding-energy distribution function we can write the energy of the system as (see section 5.13)

$$E = N\varepsilon^k + \frac{N}{2} \int_{-\infty}^{\infty} v x_B(v)\, dv, \tag{7.11.6}$$

where $x_B(v)\, dv$ is the mole fraction of molecules with binding energy between v and $v + dv$. The average potential energy of the pure liquid is identified as

$$\langle U_N \rangle_0 = \frac{N}{2} \int_{-\infty}^{\infty} v x_B(v)\, dv. \tag{7.11.7}$$

Similarly, for the conditional average $\langle U_N \rangle_s$, we can derive the analogue of (7.11.7), using the conditional distribution function $x_B(v/\mathbf{R}_s)$ instead of (7.11.7), i.e.,

$$\langle U_N \rangle_s = \frac{N}{2} \int_{-\infty}^{\infty} v x_B(v/\mathbf{R}_s)\, dv. \tag{7.11.8}$$

Since we are treating the limiting case of $N_s \to 0$, we can rewrite (7.11.5) as

$$\Delta E_s^* = \langle B_s \rangle_s + \lim_{N_s \to 0} \frac{\partial \langle U(N, N_s) \rangle}{\partial N_s}$$

$$= \langle B_s \rangle_s + \frac{N}{2} \int v[x_B(v/\mathbf{R}_s) - x_B(v)]\, dv. \tag{7.11.9}$$

Clearly, the quantity $N[x_B(v/\mathbf{R}_s) - x_B(v)]$ is the change in the average number of the v components induced by the addition of s. Therefore, the second term on the rhs of (7.11.9) may be interpreted as the contribution to ΔE_s^* due to structural changes in the solvent. For liquid water, we may further identify this general concept of "structural change" with the particular concept of structural changes as defined in section 7.6. We simply use the definition of the primitive model for water molecules and obtain, for this special case,

$$\Delta E_s^* = \langle B_s \rangle_s + [\langle U_{LJ} \rangle_s - \langle U_{LJ} \rangle_0] + \varepsilon_{HB}[\langle N_{HB} \rangle_s - \langle N_{HB} \rangle_0]. \tag{7.11.10}$$

If the LJ interaction (ε) is very small compared with the HB energy (ε_{HB}), then the first square brackets on the rhs of (7.11.10) will be negligibly small (it will be zero for a primitive model in which U_{LJ} is replaced by a hard-sphere potential). In this case, the major contribution to ΔE_s^* comes from "structural changes" induced in the solvent. Here, the term "structural changes" is used in the more special, or conventional sense as defined in section 7.6, i.e., changes in the average number of HBs in the water induced by the addition of the solute s.

Note that the contribution due to structural changes appearing in (7.11.5) appears also in the solvation entropy (7.11.3). Furthermore, when we form the combination of $\Delta E_s^* - T\Delta S_s^*$, this term cancels out. The conclusion is that any structural change in the solvent induced by the solute might affect ΔE_s^* and ΔS_s^* but will have no effect on the solvation Helmholtz energy. This conclusion has already been derived earlier in section 2.10 and in Chapter 3.

Until now we have dealt with the limit of very dilute solutions of s in water. Repeating the same derivation for any mixture of s and water, we can rewrite all the expressions above for the solvation thermodynamics of s (or of the solvent w) with a minor reinterpretation of the averages $\langle\ \rangle_0$ and $\langle\ \rangle_s$. Instead of $\langle\ \rangle_0$ in Eqs. (7.11.1), (7.11.3), and

(7.11.5), we have an average over all configurations of the system of N_s solute and N_w solvent molecules before adding the solvaton s. The conditional average $\langle\ \rangle_s$ is similarly reinterpreted as an average over all configurations of N_s and N_w molecules given a solvaton at some fixed position $\mathbf{R}_s$. The structural change induced by s either in the general sense (7.11.5) or in the particular sense of (7.11.10) should be interpreted as caused by the addition of the solvaton s to a system that may contain any number of s and w molecules.

The second case of interest is the solvation of a water molecule in pure liquid water. Again we assume that we have a T, V, N system, and we add one water molecule to a fixed position, say $\mathbf{R}_w$. The solvation Helmholtz energy is

$$\Delta A_w^* = -kT \ln\langle\exp(-\beta B_w)\rangle_0, \tag{7.11.11}$$

where $\langle\ \rangle_0$ indicates an average over all the configurations of the N_w water molecules excluding the solvaton. The corresponding entropy and energy of solvation of a water molecule are obtained by standard relationships. The results are

$$\Delta S_w^* = k \ln\langle\exp(-\beta B_w)\rangle_0 + \frac{1}{T}(\langle B_w\rangle_w + \langle U_N\rangle_w - \langle U_N\rangle_0) \tag{7.11.12}$$

and

$$\Delta E_w^* = \langle B_w\rangle_w + \langle U_N\rangle_w - \langle U_N\rangle_0, \tag{7.11.13}$$

where here the conditional average is taken with the probability distribution

$$P(\mathbf{X}^N/\mathbf{X}_w) = \frac{\exp[-\beta U_{N+1}(\mathbf{X}^{N+1})]}{\int d\mathbf{X}^N \exp[-\beta U_{n+1}(\mathbf{X}^{N+1})]}, \tag{7.11.14}$$

which is the probability density of finding a configuration $\mathbf{X}^N$ given one water molecule at a specific configuration $\mathbf{X}_w$. We note that fixing the configuration $\mathbf{X}_w$ of one of the water molecules is the same as fixing only the location of this molecule. This is so, since we can perform the integration over all orientations of this molecule and a factor $8\pi^2$ will cancel out:

$$P(\mathbf{X}^N/\mathbf{X}_w) = \frac{P(\mathbf{X}^N, \mathbf{X}_w)}{P(\mathbf{X}_w)}$$

$$= \frac{P(\mathbf{X}^N, \mathbf{R}_w)}{P(\mathbf{R}_w)}$$

$$= P(\mathbf{X}^N/\mathbf{R}_w). \tag{7.11.15}$$

Formally, Eqs. (7.11.12) and (7.11.13) are similar to the corresponding Eqs. (7.11.3) and (7.11.5). It follows that one can give also a formal interpretation to the various terms as we have done before. However, since here we have the case of pure liquid water, it is clear that the average binding energy of a water molecule is the same as the conditional average binding energy of a water molecule, i.e.,

$$\langle B_w \rangle_w = \int d\mathbf{X}^N \, P(\mathbf{X}^N / \mathbf{X}_w) B_w$$

$$= \frac{\int d\mathbf{X}^N \, P(\mathbf{X}^N, \mathbf{X}_w) B_w}{P(\mathbf{X}_w)}$$

$$= \frac{\int d\mathbf{X}^N \, P(\mathbf{X}^N, \mathbf{X}_w) B_w}{8\pi^2 / V}$$

$$= \int d\mathbf{X}_w \, d\mathbf{X}^N \, P(\mathbf{X}^N, \mathbf{X}_w) B_w$$

$$= \langle B_w \rangle_0, \qquad (7.11.16)$$

where $\langle B_w \rangle_0$ is the average binding energy of a water molecule in a system of pure water (of either N or $N + 1$ molecules). Furthermore,

$$\langle U_N \rangle_w = \frac{\int d\mathbf{X}^N \, P(\mathbf{X}^N, \mathbf{X}_w) U_N}{P(\mathbf{X}_w)}$$

$$= \frac{\int d\mathbf{X}^N \, P(\mathbf{X}^N, \mathbf{X}_w) U_N}{8\pi^2 / V}$$

$$= \int d\mathbf{X}_w \, d\mathbf{X}^N \, P(\mathbf{X}^N, \mathbf{X}_w) U_N$$

$$= \int d\mathbf{X}_w \, d\mathbf{X}^N \, P(\mathbf{X}^N, \mathbf{X}_w)(U_{N+1} - B_w)$$

$$= \langle U_{N+1} \rangle_0 - \langle B_w \rangle_0. \qquad (7.11.17)$$

Using the last equation, we can rewrite the expression for ΔE_w^* [Eq. (3.11.13)] in the simpler form

$$\Delta E_w^* = \langle B_w \rangle_0 + \frac{N+1}{2} \langle B_w \rangle_0 - \langle B_w \rangle_0 - \frac{N}{2} \langle B_w \rangle_0$$

$$= \tfrac{1}{2} \langle B_w \rangle_0. \qquad (7.11.18)$$

Thus the solvation energy of a water molecule in pure liquid water is simply half the average binding energy of a single water molecule. This result is of course more general, and applies to any liquid.

Finally, we note that the solvation volume (here in the T, P, N ensemble) of a solute s in very dilute solutions may be written as

$$\Delta V_s^* = \langle \phi_s \rangle + N_w \int \phi[x_\psi(\phi/\mathbf{R}_s) - x_\psi(\phi)] \, d\phi. \qquad (7.11.19)$$

This follows directly from the generalization of Eq. (5.13.51) for the volume of a two-component system. In (7.11.19), the solvation volume is viewed as consisting of two terms: an average Voronoi polyhedron of the solute, and a change in the volume of the solvent caused by the change in the distribution of the volumes of the Voronoi polyhedra of the solvent molecules.

7.12. APPLICATION OF THE MIXTURE MODEL APPROACH TO AQUEOUS SOLUTIONS OF SIMPLE SOLUTES

The idea that a solute changes the structure of the solvent is very old. This idea has been used to explain some puzzling observations about aqueous solutions, one of the oldest being the following: The addition of solutes such as ether or methyl acetate to water decreases the compressibility of the system, in spite of the fact that the compressibilities of these solutes by themselves are about three times larger than the compressibility of pure water. It has been postulated that water contains two species: say, monomeric water molecules and polymers of water molecules. Addition of a solute causes a shift toward the component that has a lower compressibility; hence, a qualitative "explanation" of the experimental observation is provided. Similar attempts to explain the effect of solutes on viscosity, dielectric relaxation, self-diffusion, and many other properties have been suggested in the literature.

Perhaps one of the most striking pieces of evidence that a simple solute has a significant effect on the structure of water comes from a comparison of the solvation entropy of KCl and argon in water. The corresponding values are (at 25°C)

$$\Delta S_{KCl}^* = -98.3 \text{ J K}^{-1} \text{ mol}^{-1}$$

$$2\Delta S_{Ar}^* = -130.2 \text{ J K}^{-1} \text{ mol}^{-1}. \qquad (7.12.1)$$

The fact that the solvation entropy of two argon solvatons is more negative than the solvation entropy of one K^+ solvaton and one Cl^- solvaton is quite surprising. The ions produce a strong electric field near their surface forcing a preferential orientation of the dipole moments of water molecules. This is a reasonable explanation for the negative solvation entropy of KCl. What seems quite puzzling is that the entropy of solvation of two argons is even more negative than that of KCl.

The difference between the two values in (7.12.1) corresponds roughly to the hypothetical reaction

$$K^+ + Cl^- \rightarrow Ar + Ar. \qquad (7.12.2)$$

Starting with two charged particles at fixed positions and at infinite separation from each other, we move the electron from Cl^- to the K^+ to form Cl and K atoms. Ignoring the difference in size between Cl, K, and Ar atoms, the entropy change attributed to the cancellation of the charges is negative, about -31.9 J K^{-1} mol^{-1}. Note that the enthalpy change for the same "reaction" (7.12.2) is about

$$2\Delta H_{Ar}^* - \Delta H_{KCl}^* = -22 + 696 = +674 \text{ kJ mol}^{-1}. \qquad (7.12.3)$$

Thus cancellation of the charges produce a very large *positive* enthalpy change, obviously due to the replacement of the very strong binding energies of K^+ and Cl^- by the very weak binding energy of argon. The negative entropy change might indicate that argon atoms, in spite of their weak interaction with water molecules, can produce a significant structural change in the solvent.

How does an inert atom produce such a structural change in the solvent? This effect, if true, runs against our intuition. If by "structure" we mean aggregates of HB'd water molecules, then adding any inert component should, at high enough concentrations, cause a breakdown of this structure. Therefore we expect that if such an effect exists, it must occur only for very dilute solutions. It is still not a trivial effect. In the next few subsections we shall discuss this problem from several points of view.

7.12.1. Application of a Two-Structure Model

In this section we formulate the general aspect of the application of the simplest mixture model approach to water. We shall use an exact two-structure model as introduced in section 7.9. In the following subsection, we shall illustrate the application to solutions of an interstitial model for water, and in section 7.13 we shall discuss the application of a more general MM approach to this problem.

Consider a system of N_W water molecules and N_S solute molecules at a given temperature T and pressure P. (We later confine ourselves to dilute solutions, $N_s \ll N_W$; in this section, the treatment is more general.) We henceforth assume that T and P are constant and therefore omit them from our notation.

Let N_L and N_H be the average number of L and H molecules obtained by any classification procedure (section 5.13). For the purpose of this section, we need not specify the particular choice of the two species; therefore, our treatment will be very general. For concreteness, one may think of a TSM constructed from the vector $\mathbf{x}_C$ (section 7.9), i.e.,

$$N_L = N \sum_{K=0}^{K^*} x_C(K), \qquad N_H = N \sum_{K=K^*+1}^{\infty} x_C(K), \qquad N_W = N_L + N_H. \quad (7.12.4)$$

Any extensive thermodynamic quantity can be viewed either as a function of the variables (T, P, N_W, N_s) or the variables (T, P, N_L, N_H, N_s).

Because of the equilibrium condition

$$\mu_L(T, P, N_L, N_H, N_s) = \mu_H(T, P, N_L, N_H, N_s) \qquad (7.12.5)$$

we can, in principle, solve (7.12.5) in terms of the original set of variables of the system, i.e.,

$$N_L = f(T, P, N_W, N_s), \qquad N_H = N_W - N_L. \qquad (7.12.6)$$

Consider the volume as an example of an extensive variable. The total differential of the volume (T, P constant) is

$$dV = \left(\frac{\partial V}{\partial N_S}\right)_{N_L, N_H} dN_S + \left(\frac{\partial V}{\partial N_L}\right)_{N_S, N_H} dN_L + \left(\frac{\partial V}{\partial N_H}\right)_{N_S, N_L} dN_H. \qquad (7.12.7)$$

We write

$$V_s^f = \left(\frac{\partial V}{\partial N_s}\right)_{N_L, N_H}, \qquad \bar{V}_L = \left(\frac{\partial V}{\partial N_L}\right)_{N_S, N_H}, \qquad \bar{V}_H = \left(\frac{\partial V}{\partial N_H}\right)_{N_S, N_L}, \qquad (7.12.8)$$

and rewrite relation (7.12.6) using the condition $dN_L + dN_H = 0$, as

$$dV = V_s^f \, dN_s + (\bar{V}_L - \bar{V}_H) \, dN_L. \tag{7.12.9}$$

As we have noted before, N_L is not really an independent variable, and since we have kept T, P, and N_W constant in (7.12.9), there remains only one independent variable, N_S. In fact, recognizing that N_L is a function of T, P, N_W, and N_s, through (7.12.5) or (7.12.6), we can rewrite (7.12.9) as

$$dV = V_s^f \, dN_s + (\bar{V}_L - \bar{V}_H) \left(\frac{\partial N_L}{\partial N_s}\right)_{N_W, \text{eq}} dN_s, \tag{7.12.10}$$

where now the variation is only in the number of solute molecules (we have appended the subscript "eq" to stress that this derivative is taken along the equilibrium line for the reaction $L \rightleftarrows H$).

The splitting of dV in (7.12.10) is characteristic of the application of the MM approach to the theory of solutions. It corresponds to dividing the dissolution process into two steps. First, we add dN_s moles (or molecules) to the system in the frozen-in state (i.e., keeping N_L and N_H fixed). The measurable change in volume is the first term on the rhs of (7.12.10). Next, we release the constraint imposed by fixing N_L and N_H, i.e., adding a hypothetical catalyst to our system. The system, in general, then relaxes to a new equilibrium position, and the corresponding change in volume is the second term on the rhs of (7.12.10).

The ordinary partial molar volume of the solute s is thus

$$\bar{V}_s \equiv \left(\frac{\partial V}{\partial N_s}\right)_{N_W, \text{eq}} = V_s^f + (\bar{V}_L - \bar{V}_H) \left(\frac{\partial N_L}{\partial N_s}\right)_{N_W, \text{eq}}. \tag{7.12.11}$$

A similar division can be made for other partial molar quantities. For instance, for the enthalpy, entropy, and Gibbs energy, we have

$$\bar{H}_s = \left(\frac{\partial H}{\partial N_s}\right)_{N_W} = \left(\frac{\partial H}{\partial N_s}\right)_{N_L, N_H} + (\bar{H}_L - \bar{H}_H) \left(\frac{\partial N_L}{\partial N_s}\right)_{N_W} = H_s^f + \Delta H_s$$

$$\bar{S}_s = \left(\frac{\partial S}{\partial N_s}\right)_{N_W} = \left(\frac{\partial S}{\partial N_s}\right)_{N_L, N_H} + (\bar{S}_L - \bar{S}_H) \left(\frac{\partial N_L}{\partial N_s}\right)_{N_W} = S_s^f + \Delta S_s^r$$

$$\mu_s = \left(\frac{\partial G}{\partial N_s}\right)_{N_W} = \left(\frac{\partial G}{\partial N_s}\right)_{N_L, N_H} + (\mu_L - \mu_H) \left(\frac{\partial N_L}{\partial N_s}\right)_{N_W} = \mu_s^f + \Delta \mu_s^r. \tag{7.12.12}$$

An important general result follows from the condition of chemical equilibrium $\mu_L = \mu_H$, as a result of which we have the equality

$$\mu_s = \mu_s^f. \tag{7.12.13}$$

This is a unique feature of the partial molar Gibbs energy of the solute s. We have already obtained this result in section 2.10.

One direct consequence of relation (7.12.13) is that the solubility of s does not change if we freeze in the equilibrium. This is clear since the solubility of s is governed by its chemical potential.† Another consequence of considerable importance is the so-called "entropy–enthalpy compensation." This follows directly from the equilibrium condition

$$\bar{H}_L - T\bar{S}_L = \mu_L = \mu_H = \bar{H}_H - T\bar{S}_H \qquad (7.12.14)$$

or, equivalently,

$$\Delta H_s^r = T\Delta S_s^r. \qquad (7.12.15)$$

The relaxation part of the molar enthalpy is equal to the relaxation part of $T\bar{S}_s$.

Thus far, we have not specified our components, and all the relations we have written may be applied to any TSM. The traditional interpretation of the large and negative enthalpy and entropy of solution of nonelectrolytes in water is the following: One identifies the L form with the hydrogen-bonded molecules and the H form with the nonbonded molecules. Hence, it is expected that $\bar{H}_L - \bar{H}_H$ will be negative. If, in addition, one postulates that the solute s stabilizes the L form, then we have a negative contribution from the relaxation term to the enthalpy (as well as to the entropy, by virtue of 7.12.15). In the next section, we present a simple example showing such a stabilization effect.

There are other thermodynamic quantities which, though dependent on the structural changes in the solvent, are not easily explainable in these terms. An important example is the partial molar heat capacity of the solute s, which can be expressed in either one of the following ways:

$$\bar{C}_s = \left(\frac{\partial \bar{H}_s}{\partial T}\right)_{N_W, N_S} = \left(\frac{\partial C}{\partial N_s}\right)_{T, N_W} \qquad (7.12.16)$$

(where C stands for the heat capacity of the system at constant pressure).

Using a TSM for water, one can formally write,

$$\bar{C}_s = C_s^f + (\bar{C}_L - \bar{C}_H)\left(\frac{\partial N_L}{\partial N_s}\right)_{N_W}, \qquad (7.12.17)$$

where

$$C_s^f = \left(\frac{\partial C}{\partial N_s}\right)_{N_L, N_H}. \qquad (7.12.18)$$

The difficulty of interpreting the large value of $\bar{C}_s$ within the TSM is that the quantity C itself can be split into two terms, as in section 5.13. Therefore, the quantity C_s^f already includes contributions due to structure changes in the solvent. The general expression for $\bar{C}_s$ is quite complicated and is omitted here. Instead, we present a simple example to demonstrate an important point. Suppose that the mixture of L and H is ideal (in any

---

† One should be careful with this conclusion. The freezing of the equilibrium is done at the equilibrium point for the system T, P, N_L, N_H, N_S. If on the other hand we freeze the equilibrium at any other point, say, T, P, N_L^0, N_H^0, then of course the solubility in this system is different from the solubility in any equilibrated system.

of the senses discussed in Chapter 6). Also, for simplicity, we assume that s is very dilute in water. The total heat capacity of the system can be written as (section 5.13)

$$C = C^f + (\bar{H}_L - \bar{H}_H)^2 \frac{x_L x_H N_W}{kT^2}, \qquad (7.12.19)$$

where C^f is given by

$$C^f = N_s \left(\frac{\partial H_s^f}{\partial T}\right)_{N_L, N_H, N_S} = N_L \bar{C}_L + N_H \bar{C}_H. \qquad (7.12.20)$$

The partial molar heat capacity of s in the limit $\rho_s \to 0$ is

$$\bar{C}_s^0 = \left(\frac{\partial H_s^f}{\partial T}\right)_{N_L, N_H, N_S} + (\bar{C}_L - \bar{C}_H)\left(\frac{\partial N_L}{\partial N_s}\right)_{N_W} + \frac{(\bar{H}_L - \bar{H}_H)^2}{kT^2}(x_H - x_L)\left(\frac{\partial N_L}{\partial N_s}\right)_{N_W}.$$

$$(7.12.21)$$

The first term on the rhs of (7.12.22) is essentially the contribution of the kinetic degrees of freedom of s. The second term is expected to be small since $\bar{C}_L$ and $\bar{C}_H$ are expected to be of similar magnitude. The important contribution is probably due to the last term on the rhs of (7.12.21). The sign of this term depends on the relative amounts of L and H. Thus, even when we know that s stabilizes the L form, we can still say nothing about the sign of C_s^0 in water. A detailed and explicit example is worked out in the next subsection.

7.12.2. Application of an Interstitial Lattice Model

We extend here the application of the interstitial model for water (section 7.9) to aqueous solutions of simple solutes. The merits of this model are essentially the same as those discussed in section 7.9. As before, we only outline the derivation of the various thermodynamic quantities and leave the details to the reader.

The basic assumptions of the model have been introduced in section 7.9. To adapt the model for aqueous solutions, we further assume that N_s solute molecules occupy the interstitial positions in the framework built up by the L molecules. Only one new molecular parameter is introduced in the new model, i.e., the interaction energy between the solute s and its surroundings, which we denote by E_s.

The total energy of the system is now

$$E_T = N_L E_L + N_H E_H + N_s E_s \qquad (7.12.22)$$

and the corresponding isothermal–isobaric partition function is

$$\Delta(T, P, N_W, N_s) = \sum_{N_L = N_{L_{\min}}}^{N_W} \frac{(N_0 N_L)! \exp[-\beta(N_L E_L + N_H E_H + N_s E_s + PN_L V_L)]}{N_s! N_H! (N_0 N_L - N_H - N_s)!}. \qquad (7.12.23)$$

The combinatorial factor in (7.12.23) is the number of ways in which one can place N_H H molecules and N_s solute molecules into $N_0 N_L$ holes. The summation is carried over all possible volumes of the system, which, by virtue of the assumptions of the model, is the same as a summation over all N_L. The condition $N_H \leq N_0 N_L$ for pure water is replaced by the condition

$$N_H + N_s \leq N_0 N_L. \tag{7.12.24}$$

Hence, the minimum value of N_L is $N_{L_{\min}} = (N_W + N_s)/(N_0 + 1)$.

As in the case of the one-component system, we take the maximal term in the sum (7.12.23), from which we obtain the equilibrium condition

$$\frac{N_H^*(N_0 N_L^*)^{N_0}}{(N_0 N_L^* - N_H^* - N_s)^{N_0 + 1}} = \exp[\beta(E_L - E_H + PV_L)] = K(T, P). \tag{7.12.25}$$

The "mole fractions" of empty holes, holes occupied by H-molecules, and holes occupied by s are

$$y_0 = \frac{N_0 N_L - N_H - N_s}{N_0 N_L}, \qquad y_1 = \frac{N_H}{N_0 N_L}, \qquad y_s = \frac{N_s}{N_0 N_L}. \tag{7.12.26}$$

With these mole fractions, the equilibrium condition is

$$\frac{y_1^*}{(y_0^*)^{N_0 + 1}} = K(T, P), \tag{7.12.27}$$

which has the same form as in the one-component case (Eq. (7.9.34)].

We now evaluate some thermodynamic quantities of this system. The total Gibbs energy of the system with fixed values of N_L, N_H, and N_s is

$$G(T, P, N_L, N_H, N_s) = N_L E_L + N_H E_H + N_s E_s + PN_L V_L$$
$$- kT[N_0 N_L \ln(N_0 N_L) - N_H \ln N_H - N_s \ln N_s$$
$$- (N_0 N_L - N_H - N_s) \ln(N_0 N_L - N_H - N_s)]. \tag{7.12.28}$$

As in the one-component system, this function has a minimum at a point $N_L = N_L^*$ (and $N_H = N_H^*$) which satisfies the condition (7.12.25). The mole fractions of L, H, and s are defined by

$$x_L = \frac{N_L}{N_W}, \qquad x_H = \frac{N_H}{N_W}, \qquad x_s = \frac{N_s}{N_W}. \tag{7.12.29}$$

Note that for any x_s, the addition of solute will always lead to an increase in the mole fraction of the L species. Such an effect can be ascribed to a stabilization effect on the L species by the solute molecules. In this model, the molecular reason for such a stabilization effect is quite obvious. Since we permit s to hold only interstitial sites, their presence compels the H species to vacate some of the holes and transform into L species. This statement will be given a formal proof below.

The total entropy of the system is obtained by differentiating (7.12.28) with respect to temperature

$$S = -kN_0 N_L(y_0 \ln y_0 + y_1 \ln y_1 + y_s \ln y_s). \tag{7.12.30}$$

The total volume, enthalpy, and energy of the system are readily obtainable from (7.12.28) as

$$V = N_L V_L, \qquad H = N_L E_L + N_H E_H + N_s E_s + P N_L V_L, \qquad E = H - PV. \quad (7.12.31)$$

Note that in order to obtain the values of these quantities at equilibrium, one must substitute $N_L = N_L^*$ and $N_H = N_H^*$.

The chemical potentials of the three components are

$$\mu_L = \left(\frac{\partial G}{\partial N_L}\right)_{N_H, N_S}$$

$$= E_L + P V_L - kT[N_0 \ln(N_0/N_L) - N_0 \ln(N_0 N_L - N_H - N_s)] \quad (7.12.32)$$

$$\mu_H = \left(\frac{\partial G}{\partial N_H}\right)_{N_L, N_S}$$

$$= E_H - kT[\ln(N_0 N_L - N_H - N_s) - \ln N_H] \quad (7.12.33)$$

$$\mu_s = \left(\frac{\partial G}{\partial N_s}\right)_{N_W} = \left(\frac{\partial G}{\partial N_s}\right)_{N_L, N_H}$$

$$= E_s + kT[\ln N_s - \ln(N_0 N_L - N_H - N_s)]. \quad (7.12.34)$$

Note that μ_L and μ_H are definable only for a frozen-in system (i.e., we must keep N_H and N_L constant in the respective definitions). On the other hand, the chemical potential of s is definable in both the equilibrated and the frozen-in systems; the equality between the two holds for the case in which $N_L = N_L^*$ and $N_H = N_H^*$ [i.e., when $\mu_L = \mu_H$ or, equivalently, when the equilibrium condition (7.12.25) is fulfilled].

From (7.12.32)–(7.12.34), we can compute the partial molar quantities of the various components. We first evaluate the corresponding quantities for L and H:

$$\bar{S}_L = -\left(\frac{\partial \mu_L}{\partial T}\right)_{N_L, N_H} = k N_0[\ln(N_0 N_L) - \ln(N_0 N_L - N_H - N_s)] \quad (7.12.35)$$

$$\bar{S}_H = -\left(\frac{\partial \mu_H}{\partial T}\right)_{N_L, N_H} = k[\ln(N_0 N_L - N_H - N_s) - \ln N_H] \quad (7.12.36)$$

$$\bar{H}_L = E_L + P V_L, \qquad \bar{H}_H = E_H \quad (7.12.37)$$

$$\bar{V}_L = V_L, \qquad \bar{V}_H = 0 \quad (7.12.38)$$

$$\bar{E}_L = E_L, \qquad \bar{E}_H = E_H. \quad (7.12.39)$$

Note the difference between partial molar (or molecular) quantities on the lhs of (7.12.37)–(7.12.39) and the molecular parameters of the model on the rhs.

We now turn to the corresponding quantities of the solute, definable in either the frozen-in or the equilibrated system. Before doing this, we first evaluate the following derivative:

$$\left(\frac{\partial N_L}{\partial N_s}\right)_{N_W} = -(\mu_{LL} - 2\mu_{LH} + \mu_{HH})^{-1}\left[\frac{\partial(\mu_L - \mu_H)}{\partial N_s}\right]_{N_L, N_H}$$

$$= -\frac{x_L x_H (N_0 N_L - N_H - N_s)}{kT[N_0 + x_s(N_0 x_H - x_L)]}\frac{-kT(N_0 + 1)}{N_0 N_L - N_H - N_s}$$

$$= \frac{x_L x_H (N_0 + 1)}{N_0 + x_s(N_0 x_H - x_L)}, \tag{7.12.40}$$

where, in the first step, we used a thermodynamic identity from section 5.13.10. All the derivatives on the rhs can now be evaluated in the frozen-in system by direct differentiation of (7.12.32) and (7.12.33). Although we have evaluated this quantity for a frozen-in system, at the end of our computation, we must substitute $x_L = x_L^*$ and $x_H = x_H^*$, which satisfies the equilibrium condition (7.12.25) which, in terms of the mole fractions, is written as

$$\frac{x_H^*(N_0 x_L^*)^{N_0}}{(N_0 x_L^* - x_H^* - x_s)^{N_0 + 1}} = K(T, P). \tag{7.12.41}$$

Once we have computed $\partial N_L/\partial N_s$, we can write down all the partial molar quantities of the solute (each quantity must be evaluated at the point $N_L = N_L^*$ and $N_H = N_H^*$)

$$\bar{S}_s = S_s^f + (\bar{S}_L - \bar{S}_H)\left(\frac{\partial N_L}{\partial N_s}\right)_{N_W}$$

$$= -k[\ln N_s - \ln(N_0 N_L - N_H - N_s)] + \left(\frac{\partial N_L}{\partial N_s}\right)_{N_W} k \ln K \tag{7.12.42}$$

$$\bar{H}_s = H_s^f + (\bar{H}_L - \bar{H}_H)\left(\frac{\partial N_L}{\partial N_s}\right)_{N_W}$$

$$= E_s + (E_L - E_H + PV_L)\left(\frac{\partial N_L}{\partial N_s}\right)_{N_W} \tag{7.12.43}$$

$$\bar{V}_s = V_s^f + (\bar{V}_L - \bar{V}_H)\left(\frac{\partial N_L}{\partial N_s}\right)_{N_W}$$

$$= V_L\left(\frac{\partial N_L}{\partial N_s}\right)_{N_W}. \tag{7.12.44}$$

In this model, the frozen-in part of the partial molar volume is zero (i.e., solutes in holes do not contribute to the total volume of the system); hence, all of $\bar{V}_s$ is made up of structural changes in the solvent. Note also that the entropy–enthalpy compensation law is fulfilled in this model.

As an illustrative example, consider the case in which $N_0 = 1$ and we take the limit $x_S \to 0$. The corresponding value of $\bar{H}_s$ is

$$\bar{H}_s^0 = E_s + (E_L - E_H + PV_L)2x_L^*x_H^*. \tag{7.12.45}$$

For this case, we can solve the equilibrium condition (7.12.25) and express x_L and x_H (at equilibrium) in terms of the equilibrium constant K. The result is

$$x_L^* = \frac{1}{2} + \frac{1}{2}\left(\frac{1}{4K+1}\right)^{1/2}. \tag{7.12.46}$$

Hence,

$$x_L^*x_H^* = \frac{K}{(4K+1)}, \tag{7.12.47}$$

and (7.12.45) is rewritten as

$$\bar{H}_s^0 = E_s + (kT\ln K)\frac{2K}{(4K+1)}. \tag{7.12.48}$$

An important point to be noted in (7.12.48) is that the relaxation term is independent of any property of the solute. Therefore, in principle, the two terms in (7.12.48) may be of different orders of magnitude.

As a final application of the interstitial model, consider the partial molar heat capacity of s at infinite dilution. From (7.12.45), we get

$$\bar{C}_s = \left(\frac{\partial \bar{H}_s^0}{\partial T}\right)_{P, N_W, N_s} = \frac{-2(E_L - E_H + PV_L)^2 x_L^* x_H^* (x_L^* - x_H^*)^2}{kT^2}, \tag{7.12.49}$$

which is always negative.

An important moral can be learned from this result. The traditional interpretation of the high positive partial molar heat capacity of inert gases in water is based on the following argument: If the solute s stabilizes the more structured component (here, the L form), then there is more of the "ice-like" form available for melting. From this it is concluded that the large heat capacity of pure liquid becomes even larger upon adding the solute.

The above example shows that this apparently appealing argument is false. The reason is that in general, even if s does stabilize the L form, the heat capacity does not necessarily increase. This point was also noted in subsection 7.12.1.

7.13. THE PROBLEM OF STABILIZATION OF THE STRUCTURE OF WATER BY SIMPLE SOLUTES

One of the most widely applied concepts in the study of aqueous solutions is the structural change of the solvent induced by the addition of a solute. In most cases, one measures a specific quantity for pure water and for aqueous solutions, and then a qualitative inference is made on the change in the structure of water caused by that solute.

For instance, if a solute increases the viscosity of water, it seems reasonable to attribute this effect to the increase in the structure of the solvent. This reasoning led to the introduction of the concept of "structural temperature," which is defined as follows:

Suppose one measures some physical property η for pure water and for an aqueous solution, both at the same temperature T. We denote the change in this property by

$$\Delta\eta = \eta(T, N_S) - \eta(T, N_S = 0). \tag{7.13.1}$$

Next, consider the change of the same property η for pure water caused by a change in temperature

$$\Delta\eta' = \eta(T', N_S = 0) - \eta(T, N_S = 0). \tag{7.13.2}$$

The structural temperature of the solution, with N_S moles of s at temperature T, is defined as the temperature T' for which we have the equality

$$\Delta\eta = \Delta\eta'. \tag{7.13.3}$$

The structural temperature as given here does not involve the concept of the structure of water. The idea underlying this definition is qualitatively clear. It is believed that, however we choose to define the structure of water, this quantity must be a monotonically decreasing function of the temperature. Therefore, a solute producing an effect on η in the same direction as the effect of decreasing the temperature can be said to increase the structure of water, or to decrease the structural temperature of water.

In section 7.6 we elaborated on one possible definition of the structure of water, which may also be applied to aqueous solutions of simple solutes. In this section, we discuss a more general problem. Suppose we classify molecules into quasicomponents by any one of the classification procedures. We then select one of these species and inquire about the change in its concentration upon the addition of a solute. As an example, we may choose one of the species to be the fully hydrogen-bonded molecules; hence, its concentration can serve as a measure of the structure of the solvent.

Let L and H be two quasicomponents obtained by any classification procedure. The corresponding average numbers of L and H species are N_L and N_H, respectively. We assume that the temperature and the pressure are always constant. The quantity of interest is, then, the derivative $(\partial N_L/\partial N_S)_{T,P,N_W}$. We say that the component L is *stabilized* by s if this derivative is positive, i.e., N_L increases upon the addition of the solute.

We now explore the general and exact conditions under which a stabilization of L by s occurs. We can then specialize to a particular choice of the component L and speculate on the possibility of stabilization of the structure of water by the solute s.

A convenient starting relation is the thermodynamic identity (section 5.13.10)

$$\left(\frac{\partial N_L}{\partial N_S}\right)_{N_W} = -(\mu_{LL} - 2\mu_{LH} + \mu_{HH})^{-1}(\mu_{LS} - \mu_{HS}), \tag{7.13.4}$$

where $\mu_{ij} = \partial^2 G/\partial N_i\partial N_j$. The advantage of using relation (7.13.4) is that it transforms a derivative in the equilibrated system [the lhs of (7.13.4)] into derivatives in the frozen-in system; the latter are expressible in terms of molecular distribution functions through the Kirkwood–Buff theory of solution (section 6.7).

Before we plunge into the details of the mathematical derivation, it is instructive to elaborate on the qualitative physical ideas we will be using. Suppose we start with an equilibrated system for which we have the condition

$$\mu_L(N_L, N_H, N_S) = \mu_H(N_L, N_H, N_S). \tag{7.13.5}$$

As an auxiliary device, we envisage a catalyst which, when absent from the system, causes the reaction $L \rightleftarrows H$ to become frozen in. Now, at some given equilibrium state, we remove the catalyst and add dN_s moles of the solute s. Generally, at this stage, we have the inequality

$$\mu_L(N_L, N_H, N_S + dN_S) \neq \mu_H(N_L, N_H, N_S + dN_S). \qquad (7.13.6)$$

For concreteness, suppose that at this stage we have in (7.13.6) the inequality $\mu_L < \mu_H$. Then, if we reintroduce our catalyst, water molecules will "flow" from the state of the high to the state of the low chemical potential. This means that the overall effect of adding dN_S is an increase in number of L species. Thus in order to examine the quantity $(\partial N_L/\partial N_S)_{N_W}$ at equilibrium, one can equivalently examine the change in $\mu_L - \mu_H$ in the frozen-in system. This is the content of the identity (7.13.4).

7.13.1. An Argument Based on the Kirkwood–Buff Theory

We now apply the Kirkwood–Buff theory to reexpress the rhs of (7.13.4) in terms of molecular distribution functions. The basic relation that we need is (section 6.7)

$$\mu_{\alpha\beta} = \frac{kT}{V} \frac{B^{\alpha\beta}}{|\mathbf{B}|} - \frac{\bar{V}_\alpha \bar{V}_\beta}{V \kappa_T} \qquad (7.13.7)$$

where the determinant $|\mathbf{B}|$ for the three-component system is

$$|\mathbf{B}| = \begin{vmatrix} \rho_S + \rho_S^2 G_{SS} & \rho_S \rho_L G_{SL} & \rho_S \rho_H G_{SH} \\ \rho_L \rho_S G_{LS} & \rho_L + \rho_L^2 G_{LL} & \rho_L \rho_H G_{LH} \\ \rho_H \rho_S G_{HS} & \rho_H \rho_L G_{HL} & \rho_H + \rho_H^2 G_{HH} \end{vmatrix}, \qquad (7.13.8)$$

from which we derive the various cofactors $B^{\alpha\beta}$.

From here on, we specialize to the limiting case of $\rho_S \to 0$. In a formal way, we can write the quantity (7.13.4) for any ρ_S in terms of the $G_{\alpha\beta}$. However, the general case is very complicated. Therefore we limit ourselves to very dilute solutions as, indeed, aqueous solutions of inert gases are.

At the limit $\rho_S \to 0$, the partial molar volumes $\bar{V}_L$ and $\bar{V}_H$ and the compressibility of the system tend to their values in pure water, with composition N_L and N_H, i.e.,

$$\bar{V}_L = \frac{1 + \rho_H(G_{HH} - G_{LH})}{\eta} \qquad (7.13.9)$$

$$\bar{V}_H = \frac{1 + \rho_L(G_{LL} - G_{LH})}{\eta} \qquad (7.13.10)$$

$$\kappa_T^f = \frac{\zeta}{kT\eta}, \qquad (7.13.11)$$

where ζ and η are given by (section 6.7):

$$\eta = \rho_L + \rho_H + \rho_L \rho_H(G_{LL} + G_{HH} - 2G_{LH}) \qquad (7.13.12)$$

$$\zeta = 1 + \rho_L G_{LL} + \rho_H G_{HH} + \rho_L \rho_H(G_{LL}G_{HH} - G_{LH}^2). \qquad (7.13.13)$$

Note that in (7.13.11) we have the frozen-in part of the compressibility. This is because we are now working in the frozen-in system, where N_L and N_H are assumed to behave as independent variables. Note also that $\bar{V}_L$ and $\bar{V}_H$ are definable only in the frozen-in system, and therefore there is no need to add any superscript to these symbols.

We will also need the limiting value of the frozen-in partial molar volume of the solute s. This is obtained for a system of any number of c components as follows: Using the Gibbs–Duhem relation, we find

$$0 = \sum_{\beta=1}^{c} \rho_\beta \mu_{\alpha\beta} = \frac{kT}{V|\mathbf{B}|} \sum_{\beta=1}^{c} \rho_\beta B^{\alpha\beta} - \frac{\bar{V}_\alpha}{V\kappa_T} \sum_{\beta=1}^{c} \rho_\beta \bar{V}_\beta. \tag{7.13.14}$$

Now we multiply (7.13.4) by ρ_α and sum over all the species α to obtain (noting that $\sum_{\beta=1}^{c} \rho_\beta \bar{V}_\beta = 1$)

$$\frac{kT}{V|\mathbf{B}|} \sum_{\alpha=1}^{c} \sum_{\beta=1}^{c} \rho_\alpha \rho_\beta B^{\alpha\beta} = \frac{1}{V\kappa_T}, \tag{7.13.15}$$

which provides a general expression for the isothermal compressibility of a multicomponent system. Substituting κ_T in (7.13.14), we get an expression for the partial molar volume

$$\bar{V}_\alpha = \frac{\left(\sum_{\beta=1}^{c} \rho_\beta B^{\alpha\beta}\right)}{\left(\sum_{\alpha=1}^{c} \sum_{\beta=1}^{c} \rho_\alpha \rho_\beta B^{\alpha\beta}\right)}. \tag{7.13.16}$$

For our three-component system of L, H, and s, we get from (7.13.16)

$$V_s^f = \frac{\rho_s B^{SS} + \rho_L B^{LS} + \rho_H B^{HS}}{\rho_s^2 B^{SS} + \rho_L^2 B^{LL} + \rho_H^2 B^{HH} + 2\rho_s\rho_L B^{LS} + 2\rho_s\rho_H B^{HS} + 2\rho_L\rho_H B^{LH}}. \tag{7.13.17}$$

Taking the appropriate cofactors from (7.13.8) and retaining only linear terms in ρ_S in the numerator and the denominator of (7.13.16), we get

$$\lim_{\rho_S \to 0} V_s^f = \left(\rho_S\rho_L\rho_H \begin{vmatrix} 1 + \rho_L G_{LL} & \rho_H G_{LH} \\ \rho_L G_{LH} & 1 + \rho_H G_{HH} \end{vmatrix} - \rho_L^2 \rho_H\rho_S \begin{vmatrix} G_{LS} & \rho_H G_{LH} \\ G_{HS} & 1 + \rho_H G_{HH} \end{vmatrix} \right.$$

$$+ \rho_H^2 \rho_L\rho_S \begin{vmatrix} G_{SL} & G_{SH} \\ 1 + \rho_L G_{LL} & \rho_L G_{LH} \end{vmatrix} \right)$$

$$\times \left(\rho_L^2 \rho_H\rho_S \begin{vmatrix} 1 & 0 \\ \rho_H G_{HS} & 1 + \rho_H G_{HH} \end{vmatrix} + \rho_H^2 \rho_S\rho_L \begin{vmatrix} 1 & 0 \\ G_{LS} & 1 + \rho_L G_{LL} \end{vmatrix} \right.$$

$$\left. - 2\rho_L^2 \rho_H\rho_S \begin{vmatrix} 1 & 0 \\ G_{LS} & \rho_H G_{LH} \end{vmatrix} \right)^{-1}$$

$$= \frac{\zeta - \rho_L G_{LS}(1 + \rho_H G_{HH} - \rho_H G_{LH}) - \rho_H G_{HS}(1 + \rho_L G_{LL} - \rho_L G_{HL})}{\eta}$$

$$= kT\kappa_T^f - \rho_L G_{LS}\bar{V}_L - \rho_H G_{HS}\bar{V}_H. \tag{7.13.18}$$

The last relation for the limiting value of the frozen-in partial molar volume is also the limiting value of the partial molar volume of a solute s in any two-component solvent (see section 6.17).

Next, we express the derivatives on the rhs of (7.13.4) in terms of the $G_{\alpha\beta}$. From (7.13.7), we have

$$\mu_{LS} - \mu_{HS} = \frac{kT}{V}\left[\frac{B^{LS} - B^{HS}}{|\mathbf{B}|} - \frac{V_S^f(\bar{V}_L - \bar{V}_H)}{kT\kappa_T^f}\right]. \tag{7.13.19}$$

Using the notation

$$\Delta_{LH}^S = G_{LS} - G_{HS} \tag{7.13.20}$$

and the identity

$$\bar{V}_L\bar{V}_H = \frac{\zeta - \eta G_{LH}}{\eta^2}, \tag{7.13.21}$$

we get the limiting form of (7.13.19):

$$\lim_{\rho_S \to 0}(\mu_{LS} - \mu_{HS}) = -\frac{kT}{V}\frac{\bar{V}_L - \bar{V}_H + \rho_W\Delta_{LH}^S}{\eta}, \tag{7.13.22}$$

where $\rho_W = \rho_L + \rho_H$. Note that the limiting behavior of the determinant $|\mathbf{B}|$ is

$$|\mathbf{B}| \xrightarrow{\rho_S \to 0} \rho_S\rho_L\rho_H\zeta. \tag{7.13.23}$$

Using relations (6.7.41) and (7.13.22), we get for (7.13.4)

$$\lim_{\rho_S \to 0}\left(\frac{\partial N_L}{\partial N_S}\right)_{N_W} = x_Lx_H[\eta(\bar{V}_L - \bar{V}_H) + \rho_W\Delta_{LH}^S]. \tag{7.13.24}$$

Applying relation (7.9.12), we can rewrite (7.13.24) as

$$\lim_{\rho_S \to 0}\left(\frac{\partial N_L}{\partial N_S}\right)_{N_W} = \rho_Wx_Lx_H[(G_{WH} - G_{WL}) + (G_{LS} - G_{HS})]. \tag{7.13.25}$$

Relation (7.13.25) is very general; it applies to any two-component system at chemical equilibrium and to any classification procedure we have chosen to identify the two quasicomponents.

Let us consider some general implications of (7.13.25) with regard to the conditions for the stabilization of L by s:

1. If either x_L or x_H is very small, then the whole rhs of (7.13.25) is small and we cannot get a large stabilization effect. If we choose, for instance, L to be strictly icelike molecules, then it is likely that x_L will be small; hence, a small stabilization effect will be expected.

2. If we choose two components L and H which are similar in the sense that

$$G_{LS} \approx G_{HS} \quad \text{and} \quad G_{WH} \approx G_{WL}, \tag{7.13.26}$$

then we again end up with a small stabilization effect. This is an important finding. In some *ad hoc* models it is assumed that the mixture of L and H is symmetric ideal. Such an assumption, although it does not necessarily imply extreme similarity of the two components, should be avoided.

3. Suppose that x_Lx_H is not too small and that L and H differ appreciably. The rhs of (7.13.25) will tend to zero as $\rho_W \to 0$.

All the considerations made thus far apply to any fluid. We have seen that a large stabilization effect is attainable only under very restricted conditions. Water, as one of its unique features, may conform to all the necessary conditions leading to a relatively large stabilization effect.

Note that the rhs of (7.13.25) contains two contributions to the stabilization effect. The first depends only on the properties of the solvent $G_{WH} - G_{WL}$ and not on the type of solute. The second term depends on the relative overall affinity of the solute toward the two components L and H.

We can now make a specific choice of two components for water, which seems to be the most useful choice for interpreting the thermodynamic behavior of aqueous solutions. We use the singlet distribution function $x_C(K)$, based on coordination number CN, and define the two mole fractions

$$x_L = \sum_{K=0}^{4} x_C(K), \qquad x_H = \sum_{K=5}^{\infty} x_C(K). \qquad (7.13.27)$$

Clearly, we may refer to L and H as the components of relatively low (L) and high (H) local densities. Another useful choice could be based on the singlet-distribution function for molecules with a different number of hydrogen bonds (see section 7.9). The important thing in this particular choice is that by the definition of the two components, we should have

$$\bar{V}_L - \bar{V}_H > 0 \quad \text{or, equivalently,} \quad G_{WH} - G_{WL} > 0. \qquad (7.13.28)$$

We recall that $\rho_W(G_{WH} - G_{WL})$ measures the excess of water molecules around H as compared to the excess of water molecules around L. This means that if we can find two components which are very different and for which $x_L x_H$ is close to its maximum value, then we have already guaranteed a positive term for the stabilization effect (7.13.25), which is independent of the solute. In fact, one may think of an ideal solute which does not interact with the solvent at all. In such a case, we have $G_{LS} - G_{HS} = 0$, and the whole stabilization effect becomes a property of the solvent only. It is positive for any classification for which (7.13.28) is fulfilled.

The next question concerns the sign of $G_{LS} - G_{HS}$ for simple solutes. From its definition, we have

$$\rho_S \Delta_{LH}^S = \rho_S \int_0^\infty [g_{LS}(R) - 1] 4\pi R^2 \, dR - \rho_S \int_0^\infty [g_{HS}(R) - 1] 4\pi R^2 \, dR$$

$$= \rho_S \int_0^\infty [g_{LS}(R) - g_{HS}(R)] 4\pi R^2 \, dR.$$

$$(7.13.29)$$

The quantity $\rho_S \Delta_{LH}^S$ measures the average overall excess of s molecules in the neighborhood of L relative to H. Again, from the definition of the two components in (7.13.27), we expect that an L molecule, being in a low local density region, will let more solute molecules enter its environment. Therefore Δ_{LH}^S is likely to be positive.

A somewhat different argument may be given as follows. Since both L and H are water molecules, we can assign to each the same hard-core diameter σ_W. If σ_S is the hard-core diameter of the solute, then the integral in (7.13.29) can be replaced by

$$\Delta_{LH}^{S} \approx \int_{\sigma_{WS}}^{\infty} [g_{LS}(R) - g_{HS}(R)]4\pi R^2\, dR. \qquad (7.13.30)$$

Clearly, the probability of s approaching close to an L molecule is larger than that of it approaching close to an H molecule, simply because, by definition, L has more room accessible to the solute. Therefore, we expect that within the correlation distance R_c the quantity Δ_{LH}^{S} will be positive. Another argument may be given in terms of the potential of average force. Suppose that the difference in the potential of average force between the solute and the solvent molecules is small compared with kT in the region $R \geq \sigma_{WS}$. Then we can rewrite (7.13.30) as

$$\Delta_{LH}^{S} = \int \exp[-\beta W_{LS}][1 - \exp(-\beta W_{HS} + \beta W_{LS})]4\pi R^2\, dR$$

$$\approx \beta \int \exp(-\beta W_{LS})(W_{HS} - W_{LS})4\pi R^2\, dR. \qquad (7.13.31)$$

Since H, by definition, must have more water molecules in its neighborhood than L, it is clear that at short distances, near σ_{WS}, $W_{HS}(R)$ will exhibit a repulsive behavior relative to $W_{LS}(R)$. The quantity $W_{HS}(R)$ is the work required to bring s from infinity to a distance R from H. Hence, $W_{HS}(R) - W_{LS}(R)$ is the work required to transfer an s molecule from a distance R from L to a distance R from H. If we consider the range $\sigma_{WS} \leq R \leq 2\sigma_{WS}$, then we expect this work to be positive. Furthermore, since this range of distance is expected to give the major contribution to the integral in (7.13.31), the whole term Δ_{LH}^{S} is expected to be positive as well.

Thus far, we have shown that if we choose two components in such a way so that one has a relatively low local density (L), then this component will be stabilized by the addition of a solute S. This argument applies for any fluid. More important, however, is the peculiar and probably unique coupling between low local density and strong binding energy. The latter is essential to an interpretation of the large, negative enthalpy and entropy of solution discussed in section 7.12. Consider, for example, the relaxation term of $\bar{H}_S$ [because of the compensation relation (7.12.15), the same argument applies to the relaxation part of $\bar{S}_S$].

$$\Delta H_S^r = (\bar{H}_L - \bar{H}_H)\left(\frac{\partial N_L}{\partial N_S}\right)_{N_W}. \qquad (7.13.32)$$

The following conclusion can now be drawn. For any liquid we can construct a TSM and define a low and a high local density component, as we did in (7.13.27). We have seen that for any fluid the solute s shifts the equilibrium $L \rightleftarrows H$ toward the low-density component, i.e., $\partial N_L/\partial N_S > 0$. The unique property of liquid water is that a shift in the direction $H \to L$ is also accompanied with a negative change in energy (or enthalpy). Therefore the product on the rhs of (7.13.32) will be negative for water, whereas it will be positive for any other, normal fluid. We recall that the correlation between low local density and strong binding energy was an essential ingredient in the explanation of the negative temperature dependence of the volume of water, as discussed in section 7.9. In the next subsection we shall present an exact relation between the thermal expansion coefficient and the partial molar enthalpy of a hypothetical solute.

Although we have not explicitly introduced the concept of hydrogen bonds in the present discussion, it is clear that the L component may be identified with, say, fully hydrogen-bonded molecules. In such a case, the addition of solute s is likely to enhance the formation of hydrogen bonds in the system. In sub-section 7.13.3 we discuss an approximate estimate of the amount of HBs formed by the addition of s. Other solvents, such as ammonia or hydrogen fluoride, also have strong hydrogen bonds. However, a fully hydrogen-bonded molecule is not expected to exist in a low local density; hence, a stabilization of hydrogen-bonded molecules by simple solutes is unlikely to occur in these fluids.

7.13.2. An Exact Argument for a Hypothetical Solute

In section 5.11 we derived an exact expression for the work required to create a cavity of radius $r \leq a/2$ in a liquid consisting of hard spheres with diameter a and density ρ. A cavity of radius $a/2$ can be formed by a hard solute of diameter zero. We shall refer to such a particle as a hard point. The solvation Gibbs energy of such a hard point is thus

$$\Delta G_s^* = -kT \ln(1 - \rho V_s^{EX}) \qquad (7.13.33)$$

where V_s^{EX} is the excluded volume of the hard point. Since ρV_s^{EX} is the probability of finding V_s^{EX} occupied, $1 - \rho V_s^{EX}$ is the probability of finding V_s^{EX} empty. Hence ΔG_s^* is always positive. Clearly, this relation can be applied to any liquid. V_s^{EX} is roughly the volume of the region impenetrable by the hard point. Note that the work required to create a cavity of size zero is always zero (see section 5.11); here a hard point always creates a cavity of finite size (provided that the solvent molecules can be assigned some hard-core diameter).

Relation (7.13.33) suggests that we use the hard point particle as our test solute to compare the solvation thermodynamics of this solute in different solvents. We immediately see from (7.13.33) that our test solute will be more soluble in a liquid for which the quantity ρV_s^{EX} is smaller. In other words, decreasing either the density or the size of the solvent particles causes a decrease in ΔG_s^* or an increase in solubility. Here we refer to solubility from an ideal gas phase—see for instance relation (7.10.2).

The corresponding solvation entropy and enthalpy of the hard point are

$$\Delta S_s^* = -\left(\frac{\partial \Delta G_s^*}{\partial T}\right)_p$$

$$= k \ln(1 - \rho V_s^{EX}) - \frac{kT}{1 - \rho V_s^{EX}}\left(\frac{\partial \rho}{\partial T}\right)_p \qquad (7.13.34)$$

$$\Delta H_s^* = \Delta G_s^* + T\Delta S_s^*$$

$$= \frac{-kT^2}{1 - \rho V_s^{EX}}\left(\frac{\partial \rho}{\partial T}\right)_p. \qquad (7.13.35)$$

Note that since the hard point has no "soft" interaction, the average binding energy $\langle B_S \rangle_S$ is zero (as for any hard particle of any size; see section 7.11). Therefore all of ΔH_s^* must be due to structural changes induced in the solvent. This can be interpreted in terms of a relaxation term using any method for classifying the solvent molecules into quasicomponents. The following important conclusion that can be derived from (7.13.35) is independent of any classification procedure. For normal liquids, the density always decreases with temperature, i.e., $\partial\rho/\partial T < 0$ and hence ΔH_s^* is positive. In water, we know that there exists a region between 0 and 4°C for which $\partial\rho/\partial T > 0$. Therefore, ΔH_s^* is negative. Since this must be due to structural changes in the water, we conclude that the unique temperature dependence of the density is related to the unique structural changes in the solvent.

This exact relation has been derived here for a hard point particle. For a real solute, the same conclusion was also inferred from the discussion based on the Kirkwood–Buff theory. Note also that the relaxation term in ΔH_s^* appears likewise in ΔS_s^*, and in the formation of the combination $\Delta H_s^* - T\Delta S_s^*$, it cancels out.

7.13.3. How Much Structural Change Is Induced by the Solute?

We present here an approximation similar to that used in section 7.6 to estimate the amount of structural changes induced by the solute. This estimate is based on the assumption that light and heavy water may be represented by a molecular model with a pair potential of the form (7.4.5). We further assume that light and heavy water have the same pair potential, differing only in the HB parameter ε_{HB}, which we denote by ε_D and ε_H for D_2O and H_2O, respectively. We also assume that the solute s interacts identically with both D_2O and H_2O.

Differentiating the expression for the solvation Gibbs energy

$$\Delta G_S^* = -kT\ln\langle\exp(-\beta B_S)\rangle_0 \tag{7.13.36}$$

with respect to the parameter ε_{HB} we obtain

$$\left(\frac{\partial\Delta G_S^*}{\partial\varepsilon_{HB}}\right)_{T,P,N} = \langle N_{HB}\rangle_S - \langle N_{HB}\rangle_0, \tag{7.13.37}$$

where on the rhs of (7.13.37) we have the same quantity that appeared in (7.11.10). (There the averages were in the T, V, N ensemble whereas here they are in the T, P, N ensemble.) As in section 7.6, assuming that $\varepsilon_D - \varepsilon_H$ is small enough, we can expand ΔG_S^* to first order in $(\varepsilon_D - \varepsilon_H)$ to obtain

$$\Delta G_S^*(D_2O) - \Delta G_S^*(H_2O) = (\langle N_{HB}\rangle_S - \langle N_{HB}\rangle_0)(\varepsilon_D - \varepsilon_H). \tag{7.13.38}$$

The estimate of the average change in the structure of water depends on the experimental values of the solvation Gibbs energies and an estimate of the difference $\varepsilon_D - \varepsilon_H$. Alternatively, we can use the experimental or the theoretical evidence that simple solutes do increase the structure of water to conclude that

$$\left(\frac{\partial\Delta G_S^*}{\partial(-\varepsilon_{HB})}\right)_{T,P,N} < 0. \tag{7.13.39}$$

This means that increasing the HB energy ($-\varepsilon_{HB} > 0$) will decrease ΔG_S^* (or increase the solubility of s). This finding helps to resolve the apparent puzzling fact that the solubility of small inert solute in D_2O is *larger* than in H_2O. Traditionally, the low solubility of simple solutes in water is attributed to the high degree of structure of water. If that is true, then one would expect that the solubility in D_2O will be even lower than in H_2O. The inequality (7.13.39) shows that the contrary conclusion is correct; i.e., the solubility will increase with the degree of structure. We have already seen in section 7.6 that increasing $-\varepsilon_{HB}$ will cause an increase in the structure of the water. Therefore, from (7.13.39) we can also conclude that

$$\left(\frac{\partial \Delta G_S^*}{\partial \langle N_{HB} \rangle}\right)_{P,T,N} < 0 \qquad (7.13.40)$$

for solutes s that are known to increase the structure of water.

As an example of the application of (7.13.38), we use the solvation Gibbs energy of argon in H_2O and D_2O to give, at 25°C,

$$\Delta G_S^*(D_2O) - \Delta G_S^*(H_2O) = 8.15 - 8.40 = -0.25 \text{ kJ mol}^{-1}; \qquad (7.13.41)$$

for an estimate of $\varepsilon_D - \varepsilon_H \sim 0.93 \text{ kJ mol}^{-1}$ (see section 7.6), we find

$$\langle N_{HB} \rangle_S - \langle N_{HB} \rangle_0 = 0.27. \qquad (7.13.42)$$

As in section 7.6, this value seems to be quite low. The main reason is that the first-order expansion (7.13.38) is probably unjustified.

We note that if we take larger solutes such as krypton and xenon, we obtain a negative value for $\langle N_{HB} \rangle_S - \langle N_{HB} \rangle_0$; i.e., these solutes destabilize the structure of water.

7.14. SOLVENT-INDUCED INTERACTIONS AND FORCES

In the previous four sections we have dealt with some aspects of very dilute aqueous solutions. From the formal point of view, it is sufficient to study the solvation properties of one solute s in a pure solvent. We now proceed to the next step and study a pure solvent with two solutes. In the absence of a solvent, two-particles-in-a-system determines the second virial coefficient in the density expansion of the pressure (section 1.8). Likewise, two-solute-in-a-solvent determines the second virial coefficient of the osmotic pressure (section 6.11). This quantity is expressed in terms of the pair correlation function by

$$B_2^*(T, \mu_w) = \frac{-1}{2(8\pi^2)^2 V} \int [g_{ss}^0(\mathbf{X}_1, \mathbf{X}_2) - 1] \, d\mathbf{X}_1 \, d\mathbf{X}_2, \qquad (7.14.1)$$

where g_{ss}^0 is the solute–solute pair correlation function in a pure solvent w at a given temperature T and chemical potential μ_w. In this and in the following sections we shall always assume that $g_{ss}^0 \to 1$ when the solute–solute distance is very large, $R \to \infty$. This behavior is guaranteed either by working in the open system or by taking the infinite-system limit of g_{ss}^0 when g_{ss}^0 is the closed-system pair correlation function (see section 6.7 for more details).

In general we shall be interested in the six-dimensional function $g_{ss}^0(\mathbf{X}_2)$ where $\mathbf{X}_2 = \mathbf{R}_2, \Omega_2$ are measured relative to a fixed configuration of the first solute. In (7.14.1), however, we can integrate over $\mathbf{X}_1$ and over all orientations of the second solute to obtain

$$B_2^* = -\frac{1}{2} \int_0^\infty [g_{ss}^0(R) - 1] 4\pi R^2 \, dR. \qquad (7.14.2)$$

If the system is not infinitely dilute with respect to s, then the analogue quantity is the Kirkwood–Buff integral

$$G_{ss} = \int_0^\infty [g_{ss}(R) - 1]4\pi R^2\, dR, \tag{7.14.3}$$

where now $g_{ss}(R)$ is the spatial pair correlation function for two solutes in a system containing any concentrations of s and w.

The relation between the two quantities is

$$B_2^*(T, \mu_w) = -\tfrac{1}{2}G_{ss}^0 = \lim_{\rho_s \to 0} (-\tfrac{1}{2}G_{ss}). \tag{7.14.4}$$

Instead of studying the pair correlation function we can equivalently study the corresponding potential of average force, i.e.,

$$g_{ss}(\mathbf{X}_1, \mathbf{X}_2) = \exp[-\beta W_{ss}(\mathbf{X}_1, \mathbf{X}_2)]. \tag{7.14.5}$$

We recall that g_{ss} is related to the probability of finding the two solutes at some specific configuration $\mathbf{X}_1$, $\mathbf{X}_2$. W_{ss} is the work required to bring two solute particles from fixed configurations at infinite separation to a fixed configuration $\mathbf{X}_1$, $\mathbf{X}_2$. The statistical mechanical expression for W_{ss} is

$$W_{ss}(\mathbf{X}_1, \mathbf{X}_2) = U_{ss}(\mathbf{X}_1, \mathbf{X}_2) - kT \ln \frac{\langle \exp[-\beta B_{ss}(\mathbf{X}_1, \mathbf{X}_2)]\rangle_0}{\langle \exp[-\beta B_{ss}(\infty)]\rangle_0}$$

$$= U_{ss}(\mathbf{X}_1, \mathbf{X}_2) + \delta W(\mathbf{X}_1, \mathbf{X}_2), \tag{7.14.6}$$

where $U_{ss}(\mathbf{X}_1, \mathbf{X}_2)$ is the direct pair potential for the two solutes and δW is the indirect or the solvent-induced interaction. The latter can be expressed in terms of two averages of the quantity $\exp[-\beta B_{ss}(\mathbf{X}_1, \mathbf{X}_2)]$ over all configurations of the "solvent" molecules. Note that if we have strictly two solute particles then the average $\langle\ \rangle_0$ is over all solvent molecules. However, if the system contains any number of s and w molecules, then the average $\langle\ \rangle_0$ is reinterpreted as an average over all molecules in the system except the two specific or tagged particles, 1 and 2.

As in previous sections, we found it useful to refer to a specific, or tagged, particle as a solvaton. The solvaton is a particular particle of the system (of any species) which we have chosen to place at a fixed position in order to study its solvation properties. Likewise, in (7.14.6) we have two average quantities which have the same form as the ones encountered in the study of solvation, but now we have two particles instead of one. Therefore, it is convenient to introduce here the concept of a pair of solvatons of species s. These are identical with any other molecules of the same species except that we require that they do not interact with each other. With this concept we can say that U_{ss} is the work required to bring two solutes from infinite separation to the final configuration $\mathbf{X}_1$, $\mathbf{X}_2$ in vacuum. Likewise, δW is the work required to bring the two solvatons from infinite separation to the final configuration $\mathbf{X}_1$, $\mathbf{X}_2$ within the solvent. The process could be carried out at T, V, or T, P, or T, μ_w constant.

It should be stressed that the solvaton, or tagged particle, is purely a theoretical concept. In practice we cannot follow a specific particle or pair of particles. However, theoretically we can do that. We can always think of a solvaton as an external field operating on our system. For instance, in the T, V, N ensemble, the solvation Helmholtz energy can be viewed as a ratio of two partition functions

$$\exp(-\beta \Delta A_s^*) = \frac{\int d\mathbf{X}^N \exp[-\beta U(\mathbf{X}^N) - \beta B_s(\mathbf{R}_0, \mathbf{X}^N)]}{\int d\mathbf{X}^N \exp[-\beta U(\mathbf{X}^N)]}. \tag{7.14.7}$$

In the denominator we have the configurational PF of the T, V, N system, whereas in the numerator we have the configurational PF of the same T, V, N system subjected to an "external" field of force, produced by the solvaton at $\mathbf{R}_0$.

Likewise, δW in (7.14.6) can be written in the T, V, N ensemble as

$$\exp[-\beta \delta W(\mathbf{X}_1, \mathbf{X}_2)] = \frac{\int d\mathbf{X}^N \exp[-\beta U(\mathbf{X}^N) - \beta B_{ss}(\mathbf{X}_1, \mathbf{X}_2, \mathbf{X}^N)]}{\int d\mathbf{X}^N \exp[-\beta U(\mathbf{X}^N) - \beta B_{ss}(\infty, \mathbf{X}^N)]} \tag{7.14.8}$$

where, again, the two integrals in (7.14.8) are the configurational PFs of the same T, V, N system as above subjected to an "external" field of force produced by the two solvatons, once at the configuration $\mathbf{X}_1$, $\mathbf{X}_2$, and once at some fixed configuration but at infinite separation (denoted by ∞).

In most of our study of solvent effects in Chapter 8 and in the remaining sections of this chapter, we shall focus on the quantity δW rather than on the full potential of average force. This is done mainly for the following reasons. First, the direct interaction U_{ss} is presumed to be known; therefore, any new features due to the solvent are contained in δW. Second, having excluded the direct interaction, the function δW is a smooth function of the distance, including distances that are inaccessible to real particles. Finally, as seen in (7.14.8), δW is expressible in terms of a ratio of two solvation works of a pair of particles. This suggests an approximate expression for δW in terms of solvation works of real particles.

The probability distribution corresponding to δW is defined by

$$y_{ss}(\mathbf{X}_1, \mathbf{X}_2) = \exp[-\beta \delta W(\mathbf{X}_1, \mathbf{X}_2)]$$
$$= g_{ss}(\mathbf{X}_1, \mathbf{X}_2) \exp[+\beta U_{ss}(\mathbf{X}_1, \mathbf{X}_2)]. \tag{7.14.9}$$

If ρ_s is the density of the solute s, then the probability of finding a pair of s particles at configuration $\mathbf{X}_1$, $\mathbf{X}_2$ is

$$\rho_{ss}^{(2)}(\mathbf{X}_1, \mathbf{X}_2) \, d\mathbf{X}_1 \, d\mathbf{X}_2 = \rho_s^2 g_{ss}(\mathbf{X}_1, \mathbf{X}_2) \, d\mathbf{X}_1 \, d\mathbf{X}_2. \tag{7.14.10}$$

Thus $g_{ss}(\mathbf{X}_1, \mathbf{X}_2)$ is the probability of finding the two s particles at $\mathbf{X}_1$, $\mathbf{X}_2$ relative to the probability of finding the same two particles at the same elements of volume $d\mathbf{X}_1 \, d\mathbf{X}_2$ but at infinite separation, i.e.,

$$g_{ss}(\mathbf{X}_1, \mathbf{X}_2) = \frac{\rho_{ss}^{(2)}(\mathbf{X}_1, \mathbf{X}_2) \, d\mathbf{X}_1 \, d\mathbf{X}_2}{\rho_{ss}^{(2)}(\infty) \, d\mathbf{X}_1 \, d\mathbf{X}_2}. \tag{7.14.11}$$

Likewise, the function $y_{ss}(\mathbf{X}_1, \mathbf{X}_2)$ may be interpreted as the relative probability of finding the two solvatons at $\mathbf{X}_1, \mathbf{X}_2$. This function is sometimes referred to as a cavity–cavity distribution function. This term might be misleading for two reasons. First, our solvatons exert an "external" field of force which has both attractive and repulsive parts. It is only for hard-sphere solvatons that the "external" field is equivalent to a cavity. Second, even for hard-sphere particles that are in some sense equivalent to suitable cavities, one must be careful when reference is made to the probability of finding a single or a pair of cavities. The reason for that has been discussed in section 5.10.

It is therefore safer to work with either the full potential of average force or the solvent-induced part δW. The first involves the work of bringing two real solutes from infinite separation to some final configuration. The second involves the work of the same process for a pair of solvatons (i.e., excluding the direct interaction). The corresponding probabilities were defined above without reference to cavities; i.e., the first is a relative probability of finding two real solutes at two specific configurations [Eq. (7.14.11)], the second being the relative probability for the corresponding pair of solvatons.

From now on we shall treat the extreme dilute solution of s in w. We shall drop the subscripts s whenever we discuss the potential of average force between two solute particles. Suppose we have two simple solutes s and we are interested in the solvent-induced interaction at some distance $R = \sigma$, where σ is of the order of the molecular diameter of the solute. Relation (7.14.6) can be written as

$$W(\sigma) = U(\sigma) + \Delta G^*(\sigma) - \Delta G^*(\infty)$$

$$= U(\sigma) + \Delta G^*(\sigma) - 2\Delta G_s^*, \qquad (7.14.12)$$

where $\Delta G^*(\sigma)$ is the solvation Gibbs energy of the pair of solvatons at a distance σ and $\Delta G^*(\infty)$ is the solvation Gibbs energy of the same pair but at infinite separation. The latter can be written as twice the solvation Gibbs energy of one solvaton s. This is the same as writing the average in the denominator of (7.14.6) as a product of two averages.

From Eq. (7.14.12) we have for the solvent-induced interaction

$$\delta G = W(\sigma) - U(\sigma) = \Delta G^*(\sigma) - 2\Delta G_s^*. \qquad (7.14.13)$$

This expression can be interpreted in terms of a cyclic process as depicted in Fig. 7.17.

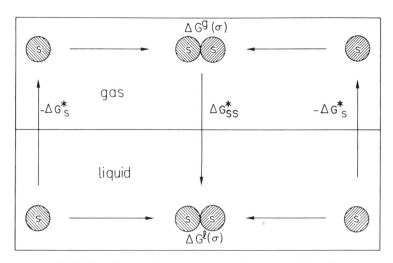

FIGURE 7.17. A cyclic process corresponding to Eq. (7.14.13).

It also suggests a way of computing δG from experimentally measurable quantities. ΔG_s^* is a measurable quantity, whereas $\Delta G^*(\sigma)$ is not. We shall discuss an approximate way of measuring δG in section 7.16.

The force acting on say particle 1 at $\mathbf{R}_1$ given that the second particle is at $\mathbf{R}_2$ is (section 6.4)

$$\mathbf{F}_1 = -\nabla_1 U(\mathbf{R}_1, \mathbf{R}_2) - \nabla_1 \delta G(\mathbf{R}_1, \mathbf{R}_2)$$

$$= -\nabla_1 U(\mathbf{R}_1, \mathbf{R}_2) - \int d\mathbf{X}_w \, \nabla_1 U_{sw}(\mathbf{R}_1, \mathbf{X}_w)\rho(\mathbf{X}_w/\mathbf{R}_1, \mathbf{R}_2). \qquad (7.14.14)$$

The solvent-induced force should be compared with the indirect force of section (3.3.6). Here, we could formally take the gradient with respect to $\mathbf{R}_1$ of the integral in the denominator of (7.14.6). But because of the homogeneity of the system, the average force on particle 1 infinitely separated from particle 2 is zero. This is different from the result obtained in section 3.3 where we had two terms in the indirect force [Eq. (3.3.75)].

The integrand on the rhs of (7.14.14) contains two factors: the forces exerted by a solvent molecule at $\mathbf{X}_w$ on the particle at $\mathbf{R}_1$ and a conditional density of solvent molecules, given two particles at $\mathbf{R}_1$, $\mathbf{R}_2$. Clearly, if $|\mathbf{R}_2 - \mathbf{R}_1|$ is very large, then the average indirect force will be zero (Fig. 7.18). Therefore we can expect to obtain significant solvent induced forces only when $|\mathbf{R}_2 - \mathbf{R}_1|$ is of the order of a few molecular diameters (Fig. 7.18). Suppose for concreteness that we have $|\mathbf{R}_2 - \mathbf{R}_1| = \sigma$. If $|\mathbf{R}_w - \mathbf{R}_1|$ is large, then the force $-\nabla_1 U_{sw}(\mathbf{R}_1, \mathbf{X}_w)$ will be negligible. Therefore the only significant contribution to the integral (7.14.14) comes when both $|\mathbf{R}_1 - \mathbf{R}_w|$ and $|\mathbf{R}_1 - \mathbf{R}_2|$ are not too large. The first condition is required to produce a large force exerted by w. The second condition is required to produce an asymmetry around the particle at $\mathbf{R}_1$.

In treating some specific examples in the next sections, we shall sometimes factor out a term that depends on the cavities of the particles and the remaining term depending on the specific interaction between the solute and the solvent. The division into the two contributions is exactly the same as in section 6.14.1. The binding energy of the solute will be written as

$$B_s = B_s^H + B_s^S, \qquad (7.14.15)$$

where B_s^H is the hard part of the binding energy of the solute s and B_s^S is the soft part of the binding energy. For a more complex solvaton, e.g., water, we shall still use B_s^S to

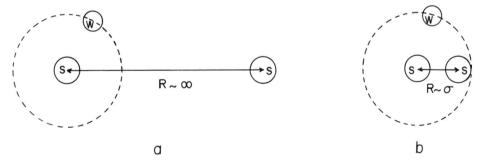

$$a \qquad\qquad\qquad\qquad\qquad b$$

FIGURE 7.18. (a) Two solute particles at infinite separation $R \sim \infty$. The solvent-induced force is zero since no water molecule can be simultaneously close to both solutes. (b) Two solutes at a short distance $R \approx \sigma$. A water molecule can be close to both solutes. Hence a solvent-induced force may be realized.

indicate the specific binding energy, say the HBing interaction with the environment. Applying (7.14.15) to each of the average quantities (7.14.6), we can write (see section 6.14.1)

$$\delta W(\mathbf{X}_1, \mathbf{X}_2) = \delta W^H(\mathbf{X}_1, \mathbf{X}_2) + \delta W^{S/H}(\mathbf{X}_1, \mathbf{X}_2)$$

$$= -kT \ln \frac{\langle \exp(-\beta B^H_{ss}) \rangle_0}{\langle \exp[-\beta B^H_{ss}(\infty)] \rangle_0} - kT \ln \frac{\langle \exp(-\beta B^S_{ss}] \rangle_H}{\langle \exp[-\beta B^S_{ss}(\infty)] \rangle_H}. \quad (7.14.16)$$

The first term on the rhs of (7.14.16) is the work required to bring the two cavities from infinite separation to the final configuration $\mathbf{X}_1$, $\mathbf{X}_2$. The second term is the conditional work corresponding to the soft part of the binding energy. A similar division applies to particles interacting through hydrogen binding, e.g., see section 7.16.

7.15. TWO SIMPLE NONPOLAR SOLUTES: HYDROPHOBIC INTERACTION

Hydrophobic (HϕO) interaction is defined here as the indirect or solvent-induced part of the potential of average force between two nonpolar or HϕO solvatons. This should be distinguished from the concept of HϕO solvation or HϕO hydration, which is concerned with the solvation properties of a single HϕO solvaton (section 7.10).

The simplest solvatons are two hard spheres in water. If we are in a T, P, N system, the corresponding work is the Gibbs energy change

$$\delta G(R) = -kT \ln \frac{\langle \exp[-\beta B^H(R)] \rangle_0}{\langle \exp[-\beta B^H(\infty)] \rangle_0}$$

$$= \Delta G^{*H}_{ss}(R) - 2\Delta G^{*H}_s. \quad (7.15.1)$$

Thus, the work required to bring two such hard spheres to a final distance R is equal to the difference in the solvation Gibbs energy of the pair ss at R (viewed as a single molecular entity) and the solvation Gibbs energy of two single hard spheres.

Since in the definition of δG we eliminated the direct hard interaction between the two solvatons, this is the same as the work required to bring two suitable cavities to a distance R. Clearly, two cavities can be brought to any distance R, including distances $R < \sigma$ that are inaccessible to the hard particles. This is the physical reason why the function $\delta G(R)$ is a smooth function of R for any $R < \sigma$. In particular, for $R = 0$, the two fused cavities at a single point are, from the point of view of the solvent, equivalent to a single cavity. Recall that a cavity is a stipulation excluding the centers of all solvent particles from a certain region. Clearly, a double stipulation on the same region is the same as one stipulation. Therefore, at $R = 0$, (7.15.1) reduces to

$$\delta G(R = 0) = \Delta G^{*H}_s - 2\Delta G^{*H}_s = -\Delta G^{*H}_s. \quad (7.15.2)$$

This result is intuitively clear. Starting with two cavities at infinite separation, $\delta G(R = 0)$ is the work required to bring the two cavities to zero separation. This is the same as the work for eliminating one cavity, or equivalently the work of desolvating a hard-sphere solvaton.

Because of this particular relation between HϕO interaction and (negative) HϕO solvation, we might say that this process is a reversal of the solubility of one hard sphere. Since the solvation of a hard sphere is large and positive, $\delta G(R = 0)$ will be large and negative (compare with some "normal" liquids). Note, however, that the result (7.15.2) is valid only for hard-sphere solvatons and only at zero separation. We shall see in section 7.18 another example where $H\phi O$ interaction is related to reversal of solvation.

For any other pair of spherical solvatons we can use the general relation (7.14.16) applied to $R = 0$ to obtain (using 7.15.2)

$$\delta G(R = 0) = \delta G^H(R = 0) + \delta G^{S/H}(R = 0)$$

$$= -\Delta G_s^{*H} + \Delta G_{ss}^{*S/H}(R = 0) - 2\Delta G_s^{*S/H}$$

$$= \Delta G_{2s}^* - 2\Delta G_s^*. \tag{7.15.3}$$

Thus, applying (7.14.16) and (7.15.1) for simple spherical solvatons, the hard part of δG behaves according to (7.15.2); on the other hand, the soft part of δG is the difference in the work of coupling twice the strength of the soft interaction minus twice the coupling work for each soft interaction. Combining the hard and the soft parts, we obtain the last form on the rhs of (7.15.3). Intuitively this result is clear. Bringing two solvatons to $R = 0$ is equivalent to eliminating the two solvatons $(-2\Delta G_s^*)$ and creating one "double" solvaton (ΔG_{2s}^*). This "double" solvaton denoted by "$2s$" has the same hard-core diameter as s, but the soft binding energy is twice as strong. Mathematically,

$$\delta G(R = 0) = -kT \ln \frac{\langle \exp(-\beta 2 B_s) \rangle_0}{\langle \exp(-\beta B_s) \rangle_0^2}. \tag{7.15.4}$$

Relation (7.15.3) suggests an approximate way of estimating the molecular quantity $\delta G(R = 0)$ in terms of experimental quantities. Ideally, if we can find two solutes having the same diameter but differing in their soft interaction as required in (7.15.4), we could have calculated $\delta G(R = 0)$ exactly. Practically, it might be possible to find such a pair of solutes.

Next we try to estimate δG at some other distance of the order of the molecular diameter of the solute. Again, we exploit the fact that δG is independent of the direct interaction between the two specific particles, so that it can be studied at any distance ρ, including $R < \sigma$.

Consider two methane molecules, which can be viewed as spheres of diameter $\sigma \sim 3.82$ Å. We now choose a particular distance $R = \sigma_1 = 1.53$ Å, which is the C—C equilibrium distance in ethane. At this particular distance, two methane molecules will have a binding energy (to the solvent) which is approximately equal to the binding energy of one ethane molecule.

Mathematically we assume that

$$B_{ss}(R = \sigma_1) \approx B_{Et} \tag{7.15.5}$$

where B_{Et} is the binding energy of ethane. Therefore, we have

$$\delta G(R = \sigma_1) = -kT \ln \frac{\langle \exp[-\beta B_{ss}(R = \sigma_1)] \rangle_0}{\langle \exp(-\beta B_s) \rangle_0^2}$$

$$\approx -kT \ln \frac{\langle \exp(-\beta B_{Et}) \rangle_0}{\langle \exp(-\beta B_{Me}) \rangle_0^2}$$

$$= \Delta G_{Et}^* - 2\Delta G_{Me}^*. \tag{7.15.6}$$

What we have achieved in (7.15.6) is an approximate measure of the molecular quantity $\delta G(\sigma_1)$ in terms of two measurable quantities: the solvation Gibbs energies of ethane

(Et) and of methane (Me). This approximation was rendered possible because of the exclusion of the direct interaction between the two methane molecules. In other words, the introduction of the two hypothetical solvatons, which are purely theoretical concepts, allowed us to reach a meaningful relation between δG and experimental quantities pertaining to real solutes, methane and ethane. The main trick is to view the ratio of the integrals in (7.15.6) as the ratio of the two PFs of *pure* water subjected to an "external" field of force produced by different solutes. The solvent recognizes the solute only through the field of force it feels. Therefore replacing two partially fused methane solvatons by one ethane molecule in (7.15.6) is approximately unnoticed by the solvent.

Figure 7.19 shows some typical values of $\delta G(\sigma_1)$ for water and some "normal" solvents. Two prominent differences between water and the other solvents are clearly conspicuous. First, $|\delta G(\sigma_1)|$ is larger in water than in the other solvents. Second, the temperature dependence is clearly negative and large in water as compared to almost temperature independence in the other solvents. Thus the quantity $\delta G(\sigma_1)$ can be useful for comparing the HϕO interaction between two simple solutes in different solvents. It does not tell us, however, what is the value of δG, say at $R \approx \sigma$.

It is known that the function $y(R) = \exp[-\beta \delta G(R)]$ is a steeply decreasing function of R for $0 \lesssim R \lesssim \sigma$. One example of this behavior was discussed in section 5.4. Therefore we expect that δG at $R \approx \sigma$ will be smaller than the values obtained at $R = \sigma_1$. In the following sections, we shall discuss some possible extensions of the same trick involved in (7.15.6) to obtain a measure of the HϕO interaction at more realistic distances as well as to improve the approximation involved in this example. In Chapter 8 we shall use the same method to study some aspects of the solvation of biochemical molecules.

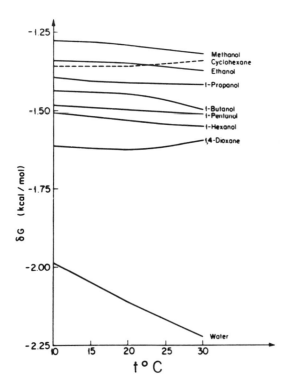

FIGURE 7.19. Values of $\delta G(\sigma_1)$, Eq. (7.15.6), as a function of temperature in various solvents.

Relation (7.15.6) might look like a simple relation between thermodynamic quantities. However, this relation cannot be obtained by purely thermodynamic reasoning. The very definition of δG involves a molecular process. The work associated with this process involves two average quantities that are as difficult to calculate as a partition function of any liquid. The role played by statistical thermodynamics was to relate these integrals to experimental quantities. In a way we let nature perform these integrals and report the results to us in terms of experimental quantities, which we then use to estimate the required molecular quantity δG. We shall encounter many other examples of this methodology in subsequent sections.

Relation (7.15.6) is also interesting from the following conceptual point of view. The potential of average force between two simple solutes is a two-particle-in-a-system property. Simple solutes such as argon or methane are extremely insoluble in water. Therefore it is difficult to measure deviations from dilute ideal behavior. For instance, to measure B_2^* would require us to study a system at very high pressures. Besides, B_2^* gives an integral over $g(R)$ or $W(R)$, not these functions themselves. Relation (7.15.6) is remarkable since it does provide an estimate of the value of δG at one distance, in terms of measurable one-particle-in-a-system properties, i.e., solvation Gibbs energies at infinite dilution.

The ratio of the average quantities can of course be computed by any of the simulation techniques based on some specific model potential for water molecules. Figure 7.20 shows one such a result obtained by Monte Carlo calculation of the potential of average force for two methane molecules in water at 25°C.

As expected, the first minimum occurs at about the distance $R \approx \sigma$, where $\sigma \approx 4.4$ Å was used as the diameter of methane. The second minimum occurs at about 6 Å. This is considerably closer than the distance one would have expected in a normal solvent (see section 6.3) i.e., a distance of about $4.4 + 3 \approx 7.4$ Å. The reason for obtaining this location of the second minimum will be further discussed in section 7.17.

The solvent-induced force between the two HϕO solvatons is (sections 6.4 and 7.14)

$$\mathbf{F}_1 = -\int \nabla_1 U_{sw}(\mathbf{R}_1, \mathbf{X}_w)\rho(\mathbf{X}_w/\mathbf{R}_1, \mathbf{R}_2)\, d\mathbf{X}_w$$

$$= -\int \nabla_1 U_{sw}(\mathbf{R}_1, \mathbf{X}_w)\rho_w g(\mathbf{X}_w|\mathbf{R}_1, \mathbf{R}_2)\, d\mathbf{X}_w, \qquad (7.15.7)$$

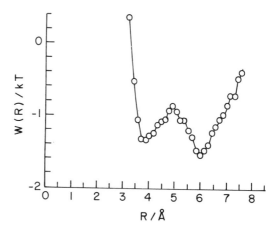

FIGURE 7.20. Potential of average force between two methane molecules in water, obtained by Monte Carlo simulation. Redrawn with changes from Ref. (1).

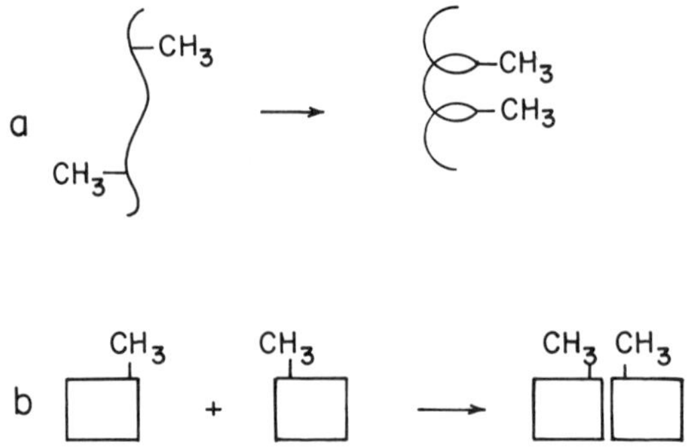

a

b

FIGURE 7.21. Two HϕO groups (methyl) are brought from a large to a close distance. (a) Intramolecular HϕO interaction. (b) Intermolecular HϕO interaction.

where ρ_w is the bulk density of water and $g(\mathbf{X}_w|\mathbf{R}_1, \mathbf{R}_2)$ is the pair correlation function between a water molecule at $\mathbf{X}_w$ and a pair of solutes at $\mathbf{R}_1$, $\mathbf{R}_2$ viewed as a single entity. Note that $g(\mathbf{X}_w|\mathbf{R}_1, \mathbf{R}_2)$ is different from the triplet correlation function $g(\mathbf{X}_w, \mathbf{R}_1, \mathbf{R}_2)$. The relation between the two is $g(\mathbf{X}_w|\mathbf{R}_1, \mathbf{R}_2) = g(\mathbf{X}_w, \mathbf{R}_1, \mathbf{R}_2)/g(\mathbf{R}_1, \mathbf{R}_2)$.

Simulation results indicate that the correlation functions between a water molecule and one or two simple solutes is of the order of two or three. This is similar to typical values in normal liquids. Furthermore, the force exerted by a water molecule on a simple solute is not expected to be much larger than the force exerted by other molecules of a similar size. Therefore, the integral in (7.15.7) does not indicate any source of strong HϕO force. This has indeed been observed by direct simulation of the force between two simple solutes. We can also estimate that this force is weak from the slope of the curve of Fig. 7.20.

Up to this point we have discussed the HϕO interaction and the corresponding force between two simple nonpolar solutes. A quantity of interest in biopolymers is the HϕO interaction and the corresponding force between two groups, say methyl groups, attached to a protein. We shall refer to this case as intramolecular HϕO interaction. Strictly, the two groups are not necessarily attached to the same polymer, but because of the presence of the polymer, the solvent induced interaction between the two groups is different from the interaction between the same groups in a pure liquid. Figure 7.21 shows two examples of such HϕO interaction between two methyl groups attached to a polymer.

The mathematical formulation of the intramolecular HϕO interaction is the same as before with one additional condition: the presence of the polymer. This condition can significantly perturb the structure of water in the surroundings of the two methyl groups. This perturbation will in turn affect the resulting induced interaction. Thus we can apply all the equations of this section, with reinterpretation of the average quantities as conditional averages.

In order to estimate the intramolecular HϕO interaction, consider a carrier molecule having two nonpolar groups, the relative positions of which can be changed. In one isomer, the two groups are far apart, in the sense that their solvation spheres do not overlap. In the second isomer, the two groups are close to each other and their solvation spheres overlap. A schematic process of this kind is depicted in Fig. 7.22. We denote the two isomers by A and B. The Gibbs energy change for the process $A \rightarrow B$ is written as

$$\Delta G(A \rightarrow B) = \Delta U(A \rightarrow B) - kT \ln \frac{\langle \exp(-\beta B_B)\rangle_0}{\langle \exp(-\beta B_A)\rangle_0}, \qquad (7.15.8)$$

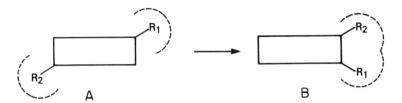

FIGURE 7.22. A schematic model in which two alkyl groups R_1 and R_2 are initially far apart in A and close to each other in B.

where ΔU is the energy change for the same process carried out in vacuum. (This includes the difference in the bond energies of the two groups in the two positions as well as direct interaction between the two groups in the isomer B. The direct interaction between the two groups in A is presumed to be small.) B_B and B_A are the total binding energies of the B and A isomers to the solvent, respectively. The averages are over all the configurations of the solvent molecules, here in the T, P, N ensemble.

The solvent-induced contribution to $\Delta G(A \rightarrow B)$ is defined by

$$\delta G(A \rightarrow B) = \Delta G(A \rightarrow B) - \Delta U(A \rightarrow B)$$

$$= -kT \ln \frac{\langle \exp(-\beta B_B) \rangle_0}{\langle \exp(-\beta B_A) \rangle_0}$$

$$= \Delta G_B^* - \Delta G_A^*. \tag{7.15.9}$$

Thus, having the solvation Gibbs energies of the isomers A and B, we can estimate the quantity $\delta G(A \rightarrow B)$.

In practical cases, A and B could be the coil and helix forms of a single polymer as in Fig. 7.21a or two proteins forming a dimer as in the case of Fig. 7.21b. In both cases we cannot measure the solvation Gibbs energies of the various polymers. This leads us to seek an approximate measure of the intramolecular HϕO interaction. To do this we first isolate the contribution of the intramolecular HϕO interaction from (7.15.9). We write the total binding energy of each isomer as

$$B_\alpha = B_{BB} + B_{R_1} + B_{R_2}, \tag{7.15.10}$$

where B_{BB} is the binding energy of the carrier (the backbone of the polymer) and B_{R_1} and B_{R_2} are the binding energies of the two radicals R_1 and R_2 in the isomer $\alpha(\alpha = A, B)$.

Using (7.15.10) for $\alpha = A, B$ in (7.15.9), we can rewrite the average quantities as (see section 6.14),

$$\delta G(A \rightarrow B) = -kT \ln \frac{\langle \exp[-\beta B_B(1, 2)] \rangle_{BB}}{\langle \exp[-\beta B_A(1, 2)] \rangle_{BB}}, \tag{7.15.11}$$

where $B_B(1, 2)$ consists of the binding energies of the two groups R_1 and R_2 in the isomer B. Similar meaning applies to $B_A(1, 2)$. The averages in (7.15.11) are now conditional averages, the condition being the presence of the backbone (BB). If in the isomer A the two groups R_1 and R_2 are far apart, as indicated in Fig. 7.22, then we can further factorize the average in the denominator of (7.15.11) as

$$\langle \exp[-\beta B_A(1,2)]\rangle_{BB} = \langle \exp[-\beta B(1)]\rangle_{BB}\langle \exp[-\beta B(2)]\rangle_{BB}, \qquad (7.15.12)$$

where each factor is related to the conditional solvation Gibbs energy of the two groups R_1 and R_2 separately.

We note the formal resemblance of (7.15.11) and (7.15.1). The difference is that now we have the HϕO interaction between two groups (instead of two solutes), and the average is over all configurations of the solvent molecules, given the backbone.

Relation (7.15.11) suggests an important approximate measure of the intramolecular HϕO interaction between two groups on a biopolymer, even when the solvation Gibbs energy of the biopolymer cannot be measured. The idea is to replace the biopolymer by a small carrier compound. This carrier should be small enough, say benzene or naphthalene, so that the solvation Gibbs energies of the various substituted molecules can be measured. It should be large enough that the condition BB in (7.15.11) is approximately the same in the model compound as in the biopolymer. We shall return to this equation in connection with the side chains of amino acids in Chapter 8. Here we give an example of a measure of an intramolecular HϕO interaction that can be obtained by using relation (7.15.9). Consider 1,2- and the 1,4-dialkylbenzene, Fig. 7.23, which we denote by ϕ_{12} and ϕ_{14}, respectively. The intramolecular HϕO interaction is

$$\delta G[(1,4) \rightarrow (1,2)] = \Delta G^*_{\phi_{12}} - \Delta G^*_{\phi_{14}}. \qquad (7.15.13)$$

Thus, by measuring the solvation Gibbs energies of ϕ_{12} and ϕ_{14}, we obtain the required indirect or solvent-induced interaction between the two alkyl groups, given the presence of the benzene ring. Strictly, $\delta G[(1,4) \rightarrow (1,2)]$ corresponds to the process of transferring one alkyl group from position 4 to position 2 with respect to an alkyl group at position 1. Table 7.5 shows some values of $\delta G[(1,4) \rightarrow (1,2)]$.

Note that the values of $\delta G[(1,4) \rightarrow (1,2)]$ are negative in H_2O and in D_2O. There is a slight increase in absolute magnitude with increasing the temperature. The values for the ethyl radicals are almost twice as large as the corresponding values for the methyl radical. Also note that the values are slightly less negative in D_2O than in H_2O.

As noted above, the quantity $\delta G[(1,4) \rightarrow (1,2)]$ actually measures the relative conditional HϕO interaction in the positions 1,2 and 1,4. We can obtain a better estimate of the conditional HϕO interaction at either the 1,4 or the 1,2 configuration by using the following reactions.

Consider reaction (a) in Fig. 7.24, where two monoalkyl benzene molecules at infinite separation from each other exchange one hydrogen at position 2 with an alkyl radical. Here we strictly bring the two alkyl groups from infinite separation, but attached to the same carrier, to the final position ϕ_{12}. The corresponding indirect work is

$$\delta G(a) = \Delta G^*_{\phi_{12}} + \Delta G^*_{\phi} - 2\Delta G^*_{\phi_1}, \qquad (7.15.14)$$

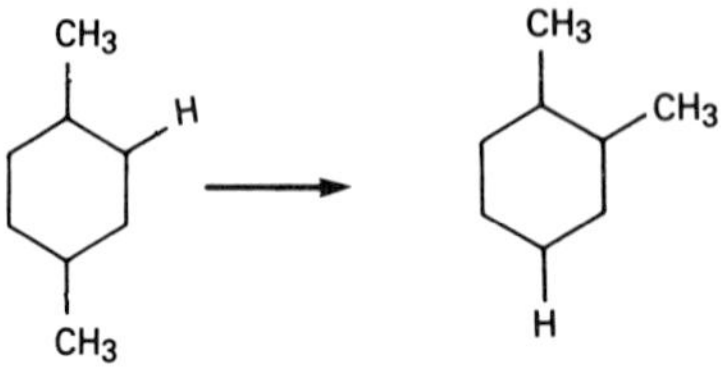

FIGURE 7.23. Model compounds used to measure the intramolecule HϕO interaction [Eq. (7.15.13)].

TABLE 7.5
Values of $\delta G[(1,4) \to (1,2)]$ Based on Eq. (7.15.13) for H_2O and D_2O and
n-Hexane

Alkyl group	Temperature (°C)	δG (cal mol^{-1})		
		H_2O	D_2O	n-Hexane
Methyl	10	-238		$+70$
	20	-300	-282	-28
Ethyl	10	-600		$+360$
	20	-616	-433	$+445$

where ϕ, ϕ_1 and ϕ_{12} stand for benzene, monoalkylbenzene, and 1,2-dialkylbenzene, respectively. Note that this is a different measure of the conditional $H\phi O$ interaction and uses different experimental data. Table 7.6 shows some values of $\delta G(a)$. The fact that these values are quite close to the values of $\delta G[(1,4) \to (1,2)]$ indicates that the two alkyl groups in the 1,4 position are nearly independent (in the sense of (7.15.12), i.e., that there is no overlapping between the solvation spheres of the two groups).

To further confirm this conclusion we perform reaction (b) in Fig. 7.24. Now, instead of (7.15.4) we have

$$\delta G(b) = \Delta G^{*}_{\phi_{14}} + \Delta G^{*}_{\phi} - 2\Delta G^{*}_{\phi_1}. \qquad (7.15.15)$$

Table 7.6 shows that these values are indeed very small, as we should expect. The two alkyl groups at the two benzene rings at infinite separation are strictly independent. The same two groups brought to the positions 1,4, are approximately independent; therefore, $\delta G(b)$ values are close to zero. Note that if we combine (7.15.15) and (7.15.14), we obtain

$$\delta G(a) - \delta G(b) = \Delta G^{*}_{\phi_{12}} - \Delta G^{*}_{\phi_{12}}, \qquad (7.15.16)$$

which is the same as the rhs of (7.15.13). Indeed, the numerical values of $\delta G[(1,4) \to (1,2)]$ should be equal to the difference of the values of $\delta G(a)$ and $\delta G(b)$.

7.16. INTERACTION BETWEEN TWO HYDROPHILIC SOLVATONS

By hydrophilic molecule (or group), we usually mean either a charged particle or particles with a permanent dipole moment. The charge–charge direct interaction is well

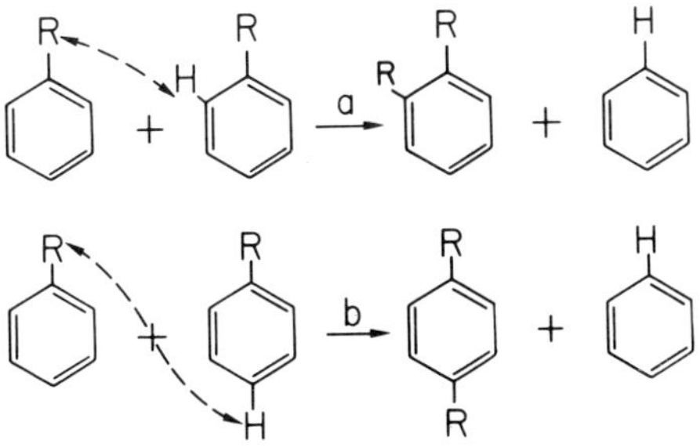

FIGURE 7.24. Two "reactions" used to estimate the intramolecular $H\phi O$ interaction. (a) corresponds to Eq. (7.15.14). (b) corresponds to Eq. (7.15.16).

TABLE 7.6
Values of $\delta G(a)$ and $\delta G(b)$ Corresponding to Eqs. (7.15.14) and (7.15.15)

Alkyl group	Temperature (°C)	$\delta G(a)$ (cal/mol)	$\delta G(b)$ (cal/mol)
Methyl	10	−205	+33
	20	−276	+24
Ethyl	10	−619	−16
	20	−609	+7

described by the Coulomb interaction of the form

$$U_{cc}(R) = \frac{z_i z_j e^2}{R}.$$ (7.16.1)

Any interaction between two polar molecules can be described in terms of the super-position of charge–charge interactions. In a solvent this interaction is modified by intro-ducing the dielectric constant of the medium ε. Thus the solvent-induced interaction can be defined as (see also section 6.12)

$$\delta W_{cc}(R) = \frac{z_i z_j e^2}{\varepsilon R} - \frac{z_i z_j e^2}{R} = \frac{-z_i z_j e^2}{R}\left(\frac{\varepsilon - 1}{\varepsilon}\right).$$ (7.16.2)

Since $\varepsilon > 1$, the effect of the medium always reduces the absolute magnitude of the charge–charge interaction. This phenomenon is well understood. The polarization induced by the charged particles on the molecules of the solvent is always such that the net interaction between the charges is weakened or screened. This is true for charged or any other polar particles.

When we say that the solvent-induced interaction between charged particles is well understood, we mean that the phenomenon can be well described by the macroscopic dielectric constant ε. This, of course, is not a molecular theory of the dielectric constant. In principle one should be able to express the dielectric constant in terms of molecular parameters, e.g., the dipole moments of the molecules.

There exists, however, an important group of molecules that can interact either between themselves or with water molecules through hydrogen bonding. These interac-tions are ubiquitous and of fundamental importance in biochemical processes. Some examples in which these interactions are involved will be presented in Chapter 8. Here, we examine the source and some characteristic features of these interactions.

The polar groups to which we shall refer as HBing groups are groups such as OH, C=O, NH, etc. We shall refer to these as functional groups (FG). Two molecules that contain such FGs will be assumed to interact through a pair potential of the form (see section 7.4)

$$U(\mathbf{X}_i, \mathbf{X}_j) = U^{H,S}(\mathbf{X}_i, \mathbf{X}_j) + U^{HB}(\mathbf{X}_i, \mathbf{X}_j),$$ (7.16.3)

where $U^{H,S}$ includes the hard (H) and soft (S) parts of the interaction and U^{HB} will be referred to as the HBing interaction. The latter will have a form similar to the one described in section 7.4 for the water–water interaction.

Solutes that have such polar groups will usually be hydrophilic, in the sense that their solubility will be relatively large in water (section 7.10.2) as compared with similar molecules with no HB capability. Formally we can think of particles for which the U^{HB} part can be switched on and off. Experimentally we may compare two molecules for which a polar group has been replaced by a nonpolar group, say ethanol and propane. Such a replacement will always cause a decrease in the solubility of that molecule in water.

We now turn to examine the question of the potential of average force, or the solvent-induced interaction between two HϕI molecules that interact either between themselves or with the solvent by a potential function of the form (7.16.13). The simplest HϕI molecules and the most important ones are water molecules themselves. We next consider the HϕI interactions between two water molecules at some particular configurations.

7.16.1. HϕI Interaction at $R_1 \approx 2.76$ Å

We recall that the radial distribution function for water has a first peak at $R_1 \sim 2.8$ Å, closely reminiscent of the oxygen–oxygen distance 2.76 Å in the ice lattice (section 7.3). The direct HBing between two water molecules is probably the most important feature of the interactions that determine the properties of liquid water. Likewise, the direct HBing between HϕI groups is of crucial importance in determining the structure and function of biomolecules such as proteins and nucleic acids.

The height of the first peak of the $g(R)$ of water at $R_1 \sim 2.8$ Å is about two. This corresponds to a minimum of the potential of average force of about

$$W(R_1) = -kT \ln g(R_1) \approx -0.4 \text{ kcal/mol}^{-1}. \qquad (7.16.4)$$

This is quite an unimpressive minimum well. However, in considering the contribution of HBing between two HϕI groups in biomolecules, we have to deal with two groups that have fixed orientation relative to each other and that are attached to some carrier or backbone. The situation is the same as in Fig. 7.21 but with HϕI groups, say hydroxyl or carbonyl, replacing the methyl group. Therefore, we should examine the conditional contribution of the solvent-induced interaction between two HϕI at some fixed relative orientation.

Consider now two water molecules brought from fixed positions and orientations at infinite separation to the final distance of $R_1 \approx 2.76$ Å and a configuration, which we denote by $\Omega_{HB}^{(1)}$, favorable to forming a HB, as in Fig. 7.25. The corresponding Gibbs energy change is

$$\Delta G(R_1, \Omega_{HB}^{(1)}) = U(R_1, \Omega_{HB}^{(1)}) + \delta G(R_1, \Omega_{HB}^{(1)}), \qquad (7.16.5)$$

where the direct interaction is mainly due to the HB energy ε_{HB}.

We are interested here in the solvent-induced contribution δG. Following similar reasoning as in section 7.14, we can write the total solvent-induced interaction as

$$\delta G = \delta G^{H,S} + \delta G^{HB/H,S}, \qquad (7.16.6)$$

where $\delta G^{H,S}$ is the solvent-induced contribution due to the process of bringing together the two solvatons when the HB part of the binding energy is switched off. The second term $\delta G^{HB/H,S}$ is the conditional solvent-induced interaction due to turning on the HBing part of the binding energy. The condition here is the hard and soft part of the water solvaton (which we have approximated as the binding energy of neon atom to water).

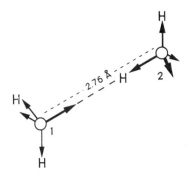

FIGURE 7.25. Two water molecules at a favorable configuration to form a direct hydrogen bond.

For HφI groups attached to a carrier, the condition should include also the presence of the carrier, e.g., the protein. The second term on the rhs of (7.16.6) can now be written as

$$\delta G^{HB/H,S}(R_1, \Omega_{HB}^{(1)}) = -kT \ln \frac{\langle \exp[-\beta B_{ww}^{HB}(R_1, \Omega_{HB}^{(1)})] \rangle_{H,S}}{\langle \exp(-\beta B_w^{HB}) \rangle_{H,S}^2}$$

$$= \Delta G_{ww}^{*HB/H,S}(R_1, \Omega_{HB}^{(1)}) - 2\Delta G_w^{*HB/H,S}, \qquad (7.16.7)$$

where $B_{ww}^{HB}(R_1, \Omega_{HB}^{(1)})$ is the HB part of the binding energy of a pair of water molecules at the specific configuration $R_1, \Omega_{HB}^{(1)}$, and B_w^{HB} is the HB part of the binding energy of a single water molecule. Note that both averages in (7.16.7) are conditional averages.

In order to estimate the quantity on the rhs of (7.16.7), all we need is the difference of the conditional solvation Gibbs energies of the pair at $R_1, \Omega_{HB}^{(1)}$, and at infinite separation. Figure 7.25 shows that at the configuration $R_1, \Omega_{HB}^{(1)}$, only six arms of the pair of water molecules are exposed to the solvent, whereas at infinite separation each water molecule is fully solvated. The conditional solvation Gibbs (or Helmholtz) energy per arm has already been estimated in section 7.7. Therefore we can write

$$\delta G^{HB/H,S}(R_1, \Omega_{HB}^{(1)}) = 6 \times (-9.42) - 2 \times 4 \times (-9.42)$$

$$= 18.8 \text{ kJ mol}^{-1} = 4.5 \text{ kcal mol}^{-1}. \qquad (7.16.8)$$

We conclude that the conditional solvent-induced interaction due to HBing with the solvent is of the order of 4.5 kcal mol^{-1}. The reason for this large and positive value is qualitatively clear. Two water molecules at infinite separation are fully solvated. When they are brought to the configuration $R_1, \Omega_{HB}^{(1)}$ to form a direct HB, each of the water molecules loses one solvated arm. Since the conditional solvation Gibbs energy per arm is about -2.25 kcal/mol, the loss of these two arms involves a net increase in Gibbs energy of about 4.5 kcal mol^{-1}.

Repeating the same arguments for two functional groups attached to a biopolymer, we can reach the same conclusion; namely, the solvent-induced interaction between two functional groups at $R_1, \Omega_{HB}^{(1)}$ is large and positive, of the order of 4.5 kcal/mol.

In estimating the overall contribution of a pair of FGs in a specific biochemical process, e.g., the association of two proteins, or protein folding, one should take into account the following two points: How many "arms" are lost in the process, and what is the direct interaction experienced by these arms in the final state. Three possible cases are the following: (1) Two FGs are transferred into the interior of a protein, where they are not involved in HBing. In this case, the loss of the solvation Gibbs energy of the FGs cannot be compensated for by the relatively weak interactions between the FGs and the interior of the protein. Such a process is clearly very improbable. In fact, it is known that in most biochemical processes the two FGs form a direct HB. We distinguish between the two cases. (2) Only two arms are lost when the two FGs form a direct HB. The loss in Gibbs energy is about 4.5 kcal/mol and the gain in direct HB is on the order of -6 kcal/mol. Therefore the net Gibbs energy change is negative, on the order of -1.5 kcal/mol; i.e., the process is favorable. This case occurs whenever the two FGs in the final product are still exposed to the solvent. For instance, in the process of formation of an α-helix, the carbonyl group has initially two arms solvated by water. On forming an α-helix only one of its arms is engaged in intramolecular HBing, the second arm being still exposed to the solvent. This process will therefore be favorable from the point of view of the HBing (direct and indirect) contribution. (3) When more than two arms of the FGs are lost, then even when forming a direct HB the net process is unfavorable. An example could be an OH and a C=O on two proteins associating to form a dimer, if these two FGs are completely removed from water when forming a direct HB. The direct energy gained is about -6 kcal/mol. The loss of Gibbs energy is now much larger, since these functional groups lose more than one arm per FG. Therefore the net Gibbs energy change will be large and positive, and the whole process will be very unfavorable.

To summarize, when two water molecules or two FGs are brought to a configuration $(R_1, \Omega_{HB}^{(1)})$ to form a direct HB, there is a gain in energy of about -6 kcal/mol. The loss of the solvation Gibbs energy could be about 4.5 kcal/mol if only two arms have been removed from the solvent and larger if more arms are removed.

We now return to the two water molecules at the configuration R_1, $\Omega_{HB}^{(1)}$ and write the three contributions to the potential of average force:

$$W(R_1, \Omega_{HB}^{(1)}) = U + \delta G^{H,S} + \delta G^{HB/H,S}. \tag{7.16.9}$$

If the direct interaction is due to HBing, we can take $U \approx \varepsilon_{HB} \sim -6.5$ kcal/mol. $\delta G^{H,S}$ is the same as the indirect part of the potential of average force between two Lennard–Jones particles at contact. From simulation calculation this value is at most on the order of $\delta G \sim -0.6$ kcal/mol, and for $\delta G^{HB/H,S}$ we take the value of (7.16.8); thus altogether we have

$$W(R_1, \Omega_{HB}^{(1)}) \sim -2.6 \text{ kcal/mol} \tag{7.16.10}$$

corresponding to

$$g(R_1, \Omega_{HB}^{(1)}) = 76. \tag{7.16.11}$$

This is a very large number compared with the experimental value of about $g(R_1) \approx 2$. The reduction from $g(R_1, \Omega_{HB}^{(1)})$ to $g(R_1)$ is brought about by the averaging over all orientations of the pair of water molecules, i.e.,

$$g(R_1) = \frac{1}{(8\pi^2)^2} \int g(R_1, \Omega_1, \Omega_2) \, d\Omega_1 \, d\Omega_2 \approx 2. \tag{7.16.12}$$

Note that this average gives equal weight to each configuration. Since the region of HBing is relatively small compared with the entire configurational space of $(8\pi^2)^2$, it is clear that this average suppresses the large value obtained at the specific configuration $\Omega_{HB}^{(1)}$ in (7.16.11). The resulting average has a normal value of about 2. We shall not carry out the average in (7.16.12). However, it is important to keep in mind that the averaging process drastically reduces the value of $g(R_1, \Omega_{HB}^{(1)})$ by a factor of about 40.

7.16.2. Hφl Interaction at $R_2 = 4.5$ Å

A second configuration of importance is related to the second-nearest neighbor configuration of water molecules in ice. We recall that the radial distribution function of water has a second peak at $R_2 = 4.5$ Å. The height of this peak is about 1.15, corresponding to a minimum value of the potential of average force of about $W(R_2) \sim 0.084$ kcal/mol. This is even less impressive than the minimum well we found at R_1. However, recalling that the experimental radial distribution function is orientation averaged, and having seen that this averaging process has a significant effect on the value of the correlation function at R_1, we can expect that upon "freezing" the orientations of the water molecules we should find correlation values far larger than the value of $g(R_2)$. The corresponding value of the potential of average force at R_2 is perhaps as important to biology as the value at R_1.

Consider again two water molecules (or Hφl groups attached to some backbone) at fixed positions and orientations at infinite separation. We bring them to a distance of 4.5 Å and orientations such that a HB bridge can be formed by a third water molecule. This is essentially the same configuration of the two next-nearest neighbors of water molecules in ice, as in Fig. 7.26. We shall denote this configuration by $(R_2, \Omega_{HB}^{(2)})$ or, for short, simply by $R_2 = 4.5$ Å. The change in the Gibbs energy for this process, carried say at P, T constant is

$$\Delta G(R_2, \Omega_{HB}^{(2)}) = U(R_2, \Omega_{HB}^{(2)}) + \delta G(R_2, \Omega_{HB}^{(2)}). \qquad (7.16.13)$$

Here, in contrast to the previous configuration at R_1, we can assume that the direct interaction is quite weak and the main contribution to ΔG comes from the solvent-induced part δG. As in the previous section, we divide δG into two parts as in (7.16.6)

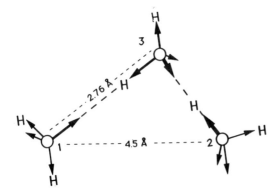

FIGURE 7.26. Two arms of two water molecules 1 and 2 (or any Hφl groups) at a distance of 4.5 Å and oriented in such a way that they can HB to a water molecule, 3.

and focus on the HBing contribution. The latter can be written as

$$\delta G^{\mathrm{HB}/H,S} = -kT \ln \frac{\langle \exp[-\beta B_{ww}^{\mathrm{HB}}(4.5)] \rangle_*}{\langle \exp(-\beta B_w^{\mathrm{HB}}) \rangle_*^2}, \qquad (7.16.14)$$

where $B_{ww}^{\mathrm{HB}}(4.5)$ is the same as in Eq. (7.16.7), but at the new configuration R_2, $\Omega_{\mathrm{HB}}^{(2)}$. The conditional average is indicated by $\langle \ \rangle_*$ where * stands for H, S if we are dealing with two water molecules or H, S and the backbone if we are considering two FGs attached to a backbone.

A theoretical estimate of $\delta G^{\mathrm{HB}/H,S}$ in (7.16.14) is quite involved (see Appendix C). We shall present here one estimate based on experimental data. Before doing that we introduce two qualitative arguments that indicate that this quantity is quite large.

First, we recall that the averaging process carried out on $g(R_1, \Omega_1, \Omega_2)$ has reduced the high value of $g(R_1, \Omega_{\mathrm{HB}}^{(1)})$ to a normal value of about 2. We expect that a similar reduction would have occurred for $g(R_2, \Omega_1, \Omega_2)$. There are two differences between the two cases. In the case of $g(R_1)$ we had large contributions from both the direct and the indirect interactions, whereas in the case of $g(R_2)$, the direct water–water interaction is probably very weak. The second difference is that in the case of $g(R_1, \Omega_{\mathrm{HB}}^{(1)})$ there are eight HBing configurations (i.e., each arm of one water molecule can form a HB with two of the arms of the second water molecule). In the case of $g(R_2, \Omega_{\mathrm{HB}}^{(2)})$, there are sixteen configurations favoring the formation of a HB'd bridge with a third molecule, as in Fig. 7.26 (i.e., any arm of one water molecule and any arm of the second water molecule can point to the third molecule). These two considerations will affect the relative values of $g(R_2, \Omega_{\mathrm{HB}}^{(2)})$ and $g(R_2)$. Qualitatively, however, we expect that by "reversing the average process"—i.e., taking a fixed configuration $(R_2, \Omega_{\mathrm{HB}}^{(2)})$—we should enhance the peak at R_2 by a factor of about 40. This corresponds to a considerable deepening of the minimum of $W(R)$ at $(R_2, \Omega_{\mathrm{HB}}^{(2)})$.

The second qualitative argument can be obtained by rewriting (7.16.14) in the form

$$\delta G^{\mathrm{HB}/H,S} = -kT \ln \frac{\langle \exp[-\beta B_{ww}^{\mathrm{HB}}(1,2; 4.5)] \rangle_*}{\langle \exp(-\beta B_w^{\mathrm{HB}}) \rangle_*^2}$$

$$= -kT \ln \frac{\langle \exp[-\beta B_w^{\mathrm{HB}}(1)] \rangle_* \langle \exp[-\beta B_w^{\mathrm{HB}}(2; 4.5)] \rangle_{*,1}}{\langle \exp(-\beta B_w^{\mathrm{HB}}) \rangle_*^2}$$

$$= -kT \ln \frac{\langle \exp[-\beta B_w^{\mathrm{HB}}(2; 4.5)] \rangle_{*,1}}{\langle \exp(-\beta B_w^{\mathrm{HB}}) \rangle_*}. \qquad (7.16.15)$$

The average in the numerator is written as a product of two averages. First we have the solvation Gibbs energy of one molecule (with the condition *) and the second is the conditional solvation Gibbs energy of the second water molecule at a distance of 4.5 Å from the first and orientation $\Omega_{\mathrm{HB}}^{(2)}$. (Here we have both the condition * and the presence of the first molecule, indicated by 1.) In the last form on the rhs of (7.16.15) we have essentially a difference in the solvation Gibbs energy of a water molecule, but the conditions are different. In one case the condition is *, in the second the condition is *,1. Clearly, because of the presence of the condition 1, the conditional density of water molecules at X_3 will be larger than in the absence of this condition. Therefore, the solvation Gibbs energy of molecule 2 given molecule 1 is expected to be more negative than the solvation Gibbs energy of a water molecule. The net result is a negative value of $\delta G^{\mathrm{HB}/H,S}$. (See also Appendix C.)

We now turn to an estimate based on experimental data. The methodology is the same as the one discussed in the preceding subsection, 7.16.1. Consider two isomers of a molecule having two FGs, as in Fig. 7.27. In reaction I we bring one functional group from position 4 to position 3 relative to a given FG at position 1. The indirect contribution to the Gibbs energy change for this reaction is

$$\delta G[(1,4) \to (1,3)] = \Delta G^*_{1,3} - \Delta G^*_{1,4}, \tag{7.16.16}$$

as in Eq. (7.15.13). Thus having the solvation Gibbs energies of the 1,3 and the 1,4 isomers, we could have estimated δG in (7.16.16). However, these data are not available. Instead, we have the differences in the solvation Gibbs energies of these compounds between water w and toluene t. From the available data we can estimate the difference

$$\delta G[(1,4) \to (1,3)]^{(w)} - \delta G[(1,4) \to (1,3)]^{(t)} = \Delta G^{*w}_{1,3} - \Delta G^{*w}_{1,4} - [\Delta G^{*t}_{1,3} - \Delta G^{*t}_{1,4}]$$

$$= -3.15 \text{ kcal/mol}, \tag{7.16.17}$$

and similarly

$$\delta G[(1,4) \to (1,2)]^{(w)} - \delta G[(1,4) \to (1,2)]^{(t)} = -1.55 \text{ kcal/mol}. \tag{7.16.18}$$

Assuming that the difference in the solvation Gibbs energies of the two isomers in toluene (t) is small, we can attribute the major part of the result (7.16.17) to the indirect interaction in water.

In order to confirm that the two FGs at positions 1,4 are far enough apart so that they may be considered independently solvated, we examine also the reaction II in Fig. 7.27. For the conversion of two (1,4) isomers into one (1,3) and phenol (1) we have

$$\delta G[2 \times (1,4) \to (1,3) + (1)] = -2.8 + 0.27 - 2 \times (0.35)$$

$$= -3.23 \text{ kcal/mol}. \tag{7.16.19}$$

Similarly, for the formation of the (1,2) isomer we have

$$\delta G[2 \times (1,4) \to (1,2) + (1)] = -1.2 + 0.27 - 2 \times (0.35) = -1.63 \text{ kcal/mol}. \tag{7.16.20}$$

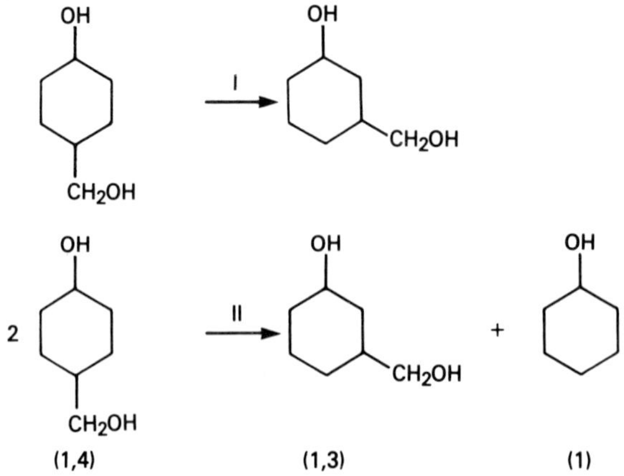

FIGURE 7.27. Model compounds used to estimate the intramolecular $H\phi I$ interaction. Process I corresponds to Eq. (7.16.17). Process II corresponds to Eq. (7.16.19).

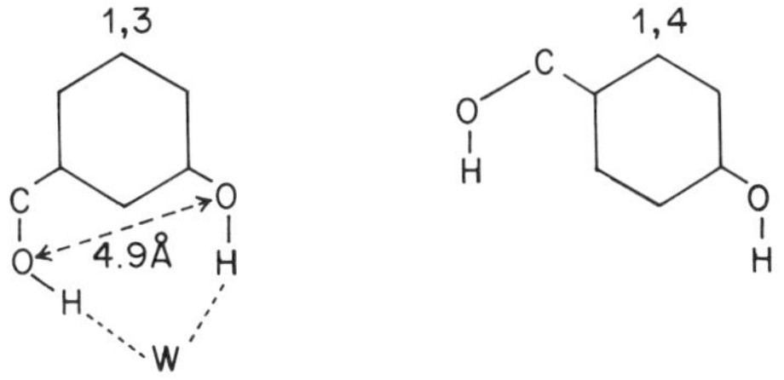

FIGURE 7.28. The distance between the two oxygens in the (1,3) isomer is about 4.9 Å, roughly the required distance to form a strong HϕI interaction. The corresponding distance in the (1,4) isomer is too large for HϕI interaction.

Thus the formation of the (1,3) isomer involves about -3.23 kcal/mol, almost the same as the value in (7.16.17). The formation of the (1,2) isomer involves as in (7.16.18) almost half of this quantity. The reason for this difference is clear. In the (1,3) position the two FGs are at a distance of about 4.9 Å (Fig. 7.28), and therefore they can form a water bridge; hence, there is an indirect contribution of about $\delta G \approx -3$ kcal/mol. In the (1,2) isomer, the distance between the two FGs is smaller, and there is partial interference between the two FGs, leading to a much lower value of δG. Note again that, as in (7.16.17) and (7.16.18), the values in (7.16.19) and (7.16.20) pertain to the difference between water and toluene. A proper estimate of the HϕI interaction should be based on the solvation Gibbs energies of the same compounds in pure water. Unfortunately, these data are not yet available.

It is of interest to note also that for the same conversion as in (7.16.17), but using n-octanol instead of water, we get a value of about -3.6 kcal/mol. The reason is that octanol, like any other alkanol, should be able to form a one-alkanol bridge between the two hydroxyl groups. Note that the alkanols can offer two acceptor arms to form a bridge between the two hydroxyl groups. Had we instead of two hydroxyl groups two carbonyl groups, then a water molecule could form a bridge between the two, but not an alkanol. This case occurs in proteins where two C=Os of the backbone can be bridged by water molecules, stabilizing intermediates in the process of protein folding (see section 8.8 for details).

Finally we note that recent Monte Carlo calculations of the potential of average force between two water molecules at a fixed relative orientation $\Omega_{HB}^{(2)}$ also indicate that there exists a minimum well of about -3 kcal/mol.

From the above considerations we conclude that there are two configurations for a pair of water molecules (or pair of HϕI groups) at which we expect large effects due to HBing. At $(R_1, \Omega_{HB}^{(1)})$, the solvent induced part δG is positive. However, in combination with a direct HB, the overall Gibbs energy change might be negative and on the order of -1.5 to -2 kcal/mol. (This is true provided that only one arm per FG is lost in the formation of the direct HB.) On the other hand, at $(R_2, \Omega_{HB}^{(2)})$, the solvent-induced part δG is negative and of the order of about -3 kcal/mol. Since the direct interaction is relatively small at this distance, δG is essentially equal to the potential of average force. We can conclude that the HϕI interaction at $(R_2, \Omega_{HB}^{(2)})$ might be as important as the total Gibbs energy change for the formation of a direct HB at $(R_1, \Omega_{HB}^{(1)})$. We therefore expect that both HϕI effects will play important roles in biochemical processes. Some possible examples are discussed in Chapter 8.

7.17. MIXED HϕO–HϕI INTERACTIONS

We discuss here some aspects of solvent-induced interactions between HϕO and HϕI molecules or groups. The simplest example is a simple nonpolar atom, say neon or argon, and a water molecule. These are "pure" HϕO and HϕI molecules. The more common cases are molecules that contain both HϕO and HϕI groups, such as alcohols, fatty acids, and proteins.

First, consider the case of a simple LJ solute, say neon or argon, denoted A, very diluted in water, w. The pair correlation function should have a first peak at $R_1 \approx (\sigma_A + \sigma_W)/2$, mainly due to the solute–solvent direct interaction. This is indeed observed in simulation by both Monte Carlo and molecular dynamics. The new feature to consider here is the orientation dependence of the pair correlation function at contact distance R_1. Simulations of aqueous solutions of simple solutes indicate that at contact distance none of the four arms of the water molecule point to the center of the nonpolar solute.

A qualitative reason for this particular orientation can easily be visualized. Water molecules tend to form a hydrogen-bonded network, leaving empty, cagelike spaces in which the solute can be accommodated. In such cages, there are no HB directions that point toward the solute molecule.

This correct pictorial argument does not provide a quantitative estimate of the extent of this preferential orientation of the water molecule around the nonpolar solute. We provide the following theoretical argument: Consider a simple solute A diluted in water w. We assume that the direct solute–solvent interaction is spherically symmetrical; that is, $U_{AW}(R)$ is a function of R only. Therefore the solute–solvent pair correlation function for this pair is written as

$$g_{AW}(1,2) = \exp[-\beta U_{LJ}(R)]\,\exp[-\beta\,\delta G(1,2)]. \tag{7.17.1}$$

In this presentation, all the orientation dependence of g_{AW} is attributed to $\delta G(1,2)$.

If a solute A approaches a water molecule w in the gaseous phase, it has the same probability of attaching to any region on the surface of w. (This is not precisely the case for real molecules. However, we make this assumption to stress the new effect of the solvent-induced interaction in producing the orientational preference discussed below.)

We now ask the question: When A approaches w in water, what is the preferential "site of binding?" To answer this question, we distinguish between two regions on the surface of a water molecule. One region consists of the four regions $\Delta\mathbf{X}_1$, $\Delta\mathbf{X}_2$, $\Delta\mathbf{X}_3$, and $\Delta\mathbf{X}_4$ (indicated by four boxes in Fig. 7.29) from which the water molecule can form a

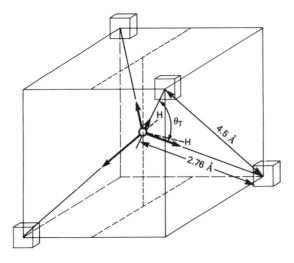

FIGURE 7.29. The four small cubes from which a water molecule can form a HB with the water molecule at the center.

HB with another water molecule. These regions will be referred to as the hydrogen-bonding regions. All the rest is referred to as the non-hydrogen-bonding region.

When A approaches w through the hydrogen-bonding region, it blocks the capacity of w to form HBs with other water molecules. This configuration is denoted by a' in Fig. 7.30. The other preferential configuration is denoted by a. Recalling the definition of δG in (7.14.6) and following the assumption that the direct interaction is spherically symmetrical, we write

$$\delta G(\infty \to a') = \Delta G^*_{Aw}(a') - \Delta G^*_{Aw}(\infty) \qquad (7.17.2)$$

$$\delta G(\infty \to a) = \Delta G^*_{Aw}(a) - \Delta G^*_{Aw}(\infty). \qquad (7.17.3)$$

Since in the configuration a all arms of w are free to form HBs to the solvent, the solvation Gibbs energy of the pair at configuration a is more negative than in the configuration a' by the amount of approximately -2.25 kcal/mol. In other words, moving the solute from configuration a to a' lowers the Gibbs energy by about 2.25 kcal/mol. From (7.17.2) and (7.17.3), we can write

$$\delta G(a \to a') = \Delta G^*_{Aw}(a') - \Delta G^*_{Aw}(a) \approx 2.25 \text{ kcal/mol}. \qquad (7.17.4)$$

In terms of probability, the difference between the two cases can be stated as follows. Suppose we sit at the center of a water molecule and count the frequency with which a solute hits the water molecule from various directions. Assuming that U_{AW} is not sensitive to the different orientations, then the ratio of the frequency of observing the two configurations a and a' is about

$$\frac{\Pr(a)}{\Pr(a')} = \frac{\exp[-\beta \Delta G^*_{Aw}(a)]}{\exp[-\beta \Delta G^*_{Aw}(a')]}$$

$$= \exp[-\beta \delta G(a' \to a)]$$

$$\sim \exp\left[\frac{+2.25}{0.6}\right]$$

$$\sim 43. \qquad (7.17.5)$$

We see that configuration a is overwhelmingly more probable than configuration a'.

Note that in (7.17.5) we compared two discrete (or "point") events: a and a'. Since the hydrogen-bonding region is considerably smaller than the non-hydrogen-bonding region, the ratio would have been much larger had we taken any of the events of type a and any events of the type a'. [If Ω_{HB} is the total solid angle of the HBing region, then the ratio (4.17.5) should be multiplied by $4\pi/\Omega_{HB}$.] We shall now use this information to predict the location of the second minimum in the potential of average force for the pair of solute molecules, i.e., for the HϕO interaction (see section 7.15).

FIGURE 7.30. Two possible orientations a and a' of a water molecule near a spherical simple solute A. (a) is the preferential orientation of the water molecule

Consider now two methane molecular (or neon, argon, etc.,) atoms in water. We have already noted that the first peak of $g_{AA}(R)$ is expected at $R_1 \approx \sigma_A$. The main reason is that $R_1 \approx \sigma_A$ corresponds to the minimum of the direct pair potential between two solute molecules. We have also seen that in normal solutions of A diluted in w the location of the second peak is expected at about $R_2 \sim \sigma_A + \sigma_W$ (section 6.3). In our case the simulated curve shown in Fig. 7.20 indicates a second minimum of $W(R)$ at about $R_2 \sim 6$ Å. This is quite different from the normal location of R_2 expected at about $2.8 + 4.4 \sim 7.2$ Å (for the parameters used in the simulated curve).

We recall that at the distances between 6 and 7 Å, the direct methane–methane interaction is negligible. Therefore, the main contribution to $g_{AA}(R_2)$ should come from the indirect interaction, i.e.,

$$g_{AA}(R_2) \approx \exp[-\beta \delta G(R_2)]. \tag{7.17.6}$$

Second, we recall our finding concerning the preferential orientation of a water molecule around a simple solute. We have seen that the ratio of the probabilities of the configurations a and a' in Fig. 7.30 is overwhelmingly in favor of a.

We now pose the following question: Having the above information in mind, where should we expect to find with high probability the second-nearest neighbors for a pair of two solutes in water? The answer is quite simple. If A is very dilute in water, then it is likely to be attached to at least one water molecule at the preferential orientation a as in Fig. 7.30. We can view this A—w pair as a "dimer." Now, if a second A approaches this "dimer," the next-nearest neighbor relation it can form with the first A occurs when it attaches to the water partner of the "dimer," again at the preferential orientation. (Note that this is not true if w is diluted in A; we are dealing here with a solution of A very diluted in w.)

In this configuration, the angle formed by the centers of the triplet A—w—A is about $\theta_T = 109.46°$, and the A—A distance would be

$$R_2 = (\sigma_A + \sigma_W) \cos \frac{180 - \theta_T}{2} = (4.4 + 2.76) \times 0.816 = 5.85 \text{ Å} \tag{7.17.7}$$

for the choice of methane and water parameters as given in Fig. 7.20.

Another possibility is that two such dimers will interact to form a HB between the two water members on each dimer. There are several possible configurations for such an event: b, c, d in Fig. 7.31. The shortest A—A distance is in configuration c, which is $R_2 = 6.15$ Å.

We can therefore conclude that the expected location of the second minimum of $W(R)$ should occur somewhere between 5.85 and 6.15 Å. Since the Monte Carlo calculation cannot distinguish between 5.85 and 6.15, we take the average value of $R_2 \sim 6$ Å, which is exactly the location of the second minimum indicated by the simulation shown in Fig. 7.20.

It should be noted that the argument given above is essentially the same as given for the preferential orientation of a water molecule around the solute. In both cases, blocking a HBing arm costs a considerable amount of Gibbs energy, about 2.25 kcal/mol.

In terms of probabilities, the water-bridged configuration, Fig. 7.31c is at least $3 \times 42.5 = 127$ more probable than that of the water-separated configuration in Fig. 7.31e. This is so because there is one possibility of attaching a second solute at the water-separated position given a first solute at the right configuration a of Fig. 7.30, but there are three possibilities of attaching the second solute at the water-bridged position.

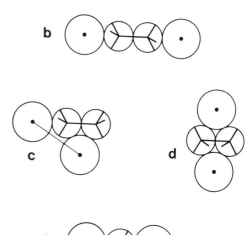

FIGURE 7.31. Various possible configurations of two simple solutes in water. (b), (c), and (d) are acceptable, but (e) violates the preferential orientation of a water molecule near a nonpolar solute.

To conclude, the $H\phi O$ interaction between two simple solutes has a first minimum at the normally expected distance of contact σ_A. It is normal since this minimum is determined mainly by the direct interactions. The second minimum occurs when the two solutes are bridged by a water molecule, and the orientation of the latter is consistent with the preferential orientation of a water molecule in contact with a simple solute. The location of the second minimum can be estimated for any σ_A from the first equality in (7.17.7).

We have discussed so far the $H\phi O$–$H\phi I$ interaction between pure $H\phi O$ and $H\phi I$ solvatons. The more frequent cases are molecules having both a $H\phi O$ and a $H\phi I$ groups, such as alcohols, carboxylic acids, and amino acids. We present here one example of the solvent-induced interaction between an $H\phi O$ group, methyl, and an $H\phi I$ group, hydroxyl, on a benzene ring. Consider the formation of 2-methylphenol from toluene and phenol, the δG for the "reaction"

$$\text{phenol} + \text{toluene} \rightarrow \text{2-methylphenol} + \text{benzene}. \qquad (7.17.8)$$

Using the same arguments as in sections 7.15 and 7.16, we estimate

$$\delta G = \Delta G^*_{\text{2-methylphenol}} + \Delta G^*_{\text{benzene}} - \Delta G^*_{\text{toluene}} - \Delta G^*_{\text{phenol}} = +0.76 \text{ kcal/mol}.$$

The fact that δG in this case is positive indicates that the methyl group, brought to within a very close proximity of the hydroxyl group, interferes with the solvation of the latter and hence makes this configuration less stable than the corresponding 4-methylphenol isomer for which we have

$$\delta G = \Delta G^*_{\text{4-methylphenol}} + \Delta G^*_{\text{benzene}} - \Delta G^*_{\text{toluene}} - \Delta G^*_{\text{phenol}} = +0.502 \text{ kcal/mol}, \quad (7.17.9)$$

which is slightly smaller than in the previous case.

7.18. GENERALIZATION TO MANY SOLUTES

In the previous four sections, we discussed the solvent-induced interactions between a pair of solvatons. These could be $H\phi O$–$H\phi O$, $H\phi I$–$H\phi I$, or a mixed $H\phi O$–$H\phi I$ pair.

The extension to any number of solvatons is quite straightforward. In this section we discuss some aspects of the solvent-induced interactions among many simple HϕO solvatons in water. An extension to more complex molecules is discussed in section 8.9.

7.18.1. An Improved Approximate Measure of the HϕO Interaction

In section 7.15 we presented an approximate measure of the HϕO interaction between two methane molecules at a distance of $R \sim 1.53$ Å. We noted also that the distance $R \sim 1.53$ Å is not accessible to real solutes, so that little can be inferred on the HϕO interaction at more realistic distances.

In this section we present a modified measure of the HϕO interaction which is based on essentially the same type of arguments as in section 7.15 but can, in principle, provide improved information. The improvement is achieved in both the nature of the approximation and the realizability of the final configuration of the solutes. The extent of the improvement depends on the availability of relevant experimental data.

As a prototype of our new measure we consider the neopentane molecule. Using the same arguments as in section 7.15, the solvent-induced interaction among five methane molecules brought to the final configuration of a neopentane molecule (see Fig. 7.32) is written as follows:

$$\delta G_5(\text{neopentane}) = \Delta G^*(\text{neopentane}) - 5\Delta G^*(\text{methane}). \tag{7.18.1}$$

This measure involves the same kind of approximation as discussed in section 7.15; i.e., the five methane molecules are brought to a very close distance of about 1.53 Å, and we neglect the contribution due to the hydrogen atoms that are lost in the formation of neopentane (see also Appendix G).

Next we consider a different process. We start with *four* methane molecules at fixed positions at infinite separation from each other. These solutes are brought (within the solvent, keeping T and P constant) to the final configuration of the four peripheral methyl groups in neopentane. These are indicated in Fig. 7.33. At this configuration we have the exact relation

$$\delta G_4(\text{agg.}) = \Delta G^*(\text{agg.}) - 4\Delta G^*(\text{methane}), \tag{7.18.2}$$

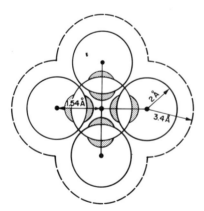

FIGURE 7.32. A schematic description of five methane molecules in the configuration of neopentane. All the inner hydrogens are indicated by the dotted areas. The boundaries of the excluded volume are indicated by the dashed curve (assuming a radius of 2 Å for the methane and 1.4 Å for the water molecule).

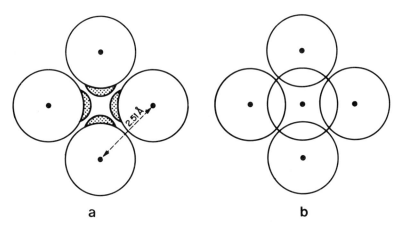

a   b

FIGURE 7.33. A schematic description of the replacement procedure corresponding to Eqs. (7.18.6) and (7.18.7). (a) Four methane molecules are brought to the position of the four peripheral methyl groups of a neopentane molecule. The four hydrogen atoms pointing toward the center are indicated by the dotted areas. The distance of closest approach between any two of the methane molecules is about 2.51 Å. (b) The four methane molecules are replaced by a single neopentane molecule. A new carbon nucleus is added, which partially compensates for the loss of the four inner hydrogen atoms.

where $\Delta G^*(\text{agg.})$ is the solvation Gibbs energy of the aggregate (agg.) viewed as a single entity. Clearly, this is not a measurable quantity. The statistical mechanical expression for $\Delta G^*(\text{agg.})$ is

$$\Delta G^*(\text{agg.}) = -kT \ln\langle \exp(-\beta B_{\text{agg.}})\rangle_0, \qquad (7.18.3)$$

where $B_{\text{agg.}}$ is the total binding energy of the aggregate to the solvent molecules; more explicitly,

$$B_{\text{agg.}} = \sum_{j=1}^{4} \sum_{i=1}^{N} U(\mathbf{R}_j, \mathbf{X}_i), \qquad (7.18.4)$$

where $U(\mathbf{R}_j, \mathbf{X}_i)$ is the solute–solvent pair potential between the jth solute at $\mathbf{R}_j$ and the ith solvent molecule at the configuration $\mathbf{X}_i$.

We notice that since the aggregate has a compact structure, no solvent molecule can penetrate into the interior of this aggregate. [Note, however, that the average $\langle\ \rangle_0$ is over all possible configurations of the solvent molecules. This also includes configurations for which solvent molecules do penetrate into the region occupied by the solute molecules. However, for each of the configurations in which a solvent molecule penetrates into this region, $B_{\text{agg.}}$ becomes very large and positive and hence $\exp(-B_{\text{agg.}}/kT)$ becomes practically zero.] In Fig. 7.32 we indicated by the dashed line the boundary of the excluded volume (assuming that the solvent is water, with a molecular diameter of 2.8 Å).

Clearly, the average in (7.18.3) gets nonzero contributions only from those configurations for which no solvent molecules penetrate into the excluded region produced by the four solute molecules. Therefore, if we insert in the center of this aggregate any particle or a group that produces a short-range field of force—such that it is not felt outside the excluded volume—the value of the average in (7.18.3) will not be affected. We exploit this fact to introduce into the center of the aggregate an "agent" that binds the four solute molecules in such a way that a real molecule is formed and for which the solvation Gibbs energy is measurable.

In our particular example we replace the aggregate of four methane molecules by one neopentane molecule. This replacement is shown schematically in Fig. 7.33. The approximation that is employed is

$$B(\text{agg.}) \approx B(\text{neopentane}). \tag{7.18.5}$$

If this is valid for all the configurations of the solvent molecules that have nonzero contribution to the average in (7.18.3), then we have the approximation

$$\Delta G^*(\text{agg.}) \approx \Delta G^*(\text{neopentane}). \tag{7.18.6}$$

Hence the exact relation (7.18.2) is transformed into the approximate, but more useful, relation

$$\delta G_4(\text{agg.}) = \Delta G^*(\text{neopentane}) - 4\Delta G^*(\text{methane}). \tag{7.18.7}$$

This should be compared with (7.18.1), which is a measure of the HϕO interaction among *five* solute particles. Here we have a measure of the HϕO interactions among *four* solutes at a configuration that is more realizable than the ones we have treated in section 7.15. Furthermore, the nature of the approximation involved in (7.18.7) is different from the one used in section 7.15. Here we have replaced the four inner hydrogens by one carbon center. In a sense we have partially compensated for the loss of the field of force produced by these hydrogens on the solvent. It is clear that had we started with four bulkier molecules, say four benzene molecules, and used the same procedure as above, we would have reached the relation

$$\delta G_4 = \Delta G^*(\text{TPM}) - 4\Delta G^*(\text{benzene}), \tag{7.18.8}$$

which measures the HϕO interaction among four benzene molecules holding the positions of the four benzyl radicals in tetraphenylmethane (TPM). In this case the boundaries of the excluded volume are quite far from the center of the aggregate. The effect of any replacement made at the center of this aggregate on the solvent becomes negligible. Clearly, the bulkier the four molecules are, the better is the replacement approximation that is used in (7.18.5) or (7.18.6).

Figure 7.34 presents some values of $\delta G_4(\text{agg.})$, defined in (7.18.7) as a function of the temperature. These values are compared with two other measures of the HϕO interaction among four methane molecules at the configuration of butane and isobutane. Note that the latter are systematically more negative than the corresponding values of $\delta G_4(\text{agg.})$. This is probably a result of the fact that the HϕO interaction becomes larger as the particles come closer together in the final configuration.

The quantity $\delta G(\text{agg.})$ in (7.18.7) measures the HϕO interaction among four methane molecules; the closest distance between any pair of molecules is about 2.51 Å, as compared to 1.5 Å between some of the pairs of methane molecules in Fig. 7.32. This is an improvement towards a more realistic configuration of solute molecules in real systems. One can make further improvement in this direction by taking four bulkier molecules, such as benzene or long-chain paraffin molecules, to form tetraalkyl or tetra-phenyl methane. In such cases the final configuration is very similar to an actually realizable configuration.

To conclude, Eqs. (7.18.7) and (7.18.8) are approximate relations between a molecular quantity, δG, and measurable quantities. These relations were obtained not for a real process, but for theoretical, on-paper, processes. The main trick is to manipulate the field of force produced by different sets of solute particles in such a way that changes in the field of force is unnoticed by the solvent. It is clear that this trick can be used to obtain

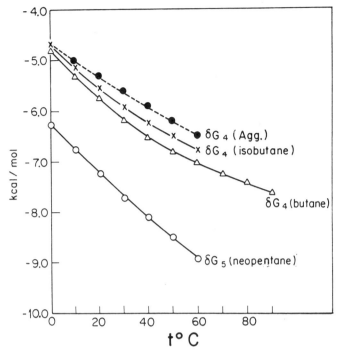

FIGURE 7.34. Values of δG_4 and δG_5 as a function of temperature for various configurations, as indicated next to each curve.

many other relations of the same kind. Improvement of the extent of approximation and of the realizability of the final configuration is limited only by the availability of relevant experimental data.

7.18.2. *HφO Interaction among Many Solute Particles Forming a Compact Aggregate in Water*

In the previous sections we presented two measures of the HφO interaction in which we made use of experimental data. In this section a partial theoretical approach to the problem is described. The basic process is the same as before. Namely, we start with m solute particles at fixed positions but at infinite separation from each other in a solvent at some given temperature T and pressure P. We then bring these particles to a close-packed configuration. More specifically, we require that the centers of all the particles be confined to a spherical region S_A, the radius of which is chosen as described below. The process is schematically written as

$$(\mathbf{R}^m = \infty) \rightarrow (\mathbf{R}^m \in S_A), \tag{7.18.9}$$

and the corresponding Gibbs energy change is

$$\Delta G_m = U_m + \delta G_m, \tag{7.18.10}$$

where U_m and δG_m are the direct and the indirect parts of the work required to carry out the process indicated in (7.18.9). Using the same argument as in section 7.18.1, we write for the indirect interaction the exact relation

$$\delta G_m = \Delta G_A^* - m\Delta G_M^*, \qquad (7.18.11)$$

where ΔG_M^* is the experimental solvation Gibbs energy of the monomer M, and ΔG_A^* is the solvation Gibbs energy of the close-packed aggregate A, viewed as a single entity.

In the previous sections we found an approximation for ΔG_A^* using experimental sources. Here, however, we appeal to the scaled particle theory (SPT) to estimate ΔG_A^*. The procedure of estimating ΔG_A^* by the SPT is the following: First, we split ΔG_A^* into two terms

$$\Delta G_A^* = \Delta G_A^*(\text{cav}) + \Delta G_A^*(\text{soft}), \qquad (7.18.12)$$

where the first term on the rhs of (7.18.12) is the work required to create a cavity of a suitable size (see below) in the solvent. The second term is due to the turning on of the soft (or attractive) part of the interaction between the aggregate A and the solvent. This is the same procedure as the one carried out in section 6.14.

We further assume that m is a large number, that the solute monomers are simple (e.g., argon, methane), and that the aggregate A has a spherical shape and consists of closely packed monomers. Following these assumptions, we expect that the soft part of the field of force of A will originate essentially from those molecules that are in direct contact with the solvent, i.e., the molecules that form the surface of the aggregate A. If the number of monomers m is large, the contribution of ΔG_A^* (soft) to ΔG_A^* becomes small compared to ΔG_A^* (cav). Thus for sufficiently large m we use the approximation

$$\Delta G_A^* \approx \Delta G_A^*(\text{cav}) \qquad (7.18.13)$$

where $\Delta G_A^*(\text{cav})$ may be computed from the SPT. Note that for a hard-sphere solute $\Delta G_A^*(\text{soft})$ is zero and (7.18.13) is an equality. We therefore expect that for a simple solute such as methane, (7.18.13) is a good approximation.

To proceed we need to estimate the size of the appropriate cavity in which the aggregate is to be accommodated. Let σ_M be the effective hard-core diameter of methane, which we take as equal to the Lennard–Jones diameter of methane $\sigma_M = 3.82$ Å. If m solutes of diameter σ_M are packed compactly in such a way that they form a sphere of diameter σ_A, it is well known that the ratio of the volume of the m particles to the volume of the sphere S_A is

$$\frac{m\pi\sigma_M^3/6}{\pi\sigma_A^3/6} = 0.7405. \qquad (7.18.14)$$

From (7.18.14) we may eliminate σ_A:

$$\sigma_A = \left(\frac{m\sigma_M^3}{0.7405}\right)^{1/3}. \qquad (7.18.15)$$

Let the diameter of the solvent molecules be σ_S; then the radius of the cavity produced by the aggregate A is given by

$$R_{cav} = \frac{\sigma_A + \sigma_S}{2}. \qquad (7.18.16)$$

The situation is schematically depicted in Fig. 7.35. Once we have the radius R_{cav}, the molecular diameter, and the number density of the solvent, we can use the SPT to estimate $\Delta G_A^*(\text{cav})$.

Thus in essence we have replaced the exact result in (7.18.11) by the approximate relation

$$\delta G_m \approx \Delta G_A^*(\text{cav}) - m\Delta G_M^*, \qquad (7.18.17)$$

where ΔG_M^* is taken from experimental sources and $\Delta G_A^*(\text{cav})$ is computed from the SPT. For the following numerical examples we use methane, with a molecular diameter of $\sigma_M = 3.82$ Å, as our monomer. The solvents used in these illustrations are water (and heavy water) with $\sigma_S = 2.90$ Å, methanol with $\sigma_S = 3.69$ Å, ethanol with $\sigma_S = 4.34$ Å, and cyclohexane with $\sigma_S = 5.63$ Å. These are effective hard-core diameters of the molecules that have been used in the application of the SPT. It should be borne in mind, however, that for nonspherical molecules, σ_S has no clear-cut physical meaning, as in the case of simple spherical molecules.

In addition to the effective molecular diameter, the solvents are characterized by their number densities at the given temperature and pressure. This information is sufficient for the computation of the Gibbs energy change associated with the formation of a cavity $\Delta G_A^*(\text{cav})$ and hence the computation of δG_m through relation (7.18.17).

The corresponding entropy and enthalpy of the process (7.18.9) are

$$\delta S_m = -\frac{\partial \delta G_m}{\partial T} = \Delta S_A^*(\text{cav}) - m\Delta S_M^* \qquad (7.18.18)$$

and

$$\delta H_m = \delta G_m + T\delta S_m$$
$$= \Delta H_A^*(\text{cav}) - m\Delta H_m^*, \qquad (7.18.19)$$

where again we use the SPT to compute $\Delta S_A^*(\text{cav})$ and $\Delta H_A^*(\text{cav})$ but use experimental sources for ΔS_M^* and ΔH_M^*.

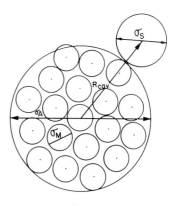

FIGURE 7.35. A cavity of radius R_{cav} is formed by an aggregate of diameter σ_A in a solvent; the diameter of the solvent molecules is σ_S.

In using the SPT to compute $\Delta S_A^*(\text{cav})$ and $\Delta H_A^*(\text{cav})$, we must also consider the temperature dependence of the effective hard-core diameter of the solvent molecules. (Only for hard-sphere particles is the molecular diameter, by definition, strictly temperature-independent.) As we have noted above, the effective diameter, especially for non-spherical molecules, is not a uniquely defined quantity, and clearly the same is true for its temperature dependence. Several procedures have been suggested in the literature to obtain this temperature dependence, but none is satisfactory from the theoretical point of view. This fact is another difficulty of the SPT when applied to complex solvents such as water, methanol, and ethanol.

Note that δS_m as defined in (7.18.18) is the same as the total entropy change for the process indicated in (7.18.9). On the other hand, δH_m is only the indirect part of the enthalpy change that corresponds to this process. The relation between the total and the indirect enthalpy changes is

$$\Delta H_m[(\mathbf{R}^m = \infty) \rightarrow (\mathbf{R}^m \in S_A)] = U_m + \delta H_m, \tag{7.18.20}$$

which should be compared with relation (7.18.10).

In Fig. 7.36 we present some computed values of δG_m as a function of m, the number of monomers. It is clear that as m becomes large enough (say $m \gtrsim 100$), the values of δG_m in water (and heavy water) become large and negative. In methanol, ethanol, and cyclohexane, the corresponding values are either positive or slightly negative. These results may be easily transformed into ratios of probabilities. Suppose we take two solvents, say water and methanol, both at the same temperature T and pressure P. Also, we assume that the solute M forms a very dilute solution in these two solvents, in such a way that the number density ρ_M is the same in the two solvents; i.e., $\rho_M(\text{in water}) = \rho_M(\text{in methanol})$. ($\rho_M$ is the number of solute molecules per unit volume of the solvent.) In such a solution we may ask what the probability is of finding a close-packed aggregate containing m solute molecules. Clearly, if we have given a precise configuration $\mathbf{R}^m = \mathbf{R}_1, \ldots, \mathbf{R}_m$ to these particles, then the probability of its occurrence is zero. However, the ratio of such probabilities in two solvents is a finite quantity. This ratio is given by

$$\eta = \frac{P_m(\text{water})}{P_m(\text{methanol})}$$

$$= \exp[-\beta\delta G_m(\text{water}) - \beta\delta G_m(\text{methanol})] \tag{7.18.21}$$

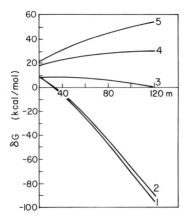

FIGURE 7.36. Values of δG_m as a function of the number of monomers m for different solvents. The solute is methane and the solvents are (1) H_2O; (2) D_2O; (3) methanol; (4) ethanol; and (5) cyclohexane. All values are for atmospheric pressure at $t = 30°C$.

As an example, we choose $m = 100$ and compute the ratio η at three temperatures. The results are

$$\eta(t = 10°C) \approx 4 \times 10^{32}, \qquad \eta(t = 30°C) \approx 5 \times 10^{53}, \qquad \eta(t = 60°C) \approx 2 \times 10^{57}.$$

$$(7.18.22)$$

These results indicate that in a dilute solution of the monomers M, as described above, the probability of finding a close-packed aggregate in water is far larger than the corresponding probability in methanol. Furthermore, this ratio becomes larger as the temperature increases in the range of temperatures of, say, $0 \lesssim t \lesssim 80°C$.

We have seen in section 7.15 that the pairwise HϕO interaction between methane molecules is stronger in H_2O as compared to D_2O. From Fig. 7.36 we see that the HϕO interaction among m solute particles in D_2O is weaker than in H_2O, in conformity with the behavior of pairwise HϕO interaction.

Figure 7.37 shows the variation with m of the three quantities involved in (7.18.17). We see that, for small m, values of $\Delta G_A^*(cav)$ and $m\Delta G_M^*$ are of comparable magnitude. As m increases, it is clear that the term $m\Delta G_M^*$ becomes the dominating one. This means that for large values of m, our computed results depend on the experimental rather than on the theoretical source. This conclusion may also be understood on intuitive grounds. To see this, we make a distinction between two kinds of monomers that build up our aggregate; let m_S be the number of solute monomers that form the surface of the aggregate (i.e., those that are in contact with the solvent) and m_I be the number of solute monomers that are in the interior of the aggregate (i.e., those that are surrounded by other solute monomers only).

Thus the overall process of aggregation

$$m \text{ monomers} \rightarrow \text{aggregate} \qquad (7.18.23)$$

may be viewed as being split into two parts:

$$m_S \text{ monomers} \rightarrow \text{surface of the aggregate} \qquad (7.18.24)$$

$$m_I \text{ monomers} \rightarrow \text{interior of the aggregate.} \qquad (7.18.25)$$

Clearly, for very large aggregates, the number of surface particles may be neglected with respect to the number of interior particles. This means that the reaction (7.18.25) will dominate the overall process (7.18.23). Hence the Gibbs energy change of the overall process will be determined by the Gibbs energy of transferring m_I monomers from the

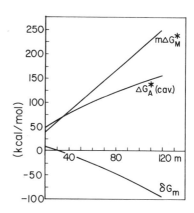

FIGURE 7.37. Values of δG_m, $\Delta G_A^*(cav)$, and $m\Delta G_M^*$ as a function of the number of monomers m in water at $P = 1$ atm and $t = 30°C$.

solvent into the interior of the aggregate. Furthermore, since we have eliminated the direct solute–solute interaction in the definition of δG_m, the process (7.18.25) is the same as transferring m_l solutes from the liquid to the gas. Hence, this process is approximately represented by $-m\Delta G_M^*$.

The above considerations are valid for very large m, in which case δG_m becomes essentially equal to m times $-\Delta G_M^*$. As we see from Fig. 7.37, for m on the order of 100, both terms in (7.18.17) contribute to δG_m; hence for such a size of aggregate we are still far from the limiting behavior that we mentioned above. This means that δG_m, with m on the order of 100, is more relevant to an aggregation process than to a mere reversal of the solubility of the monomer.

Figure 7.38a and b present the values of δS_m and δH_m as a function of the number of monomers m. These computations are based on the assumption that the diameter of the solvent molecules is temperature independent (see below).

The curves clearly indicate that both the entropy and the enthalpy changes associated with the process of aggregation in water are larger than the corresponding values in methanol, ethanol, and cyclohexane. These results are in complete agreement with the results obtained for $\delta G_2(\sigma_1)$ using the model of section 7.15.

It must be noted, however, that the question of which temperature dependence for the molecular diameter one should employ in these computations is not yet settled. Only for hard spheres is the diameter of the particles a well-defined quantity and temperature independent. For simple fluids, say argon, methane, etc., one may reasonably argue that the effective hard-core diameter should be a decreasing function of the temperature. The physical idea is that as one increases the temperature, the kinetic energy of the particles increases. Hence, on the average, interparticle collisions would lead to more extensive penetration into the repulsive region of the pair potential for the two particles. Indeed, it has been demonstrated that if such a negative temperature dependence of σ_S is adopted, then one can obtain a good agreement between the prediction from the SPT and experimental results.

The situation is far more complicated for nonspherical or more complex solvent molecules. In the first place, the very concept of a hard-core diameter is not a well-defined

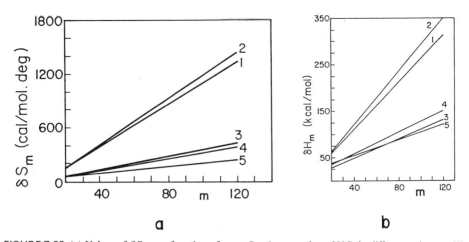

FIGURE 7.38. (a) Values of δS_m as a function of m, at $P = 1$ atm and $t = 30°$C, in different solvents: (1) H_2O; (2) D_2O; (3) methanol; (4) ethanol; (5) cyclohexane. In all the calculations the molecular diameter of the solvent is taken to be temperature-independent. (b) Values of δH_m for the same process.

quantity. For water, for instance, one may conveniently choose the effective diameter of the water molecule as the location of the first peak in the radial distribution function $g(R)$ for pure water. If we adopt this definition, we find that there exists a small positive temperature dependence of the molecular diameter of water. The rationale for this behavior is quite simple. In liquid water at room temperature most of the water molecules are engaged in hydrogen bonds. The optimal distance for a hydrogen bond is about 2.76 Å, well within the effective hard-core diameter assigned to a water molecule, about 2.8 Å. As we increase the temperature, we should consider at least two competing effects. On the one hand, we have the kinetic effect that was described above, which tends to decrease the effective hard-core diameters of free water molecules. On the other hand, hydrogen-bonded pairs are broken as we increase the temperature: hence, fewer pairs of molecules will be found at the relatively short distance of 2.76 Å. This tends to increase the effective hard-core diameters of the bonded molecules. Indeed it has been found that if one takes a positive temperature dependence for σ_S, the computed results from the SPT are more consistent with experimental findings.

REFERENCE

1. G. Ravishanker, M. Mezei, and D. L. Beveridge, *Faraday Sym. Chem. Soc.* **17**, 79 (1982).

SUGGESTED READINGS

A good survey on the outstanding properties of water is

D. Eisenberg and W. Kauzmann, *The Structure and Properties of Water* (Oxford University Press, Oxford, 1969).
More detailed information about many aspects of liquid water and aqueous solutions is provided in

F. Franks (ed.), *Water, A Comprehensive Treatise* (Plenum Press, New York, 1972–79).
Some theoretical and computational aspects of liquid water are reviewed in

F. H. Stillinger, *Science* **209**, 451 (1980).
D. L. Beveridge, M. Mezei, P. K. Mehrotra, F. T. Marchese, G. Ravishanker, T. Vasu, and S. Swaminathan, *Adv. Chem. Ser.* **204**, 297 (1983).
Specialized treatments of hydrophobic interactions are provided in

C. Tanford, *The Hydrophobic Effect: Formation of Micelles and Biological Membranes* (Wiley Interscience, New York, 1973).
A. Ben-Naim, *Hydrophobic Interactions* (Plenum Press, New York, 1980).
And a more recent review is

G. Ravishanker and D. L. Beveridge, in *Theoretical Chemistry of Biological Systems* (G. Naray-Szabo, ed.) (Elsevier Science Pub, Amsterdam, 1986).

On hydrophilic interactions, see

A. Ben-Naim, *J. Chem. Phys.* **90**, 7412 (1989).
A. Ben-Naim, *Biopolymers* **29**, 567 (1990).

8

Solvent Effects on Processes in Aqueous Solutions

8.1. INTRODUCTION

Most of the processes treated in Chapters 2 and 3 and part of 4 were assumed to take place in the vacuum. In reality these processes are usually carried out in a solvent. This is certainly true for biochemical processes such as the binding of oxygen to hemoglobin, the binding of substrate to enzymes, the helix–coil transition, etc. In a simple solvent it is often assumed that the theory itself may be retained except for an appropriate rescaling of the parameters involved. This, in general, is not true for complex solvents such as water. Here, the presence of a solvent might change not only the parameters involved in a specific theory, but also the theory itself. For instance, the PF of a simple Ising model with two-state units and nearest-neighbor interactions can be represented by a 2×2 matrix. The same model in a solvent could be affected in two ways. First, if the assumptions of the model can be retained, then only the parameters of the model are changed, e.g., the energies of each state or the interaction energies between nearest neighbors. Second, when the assumptions of the model cannot be retained, then the theory itself must be modified. For instance, if the solvent introduces next-nearest-neighbor correlations then the 2×2 matrix will not be sufficient to represent the PF of the system. Higher-order matrices would be necessary to treat the same model in a solvent. In more extreme cases, the 1-D model itself must be abandoned.

In this chapter we shall rederive some of the theories dealt with in the previous chapters and see where the solvent might modify the theory. We shall also discuss some new processes where the solvent plays a crucial role: the association and folding of proteins, the formation of micelles, or the aggregation of biological molecules.

There are essentially two classes of effects that the solvent can have on a given process. These can be demonstrated by the following example. Consider an equilibrium system where ρ_i is the equilibrium concentration of a species i. This can be a ligand, a polymer with a specific number of ligands, or a one-dimensional system with a specific configuration. The equilibrium concentration $(\rho_i)_{eq}$ is related to the chemical potential of that species, in the ideal gas phase by

$$\mu_i^{i.g.} = kT \ln(\rho_i)_{eq}^g \Lambda_i^3 q_i^{-1}. \tag{8.1.1}$$

In the liquid phase, the relation between the chemical potential and the equilibrium density of i can be modified in two ways. First, there is the coupling work of i against the entire solvent, which we denote by $W(i|l)$. Second, the solvent molecules can perturb the internal degrees of freedom of the solvated molecule. The latter is essentially a quantum-mechanical problem. We shall always assume that the effect of the solvent on the electronic, vibrational–rotational PF of a single molecule is negligibly small. Therefore the chemical potential of the species i in the liquid l will be written as

$$\mu_i^l = W(i|l) + kT \ln(\rho_i)_{eq}^l \Lambda_i^3 q_i^{-1}. \tag{8.1.2}$$

Thus, if we measure the equilibrium density $(\rho_i)_{eq}$ with respect to the same reference state, say a pure solid i, then the equilibrium densities $(\rho_i)_{eq}^g$ and $(\rho_i)_{eq}^l$ will be different only because of the coupling work $W(i|l)$. The internal PF q_i is assumed to be the same in (8.1.1) and (8.1.2).

If the molecule i has internal rotational degrees of freedom, then each conformation of i will be treated as a different species having different coupling work. This will effectively account for the effect of the solvent on the equilibrium distribution of the conformers. Proteins are examples of such molecules. These are treated in the subsequent sections of this chapter.

8.2. FROM PAIR POTENTIAL TO POTENTIAL OF AVERAGE FORCE

As our first example of solvent effects, we consider the process of bringing two particles from infinite separation to some close distance. In a vacuum the corresponding work is the (direct) pair potential. The work involved in carrying out the same process in a solvent, say at a given temperature and pressure, is the Gibbs energy change or the corresponding potential of average force.

In Fig. 5.8 we have demonstrated how the pair potential $U(R)$ for two Lennard–Jones particles is transformed into the potential of average force when the density of the surroundings of the pair of particles increases. We noticed that the first minimum of the potential of average force becomes deeper as we increase the density. Also, new minima appear at $R_2 \approx 2\sigma$, $R_3 \approx 3\sigma$, etc.

As a concrete example, consider argon, with a pair potential of the form given in Fig. 1.10. The work required to bring two argon atoms from infinity to a distance of $R_1 \approx \sigma \approx 3.4 \, \text{Å}$ is about

$$U_{AA}(R_1 \approx 3.5 \, \text{Å}) \approx -0.24 \, \text{kcal/mol}. \tag{8.2.1}$$

The work involved in the same process carried out in liquid argon can be obtained from the experimental radial distribution function. From Fig. 7.6 we estimate (assuming that $g(R)$ has the same form at room temperature)

$$W_{AA}(R_1 \approx 3.4 \, \text{Å, in liquid argon}) \approx -0.73 \, \text{kcal/mol}. \tag{8.2.2}$$

We see that the work required to perform the same process in the liquid is almost three times larger than in vacuum. Thus the "solvent" deepens the minimum well. In terms of force we can say that, approaching the minimum at $R_1 \approx 3.4 \, \text{Å}$, the solvent-induced force is attractive. It acts to enhance the attraction in that region (say between $R \approx 3.4 \, \text{Å}$ and $R \approx 5 \, \text{Å}$).

Although we do not have the potential of average force between two argon atoms in water, we can estimate the relative reduction in the work from the simulated results for two methanes in water. Based on the Lennard–Jones parameter $\varepsilon \approx -0.3 \, \text{kcal/mol}$ for methane, and the first minimum of $W_{AA}(R)$ from Fig. 7.20, which is about $-0.8 \, \text{kcal/mol}$, we can conclude that had we carried out the same process for two argon atoms in water, we would have obtained roughly the same value as for two argon atoms in liquid argon [Eq. (8.2.2)].

The results (8.2.1) and (8.2.2) can be translated into a ratio of the radial distribution functions as follows:

$$y_{AA}(R_1) = \frac{g_{AA}(R_1, \text{ in liquid argon})}{g_{AA}(R_1, \text{ in vacuum})} = \frac{\exp(-\beta W_{AA})}{\exp(-\beta U_{AA})} = \frac{77}{4.2} = 18. \tag{8.2.3}$$

The pair correlation function for two argon atoms at contact is about 18 times *larger* in the liquid (this is at $T = 84$ K and 0.7 atm) than in a vacuum at the same temperature. Assuming that for water at room temperature the value of $W_{AA}(R_1)$ is about -0.8 kcal/mol, we estimate

$$y_{AA}(R_1) = \frac{g_{AA}(R_1, \text{ in water})}{g_{AA}(R_1, \text{ in vacuum})} = \frac{3.8}{1.5} = 2.5. \tag{8.2.4}$$

In both cases, the solvent enhances the first peak of $g(R_1)$ in the liquid relative to the vacuum. The main reason for the different values in (8.2.3) and (8.2.4) results from the difference in the temperatures at which $y(R_1)$ was evaluated.

The change from vacuum to liquid is dramatically different when we consider the interaction between two water molecules. In vacuum, the work required to bring two water molecules from infinite separation to a fixed configuration so that they can form a HB is about

$$U_{ww}(R_1, \Omega_{HB}^{(1)}) \approx -6.5 \text{ kcal/mol.} \tag{8.2.5}$$

The configuration $(R_1, \Omega_{HB}^{(1)})$ is the same as the one we used in section 7.16; i.e., $R_1 \sim 2.8$ Å and $\Omega_{HB}^{(1)}$ corresponds to a specific orientation of a water molecule with respect to forming a HB with one arm of another water molecule.

The same process carried out in liquid water involves (see section 7.16) the work

$$W_{ww}(R_1, \Omega_{HB}^{(1)}) = U_{ww}(R_1, \Omega_{HB}^{(1)}) + \delta G_{ww}(R_1, \Omega_{HB}^{(1)})$$

$$\approx -6.5 + 4.5 - 0.6 \approx -2.6 \text{ kcal/mol.} \tag{8.2.6}$$

The work required to perform the same process in the liquid is almost three times *smaller* than in the vacuum. This is in sharp contrast with the effect we observed for liquid argon. As we have seen in section 7.16, the main reason for this effect is the loss of the solvation Gibbs energy of the two arms involved in the formation of a HB. In (8.2.6) we took 4.5 kcal/mol for the loss of the solvation Gibbs energy of the two arms and -0.6 kcal/mol to account for the hard and soft contributions to δG [see Eq. (7.16.6)]. In terms of force, two water molecules approaching each other at the configuration $\Omega_{HB}^{(1)}$ experience a weakened force, compared to the force in the vacuum. Again, this is in sharp contrast to the case of argon in liquid argon.

The figures in Eqs. (8.2.5) and (8.2.6) can be translated into a ratio of pair correlation functions as in (8.2.3), so that we find

$$y_{ww}(R_1, \Omega_{HB}^{(1)}) = \frac{g_{ww}(R_1, \Omega_{HB}^{(1)}, \text{ in water})}{g_{ww}(R_1, \Omega_{HB}^{(1)}, \text{ in vacuum})} = \frac{76}{5 \times 10^4} = 1.5 \times 10^{-3}. \tag{8.2.7}$$

Here, in contrast to the case of argon, the pair correlation function is reduced by a few orders of magnitude in passing from the vacuum to liquid water.

Some care must be exercised in interpreting the results (8.2.7) and (8.2.3) in terms of probabilities. The reason is that the probability is obtained from $g(R)$ by multiplying by $\rho^2 d\mathbf{R}$. The number densities are obviously different in the two phases. However, suppose we take two tagged argon molecules in a given volume V in the vacuum. Then we fill the entire volume with untagged argon atoms. In this case, the quantity in (8.2.3) is the ratio of the probabilities of finding the two tagged particles in the two phases. A similar interpretation applies to the quantity y_{ww} in (8.2.7) for two tagged water molecules.

In considering the pair correlation of water, we have chosen a specific relative orientation of the two water molecules. In applying these results to HϕI groups attached to biopolymers, we shall indeed need to consider a *fixed* relative orientation of the two groups. (We shall encounter such examples in the following sections.) The experimental pair correlation function of water is the orientational average of the function $g_{ww}(R_1, \Omega_1, \Omega_2)$. In order to obtain $\bar{g}_{ww}(R_1)$ we can either first transform the pair of water molecules into the liquid and then perform the average over all orientations (section 7.16), or we can first perform the average and then transfer the pair into the liquid.

The orientational average is

$$\bar{g}_{ww}(R_1) = \frac{1}{(8\pi^2)^2} \int g_{ww}(R_1, \Omega_1, \Omega_2)\, d\Omega_1\, d\Omega_2. \tag{8.2.8}$$

Here we integrate over all orientations of the two water molecules, leaving only the distance R_1 fixed (note that one angle in Ω_1, Ω_2 is redundant, since only five angles are sufficient to determine the relative orientation of the two molecules). Estimating this average for one model potential for a pair of water molecules gives a value of about

$$\bar{g}_{ww}(R_1, \text{in vacuum}) \approx 140. \tag{8.2.9}$$

This is reduced to a value of about two when transformed into the liquid—again, a drastic reduction due to the presence of the solvent.

8.3. CHEMICAL REACTION

We consider here the simplest chemical reaction treated in section 2.4.1, i.e., particles having two energy levels E_A and E_B with degeneracies ω_A and ω_B, respectively. We use the same notation as in section 2.4. The conversion reaction

$$A \rightleftarrows B \tag{8.3.1}$$

is now taking place in a solvent. For simplicity, we assume that the solvent is pure water w and that A and B are very dilute in the solvent. The generalization to any mixture of solvents, including high concentrations of A and B, is quite straightforward.

The condition of chemical equilibrium is

$$\mu_A = \mu_B. \tag{8.3.2}$$

Here, the expressions for the chemical potentials of A and B are:

$$\mu_A = W(A|w) + kT \ln \rho_A \Lambda_A^3 q_A^{-1} \tag{8.3.3}$$

$$\mu_B = W(B|w) + kT \ln \rho_B \Lambda_B^3 q_B^{-1}. \tag{8.3.4}$$

Note that q_A and q_B are the same quantities as the ones defined in section 2.4. There are only two internal energy levels, and these are measured relative to a common zero energy.

Note also that in this case, $\Lambda_A^3 = \Lambda_B^3$. The equilibrium constant is now

$$K = \left(\frac{\rho_B}{\rho_A}\right)_{eq}$$

$$= \frac{q_B}{q_A} \exp\{-\beta[W(B|w) - W(A|w)]\}$$

$$= \frac{\omega_B}{\omega_A} \exp[-\beta(E_B - E_A)] \exp\{-\beta[W(B|w) - W(A|w)]\}. \tag{8.3.5}$$

The modification of the equilibrium constant, Eq. (2.4.45), is in the addition of the new factor depending on the difference of the coupling work of B and A to the solvent. Since we always assume that the internal degrees of freedom are unaffected by the solvent, the difference in the coupling work is the same as the difference in the solvation Helmholtz or Gibbs energies depending on whether we are working in the T, V, N or T, P, N ensemble.

For concreteness, suppose we are working in the T, V, N ensemble then

$$K = \frac{\omega_B}{\omega_A} \exp[-\beta(E_B - E_A)] \exp - \beta(\Delta A_B^* - \Delta A_A^*)]$$

$$= \frac{\omega_B}{\omega_A} \exp[-\beta(E_B - E_A)] \exp[-\beta\delta A(A \to B)]. \tag{8.3.6}$$

In the last form on the rhs of (8.3.6), the new factor is the solvent-induced contribution to the conversion between the two isomers, $A \to B$. In the T, P, N ensemble, δG replaces δA in (8.3.6).

The standard Helmholtz energy change and the corresponding entropy and energy changes are

$$\Delta A^0 = -kT \ln K = -kT \ln \frac{\omega_B}{\omega_A} + (E_B - E_A) + (\Delta A_B^* - \Delta A_A^*) \tag{8.3.7}$$

$$\Delta S^0 = -\frac{\partial \Delta A^0}{\partial T} = k \ln \frac{\omega_B}{\omega_A} + (\Delta S_B^* - \Delta S_A^*) \tag{8.3.8}$$

$$\Delta E^0 = \Delta A^0 + T\Delta S^0 = (E_B - E_A) + (\Delta E_B^* - \Delta E_A^*). \tag{8.3.9}$$

These should be compared with Eqs. (2.4.43), (2.4.46), and (2.4.47). In section 2.4.3, we noted that in the simplest two-energy-level problem there is a clear-cut separation between the difference in the energy levels, $E_B - E_A$, that appeared in ΔE^0 in (2.4.47), and the ratio of the degeneracies ω_B/ω_A that appeared in ΔS^0 in (2.4.46). Here, the standard energy and entropy of the reaction are "coupled" in the sense that their values are no longer independent.

This is clearly seen when we write the expression for the solvation thermodynamic quantities in the form (section 7.11)

$$\Delta A_\alpha^* = -kT \ln\langle\exp(-\beta B_\alpha)\rangle_0 = W(\alpha|w) \tag{8.3.10}$$

$$\Delta S_\alpha^* = k \ln\langle\exp(-\beta B_\alpha)\rangle_0 + \frac{1}{T}(\langle B_\alpha\rangle_\alpha + \langle U_N\rangle_\alpha - \langle U_N\rangle_0) \tag{8.3.11}$$

$$\Delta E_\alpha^* = \langle B_\alpha\rangle_\alpha + \langle U_N\rangle_\alpha - \langle U_N\rangle_0, \tag{8.3.12}$$

where α is either A or B. Using Eqs. (8.3.10)–(8.3.12) in (8.3.7)–(8.3.9), we write

$$\Delta A^0 = -kT \ln \frac{\omega_B}{\omega_A} + (E_B - E_A) + [W(B|w) - W(A|w)] \qquad (8.3.13)$$

$$T\Delta S^0 = kT \ln \frac{\omega_B}{\omega_A} + [-W(B|w) + W(A|w)]$$
$$+ (\langle B_B \rangle_B - \langle B_A \rangle_A) + (\langle U_N \rangle_B - \langle U_N \rangle_A) \qquad (8.3.14)$$

$$\Delta E^0 = E_B - E_A + (\langle B_B \rangle_B - \langle B_A \rangle_A) + (\langle U_N \rangle_B - \langle U_N \rangle_A). \qquad (8.3.15)$$

The last two equations clearly show how ΔS^0 and ΔE^0 become coupled in the presence of a solvent. Both depend on the difference in the conditional average binding energies of A and B, as well as on the difference in the structural changes induced by A and B on the solvent. These two quantities exactly compensate for each other on forming the combination $\Delta A^0 = \Delta E^0 - T\Delta S^0$.

From the general Eqs. (8.3.13) to (8.3.15) we can make the following two conclusions. First, the equilibrium constant in an ideal-gaseous phase is determined by the competition between the *energy difference* and the *degeneracy ratio* of the two isomers. In solution, the equilibrium constant generally changes when the solvation Helmholtz energies of A and B are different. Second, the standard entropy and energy of the reaction in the absence of the solvent depend only on the ratio of the degeneracies and the difference in the energy levels, respectively. The modifications in ΔS^0 and ΔE^0 are more complicated than in the case of ΔA^0; ΔS^0 includes, now, the difference in the coupling work of A and B, the difference in the average binding energies of A and B, and the difference in the structural changes induced by A and B. The last two factors can contribute significantly to both ΔS^0 and ΔE^0, but they cancel out and do not contribute to ΔA^0.

The relation between ΔE^0 and $T\Delta S^0$ can be written as

$$\Delta E^0 = (E_B - E_A) + \Delta E^r \qquad (8.3.16)$$

$$T\Delta S^0 = kT \ln \frac{\omega_B}{\omega_A} + W(A|w) - W(B|w) + T\Delta S^r \qquad (8.3.17)$$

$$\Delta E^0 = \left[(E_B - E_A) - kT \ln \frac{\omega_B}{\omega_A} + W(B|w) - W(A|w) \right] + T\Delta S^0, \qquad (8.3.18)$$

where ΔE^r and $T\Delta S^r$ may be referred to as the relaxation terms in ΔE^0 and $T\Delta S^0$, respectively (see section 2.10). From (8.3.14) and (8.3.15) it follows that

$$\Delta E^r = T\Delta S^r. \qquad (8.3.19)$$

If ΔE^r in (8.3.16) and $T\Delta S^r$ in (8.3.17) dominates the values of ΔE^0 and $T\Delta S^0$, respectively, then the entire expression in the square brackets in (8.3.18) is small relative to the values of ΔE^0 and $T\Delta S^0$. Therefore a plot of the experimental values of ΔE^0 against ΔS^0 for various reactions (8.3.1) might give an approximate straight line with a slope of T. This is indeed observed experimentally for some processes in aqueous solutions. This is sometimes referred to as the entropy–enthalpy (or energy) compensation. One must distinguish, however, between the *exact* compensation in (8.3.19) and the *approximate* compensation between the experimental quantities ΔE^0 and $T\Delta S^0$ in (8.3.18).

8.4. SIMPLE LANGMUIR ISOTHERMS IN SOLUTION

In this section we treat two simple Langmuir isotherms. The adsorbing polymers as well as the ligand are now immersed in a liquid. Formally the binding isotherm looks the same as that of the vacuum examples treated in Chapter 2. However, because of solvation effects, the thermodynamics of the binding process could change considerably when the system is in a solvent. We shall first generalize the simplest Langmuir isotherm, then proceed to the case when the polymer can attain one of two conformational states. The latter will be used in the study of allosteric effects in a solvent in section 8.5.

8.4.1. Langmuir Isotherm in Solution with No Conformational Changes

We generalize the adsorption model of section 2.8. Most of the assumptions of the simple Langmuir model are still retained, except for two: the sites are not localized, and they are solvated by the solvent (the solvent may be any mixture of liquids, but for simplicity we assume that it is a one-component liquid, denoted by w). The simplest way to obtain the modified binding isotherm is to use the thermodynamic approach of section 2.9.1 with the appropriate modification introduced into the chemical equilibrium condition in solution, as in section 8.3.

Let S_0 and S_1 be the empty and occupied polymer in a solution containing ligand A. The adsorption of A on S_0 is viewed as a chemical reaction (section 2.9.1) of the form

$$S_0 + A \rightleftarrows S_1. \tag{8.4.1}$$

The condition of chemical equilibrium is

$$\mu_{S_0} + \mu_A = \mu_{S_1}. \tag{8.4.2}$$

Using the general expression for the chemical potential for each of the three species [see Eq. (8.3.3)], we rewrite (8.4.2) as

$$W(S_0|w) + kT \ln \rho_{S_0}\Lambda_{S_0}^3 Q_{S_0}^{-1} + kT \ln \lambda_A = W(S_1|w) + kT \ln \rho_{S_1}\Lambda_{S_1}^3 Q_{S_1}^{-1}. \tag{8.4.3}$$

We shall use Q_α for the internal PF of a single polymer and reserve q_A for the PF of a single ligand.

From (8.4.3) we obtain the ratio of the average number (or density) of the polymer in the occupied and the empty state, at equilibrium.

$$\frac{\rho_{S_1}}{\rho_{S_0}} = \lambda_A \frac{\Lambda_{S_0}^3 Q_{S_1}}{\Lambda_{S_1}^3 Q_{S_0}} \exp\{-\beta[W(S_1|w) - W(S_0|w)]\}$$

$$= \lambda_A Y_A. \tag{8.4.4}$$

In (8.4.4) we have defined the new binding constant Y_A. The fraction of occupied sites is

$$\theta = \frac{\rho_{S_1}}{\rho_{S_0} + \rho_{S_1}}. \tag{8.4.5}$$

From (8.4.4) and (8.4.5) we obtain the required generalized binding isotherm

$$\theta = \frac{Y_A \lambda_A}{1 + Y_A \lambda_A}, \tag{8.4.6}$$

which is formally the same as Eq. (2.8.23), except for the new definition of the constant Y_A in (8.4.4). In most practical cases the polymer is very large compared with the ligand A (say, myoglobin and oxygen). In such a case we can assume that the mass as well as the moments of inertia of the polymer do not change appreciably on binding of the ligand A. In this case, the momentum and the rotational PF of S_0 and S_1 are approximately the same; hence, we write

$$Q_{S_1} = q_A Q_{S_0} \exp(-\beta U), \tag{8.4.7}$$

where q_A is the internal PF of A. The approximate binding constant is now

$$Y_A = q_A \exp\{-\beta[U + W(S_1|w) - W(S_0|w)]\}. \tag{8.4.8}$$

This should be compared with Eq. (2.8.20). The modification here is the difference in the coupling work of S_0 and S_1 to the solvent. The adsorption isotherm (8.4.6) is still very general in the sense that A can come either from the solution or from a gaseous phase. If the ligands are in the same solution with density ρ_A, then λ_A is [compare with (2.8.25)]

$$\lambda_A = \rho_A \Lambda_A^3 q_A^{-1} \exp[\beta W(A|w)]. \tag{8.4.9}$$

Note that this relation is valid for any concentration ρ_A. In general, $W(A|w)$ is the coupling work of A to the solvent; the solvent may include any number of A molecules. In practical applications, however, the ligand is presumed to be very dilute in the solvent w. In that case $W(A|w)$ is independent of ρ_A; hence, λ_A is proportional to ρ_A. This is essentially the ideal dilute limit.

In terms of the density of the ligand the new isotherm can be written as

$$\theta = \frac{K'\rho_A}{1 + K'\rho_A}, \tag{8.4.10}$$

where the new binding constant is defined by

$$K' = \Lambda_A^3 \exp(-\beta U) \exp\{-\beta[W(S_1|w) - W(S_0|w) - W(A|w)]\}. \tag{8.4.11}$$

This should be compared with (2.8.16). The generalization of K' for the Langmuir isotherm of section 2.8.2 involves the new quantity

$$\delta G(A + S_0 \rightarrow S_1) = W(S_1|w) - W(S_0|w) - W(A|w), \tag{8.4.12}$$

which is exactly the indirect work required to bring A from an infinite position relative to S_0 to the final binding site to form S_1 (here in the T, P, N ensemble). Combining (8.4.11) with (8.4.12), we can write

$$K' = \Lambda_A^3 \exp[-\beta \Delta G(A + S_0 \rightarrow S_1)], \tag{8.4.13}$$

where $\Delta G(A + S_0 \rightarrow S_1)$ is the total change in the Gibbs energy for the same process as above. If we are working in the T, V, N ensemble, then we need to take the Helmholtz energy in (8.4.12) and (8.4.13). In the absence of a solvent, $\delta G = 0$ and (8.4.13) reduces to (2.8.16).

The second case of interest is when A comes from an ideal gaseous phase, say oxygen at a partial pressure P_A. In this case

$$\lambda_A = \lambda_A^{\text{i.g.}} = \frac{P_A \Lambda_A^3 q_A^{-1}}{kT}. \tag{8.4.14}$$

Substituting in (8.4.6) and defining the new constant

$$K = \frac{\Lambda_A^3}{kT} \exp\{-\beta[U + W(S_1|w) - W(S_0|w)]\}. \tag{8.4.15}$$

We can rewrite the isotherm in terms of the partial pressure P_A as

$$\theta = \frac{KP_A}{1 + KP_A}. \tag{8.4.16}$$

Compare (8.4.16) with (2.8.17). Note that K contains the work required to bring the ligand A from a fixed point in the vacuum to the binding site. The coupling work of the free ligand A does not appear in (8.4.15).

A particular simple case occurs when A is *absorbed* into the polymer (for instance, when the site is in the interior of the protein). In that case, the coupling work of S_0 and S_1 are nearly the same. (Note that in this model we assume that A does not perturb the polymer.) Hence the constant K reduces to

$$K = \frac{\Lambda^3}{kT} \exp(-\beta U), \tag{8.4.17}$$

which is the same as in (2.8.17). Thus, although the polymers are in a solution, the binding isotherm is identical with (2.8.17). If, on the other hand the ligand is in the liquid phase, then we use the isotherm (8.4.10) with the new binding constant

$$K'' = \Lambda_A^3 \exp(-\beta U) \exp[\beta W(A|w)]. \tag{8.4.18}$$

Here the constant K'' involves the loss of the solvation work of A and the gain of the binding energy of A to the site. We see that the various binding isotherms in a solvent have the same formal appearance as in section 2.8 except for the proper modification of the binding constants, required by the presence of the solvent.

8.4.2. Langmuir Isotherm with Conformational Changes

We now generalize the model treated in section 2.10, but this time the polymers and the ligands are in the liquid phase. Again, the simplest way to obtain the generalized adsorption isotherm as well as other thermodynamic quantities of the system is to view the adsorption process as a chemical reaction, exactly as in the previous section, i.e.,

$$S_0 + A \rightleftarrows S_1. \tag{8.4.19}$$

Formally, all the equations of the previous example apply here. The general result (8.4.10) is the same, with the reinterpretation of the coupling work $W(S_1|w)$ and $W(S_0|w)$ in terms of the coupling work of the two conformations. Also, the binding energy U is reinterpreted as an average in the same way as in section 2.10.

Again we assume that the polymer can be in one of two conformational states in equilibrium

$$L \rightleftarrows H. \tag{8.4.20}$$

Normally, these two states are characterized by their Helmholtz (or Gibbs) energy levels (see section 2.4.4) which include many quantum-mechanical energy levels. However, as in the case of section 2.10, in order to stress the new appearance of Helmholtz (or Gibbs) energies, we shall assume that L and H are characterized by their energy levels E_L and E_H only.

The chemical equilibrium condition for the empty polymers is

$$\mu_{L_0} = \mu_{H_0} \qquad (8.4.21)$$

or, equivalently,

$$W(L_0|w) + kT \ln \rho_{L_0} Q_{L_0}^{-1} = W(H_0|w) + kT \ln \rho_{H_0} Q_{H_0}^{-1}, \qquad (8.4.22)$$

which gives the equilibrium constant

$$K^* = \frac{\rho_{H_0}}{\rho_{L_0}}$$

$$= \frac{Q_{H_0}}{Q_{L_0}} \exp\{-\beta[W(H_0|w) - W(L_0|w)]\}$$

$$= \exp[-\beta(E_H - E_L)] \exp[-\beta \delta G(L \to H)]. \qquad (8.4.23)$$

This should be compared with (2.10.9). In the general case, Q_{H_0} and Q_{L_0}, the PFs of the *empty* L and H, could be different in their rotational–vibrational PFs (but not the momentum PF). Here we have assumed, as in section 2.10, that the two states L and H are characterized only by their energy levels E_L and E_H. The new factor that appears in (8.4.23) is the indirect work, here the Gibbs energy change, for the conversion of an empty L into an empty H.

The corresponding mole fractions of the empty polymers are

$$x_L^* = \frac{1}{1 + K^*}, \qquad x_H^* = \frac{K^*}{1 + K^*}, \qquad (8.4.24)$$

which are formally the same as (2.10.11) and (2.10.12), with the reinterpretation of the equilibrium constant as in (8.4.23).

Another way of writing the equilibrium constant K^* is by defining the Gibbs energy levels

$$G_{H_0} = E_H + W(H_0|w) \qquad (8.4.25)$$

$$G_{L_0} = E_L + W(L_0|w), \qquad (8.4.26)$$

in terms of which

$$K^* = \exp[-\beta(G_{H_0} - G_{L_0})]. \qquad (8.4.27)$$

In order to obtain the generalization of the binding constant Y_A in (8.4.6) we must find the relation between the quantities G_H and G_L and the corresponding average quantity of the polymer. This relation has been obtained in section 6.15 for the case of two isomers in solution. To obtain the required relation we use the identities

$$\mu_{S_0} = \mu_{L_0} = \mu_{H_0} \qquad (8.4.28)$$

and

$$\mu_{S_1} = \mu_{L_1} = \mu_{H_1}, \qquad (8.4.29)$$

from which we obtain (see section 6.15 for details)

$$Q_{S_0} \exp[-\beta W(S_0|w) = Q_{L_0} \exp[-\beta W(L_0|w)] + Q_{H_0} \exp[-\beta W(H_0|w)] \quad (8.4.30)$$

and

$$Q_{S_1} \exp[-\beta W(S_1|w)] = Q_{L_1} \exp[-\beta W(L_1|w) + Q_{H_1} \exp[-\beta W(H_1|w)]. \quad (8.4.31)$$

These can be substituted into the equilibrium condition (8.4.3) to obtain the generalization of (8.4.4). Using again the assumption that the polymer is much larger than the ligand, we have

$$\frac{\rho_{S_1}}{\rho_{S_0}} = \lambda_A \frac{Q_{S_1} \exp[-\beta W(S_1|w)]}{Q_{S_0} \exp[-\beta W(S_0|w)]}$$

$$= \lambda_A q_A \frac{\exp\{-\beta[E_L + U_L + W(L_1|w)]\} + \exp\{-\beta[E_H + U_H + W(H_1|w)]\}}{\exp\{-\beta[E_L + W(L_0|w)]\} + \exp\{-\beta[E_H + W(H_0|w)]\}}$$

$$= \lambda_A q_A \frac{\exp[-\beta(G_{L_1} + U_L)] + \exp[-\beta(G_{H_1} + U_H)]}{\exp(-\beta G_{L_0}) + \exp(-\beta G_{H_0})}$$

$$= \lambda_A q_A \{x_L^* \exp[-\beta(U_L + G_{L_1} - G_{L_0})] + x_H^* \exp[-\beta(U_H + G_{H_1} - G_{H_0})]\}$$

$$= \lambda_A Y_A, \quad (8.4.32)$$

where we have defined the new binding constant Y_A, which is a generalization of (8.4.8). If the solvent is absent, then this constant reduces to average binding constant of section 2.10. Note that $G_{L_1} - G_{L_0}$ is essentially the difference in the coupling work of S_0 and S_1. If the ligand is absorbed into the polymer, then $G_{L_1} - G_{L_0} = G_{H_1} - G_{H_0} = 0$. The average in (8.4.32) is still modified from the average in section 2.10 [see Eq. (2.10.35)] by the new mole fractions x_L^* and x_H^* of the empty polymers, but solvated by the solvent.

A somewhat simpler form of the binding constant can be obtained by generalizing (8.4.10) instead of (8.4.6). In that case, the generalization of K' is

$$K' = \Lambda_A^3 \{x_L^* \exp[-\beta \Delta G(A + S_{L_0} \rightarrow S_{L_1})] + x_H^* \exp[-\beta \Delta G(A + S_{H_0} \rightarrow S_{H_1})]\}, \quad (8.4.33)$$

where

$$\Delta G(A + S_{H_0} \rightarrow S_{H_1}) = U_H + \delta G(A + S_{H_0} \rightarrow S_{H_1}) \quad (8.4.34)$$

$$\Delta G(A + S_{L_0} \rightarrow S_{L_1}) = U_L + \delta G(A + S_{L_0} \rightarrow S_{L_1}). \quad (8.4.35)$$

The last two quantities may be interpreted as the potential of average force (here in the T, P, N ensemble) between A and the polymer (either in the state L or H) at the binding site.

The binding isotherm is now

$$\theta = \frac{K' \rho_A}{1 + K' \rho_A}, \quad (8.4.36)$$

where K' is given by (8.4.33) instead of (8.4.11). If there is only one state, then (8.4.33) reduces to (8.4.11). If there is no solvent, then (8.4.33) reduces to

$$K' = \Lambda_A^3 [x_L^0 \exp(-\beta U_L) + x_H^0 \exp(-\beta U_H)], \quad (8.4.37)$$

where now the expression within the square brackets in (8.4.37) is the same as K in section 2.10.2.

Another quantity of interest in this model is the extent of conformational change induced by the adsorption of the ligand. This question was given an exact answer in section 2.10.3 for the vacuum case. One can repeat the fairly lengthy derivation of the same quantity in solution. This is unnecessary, however. A shortcut to obtain this and other analogous results is to write the equilibrium constant for the fully occupied polymer as

$$K^{**} = \exp[-\beta(G_{H_1} - G_{L_1})], \tag{8.4.38}$$

where the new energy levels are now [see (8.4.25), (8.4.26)]

$$G_{H_1} = E_H + U_H + W(H_1|w) \tag{8.4.39}$$

$$G_{L_1} = E_L + U_L + W(L_1|w). \tag{8.4.40}$$

The ratio between the two equilibrium constants K^{**} and K^* defines the new parameter

$$h^* = \frac{K^{**}}{K^*} = \exp[-\beta\Delta G(A + S_{L_0} \to S_{L_1}) + \beta\Delta G(A + S_{H_0} \to S_{H_1})]. \tag{8.4.41}$$

The last relation is essentially the same as the relation between K and Kh of section 2.10. The differential conformational change is thus

$$d_L^* = \left(\frac{\partial x_L^*}{\partial\theta}\right)_{T,P,M,N_w} = \frac{K^*(1-h^*)}{(1+K^*)(1+h^*K^*)} = \frac{(1-h^*)x_L^*x_H^*}{x_L^* + h^*x_H^*}, \tag{8.4.42}$$

which is formally the same as (2.10.69), but with the reinterpretation of the quantities K^*, h^*, x_L^*, and x_H^*.

The $L \rightleftarrows H$ equilibrium constants K^* and K^{**} are now determined not only by E_L and E_H, or by $E_L + U_L$ and $E_H + U_H$ as in the vacuum case, but are modified by the solvation Gibbs energy of the various species. Similarly, h defined in section 2.10 depends only on the difference $U_H - U_L$. Here, h^* depends on the difference in the Gibbs energy changes for bringing the ligand from infinite separation to the site on L or on H, respectively.

As an illustration of the above two modifications, consider first the case that the two conformations have almost the same energy $E_L \approx E_H$. In this case the binding of a ligand to such a polymer in the vacuum would cause a differential conformational change as in (2.10.69), with $K = 1$. However, this conformational change will not involve any contribution to the entropy and energy of the binding process [see (2.10.80), (2.10.81), (2.10.83), and (2.10.85)]. The situation could be markedly different when the same polymer is in water. For instance, a small change in the location of two functional groups, say C=O and NH, could bring these to form a strong HϕI interaction. In such a case, although $E_L \approx E_H$, the difference $G_H - G_L$ could be on the order of 3 kcal/mol per one HϕI interaction. The situation is schematically illustrated in Fig. 8.1, where a small rotation about one bond can produce a large change in the solvation of the protein.

The second example involves a ligand for which $U_L \approx U_H$. This will lead to $d_L = 0$ in the vacuum example of section 2.10. The same system placed in water could produce a large conformational change due to the difference in δG for the H and the L form [Eqs. (8.4.34) and (8.4.35)]. To illustrate this case, we use the same example as in Fig. 8.1, but now we assume that the binding site for the ligand A is in the same region where a HϕI interaction exists in the L form but not in the H form. (Note that in the previous example we were concerned only with the difference between the solvation Gibbs

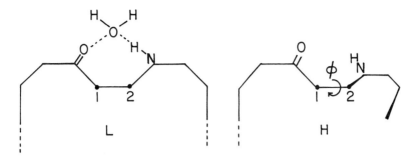

FIGURE 8.1. Two conformations of a protein. In the L form a strong HϕI interaction exists. A small conformational change, resulting from a rotation about a single bond (1—2), can eliminate this interaction and hence increase the Gibbs energy of solvation by about 3 kcal/mol.

energies of L and H forms. The binding site could be any place in the polymer. Here we are concerned with the difference in the solvation Gibbs energy caused by the binding of the ligand A.)

Clearly, if A blocks the possibility of formation of a HϕI interaction in the L form, then $\delta G(A + S_{L_0} \rightarrow S_{L_1})$ will be of about $+3$ kcal/mol more positive than $\delta G(A + S_{H_0} \rightarrow S_{H_1})$, where the ligand is presumed not to interfere with the solvation of the C=O and the NH groups. Therefore, although $U_L \approx U_H$, the modified constant h^* in (8.4.41) could induce a large conformational change. In this particular example, it would favor the H form.

8.5. ALLOSTERIC SYSTEMS IN SOLUTION

In the previous section we generalized two of the Langmuir models to include solvent effects on the binding isotherm. The formal generalization of all the models treated in Chapter 3 to include solvent effect is quite straightforward. We shall focus here only on the new quantity that characterizes the allosteric models: the ligand–ligand correlation functions. In section 3.5.4 we mentioned two possible sources of solvent effects on the correlation function, hence on the extent of cooperativity of the system. These are, indirect correlation between ligands mediated by the solvent only and indirect correlation mediated by both the polymer and the solvent. We shall study these two effects separately in the following two subsections.

8.5.1. Ligand–Ligand Correlation Mediated by the Solvent

We use the model of section 3.1 to introduce the ligand–ligand correlation mediated by the solvent. We use here the same notation as in Chapter 3. Direct correlation due to direct interaction is denoted by S. Indirect correlation mediated by the polymer only was denoted by y. In this section we introduce a new source of indirect correlation between the ligands. In order to stress this new correlation due to the solvent, we have chosen the model of section 3.1, i.e., no conformational change, hence $y = 1$.

The pair correlation function for two ligands on the same polymer is related to the work, here simply the energy, required to convert two singly occupied polymers into one empty and one doubly occupied polymer. The reaction is symbolically written as [see (3.1.22)]

$$2(0, 1) \rightarrow (1, 1) + (0, 0) \tag{8.5.1}$$

or, in the notation of this chapter,

$$2S_1 \rightarrow S_2 + S_0.$$

(Note that S without a subscript is the direct correlation as defined in Chapter 3; on the other hand, S_n denotes a polymer with n ligands. Thus S_0 is the empty polymer which is also denoted by $\mathbf{P}$, S_1 the singly occupied polymer, etc.) In this particular model, the corresponding work is simply the direct ligand–ligand interaction energy U_{12}.

The simplest way to obtain the corresponding Gibbs (or Helmholtz) energy change for the same process (8.5.1) in the liquid phase, keeping P, T constant (or T, V constant) is to appeal to the diagram in Fig. 8.2. Instead of carrying out the process in the liquid, we first remove the two polymers from the liquid to the gaseous phase, perform the process (8.5.1), and reintroduce the products into the liquid. The balance of the Gibbs energy for the cyclic process is

$$\Delta G^l(2S_1 \rightarrow S_2 + S_0) = -2\Delta G_{0,1}^* + \Delta G^g(2S_1 \rightarrow S_2 + S_0) + \Delta G_{0,0}^* + \Delta G_{1,1}^*, \tag{8.5.2}$$

where ΔG^l and ΔG^g are the Gibbs energy changes for process (8.5.1) in the liquid and gaseous phases, respectively, and the quantities ΔG_α^* are the solvation Gibbs energies of the various species α as indicated. In this particular case

$$\Delta G^g(2S_1 \rightarrow S_2 + S_0) = U_{12}. \tag{8.5.3}$$

Hence the solvent contribution to ΔG^l is defined by

$$\delta G(2S_1 \rightarrow S_2 + S_0) = \Delta G^l(2S_1 \rightarrow S_2 + S_0) - \Delta G^g(2S_1 \rightarrow S_2 + S_0)$$
$$= \Delta G_{1,1}^* + \Delta G_{0,0}^* - 2\Delta G_{0,1}^*. \tag{8.5.4}$$

The total ligand–ligand pair correlation function is now, in generalization of (3.1.24),

$$g^0(1, 1) = Sy'(1, 1), \tag{8.5.5}$$

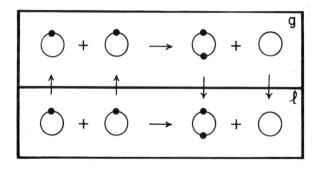

FIGURE 8.2. A cyclic process corresponding to Eq. (8.5.2). Instead of performing the reaction (8.5.1) in the liquid phase, we first transform the two polymers (0.1) to the gaseous phase. We perform the conversion (8.5.1) and reintroduce the new polymers (0, 0) and (1, 1) into the liquid.

where y' is referred to as the indirect ligand–ligand correlation function mediated by the solvent. Formally, this is defined as

$$y'(1, 1) = \exp[-\beta(\Delta G_{1,1}^* + \Delta G_{0,0}^* - 2\Delta G_{0,1}^*)]. \qquad (8.5.6)$$

In order to focus on the origin of this new correlation we rewrite each of the solvation Gibbs energies as

$$\Delta G_{1,1}^* = -kT \ln\langle \exp[-\beta B(1, 1)]\rangle_0$$
$$= -kT \ln\langle \exp(-\beta B_{\mathbf{P}} - \beta B_{A,A})\rangle_0$$
$$= -kT \ln\langle \exp(-\beta B_{\mathbf{P}})\rangle_0\langle \exp(-\beta B_{A,A})\rangle_{\mathbf{P}}$$
$$= \Delta G_p^* + \Delta G_{A,A/\mathbf{P}}^*, \qquad (8.5.7)$$

where the first average is over all configurations of the solvent molecules in the T, P, N ensemble. In the second form on the rhs of (8.5.7) we have separated the total binding energy to the solvent $B(1, 1)$ into two terms; the binding energy of the polymer $B_{\mathbf{P}}$, which is the same as $B(0, 0)$, and the binding energy of the two ligands at the sites on the polymer. On the third form we factorized the average into two factors. One is related to the solvation Gibbs energy of the empty polymer (this is the same as $\Delta G_{0,0}^*$) and the second is the conditional solvation Gibbs energy of the pair of ligands, given the polymer. Similarly, for $\Delta G_{0,0}^*$ and $\Delta G_{1,0}^*$ we have

$$\Delta G_{0,0}^* = \Delta G_{\mathbf{P}}^* \qquad (8.5.8)$$

$$\Delta G_{0,1}^* = \Delta G_{\mathbf{P}}^* + \Delta G_{A/\mathbf{P}}^*. \qquad (8.5.9)$$

Since $\Delta G_{\mathbf{P}}^*$ appears in (8.5.7), (8.5.8), and (8.5.9), it is canceled out in (8.5.6). The final expression for $y'(1, 1)$ is thus

$$y'(1, 1) = \exp[-\beta(\Delta G_{A,A/\mathbf{P}}^* - 2\Delta G_{A/\mathbf{P}}^*)]$$
$$= \frac{\langle \exp(-\beta B_{A,A})\rangle_{\mathbf{P}}}{\langle \exp(-\beta B_A)\rangle_{\mathbf{P}}^2}. \qquad (8.5.10)$$

This is almost the same as the solvaton–solvaton pair correlation function in liquids (see, for example, section 7.15). The only difference is the appearance of the condition $\mathbf{P}$, i.e., the presence of the polymer.

Thus, for the model of section 3.1, when transferred into the solvent, there are two sources for ligand–ligand correlation, and hence cooperativity. One is the direct ligand–ligand interaction, and the second is a solvent-induced correlation, which is essentially the same as the solvent-induced correlation between two solvatons in any liquid, except for the presence of the polymer. Since we know that the pair correlation function in liquids has a range of a few molecular diameters, we can expect that $y' \neq 1$ even when the ligand–ligand distance is larger than the range of the direct interaction. Note that the direct and indirect cooperativity could be independently zero, positive, or negative.

As an example, suppose that the two ligands are far enough apart so that direct interaction is negligible (i.e., $S = 1$) but they are within the range of the indirect correlation. Figure 8.3 shows two possible cases. In one case the HϕI–HϕI interaction produces a positive cooperativity, which could be quite large, on the order of $y' \approx \exp(5) \approx 150$. In the second, a H$\phi$I–H$\phi$O interaction produces a negative cooperativity. In this case, the methyl group simply interferes with the solvation of the hydroxyl group on the second ligand—(see example in section 7.17). Thus, in this particular model, the binding isotherm

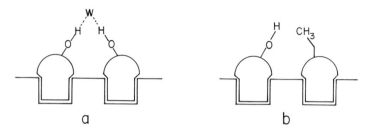

FIGURE 8.3. (a) Two ligands forming a strong HϕI interaction, hence contributing to the positive cooperativity, $y' > 1$ in Eq. (8.5.10). (b) Two ligands contributing negatively to the cooperativity of the system. (Note that the two ligands in each case are identical; only the correlated groups are shown in the figure.)

will be a simple Langmuir type in the gaseous phase (no direct interactions). Placed in water, the same model could be cooperative (either positively or negatively); i.e., the binding of one ligand strongly affects the probability of the binding of a second ligand.

This effect is not expected to be significant if the two ligands are very far apart, or if they are buried in the polymer so that they are not exposed to the solvent. In the first case, the indirect correlation is negligible; in the second case, it is the condition (**P** in 8.5.10) that will preclude the solvent molecules to reach the ligands. In the next subsection we shall encounter a different solvent-mediated effect which could be operative even when the two ligands are very far apart. We conclude this section by generalizing the relation between the Gibbs energy change for binding on first and second sites.

The Gibbs energy change for binding to the first site, i.e., for the process

$$A + (0, 0) \rightarrow (1, 0) \tag{8.5.11}$$

can be obtained by using a similar cyclic process to that in Fig. 8.2, the result being

$$\Delta G^{*(1)} = U + \Delta G^{*}_{1,0} - \Delta G^{*}_{0,0} - \Delta G^{*}_{A}$$

$$= U + \Delta G^{*}_{A/\mathbf{P}} - \Delta G^{*}_{A}. \tag{8.5.12}$$

Similarly, for binding to the second site, i.e., for the process

$$A + (1, 0) \rightarrow (1, 1) \tag{8.5.13}$$

we have

$$\Delta G^{*(2)} = U + U_{12} + \Delta G^{*}_{1,1} - \Delta G^{*}_{1,0} - \Delta G^{*}_{A} \tag{8.5.14}$$

$$= U + U_{12} + \Delta G^{*}_{A,A/\mathbf{P}} - \Delta G^{*}_{A/\mathbf{P}} - \Delta G^{*}_{A}. \tag{8.5.15}$$

Hence, for $\Delta G^{I}(2S_1 \rightarrow S_2 + S_0)$, we have

$$\Delta G^{I}(2S_1 \rightarrow S_2 + S_0) = \Delta G^{*(2)} - \Delta G^{*(1)}$$

$$= U_{12} + \Delta G^{*}_{A,A/\mathbf{P}} - 2\Delta G^{*}_{A/\mathbf{P}} \tag{8.5.16}$$

or, equivalently,

$$\exp(-\beta \Delta G^{*(2)}) = S y'(1, 1) \exp(-\beta \Delta G^{*(1)}), \tag{8.5.17}$$

which is formally the same as (3.1.28), but with the additional indirect correlation $y'(1, 1)$.

8.5.2. Ligand–Ligand Correlation Mediated by the Polymer and the Solvent

In this section we use the model of section 3.2, but in a solvent. Again, to minimize the complexity of the problem and in order to focus on a new phenomenon, we assume that the two ligands are very far apart so that both direct correlation and indirect correlation of the type discussed in section 8.5.1 are negligible; i.e., $S = 1$ and $y'(1, 1) = 1$.

In section 3.2 we studied the indirect correlation $y(1, 1)$ arising from conformational equilibrium $L \rightleftarrows H$. We also found that in this model we always have positive cooperativity in the sense $y(1, 1) > 1$ (unless $h = 1$ or $K = 0$).

To study the same model in a solvent, we again use the cyclic process as in Fig. 8.2, but now the species are denoted by S_0, S_1, and S_2 for the empty, singly occupied, and doubly occupied polymer. Each of these can be either L or H. We shall use the notation L_i and H_i ($i = 0, 1, 2$) for the polymer in state L or H having i ligands.

The pair correlation function is related to the work associated with the process

$$2(0, 1) \rightarrow (1, 1) + (0, 0)$$

or, equivalently,

$$2S_1 \rightarrow S_2 + S_0, \qquad (8.5.18)$$

which is the same as (8.5.1) except that now each of the species S_0, S_1, and S_2 can be in one of two states, L or H. We can use the same cyclic process as in Fig. 8.2 to obtain the solvent contribution to the correlation function. As in Eq. (8.5.2), the Gibbs energy change for the process $(2S_1 \rightarrow S_2 + S_0)$ is

$$\Delta G^l(2S_1 \rightarrow S_2 + S_0) = \Delta G^g(2S_1 \rightarrow S_2 + S_0) + \Delta G_{1,1}^* + \Delta G_{0,0}^* - 2\Delta G_{1,0}^*. \quad (8.5.19)$$

In section 8.5.1, the quantity ΔG^g was simply the interaction *energy* [Eq. (8.5.3)]. By passing to the liquid, the solvation of the species S_0, S_1, and S_2, has converted ΔG^g into a Gibbs energy change [see Eq. (8.5.2)]. In our case, ΔG^g in (8.5.19) is already a Gibbs energy change due to the existence of conformational equilibrium (section 3.2), and by adding the solvation Gibbs energies we add the solvent contribution to the Gibbs energy of the process, which in our case is

$$\delta G(2S_1 \rightarrow S_2 + S_0) = \Delta G_{1,1}^* + \Delta G_{0,0}^* - 2\Delta G_{1,0}^*. \quad (8.5.20)$$

Each of the solvation Gibbs energies in (8.5.20) can be expressed as a combination of the solvation Gibbs energies of the L and H species. The relation for each species S_α ($\alpha = 0, 1, 2$) is (see section 6.15) is

$$\exp[-\beta \Delta G^*(S_\alpha)] = x_L^{(\alpha)} \exp[-\beta \Delta G^*(L_\alpha)] + x_H^{(\alpha)} \exp[-\beta \Delta G^*(H_\alpha)], \quad (8.5.21)$$

where $x_L^{(\alpha)}$ are the mole fractions of the L_α species in the gaseous phase, i.e.,

$$x_L^{(0)} = x_L^0 = \frac{1}{1 + K}$$

$$x_L^{(1)} = \frac{1}{1 + Kh} \qquad (8.5.22)$$

$$x_L^{(2)} = \frac{1}{1 + Kh^2}.$$

For each of the solvation Gibbs energies on the rhs of (8.5.21) we factorize as in (8.5.7). For example

$$\exp[-\beta\Delta G^*(L_2)] = \langle\exp[-\beta B(L_2)]\rangle_0$$

$$= \langle\exp(-\beta B_L)\rangle_0\langle\exp(-\beta B_{A,A})\rangle_L. \qquad (8.5.23)$$

The total ligand–ligand pair correlation function, see (8.5.19) and (8.5.20) is

$$g^0(1, 1) = \exp[-\beta\Delta G'(2S_1 \to S_2 + S_0)]$$

$$= y(1, 1)y''(1, 1), \qquad (8.5.24)$$

where $y(1, 1)$ is the indirect correlation function resulting from the conformational changes, as in section 3.2. The new factor $y''(1, 1)$ includes both ligand–ligand correlation of the type discussed in section 8.5.1 and a new effect due to the difference of the solvation Gibbs energies of the L and H form. In order to isolate this effect, suppose that $y(1, 1) = 1$. Furthermore, suppose that the ligands are so far apart that even the ligand–ligand correlation mediated by the solvent is negligible. This is equivalent to the factorizations

$$\langle\exp(-\beta B_{A,A})\rangle_L = \langle\exp(-\beta B_A)\rangle_L^2 \qquad (8.5.25)$$

$$\langle\exp(-\beta B_{A,A})\rangle_H = \langle\exp(-\beta B_A)\rangle_H^2; \qquad (8.5.26)$$

i.e., y' of section 8.5.1 is unity for each of the species [see (8.5.10)].

With these assumptions the remaining indirect correlation function is

$$y''(1, 1) = \exp[-\beta\delta G(2S_1 \to S_2 + S_0)]$$

$$= \exp(-\beta\Delta G_{1,1}^* - \beta\Delta G_{0,0}^* + 2\beta\Delta G_{1,0}^*). \qquad (8.5.27)$$

Each of the solvation quantities in (8.5.27) can be further expressed, as in (8.5.21) and (8.5.23). The general result is quite complicated. In order to focus on the new correlation, we choose a particularly simple example. Suppose that $h = 1$ (i.e., $U_L = U_H$) and $K = 1$ (i.e., $E_L - E_H$), in which case $y(1, 1) = 1$ (section 3.2). In this case also $x_L^{(\alpha)} = x_H^{(\alpha)} = \frac{1}{2}$ and (8.5.27) reduces to a simple form.

We denote by

$$h'' = \frac{\langle\exp(-\beta B_A)\rangle_H}{\langle\exp(-\beta B_A)\rangle_L} \qquad (8.5.28)$$

and

$$K'' = \frac{\langle\exp(-\beta B_H)\rangle_0}{\langle\exp(-\beta B_L)\rangle_0}. \qquad (8.5.29)$$

Then for the particular assumptions made above, (8.5.27) reduces to

$$y''(1, 1) = \frac{(1 + K'')[1 + K''(h'')^2]}{[1 + h''K'']^2}. \qquad (8.5.30)$$

This is formally the same as $y(1, 1)$ in (3.2.30), with the new interpretation of the h and K of section 3.2. Note that K'' and h'' may be obtained from K^* [Eq. (8.4.27)] and h^*

[Eq. (8.4.41)] by eliminating h and K of section 3.2; i.e.,

$$K^* = \exp[-\beta(G_{H_0} - G_{L_0})]$$
$$= \exp[-\beta(E_H - E_L)] \exp[-\beta\delta G(L_0 \to H_0)]$$
$$= KK'' \tag{8.5.31}$$

and

$$h^* = \exp[-\beta(U_H - U_L)] \exp[-\beta\delta G(A + S_{H_0} \to S_{H_1}) + \beta\delta G(A + S_{L_0} \to S_{L_1})]$$
$$= h \exp[-\beta(\Delta G_{H_1}^* - \Delta G_{H_0}^*)] \exp[-\beta(\Delta G_{L_1}^* - \Delta G_{L_0}^*)]$$
$$= h \exp[-\beta(\Delta G_{A/H}^* - \Delta G_{A/L}^*)] = hh''. \tag{8.5.32}$$

Note that from (8.5.30) it follows as in section 3.2 that $y''(1, 1) = 1$ if either $h'' = 1$ or $K'' = 0$. For all other cases, $y''(1, 1) > 1$; i.e., this effect produces positive cooperativity.

Thus, the solvent can produce a positive ligand–ligand correlation when there is a finite difference in the solvation Gibbs energies of L_0 and H_0, and if the conditional solvation of the ligand A, given H, differs from the conditional solvation of A given L. As we noted in the preceding section, this correlation could be large, even when the ligands are very far apart (so that they are uncorrelated either directly, $S = 1$, or indirectly through the solvent, $y' = 1$). Therefore, in considering the cooperativity of multisubunit systems such as hemoglobin or regulatory enzymes, one must take into consideration also the solvation properties of both the empty and the occupied polymers.

As a final comment we note that the extent of cooperativity in the vacuum examples treated in Chapter 3 could be varied by changing the temperature. The same system in a solvent has a new degree of freedom, i.e., one can change the extent of the cooperativity of a system by changing the properties of the solvent. For instance, the addition of various solutes could change the cooperativity to such an extent as to cause loading or unloading of oxygen onto or from hemoglobin. In this sense, the solvent effect can be viewed as a generalized allosteric effect. In the simple allosteric mechanism we have looked at the *binding* of specific effectors on specific sites, to cause a conformational change—which in turn was noticed at a distant site. In the case of solvent effects, the binding of specific effectors on specific sites is replaced by the total *interaction* of the polymer with all its surrounding solvent molecules. In real systems, both specific effectors and solvent effects combine to determine the behavior of biomolecules.

8.6. ONE-DIMENSIONAL MODELS IN A SOLVENT

In Chapter 4 we studied several one-dimensional (1-D) systems. All of those systems consisted of interacting particles. The simplicity of the systems, which led to their solvability, was due to the fact that all the interactions were along the 1-D system. We also noted in Chapter 4 that there are real systems that can be approximately viewed as 1-D systems. These range from simple polymers to proteins and to nucleic acids. In reality, most of these systems are in a solution, not in vacuum. Specifically, proteins and nucleic acids are always in aqueous solution. Though it is still true that their main feature changes along a 1-D line, e.g., the sequence of amino acids in protein or the sequence of bases in DNA, their properties are affected by the presence of the solvent. These effects could be moderate, requiring only a minor rescaling of the energy levels of the units, or could have a major change on the behavior of the 1-D system. The latter is probably the case in systems of biological interest in aqueous solutions.

8.6.1. The General Modification of the PF of a 1-D System in Solution

We consider the simple Ising model of M units, each of which can be in either of two states. The configurational PF of such a system is

$$Z(T, M) = \sum_{\mathbf{s}} \exp[-\beta E(\mathbf{s})], \tag{8.6.1}$$

where $\mathbf{s} = s_1, \ldots, s_M$ is a specific configuration of the 1-D system and $E(\mathbf{s})$ is the corresponding energy level. This could include interaction between the units as well as interaction with an external field. For simplicity and in order to stress the new effects induced by the solvent, we assume that $E(\mathbf{s})$ consists of only nearest-neighbors interactions; i.e.,

$$E(\mathbf{s}) = \sum_{i=1}^{M} U(s_i, s_{i+1}), \tag{8.6.2}$$

where we identify the $(M + 1)$th unit with the first unit. With these assumptions we have seen that $Z(T, M)$ may be written as

$$Z(T, M) = \lambda_{\max}^M, \tag{8.6.3}$$

where $\lambda_{\max}$ is the largest eigenvalue of the matrix $\mathbf{P}$ defined by its elements

$$P_{kl} = \exp[-\beta U(k, l)]. \tag{8.6.4}$$

The probability of finding the system in any specific configuration is

$$\Pr(\mathbf{s}) = \frac{\exp[-\beta E(\mathbf{s})]}{Z(T, M)}. \tag{8.6.5}$$

We now insert the 1-D system in a solvent w, consisting of N molecules at some temperature T and volume V. Thus in effect we have a T, V, N system with one "solute" inserted at some fixed position and orientation. This solute can be viewed as having many isomers, each characterized by its configuration $\mathbf{s}$.

The Helmholtz energy of the entire system is

$$\exp[-\beta A(T, V, N; 1)] = Q(T, V, N; 1)$$

$$= \frac{q_w^N}{\Lambda_w^{3N} N!} \sum_{\mathbf{s}} \exp[-\beta E(\mathbf{s})] \int d\mathbf{X}^N \exp\{-\beta[U(\mathbf{X}^N) + B(\mathbf{s}, \mathbf{X}^N)]\}$$

$$= \frac{q_w^N}{\Lambda_w^{3N} N!} Z(N, M), \tag{8.6.6}$$

where $U(\mathbf{X}^N)$ is the total interaction energy among the solvent molecules at a specific configuration $\mathbf{X}^N$ and $B(\mathbf{s}, \mathbf{X}^N)$ is the total interaction energy between the 1-D system and all solvent molecules at a specific configuration $(\mathbf{s}, \mathbf{X}^N)$. $Z(N, M)$ is the configurational PF of the system (T, V are suppressed in this notation). The Helmholtz energy of the pure solvent is

$$\exp[-\beta A(T, V, N)] = Q(T, V, N)$$

$$= \frac{q_w^N}{\Lambda_w^{3N} N!} \int d\mathbf{X}^N \exp[-\beta U(\mathbf{X}^N)]$$

$$= \frac{q_w^N}{\Lambda_w^{3N} N!} Z(N). \qquad (8.6.7)$$

From (8.6.6) and (8.6.7) we obtain the pseudochemical potential of the 1-D system

$$\exp(-\beta\mu^*) = \frac{Q(T, V, N; 1)}{Q(T, V, N)}$$

$$= \sum_s \exp[-\beta E(s)] \int d\mathbf{X}^N P_0(\mathbf{X}^N) \exp[-\beta B(s, \mathbf{X}^N)]$$

$$= \sum_s \exp[-\beta E(s)] \langle \exp[-\beta B(s)] \rangle_0$$

$$= \sum_s \exp\{-\beta[E(s) + W(s|w)]\}, \qquad (8.6.8)$$

where $W(s|w)$ is the coupling work of the 1-D system being at a specific configuration s, to the solvent. We now define the Helmholtz energy level, or the pseudochemical potential of each individual configuration s, by

$$\mu^*(s) = E(s) + W(s|w), \qquad (8.6.9)$$

and rewrite (8.6.8) as

$$Z^*(M; \rho) = \exp(-\beta\mu^*)$$

$$= \sum_s \exp[-\beta\mu^*(s)]$$

$$= \frac{Z(N, M)}{Z(N)}, \qquad (8.6.10)$$

which is formally a generalization of Eq. (6.15.13) of section 6.15.

The quantity $Z^*(M; \rho)$ is the effective configurational partition function of the 1-D system; $\rho = N/V$ is the density of the solvent. If $\rho \to 0$, or if there are no interactions between the 1-D system and the solvent [(i.e., when $B(s, \mathbf{X}^w) = 0$), then $Z^*(M; \rho)$ reduces to $Z(T, M)$ in (8.6.1)].

The probability of finding the 1-D system in any specific configuration s is now

$$\Pr^*(s) = \frac{\exp[-\beta\mu^*(s)]}{Z^*(M; \rho)}$$

$$= \frac{\exp[-\beta E(s)] \exp[-\beta W(s|w)]}{Z^*(M; \rho)}$$

$$= \Pr(s)y(s), \qquad (8.6.11)$$

where the vacuum probability Pr(s) is given in (8.6.5) and the solvent contribution to the probability is defined by

$$y(\mathbf{s}) = \frac{Z(T, M) \exp[-\beta W(\mathbf{s}|w)]}{Z^*(M; \rho)}. \tag{8.6.12}$$

Let $\mathbf{s}_0$ be any specific configuration of the 1-D system (this can be chosen as the configuration for which $E(\mathbf{s}_0) = 0$). Then

$$\frac{y(\mathbf{s})}{y(\mathbf{s}_0)} = \exp[-\beta W(\mathbf{s}|w) + \beta W(\mathbf{s}_0|w)]$$

$$= \exp[-\beta \delta A(\mathbf{s}_0 \rightarrow \mathbf{s})], \tag{8.6.13}$$

where $\delta A(\mathbf{s}_0 \rightarrow \mathbf{s})$ is the indirect or solvent-induced work (here in the T, V, N ensemble) to make the transition $\mathbf{s}_0 \rightarrow \mathbf{s}$. For M solute particles, we normally choose the reference configuration where all particles are at infinite separation from each other. In that case, the corresponding $y(\mathbf{s})$ would be related to the indirect work required to bring the M particles from infinite separation to the final configuration $\mathbf{s}$.

The simplest case of a solvated 1-D system occurs when the coupling work $W(\mathbf{s}|w)$ is approximately the same for all configurations $\mathbf{s}$. Two examples are shown in Fig. 8.4. In a, the solvent consists of hard-spheres particles and each unit contributes the same excluded volume with respect to the solvent independently of its state (say up or down spin). In this case $W(\mathbf{s}|w)$ is simply the work required to create a cavity in the solvent to accommodate the 1-D system. The Helmholtz energy levels (8.6.9) are modified by a constant

$$\mu^*(\mathbf{s}) = E(\mathbf{s}) + W(\text{cavity}). \tag{8.6.14}$$

In this case $W(\text{cavity})$ is independent of $\mathbf{s}$; therefore, from (8.6.10), we have

$$Z^*(M; \rho) = Z(T, M) \exp[-\beta W(\text{cavity})]; \tag{8.6.15}$$

hence,

$$y(\mathbf{s}) = 1.$$

I.e., there is no effect of the solvent on the probability distribution of the configurations. Equation (8.6.15) means that the partition function $Z(T, M)$ of the vacuum model, is modified by the constant parameter $W(\text{cavity})$.

In the second example, shown in Fig. 8.4b, the solvation of each unit depends on its state. A unit in a vertical state can form HBs with solvent molecules. A unit in a

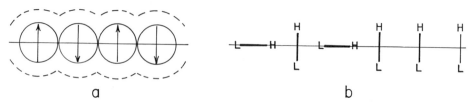

FIGURE 8.4. Two 1-D systems in a solvent. (a) The excluded volume is the same for all configurations $\mathbf{s}$. (b) The solvation of each unit depends on its state, but if the units are far apart, they are independently solvated.

horizontal state cannot. The units are assumed to be far apart, hence independently solvated in the following sense. We rewrite the coupling work defined in (8.6.8) as

$$\exp[-\beta W(\mathbf{s}|w)] = \langle \exp[-\beta B(\mathbf{s})]\rangle_0$$

$$= \langle \exp[-\beta B^H(\mathbf{s})]\rangle_0 \left\langle \exp\left[-\beta \sum_{i=1}^{M} B_i\right]\right\rangle_H$$

$$= \exp[-\beta W(\text{cavity})] \prod_{i=1}^{M} \exp[-\beta \Delta A^*(i/H)], \qquad (8.6.16)$$

where we assumed that the total binding energy of the 1-D system can be written as a sum of the hard (H) part and a specific binding energy, say hydrogen bonding; i.e.,

$$B(\mathbf{s}) = B^H(\mathbf{s}) + \sum_{i=1}^{M} B_i. \qquad (8.6.17)$$

If the units are far apart, then the conditional average $\langle \exp[-\beta \sum_{i=1}^{M} B_i]\rangle_H$ can be factored into a product of M averages, each of which corresponds to the conditional solvation Helmholtz energy of one unit.

The Helmholtz energy levels are now

$$\mu^*(\mathbf{s}) = E(\mathbf{s}) + W(\text{cavity}) + \sum_{i=1}^{M} \Delta A^*(i/H). \qquad (8.6.18)$$

Note that in this case, although the units are independently *solvated*, the conditional solvation Helmholtz energy of each unit can either be dependent or independent of the state of the unit (Fig. 8.5). It is assumed that $W(\text{cavity})$ is independent of $\mathbf{s}$.

If $\Delta A^*(i/H)$ are independent of the state of the ith unit, then $\sum_{i=1} \Delta A^*(i/H)$ is independent of the configuration, and therefore $\mu^*(\mathbf{s})$ in (8.6.18) depends on $\mathbf{s}$ only through the energies $E(\mathbf{s})$. Hence

$$Z^*(M;\rho) = Z(T, M) \exp\left[-\beta W(\text{cavity}) - \beta \sum_{i=1}^{M} \Delta A^*(i/H)\right], \qquad (8.6.19)$$

and therefore

$$y(\mathbf{s}) = 1;$$

i.e., the solvent has no effect on the probability distribution $\Pr(\mathbf{s})$. The more important case occurs when $\Delta A^*(i/H)$ does depend on the state s_i, (e.g., Fig. 8.5b), but not on the

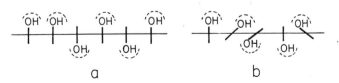

a   b

FIGURE 8.5. The conditional solvation is independent of (a) or dependent on (b) the state of the units. The cavity work is the same for each configuration.

state of the other units: for instance, when the units are far apart, but the conditional solvation is different for different configurations s_i. In this case we have

$$Z^*(M; \rho) = \exp[-\beta W(\text{cavity})] \sum_{\mathbf{s}} \exp\left[-\beta E(\mathbf{s}) - \beta \sum_{i=1}^{M} \Delta A^*(s_i/H)\right], \quad (8.6.20)$$

and therefore

$$\text{Pr}^*(\mathbf{s}) = \text{Pr}(\mathbf{s})y(\mathbf{s}),$$

where

$$y(\mathbf{s}) = \frac{Z(T, M) \exp\left[-\beta \sum_{i=1}^{M} \Delta A^*(s_i/H)\right]}{\sum_{\mathbf{s}} \exp\left[-\beta E(\mathbf{s}) - \beta \sum_{i} \Delta A^*(s_i/H)\right]}. \quad (8.6.21)$$

In this case the solvent will affect the probability distribution of the configurations·s, giving preference to those configurations which are better solvated, i.e., configurations for which $\sum_i \Delta A^*(s_i/H)$ is more negative.

Formally, this system can be viewed as a 1-D system subjected to an "external field" which operates on each of the units individually. The effective PF, $Z^*(M; \rho)$ is formally the same as that of the simple Ising model with an additional interaction with the external field. Therefore, the theory of the 1-D system remains unchanged, except for the rescaling of the energy levels into Helmholtz energy levels. The latter depends on the activity or the density of the solvent. Similarly, in the T, P, N system the energy levels turn into Gibbs energy levels.

Another case where the theory does not change occurs when the conditional solvation Helmholtz energy of all the units cannot be factorized as in (8.6.16), but can be factorized into nearest-neighbor contributions; i.e.,

$$\left\langle \exp\left[-\beta \sum_{i=1}^{M} B_i\right] \right\rangle_H = \prod_{i=1}^{M} \exp[-\beta \delta A(s_i, s_{i+1})]. \quad (8.6.22)$$

In that case,

$$Z^*(M; \rho) = \exp[-\beta W(\text{cavity})] \sum_{\mathbf{s}} \exp\left\{-\beta \sum_{i} [U(s_i, s_{i+1}) + \delta A(s_i, s_{i+1})]\right\}. \quad (8.6.23)$$

In this case, the theory of the Ising model can still be retained with a rescaling of the pairwise interaction energies, i.e., besides the factor $W(\text{cavity})$ we must replace $U(s_i, s_{i+1})$ with $U(s_i, s_{i+1}) + \delta A(s_i, s_{i+1})$. This is what is usually assumed in the theory of polymers, proteins, etc., in a solvent.

In real systems such as proteins or nucleic acids, one cannot assume that the units are independently solvated, nor that the units are correlated up to nearest neighbors only. This will in general require a modification of the theory itself, not only a rescaling of the parameters in the simple Ising model. In simple cases one can argue that the solvent introduces next-nearest-neighbor interactions; this will lead to a representation of the PF by a higher-order matrix. In more complex systems, long-range correlations between distant units could invalidate the very application of the 1-D model. This is the case in the protein folding process. Here, even the cavity work depends collectively on the configuration of the entire system. In addition, strong correlation between functional groups on distant locations along the sequence of amino acids should lead to a profound effect on the behavior of the system. This requires a different theory, some aspects of which will be discussed in sections 8.6.3 and 8.8.

8.6.2. 1-D "Water" in Liquid Water

In section 4.5.4 we worked out a 1-D "water" system. The essential ingredient of that model was the double-square-well potential with the additional requirement that the depth of the more distant well be deeper than the shorter-range well. We have seen that this feature of the interaction led to a minimum in the volume, or length, of the system, as a function of the temperature—a behavior akin to that of real liquid water. However, a double-well pair potential of this kind does not exist in reality.

We shall now see that by inserting a similar type of model into liquid water we can get a potential of average force which has essentially the double-well character of the model of section 4.5.4. Thus, although the 1-D model itself is still an artificial model, the effective pair interaction has a real character, originating from the solvent.

As in section 4.5.4, we consider a 1-D fluid of particles having the following properties: Each particle has four selected directions along which it can form hydrogen bonds. We can distinguish between the two donor and the two acceptor directions, but for the purpose of the present demonstration this distinction is not important. The four unit vectors are in the same plane, and they have a fixed orientation with respect to the 1-D line of the system, as shown in Fig. 8.6. The particles are free to move along the 1-D line as in the model of section 4.5.4. Since their relative orientation is fixed, they are precluded from forming direct HBs between themselves. The pair potential between nearest neighbors is assumed to be a LJ type of potential, which we denote by $U(R)$. This 1-D fluid

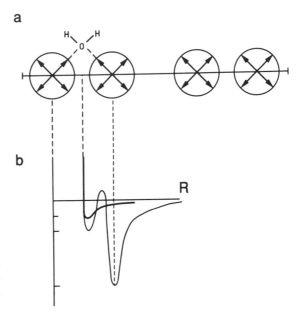

FIGURE 8.6. A 1-D "water" in liquid water. A pair of units at a favorable distance can form a strong HϕI interaction. The potential of average force [light curve shown in (b)], has a first minimum at about the contact distance σ but a deeper minimum at the distance of 4.5 Å, where HϕI interaction is the strongest.

has a T, L, M partition function of the form [see Eq. (4.5.6)], which we write in a shorthand notation as

$$Q(T, L, M) = \frac{1}{\Lambda^M} \int dx^M \exp[-\beta U(x^M)].$$ (8.6.24)

We have already seen that this model will have a normal thermal expansion coefficient; i.e. [see Eq. (4.5.62)],

$$\left(\frac{\partial \bar{L}}{\partial T}\right)_{M,P} > 0.$$ (8.6.25)

Note that we use M for the number of particles in the 1-D system, and P is the 1-D pressure. We also use $x = x_1, \ldots, x_M$ for the configuration of the particles in the 1-D system. In (8.6.24) we have already assumed that the particles are ordered; i.e., $x_1 < x_2 < \cdots x_M$.

We now insert this 1-D system into 3-D liquid water at T, V, N. The PF of the pure water is $Q(T, V, N)$ and the PF of the combined system is denoted by $Q(T, V, N; L, M)$. The analogue of the pseudochemical potential of the 1-D system is defined as in (8.6.8) by

$$\exp(-\beta \mu^*) = \frac{Q(T, V, N; L, M)}{Q(T, V, N)}$$

$$= \frac{\frac{1}{\Lambda^M} \int dx^M \exp[-\beta U(x^M)] \int dX^N \exp[-\beta U(X^N) - \beta B(X^N, x^M)]}{\int dX^N \exp[-\beta U(X^N)]}$$

$$= \frac{1}{\Lambda^M} \int dx^M \exp[-\beta U(x^M)]\langle \exp[-\beta B(x^M)]\rangle_0$$

$$= \frac{1}{\Lambda^M} \int dx^M \exp[-\beta U(x^M) - \beta W(x^M | w)]$$

$$= Q^*(T, L, M; \rho).$$ (8.6.26)

Here $B(X^N, x^M)$ is the total binding energy of the 1-D system at x^M to all solvent molecules at X^N. The corresponding coupling work is denoted by $W(x^M | w)$. The resulting quantity may be viewed as the effective PF of the 1-D system in the solvent. This is the analogue of $Z^*(M; \rho)$ of section 8.6.1. Here, Q^* is also a function of the density of the solvent $\rho = N/V$.

In the 1-D water treated in section 4.5.4, the canonical PF (8.6.24) was converted to a simple form [Eqs. (4.5.11) and (4.5.13)] by transforming variables and applying the convolution theorem. The generalized PF (8.6.26) includes the new factor $W(x^M | w)$ which in general is not pairwise additive. Therefore, in general, transformation to the $(T, P, M; \rho)$ PF does not lead to a simple PF. (Note that here, P is still the 1-D analogue of the pressure. The entire system is at a constant volume V.)

The probability density of finding a specific configuration $\mathbf{x}^M$ is (in analogy with 8.6.11)

$$\text{Pr*}(\mathbf{x}^M) = \frac{\exp[-\beta U(\mathbf{x}^M)]\exp[-\beta W(\mathbf{x}^M|w)]}{\int d\mathbf{x}^M \exp[-\beta U(\mathbf{x}^M)]\exp[-\beta W(\mathbf{x}^M|w)]}$$

$$= \text{Pr}(\mathbf{x}^M)y(\mathbf{x}^M), \qquad (8.6.27)$$

where $\text{Pr}(\mathbf{x}^M)$ is the probability density of finding the 1-D model in configuration $\mathbf{x}^M$ in the absence of the solvent, and $y(\mathbf{x}^M)$ is given, by analogy with (8.6.12), by

$$y(\mathbf{x}^M) = \frac{\exp[-\beta W(\mathbf{x}^M|w)] \int d\mathbf{x}^M \exp[-\beta U(\mathbf{x}^M)]}{\int d\mathbf{x}^M \exp[-\beta U(\mathbf{x}^M) - \beta W(\mathbf{x}^M|w)]}. \qquad (8.6.28)$$

Again, we choose a particular configuration, which we denote by $\mathbf{x}_0^M$, and rewrite (8.6.28) as

$$y(\mathbf{x}^M) = y(\mathbf{x}_0^M)\exp[-\beta W(\mathbf{x}^M|w) + \beta W(\mathbf{x}_0^M|w)]$$

$$= y(\mathbf{x}_0^M)\exp[-\beta\delta A(\mathbf{x}_0^M \rightarrow \mathbf{x}^M)], \qquad (8.6.29)$$

which is the analogue of (8.6.13). δA is the indirect work required to change from the reference configuration $\mathbf{x}_0^M$ to $\mathbf{x}^M$. This, in general, cannot be viewed as a pairwise additive quantity.

We now return to the effective partition function (8.6.26) and assume that δA can be written as pairwise additive quantities and, in addition, we assume that there exists a reference configuration $\mathbf{x}_0^M$ for which the M particles are independently solvated. Then

$$W(\mathbf{x}^M|w) = W(\mathbf{x}_0^M|w) + \delta A(\mathbf{x}_0^M \rightarrow \mathbf{x}^M)$$

$$= \sum_{i=1}^M W(i|w) + \sum_{i=1}^M \delta A(x_i, x_{i+1}), \qquad (8.6.30)$$

where $W(i|w)$ is the coupling work of the individual particles and $\delta A(x_i, x_{i+1})$ is the indirect work required to bring the ith and $(i+1)$th particles from a configuration where they are independently solvated (normally when $x_{i+1} - x_i \rightarrow \infty$) to the final configuration x_i, x_{i+1}. With the assumptions made in (8.6.30), the effective PF of (8.6.26) can be written as

$$Q^*(T, L, M; \rho) = \frac{q^M}{\Lambda^M} \int d\mathbf{x}^M \exp\left[-\beta U(\mathbf{x}^M) - \beta \sum_i \delta A(x_i x_{i+1})\right], \qquad (8.6.31)$$

where $q = \exp[-\beta W(i|w)]$ may be viewed as an effective "internal" PF of one unit. Equation (8.6.31), with the assumption of pairwise additivity for both the direct [Eq. (4.5.1)] and the indirect (8.6.30) interaction has now the same formal form as (4.5.6). Therefore the convolution theorem may be applied to obtain the analogue of (4.5.11), i.e.,

$$\Delta(T, P, M; \rho) = \frac{q^M}{\Lambda^M}\left[\int d\mathbf{R} \exp[-\beta PR - \beta W(R)]\right]^M, \qquad (8.6.32)$$

where $W(R) = U(R) + \delta A(R)$ is the potential of average force between the two consecutive units. If the units are oriented in such a way that they can form HBs with the solvent as depicted in Fig. 8.6a, then $\delta A(R)$ has a deep well centered at about 4.5 Å of a depth of about -3 kcal/mol.

On the other hand, $U(R)$ (heavy line in Fig. 8.6b), for particles of neon size will have a relatively shallow well at about 2.8 Å. Schematically, the potential of average force will have the form, shown as a light curve in Fig. 8.6b. Repeating the same arguments as in section 4.5.4, we should obtain a minimum in the $\bar{L}(T)$ curve as the one obtained in the vacuum model. (Note that the solvation of the individual units contained in q^M does not affect the equation of state of the 1-D system.)

Although both this and the previous model treated in section 4.5.4 are rather artificial models, the present one has one realistic feature which is relevant to liquid water. This is the second, deeper potential well which arises from the hydrophilic interactions through the solvent.

We can conclude that even with the simplifying assumptions made in (8.6.30), the transition from the vacuum to the liquid case produces a new phenomenon. Here we demonstrate the transition from a monotonic $\bar{L}(T)$ curve to a curve having a minimum average length at some temperature. The simplifying assumption made in (8.6.30) allowed us to use the same theory as in section 4.5.4. In the more general case, when these assumptions cannot be made, the theory itself must be modified to account for longer-range correlations.

8.6.3. The Helix–Coil Transition in a Solvent

In section 4.7.2, we developed the theory of helix–coil transition in the vacuum. We have seen that the theory is based on the reduction of the entire two-dimensional configurational space of a single amino acid residue, $\phi_i\psi_i$ into three coarse states: helix (H), coiled (C), and impossible (I). These roughly correspond to the α-helix, the β-strand, and the inaccessible regions in the Ramachandran maps. Since the I states involve large positive energies, their probability of occurrence is very small and therefore can be ignored. Thus, for a linear system of M amino-acid residues, each microstate of the system is translated into a sequence of macrostates, symbolically

$$\phi_1\psi_1 \cdots \phi_M\psi_M \to HHCHCHHC. \qquad (8.6.33)$$

This classification of states leads to a simplification of the configurational partition into a sum of terms of the form

$$uuvwvuuv \cdots , \qquad (8.6.34)$$

where

$$u = \Omega_C \exp(-\beta U_C^*)$$
$$v = \Omega_H \exp(-\beta U_C^*) \qquad (8.6.34)$$
$$w = \Omega_H \exp(-\beta U_{HB}),$$

where Ω_C and Ω_H are the configurational range of the coiled and helix states, roughly corresponding to the areas of the β-strand and α-helix in the Ramachandran plot; U_C^* is a typical energy of the coiled state; and U_{HB} is the typical energy of the helical state, which is essentially the hydrogen-bond energy.

The helix–coil transition as a function of temperature is a result of the competition between the difference in energies and the ratio of degeneracies. In terms of the quantities u, v, and w, we write

$$\frac{u}{v} \approx \frac{\Omega_C}{\Omega_H} \gg 1 \tag{8.6.35}$$

$$\frac{w}{v} \approx \exp[-\beta(U_{HB} - U_C^*)] \gg 1. \tag{8.6.36}$$

Thus u/v is essentially the ratio of the degeneracies of the two states, and w/v depends essentially on the difference in the energies of the two states. A simplified 1-D model of this kind was treated in section 4.4.2.

The situation becomes markedly different if the same model is inserted in water. Although the reduction from the detailed microstates to the coarser macrostates, as accomplished in (8.6.33), can be retained, one cannot represent the contribution to the PF of a specific sequence of Hs and Cs by a term of the form (8.6.34).

The reason is as follows. Consider first the simplest case of polyglycine. The effective PF of the polymer in a solvent, is, in analogy with (8.6.26),

$$Z^*(M; \rho) = \frac{Q(T, V, N; M)}{Q(T, V, N)} = \int d\mathbf{\Omega}^M \exp[-\beta U(\mathbf{\Omega}^M) - \beta W(\mathbf{\Omega}^M | w)], \tag{8.6.37}$$

where $\mathbf{\Omega}^M$ stands for a specific configuration $\phi_1 \psi_1 \cdots \phi_M \psi_M$ of the entire system of M units and $W(\mathbf{\Omega}^M | w)$ is the coupling work of the polymer at the specific configuration $\mathbf{\Omega}^M$.

In the vacuum case $W(\mathbf{\Omega}^M | w) = 0$ and $Z^*(M; \rho)$ reduces to the PF of the system treated in section 4.7.3. The transition to a trace of a 4×4 matrix was rendered possible due to some simplifying assumptions made on the total potential energy $U(\mathbf{\Omega}^M)$ [see Eq. (4.7.10)]. In a solvent $W(\mathbf{\Omega}^M | w)$ cannot be written in a way similar to Eq. (4.7.10).

For simplicity, suppose that the total interaction energy of polyglycine with a water molecule can be written as

$$U(\mathbf{\Omega}^M, \mathbf{X}_i) = U^H(\mathbf{\Omega}^M, \mathbf{X}_i) + \sum_{k=1}^{M} U_k^{HB}(\mathbf{\Omega}^M, \mathbf{X}_i), \tag{8.6.38}$$

where U^H is the hard part of the interaction and $U_k^{HB}(\mathbf{\Omega}^M, \mathbf{X}_i)$ is the HB interaction between the kth functional group (here the C=O and NH of the backbone only), and a water molecule (we neglect the soft interaction).

The total binding energy of the polymer to the solvent is thus

$$B(\mathbf{\Omega}^M) = B^H(\mathbf{\Omega}^M) + \sum_{k=1}^{M} B_k^{HB}(\mathbf{\Omega}^M). \tag{8.6.39}$$

Hence the coupling work of the polymer at a specific configuration $\mathbf{\Omega}^M$ is, in analogy to (8.6.16):

$$\exp[-\beta W(\mathbf{\Omega}^M | w)] = \langle \exp[-\beta B(\mathbf{\Omega}^M)] \rangle_0$$

$$= \langle \exp[-\beta B^H(\mathbf{\Omega}^M)] \rangle_0 \left\langle \exp\left[-\beta \sum_{k=1}^{M} B_k^{HB}(\mathbf{\Omega}^M)\right] \right\rangle_H, \tag{8.6.40}$$

where the first factor on the rhs of (8.6.40) is the work required to create a cavity for the polymer at Ω^M and the second factor is a conditional average over all configurations of the solvent molecules, given that the hard interaction has already been turned on.

We have seen (section 5.10) that the first term on the rhs of (8.6.40) is related to the work required to create a cavity equal to the excluded volume with respect to the solvent. (Note that this is different from the excluded volume effect due to repulsions within the polymer itself. This has been taken into account by restricting ourselves to the accessible regions in the Ramachandran map.) Since different conformations Ω^M have different excluded volumes, and since the excluded volume is a very complicated function of the configuration Ω^M, we cannot further factor the quantity $\langle \exp[-\beta B^H(\Omega^M)]\rangle_0$ into single, pair, triplet, etc., contributions. The reason is that the polymer actually occupies the three dimensional space. The excluded volume of each configuration Ω^M collectively depends on the configuration of the entire polymer. This in itself invalidates the main assumption made in section 4.7.3, i.e., that the system can be viewed as a 1-D system.

Furthermore, suppose we take some average work of creating a cavity—average over all the configurations of the polymer—and replace the first factor on the rhs of (8.6.40) by a constant, as we did in (8.6.16). We still have to face the problem of factorizability of the second term on the rhs of (8.6.40). This term includes the conditional solvation, Gibbs or Helmholtz energies of the $C{=}O$ and NH groups and $H\phi I$ interactions between such groups. As an example, two successive units in the helical state could bring the $C{=}O$ of the ith unit to a distance of about 4–5 Å from the $C{=}O$ of the $(i+2)$th unit so that a strong $H\phi I$ interaction becomes operative (see Fig. 8.33). This in itself will invalidate the assignments made in section 4.7.3 for the eight configurations of a triplet of consecutive units. Note also that the direct HB energy formed by the ith and $(i+4)$th units that we denoted by U_{HB} in (8.6.34) will now be converted to Gibbs or Helmholtz energy change and will be considerably reduced owing to the loss of the solvation of the two "arms" of the $C{=}O$ and NH involved in this bond.

In a more realistic polypeptide, both the cavity work as well as the specific effects of the side chains depend not only on the configuration Ω^M but also on the specific sequence of side chains. Here, we must take into account $H\phi I - H\phi I$, $H\phi I - H\phi O$, and $H\phi O - H\phi O$ interactions between units five or more units apart. The number and extent of each of these interactions depend on the specific sequence. Clearly, this calls for a different theory for a different sequence.

8.7. PROTEIN–PROTEIN ASSOCIATION AND MOLECULAR RECOGNITION

In this section we treat one of the most fundamental processes in biochemistry, the association between two macromolecules. A second important process, the folding of a protein, is discussed in section 8.8. Other, more complex processes may be viewed as combinations of these two fundamental processes.

The protein–protein association is extremely complicated. Even in the vacuum we have to take into account interaction among many different amino-acid residues. In a solvent, the process becomes far more difficult to treat theoretically. Here, we have to account for direct interactions as well as indirect interactions mediated through the solvent.

Given two proteins of known structure, one can list all possible interactions between the groups on their surfaces. This list will tell us how the proteins are most likely to bind to each other. Statistical thermodynamics plays a relatively minor role in studying this problem. It can provide us with an approximate estimate of the translational and rotational PFs of the proteins before and after the association (assuming the vibrational degrees of freedom are unaffected by the process) and therefore an estimate of the overall standard Gibbs energy of the process. The main input into the theory is the interaction energy between the two monomers.

The same process in a solvent also requires, in addition to all the information needed for the vacuum case, the changes in the solvation Gibbs energies of the two proteins upon association. It is here that statistical thermodynamics plays a major role. As we shall see in this section, statistical thermodynamics not only provides a way of classifying all possible contributions to the standard Gibbs energy of the process, but also suggests a method of studying each contribution separately, using simple model compounds. In effect, the theory provides a method of dissecting an extremely complicated problem into smaller, more manageable problems. The latter, though still difficult to solve theoretically, can be studied by experimental means.

Throughout this section, we shall emphasize specific effects that originate from the solvent—here, water. We shall defer to section 8.10 the study of the effect of added solutes on the thermodynamics of the association process.

8.7.1. Formal Separation of the Solvent Effects

We consider the association process

$$\mathbf{P} + \mathbf{L} \rightleftharpoons \mathbf{PL} \tag{8.7.1}$$

where $\mathbf{P}$ and $\mathbf{L}$ are referred to as a protein and a ligand respectively, but the treatment is very general and could apply to any association between two macromolecules in a solution.

The chemical equilibrium condition is

$$\mu_{\mathbf{P}} + \mu_{\mathbf{L}} = \mu_{\mathbf{PL}}, \tag{8.7.2}$$

where for each species $\alpha (=\mathbf{P}, \mathbf{L} \text{ or } \mathbf{PL})$ we write

$$\mu_\alpha = W(\alpha|w) + kT \ln Q_\alpha^{-1} \rho_\alpha, \tag{8.7.3}$$

where $W(\alpha|w)$ is the coupling work of the solute α to the solvent w (the solvent could be pure water or any mixture of solvents containing any concentrations of $\mathbf{P}$, $\mathbf{L}$, and $\mathbf{PL}$. For simplicity we shall assume that we have a pure one-component solvent). Q_α is the internal PF of the species α. We have included in Q_α the momentum PF, Λ_α^3, and the rotational–vibrational and electronic PFs of the α species.

In general, each of the species $\mathbf{P}$, $\mathbf{L}$, and $\mathbf{PL}$ can assume many conformations. In this section we treat only one conformation of each species. We shall return to the averaging process over all possible conformations in section 8.7.5. For the moment, Q_α does not include the contributions due to internal rotations (or conformations). For the complex $\mathbf{PL}$ we also write

$$\mu_{\mathbf{PL}} = W(\mathbf{PL}|w) + \Delta U(\mathbf{PL}) + kT \ln Q_{\mathbf{PL}}^{-1} \rho_{\mathbf{PL}}, \tag{8.7.4}$$

where $\Delta U(\mathbf{PL})$ is the direct interaction between $\mathbf{P}$ and $\mathbf{L}$ in the complex $\mathbf{PL}$. Note that we assume here a specific configuration of the dimer $\mathbf{PL}$. The chemical equilibrium for the reaction (8.7.1) is thus

$$K^l = \frac{\rho_{PL}}{\rho_L \rho_P}$$

$$= \frac{Q_{PL}}{Q_L Q_P} \exp[-\beta \Delta U - \beta W(\mathbf{PL}|w) + \beta W(\mathbf{P}|w) + \beta W(\mathbf{L}|w)]$$

$$= \exp(-\beta \Delta G^{0l}), \tag{8.7.5}$$

where ΔG^{0l} is the thermodynamic *standard Gibbs energy* of the reaction in the liquid.

The same reaction (8.7.1), with the same assumed conformations, in an ideal-gas phase would have an equilibrium constant

$$K^g = \frac{Q_{PL}}{Q_L Q_P} \exp(-\beta \Delta U) = \exp[-\beta \Delta G^{0g}], \tag{8.7.6}$$

where ΔG^{0g} is the standard Gibbs energy for the same reaction in the ideal-gas phase.

The solvent contribution to the Gibbs energy change is now defined by

$$\delta G(\mathbf{P} + \mathbf{L} \rightarrow \mathbf{PL}) = \Delta G^{0l} - \Delta G^{0g}$$

$$= W(\mathbf{PL}|w) - W(\mathbf{P}|w) - W(\mathbf{L}|w). \tag{8.7.7}$$

Clearly, in the absence of the solvent all of the coupling-work terms are zeros and $\delta G = 0$.

In this section we shall mainly focus on the δG of the association reaction. We shall also compare it with ΔU to get some estimates of the order of magnitude of the different contributions to ΔG^{0l}. It is convenient to define a new binding Gibbs energy by

$$\Delta G_B = \Delta U + \delta G, \tag{8.7.8}$$

which is simply the Gibbs energy change for the process of bringing $\mathbf{P}$ and $\mathbf{L}$ from fixed positions and orientations at infinite separation to a final fixed position and orientation of the complex $\mathbf{PL}$. The process is carried out within the liquid at some given pressure and temperature. This is equivalent to freezing the translational and rotational degrees of freedom of the species $\mathbf{P}$, $\mathbf{L}$, and $\mathbf{PL}$.

In order to further analyze the content of the coupling works that appear in (8.7.7), we recall that for each solute $\alpha = \mathbf{P}$, $\mathbf{L}$, or $\mathbf{PL}$, we have

$$\exp[-\beta W(\alpha|w)] = \langle \exp(-\beta B_\alpha) \rangle_0, \tag{8.7.9}$$

where B_α is the total binding energy of the solute α to all the solvent molecules, and the average is over all configurations of the solvent molecules in the T, P, N ensemble. Thus

$$B_\alpha(\mathbf{X}_\alpha, \mathbf{X}^N) = \sum_{i=1}^{N} U(\mathbf{X}_\alpha, \mathbf{X}_i), \tag{8.7.10}$$

where $U(\mathbf{X}_\alpha, \mathbf{X}_i)$ is the pair interaction between the solute α at $\mathbf{X}_\alpha$ and a solvent molecule at $\mathbf{X}_i$. Note that no pairwise additivity is presumed for the total interaction energy among the solvent molecules. In (8.7.9) we wrote B_α and not $B_\alpha(\mathbf{X}_\alpha)$ since the value of the average is independent of the specific choice of position and orientation of the solute α. This is true only if α has a fixed conformation. If α has many conformations, another average is required (see section 8.7.5).

There are several ways of further separating each of the pair interactions $U(\mathbf{X}_\alpha, \mathbf{X}_i)$ into various components. The two most convenient ones are the following:

$$U(\mathbf{X}_\alpha, \mathbf{X}_i) = U^H + U^S + U^{HB} + U^C \qquad (8.7.11)$$

$$U(\mathbf{X}_\alpha, \mathbf{X}_i) = U^{BB} + \sum_{k=1}^{M} U(k, \mathbf{X}_i). \qquad (8.7.12)$$

In the first form we divide the total interaction between α and a water molecule into different types of interactions, e.g., the hard (H), soft (S), hydrogen-bonding (HB), or charge–dipole (C) interaction. In the second form, we "cut off" all the M side chains and view the interaction $U(\mathbf{X}_\alpha, \mathbf{X}_i)$ as consisting of two parts: one originating from the backbone (BB), and the second originating from the side chains. The latter is further written as a sum of contributions from each individual side chain. The two alternative views of $U(\mathbf{X}_\alpha, \mathbf{X}_i)$ have already been used in sections 6.14.1 and 6.14.2 for small solutes. Note that in the second choice, (8.7.12) can be further divided into components, e.g.,

$$U^{BB} = U^{BB,H} + U^{BB,S} + U^{BB,HB} + \cdots. \qquad (8.7.13)$$

The specific choice we shall make in the following sections of this chapter will depend on the particular problem we study and on the availability of relevant experimental data, which we shall need for estimating the various contributions to $W(\alpha|w)$. When using (8.7.12) for globular proteins, the sum over k extends only to the side chains that are exposed to the solvent.

Suppose we use the division (8.7.12); then Eq. (8.7.9) can be rewritten as

$$\exp[-\beta W(\alpha|w)] = \left\langle \exp\left[-\beta B_\alpha^{BB} - \beta \sum_k B_\alpha(k) \right] \right\rangle_0$$

$$= \langle \exp(-\beta B_\alpha^{BB}) \rangle_0 \left\langle \exp\left[-\beta \sum_k B_\alpha(k) \right] \right\rangle_{BB}, \qquad (8.7.14)$$

where

$$B_\alpha^{BB} = \sum_{i=1}^{N} U^{BB}(\mathbf{X}_\alpha, \mathbf{X}_i) \qquad (8.7.15)$$

$$B_\alpha(k) = \sum_{i=1}^{N} U(k, \mathbf{X}_i). \qquad (8.7.16)$$

The factoring in (8.7.14) corresponds to splitting the solvation process into two parts, namely,

$$\Delta G_\alpha^* = \Delta G_\alpha^{*BB} + \Delta G_\alpha^{*SC/BB}, \qquad (8.7.17)$$

where ΔG_α^* is the solvation Gibbs energy of the entire solute α, ΔG_α^{BB} is the solvation Gibbs energy of the backbone, and $\Delta G_\alpha^{*SC/BB}$ is the conditional solvation Gibbs energy of all the side chains (SC) given the backbone. Note that (8.7.17) is exact provided that we accept representation (8.7.12). It does not involve any assumption on group additivity of the solvation Gibbs energy (see also section 6.14.2).

If all the side chains are far apart from each other, then one can further factor the second average on the rhs of (8.7.14) as

$$\left\langle \exp\left[-\beta \sum_k B_a(k) \right] \right\rangle_{BB} = \prod_k \langle \exp[-\beta B_a(k)] \rangle_{BB}, \qquad (8.7.18)$$

which corresponds to the sum

$$\Delta G_a^{*SC/BB} = \sum_k \Delta G_a^{*k/BB}, \qquad (8.7.19)$$

where $\Delta G_a^{*k/BB}$ is the conditional solvation Gibbs energy of the kth side chain of the solute a, given the backbone BB. The factorization (8.7.18), or equivalently (8.7.19), is in general unjustified. This will be discussed further in the subsequent sections.

Using the expression (8.7.14) or (8.7.17) for each of the species involved in the association reaction, we can rewrite (8.7.7) as

$$\delta G(P + L \to PL) = \Delta G_{PL}^{*BB} - \Delta G_P^{*BB} - \Delta G_L^{*BB} + \Delta G_{PL}^{*SC/BB} - \Delta G_P^{*SC/BB} - \Delta G_L^{*SC/BB}$$

$$= \delta G^{BB} + \delta G^{SC/BB}. \qquad (8.7.20)$$

A similar expression can be written if we use the representation (8.7.11) instead of (8.7.12).

Thus, the total solvent-induced effect on the association process is written in (8.7.20) as consisting of two terms: the contribution of the backbone and the contribution of all the side chains. Had we used (8.7.11), the analogue of (8.7.20) would be

$$\delta G(P + L \to PL) = \delta G^H + \delta G^{S/H} + \delta G^{HB/H,S} + \cdots, \qquad (8.7.21)$$

where δG^H is the contribution due to the hard part, $\delta G^{S/H}$ is the contribution due to the soft interaction given the hard part, $\delta G^{HB/H,S}$ is the hydrogen-bonding contribution, etc.

8.7.2. Classification of the Various Contributions to δG

In the following we shall be using a separation of the pair potential function, either (8.7.11) or (8.7.12), by writing

$$U(\mathbf{X}_a, \mathbf{X}_i) = U^H(\mathbf{X}_a, \mathbf{X}_i) + U^S(\mathbf{X}_a, \mathbf{X}_i) + \sum_{k=1}^{M} U^{FG}(k, \mathbf{X}_i), \qquad (8.7.22)$$

where U^H could be either the hard part of the backbone as in (8.7.12) or the hard part of the entire molecule as in (8.7.11). Similar meaning applies to U^S. U^{FG} is a specific interaction due to a specific functional group (FG). If we use (8.7.11), then a FG is only the HB or the C part of the interaction, if we use (8.7.12), then a FG is the entire side chain. Thus in (8.7.22) we do not commit ourselves to a specific separation of the pair potential. The total solvent-induced interaction between P and L is written as

$$\delta G(P + L \to PL) = \delta G^H + \delta G^{S/H} + \delta G^{FG/H,S}. \qquad (8.7.23)$$

The three terms in (8.7.23) correspond to the following three hypothetical processes: Suppose we switch off the soft and all FG interactions. We are left with the two hard particles corresponding to the solutes P and L. We bring these two hard particles to form the hard complex PL. The corresponding contribution to δG is δG^H. The second term on the rhs of (8.7.23) is the contribution to δG due to turning on the soft interaction between the solutes and all water molecules. The last term $\delta G^{FG/H,S}$ corresponds to turning on the specific FGs. We can now further classify the various FGs according to their nature and according to the role they play in the association process. We start with a classification of all FGs into three groups. This classification pertains to a specific association process by two specific solutes $\mathbf{P}$ and $\mathbf{L}$. Note that in (8.7.22) the sum over k is over all FGs of the solute α that are exposed to the solvent. The hard part includes all those side chains that are buried within α. We distinguish between three groups of FGs:

a. The E group

The E (for external) group consists of all FG the solvation properties of which are unchanged in the process (8.7.1). Intuitively, any FG that when switched off in the initial state (separated $\mathbf{P}$ and $\mathbf{L}$) and switched off in the final state (the complex $\mathbf{PL}$) does not affect δG will belong to the E group. Clearly, these FGs are likely to be external to the scene where changes do occur in the association process (see Fig. 8.7).

The formal definition of a FG in the E group in terms of the solvation Gibbs energy is

$$\Delta G_{\mathbf{P}}^{*k/H,S} = \Delta G_{\mathbf{PL}}^{*k/H,S} \tag{8.7.24}$$

$$\Delta G_{\mathbf{L}}^{*k/H,S} = \Delta G_{\mathbf{PL}}^{*k/H,S}. \tag{8.7.25}$$

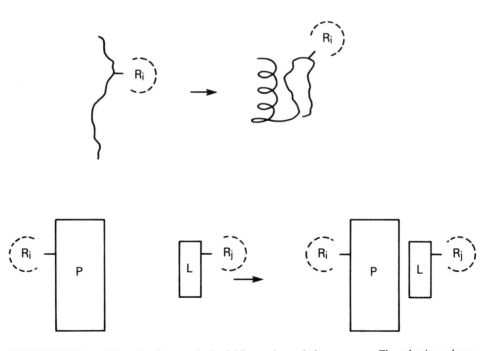

FIGURE 8.7. External functional groups in the folding and association processes. The solvation spheres around these groups are unchanged in the processes.

More specifically, if the conditional solvation Gibbs energy of the kth FG in the solute **P** is the same as in the solute **PL**, then k belongs to the E group. Similarly, (8.7.25) applies to the solute **L**.

b. The I group

The I (for internal) group consists of all FGs whose solvation in the initial state (separated **P** and **L**) is completely lost in the process. By completely lost we mean that in the final state (**PL**) these FGs are not exposed to the solvent; i.e., they are buried in the binding domain between **P** and **L** in the complex **PL**. The loss of solvation of such a FG is illustrated in Fig. 8.8. The formal requirement in terms of solvation Gibbs energy is

$$\Delta G_{\mathbf{P}}^{*k/H,S} \neq 0 \quad \text{and} \quad \Delta G_{\mathbf{PL}}^{*k/H,S} = 0 \tag{8.7.26}$$

or

$$\Delta G_{\mathbf{L}}^{*k/H,S} \neq 0 \quad \text{and} \quad \Delta G_{\mathbf{PL}}^{*k/H,S} = 0. \tag{8.7.27}$$

c. The J group

The J (for joint, or boundary) group consists of all FGs that do not belong to either the E or to the I group. These FGs are characterized by an intermediate change in their solvation properties—intermediate between no change as in the E group or total loss as in the I group. Qualitatively, the J group consists of all FGs whose solvation environment changes in the process (8.7.1). Examples are FGs that are fully exposed to the solvent in one state and partially buried in the second state. The more important case occurs when

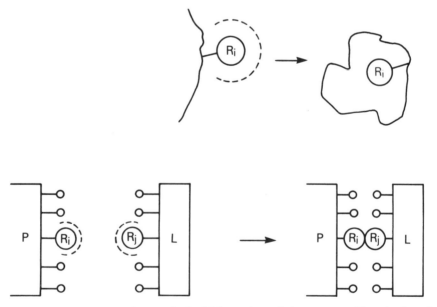

FIGURE 8.8. Internal functional groups in the folding and association processes. The solvating spheres around these groups are completely lost in the processes.

the FGs are independently solvated in one state but their solvation becomes correlated or dependent in the other state. An example of two groups in the joint region between **P** and **L** is shown in Fig. 8.9.

The formal requirements in terms of the solvation Gibbs energy are

$$\Delta G_{\mathbf{P}}^{*k/H,S} \neq \Delta G_{\mathbf{PL}}^{*k/H,S} \tag{8.7.28}$$

or

$$\Delta G_{\mathbf{L}}^{*k/H,S} \neq \Delta G_{\mathbf{PL}}^{*k/H,S}. \tag{8.7.29}$$

With this we have completed the classification of FGs according to the extent of change that occurs in their solvation environment. It should be noted that this classification pertains to a specific process. If **L** can bind to **P** on two different sites, then a different classification pertains to each mode of binding. Similarly, a different classification into the three groups E, I, and J pertains to the folding process discussed in section 8.8.

We have started with M FGs in the sum on the rhs of (8.7.22) and classified these into three groups. Denoting by M_E, M_I, and M_J the number of FGs in the E, I, and J groups, respectively, then

$$M_E + M_I + M_J = M, \tag{8.7.30}$$

and (8.7.22) can be written as

$$U(\mathbf{X}_a, \mathbf{X}_i) = U^H + U^S + \sum_{k \in E} U^{\mathrm{FG}}(k, \mathbf{X}_i) + \sum_{k \in I} U^{\mathrm{FG}}(k, \mathbf{X}_i) + \sum_{k \in J} U^{\mathrm{FG}}(k, \mathbf{X}_i), \tag{8.7.31}$$

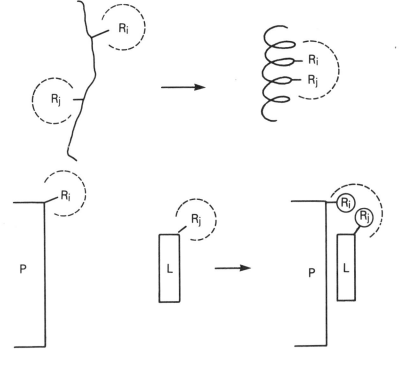

FIGURE 8.9. Two functional groups in the joint group for the folding and association processes.

and the corresponding binding energy between α and the solvent is

$$B_\alpha = B_\alpha^H + B_\alpha^S + \sum_{k \in E} B_k + \sum_{k \in I} B_k + \sum_{k \in J} B_k$$
$$= B_\alpha^H + B_\alpha^S + B_E + B_I + B_J, \tag{8.7.32}$$

where B_E is the total binding energy due to all FGs in the E group. Similar meanings apply to B_I and B_J.

The solvation Gibbs energy of the entire solute α is now written as (see, for instance, 8.7.17 and 8.7.19)

$$\Delta G_\alpha^* = \Delta G_\alpha^{*H} + \Delta G_\alpha^{*S/H} + \Delta G_\alpha^{*I/H,S} + \Delta G_\alpha^{*J/H,S,I} + \Delta G_\alpha^{*E/H,S,I,J}. \tag{8.7.33}$$

Note that Eq. (8.7.33) corresponds to turning on successively the binding energies due to different sources. In each step the condition consists of all binding energies that have been previously turned on. For instance, $\Delta G_\alpha^{*J/H,S,I}$ means that we turn on all the FGs in J, given that H, S, and I are already turned on. Clearly, we can turn on the various contributions in a different order; only one of these is chosen in (8.7.33).

Applying (8.7.33) to $\alpha = P$, L, and PL, we can rewrite (8.7.23) as

$$\delta G(P + L \rightarrow PL) = \delta G^H + \delta G^{S/H} + \delta G^{I/H,S} + \delta G^{J/H,S,I} + \delta G^{E/H,S,I,J}. \tag{8.7.34}$$

From the definition of the E group, it follows that each FG in this group has the same contribution to the solvation Gibbs energy in the initial and the final states of the process. Thus, from Eqs. (8.7.24) and (8.7.25) it is clear that the FGs in the group E will not contribute to δG. In other words, since

$$\Delta G_{PL}^{*E/H,S,I,J} = \Delta G_P^{*E/H,S,I,J} + \Delta G_L^{*E/H,S,I,J}, \tag{8.7.35}$$

it follows that

$$\delta G^{E/H,S,I,J} = 0. \tag{8.7.36}$$

Each of the terms in (8.7.35) corresponds collectively to the entire contribution of the E group of a specific solute (P, L, or PL) to the solvation Gibbs energy. If all the M_E FGs were independent, then each of the terms in (8.7.35) could be written as a sum of terms of the type (8.7.24) or (8.7.25), e.g.,

$$\Delta G_L^{*E/H,S} = \sum_{k \in E} \Delta G_L^{*k/H,S}. \tag{8.7.37}$$

In this case, the same individual terms appear on the two sides of (8.7.35); hence, Eq. (8.7.36) follows. However, even when the M_E FGs are dependent, we can choose to turn on these FGs in a specific order. In this case, instead of (8.7.37) we shall write

$$\Delta G_L^{*E/H,S} = \sum_{k \in E} \Delta G_L^{*k/H,S,1,2,...,k-1}. \tag{8.7.38}$$

At each stage we turn on one FG, say the kth, given that H, S and all other $k - 1$ FGs have already been turned on. Writing similar expression for P and PL leads to cancellation of each individual term pertaining to one FG.

Returning to (8.7.34), we can write the explicit form of each of the contributions on the rhs in terms of the corresponding solvation Gibbs energies.

$$\delta G^H = \Delta G_{PL}^{*H} - \Delta G_P^{*H} - \Delta G_L^{*H} \tag{8.7.39}$$

$$\delta G^{S/H} = \Delta G_{PL}^{*S/H} - \Delta G_P^{*S/H} - \Delta G_L^{*S/H} \tag{8.7.40}$$

$$\delta G^{I/H,S} = - \Delta G_P^{*I/H,S} - \Delta G_L^{*I/H,S} \tag{8.7.41}$$

$$\delta G^{J/H,S,I} = \Delta G_{PL}^{*J/H,S,I} - \Delta G_P^{*J/H,S,I} - \Delta G_L^{*J/H,S,I}. \tag{8.7.42}$$

Note that none of the ΔG^{*E} terms appear in the above equations. Also note that, by definition, $\Delta G_{PL}^{*I/H,S}$ does not appear in (8.7.41). This follows from (8.7.26) and (8.7.27).

Equations (8.7.39)–(8.7.42) apply to the entire groups I and J. We shall next turn to examine the contribution due to individual FGs. In general, it is not possible to write, say, $\Delta G_P^{I/H,S}$ as a sum of the solvation Gibbs energies of all the FGs in the I group of P. Therefore we shall first assume that, in the initial state (P and L separated), all FGs are independently solvated and that, in the final state (PL), the solvation of the FGs in the J group of P becomes correlated with the FGs in the J group of L. We also assume at this stage that only pairwise correlations exist, but not higher correlations. All these assumptions can be relaxed to take into account higher-order correlations between FGs. Also, for simplicity, we neglect partially buried FGs. These can always be viewed and counted separately as different FGs.

With these assumptions we rewrite (8.7.41) and (8.7.42) as

$$\delta G^{I/H,S} = - \sum_{k \in I} \Delta G_P^{*k/H,S} - \sum_{k \in I} \Delta G_L^{*k/H,S} \tag{8.7.43}$$

$$\delta G^{J/H,S,I} = \sum_{k,l} \Delta G_{PL}^{*k,l/H,S,I} - \sum_{k} \Delta G_P^{*k/H,S,I} - \sum_{l} \Delta G_L^{*l/H,S,I}$$

$$= \sum_{k,l} \delta G^{k,l/H,S,I}. \tag{8.7.44}$$

The quantity $\delta G^{k,l/H,S,I}$ on the rhs of (8.7.44) is exactly the same as the conditional $H\phi O$ or $H\phi I$ interaction between two FGs k and l, one on P and the other on L, in the complex PL. It has the form of a correlation function in the sense of

$$\delta G^{k,l/H,S,I} = -kT \ln \left\{ \frac{\langle \exp[-\beta(B_k + B_l)] \rangle_{H,S,I}}{\langle \exp(-\beta B_k) \rangle_{H,S,I} \langle \exp(-\beta B_l) \rangle_{H,S,I}} \right\}$$

$$= -kT \ln y(k, l), \tag{8.7.45}$$

where $y(k, l)$ is the conditional correlation function between the FGs k, l given the conditions H, S, I.

The total solvent induced contribution δG can now be written as

$$\delta G = \delta G^H + \delta G^{S/H} + \left[- \sum_{k \in I} \Delta G_P^{*k/H,S} - \sum_{k \in I} \Delta G_L^{*k/H,S} \right] + \sum_{k,l} \delta G^{k,l/H,S,I}. \tag{8.7.46}$$

This division into four terms corresponds to a classification of all solvent effects into four major terms. The first is due to the change of the excluded volume with respect to the solvent upon the formation of PL. The second is due to the loss of the soft interaction with the solvent caused by transferring part of the surfaces of P and L into the binding domain between P and L in the complex PL. The third contribution is also due to the loss of the conditional solvation of the specific FGs that are in the *I* group. The last term is due to FGs that are independent in the initial state but become pairwise correlated in the final state.

Clearly, the contribution of each FG depends on the nature of that FG. Different side chains will have different solvation properties; therefore the sums over k and over k, l in (8.7.46) include many different terms depending on the nature of the FGs. In the next section we shall turn to estimating representatives of each of these contributions.

At present, it is impractical to consider all different possible side chains. There are too many of these. Therefore, we shall focus on only two classes of FGs, the HϕO and the HϕI types. Even with this broad classification we have seven different types of contributions to δG. These are δG^H, $\delta G^{S/H}$, loss of solvation of HϕO FGs, loss of solvation of HϕI FGs, and correlation between HϕO–HϕO, HϕI–HϕI, or HϕO–HϕI FGs.

8.7.3. Methods of Estimating the Various Contributions to δG

The calculation of each of the contributions to δG in (8.7.46) is extremely complicated. Each quantity involves an average over all configurations of the solvent molecules. In this section we indicate possible ways of estimating these contributions.

First, δG^H is given by

$$\delta G^H = \Delta G_{PL}^{*H} - \Delta G_P^{*H} - \Delta G_L^{*H}. \qquad (8.7.47)$$

Each of the quantities on the rhs of (8.7.47) is the work required to create a cavity corresponding to one of the three solutes **PL**, **P**, or **L**. For spherical solutes, the best way to estimate these quantities is by the scaled-particle theory. For large and irregularly shaped solutes, say proteins, nucleic acids, etc., we can use the approximate expression for the cavity work (see section 5.10)

$$\Delta G_\alpha^{*H} = P V_\alpha^{EX}, \qquad (8.7.48)$$

where P is the pressure and V_α^{EX} is the volume of the cavity, i.e., the volume excluded to solvent molecules by the hard repulsive interaction of the solute α. Using (8.7.48) for each of the solutes in (8.7.47), we can estimate

$$\delta G^H \approx P(V_{PL}^{EX} - V_P^{EX} - V_L^{EX}) < 0. \qquad (8.7.49)$$

Since the excluded volume of the complex **PL** is always smaller than the excluded volume of the two monomers at infinite separation, δG^H is predicted to be negative, and is roughly proportional to the volume of the overlapping excluded regions of **P** and **L** (Fig. 8.10).

Note that this estimate is not valid for small cavities. However, when it is valid, it will be roughly the same for different liquids provided that the diameter of the solvent molecules is roughly the same as in water. If the solvent molecules are of larger diameter, then V_α^{EX} is larger than in water, and the loss of the excluded volume contained in δG^H changes accordingly. Thus we conclude that δG^H is likely to be negative and not very specific to either the solvent or the specific distribution of FGs on the surfaces of **P** and **L**.

The soft part, $\delta G^{S/H}$, is the contribution due to turning on the solute–solvent soft interaction. Here by soft we mean van der Waals or Lennard–Jones types of interaction, but not hydrogen bonding or charge–dipole interactions. Since we have excluded all

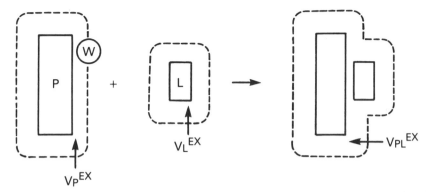

FIGURE 8.10. The excluded volume, indicated by the dashed curves, of the two monomers is reduced upon association.

specific or FG interactions, we can view the resulting solute as a large nonpolar solute. The interaction between the solute and the water molecules depends roughly on the surface area of the solute, i.e.,

$$\Delta G_\alpha^{*S/H} \approx - \varepsilon \mathscr{A}_\alpha \qquad (8.7.50)$$

where $\mathscr{A}_\alpha$ is the surface area of the solute α and ε is an energy parameter, of the order of magnitude of the Lennard–Jones parameter for the neon–methane interaction. Note that the last quantity is not related to the surface tension of the solvent. A Gibbs energy contribution proportional to the area $\mathscr{A}_\alpha$ would be added to ΔG_α^{*H} in (8.7.48) if we were to include surface tension effects (see section 5.11).

Using (8.7.50) for each of the three solutes, we can estimate

$$\delta G^{S/H} = \Delta G_{PL}^{*S/H} - \Delta G_P^{*S/H} - \Delta G_L^{*S/H} \sim - \varepsilon [\mathscr{A}_{PL} - \mathscr{A}_P - \mathscr{A}_L] > 0. \qquad (8.7.51)$$

Since the total surface area decreases in the process, we expect that this contribution will be positive, i.e., the formation of **PL** involves the loss of part of the soft solvation Gibbs energy of the two solutes. Again, for large solutes we do not expect that this quantity will be specific to liquid water. Both of the quantities δG^H and $\delta G^{S/H}$ can therefore be estimated by knowing the volumes and the surface areas of the three solutes **P**, **L**, and **PL**.

The quantities that are more difficult to estimate are the specific conditional solvation Gibbs energies of the various FGs. These are also the more important quantities in determining the specificity of the process; e.g., the molecular recognition (section 8.7.4) or the folding pathway of a protein (section 8.8). Since there exists no theory that can be used to calculate these quantities, we appeal to a theory that transforms our problem into a more manageable one.

We recall the definition of the conditional solvation Gibbs energy of a FG, k, in a polymer α:

$$\Delta G_\alpha^{*k/H,S} = \Delta G_{\alpha+k}^* - \Delta G_\alpha^*. \qquad (8.7.52)$$

Some care should be exercised in interpreting the quantities written in Eq. (8.7.52). Until now we have used α to denote any one of the solutes **P**, **L**, or **PL**. Here the notation is slightly different. We want to make the distinction between the solvation Gibbs energy of the solute α with and without the kth FG. We could have used either $\alpha + k$ and α, or α and $\alpha - k$ to make this distinction. We have chosen the first of these notations in (8.7.52). Thus on the rhs of (8.7.52) we have the solvation Gibbs energy of the polymer $\alpha + k$ and of the same polymer without k. Recall that a FG may be either a real group (e.g., methyl, hydroxyl, etc.) or only a part of the interaction energy (say the charge–dipole or HB interaction), depending on whether we choose the separation (8.7.11) or (8.7.12).

The quantity $\Delta G_\alpha^{*k/H,S}$ is the indirect work required to bring the kth FG from a fixed position in the vacuum to its final position next to the polymer α or simply to turn on the kth FG, given the appropriate conditions. Clearly for a large polymer α neither of the quantities on the rhs of (8.7.52) is measurable. Therefore the conditional quantity $\Delta G_\alpha^{*k/H,S}$ cannot be estimated directly from the rhs of (8.7.52). It is at this point that statistical thermodynamics comes to our aid. It does so by transforming Eq. (8.7.52) into an approximate equation which can be used to estimate the required quantity.

The reasoning is as follows: Consider the conditional quantity $\Delta G_\alpha^{*k/H,S}$ defined by

$$\exp(-\beta \Delta G_\alpha^{*k/H,S}) = \langle \exp(-\beta B_k) \rangle_{H,S}. \tag{8.7.53}$$

This is a conditional average of the quantity $\exp(-\beta B_k)$ over all configurations of the solvent molecules in the T, P, N ensemble, given H, S (or in the more general case given the solute α, excluding the kth FG).

Clearly, this quantity depends on the binding energy B_k and on the distribution of solvent molecules around this FG. The situation is schematically depicted in Fig. 8.11. It is also clear that the solvation of the entire solute α (or $\alpha + k$) is not necessary for estimating the quantity (8.7.53). This suggests that if we find two model compounds $\alpha' + k$ and α' such that the environment around the group k is similar to the environment

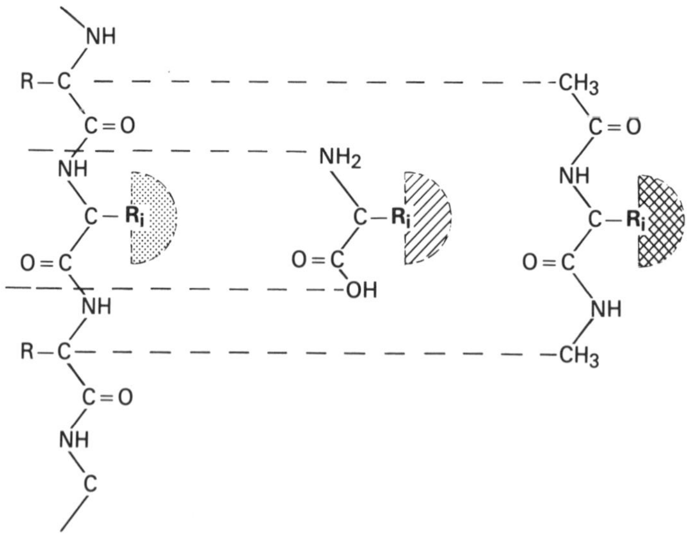

FIGURE 8.11. Schematic illustration of the conditional solvation of a functional group R_i. [R_i is referred to as the kth FG in Eq. (8.7.53).] On the lhs, R_i is attached to the entire protein. On the rhs, the same group R_i is attached to two model compounds that resemble the surroundings of R_i in the original protein.

of the same group in the original molecule $\alpha + k$, then we can approximate Eq. (8.7.52) by

$$\Delta G_{\alpha}^{*k/H,S} \approx \Delta G_{\alpha'}^{*k/H,S} = \Delta G_{\alpha'+k}^{*} - \Delta G_{\alpha'}^{*}. \tag{8.7.54}$$

The difference between (8.7.52) and (8.7.54) is that in the latter we have experimental quantities on the rhs of the equation.

Clearly, the model compounds α' and $\alpha' + k$ should be small enough so that $\Delta G_{\alpha'+k}^{*}$ and $\Delta G_{\alpha'}^{*}$ are measurable quantities. They should be large enough so that the environment surrounding the kth FG is approximately the same in $\alpha + k$ as in $\alpha' + k$. If these conditions can be met, then (8.7.54) can be used to estimate the required quantity $\Delta G_{\alpha}^{*k/H,S}$. In Appendices G and H we further elaborate on the nature of this approximation and give some estimated quantities based on Eq. (8.7.54).

The same methodology can be applied to estimate the correlation quantities $\delta G^{k,l/H,S,I}$. In fact, these are exactly the intramolecular, solvent-induced HϕO or HϕI interactions discussed in sections 7.15 and 7.16. For any pair of FGs k and l we write

$$\delta G^{k,l/C} = \Delta G_{\mathbf{PL}}^{*k,l/C} - \Delta G_{\mathbf{P}}^{*k/C} - \Delta G_{\mathbf{L}}^{*l/C}, \tag{8.7.55}$$

where the condition is denoted by C. This can be either H, S, I as in (8.7.46), or any other condition, depending on which of the divisions (8.7.11) or (8.7.12) we use. Again the quantities on the rhs of (8.7.55) cannot be measured for large molecules. But using the same arguments as above, we can appeal to small model compounds to estimate the required quantities $\delta G^{k,l/C}$. Some estimates have already been presented in sections 7.15 and 7.16.

In writing (8.7.46) we assumed that all FGs were independently solvated in the initial state (separated $\mathbf{P}$ and $\mathbf{L}$) and that only pairwise correlations exist in the final state $\mathbf{PL}$. Having information on the various conditional solvation Gibbs energies, we can relax these requirements and take into account both correlations in the initial state and higher-order correlations in the final state. This procedure should be carried out for specific solutes and for a specific process. We shall discuss a few simple representative processes in the next subsection.

8.7.4. Selection of Specific Binding Site: Molecular Recognition

In our discussion of the association process Eq. (8.7.1), we have assumed that the configuration of the final state, $\mathbf{PL}$, is known. The same assumption has been made in the treatment of the binding of ligands to their sites. In this section we address ourselves to the question of how the ligand selects the binding site or, more generally, what is the most likely configuration of the final state $\mathbf{PL}$. If the ligand is relatively small and the polymer P can offer several binding sites, then the question is which site is more likely to be chosen—this is essentially the question of molecular recognition. A more general question arises when both $\mathbf{P}$ and $\mathbf{L}$ can offer different binding sites—or binding surfaces. We present here a few examples in which there are several modes of binding or association and focus on the role of the solvent in influencing or even determining the choice of the most probable configuration of the final product $\mathbf{PL}$. For simplicity, we still assume that $\mathbf{P}$, $\mathbf{L}$, and $\mathbf{PL}$ have one conformation each. The more realistic cases are those in which each of the solutes can be in many conformational states. These cases require averaging over all possible conformations, before and after the association process, and will be further discussed in section 8.7.5.

In each of the following examples we assume that there are only two modes of association. Also, for simplicity, we select a solute containing only a few kinds of FG, say methyl to represent a HϕO FG and hydroxyl or carbonyl to represent a HϕI FG. Although we shall be mainly interested in the solvent-induced effect δG, we shall also use the direct interaction ΔU in order to compare the change in the mode of binding when passing from vacuum to the solvent. In an actual calculation of the equilibrium constant [Eq. (8.7.5)], we must also consider the contributions of the internal partition functions, Q_L, Q_P, and Q_{PL}. For simplicity we assume that these are approximately the same for the different modes of association. The numerical values given in the following examples are only order-of-magnitude estimates.

1. We consider first the association between two rigid linear molecules having alternating HϕO and HϕI FGs, Fig. 8.12. There are infinitely many possible configurations for the dimer. In a vacuum, the most probable configuration is the one for which the direct interaction energy ΔU in Eq. (8.7.8) is minimal (or the most negative). In this example, it is clear that the dimer in which the maximum number of HBs is formed is the most probable one. For example, if we have six FGs capable of forming direct HBs, we can estimate

$$\Delta G_B = \Delta U \approx -6 \times 6.5 = -39 \text{ kcal/mol.} \tag{8.7.56}$$

For comparison, the direct interaction energy for the dimer in which the HϕO groups face each other (Fig. 8.12b) is about

$$\Delta G_B = \Delta U \approx -6 \times 0.5 = -3 \text{ kcal/mol,} \tag{8.7.57}$$

where we have chosen -6.5 kcal/mol and about -0.5 kcal/mol for the direct HB and the direct van der Waals interaction between the two methyl groups.

Comparing these particular configurations, we can estimate the relative probabilities of the two dimers (Fig. 8.12) in a vacuum, at room temperature ($kT \approx 0.6$ kcal/mol) as

$$\frac{\text{Pr}(\text{H}\phi\text{O--H}\phi\text{O})}{\text{Pr}(\text{H}\phi\text{I--H}\phi\text{I})} = \exp\left(-\frac{36}{0.6}\right) = 8.7 \times 10^{-27}. \tag{8.7.58}$$

As expected, in a vacuum the HϕI–HϕI configuration is overwhelmingly more probable than the HϕO–HϕO configuration.

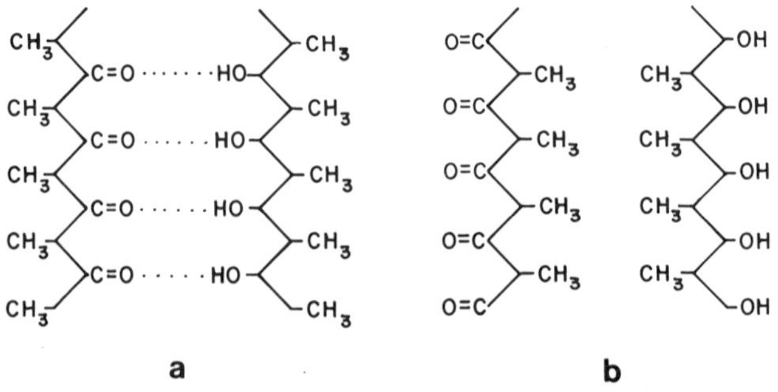

FIGURE 8.12. Two rigid linear molecules having alternating HϕO and HϕI groups. The association in the HϕI–HϕI mode (a) and in the HϕO–HϕO mode (b) are compared in Eqs. (8.7.56) and (8.7.57).

We now take the same two configurations and insert the system in water. In order to calculate the solvent contribution we shall use the division (8.7.11) for the solute–solute interaction, i.e.,

$$U(\mathbf{X}_\alpha, \mathbf{X}_i) = U^H + U^S + U^{HB}. \tag{8.7.59}$$

Again we compare the two configurations in Fig. 8.12. The direct interactions are the same as in Eqs. (8.7.56) and (8.7.57). For the indirect interactions, we assume that the FGs are independent, and therefore [see Eq. (8.7.46)]

$$\delta G = \delta G^H + \delta G^{S/H} - \Delta G^{*HB/H,S}, \tag{8.7.60}$$

where $\Delta G^{*HB/H,S}$ is the conditional solvation Gibbs energy of all FGs.

Thus we first switch off all the soft and HB interactions with the solvent. The hard contribution is

$$\delta G^H = \Delta G^{*H}_{PL} - \Delta G^{*H}_P - \Delta G^{*H}_L. \tag{8.7.61}$$

This may be estimated by the excluded volume of the solutes using the approximation (8.7.49). However, if we are interested only in comparing the relative probabilities of the two configurations, we need not estimate δG^H for each configuration. Instead, we can assume that the excluded volume of PL is approximately the same for the two configurations and that therefore δG^H will not affect the relative probabilities. Similarly, when we turn on the soft interactions, we have

$$\delta G^{S/H} = \Delta G^{*S/H}_{PL} - \Delta G^{*S/H}_P - \Delta G^{*S/H}_L. \tag{8.7.62}$$

If the interaction between the hydroxyl or the carbonyl groups with water can be written as in Eq. (8.7.11), then turning on the soft part of the interaction will again be approximately the same for the two configurations. Therefore, $\delta G^{S/H}$ will approximately not affect the probability ratio. The only remaining quantity is the loss of the conditional HBing interaction. Assuming that each FG participating in direct HBing in the configuration (Fig. 8.12a) loses one arm, we have altogether a loss of twelve arms. We have estimated in section 7.7 (see also Appendix C) that the conditional solvation Gibbs energy per arm is about −2.25 kcal/mol. Thus the relative probability of the two configurations in water is now modified into

$$\frac{Pr^*(H\phi O\text{–}H\phi O)}{Pr^*(H\phi I\text{–}H\phi I)} = 8.7 \times 10^{-27} \exp(12 \times 2.25/0.6)$$

$$= 8.7 \times 10^{-27} \times 3.5 \times 10^{19}$$

$$= 3 \times 10^{-7}. \tag{8.7.63}$$

We see that the probability ratio is still in favor of the $H\phi I$–$H\phi I$ configuration a (Fig. 8.12) but this has been considerably reduced by the solvent. The reason for this reduction is essentially the same as the reduction of the first peak of the water–water pair correlation function discussed in section 8.2.

To summarize, the configuration in Fig. 8.12a in the vacuum is more favorable, mainly due to the strong direct HBing between the two monomers. In water, the direct interaction is still the same, but the indirect or solvent-induced effect tends to weaken the overall binding Gibbs energy. The reason is the loss of one arm per FG. Since the direct HB energy is larger (in absolute magnitude) than the loss of the conditional Gibbs energy of solvation of two arms, we find that the $H\phi I$–$H\phi I$ mode of binding is still the more favorable in water.

The same reasoning leads to a different result if more than one arm per FG is lost in the association process. This is demonstrated in the next example.

2. We extend the previous example by taking instead of two rigid linear molecules, two planes or two cubes. The preceding example is related to the formation of secondary structures in protein (section 8.8). This present model is more relevant to protein–protein association. We model a protein by a cube (Fig. 8.13). We assume that three surfaces contain $H\phi O$ groups, say methyl groups, and three surfaces contain $H\phi I$ groups. These groups are assumed to be independently solvated, and they are spaced in such a way that when two $H\phi I$ surfaces approach each other, each FG on one cube can form one HB with a FG on the second cube.

Again we use the separation (8.7.11) for the total interaction between a solute and a water molecule. Most of the arguments are the same as in the previous example. The probability ratio in a vacuum is again in favor of the $H\phi I$–$H\phi I$ mode; the reason is exactly the same as before. We estimate the probability ratio per pair of FGs as

$$\frac{\Pr(H\phi O - H\phi O)}{\Pr(H\phi I - H\phi I)} = \exp\left(-\frac{6.5 - 0.5}{0.6}\right) = 4.5 \times 10^{-5}. \tag{8.7.64}$$

Again, both δG^H and $\delta G^{S/H}$ are expected to be approximately the same for the two configurations. The only difference between this and the previous example arises from the loss of the conditional solvation Gibbs energy. In the previous example, each pair of FGs forming a direct HB contribute about -6.5 kcal/mol due to direct HB and about $+4.5$ kcal/mol due to the loss of the solvation of two arms. This still leaves a *negative* net Gibbs energy of about -2 kcal/mol per pair of FGs. The situation changes drastically if just one more arm is lost in the formation of the dimer. In real cases, the FGs, say hydroxyl or carbonyl, are completely removed from the water into the binding region. Since each of these FGs has more than one arm, the balance of Gibbs energy per pair of FGs turns from negative to positive.

As an example, consider one pair of OH and C=O forming a direct HB. Again, the energy gained is about -6.5 kcal/mol. But the Gibbs energy loss is about -6.7 kcal/mol for the OH and about -5.2 kcal/mol for the C=O (see Appendix H). This is roughly equivalent to a loss of five arms, which is equivalent to a *positive* Gibbs energy change of about $\delta G \approx 11.9$ kcal/mol per pair of FGs. In terms of probability ratio, Eq. (8.7.64) is modified into

$$\frac{\Pr^*(H\phi O - H\phi O)}{\Pr^*(H\phi I - H\phi I)} = 4.5 \times 10^{-5} \exp(11.9/0.6) = 1.8 \times 10^4. \tag{8.7.65}$$

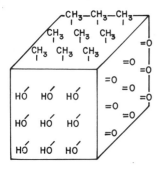

FIGURE 8.13. A cube having $H\phi I$ and $H\phi O$ surfaces as a model for a protein.

Thus, in contrast to the previous example where the $H\phi I$–$H\phi I$ configuration was still the preferable one in water, in the present example the preference turns in favor of the $H\phi O$–$H\phi O$ mode. This phenomenon is very common in protein binding and protein–protein association. The preference of the $H\phi O$–$H\phi O$ mode of binding is commonly attributed erroneously to $H\phi O$ interaction. An analysis of the various factors involved in determining the probability ratio shows that the main reason for the $H\phi O$–$H\phi O$ mode of association is the large and negative solvation Gibbs energies of the $H\phi I$ groups, which are more than an order of magnitude larger than the corresponding solvation Gibbs energy of a $H\phi O$ group. Thus two proteins, when given the possibility to associate, will prefer to choose the $H\phi O$–$H\phi O$ mode not because of strong $H\phi O$ interactions, but because of the very strong reluctance of the $H\phi I$ groups to lose their solvation. It should be noted that this and the previous examples were intentionally designed to eliminate the $H\phi O$ interaction. One can legitimately refer to both δG^H and $\delta G^{S/H}$ as $H\phi O$ interactions in the sense of section 7.15. In these examples, both of these quantities are nearly the same for the two modes of association and therefore do not affect the probability ratio. The only factor that does affect the probability ratio is the loss of the conditional solvation Gibbs energy of the $H\phi I$ groups. In both cases, this factor contributes positively to the overall δG. The only difference between the two examples is that in the first case, the positive contribution to δG is not enough to compensate for the negative direct HB energy. In the second case, the positive contribution to δG is more than enough to compensate for the negative direct HB energy. Therefore, the preferential mode of association cannot be attributed to $H\phi O$ interactions.

To sharpen the above argument, we shall now change slightly the distribution of the FGs on the surface of the cube and observe a dramatic change in the probability ratio. In the model of a protein depicted as a cube in Fig. 8.13 we assumed that the FGs on the surface are independently solvated. Therefore, the loss of the solvation Gibbs energies of the FGs was additive. Since each of the C=O and OH forming a direct HB have lost more than one arm, we got a large probability ratio in favor of the $H\phi O$–$H\phi O$ model in Eq. (8.7.65).

Now suppose that we change the distance between the members of one pair of FGs on each surface in such a way that one pair of OHs on one cube and one pair of C=Os on the second cube are at a distance of about 4.5 Å, so that they can form a strong $H\phi I$ interaction of about -3 kcal/mol when P and L are in isolation. Thus the conditional Gibbs energy of solvation of a pair of C=Os and a pair of OHs in the isolated P and L is about (see Appendix H)

$$\Delta G^*(\text{pair of C=O at 4.5 Å}) \approx -2 \times 5.2 - 3 = -13.4 \text{ kcal/mol} \qquad (8.7.66)$$

$$\Delta G^*(\text{pair of OH at 4.5 Å}) \approx -2 \times 6.7 - 3 = -16.4 \text{ kcal/mol}. \qquad (8.7.67)$$

When these two pairs are brought to the final position PL, the total loss of the solvation Gibbs energy is about $+29.8$ kcal/mol. Taking $29.8/2 = 14.9$ kcal/mol as the loss per one pair of C=O and OH forming a direct HB in PL, we have here an excess loss of 3 kcal/mol compared with 11.9 kcal/mol used in (8.7.65) (for the independent FGs). Therefore, the probability ratio in (8.7.65) will now be multiplied by a factor of $\exp(3/0.6) \approx 148$ in favor of the $H\phi O$–$H\phi O$ mode. We stress that this enhancement of the probability ratio was achieved only by changing the distances of one pair of the $H\phi I$ group. Any effect due to the $H\phi O$ groups was not changed. Therefore, this enhancement certainly cannot be attributed to $H\phi O$ interactions.

A more dramatic effect can be observed if we further change the distance between the HϕI groups. Instead of bringing the pair of C=Os and the pair of OHs to a distance of 4.5 Å, we now chose a smaller distance, in such a way that the two C=Os are so close to each other that they interfere with the solvation of each other. A schematic illustration is shown in Fig. 8.14. In such a configuration, the C=O and the OH cannot use all their arms to form HBs with the solvent. Assuming that each of these groups can be solvated through only one arm, then the loss of the solvation Gibbs energy per pair of C=O and OH will now be +4.5 kcal/mol. This will be less than enough to compensate for the direct HB energy of −6.5 kcal/mol. Therefore the probability ratio will now turn in favor of the HϕI–HϕI mode, exactly as in example a of Fig. 8.12.

To summarize, the probability ratio can be changed in favor of either the HϕO–HϕO mode or the HϕI–HϕI, simply by changing the distribution of HϕI groups on the surface of the protein. The HϕO groups play a minor role in the determination of this probability ratio.

The formation of a direct HB between two functional groups, either intramolecularly or intermolecularly, is often presented as a stoichiometric "reaction" of the form

$$E-OH\cdots H_2O + S-C=O\cdots H_2O \rightarrow E-OH\cdots O=C-S + H_2O\cdots H_2O,$$

where an enzyme (E) and a substrate (S) are initially solvated by water. When they form a direct HB, they release the water molecules to form HBs between themselves.

This presentation suggests that there is an overall balance in the number of HBs formed and destroyed on the two sides of the equation, leading to the conclusion that HBs do not contribute to the Gibbs energy (or to the enthalpy) of the reaction. But this presentation is misleading for two reasons. First, the loss on the lhs is the *Gibbs energy* (or the enthalpy, the entropy, etc.) of *solvation* of one OH and one C=O. The gain on the rhs is the HB *energy*. Second, the fate of the water molecules released in the reaction is taken implicitly into account when considering the solvation Gibbs energies of E, S, and ES. There is no need to consider explicitly the formation of water–water HB. The reason for this is essentially the same as the one given in section 7.11.

3. We now consider the solvent effect on molecular recognition. Let **L** be a relatively small ligand and **P** a large polymer. We assume that **L** can bind to several distinguishable sites of **P**. For each site i, the association reaction is

$$\mathbf{P} + \mathbf{L} \rightarrow \mathbf{LP}_i, \tag{8.7.68}$$

where $\mathbf{LP}_i$ is the complex of **L** and **P** at the specific site i. The selection of a specific binding site is a ubiquitous phenomenon in biochemical processes. These range from the adsorption of fatty acids or drugs by plasma proteins to the binding of a substrate to an enzyme or the binding of proteins (such as a repressor) to nucleic acids, etc.

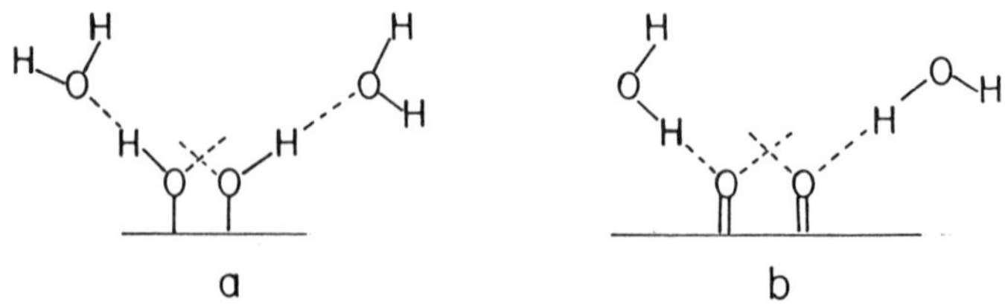

FIGURE 8.14. Two hydroxyl groups (a) or two carbonyl groups (b) brought to a very short distance apart so that their solvation spheres overlap. The result is the loss of the solvation of a pair of arms.

In all of these processes, the selection of the binding site is highly specific. How does the ligand **L** know how to make the selection of a particular site? This is essentially the question of molecular recognition in biochemistry. The classical answer to this question is given within the so-called "lock and key" model. This metaphoric model requires that the two partners involved in the binding fit tightly in the complex **LP**$_i$ (Fig. 8.15a).

Clearly, a geometrical fit is equivalent to maximum direct interaction through weak van der Waals interactions. A somewhat more general requirement that does not require geometrical fit is that **L** and **P** possess a complementary pattern of FGs such as is depicted in Fig. 8.15b. Another variation of the same model is the "induced-fit" model. Here, the fitting between **L** and **P** is produced after the binding process has been initiated. This case may be treated in the context of conformational changes in the protein (section 8.7.5). There is also a possibility that the rotational PF of the various complexes **LP**$_i$ would influence the preference for a given site i. This is probably a small effect and will be neglected.

In all of these models the criterion for choosing the right site depends on the *direct* interaction between **L** and **P**$_i$. The most probable site is the site j for which

$$\Delta U(j) = \min_i \Delta U(i), \tag{8.7.69}$$

where $\Delta U(i)$ is the direct interaction between **L** and **P** at the site i.

The situation is markedly different in solution, where the criterion should be replaced by

$$\Delta G_B(j) = \min_i \Delta G_B(i)$$

$$= \min_i [\Delta U(i) + \delta G(i)]. \tag{8.7.70}$$

The presence of the solvent can completely change the preference for the selected site. Here we present one example of such a switching of the preferential binding site, which is due to one type of contribution to $\delta G(i)$, namely, loss of conditional solvation of FG in the I group. In the next example we show how FGs in the J group can also produce such a shift in the selection of binding sites.

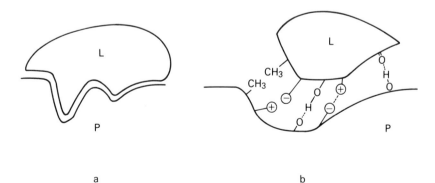

a   b

FIGURE 8.15. Lock and key models for recognition. (a) Geometrical fit. (b) Complementary pattern of functional groups.

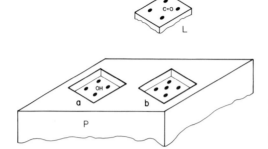

FIGURE 8.16. Two binding sites on the polymer **P**. In site a there is an OH at the center. This is replaced by a methyl group (dark points) in site b.

Consider two binding sites depicted as a and b in Fig. 8.16. The ligand has a complementary pattern of FG. The pattern of the binding surface is shown in the figure. The only difference between the two sites is the replacement of one methyl group by a polar group. In the absence of solvent, the ligand **L** will bind to the site a. The preference for site a is due to the direct HB between the OH on site a and the C=O on the ligand **L**. The relative probability of binding to a and to b is

$$\frac{\Pr(a)}{\Pr(b)} = \exp[-\beta U(a) + \beta U(b)]$$

$$\approx \exp\left(\frac{6.5 - 0.5}{0.6}\right)$$

$$= 2.2 \times 10^4, \tag{8.7.71}$$

where we assume that the HB energy is about $\varepsilon_{HB} = -6.5\,\text{kcal/mol}$ and the energy for the interaction of the C=O with the methyl group is about $-0.5\,\text{kcal/mol}$. Note that all interactions between **L** and **P** are assumed to be the same except for the interaction between the C=O and OH in a or the C=O and the methyl in b.

The situation changes drastically when the entire system is in water. Again, assuming all other things being equal, the relative probability is now

$$\frac{\Pr^*(a)}{\Pr^*(b)} = \exp[-\beta U(a) + \beta U(b)]\exp[-\beta\delta G(a) + \beta\delta G(b)]$$

$$= 2.2 \times 10^4 \exp[-\beta(-\Delta G_{\mathbf{P}}^{*OH/a} + \Delta G_{\mathbf{P}}^{*CH_3/b})]$$

$$= 2.2 \times 10^4 \exp\left(\frac{-6.7 - 0.3}{0.6}\right)$$

$$= 0.19. \tag{8.7.72}$$

Thus, the solvent causes a switch of preference from a (in vacuum) to b. Note that the loss of the conditional solvation Gibbs energy for the C=O of the ligand is the same for binding to a or to b. Therefore, the solvent contribution to the probability ratio comes only from the loss of the solvation of the OH (in site a) or of the CH$_3$ (in site b). The reason for this switch is the reluctance of site a to lose the solvation of the OH group as well as the eagerness of site b to lose the solvation of its central methyl group. Both contribute negatively to $\delta G(a \to b)$, but note that the HϕI contribution is more than twenty times larger than the HϕO contribution to this switch.

4. We now turn to molecular recognition through indirect HϕI interaction. In all of the examples given above, we have stressed one aspect of the contribution to δG, the loss of the solvation of a HϕO or HϕI groups in the I group [see Eq. (8.7.46)]. We now demonstrate that correlation between FG in the J group can also make selective preference for a binding site.

Consider Fig. 8.17 where we have two identical sites A and B. From the point of view of the lock and key model, the ligand **L** would have exactly the same probability to bind to either A or to B. In Fig. 8.17 we have depicted a perfect *geometrical* fit between the ligand and the sites, but the same argument will hold if the fit is produced by a complementary pattern of FG. The important point is that the sites A and B provide the same binding *energy* to the ligand **L**.

Now suppose that both **L** and **P** have carbonyl groups in the J region. These are sufficiently far apart so that they do not interact directly. In other words, the presence of these FG groups does not change the direct interaction $U(A) = U(B)$. In the vacuum, the probability ratio is therefore

$$\frac{Pr(A)}{Pr(B)} = 1. \tag{8.7.73}$$

In water, one strong HϕI interaction can be formed by the two carbonyl groups. This in itself will change the probability ratio into

$$\frac{Pr^*(A)}{Pr^*(B)} = \exp[-\beta\delta G(B \to A)]$$

$$\approx \exp\left(\frac{3}{0.6}\right)$$

$$= 150. \tag{8.7.74}$$

Thus the two sites that are identical from the point of view of the lock and key model become unidentical with preference towards one of the sites. This preference is due specifically to the HϕI interaction. In protein association or in protein binding to DNA, there are many possibilities for such HϕI interaction. If there are m such pairs, then the probability ratio is about 150^m, a huge number even for small m.

Similar effect could be obtained by HϕO groups. Suppose that in the same situation as in Fig. 8.17 we have two methyl groups instead of the two carbonyl group. The corresponding probability ratio would be on the order of

$$\frac{Pr^*(A)}{Pr^*(B)} \sim \exp[-\beta\delta G(B \to A)] \sim \exp(1) \approx 3. \tag{8.7.75}$$

Clearly the HϕI effect in (8.7.74) is much larger than the HϕO effect in (8.7.75).

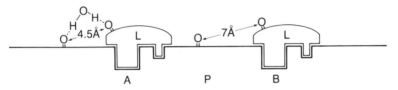

FIGURE 8.17. Two identical sites from the point of view of the lock and key model become unidentical in water due to HϕI interaction.

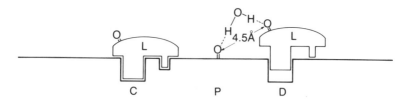

FIGURE 8.18. Two different sites on a polymer. C is preferable by direct interaction (lock and key model). In water the preference may switch to D due to HϕI interaction.

A second example is shown in Fig. 8.18. Here **L** fits better into the "lock" C. Therefore, in the vacuum, there is a preference for binding at the site C. This preference might be switched to D if there existed HϕI interaction as indicated in Fig. 8.18D.

In the examples discussed in (3) and (4), above, we observed a solvent induced preference of a binding site. In (3), the preference is due to FGs in the I group. In (4) it was due to FGs in the J group. In any real case both of these effects as well as the direct interaction operate concertedly to determine the most probable binding site.

8.7.5. Averaging over All Conformations of **P**, **L**, and **PL**

We started in section 8.7.1 with the consideration of rigid molecules; i.e., we selected one conformation for each of the solutes involved in the binding process Eq. (8.7.1). Let α be one of the solutes **L**, **P**, or **PL**, and let α_i be one specific conformation of α. The equilibrium condition requires that

$$\mu_\alpha = \mu_{\alpha_i} \qquad \text{for all } i. \tag{8.7.76}$$

Here μ_α is the chemical potential of the solute α and μ_{α_i} the chemical potential of a specific conformer α_i.

In order to obtain the thermodynamic standard Gibbs energy of reaction (8.7.1) (except that now **P**, **L**, and **PL** are understood to attain all possible conformations), we write in thermodynamic notation

$$\mu_\alpha^{0l} + kT \ln \rho_\alpha = \mu_{\alpha_i}^{0l} + kT \ln \rho_{\alpha_i} \tag{8.7.77}$$

or, equivalently,

$$\exp(-\beta\mu_\alpha^{0l}) = \sum_i \exp(-\beta\mu_{\alpha_i}^{0l}). \tag{8.7.78}$$

The standard Gibbs energy of the reaction is therefore

$$\Delta G^{0l} = \mu_{PL}^{0l} - \mu_P^{0l} - \mu_L^{0l}, \tag{8.7.79}$$

where each of the quantities on the rhs of (8.7.79) is expressed in terms of the sum over all possible conformations as in (8.7.78).

In order to extract the solvent-induced contribution, we write the analogues of (8.7.78) and of (8.7.79) for an ideal-gas phase

$$\exp(-\beta\mu_\alpha^{0g}) = \sum_i \exp(-\beta\mu_{\alpha_i}^{0g})$$

$$= \sum_i \exp[-\beta U^*(\alpha_i)]Q_i, \tag{8.7.80}$$

where $U^*(\alpha_i)$ is the intramolecular energy of the conformation α_i relative to some common zero energy. The relation between μ_a^{0l} and μ_a^{0g} is thus

$$
\begin{aligned}
\exp(-\beta\mu_a^{0l}) &= \sum_i \exp(-\beta\mu_{\alpha_i}^{0l}) \\
&= \sum_i \exp[-\beta W(\alpha_i|w)] \exp[-\beta U^*(\alpha_i)]Q_i \\
&= \exp(-\beta\mu_a^{0g}) \sum_i \exp[-\beta W(\alpha_i|w)] Y_{\alpha_i},
\end{aligned}
\tag{8.7.81}
$$

where Y_{α_i} is the mole fraction of the conformer α_i in the ideal-gas phase,

$$
Y_{\alpha_i} = \frac{Q_i \exp[-\beta U^*(\alpha_i)]}{\sum_j Q_j \exp[-\beta U^*(\alpha_j)]}.
\tag{8.7.82}
$$

Writing the same expression for **P**, **L**, and **PL**, we obtain the required relation

$$
\exp(-\beta\Delta G^{0l}) = \exp(-\beta\Delta G^{0g}) \exp(-\beta\delta G),
\tag{8.7.83}
$$

where the solvent contribution is defined by

$$
\exp(-\beta\delta G) = \frac{\langle\exp[-\beta W(\mathbf{PL}|w)]\rangle_{\text{conf}}}{\langle\exp[-\beta W(\mathbf{P}|w)]\rangle_{\text{conf}}\langle\exp[-\beta W(\mathbf{L}|w)]\rangle_{\text{conf}}}.
\tag{8.7.84}
$$

The symbol $\langle\ \rangle_{\text{conf}}$ is an average over all conformations relevant to the particular solute, with the probability distribution Y_{α_i} appropriately chosen for each solute.

Equation (8.7.84) means that when the solutes **P**, **L**, and **PL** can have different conformations, we have to first compute the coupling work for each conformer (which is itself an average over all configurations of the solvent molecules), then we average over all conformations of that species. In the case of the dimer **PL**, we include in the term "conformations" all the configurations of **PL** that we recognize as belonging to the dimer.

When applying these averages, one should bear in mind that the characterization of the various contributions to δG and even the classification of FGs into various groups (E, I, and J) depend on the particular conformation. A FG can be in group E in one conformer and in group I in a different conformer. Therefore, the general expression for δG is useful only when the averages over all conformations do not include widely different conformations.

For instance, suppose that **P** and **L** are rigid, but there are two binding sites on **P**, say A and B as in Fig. 8.17. Formally, the two complexes $\mathbf{LP}_A$ and $\mathbf{LP}_B$ could be viewed as two conformations of **PL**, and Eq. (8.7.84) for δG would apply. However, it is clear that all FG of **P** that belong to the I group in the complex $\mathbf{LP}_A$ are in the E group in the complex $\mathbf{LP}_B$, and *vice versa*. Therefore, in this case the classification into groups E, I and J is not useful. A better procedure would be to classify $\mathbf{LP}_B$ and $\mathbf{LP}_A$ as different species and take averages within the restricted range of conformations belonging to $\mathbf{LP}_A$ and $\mathbf{LP}_B$, respectively.

Another example of importance is the induced-fit model. Suppose that **L** is rigid and **P** can be in two conformations, A and B. Again suppose that the equilibrium constant for the conversion

$$
A \rightleftarrows B
\tag{8.7.85}
$$

is in favor of, say, B, such that before the binding of $\mathbf{L}$ we have

$$x_B^0 \gg x_A^0. \tag{8.7.86}$$

Now suppose that the binding of $\mathbf{L}$ on A is much stronger than on B, i.e., $U_A \ll U_B$. Then, upon binding, the equilibrium concentration might change in favor of A; i.e.,

$$x_B' \ll x_A'. \tag{8.7.87}$$

This case may be referred to as the induced-fit model, shown in Fig. 8.19. Again one could have applied δG in (8.7.74) and included A and B as two conformations of $\mathbf{P}$. A more useful procedure would be to classify $\mathbf{P}_A$ and $\mathbf{P}_B$ as two species and treat these as having different binding energies, exactly as we have done in section 2.10 for the vacuum case or in section 8.4.2 in a solvent.

8.7.6. Solvent-Induced Forces between Macromolecules

The solvent-induced forces operating between any two molecules were discussed in sections 6.4 and 7.14. Here we extend the treatment for two macromolecules, denoted by $\mathbf{L}$ and $\mathbf{P}$. These could be two proteins, two DNA molecules, two membranes, or any two walls containing HϕI groups on their surfaces. The emphasis here is on the origin of strong and relatively long-range forces induced specifically by water as a solvent.

Consider the force on $\mathbf{L}$ at $\mathbf{X_L}$ given $\mathbf{P}$ at $\mathbf{X_P}$ in water at a given temperature and pressure. The general expression for the indirect force or solvent-induced (SI) force on $\mathbf{L}$ is

$$\mathbf{F}_{\mathbf{L}}^{(\mathrm{SI})} = \int -\nabla_{\mathbf{L}} U(\mathbf{X_L}, \mathbf{X}_w)\rho(\mathbf{X}_w/\mathbf{X_L}, \mathbf{X_P})\, d\mathbf{X}_w, \tag{8.7.88}$$

where $\nabla_{\mathbf{L}} U(\mathbf{X_L}, \mathbf{X}_w)$ is the gradient of U with respect to $\mathbf{X_L}$. This, in general, includes both linear forces as well as rotational torques on $\mathbf{L}$. (Note that here we assume that $\mathbf{L}$ has a fixed conformation. Forces that change the conformation are discussed in section 8.8.)

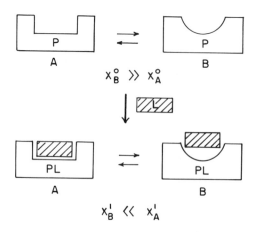

FIGURE 8.19. A conformational equilibrium between A and B. In the absence of ligand $x_B^0 \gg x_A^0$. In the presence of ligand the equilibrium shifts towards A, so that $x_B' \ll x_A'$.

In general, $\nabla_L U$ is a six-dimensional vector. Here we shall discuss only one component of the force, the component perpendicular to the surfaces of **L** and **P** as shown in Fig. 8.20A. Since

$$U(\mathbf{X_L}, \mathbf{X}_w) = U^H + U^S + \sum_k U(k, \mathbf{X}_w), \qquad (8.7.89)$$

where U^H, U^S, and $U(k, \mathbf{X}_w)$ are the hard, soft, and specific FG interaction between **L** and a water molecule, the indirect force can also be written as a sum of three terms

$$\mathbf{F}_L^{(SI)} = \mathbf{F}^H + \mathbf{F}^S + \sum_k \mathbf{F}_k. \qquad (8.7.90)$$

Both $\mathbf{F}^H$ and $\mathbf{F}^S$, when considered per unit surface area of **L**, are expected to be on the same order of magnitude as in the case of a simple small solute **L**. By unit area of **L** we mean here an area on an order of magnitude of the cross-section area of, say, a methane molecule. We have already seen in section 7.14 that a strong solvent-specific force can be expected if both **L** and **P** contain FGs that can form HBs with the solvent.

We explore various possibilities of such HϕI forces in the following examples. In all of the following examples we assume that the surfaces of **L** and **P** are parallel. Each surface contains FGs, either C=O or OH, and we shall be interested in the force perpendicular to the surface of **L**.

1. The simplest case occurs when all the FGs on the surfaces of **L** and **P** are far apart and therefore independently solvated. In this case the force acting on each FG can be treated separately, then summed over all FGs as in (8.7.90).

We have already seen that the largest contribution to the integrand of (8.7.88) comes when a water molecule is at a configuration $\mathbf{X}_w$ such that it can form a HB with one FG on **L**, say FG 1 in Fig. 8.20A, and that the conditional density at $\mathbf{X}_w$, given $\mathbf{X_L}$, $\mathbf{X_P}$, is high. The latter is achieved when **P** also provides a FG that produces a high local density at $\mathbf{X}_w$.

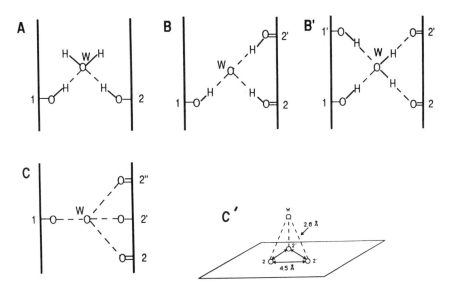

FIGURE 8.20. Configurations for which a one-water bridge can be formed between **L** and **P**. (**P** is the right-hand-side wall, and **L** the left-hand-side wall.) A cross-section of the surface of **P** is shown in C'.

In Fig. 8.20A we show the simplest configuration of the two solutes, **P** and **L**, shown as two walls each containing one FG. This case produces a HφI force identical with the force discussed in section 7.14. We shall refer to this force as originating from a one-water bridge. The strong force operating on FG 1 is a result of the strong force exerted by a water molecule at X_w multiplied by a relatively high conditional density of water molecules at X_w. Some estimates of the conditional densities in the surroundings of FGs are given in Appendix D.

Assuming that the main contribution to the force comes from the configuration where w forms simultaneous HBs with both **L** and **P**, we rewrite (8.7.88) as

$$F_L^{(SI)} \approx -\nabla_L U(\mathbf{X_L}, \mathbf{X}_w)\rho(\mathbf{X}_w/\mathbf{X_L}, \mathbf{X_P})\Delta\mathbf{X}_w$$

$$= -\nabla_L U(\mathbf{X_L}, \mathbf{X}_w)g(\mathbf{X}_w|\mathbf{X_L}, \mathbf{X_P})\frac{\rho_w\Delta\mathbf{X}_w}{8\pi^2}. \tag{8.7.91}$$

Here, $\rho_w\Delta\mathbf{X}_w/8\pi^2$ is the probability of finding a water molecule in a specific configuration $\mathbf{X}_w$ within the element $\Delta\mathbf{X}_w = \Delta\mathbf{R}_w\Delta\mathbf{\Omega}_w$. $g(\mathbf{X}_w|\mathbf{X_L}, \mathbf{X_P})$ is the pair correlation function between w and **L** and **P** viewed together as a single entity at a configuration $(\mathbf{X_L}, \mathbf{X_P})$. Note that this is different from the triplet correlation function $g(\mathbf{X}_w, \mathbf{X_L}, \mathbf{X_P})$. The relation between the two follows from the identity

$$\rho(\mathbf{X}_w/\mathbf{X_L}, \mathbf{X_P}) = \frac{\rho(\mathbf{X}_w, \mathbf{X_L}, \mathbf{X_P})}{\rho(\mathbf{X_L}, \mathbf{X_P})}$$

or, equivalently,

$$g(\mathbf{X}_w|\mathbf{X_L}, \mathbf{X_P}) = \frac{g(\mathbf{X}_w, \mathbf{X_L}, \mathbf{X_P})}{g(\mathbf{X_L}, \mathbf{X_P})}. \tag{8.7.92}$$

If we have m independently solvated pairs of FGs as 1 and 2 in Fig. 8.20A, the force exerted on **L** will be simply m times the force due to one pair.

2. A stronger force produced by a one-water bridge can be obtained if either **L** or **P** have a pair of FGs at a distance of about 4.5 Å to form a strong HφI interaction. The situation is schematically depicted in Fig. 8.20B and B'. For instance, in B' **L** and **P** each contain a pair of such FGs. Note that in this case we must have two donor and two acceptor FGs to satisfy the HBing capability of the two pairs of arms of a water molecule.

Applying Eq. (8.7.88) to this case, we see that the force exerted by a water molecule at X_w perpendicular to **L** is roughly twice the force we had in case A. However, the local density at X_w is much larger than in case A, about $(1.96 \times 10^4)\rho_w/8\pi^2$ in case B, and $(7.5 \times 10^4)\rho_w/8\pi^2$ in case B' (see Appendix D). This configuration produces a large integrand in (8.7.88), hence a very strong force.

The largest force produced by a one-water bridge is shown schematically in Fig. 8.20C. Here, **L** has one FG but **P** has three FGs at (2, 2', 2'') at a distance of about 4.5 Å from each other (Fig. 8.20C'); i.e., these three FGs are triply correlated by HφI interactions (see Appendix C). In this case, the force $\nabla_L U(\mathbf{X_L}, \mathbf{X}_w)$ is the same as in case A, although acting in a different direction, but the local density at X_w is about the same as in case B'.

With the two cases B' and C, we exhausted all four arms of a single water molecule and therefore all possible one-water bridges. Note that the range of the force is the shortest at C, larger at B, and largest at A. In order to obtain longer-range forces we now have to consider two-water bridges.

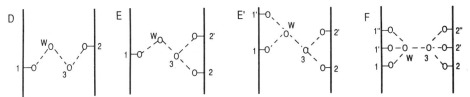

FIGURE 8.21. Configurations for which two-water bridges can be formed between **L** and **P**.

3. The simplest case of a two-water bridge is shown in Fig. 8.21D. Here again, the force exerted on 1 by a water molecule at $\mathbf{X}_w$ is as in case A. However, the local density of water molecules at $\mathbf{X}_w$ is enhanced by the presence of FG 1 as in case A, but is affected by 2 through an intermediate water molecule at $\mathbf{X}_3$. To see this we use the following identity

$$\rho(\mathbf{X}_w/\mathbf{X}_\mathbf{L}, \mathbf{X}_\mathbf{P}) = \frac{\rho(\mathbf{X}_w, \mathbf{X}_\mathbf{L}, \mathbf{X}_\mathbf{P})}{\rho(\mathbf{X}_\mathbf{L}, \mathbf{X}_\mathbf{P})}$$

$$= \int \frac{\rho(\mathbf{X}_w, \mathbf{X}_3, \mathbf{X}_\mathbf{L}, \mathbf{X}_\mathbf{P})}{\rho(\mathbf{X}_\mathbf{L}, \mathbf{X}_\mathbf{P})} d\mathbf{X}_3$$

$$= \int \rho(\mathbf{X}_w/\mathbf{X}_3, \mathbf{X}_\mathbf{L}, \mathbf{X}_\mathbf{P})\rho(\mathbf{X}_3/\mathbf{X}_\mathbf{L}, \mathbf{X}_\mathbf{P}) d\mathbf{X}_3. \qquad (8.7.93)$$

Suppose that we take the most favorable configuration of the system, as shown in Fig. 8.21D, so that the integrand in (8.7.93) is the largest. Then we have approximately, as in (8.7.91),

$$\mathbf{F}_\mathbf{L}^{(SI)} = -\nabla_\mathbf{L} U(\mathbf{X}_\mathbf{L}, \mathbf{X}_w)\rho(\mathbf{X}_w/\mathbf{X}_\mathbf{L}, \mathbf{X}_\mathbf{P}), \Delta\mathbf{X}_w$$

$$= [-\nabla_\mathbf{L} U(\mathbf{X}_\mathbf{L}, \mathbf{X}_w)\rho(\mathbf{X}_w/\mathbf{X}_3, \mathbf{X}_\mathbf{L}, \mathbf{X}_\mathbf{P})\Delta\mathbf{X}_w]\rho(\mathbf{X}_3/\mathbf{X}_\mathbf{L}, \mathbf{X}_\mathbf{P})\Delta\mathbf{X}_3. \qquad (8.7.94)$$

The force in the square brackets will have roughly the same value as in case A, but now this force is multiplied by the probability that a water molecule can be found within the range of configurations $\Delta\mathbf{X}_3$. This probability is quite small, about 0.06 [see Appendix D, Eq. (D.7)], hence we do not expect a strong force in this case.

4. The next two-water-bridge configuration is shown in Fig. 8.21E. The same equation (8.7.91), when applied to this case, produces a stronger force than in case D. The reason is that the two FGs at 2 and 2′, both belonging to **P**, produce a higher local density at $\mathbf{X}_3$. This, in turn, together with 1 on **L**, produces a high local density at $\mathbf{X}_w$.

A variety of two-water bridges can be conceived of and the one producing the strongest force is depicted in Fig. 8.21F. Here, the water molecule at $\mathbf{X}_w$ exerts a force on three FGs on **L** (1, 1′, and 1″). The local density at $\mathbf{X}_3$ is now the same as in case B. In configuration F we have exhausted all eight arms of the two water molecules at $\mathbf{X}_w$ and $\mathbf{X}_3$. Therefore this case produces the strongest force for the two-water bridges.

5. The next step is the case of three-water bridges. It is clear that if we take a linear chain of water molecules as depicted in Fig. 8.21D, but with one more molecule, the force obtained will be even weaker than the one obtained in case D. This can be seen by

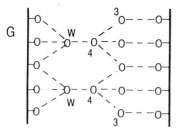
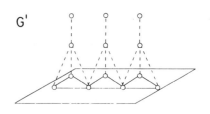

FIGURE 8.22. Three-water bridges between **L** and **P**. G' shows a cross section of the surface of **P**.

extension of the argument given in (8.7.93) and (8.7.94). The extension of the identity (8.7.93) for this case is,

$$\rho(\mathbf{X}_w/\mathbf{X_L}, \mathbf{X_P}) = \frac{\rho(\mathbf{X}_w, \mathbf{X_L}, \mathbf{X_P})}{\rho(\mathbf{X_L}, \mathbf{X_P})}$$

$$= \int \rho(\mathbf{X}_w/\mathbf{X}_3, \mathbf{X}_4, \mathbf{X_L}, \mathbf{X_P})$$

$$\times \rho(\mathbf{X}_3/\mathbf{X}_4, \mathbf{X_L}, \mathbf{X_P})\rho(\mathbf{X}_4/\mathbf{X_L}, \mathbf{X_P}) \, d\mathbf{X}_3 \, d\mathbf{X}_4. \qquad (8.7.95)$$

Equation (8.7.95) can now be substituted into (8.7.91) to obtain the analogue of (8.7.94), which for this case is

$$\mathbf{F_L^{(SI)}} = [-\nabla_{\mathbf{L}} U(\mathbf{X_L}, \mathbf{X}_w)\rho(\mathbf{X}_w/\mathbf{X}_3, \mathbf{X}_4, \mathbf{X_L}, \mathbf{X_P})\Delta\mathbf{X}_w]$$

$$\times \rho(\mathbf{X}_3/\mathbf{X}_4, \mathbf{X_L}, \mathbf{X_P})\Delta\mathbf{X}_3\rho(\mathbf{X}_4/\mathbf{X_L}, \mathbf{X_P})\Delta\mathbf{X}_4. \qquad (8.7.96)$$

The force in the square brackets in (8.7.96) is roughly the same as in (8.7.91) for case A, but now it is multiplied by the conditional probability of finding a water molecule within $\Delta\mathbf{X}_3$ and by the conditional probability of finding a water molecule within $\Delta\mathbf{X}_4$. Clearly, if these two probabilities are small, the force obtained for a three-water bridge will be small. In order to obtain a relatively strong force we need to guarantee a relatively high probability of finding a water molecule both in $\Delta\mathbf{X}_3$ and in $\Delta\mathbf{X}_4$. One such case is shown in Fig. 8.22G.

The extension of the above arguments to obtain strong forces at longer distances is quite clear. At each step we can add one water molecule to the bridge—but by doing so we also decrease the local density at $\mathbf{X}_w$. For two walls containing functional groups at the correct distances and orientations (as depicted in Fig. 8.22G'), ideally two surfaces of ice, there will be many possible water bridges that will transmit the force between the two walls. Such forces between large bodies such as micas have indeed been measured. Similar forces based on HϕI groups are operative between proteins, nucleic acids, or biological membranes.

8.8. PROTEIN FOLDING

One of the most intriguing problems in biochemistry is how and why nascent proteins fold into a precise three-dimensional (3-D) structure. It is known that some proteins, the best known being ribonuclease A, can fold spontaneously into its native folded form in a proper aqueous environment.

Ribonuclease A has about 130 amino-acid residues. In the unfolded form, obtained by treatment with denaturation agents, it can rotate about some 2×130 single bonds (not including rotations within the side chains). It is clear that the number of configurations available to the protein is extremely large. A rough estimate shows that even if we allow only a small number of discrete rotations about each single bond along the main backbone, we obtain an enormously large number of conformations. This has led to the conclusion that if the proteins were to move randomly in the $2M$ configurational space $(\phi_1 \psi_1 \cdots \phi_M \psi_M)$, it would have taken a fantastically long time to reach the native configuration.

The fact is that proteins do fold in a relatively short time, of the order of seconds or minutes. Therefore there must be some efficient route to the folding process. The fact that the protein does fold spontaneously has led to the conclusion that the information contained in the 1-D sequence of amino acids is sufficient to determine the ultimate 3-D structure of the protein. But how does the protein know how to translate this 1-D information into the 3-D structure? This is often described as the "3-D code" that is embedded in the sequence. Even if there is such a code, there must be some mechanism that deciphers it and translates it into executable orders. The situation may be likened to a book containing all the information required to construct a building. Such a building would never be constructed spontaneously unless some agent translates the information contained in the book into a series of executable orders.

It is unlikely that there is a one-to-one 3-D code. The fact that there are many different sequences resulting in the formation of the same 3-D structure means that if such a code exists, it must be very redundant. Also, the structure of the native protein is not unique. Proteins always experience structural fluctuations; therefore, a given sequence of amino acids does not determine a unique set of coordinates for the folded form of the protein.

Yet the unfolded 1-D chain does fold into a functionally active unit, which maintains a reasonably well defined structure, within certain limits. It does so in a very short time, far shorter than any estimated time that would have been required by the random selection of a folding pathway. This means that there exists not only a thermodynamic force leading from the unfolded to the folded state, but there must be some dynamical forces, at least in part of the folding pathway, that direct the protein to move toward the final product.

As in the case of association between proteins, interactions and forces can be classified either as direct, arising from the protein themselves, or as indirect, arising from the solvent. Again, we note that the folding process is not a statistical mechanical problem. Each specific sequence has a specific folding pathway, leading to a specific target. Statistical mechanics enters only in the elucidation of the indirect forces and suggests ways of studying these forces through model compounds. These indirect forces are statistical in nature since they involve averaging over the configurations of all solvent molecules.

Having a given sequence of amino acids, one can in principle compute the force acting on each of its nuclei at any moment, and therefore in principle one can follow its folding trajectory under its own forces. This is a very complex problem. One needs to write the potential function for at least some $3M$ coordinates, where M is the number of residues, and compute the forces at any given configuration. Clearly, different sequences will have different trajectories.

The solvent-induced forces are an "order of magnitude" more difficult to study than the direct forces. Here we need to take into account protein–solvent interactions as well as solvent–solvent interactions.

As in the case of protein association, the solvent-induced forces can profoundly alter the specificity of the process. In the association process, we have seen how the solvent may change or even determine the selection of the specific binding mode. In the folding process the solvent can affect both the specificity of the trajectory as well as the ultimate specific 3-D structure.

There is one essential difference between the association and the folding processes, as far as the involvement of the solvent is concerned. In the first, the force operating between the two solutes, say **P** and **L**, are of relatively short range, and therefore come into play when the two solutes are very close to each other. In the folding process, no matter what the initial state is, there are always forces both direct and indirect that affect the motion of the protein toward the target product.

We shall first examine the thermodynamic forces for the overall folding process. We shall see that a classification of FGs into three groups E, I, and J can be made in the same way as in the association process.

8.8.1. Formal Separation of Solvent Effects

We consider the transition

$$\mathbf{U} \rightleftarrows \mathbf{F}, \tag{8.8.1}$$

where **U** denotes one specific conformer of the unfolded protein, say the fully extended all-trans protein. **F** denotes one conformation of the folded protein, say the specific structure of the protein as determined by X-ray diffraction. As in section 8.7.1, we first study a transition between two specific conformers, and later modify the treatment as required due to the existence of many conformational states.

The condition of chemical equilibrium is

$$\mu_{\mathbf{U}} = \mu_{\mathbf{F}}$$

or, equivalently,

$$
\begin{aligned}
K^l &= \left(\frac{\rho_{\mathbf{F}}}{\rho_{\mathbf{U}}}\right)_{\text{eq}} \\
&= \frac{Q_{\mathbf{F}}}{Q_{\mathbf{U}}} \exp[-\beta \Delta U - \beta W(\mathbf{F}|w) + \beta W(\mathbf{U}|w)] \\
&= \frac{Q_{\mathbf{F}}}{Q_{\mathbf{U}}} \exp[-\beta \Delta U - \beta \delta G(\mathbf{U} \to \mathbf{F})] \\
&= \exp(-\beta \Delta G^{0l}).
\end{aligned}
\tag{8.8.2}
$$

The symbols in (8.8.2) are similar to the symbols in (8.7.5). ΔU is the energy change for the transition (8.8.1) in vacuum, $W(\alpha|w)$ is the coupling work of the solute α (either **U** or **F**) to the solvent w (which may be pure liquid or any mixture of liquids), and ΔG^{0l} is the thermodynamic standard Gibbs energy change for the specific reaction (8.8.1). In (8.7.5) we also included in Q_α the momentum PF of each solute. Here, the momentum PFs of **U** and **F** are the same and therefore cancel out in (8.8.2). The solvent-induced contribution to ΔG^{0l} is defined as in (8.7.7) by

$$\delta G(\mathbf{U} \to \mathbf{F}) = \Delta G^{0l} - \Delta G^{0g}, \tag{8.8.3}$$

where ΔG^{0g} is the standard Gibbs energy of the same reaction in an ideal-gas phase, i.e.,

$$K^g = \frac{Q_F}{Q_U} \exp(-\beta \Delta U) = \exp(-\beta \Delta G^{0g}). \qquad (8.8.4)$$

The solvent contribution defined in (8.8.2) can be expressed as a ratio of two average quantities; i.e.,

$$\exp[-\beta \delta G(U \to F)] = \frac{\langle \exp(-\beta B_F) \rangle_0}{\langle \exp(-\beta B_U) \rangle_0}, \qquad (8.8.5)$$

where B_F and B_U are the binding energies of F and U to the solvent at some *specific* configuration X^N. The average $\langle \ \rangle_0$ is over all configurations of the solvent molecules, here in the T, P, N ensemble. Formally, this expression is analogous to the corresponding expression for the association process, which may be written symbolically as

$$\exp[-\beta \delta G(P + L \to PL)] = \frac{\langle \exp(-\beta B_{PL}) \rangle_0}{\langle \exp[-\beta(B_P + B_L \text{ at } \infty)] \rangle_0}, \qquad (8.8.6)$$

where in both (8.8.5) and (8.8.6) we have a ratio of two averages, one pertaining to the final state (F or PL) and the second pertaining to the initial state (U, or P and L at infinite separation). Similar expressions were discussed in connection with the problem of $H\phi O$ and $H\phi I$ interactions in section 7.14.

In order to proceed with the analysis of the various contributions to $\delta G(U \to F)$, we use the same division of the solute–solvent interaction as in (8.7.11) or (8.7.12). We shall find both of these useful, depending on how we want to estimate the overall value of δG and on the availability of relevant experimental data (see section 8.8.2). In the following we shall use the notation of (8.7.22), i.e.,

$$U(X_\alpha, X_i) = U^H(X_\alpha, X_i) + U^S(X_\alpha, X_i) + \sum_{k=1}^{M} U^{FG}(k, X_i), \qquad (8.8.7)$$

where by FG we mean either an entire side chain as in (8.7.12) or a specific part of the interaction, such as an HB interaction, as in (8.7.13).

The classification of all FGs into three groups is essentially the same as in section 8.7.2. Briefly, a FG belongs to the E group if

$$\Delta G_U^{*k/H,S} = \Delta G_F^{*k/H,S}. \qquad (8.8.8)$$

Compare with (8.7.24) and (8.7.25).

A FG belongs to the I group if

$$\Delta G_U^{*k/H,S} \neq 0 \quad \text{and} \quad \Delta G_F^{*k/H,S} = 0. \qquad (8.8.9)$$

This should be compared with the conditions (8.7.26) and (8.7.27).

All FG that were not classified either as in E or in I are referred to as belonging to the J group. These FG have different solvation Gibbs energy in the two states U and F, i.e.,

$$\Delta G_U^{*k/H,S} \neq \Delta G_F^{*k/H,S}. \qquad (8.8.10)$$

The difference could occur from partial burial of the FG or correlation with other FGs adjacent to k.

Note that the classification into groups pertains to the specific choice of the conformers in (8.8.1). A major difficulty of the protein-folding problem is related to the change in the classification of FGs into groups along the folding pathway. We shall return to this problem in section 8.8.4.

As in section 8.7.2, we write the total binding energy of a solute α (either **U** or **F**) to the solvent as (see 8.7.32)

$$B_\alpha = B_\alpha^H + B_\alpha^S + \sum_{k=1}^{M} B_k$$

$$= B_\alpha^H + B_\alpha^S + B_E + B_I + B_J, \tag{8.8.11}$$

where the meaning of the symbols in (8.8.11) is the same as in (8.7.32).

The solvation Gibbs energy of **U** and **F** can now be written as (see section 8.7.33)

$$\Delta G_U^* = \Delta G_U^{*H} + \Delta G_U^{*S} + \Delta G_U^{*I/H,S} + \Delta G_U^{*J/H,S,I} + \Delta G_U^{*E/H,S,I,J} \tag{8.8.12}$$

$$\Delta G_F^* = \Delta G_F^{*H} + \Delta G_F^{*S} + \Delta G_F^{*J/H,S} + \Delta G_F^{*E/H,S,I}. \tag{8.8.13}$$

Note that since the FGs in the I group lose their solvation, ΔG_F^* does not include the term $\Delta G_F^{*I/H,S}$.

Using the definition of $\delta G(\mathbf{U} \to \mathbf{F})$ in (8.8.5), we can now write the general expression for the solvent contribution in terms of the various ingredients as

$$\delta G(\mathbf{U} \to \mathbf{F}) = \delta G^H + \delta G^{S/H} + \delta G^{I/H,S} + \delta G^{J/H,S,I}, \tag{8.8.14}$$

where the contributions of the FGs in the E group cancel out. This should be compared with Eq. (8.7.34). In terms of the solvation Gibbs energies, the four contributions to δG are

$$\delta G^H = \Delta G_F^{*H} - \Delta G_U^{*H} \tag{8.8.15}$$

$$\delta G^{S/H} = \Delta G_F^{*S/H} - \Delta G_U^{*S/H} \tag{8.8.16}$$

$$\delta G^{I/H,S} = -\Delta G_U^{*I/H,S} \tag{8.8.17}$$

$$\delta G^{J/H,S,I} = \Delta G_F^{*J/H,S,I} - \Delta G_U^{*J/H,S,I}. \tag{8.8.18}$$

As in section 8.7.2, the first two contributions are due to the entire molecule. δG^H is the change in the cavity work for the process $\mathbf{U} \to \mathbf{F}$. $\delta G^{S/H}$ is the contribution due to turning on the soft interactions. The last two contributions, (8.8.17) and (8.8.18), were written here for the entire I and J groups. They can be further divided into the contributions of the individual FGs, as we did in section 8.7.2. If all FGs in the **U** form were independent, then $\delta G^{I/H,S}$ would include the loss of the solvation of all FGs in the I group. The term $\delta G^{J/H,S,I}$ would then include all the correlations between FGs that come close to each other in the **F** form. The assumption of independence of the FGs is unlikely to be correct for the **U** form. Specifically, the C=O and NH FGs on the i and $i+1$ amino acids could come to a distance of about 5 Å and therefore will be strongly correlated by HϕI interactions.

It is impossible to analyze the multitude of different contributions due to $\delta G^{I/H,S}$ and $\delta G^{J/H,S,I}$ for a "general" protein. The practical approach is to study a specific protein for which both the sequence of amino acids and its crystallographic structure are known. For such a specific protein, the classification of all FGs into the three groups E, I, J, can easily be made by following the fate of each FG in the folding process. Furthermore, we can also subclassify all the FGs in each group I and J (the FGs in E do not contribute to δG and therefore can be ignored), according to whether they are independent, correlated with one or more FGs, partially buried, etc. Having this detailed information on one specific protein, one can then assess the relative contribution of the various factors to δG. Unfortunately, this information is not available for any protein. Therefore, in the next section we shall treat a hypothetical protein consisting of only two kinds of side chains; HϕO and HϕI.

8.8.2. Methods of Estimating the Various Contributions to δG

We consider a polypeptide consisting of alanine and serine residues only. Suppose we start with the fully extended, all-trans U form and suppose that we know the detailed structure of the final form F. There are two ways to estimate the various contributions to δG, according to whether we use the division of the solute–solvent interaction as in (8.7.11) or as in (8.7.12).

The simpler choice is (8.7.11). We write the total binding energy of U and F as

$$B_U = B_U^H + B_U^S + \sum B_k^{HB} \tag{8.8.19}$$

$$B_F = B_F^H + B_F^S + \sum B_k^{HB}. \tag{8.8.20}$$

The first contribution is δG^H, defined in (8.8.15). This is due to the change of the cavity produced by U and F. Again, using an argument similar to the one used in Eq. (8.7.49), we expect that for large proteins δG^H will roughly depend on the excluded volume change, i.e.,

$$\delta G^H \approx P(V_F^{EX} - V_U^{EX}) < 0. \tag{8.8.21}$$

This is based on the assumption that the work of creating a large cavity is given by (8.7.48), where P is the pressure of the system. This reduction in the excluded volume contributes negatively to δG^H. Similarly, the soft part of the interaction, $\delta G^{S/H}$, defined in (8.8.16), is expected to be positive. The argument is similar to the one given in (8.7.51). The exact expression for $\delta G^{S/H}$ is

$$\exp(-\beta \delta G^{S/H}) = \frac{\langle \exp(-\beta B_F^S)\rangle_H}{\langle \exp(-\beta B_U^S)\rangle_H}. \tag{8.8.22}$$

Assuming that the soft interaction is weak, we can approximate each of the average quantities in (8.8.22) by

$$\langle \exp(-\beta B_a^S)\rangle_H \approx 1 - \langle \beta B_a^S\rangle_H \tag{8.8.23}$$

or, equivalently [see (8.7.50)],

$$\Delta G_a^{*S/H} \approx -\varepsilon \mathscr{A}_a \tag{8.8.24}$$

$$\delta G^{S/H} \approx \langle B_F^S\rangle_H - \langle B_U^S\rangle_H \approx -\varepsilon[\mathscr{A}_F - \mathscr{A}_U] > 0, \tag{8.8.25}$$

where $\varepsilon > 0$ is the Lennard–Jones parameter for the pair interaction between a methyl or methylene group and neon. The result (8.8.25) is qualitatively clear. In the U form there are more contacts between the protein and the solvent. In the F form, less surface area is exposed to the solvent. Therefore, in the process of folding we lose part of the negative interaction with the solvent; hence, the resulting contribution to $\delta G^{S/H}$ is positive. We have assumed in (8.8.23) and (8.8.25) that in the U form, all the side chains are nearly independent (these are the side chains of alanine and the CH_2OH side chain of serine, excluding the HB part of the interaction). On the other hand, some of these side chains may come close together so that their solvations will be correlated. For each correlated pair there would be a negative intramolecular $H\phi O$ interaction (see section 7.15), of about -0.5 kcal/mol.

Having taken care of the hard and the soft interactions, the only interactions left are the solute–solvent HBing capacity of the OH of serine as well as the $C=O$ and NH of the backbone. Even in this simple polypeptide we have to consider a multitude of very large effects which will determine both the folding pathway and the final product. Some of these are:

1. Loss of the conditional solvation Gibbs energy of each independent FG in the I group. For the OH we have an estimate of about $+6.7$ kcal/mol; for a $C=O$ group we have an estimate of about $+5.2$ kcal/mol (see Appendix H). We do not have an estimate for the NH group—but assuming that it can be solvated by one arm, the loss of the solvation Gibbs energy will be at least $+2.25$ kcal/mol. All these are much larger than the corresponding loss of the conditional solvation Gibbs energy of a methyl group which is about -0.5 kcal/mol.

2. Loss of pairs of FGs in the I group that are correlated in the U form. Suppose that k and l are two such FGs at the correct distance to form a strong $H\phi I$ interaction. The corresponding loss of the solvation Gibbs energy of such a pair is about $+7.7$ kcal/mol (see Appendix C).

3. Loss of triplet of FGs in the I group which are correlated in the U form. If this triplet has a strong $H\phi I$ interaction as described in Appendix C, then the corresponding loss of Gibbs energy is on the order of $+13.8$ kcal/mol.

4. Two FGs that are independent in the U form but become correlated in the F form. This could contribute about -3 kcal/mol per pair of $H\phi I$ groups (see Appendix C), and about -0.3 kcal/mol per pair of $H\phi O$ groups.

5. Triplet correlations formed by three FGs that are initially independent in the U form but are correlated in the F form. This could contribute about -7.0 kcal/mol (see Appendix C).

6. Triplet correlations formed by one pair of FGs and one single FG, such that in the U form the pair of FGs are correlated by $H\phi I$ interactions, but the pair as a unit is independent of the single FG. If these three FGs are triply correlated in the F form, then, there is a contribution of about -4 kcal/mol to δG.

7. Quadruplet correlation formed by four FGs that are initially independent will contribute about -11.3 kcal/mol (see Appendix C).

There are many more cases involving different distances between the FGs as well as different kinds of FGs that change their extent of correlations in reaction (8.8.1).

The above analysis demonstrates how complicated it is to build up an inventory of all possible solvent effects even for a very simple protein. Nevertheless, for a specific protein of known sequence and structure, the relative contributions of each specific residue can be estimated based on experimental data on small model compounds (see Appendix G). With an imaginative choice of such models it is in principle possible to estimate the contribution of all the conditional solvation Gibbs energies of single FGs, pair-correlated FGs, and triplet-correlated FGs. As we have seen above, the overall value of δG would be the balance between many positive and negative contributions of the various FGs as well as the backbone.

The analysis carried out in this section was based on the division of the solute–solvent potential as in (8.7.11). A similar analysis may be carried out based on (8.7.12). Both of these are equivalent. The choice of one or another or of a combination of the two will depend on the availability of experimental and theoretical information on the various contributions included in the terms (8.8.15)–(8.8.18).

In this section we have dealt with the overall reaction (8.8.1) from the initial to the final form. We have chosen one representative conformation of each form. For a real protein, we need to take proper averages, the same as in section 8.7.5, over all possible conformations recognized as belonging to the **U** and **F** forms. As was pointed out in section 8.7, such an average process is useful whenever the average is over a group of conformations for which the classification of FGs into the three groups E, I, and J does not change.

In the process of protein folding, the classification we have done refers to one conformer **U** and one conformer **F**. Clearly, different conformations belonging to the denatured protein have different classifications of the FGs. Therefore we must subdivide all conformations into subgroup, which are the intermediary folding structures. This leads us to examine the forces and the Gibbs energies of the intermediates in the folding process.

8.8.3. Force in Protein Folding

In sections 7.14 and 8.7.6 we discussed the average force operating on a particle that originates from the presence of the solvent. We generalized this concept for a solution containing a single macromolecule.

First, consider a macroscopic process, e.g., expansion of N molecules in an ideal-gas phase from volume V to $2V$. In this process $\Delta G < 0$ and we say that there is a *thermodynamic force* driving the system from the initial state i to the final state f.

$$\Delta G = G_f - G_i < 0. \qquad (8.8.26)$$

If the system is macroscopically large, then once we reach the equilibrium state, the thermodynamic force becomes zero. We know also that even at equilibrium there are always fluctuations around the final state. For very large N, these fluctuations are relatively small, and are usually ignored in thermodynamics.

If N is small, then repeating the same experiment as above, we shall again find that the system approaches an equilibrium state, but now the fluctuations about the equilibrium state are relatively large. As an extreme case of a small system, let $\mathbf{R}_1$ and $\mathbf{R}_2$ be the positions of two spherical solutes in a solvent. The potential of average force between the two solute particles looks like the curve in Fig. 8.23. Suppose that we could move the two solutes along the line connecting their center. The entire system includes both

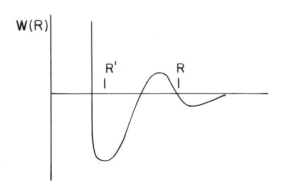

FIGURE 8.23. A schematic form of the potential of average force between two solutes. The two solutes at R will initially move away from each other. In spite of the thermodynamic force towards R'.

the solvent and the two solutes. We now focus on the two solutes only, viewing these as our small system.

Let $R = |\mathbf{R}_2 - \mathbf{R}_1|$ be the scalar distance between the two solute particles. For any two selected distances R' and R we can write the Gibbs energy of the system $G(R')$ and $G(R)$, respectively. If $G(R') - G(R) < 0$, we can say that there is a *thermodynamic force* leading from R to R'. This does not mean that the actual force between the two solute particles at R is attractive. In Fig. 8.23 we indicate by vertical lines two points R and R' for which we have

$$\Delta G = G(R') - G(R) < 0. \tag{8.8.27}$$

The actual or *dynamic* force at R is repulsive; i.e., if we start with two particles at R and release them, on the average they will initially move away from each other, in spite of the fact that the thermodynamic force is in favor of R' such that $R' < R$.

It is clear that in order to find the actual force operating at a specific configuration, we need to examine the change in G resulting from an infinitesimal change in the configuration. It is true that there is a large probability for the system to be at the configuration of lowest Gibbs energy, say R' in Fig. 8.23. However, because our system is small, there will be large fluctuations even after reaching the location of minimum Gibbs energy.

We now generalize the concept of thermodynamic and dynamic forces for a macromolecule. Let $\mathbf{R}^M = \mathbf{R}_1 \cdots \mathbf{R}_M$ be a specific configuration of a polymer having M nuclei. If the polymer is inserted in a solvent of N molecules at a given T and P, then the Gibbs energy of the system is given by

$$\exp[-\beta G(\mathbf{R}^M)] = \Delta(T, P, N; \mathbf{R}^M), \tag{8.8.28}$$

where $\Delta(T, P, N; \mathbf{R}^M)$ is the T, P, N PF of the system having one polymer at a specific configuration $\mathbf{R}^M$.

For any two configurations of the polymer $\mathbf{R}^M$ and $\mathbf{R}'^M$, the difference in the Gibbs energy (keeping T, P, N constant) is

$$G(\mathbf{R}'^M) - G(\mathbf{R}^M) = -kT \ln \left\{ \frac{\int dV \exp(-\beta PV) \int d\mathbf{X}^N \exp[-\beta U(\mathbf{R}'^M, \mathbf{X}^N)]}{\int dV \exp(-\beta PV) \int d\mathbf{X}^N \exp[-\beta U(\mathbf{R}^M, \mathbf{X}^N)]} \right\}, \tag{8.8.29}$$

where

$$U(\mathbf{R}^M, \mathbf{X}^N) = U(\mathbf{R}^M) + U(\mathbf{X}^N) + B(\mathbf{R}^M, \mathbf{X}^N). \tag{8.8.30}$$

$B(\mathbf{R}^M, \mathbf{X}^N)$ is the total binding energy of the polymer at $\mathbf{R}^M$ to the solvent at a specific configuration $\mathbf{X}^N$. Note that the entire system is macroscopic, but we are following the configurations of the polymer only.

The change in Gibbs energy can also be written as

$$\Delta G(\mathbf{R}^M \rightarrow \mathbf{R}'^M) = \Delta U(\mathbf{R}^M \rightarrow \mathbf{R}'^M) + \delta G(\mathbf{R}^M \rightarrow \mathbf{R}'^M), \tag{8.8.31}$$

where

$$\exp[-\beta \delta G(\mathbf{R}^M \rightarrow \mathbf{R}'^M)] = \frac{\langle \exp[-\beta B(\mathbf{R}'^M)] \rangle_0}{\langle \exp[-\beta B(\mathbf{R}^M)] \rangle_0}. \tag{8.8.32}$$

As in the simple example of two particles, whenever $\Delta G(\mathbf{R}^M \to \mathbf{R}'^M)$ is negative, we say that there is a thermodynamic force leading from $\mathbf{R}^M$ to $\mathbf{R}'^M$. Again, this does not mean that at the starting point $\mathbf{R}^M$, the polymer, will move directly to the final configuration $\mathbf{R}'^M$.

If the sign of ΔG is negative for any finite change $\mathbf{R}^M \to \mathbf{R}'^M$, we cannot say anything about the direction in which the polymer will move. To do this we need to examine the change of $G(\mathbf{R}^M)$ caused by an infinitesimal change in $\mathbf{R}^M$.

For instance, if we are interested in the force acting on nucleus i at $\mathbf{R}_i$, keeping all $\mathbf{R}_j$ ($j \neq i$) fixed, we have to take the gradient of $G(\mathbf{R}^M)$ with respect to $\mathbf{R}_i$. This is the same as examining the slope at the point R in Fig. 8.23.

Thus the average force on the nucleus i is

$$-\mathbf{F}_i(\mathbf{R}_i)\, d\mathbf{R}_i = G(\mathbf{R}_i + d\mathbf{R}_i) - G(\mathbf{R}_i)$$

$$= [\nabla_i U(\mathbf{R}^M) + \nabla_i \delta G(\mathbf{R}^M)]\, d\mathbf{R}_i. \qquad (8.8.33)$$

Note that here $d\mathbf{R}_i$ is an infinitesimal vector, not an element of volume as used in other parts of this book. This is essentially the same expression for the average force as was discussed in section 8.7.6, but now we are dealing with an intramolecular force. The force in (8.8.33) has two components. One results from the polymer itself. If we know the internal energy of the molecule for each configuration $\mathbf{R}^M$, we can obtain the force acting on any nucleus i. The second force originates from the solvent. Since

$$\delta G(\mathbf{R}^M) = \Delta G^*(\mathbf{R}^M) - \Delta G^*(\infty), \qquad (8.8.34)$$

the solvent-induced force is given by

$$\mathbf{F}_i^{(SI)}(\mathbf{R}_i) = -\nabla_i \delta G(\mathbf{R}^M)$$

$$= -\nabla_i \Delta G^*(\mathbf{R}^M). \qquad (8.8.35)$$

Thus for any initial configuration of the polymer $\mathbf{R}^M$, there is a force exerted on i if and only if a change in $\mathbf{R}_i$ causes a change in the solvation Gibbs energy of the polymer (all $\mathbf{R}_j$, $j \neq i$, and T, P, N are constant).

Taking the gradient with respect to $\mathbf{R}_1$ and following exactly the same steps as in sections 6.4 and 7.14, we obtain

$$\mathbf{F}_1^{(SI)}(\mathbf{R}_1) = -\nabla_1 \Delta G^*(\mathbf{R}^M)$$

$$= \nabla_1 \{kT \ln\langle \exp[-\beta B(\mathbf{R}^M)]\rangle_0\}$$

$$= -\int \nabla_1 U(\mathbf{R}_1, \mathbf{X}_w) \rho(\mathbf{X}_w/\mathbf{R}^M)\, d\mathbf{X}_w. \qquad (8.8.36)$$

This is essentially the same result as (7.14.14), except that here the condition $\mathbf{R}^M$ replaces the condition $\mathbf{R}_1$, $\mathbf{R}_2$ in (7.14.14). The analysis of the various possible contributions to the solvent-induced force is therefore the same as we have performed in section 7.14.

In order to obtain a large solvent-induced force we need both factors in the integrand to be large. This is obtained if the water exerts a strong force on 1 [i.e., $\nabla_1 U(\mathbf{R}_1, \mathbf{X}_w)$ is large] and at the same time the condition $\mathbf{R}^M$ produces a large excess local density of water molecules at $\mathbf{X}_w$ (see section 7.14 for more details).

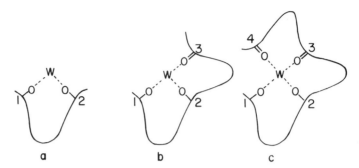

FIGURE 8.24. Pair, triplet, and quadruplet functional groups in favorable configurations to form strong HφI interactions.

We consider here four possible cases:

1. If 1 is a HφO group and the neighboring groups at $\mathbf{R}^M$ are also HφO, then both the force $\nabla_1 U(\mathbf{R}_1, \mathbf{X}_w)$ and the local density $\rho(\mathbf{X}_w/\mathbf{R}^M)$ are expected to be "normal." (By normal we mean here that the forces and the local densities are as in an a polar solvent.) In this case we cannot expect large solvent-induced forces.

2. If 1 is HφO and the neighboring groups at $\mathbf{R}^M$ are HφI, then the force is still normal but the density of water molecules might be large (see Appendix D).

3. If 1 is HφI and the neighboring groups at $\mathbf{R}^M$ are HφO, then the force is large but the local density is normal.

4. If both 1 and its neighboring groups at $\mathbf{R}^M$ are HφI, then both the force and the local density might be large.

We therefore conclude that the strongest possible solvent-induced forces are expected to operate on HφI groups that also have in their environment HφI groups in appropriate locations and orientations to provide a large local density of solvent molecules (see Appendix D).

Note that by "neighboring groups at $\mathbf{R}^M$" we mean those groups $j \neq 1$ which at the specific configuration $\mathbf{R}^M$ are close enough to 1 so that they can influence the local density of the solvent at $\mathbf{X}_w$. These groups are not necessarily close to 1 along the sequence of the polypeptide. The number of HφI neighboring groups at $\mathbf{R}^M$ could be one, two, or three (Fig. 8.24). These, in favorable configurations, could produce a very large local density of solvent molecules at $\mathbf{X}_w$ (see Appendix D) and therefore also very strong forces. We have seen in section 8.7 how one can obtain strong as well as long-range forces mediated by the solvent. The same applies in the case of intramolecular forces. These, together with the direct force will determine the overall trajectory of the protein, subjected to fluctuations due to the fact that the protein is a small system. We shall next examine some specific effects of the solvent-induced forces along the folding pathway of the protein.

8.8.4. The Solvent Effect on the Specificity of the Protein-Folding Pathway

In the association process treated in section 8.7, we have seen that the selection of a specific binding site depends on the particular distribution of FGs on the surface of the solutes. The analogue of the specific binding site in the case of protein folding is the specific three-dimensional structure of the **F** form. The problem of association differs, however, from the problem of folding in one important sense. In the case of association, there could be strong *thermodynamic force* to form a complex **PL**. But as long as **P** and **L** are far apart, there is no net *dynamical force* that directs the two monomers toward each other. The forces, both direct and indirect, come into play only when **P** and **L** reach a distance of a few molecular diameters.

In the case of folding, we start with an unfolded form **U** and suppose that there is a strong thermodynamic force towards the final form **F**. In contrast to the case of association, a dynamical force, both direct and indirect, operates on the protein throughout the entire process leading from **U** to **F**. This fact affects the specificity of both the dynamics of the process (selection of preferential pathways) and the specificity of the end product.

To appreciate the significance of the different cases, we consider a simple analogue. Let inert molecules, A, in an ideal-gas phase be distributed in two compartments i and f. At equilibrium, the ratio of the number of A molecules in the two compartments is simply (Fig. 8.25)

$$\frac{N_f}{N_i} = \frac{V_f}{V_i},\tag{8.8.37}$$

where V_i and V_f are the volumes of the two compartments i and f, respectively. (At this stage, i and f are arbitrary indices; in later examples, we shall use these indices to indicate initial and final states.)

We now introduce a partition between the two compartments and fill the two compartments with different solvents. The partition allows the free passage of A but is impermeable to the two solvents, which we denote also by i and f. The new equilibrium ratio of A in the two phases is now

$$\frac{N_f^*}{N_i^*} = \frac{V_f}{V_i} \exp[-\beta W(A|f) + \beta W(A|i)]$$

$$= \frac{N_f}{N_i} \exp[-\beta \delta G(i \to f)].\tag{8.8.38}$$

Here $\delta G(i \to f)$ is the solvents' contribution to the thermodynamic force. Suppose that we start with the ideal-gas distribution as in Eq. (8.8.37) and add the two solvents i and f. If $\delta G(i \to f) < 0$, then we have

$$\frac{N_f^*}{N_i^*} > \frac{N_f}{N_i}.\tag{8.8.39}$$

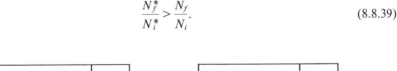

FIGURE 8.25. Two compartments i and f initially containing ideal gases (left). The two compartments are filled with different solvents, producing a new driving force that changes the distribution of particles from N_i, N_f to N_i^*, N_f^*.

Since the total number of A molecules is fixed, i.e.,

$$N_A = N_i + N_f = N_i^* + N_f^*, \qquad (8.8.40)$$

it follows that

$$N_i^* < N_i$$

or

$$N_f^* > N_f. \qquad (8.8.41)$$

This means that by adding the two solvents i and f we have introduced a new driving force. Therefore, A molecules will flow from i to f due to the thermodynamic force $\delta G < 0$, which here is due to the change of the solvation of A in favor of the solvent f. The resulting new distribution of A molecules will now be N_i^* and N_f^*.

Now suppose that we start with the initial distribution N_i and N_f, add the solvents into the two compartments, and follow the trajectory of one *specific* A molecule. As long as the A molecule is in one of the pure solvents, either i or f, it experiences no net force toward one phase or another. The molecule moves randomly until it reaches the boundary between the two phases. It is only within the boundary region that the A molecule experiences a solvent-induced force. Reaching the boundary from i, the molecule will experience attraction toward f. Reaching the boundary from f, it will be repelled from the boundary. The reason is that only in the boundary region is there a gradient in the Gibbs energy $G(\mathbf{R}_A)$, where $\mathbf{R}_A$ is the location of A caused by the difference in the solvation Gibbs energy of A. This is shown schematically in Fig. 8.26.

We next extend the model of Fig. 8.26 to the association of two solutes $\mathbf{P}$ and $\mathbf{L}$. Let a bold point in the ith compartment represent the configuration of $\mathbf{P}$ and $\mathbf{L}$ in the twelve-dimensional configurational space $(\mathbf{X_L}, \mathbf{X_P})$. The density of the dots in each phase represents the solvation Gibbs energy of the pair of solutes $\mathbf{L}$ and $\mathbf{P}$. The higher the density of points, the more negative the solvation Gibbs energy. The two solutes $\mathbf{P}$ and $\mathbf{L}$ moving at large separation from each other do not experience any change in their solvation—this is the same as saying that there is no force acting between them. A solvent-induced force becomes operative only when the solvation of the pair changes. This occurs only when they come close to each other. This situation is represented by the denser

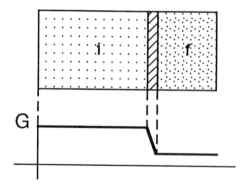

FIGURE 8.26. Two homogeneous phases i and f. The density of the dots represents the solvation Gibbs energy of a solute A. The Gibbs energy of the system $G(\mathbf{R}_A)$ drops when A passes from left to right. This produces a force toward f.

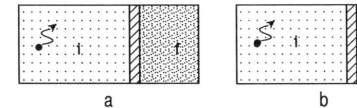

a   b

FIGURE 8.27. A bold point moves in the configurational space of **P** and **L**. The density of dots represents the solvation Gibbs energy of the pair at $\mathbf{X_L}$, $\mathbf{X_P}$. (a) The region f represents a single configuration of a dimer. (b) The region f represents two possible dimers.

region denoted by f in Fig. 8.27*a*. If there is a particularly high-density region, then the system will be attracted to that region. This corresponds to the selection of a particular binding site. In Fig. 8.27*b* we indicated two highly dense regions corresponding to two possible binding sites.

Clearly, if we start with two solutes **P** and **L**, very diluted in a solvent, it will take a very long time to form a dimer. This corresponds to starting with a bold point in a very large configurational space i in Fig. 8.27. No matter how large the thermodynamic force δG is, it might take a very long time to reach the final region f (i.e., the region corresponding to the dimer **PL**). Note again that in the above considerations we have focused on the solvent effect only. In reality, one must follow the entire Gibbs energy $G(\mathbf{X_L}, \mathbf{X_P})$ as a function of the configuration $\mathbf{X_L}$, $\mathbf{X_P}$.

The situation is considerably different in the case of protein folding. Again we can use the bold point as in Fig. 8.27 to represent the configuration of the entire molecule $\mathbf{R}^M$. (Here we use $\mathbf{R}^M$ for the configuration of all nuclei. We could also use the rotation angles ϕ, ψ to describe the configuration of the protein.)

We also make use again of the density of dots in the configurational space to represent the solvation Gibbs energy of the entire protein. Starting from an initial configuration i (this corresponds to the unfolded state U), the system will move in the $3M$-dimensional space. If $\delta G(U \rightarrow F)$ is negative, then there is a solvent induced thermodynamic force to proceed from U to F. However, if M is large and if the motion were random, then it would have taken a very long time to reach the final state F. Knowing, however, that proteins fold spontaneously in a relatively short time and that the presence of the solvent (aqueous solution) is essential for both the folding and the stability of the protein, there must therefore be some solvent-induced forces that direct the protein into a relatively few preferential folding pathways.

The fundamental difference between the present case and the association process is that a solvent-induced force is operative at any point in the configurational space of the protein. The analogue of the model of Fig. 8.27 is that the bold point representing $\mathbf{R}^M$ moves initially not in a homogeneous configurational space (represented by the uniform density in Fig. 8.27), but in a configurational space which is very inhomogeneous. This means that there is a dynamic force operating on the polymer at every configuration $\mathbf{R}^M$. What are the ingredients of these solvent-induced forces? Any time $\mathbf{R}^M$ changes in such a way that a HϕO group is removed from the solvent, there is a corresponding decrease in G. When two such HϕO groups come close to each other so that they become correlated, we again get a decrease in G. In both cases the effect is relatively small, of the order of about $-0.5\,\text{kcal/mol}$. Stronger effects on G occur when a HϕI group changes its solvation. An OH group that is excluded from the solvent will cause an increase in G of

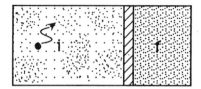

FIGURE 8.28. A bold point moving in the configurational space of the protein. The density of dots represents the solvation Gibbs energy of the protein at $\mathbf{R}^M$. The unfolded protein moves in configurational space which is highly inhomogeneous in the density of dots. The final state f is represented by a high-density region.

about $+6.7$ kcal/mol. (Note that, in general, part of this is compensated for by the formation of direct HB but this belongs to the direct intramolecular interaction.) When two HϕI groups come close to each other they might cause a decrease in G of about -3 kcal/mol. Three such HϕI may contribute up to -7 kcal/mol, and so forth.

We therefore expect that the bold point representing $\mathbf{R}^M$ will move in a highly inhomogeneous configurational space. This is schematically represented in Fig. 8.28. If the distribution of the dots in Fig. 8.28 were completely random, then it would again take a very long time to reach the final state f from an initial state i. However, if the pattern of the density of the dots changes in such a way that on the average there is an increase in the density of dots toward f, then, although the motion of the bold point is still random, there will be a general bias toward f. Two examples are shown in Figs. 8.29a and 8.29b. In the first case, in the region i, there are some "tunnels" with high densities of dots. These tunnels correspond to regions in which the solvation Gibbs energy of the polymer is particularly low. The polymer, when reaching such a tunnel, will be partially trapped there, trapped in the sense that there is a low probability of leaving the tunnel. Equivalently, there is force that attracts the point to enter the tunnel but is repulsive to its exit.

Even when the density of dots within the tunnel is constant, it is clear that the average time required to proceed from i to f is reduced compared with the case in Fig. 8.28. Starting from any point in i, the system wanders randomly until it reaches one of these tunnels. Once inside, it stays there with high probability. Its motion is still random, but it spends most of the time within the tunnel. This effectively reduces the dimensionality of the configurational space in which it moves. Therefore, though moving randomly within the tunnel, the system goes from i to f in a shorter time. Of course we can think of a more intricate system of tunnels within tunnels, where in each step the system moves from one density to a higher density. This would further accelerate the motion toward f. An extreme case is shown in Fig. 8.29b, where the density changes gradually throughout the entire configurational space. Here, there is a constant bias to move into a region of higher density; hence, the motion from i to f is considerably accelerated.

It should be noted that in both examples given above, Figs. 8.29a and b, the motion of the system is always random, and all possible trajectories are allowable. However, if

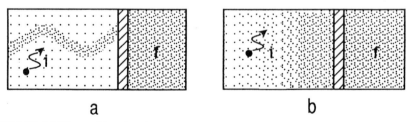

a   b

FIGURE 8.29. (a) The configurational space i has a tunnel of relatively high density of dots. (b) The configurational space has a distribution of densities, increasing toward f.

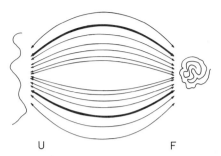

FIGURE 8.30. Schematic description of all possible pathways from **U** to **F**. The preferential pathways are indicated by the bold arrows.

there are some density gradients favoring the motion toward f, the overall random motion will be biased toward f. In other words, there are trajectories that have higher probability, and therefore the system will be "drained" with high probability through the trajectories along which the density gradients are large. These are the preferential pathways of protein folding. In Fig. 8.30 we indicate with double-arrowed lines all possible pathways connecting U and F. The bold lines indicate the preferential pathways. Although *all* of the pathways are accessible, the protein will move along the preferential pathways with high probability. In the next section we discuss the molecular origin of these tunnels for a real protein. We conclude this section by a comment on the existence of a configuration of local or global minimum Gibbs energy.

We first formulate the question in terms of a potential of average force for a pair of particles. Suppose that $W(R) = G(R) - G(\infty)$ has two minima, at R_1 and R_2 (Fig. 8.31), such that

$$G(R_2) < G(R_1). \tag{8.8.42}$$

Clearly, starting from any distance $R(R > R_1)$, there is a *thermodynamic force* toward R_2. There is also a thermodynamic force toward R_1. Therefore, from any initial point $R > R_1$ the system must first reach the minimum at R_1. If the barrier between the two minima is very large compared with kT, then the chances of arriving at R_2 will be very small. Thus, in spite of the thermodynamic force leading from R_1 to R_2, the system effectively stays at R_1. In extreme cases, the minimum at R_1 is practically the equilibrium

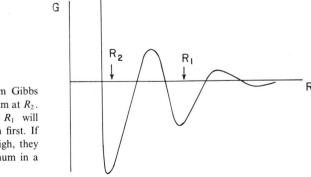

FIGURE 8.31. A local minimum Gibbs energy at R_1 and a global minimum at R_2. Two particles initially at $R > R_1$ will always reach the local minimum first. If the barrier to reach R_2 is too high, they may not reach the global minimum in a practical time span.

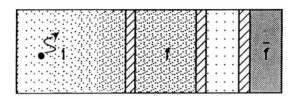

FIGURE 8.32. An extension of Figs. 8.29 and of 8.31. The protein starting in i will always reach f first. It may not reach the global minimum at $\bar{f}$ if the barrier to cross from f to $\bar{f}$ is too high.

state of the system. As always, there will be large fluctuations about R_1, but if these fluctuations are not large enough to cross to R_2, the system will never reach the global minimum.

The same situation can occur for the protein. Figure 8.32 is an extension of Fig. 8.29b. The details of the regions i and f are the same. In addition, we now have another region denoted by $\bar{f}$. Here the density of dots is very high, corresponding to a large negative solvation Gibbs energy. However, if the region $\bar{f}$ is separated from f by a very-low-density region, then starting from any initial state in i, the system must first reach the final region f. Although there exists a thermodynamic force toward the region $\bar{f}$, the system in practice stays in f. Any attempt to cross from f into the separating zone toward $\bar{f}$ induces a strong force to bring the system back into f. In extreme cases, therefore, although $\bar{f}$ has a lower Gibbs energy, the system may never reach there in practice. Again, for all practical purposes the system will stay in f. The question of whether the native protein is a state of local or global minimum Gibbs energy may be an interesting theoretical problem. Practically, however, this question may not be relevant to biology. If the native state is reached from any initially denatured state and if it is trapped there long enough on the time scale of biological processes, then it can be considered the practical equilibrium state. In this section we have stressed the solvent-induced contribution to the Gibbs energy. In any real process, one should consider both direct and indirect contributions to the Gibbs energy.

8.8.5. Possible Solvent-Induced Effects on the Formation of the α Helix

In section 4.7 we treated the helix coil transition in the vacuum. It was assumed that the main thermodynamic force to form the helix is the strong HB between the $C\!=\!O$ of the kth residue and the NH of the $(k + 4)$th residue. Because of some simplifying assumptions it was possible to treat the helix–coil transition as a 1-D problem. In section 8.6.3 we pointed out how the insertion of a 1-D model in water might invalidate the theory. We now discuss one possible solvent-induced effect on the selection of a preferential folding pathway—here, a transition from coil to helix.

Consider polyglycine or polyalanine. Let us start with an initial U form, which is the fully extended all-trans conformation. The polypeptide starts a motion under the forces originating both from the protein itself and from the solvent. We focus on the latter only. Each change of configuration will in general result in change in $G(\mathbf{R}^M)$. Some changes will increase G, some will decrease it.

Consider first the system in the vacuum. In order to form the first intramolecular HB between the k and the $k + 4$ residues, the protein has to move through many possible configurations until k and $k + 4$ residues are favorably oriented to form such a HB. The lowering of the energy by ε_{HB} gives a temporary stability to this structure until further residues come to favorable configurations to form other direct HBs. When the system is in a solvent, many new causes can change the Gibbs energy of the system. We point out two possible intermediates that are partially stabilized before the formation of a direct HB. One occurs when the two C=O's of the k and the $k + 2$ residues come to a distance of about 4.5 Å from each other and are oriented roughly in the same direction. This configuration in a vacuum will have no special preference. However, in water, the two C=O's can form a strong HϕI interaction, giving rise to a lowering of the Gibbs energy by about -3 kcal/mol. This means that formation of one such interaction moves the system into a high-density region in the configurational space of the polymer (Fig. 8.29). Furthermore, when the C=O of the $k + 1$ residue is similarly oriented towards the C=O of the $k + 3$ residue, another HϕI interaction can further lower the Gibbs energy by -3 kcal/mol. This is shown schematically in Fig. 8.33. Finally, $k + 2$ and $k + 4$ can form another HϕI interaction. This brings the NH group of the $k + 4$ residue to a favorable configuration to form a direct HB with the C=O of the kth residue. Note that in contrast to the vacuum case, the formation of a direct HB involves only about $-6.5 + 4.5 \approx -2$ kcal/mol. The loss of the solvation Gibbs energy of the two arms (of NH and of C=O) costs about $+4.5$ kcal/mol.

The overall picture is the following. In moving from the random coil to the helix coil there are many possible trajectories. However, in water, there are at least two "stations" or intermediates that are stabilized by HϕI interaction. Therefore there will be a higher probability for the system to pass through these stations, and therefore the overall average time required to reach the helix configuration will be shortened.

The same story can be translated into the language of forces. Again, starting with a random coil, there will be initially an almost random motion (from the point of view of the solvent-induced forces). When the kth and $k + 2$ carbonyls come to a distance of slightly above, say, 5 Å, they will be attracted by one-water-bridge forces to the point of lowest pairwise HϕI interaction. This will temporarily stabilize the configuration, since any attempt to break the HϕI interaction will cost about $+3$ kcal/mol. This is equivalent to entering a tunnel in the configurational space, Fig. 8.29a, where our bold point will

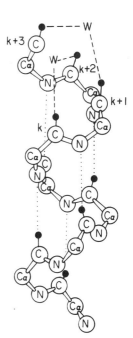

FIGURE 8.33. A segment of a growing helix. The dotted lines indicate HBs between NH and C=O four residues apart. The dashed lines indicate HϕI interaction by means of a one-water bridge, connecting two next-nearest-neighbor carbonyl groups.

be partially trapped. Next, random motion of other residues (corresponding to random motion in the first tunnel) will bring the $k + 3$ carbonyl to close separation where it will be attracted by HϕI forces exerted by the $k + 1$ carbonyl. This is equivalent to entering a tunnel within a tunnel and so forth.

Note that in this sequence of events, we have used only the solvent and carbonyl groups of the backbone to fold the coil into the helix form. This sequence of events gives preference to folding along a particular pathway with high probability. It is specific to water as a solvent. For instance, an alcohol can also form three HBs by using three of its arms. Therefore, in principle, alcohol can produce strong forces analogous to the one-water bridge, but in our particular example we have used the carbonyl groups for the HϕI interaction. This requires a solvent molecule that can offer two donor arms. Water can, but alcohols cannot, offer that.

8.9. AGGREGATION AND SELF-ASSEMBLY IN AQUEOUS SOLUTIONS

The association process discussed in section 8.7 can be easily extended to the association or aggregation of any number of monomers. The aggregation of biomolecules to form larger structures is very common in biological systems. Examples are the association of subunits to form a multisubunit enzyme, the formation of membranes, the recombination of both the small and large subunits of ribosomes, the assembly of subunits and RNA to form the tobacco mosaic virus, and many others.

All of these processes are highly specific in terms of the mode of packing as well as in the environment required for their formation. The principles that determine the driving forces for these processes are essentially the same as in the case of the association of two solutes discussed in section 8.7. However, when large solutes assemble to form a single aggregate, the driving force for the assembly process is in general not simply linear in the number of units. This gives rise to a phenomenon that closely resembles a phase transition. We shall discuss in some detail only one case in this section—the formation of micelles. At the end of this section we shall point out how the solvent can contribute to the specificity of the structure of the aggregates of biological molecules.

8.9.1. Formation of Micelles in Aqueous Solutions

The basic experimental observation is the following. A surface-active molecule usually contains a polar "head" group and a nonpolar "tail." Such molecules are known to reduce the surface tension of water when they are added to water—hence the term "surface-active" or "surfactants." The main reason they do so is their tendency (more precisely, the tendency of their nonpolar part) to avoid contact with water and to seek a nonaqueous environment. These molecules are sometimes referred to as amphiphiles, but more recently the term amphipaths has been found more appropriate to describe the contrasting behavior of the two parts of the same molecule toward the solvent.

When the concentration of the surfactant is gradually increased, one observes systematic deviations from the behavior of ideal dilute solution. This phenomenon may be ascribed to the formation of small aggregates, e.g., dimer, trimer, and so on, of solute molecules. This is a common phenomenon shared by many concentrated solutions. What makes aqueous surfactant solutions so remarkable is that over some small concentration range one finds an abrupt change in the properties of the solution. This phenomenon is ascribed to the formation of larger aggregates of solute molecules known as *micelles*. The concentration (or better the range of concentrations) at which this turning point occurs is referred to as the critical micelle concentration (CMC).

Many physical properties may be followed in order to determine the CMC. The most common are surface tension and conductivity of the solution. Experimental evidence indicates that below the CMC micelles are not formed (or at least are undetectable by experimental means). Above the CMC it has been established that most of the added surfactant is used to build up more micelles. The concentration of the monomeric solute remains fairly constant. If it were exactly constant, then we should have a phenomenon similar to a phase transition at the CMC.

We now turn to the thermodynamic description of surfactant solutions. First we note that if the micelles are viewed as a separate phase, then the chemical potential of the surfactant S in the two phases is (assuming dilute ideality of the aqueous solution)

$$\mu_S(\text{in micelle}) = \mu_S(\text{in water}) = \mu_S^{0\rho} + kT \ln \rho_S. \qquad (8.9.1)$$

If $\mu_S(\text{in micelle})$ is treated as the chemical potential of a pure phase, then one would have predicted that ρ_S is constant and equal to the CMC. Hence,

$$\Delta G^0 = \mu_S(\text{in micelle}) - \mu_S^{0\rho}$$
$$= kT \ln(\text{CMC}), \qquad (8.9.2)$$

where ΔG^0 could be interpreted as a standard Gibbs energy for the transfer of S from water into the micellar "phase." However, the fact that the CMC is not a singular point and the fact that the solute concentration does not remain strictly constant above the CMC are sufficient reasons to abandon the phase-separation model of micellar solutions.

The more realistic approach is to assume a sequential series of association reactions of the form

$$nM \rightleftarrows A_n, \qquad n = 2, 3, \dots, \qquad (8.9.3)$$

where A_n is an aggregate consisting of n monomers of M. One can also make a distinction between aggregates having the same n but differing in their shape. However, these can always be viewed as different conformations of the same aggregate A_n.

A further assumption that is customarily made for these solutions is that they form an associated ideal dilute solution. This means that the solution, when viewed as a two-component system of water w and surfactant S, is not an ideal dilute solution. The deviation from ideal behavior is a result of the solute–solute correlations. However, if these correlations are of short range, then any n-tuplet of solutes within this range is identified as an aggregate and is assigned the symbol A_n. Thus, by definition, all the solute–solute correlations are contained within the various aggregates A_n. This allows us to view the system as a multicomponent mixture $w, M, A_2, A_3, \dots, A_n$ and, to a good approximation, to ignore the correlations between these species. This is the basic argument that leads to the idea of an associated ideal dilute solution.

We now write for each species the corresponding chemical potential as

$$\mu_M = W(M|w) + kT \ln \rho_M \Lambda_M^3 q_M^{-1} \qquad (8.9.4)$$

$$\mu_{A_n} = U(A_n) + W(A_n|w) + kT \ln \rho_n \Lambda_{A_n}^3 Q_{A_n}^{-1}, \qquad (8.9.5)$$

where q_M is the internal PF of the monomer and Q_{A_n} is the internal PF of the aggregate. We have denoted by $U(A_n)$ the direct or the internal interactions between the monomers within the aggregates. Note also that the assumption of an associate ideal solution is equivalent to the assumption that the coupling work in (8.9.4) and (8.9.5) is against the pure solvent.

The condition of chemical equilibrium requires that

$$\mu_{A_n} = n\mu_M \qquad \text{for } n = 2, 3, \ldots. \tag{8.9.6}$$

As in the case of proteins, Eq. (8.9.6) can be viewed either as an equality between the chemical potentials of specific conformers or as an equality between the chemical potentials of the species A_n and M. The latter are obtained by averaging over all possible conformations as in section 8.7.5. Within the assumption of an associate ideal dilute solution, the total density of the species is

$$\rho_C = \sum_{n=1} \rho_n = \rho_M + \rho_2 + \rho_3 + \cdots, \tag{8.9.7}$$

where ρ_i is the number density of the aggregate $A_n(\rho_1 = \rho_M)$. Since there are no correlations between the aggregates, the experimental osmotic pressure is determined by

$$\beta \pi_{\text{exp}} = \rho_C = \sum \rho_n. \tag{8.9.8}$$

On the other hand, the theoretical or hypothetical osmotic pressure (π_{th}) that we would have calculated if the system had been a truly ideal dilute solution is

$$\beta \pi_{\text{th}} = \rho_T = \sum_{n=1} n\rho_n, \tag{8.9.9}$$

where ρ_T is the total density of the surfactant in the solution.

From the ratio of the last two quantities, we obtain

$$\frac{\pi_{\text{th}}}{\pi_{\text{exp}}} = \frac{\sum n\rho_n}{\sum \rho_n} = \langle n \rangle, \tag{8.9.10}$$

where $\langle n \rangle$ is the average aggregate size. This can be obtained from the experimental quantity π_{exp} (provided the assumption of an associated ideal dilute solution is valid) and the total density ρ_T.

From (8.9.4) and (8.9.6) we can write the density of the aggregate A_n as

$$\rho_n = \frac{Q_{A_n} \Lambda_M^{3n}}{Q_M^n \Lambda_{A_n}^3} \rho_M^n \exp[-\beta U(A_n) - \beta \delta G(A_n)], \tag{8.9.11}$$

where

$$\delta G_n = \delta G(A_n) = W(A_n | w) - nW(M | w) \tag{8.9.12}$$

is the solvent contribution to the total Gibbs energy of formation of the aggregate A_n. In thermodynamic notation, Eq. (8.9.11) is written as:

$$\rho_n = \rho_M^n \exp[-\beta \Delta G^0(A_n)]. \tag{8.9.13}$$

This, together with relation (8.9.9), determines the distribution of micellar sizes. In order to determine the micelle size distribution, one must first know all $\Delta G^0(A_n)$, then solve Eq. (8.9.9) for ρ_M, i.e., the implicit equation

$$\rho_T = \rho_M + \sum_{n=2} \rho_M^n \exp[-\beta \Delta G^0(A_n)], \tag{8.9.14}$$

which gives ρ_M as a function of the total density ρ_T. This is used in (8.9.13) to obtain

ρ_n for each n. The dependence of ρ_n on n is very complicated. Each of the factors in (8.9.11) depends on n. The dependence on n is quite simple for most of the factors in (8.9.11) except $\delta G(A_n)$. More specifically it is $W(A_n|w)$ the coupling work of A_n to the solvent, that is difficult to predict theoretically. There have been many attempts to predict the size distribution as well as the most probable shapes of the micelles (e.g., spherical, oblate or disklike spheroids, prolate or rodlike spheroids, etc.). All of these involve some *ad hoc* assumptions on the dependence of $\Delta G^0(A_n)$ on n.

We next turn to the question of why micelles form and what is the molecular reason for the abrupt change that we observe at the CMC. To answer this, we rewrite (8.9.11) or (8.9.13) as a series of equilibrium constants as follows

$$\rho_n = \rho_M^n K_n$$
$$= \rho_M^n K_n^{i.g} \exp[-\beta U(A_n) - \beta \delta G(A_n)]$$
$$= \rho_M^n K_n^{i.g} K_n^{int.}, \qquad (8.9.15)$$

where K_n is simply $\exp[-\beta \Delta G^0(A_n)]$ of Eq. (8.9.13) and $K_n^{i.g}$ and $K_n^{int.}$ are defined in Eq. (8.9.15). $K_n^{int.}$ depends on all the interactions in the system. It is simple to demonstrate that if we choose K_n to be small for $n = 2, 3, \ldots, n^*$ up to some n^* and then it is abruptly increased either for just one value of $n = n^*$ or for a few ns $n \geq n^*$, we get a sharp transition when we plot ρ_M as a function of ρ_T. Initially, ρ_M is linear with ρ_T and then turns sharply and remains nearly constant upon further increase of ρ_T. This is shown in Fig. 8.34. This finding suggests that in aqueous solutions, there is a certain aggregate size n^* above which K_n changes abruptly. A glance at (8.9.11) or (8.9.15) shows that the factor $K_n^{i.g}$ is a decreasing function of n; i.e., if there are no interactions at all, the formation of clusters of size n will be less probable as n increases. On the other hand, the quantity $U(A_n) + \delta G(A_n)$ is negative; i.e., $K_n^{int.}$ increases with n. Therefore it is possible that a product of a strongly decreasing $(K_n^{i.g})$ and a strongly increasing $(K_n^{int.})$ function of n will give rise to a function that has a maximum at some $n = n^*$. All the factors included in K_n, except $W(A_n|w)$, should change in a regular manner with n, even when we consider all possible conformations of the aggregates.

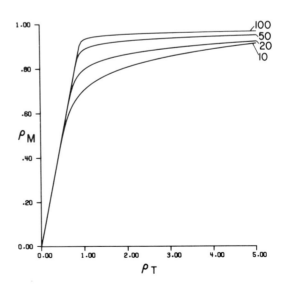

FIGURE 8.34. The dependence of the monomer concentration ρ_M on the total concentration ρ_T for a system in which $K_n = 0$ for $n < n^*$ and $K_n = 1$ for $n = n^*$. The various curves correspond to different choices of n^*, as indicated next to each curve.

The only quantity that can be suspected of having an irregular dependence on n is $W(A_n|w)$, or $\delta G(A_n)$.

A convenient way to examine the factors involved in the solvent contribution δG is to separate the contributions due to the hydrophilic "head" groups h and the hydrophobic "tail" groups t. This can be done neatly for this process as follows: For each monomer we write

$$W(M|w) = \Delta G_M^* = \Delta G_M^{*t} + \Delta G_M^{*h/t} \qquad (8.9.16)$$

or, equivalently,

$$\Delta G_M^* = \Delta G_M^{*h} + \Delta G_M^{*t/h}. \qquad (8.9.17)$$

In (8.9.16) the solvation Gibbs energy of M is divided into two parts. We transfer first the tail group and then the head group, given the condition that the tail group has already been solvated. We can make the same transfer in a different order, as in Eq. (8.9.17), but the first seems to be more useful for the present case. Note that both (8.9.16) and (8.9.17) are exact and do not involve the assumption of group additivity (see also section 8.14). Similarly, for any A_n we write

$$W(A_n|w) = \Delta G_{A_n}^* = \Delta G_{A_n}^{*t} + \Delta G_{A_n}^{*h/t}. \qquad (8.9.18)$$

Thus the solvent contribution δG_n can be rewritten as

$$\begin{aligned}\delta G_n &= \delta G_n^t + \delta G_n^{h/t} \\ &= [\Delta G_{A_n}^{*t} - n\Delta G_M^{*t}] + [\Delta G_{A_n}^{*h/t} - n\Delta G_M^{*h/t}],\end{aligned} \qquad (8.9.19)$$

where δG_n^t is the indirect work required to form an aggregate A_n of a specific size and shape (i.e., conformation) in the liquid. $\delta G_n^{h/t}$ is the corresponding conditional work to transfer the heads (i.e., $-\Delta G_M^{*h/t}$) from the monomers onto their final locations on the aggregate (i.e., $\Delta G_{A_n}^{*h/t}$). Figure 8.35 shows schematically the various ingredients of δG_n^t and $\delta G_n^{h/t}$. Note that in this diagram we consider a specific conformation of both monomers and the micelles; eventually one must take the appropriate average over all possible conformations, as in section 8.7.5.

The first step involves only the HϕO interaction between the tails. Therefore for any n, and for any possible shape of the aggregate, we should expect that $\delta G_n^t < 0$. The second step involves the transfer of n "conditional heads"—i.e., heads from the n monomers—to the aggregate. The value of $\delta G_n^{h/t}$ depends on the shape of the particular aggregates. For small aggregates, say $n = 2$ if the heads are charged, the most probable configuration of the dimer is the one in which the heads are far apart. In this case, $\delta G_n^{h/t}$ would be nearly zero. For nonionic surfactants, indirect head–head correlation could make $\delta G_n^{h/t}$ negative. For instance, the two head groups of carboxylic acid could come to within the distance so that they can form a strong HϕI interaction through a one-water bridge. Under favorable conditions, even two one-water bridges could be formed, giving rise to a large negative value of the order of $\delta G^{h/t} \approx -6$ kcal/mol per pair of heads (Fig. 8.36). (For some small carboxylic acids it has been proposed that the carboxylic groups form a direct HB. This will contribute about $\delta G^{h/t} \approx 4.5$ kcal/mol due to the loss of the solvation Gibbs energy of the two arms involved in the formation of the HB. Therefore, it seems that the formation of either one or two one-water bridges will be more likely. The formation of direct HBs between the heads is even less likely for larger aggregates.)

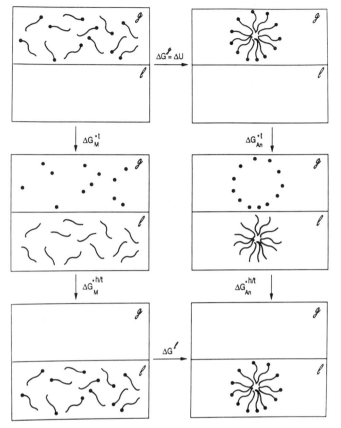

FIGURE 8.35. Schematic procedure of breaking the process of micelle formation into two parts corresponding to the two terms in Eq. (8.9.19).

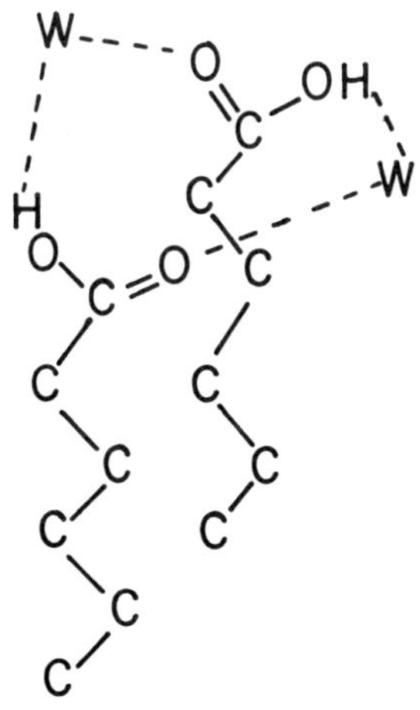

FIGURE 8.36. Two carboxylic acids in a configuration favorable to the formation of two one-water bridges between the two head groups.

We define the solvent contribution per monomer $\delta G^{(1)}$ by

$$\delta G^{(1)} = \frac{\delta G_n}{n} = \frac{W(A_n|w)}{n} - W(M|w). \qquad (8.9.20)$$

Since $W(M|w)$ is a constant independent of n, the only quantity that might possibly be responsible for the irregular dependence on n is $W(A_n|w)$. For small aggregates, say $n \le 10$, the transfer of the tails in the first step is likely to give rise to a relatively small negative contribution to $\delta G^{(1)}$. The reason is that the tails that are fully exposed to the solvent in the monomers become only partially exposed to the solvent in the small aggregate. In this case the HϕO *interaction* is not expected to be large. For a sufficiently large n, say $n \ge n^*$, the aggregates can pack in a new way such that the tails become almost unexposed to the solvent. Thus we have a transition from HϕO *interactions*, among the tails of a small aggregate (say $n < n^*$), to an almost total loss of the HϕO *solvation* of the tails in the larger aggregate. This transition involves an abrupt change in $\delta G'$, which in turn causes an abrupt change in K_n.

Of course the precise way K_n changes with n would depend on the shape of the aggregate. The rotational PF of A_n and the internal interactions $U(A_n)$ also depend on that shape. However, for whatever shapes (spherical, rodlike, or disklike) it is clear that packing of the tails in such a way that only a small fraction of their surface is exposed to the solvent gives rise to a large and negative value of $\delta G'$.

The fact that the CMC decreases with the chain length of the tail and the fact that it also decreases (at least over some range) with increase of temperature suggest that the main reason for the abrupt change of K_n as a function of n is due to the loss of the HϕO solvation of the tails.

In the next step we transfer the "conditional heads" to their ultimate destination on the surface of the micelles. If the head groups on the micelles were far apart, hence independently solvated, then there should not be much change in the environment of a given head next to a monomer or next to the micelles. In this case, we should expect that $\delta G^{h/t} \approx 0$. However, large negative contributions due to HϕI interaction between the heads could not be excluded. As the aggregates become larger, the curvature of the micelles decreases and the distance between the head groups on the surface of the micelles become smaller. This can bring the head groups to favorable distances to form pairwise HϕI as well as triple HϕI interaction. Recall that a carboxylic head group has five arms that can form HBs. If each head can form only three one-water bridges with three neighbor heads, this would contribute at least -3 kcal/mol per head group. This contribution is of comparable magnitude to the loss of the solvation Gibbs energy of a hydrocarbon of length $n_c = 10$, which is about -3.2 kcal/mol.

We have started with the assumption that the micellar solution is an associated ideal dilute solution. When the aggregates are very large, deviations from this ideality cannot be ignored. Formally one can retain Eq. (8.9.11) but reinterpret δG_n in (8.9.12) in terms of the solvation Gibbs energies of A_n and M not in pure water, but in a mixture of various aggregates.

If the system is open with respect to the solvent (osmotic system), then the first-order deviation from the ideality for, say, $W(A_n|w)$ is

$$W(A_n|\text{solvent}) = (A_n|\text{pure } w) + 2B_2^*(T, \mu_w)\rho_C + \cdots, \qquad (8.9.21)$$

where B_2^* is an average second virial coefficient of the osmotic pressure (average in the sense of section 1.8, but here over all solute species).

The value of B_2^* can be either positive or negative depending on the size of the micelles and on the extent of attractive interaction between the micelles. For example, for two micelles of equal size, we can write approximately

$$B_2^* = -\frac{1}{2}\int[g(R)-1]4\pi R^2\,dR$$

$$\approx \int_0^{\sigma} 2\pi R^2\,dR - \int_{\sigma}^{\infty}[g(R)-1]2\pi R^2\,dR. \qquad (8.9.22)$$

The first term on the rhs of (8.9.22) is positive and proportional to the size of the micelles. The second term depends on the micelle–micelle correlation function. Since the surfaces of the micelles consist of HϕI groups, it is possible that strong HϕI correlation will give rise to a large negative contribution to B_2^*. Therefore B_2^* could be either positive or negative depending on the relative magnitude of the two terms on the rhs of (8.9.22). Note that Eq. (8.9.21) is valid only for osmotic systems; i.e., T, μ_w are constants. Normally, experiments are carried in a system under T, P constants, in which case the first-order correction is different from (8.9.21) (see section 6.9).

8.9.2. Solubilization

Aqueous micellar solutions provide some interesting theoretical problems, such as the prediction of the micelle size distribution, the most stable shapes of the micelles and the elucidation of their behavior at the CMC. From the practical point of view, the most important aspect of micellar solutions is their capability to solubilize solutes that are very sparingly soluble in pure water. This phenomenon occurs naturally in biological systems and has been exploited in many applications, as for example in the pharmaceutical and detergent industries.

The basic experimental observation is the following. We take a nonpolar solute, such as naphthalene, which is sparingly soluble in water (about 2.55×10^{-4} mol/liter) and dissolve it in an aqueous surfactant solution. One finds that in the premicellar region the solubility of the naphthalene changes very slowly with the concentration of the surfactant. At the CMC an abrupt change in the solubility of the solute is observed. Typical behavior is illustrated in Fig. 8.37. The interpretation of this phenomenon is quite simple. Once

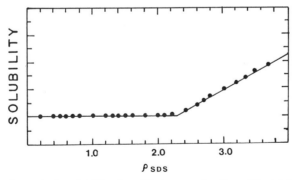

FIGURE 8.37. The change in the solubility of naphthalene as a function of the total concentration of the surfactant (sodium dodecylsulfate).

micelles are formed, they provide some pockets of nonpolar environments in which the nonpolar solute might enter. Since these solutes prefer the nonpolar environment of the interior of the micelles, they preferentially dissolve into the micelles. This explains the sharp increase in the solubility of these solutes in such solutions.

The theoretical treatment of these systems is similar to the treatment of the water–surfactant solution. Let α be the sparingly soluble solute. The chemical potential of α in pure water (w) and in the aqueous surfactant solution (l) are given by

$$\mu_\alpha^l = W(\alpha|l) + kT \ln \rho_\alpha^l \Lambda_\alpha^3 q_\alpha^{-1} \qquad (8.9.23)$$

$$\mu_\alpha^w = W(\alpha|w) + kT \ln \rho_\alpha^w \Lambda_\alpha^3 q_\alpha^{-1}. \qquad (8.9.24)$$

If the solubility is measured with respect to, say, pure solid α, then the ratio of the solubilities of α in the two phases l and w is given by the condition

$$(\rho_\alpha^l/\rho_\alpha^w)_{eq} = \exp[-\beta W(\alpha|l) + \beta W(\alpha|w)]$$

$$= \exp(-\beta \Delta W), \qquad (8.9.25)$$

where

$$\Delta W = W(\alpha|l) - W(\alpha|w) \qquad (8.9.26)$$

is the difference in the solvation Gibbs energy of α in the two phases. The more negative ΔW, the larger the solubility ratio $\rho_\alpha^l/\rho_\alpha^w$. The solubilization is sometimes defined as the ratio

$$\text{Solubilization} = \frac{\rho_\alpha^l - \rho_\alpha^w}{\rho_\alpha^w} = \frac{\rho_\alpha^l}{\rho_\alpha^w} - 1. \qquad (8.9.27)$$

If the system is a strictly ideal dilute solution with respect to all solutes, then, by definition $W(\alpha|l)$ must be equal to $W(\alpha|w)$, and there will be no solubilization. This is what is observed in the premicellar region, where the solubility of α in the solution l is almost constant and equal to the solubility in pure w up to the CMC.

Above the CMC we cannot ignore the correlation between the solute α and the micelles. For simplicity, assume that we have only one kind of micelle, say with fixed $n = n^*$, and suppose that the system is osmotic, i.e., open with respect to the solvent. In such a system, the first-order deviation of the solvation Gibbs energy of α is given by

$$W(\alpha|l) = W(\alpha|w) + 2B_2^*(T, \mu_w)\rho_n + \cdots, \qquad (8.9.28)$$

where B_2^* is the α-micelles second virial coefficient, defined by

$$B_2^* = -\frac{1}{2} \int_0^\infty [g_{\alpha,A_n}(R) - 1] 4\pi R^2 \, dR \qquad (8.9.29)$$

and ρ_n is the density of the micelles of size n.

Since α can penetrate the micelle A_n, the pair correlation function g_{α,A_n} obtains a large positive contribution from the region $R < \sigma = (\sigma_{A_n} + \sigma_\alpha)/2$. To see this, we write g_{α,A_n} in terms of the potential of average force as

$$g_{\alpha,A_n}(R) = \exp[-\beta W_{\alpha,A_n}(R)]. \qquad (8.9.30)$$

Note that here W is the potential of average force, whereas in (8.9.26) it is the coupling work of, say, α to w.

For sufficiently small R, the solute α is in the interior of the micelle; hence, we write for $R < \sigma$

$$W_{\alpha,A_n}(R) = U_{\alpha,A_n}(R) + \delta G_{\alpha,A_n}, \qquad (8.9.31)$$

where U is the direct interaction between α and the interior of the micelle at some point $R < \sigma$. δG is the indirect work to bring α from infinite separation into the micelle. Since the solvation Gibbs energies of the micelle with and without the solute are nearly equal (the surface of the micelle exposed to the solvent is assumed to be unaffected by the inclusion of the solute α), it follows that δG is essentially $-\Delta G_{\alpha}^{*}$. Recall that the solute is sparingly soluble in water; therefore ΔG_{α}^{*} is a large and positive quantity. It follows that $W_{\alpha,A_n}(R)$ is large and negative for $R < \sigma$; this in turn gives rise to a large negative B_2^{*} in (8.9.29). Thus, to first-order deviation (8.9.28), the quantity ΔW defined in (8.9.26) is large and negative. This is the phenomenon of solubilization.

The argument above can be repeated for aggregates of any size or shape. Thus the main reason for the solubilization phenomenon is the loss of the large positive solvation Gibbs energy of the solute α upon penetrating into the micelle.

Returning to the micelle size distribution, we can estimate how this distribution changes upon the addition of solute α. Suppose again that the two-component system of surfactant and water behaves as an associated ideal dilute solution for which (8.9.11) and (8.9.13) are valid. By adding the solute α we can define the new species $A(k, n)$, i.e., an aggregate with n monomers and k solute molecules α. Since the strong correlations between α and A_n are within the species $A(k, n)$, the assumption of associate ideality can be retained for all the species $A(k, n)$. Hence the analogue of (8.9.11) is now

$$\rho_n = \sum_k \rho_{n,k} = \sum_k \frac{Q_{A(k,n)} \Lambda_M^{3n}}{Q_M^n \Lambda_{A(k,n)}^3} \rho_M^n \exp[-\beta U(A(k, n)) - \beta \delta G(A(k, n))], \qquad (8.9.32)$$

where $\rho_{n,k}$ is the density of the species $A(k, n)$ and other quantities have their obvious meaning as in (8.9.11). Note however that in this case $U(A(k, n))$ includes the direct interactions between all the monomers as well as with the k solutes within the micelle. The difference between $U(A(k, n))$ and $U(A_n)$ in (8.9.11) is essentially k times the interaction between the solute and the interior of the micelle. (After averaging over all possible conformations of the micelles, these energies turn into Gibbs energies.)

On the other hand, $\delta G(A(k, n))$ in (8.9.32) is almost the same as $\delta G(A_n)$ in (8.9.11). The reason is that the coupling work of M and A_n to the solvent are nearly independent of the extent of occupation of the micelles by α. If the micelles are very large compared with the solute α, and if the interaction of α with the micelle is weak, then all the quantities on the rhs of (8.9.32) are approximately independent of k; hence the micelle size distribution is almost unaffected by α. This is indeed observed for simple nonpolar solutes. For larger or polar solutes, the effect of the solutes on the distribution could be significant.

8.9.3. Self-Assembly of Macromolecules

The generalization of the process of association of two solutes discussed in section 8.7 to many solutes is straightforward.

It is easy to generalize the lock and key model to any number of macromolecules. For instance, in Fig. 8.38a the subunit A recognizes B, B recognizes C, C recognizes D, and D recognizes A. Therefore, from the point of view of the lock and key model, there is one way of assembling the four subunits which has the highest probability. Here the preference for this particular mode of assembly is due to the direct interaction between the subunits. (In Fig. 8.38a the fit is achieved by maximizing the van der Waals interaction between the surfaces, i.e., by geometrical fit. However, the same effect can be achieved by having complementary FGs on the surfaces such that the direct interaction between the subunits leads to a unique mode of assembling.)

As in the association of two proteins, the problem of finding the specific mode of packing due to direct interactions is not a statistical mechanical problem. Knowing the surfaces of the monomers, we can tell by inspection which is the most probable mode of packing. For many subunits, the problem is very much like fitting the pieces into a jigsaw puzzle.

The situation is quite different when both direct and indirect interactions are operative. In a solvent there are many ways by which the solvation effects could affect the mode of assembly of the subunits. These are the same as in the case of molecular recognition of two subunits, discussed in section 8.7. Here we point out one example that is specific to the solvent. Suppose that we have four cubes that do not have any preference for a particular mode of packing in the vacuum (Fig. 8.38b). Therefore these cubes will have roughly the same probability of associating in any mode. In a solvent, strong HϕI interaction could dictate a highly specific mode of assembly. For instance, in Fig. 8.38b subunits A and B recognize each other by a specific pattern of HϕI groups in the J group (i.e., those FGs that do not interact directly, but only through the solvent). Likewise, C and D recognize each other by a different pattern of HϕI groups. Finally the pairs AB and CD recognize each other by forming pairwise or perhaps even triple HϕI interactions. Since each HϕI interaction through a one-water bridge contributes about -3 kcal/mol to the binding Gibbs energy, then for the particular mode of assembly shown in Fig. 8.38b, the solvent contributes a factor on the order of $\exp(-\beta\delta G) \sim \exp(8 \times 3/0.6) \approx 2 \times 10^{17}$. This example shows how the solvent, by itself, can give rise to a specific mode of packing. It can also switch preference—as in example (4) of section 8.7. For instance, suppose that the same subunits A, B, C, and D in Fig. 8.38a, have in addition to their geometrical

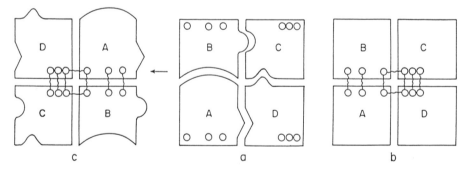

c a b

FIGURE 8.38. (a) Four subunits assembling by direct interactions (lock and key model). This configuration is favored in the vacuum. (b) For subunits assembling by indirect interactions. The open circles representing HϕI groups. The HϕI interactions are indicated by wiggled lines. (c) Same four subunits as in (a). In a solvent a different configuration might be preferred.

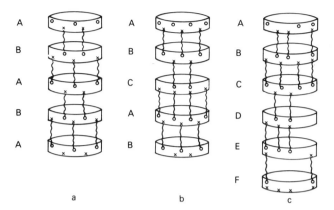

FIGURE 8.39. Various ways of stacking subunits. The recognition here is achieved by means of HϕI interactions between function groups in the J regions indicated by x and o.

fit also some HϕI groups distributed as in Fig. 8.38a, but note that these FGs do not interact directly; therefore, in vacuum the configuration of Fig. 8.38a is still the most probable. In a solvent, however, the preference might switch over to a new configuration as indicated in Fig. 8.38c.

The same ideas can be generalized to the self-assembly of any number of identical or different subunits to form macrostructures having a specific mode of packing. Again we note that it is relatively easy to see how such a specific mode of assembly can arise by direct interaction. For instance, one surface of A can recognize (by direct interactions) a surface on B, another surface of B can recognize a surface of C, and so forth. It is more subtle to show that a pattern of recognition can also originate from the solvent.

Figure 8.39 shows three cases of such self-assembly by the stacking of subunits. Here the binding surfaces, i.e., the surfaces that come into contact, are made featureless; hence, there will be no specific recognition in the vacuum. In a solvent, due to a specific distribution of HϕI groups in the J group, subunit A can recognize B, B can recognize C, and so forth. This pattern of recognition will be highly specific to the solvent. Small changes in the solvent can produce large effects on the extent and on the specificity of these processes (see section 8.10). Hence, a mechanism of regulating these processes by changing the properties of the solvent becomes feasible. Clearly, such a mechanism will have an evolutionary advantage over the conventional lock and key mechanism, based on direct interactions.

In any real process, both direct and indirect effects will combine concertedly to determine the ultimate product of an assembly process. It is inconceivable that nature would not have exploited the solvent-induced effects to gain control and to regulate vital biological processes.

8.10. SOLUTE EFFECTS ON PROCESSES IN AQUEOUS SOLUTIONS

In this chapter we have studied the modifications required to be made in the theory of some biochemical processes when a solvent is present. In this sense, this chapter may be viewed as a condensed version of earlier chapters such as 2, 3, and 4. This section is concerned with the further modifications needed to be introduced into the theory in order to include solute effects on the same processes.

Solute effects on almost any process discussed earlier in this chapter have been investigated extensively. They are important since most biochemical processes do take place in aqueous solutions, not in a pure solvent.

In order to investigate solute effects, we could have repeated all the treatments of previous examples, e.g., binding isotherms, the helix–coil transition, protein folding, etc., and made the required modifications so as to take into account the solute effect. This is not necessary, however, for two reasons.

First, in our treatment of the solvent effects throughout this chapter we have noted several times that the solvent could be either a pure liquid, say water, or any mixture of liquids, including any number and concentration of solutes. Therefore the modifications that were carried out in this chapter already include, implicitly, the solute and the solvent effects. In every place where a coupling work was written as $W(\alpha|w)$, one should reinterpret w not as pure solvent, but as the entire solvent, including any number of components.

Second, if we insist on studying the specific effects caused by the addition of a specific solute, we need not repeat the treatment of all the previous examples. The reason is that we already know that the general modification from the *vacuum theory* to the *solution theory* involves the solvent-induced quantity δG (if we are working in the T, P, N system, which in practice is the most important system). We also know that δG can always be expressed as a combination of solvation Gibbs energies. For instance, for the association and the folding of proteins, we have

$$\delta G(\mathbf{P} + \mathbf{L} \to \mathbf{PL}) = \Delta G^*_{\mathbf{PL}} - \Delta G^*_{\mathbf{P}} - \Delta G^*_{\mathbf{L}} \tag{8.10.1}$$

$$\delta G(\mathbf{U} \to \mathbf{F}) = \Delta G^*_{\mathbf{F}} - \Delta G^*_{\mathbf{U}}. \tag{8.10.2}$$

Therefore, in order to study specific solute effects on δG of any process, it is sufficient to examine the effect of the solute on the solvation Gibbs energies of the various species involved in (8.10.1) and (8.10.2).

The solvation Gibbs energy of a given species in itself is an important quantity that determines the solubility of that species. Therefore we start in the next section with an analysis of the various factors that determine the solubility of a biomolecule; then we examine how each of these factors might be affected by the addition of a solute.

8.10.1. Solubility

The solubility of a species α in a solvent w is determined with respect to some reference state of α. For example, if α is solid in its pure state at the same temperature and pressure of the solution, then the solubility of α in the solvent w is determined by the equilibrium condition

$$\mu_\alpha^{\text{solid}}(P, T) = \mu_\alpha^w(P, T, \rho_\alpha^w), \tag{8.10.3}$$

where ρ_α^w is the equilibrium concentration, say in moles per unit volume, of α in the solvent w. Here the solvent could be a mixture of any number of components. For simplicity, we start with pure water.

In all our discussions in this section we shall assume that the reference state—the pure α—is unchanged. We shall examine the factors that determine the solubility of α as well as the change of solubility caused by the addition of a solute s, all with respect to the same reference state.

In general, the chemical potential of α in the solvent w is given by

$$\mu_\alpha^{\text{solid}}(P, T) = W(\alpha\,|\,w + \alpha) + kT \ln \rho_\alpha \Lambda_\alpha^3 q_\alpha^{-1}, \tag{8.10.4}$$

where $W(\alpha\,|\,w + \alpha)$ is the coupling work of one α-solvaton against a system containing both w and α. Therefore, in order to solve for ρ_α at equilibrium we need to solve the implicit equation (8.10.4). This is in general a very complicated equation. On the other hand, if we know that the solubility of α is very low, then $W(\alpha\,|\,w + \alpha)$ does not depend on ρ_α. This is the limit of the ideal dilute solution. In this case the equilibrium value of ρ_α, i.e., the solubility of α in w, is given by

$$\rho_\alpha^w = \exp[\beta\mu_\alpha^{\text{solid}} - \beta W(\alpha\,|\,w)]. \tag{8.10.5}$$

Thus the more negative $W(\alpha\,|\,w)$ is, the larger the solubility ρ_α^w.

For simple solutes such as argon or simple hydrocarbons, there are essentially two components that determine the solubility. These are the hard and the soft parts of the solvation Gibbs energy, i.e.,

$$W(\alpha\,|\,w) = \Delta G_\alpha^* = \Delta G_\alpha^{*H} + \Delta G_\alpha^{*S/H}. \tag{8.10.6}$$

The relative contribution of the two terms on the rhs of (8.10.6) for simple solutes in water has been discussed in Chapter 7.

For complex solutes such as proteins, we can add a term which depends on the specific FGs and on their specific distribution on the surface of α. By the surface of α we mean all the FGs that are exposed to the solvent. Thus, instead of (8.10.6) we have

$$W(\alpha\,|\,w) = \Delta G_\alpha^{*H} + \Delta G_\alpha^{*S/H} + \sum_k \Delta G_\alpha^{*k/S,H}, \tag{8.10.7}$$

where the sum is over all the FGs exposed to the solvent. Note that this sum should be understood in the sense of Eq. (8.7.38); i.e., it is important to specify the order in which we add the kth FGs given that other FGs have already been turned on.

Let α be a globular protein that can be in two conformational forms A and B in equilibrium. The solvation Gibbs energy of the protein is related to the solvation Gibbs energies of A and B by

$$\exp(-\beta\Delta G_\alpha^*) = x_A \exp(-\beta\Delta G_A^*) + x_B \exp(-\beta\Delta G_B^*), \tag{8.10.8}$$

where x_A and x_B are the mole fractions of A and B as determined in the absence of the solvent (see section 6.15). Suppose that A and B are two very similar conformations, the only difference between which is that two FGs, say 1 and 2, that are independently solvated in A become correlated in B. A schematic illustration of the conformational change is shown in Fig. 8.40.

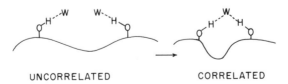

UNCORRELATED   CORRELATED

FIGURE 8.40. Two functional groups independently solvated in one conformation become correlated as a result of a small conformational change.

For this change we can assume that the solvation of the hard and the soft, as well as of all FGs except this particular pair (say 1 and 2), is not affected by the conformational change. Therefore

$$\Delta G_B^* - \Delta G_A^* = \delta G(A \to B)$$

$$= \delta G(1, 2/\alpha). \tag{8.10.9}$$

Thus for a small conformational change the difference between the solvation Gibbs energies of A and B is simply the conditional indirect or solvent-induced interaction between the two FGs, given α.† The ratio of the solubility of A and B is therefore

$$\frac{\rho_B^w}{\rho_A^w} = \exp(-\beta \Delta G_B^* + \beta \Delta G_A^*)$$

$$= \exp[-\beta \delta G(1, 2/\alpha)]. \tag{8.10.10}$$

We consider the following examples of the groups 1 and 2.

1. For two methyl groups, $\delta G \approx -0.3 \, \text{kcal/mol}$, hence

$$\frac{\rho_B^w}{\rho_A^w} \approx \exp(0.3/0.6) = 1.65. \tag{8.10.11}$$

2. For two ethyl groups, $\delta G \approx -0.6 \, \text{kcal/mol}$, hence

$$\frac{\rho_B^w}{\rho_A^w} \approx \exp(0.6/0.6) = 2.7. \tag{8.10.12}$$

3. For two hydroxyl groups brought to a distance of 4.5 Å and the correct configuration to form a strong HφI interaction, $\delta G \approx -3 \, \text{kcal/mol}$; hence,

$$\frac{\rho_B^w}{\rho_A^w} \approx \exp(3/0.6) = 148. \tag{8.10.13}$$

4. For two hydroxyl groups that are initially in a configuration to form a strong HφI interaction as in example (3), brought to a very close distance so that this HφI interaction is lost. In that case $\delta G \approx +3 \, \text{kcal/mol}$; hence,

$$\frac{\rho_B^w}{\rho_A^w} \approx \exp(-3/0.6) = 6.7 \times 10^{-3}. \tag{8.10.14}$$

We see that a small change in the conformation can cause a large change in the solvation Gibbs energy of the protein, hence of the solubility of the protein. It is well known that one of the most outstanding properties of proteins is their solubility. A polypeptide of a given number and kinds of residues can have a very large number of different sequences. Only a very small number of these will be soluble in water. Obviously, soluble proteins that do occur in biological systems are those that have survived the evolutionary selection. They would not have survived if they were not soluble. As we have seen above, the most dramatic effect on solubility involves changes in HφI interactions. Such changes can be induced by a small mutation. For instance, a carboxylic group of aspartic acid on the

---

† Recall that the same relation (8.10.9) was used in sections 7.14–7.16 to study the correlation δG from available information on ΔG_A^* and ΔG_B^*.

surface of the protein can form under favorable conditions up to five HϕI interactions. If a mutation changes the aspartic acid into glutamic acid, i.e., the new protein differs from the original type by adding only one carbon atom to the side chain (see Fig. 8.41), the resulting change in solubility could be on the order of $(148)^{-5} \approx 1.4 \times 10^{-11}$. Note that this change has retained the overall polarity and structure of the entire protein. Similarly, a small conformational change can break all HϕI interactions leading to the same change in solubility.

In actual biological systems, changes in solubility can be induced by changing the environment of the protein, i.e., by changing the concentration of one or more solutes. We turn next to study the factors involved in the effect of solutes on the solvation Gibbs energy of a protein.

8.10.2. Solute Effect on the Solvation Gibbs Energy of a Molecule with a Fixed Conformation

Let α be any solute forming a dilute solution in a pure solvent w. We are interested in the change in the solvation Gibbs energy caused by the addition of a small quantity of solute s. We assume here that α has a fixed conformation; α can be a simple solute such as argon, benzene, or a single conformer of a large polymer.

The chemical potential of α in the system is

$$\mu_\alpha(T, P, N_w, N_\alpha, N_s) = \Delta G_\alpha^* + kT \ln \rho_\alpha \Lambda_\alpha^3 q_\alpha^{-1}. \qquad (8.10.15)$$

We shall be interested in the first-order effect of a nonionic solute s on ΔG_α^* in a system at constant P, T, N_w. (The case of an osmotic system is simpler to examine. This may

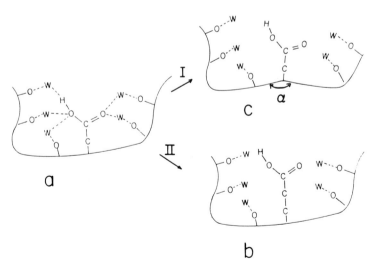

FIGURE 8.41. (a) An aspartic acid residue [shown in the center of (a)] favorably oriented to form five HϕI interactions. These interactions may be eliminated either by a small conformational change (c), or by a mutation (b).

be the case in a living cell. However, in laboratory experiments, we usually carry out measurements in a T, P, N_w system.)

$$\Delta G_\alpha^*(T, P, N_w; m_s) = \Delta G_\alpha^*(T, P, N_w) + \left(\frac{\partial \Delta G_\alpha^*}{\partial m_s}\right)_{T,P} dm_s, \qquad (8.10.16)$$

where $m_s = N_s/N_w$.

The general expression for the derivative of the chemical potential of α with respect to N_s is

$$\left(\frac{\partial \mu_\alpha}{\partial N_s}\right)_{T,P,N_w,N_\alpha} = \frac{\partial \Delta G_\alpha^*}{\partial N_s} - \frac{kT}{V} \bar{V}_s, \qquad (8.10.17)$$

where V is the average volume of the system and $\bar{V}_s$ is the partial molar volume of s. Note that we assumed that the internal partition function of α is not affected by addition of s.

Using the Kirkwood–Buff theory, we have (see Appendix J)

$$\left(\frac{\partial \mu_\alpha}{\partial N_s}\right)_{T,P,N_w,N_\alpha} = \frac{kT|E(\alpha, A)|}{\rho_\alpha \rho_s V|D|}, \qquad (8.10.18)$$

where

$$E(\alpha, A) = \begin{vmatrix} 1 & 0 & 1 \\ G_{\alpha s} & 1 & G_{ws} \\ G_{\alpha w} & 1 & G_{ww} + \rho_w^{-1} \end{vmatrix} \qquad (8.10.19)$$

and

$$|D| = \begin{vmatrix} 0 & 1 & 1 & 1 \\ 1 & G_{\alpha\alpha} + \rho_\alpha^{-1} & G_{\alpha s} & G_{\alpha w} \\ 1 & G_{\alpha s} & G_{ss} + \rho_s^{-1} & G_{sw} \\ 1 & G_{\alpha w} & G_{sw} & G_{ww} + \rho_w^{-1} \end{vmatrix}. \qquad (8.10.20)$$

We are interested in the limit as $\rho_s \to 0$ and $\rho_\alpha \to 0$. At this limit we have, from (8.10.18) and (8.10.19),

$$\left(\frac{\partial \Delta G_\alpha^*}{\partial m_s}\right)_{T,P} = N_w \left(\frac{\partial \Delta G_\alpha^*}{\partial N_s}\right)$$

$$= N_w \left[\frac{\partial \mu_\alpha}{\partial N_s} + \frac{kT}{V} \bar{V}_s\right]. \qquad (8.10.21)$$

Using Eq. (8.10.20) at the limit of $\rho_s \to 0$ and $\rho_\alpha \to 0$, we find the final result

$$\left(\frac{\partial \Delta G_\alpha^*}{\partial m_s}\right)_{T,P} = kT\rho_w(G_{\alpha w}^{oo} - G_{\alpha s}^{oo}), \qquad (8.10.22)$$

where the superscripts oo indicate the double limit of $\rho_s \to 0$, $\rho_\alpha \to 0$.

In an osmotic system the first-order analogue of (8.10.18) would be

$$\Delta G_a^*(T, V, \mu_w; \rho_s) = \Delta G_a^*(T, V, \mu_w) - kTG_{as}^{\infty\infty}\rho_s + \cdots. \tag{8.10.23}$$

In the T, P, N system, the first order effect of s on the solvation of α depends on the difference in the affinity of α towards the solvent and towards the solute s. This may be rewritten as

$$\left(\frac{\partial \Delta G_a^*}{\partial m_s}\right)_{T,P} = kT\rho_w \left\{ \int_0^\infty [g_{aw}(R) - g_{as}(R)]4\pi R^2\, dR \right\}. \tag{8.10.24}$$

To further analyze the rhs of (8.10.24), we recall that the main contribution to the pair correlation function comes from a small region around the first peak. Let us denote by R_1 the location of the first peak of $g(R)$.

Suppose first that α, w, and s are all simple solutes of about the same diameter σ. Then $R_1 \approx \sigma$ and we write

$$g_{aw}(R_1) - g_{as}(R_1) = g_{aw}(R_1)\left[1 - \frac{g_{as}(R_1)}{g_{aw}(R_1)}\right]$$

$$= g_{aw}(R_1)\{1 - \exp[-\beta\Delta G(aw \to as)]\}$$

$$= g_{aw}(R_1)\{1 - \exp[-\beta\Delta U(aw \to as) - \beta\delta G(aw \to as)]\}. \tag{8.10.25}$$

In (8.10.25) we denoted by $\Delta G(aw \to as)$ the Gibbs energy change for the formation of the pair aw at R_1 from the pair as at R_1.

We see that there are two main contributions that determine the effect of the solute s on the solvation (hence the solubility) of α. First, the difference in the direct interaction between the pairs as and aw. For instance, if s interacts more strongly with α compared with the interaction between w and α, i.e., if $\Delta U(aw \to as) < 0$, then this effect will give a negative contribution to the integrand in (8.10.24), and therefore the solvation Gibbs energy of α becomes more negative (or less positive) upon the addition of s. This is equivalent to the increase in the solubility of α.

The second contribution is the indirect effect $\delta G(aw \to as)$, which measures the extent of change of the solvation Gibbs energy of the pair aw relative to the pair as at R_1.

If the solute α is a polymer and w and s differ in their sizes then we still can identify the direct and indirect effects as in the simple case, but now we must consider these effects at different distances and orientations between α and w and α and s. Although the integral in (8.10.24) requires only the angle averaged pair correlation functions, the following analysis is made for the full-angle-dependent pair correlation functions. Thus instead of (8.10.25), we write (8.10.22) as

$$\left(\frac{\partial \Delta G_a^*}{\partial m_s}\right)_{T,P} = \frac{kT\rho_w}{8\pi^2}\left\{\int [g_{aw}(\mathbf{X}_w) - 1]\, d\mathbf{X}_w - \int [g_{as}(\mathbf{X}_s) - 1]\, d\mathbf{X}_s\right\}. \tag{8.10.26}$$

Thus in (8.10.26) we have fixed the orientation of α and the integration is over all configurations of w in the first integral and of s in the second. For a given polymer α in a given solvent w the first integral $G_{aw}^{\infty\infty}$ is constant. We can use the second integral to compare different solutes. We write

$$g_{as}(\mathbf{X}_s) = \exp[-\beta U(\mathbf{X}_s) - \beta\delta G(\mathbf{X}_s)], \tag{8.10.27}$$

where $U(\mathbf{X}_s)$ and $\delta G(\mathbf{X}_s)$ are the direct and indirect work required to bring s from infinite separation from α to a fixed configuration $\mathbf{X}_s$ relative to the configuration of α. Each of these can be further divided into separate contributions due to the hard, soft, and specific functional groups on both α and s. In general, one must consider all these effects, then integrate over all configurations of s to estimate the contribution of $G_{\alpha s}^{\infty}$. This is a fairly complicated task even when s is a simple solute. Note also that the binding of s on α is a special case where the interaction between α and s can be treated as a discrete event. We have discussed so far only the limiting case $\rho_s \to 0$ and $\rho_\alpha \to 0$ when the solvent consists of one component only. All of these limitations can be removed. Formally, we can write the general expression for the required quantity as in (8.10.22)

$$\left(\frac{\partial \Delta G_\alpha^*}{\partial m_s}\right)_{T,P,N} = kT\rho_w \left[\frac{|E(\alpha, A)|}{\rho_s \rho_\alpha |D|} + \bar{V}_s\right], \tag{8.10.28}$$

where all the quantities $|E(\alpha, A)|$, $|D|$, and $\bar{V}_s$ may be expressed in terms of all the G_{ij}s in the system through an extension of the Kirkwood–Buff theory (see Appendix J). The general expression is fairly complicated. For the general case of a three-component system at arbitrary concentrations of α, w, and s, we have to consider the relative magnitude of all six affinities $G_{\alpha\alpha}$, $G_{\alpha w}$, $G_{\alpha s}$, G_{ss}, G_{sw}, and G_{ww}. For this case the explicit expression for (8.10.28) may be obtained by using (8.10.19) and (8.10.20).

The above treatment is valid for a solute α having a fixed conformation. For instance α can be a small rigid molecule. It can also be applied to a polymer, having a range of conformational changes all of which have the same affinity towards s and w. If, however, there are conformations that differ appreciably in their affinities towards s and w, then we must treat each of these separately. We therefore next treat an example where there are two conformations only (either two rigid molecules or two conformations of protein each of which is allowed a narrow range of conformational change).

8.10.3. Solute Effect on the Solvation Gibbs Energy of a Molecule Having Two Conformations

We return to the example discussed at the beginning of this section. Let A and B be two conformations of the protein α. The solvation Gibbs energy of the protein is given by (8.10.8). We can now ask two related questions: (1) What is the effect of a solute s on the equilibrium concentrations of A and B? (2) What is the effect of the solute s on the solubility of the protein α?

To answer the first question, we can use the general identity derived in section 5.13, which for our case is written as

$$\left(\frac{\partial \bar{N}_A}{\partial N_s}\right)_{P,T} = -\frac{\mu_{As} - \mu_{Bs}}{\mu_{AA} - 2\mu_{AB} + \mu_{BB}}, \tag{8.10.29}$$

where $\mu_{\alpha\beta} = (\partial \mu_\alpha / \partial N_\beta)_{P,T,N_j (j \neq \beta)}$. On the lhs of (8.10.29) we have a derivative of the average number of the A conformers with respect to N_s along the equilibrium line, i.e., when the equilibrium $A \rightleftarrows B$ is maintained during the addition of N_s (at T, P and $N_A + N_B$ constant). All the derivatives on the rhs of (8.10.29) are in the frozen-in system, i.e., where the transition between A and B is forbidden. Since the denominator is always positive and is a property of the system of A and B in the absence of s, we need to examine only the numerator. The two quantities μ_{As} and μ_{Bs} can be expressed in terms of the Kirkwood–Buff integrals (see also Appendix J). Hence relation (8.10.29) can provide an answer to our first question.

The second question is more important from the practical point of view. As before, assuming that s does not affect the internal properties of A and B, we can take the derivative of (8.10.8) with respect to m_s. The mole fractions x_A and x_B do not depend on m_s. Taking the derivative of (8.10.8) we get (P, T constant)

$$\exp(-\beta \Delta G_\alpha^*) \frac{\partial \Delta G_\alpha^*}{\partial m_s} = x_A \exp(-\beta \Delta G_A^*) \frac{\partial \Delta G_A^*}{\partial m_s} + x_B \exp(-\beta \Delta G_B^*) \frac{\partial \Delta G_B^*}{\partial m_s} \quad (8.10.30)$$

or, equivalently,

$$\frac{\partial \Delta G_\alpha^*}{\partial m_s} = x_A^* \frac{\partial \Delta G_A^*}{\partial m_s} + x_B^* \frac{\partial \Delta G_B^*}{\partial m_s}, \quad (8.10.31)$$

where x_A^* and x_B^* are the mole fractions of A and B in the solution (P, T, N_w, N_α, N_s). At the limit $\rho_\alpha \to 0$ and $\rho_s \to 0$, we can write the generalization of (8.10.22) for this case as

$$\left(\frac{\partial \Delta G_\alpha^*}{\partial m_s} \right)_{T,P} = kT \rho_w [x_A^*(G_{Aw}^{\infty} - G_{As}^{\infty}) + x_B^*(G_{Bw}^{\infty} - G_{Bs}^{\infty})]. \quad (8.10.32)$$

Thus, in order to examine the initial effect of s on the solvation of α (and hence on its solubility), we need the two differences of the affinities $G_{Aw}^{\infty} - G_{As}^{\infty}$ and $G_{Bw}^{\infty} - G_{Bs}^{\infty}$. In other words, we have to examine the relative affinities of s and w toward A and toward B.

For the more general case, when the concentrations of α and s are arbitrary, we must use the more general relation (8.10.28) applied for A and B, respectively. Note that if the conformations A and B are very similar, so that $G_{Aw}^{\infty} = G_{Bw}^{\infty}$ and $G_{As}^{\infty} = G_{Bs}^{\infty}$, then Eq. (8.10.32) reduces to the particular case of the previous section: Eq. (8.10.22).

The generalization to many conformations (discrete or continuous) is now straightforward.

8.10.4. Conclusion

We have seen that even at the limit of very dilute solutions (with respect to both α and s) the examination of the solute effect on the solvation Gibbs energy of α is very complicated. If α has only one conformation (i.e., it is either a rigid small molecule or a protein with a narrow range of accessible conformations), we need to know how the solute s interacts directly with all the specific groups on the polymer (this includes binding to a fixed site on α), as well as how it interacts indirectly, mediated through the solvent. If there are several conformations, then all these should be examined for each of the conformations. Therefore, from an experimental study of the effect of a solute, say urea, on the solubility of a protein, it is in general impossible to infer the molecular events that are responsible for any particular effect. A solute such as urea interacts both directly and indirectly with all the FGs on the surface of α. The average of all these interactions enters into the experimental quantity given in (8.10.22) or in (8.10.32).

In spite of the enormous complexity, the study of the effect of various solutes on the solvation of biomolecules such as proteins, nucleic acids, glycoproteins, polysaccharides, etc., is important. The knowledge of such solute effects can give us a method of controlling biochemical processes such as enzymatic activity, binding of proteins to DNA, etc. No doubt, nature uses these same methods to control and regulate its own processes.

SUGGESTED READINGS

General introductions to protein structure and properties are

T. E. Creighton, *Proteins, Structure and Molecular Principles* (W. H. Freeman and Co., New York, 1984).

G. E. Schulz and R. H. Schirmer, *Principles of Protein Structure* (Springer Verlag, Berlin, 1978).

C. R. Cantor and P. R. Schimmel, *Biophysical Chemistry* (W. H. Freeman and Co., San Francisco, 1980).

R. Jaenicke, *Prog. Biophys. Mol. Biol.* **49**, 117 (1987).

A recent review of the theoretical and computational aspects of proteins is

C. L. Brooks, M. Karplus, and B. M. Pettit, *Proteins: A Theoretical Perspective of Dynamics, Structure and Thermodynamics* (John Wiley and Sons, New York, 1988).

Some Geometries Involving Hydrogen Bonding

The fundamental tetrahedral geometry around an oxygen atom in ice is shown in Fig. 7.2. The nearest neighbor's distance is 2.76 Å and the second-nearest neighbor's distance is

$$2 \times 2.76 \times \sin\left(\frac{\theta_T}{2}\right) = 4.507 \text{ Å}, \tag{A.1}$$

where $\theta_T = 109.46°$ is the characteristic tetrahedral angle defined by

$$\cos\theta_T = -1/3. \tag{A.2}$$

If a is the edge of the cube containing the four nearest-neighbor oxygens as in Fig. 7.2, then

$$a = \frac{4.507}{2^{1/2}} = 3.19 \text{ Å} \tag{A.3}$$

and the long diagonal of the cube is

$$c = 2 \times 2.76 = 5.52 \text{ Å}. \tag{A.4}$$

The distance between the central oxygen and one of the faces of the cube is

$$d = a/2 = 1.59 \text{ Å}. \tag{A.5}$$

The distance between the central oxygen and the plane containing three of the nearest-neighbor oxygens is

$$e = 0.92 \text{ Å}. \tag{A.6}$$

In section 7.5 we calculated the configurational range for a single hydrogen bond, based on data from the second virial coefficient:

$$\frac{\Delta \mathbf{X}}{8\pi^2} \approx \frac{298}{\exp(-\beta\varepsilon_{\text{HB}}) - 1}. \tag{A.7}$$

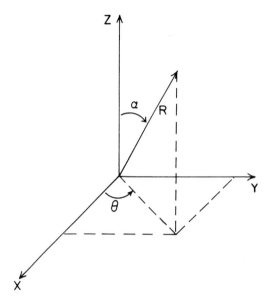

FIGURE A.1. Polar angles α and θ for the orientation of one arm of a water molecule relative to the O—O axis, chosen as the z axis.

The estimate of ΔX depends on the choice of the HB energy. For a choice of $\varepsilon_{HB} = -6.5\ \text{kcal/mol}$, we find

$$\frac{\Delta X}{8\pi^2} = 5.1 \times 10^{-3}\ \text{cm}^3/\text{mol}. \tag{A.8}$$

In some calculations we might need to estimate the separate regions of location and orientation, $\Delta X = \Delta R\ \Delta \Omega$.

For a single HB we choose the O—O line as the z axis; then the orientation of the specific arm of one water molecule that engages in a HB with a second molecule is defined in terms of the two polar angles θ and α relative to this axis (Fig. A.1). In addition, the molecule has a full rotational range of 2π about the O—O axis. Therefore an estimate of $\Delta \Omega$ for the allowable configurations of one arm relative to a second arm with a fixed orientation is

$$\Delta \Omega = \int_0^{2\pi} d\psi \int_0^{20\pi/180} \sin \alpha\ d\alpha \int_0^{2\pi} d\theta = 4\pi^2 6 \times 10^{-2} = 2.38, \tag{A.9}$$

where we arbitrarily chose 20° for the allowable angle between the direction of the arm and the z axis (Fig. A.1).

From (A.9) and (A.8) we can estimate that

$$\Delta R = \Delta x\ \Delta y\ \Delta z = 0.17\ \text{cm}^3/\text{mol} = 0.28\ \text{Å}^3. \tag{A.10}$$

Choosing polar coordinates again, we express ΔR as

$$\Delta R = R^2\ dR \int_0^{20\pi/180} \sin \alpha\ d\alpha \int_0^{2\pi} d\phi = 2.88\ dR. \tag{A.11}$$

We find

$$dR \approx 0.1\ \text{Å}, \tag{A.12}$$

which is the linear allowable variation in the O—O distance.

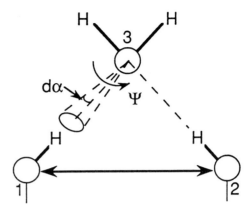

FIGURE A.2. A water molecule forming a HB with arm 1 can rotate by 2π about the O—O axis. If it is required to form a second HB with another arm 2, the angle of rotation Ψ has to be restricted to about $\Delta\Psi \approx 20°$.

The value of $\Delta X/8\pi^2$ given in (A.8) is for the formation of one HB by a pair of arms. If we require that a water molecule form simultaneous HBs with two arms as in Fig. A.2, we need to restrict the rotation over the angle ψ. In Eq. (A.9) we allowed the full rotational range of 2π, but now we restrict it to about 20°, and we obtain

$$\Delta X' = \frac{20}{360}\Delta X. \qquad (A.13)$$

In section 8.7 and in Appendices C and D we require a water molecule to form simultaneously three HBs with three arms. The corresponding region is denoted by $\Delta X''$. The requirement on the configuration $\Delta X''$ is slightly more stringent in this case, but one can take the value of $\Delta X'$ as an approximation for $\Delta X''$.

On the Extent of Independence of the Conditional Solvation Gibbs Energies of the Four Arms of a Water Molecule

In section 7.7 we assumed that the binding energies along the four arms of a water molecule are independent; hence, we wrote Eq. (7.7.6):

$$\left\langle \exp\left(-\beta \sum_{l=1}^{4} B_l^{HB}\right)\right\rangle_{LJ} = \prod_{l=1}^{4} \langle \exp(-\beta B_l^{HB})\rangle_{LJ}. \tag{B.1}$$

To what extent is this assumption justified? To answer this question, we analyze the quantity on the lhs of (B.1) using an argument similar to the one used in section 7.7. For notational simplicity, we assume that all arms are equivalent (the same conclusion applies when we distinguish between the donor and acceptor arms):

$$\left\langle \exp\left(-\beta \sum_{l=1}^{4} B_l^{HB}\right)\right\rangle_{LJ} = \int d\mathbf{X}^N P(\mathbf{X}^N/LJ) \exp\left(-\beta \sum_l B_l^{HB}\right)$$

$$= 4 \sum_{i=1}^{N} \int_{\text{one}} + 6 \sum_{i \neq j} \int_{\text{two}} + 4 \sum_{i \neq j \neq k} \int_{\text{three}}$$

$$+ \sum_{i \neq j \neq k \neq h} \int_{\text{four}} + \int_{\text{none}}, \tag{B.2}$$

where $P(\mathbf{X}^N/LJ)$ is the conditional distribution function for all "solvent" molecules given that the LJ part of the potential of our central (or the solvaton) molecule is turned on. The first term corresponds to the configurational space from which one arm is hydrogen bonded. There are four specific arms and N specific water molecules that can do that. The indication "one" means that the central molecule is hydrogen bonded to only one specific water molecule. Similarly, there are six possibilities for forming two bonds along two arms, there are four possibilities for forming three bonds and one possibility for forming four bonds. The last term corresponds to the case where no bonds are formed. This last integral is the probability of finding the central molecule unbound. Hence,

$$\int_{\text{none}} d\mathbf{X}^N P(\mathbf{X}^N/LJ) = 1 - 4 \sum_{i=1}^{N} \int_{\text{one}} - 6 \sum_{i \neq j} \int_{\text{two}} - 4 \sum_{i \neq j \neq k} \int_{\text{three}} - \sum_{i \neq j \neq k \neq h} \int_{\text{four}}. \tag{B.3}$$

Since each arm can form at most one HB at a time, the integrals in (B.2) can be simplified considerably; for instance,

$$4 \sum_{i=1}^{N} \int_{\text{one}} d\mathbf{X}^N P(\mathbf{X}^N/\text{LJ}) \exp[-\beta U^{\text{HB}}(1, \mathbf{X}_i)]$$

$$= 4 \int_{\text{one}} \exp[-\beta U^{\text{HB}}(1, \mathbf{X}_1)] \rho^{(1)}(\mathbf{X}_1/\text{LJ}) \, d\mathbf{X}_1, \qquad (\text{B.4})$$

where $U^{\text{HB}}(1, \mathbf{X}_1)$ is the HB interaction between arm 1 and a water molecule at $\mathbf{X}_1$. Similarly, for the other terms in (B.2), we have

$$6 \sum_{i \neq j} \int_{\text{two}}$$

$$= 6 \sum_{i \neq j} \int_{\text{two}} \exp[-\beta U^{\text{HB}}(1, \mathbf{X}_1) - \beta U^{\text{HB}}(2, \mathbf{X}_2)] \rho^{(2)}(\mathbf{X}_1, \mathbf{X}_2/\text{LJ}) \, d\mathbf{X}_1 \, d\mathbf{X}_2, \qquad (\text{B.5})$$

where $\rho^{(2)}(\mathbf{X}_1, \mathbf{X}_2/\text{LJ})$ is the conditional pair distribution function. A similar expression applies to the third and fourth terms in (B.2). Combining these with (B.3), we have

$$\left\langle \exp\left[-\beta \sum_{l} B^{\text{HB}}\right] \right\rangle_{\text{LJ}}$$

$$= 1 + 4 \int_{\text{one}} \{\exp[-\beta U^{\text{HB}}(1, \mathbf{X}_1)] - 1\} \rho^{(1)}(\mathbf{X}_1/\text{LJ}) \, d\mathbf{X}_1$$

$$+ 6 \int_{\text{two}} \{\exp[-\beta U^{\text{HB}}(1, \mathbf{X}_1) - \beta U^{\text{HB}}(2, \mathbf{X}_2)] - 1\} \rho^{(2)}(\mathbf{X}_1, \mathbf{X}_2/\text{LJ}) \, d\mathbf{X}_1 \, d\mathbf{X}_2$$

$$+ 4 \int_{\text{three}} \left\{\exp\left[-\beta \sum_{l=1}^{3} U^{\text{HB}}(l, \mathbf{X}_l)\right] - 1\right\} \rho^{(3)}(\mathbf{X}_1, \mathbf{X}_2, \mathbf{X}_3/\text{LJ}) \, d\mathbf{X}_1 \, d\mathbf{X}_2 \, d\mathbf{X}_3 \qquad (\text{B.6})$$

$$+ \int_{\text{four}} \left\{\exp\left[-\beta \sum_{l=1}^{4} U^{\text{HB}}(l, \mathbf{X}_l)\right] - 1\right\} \rho^{(4)}(\mathbf{X}_1, \mathbf{X}_2, \mathbf{X}_3, \mathbf{X}_4/\text{LJ}) \, d\mathbf{X}_1 \, d\mathbf{X}_2 \, d\mathbf{X}_3 \, d\mathbf{X}_4.$$

We now assume, first, that there is only a very small region from which a water molecule can form a HB using any one of its arms. Second, two of such hydrogen-bonding regions are at a distance of about 4.5 Å (see Fig. 7.2). At this distance, there is only a very weak correlation between two water molecules. (Note that this is true only when the LJ condition applies. If the condition is a fully switched-on water molecule, then there would be a strong correlation between two molecules occupying two such regions.)

Thus we write, for the first integral on the rhs of (B.6),

$$\int \{\exp[-\beta U^{\text{HB}}(1, \mathbf{X}_1)] - 1\} \rho^{(1)}(\mathbf{X}_1/\text{LJ}) \, d\mathbf{X}_1 \approx \{\exp(-\beta \varepsilon_{\text{HB}}) - 1\} \rho^{(1)}(\mathbf{X}_w/\text{LJ}) 2\Delta\mathbf{X},$$

$$(\text{B.7})$$

where $\Delta\mathbf{X}$ is the configurational volume from which a water molecule can form one HB to arm 1. The factor 2 enters since each water molecule can offer two arms for HBing with arm 1.

For the second integral on the rhs of (B.6), we write

$$\int_{two} \{\exp[-\beta U^{HB}(1, \mathbf{X}_1) - \beta U^{HB}(2, \mathbf{X}_2)] - 1\}\rho^{(2)}(\mathbf{X}_1, \mathbf{X}_2/LJ)\, d\mathbf{X}_1\, d\mathbf{X}_2$$

$$\approx [\exp(-\beta 2\varepsilon_{HB}) - 1]\rho^{(2)}(\mathbf{X}_1, \mathbf{X}_2/LJ)(2\Delta\mathbf{X})^2 \approx [\exp(-\beta\varepsilon_{HB})\rho^{(1)}(\mathbf{X}_w/LJ)2\Delta\mathbf{X}]^2.$$

(B.8)

On the last form on the rhs of (B.8) we have assumed that the conditional correlation between $\mathbf{X}_1$ and $\mathbf{X}_2$ is negligible; i.e., we write

$$\rho^{(2)}(\mathbf{X}_1, \mathbf{X}_2/LJ) = \rho^{(1)}(\mathbf{X}_1/LJ)\rho^{(1)}(\mathbf{X}_2/LJ)g(\mathbf{X}_1, \mathbf{X}_2/LJ).$$

(B.9)

Since $R_{12} = 4.5$ Å, and since we know that the experimental (orientation-averaged) pair correlation function at $R_{12} = 4.5$ Å is about 1.15, we can assume that the conditional pair correlation function might be even smaller than 1.15. Therefore we have neglected this correlation in (B.8). Similarly, for the triplet and the quadruplet correlations we write

$$\rho^{(3)}(\mathbf{X}_1, \mathbf{X}_2, \mathbf{X}_3/LJ) = \rho^{(1)}(\mathbf{X}_1/LJ)\rho^{(1)}(\mathbf{X}_2/LJ)\rho^{(1)}(\mathbf{X}_3/LJ)$$

(B.10)

$$\rho^{(4)}(\mathbf{X}_1, \mathbf{X}_2, \mathbf{X}_3, \mathbf{X}_4/LJ) = \rho^{(1)}(\mathbf{X}_1/LJ)\rho^{(1)}(\mathbf{X}_2/LJ)\rho^{(1)}(\mathbf{X}_3/LJ)\rho^{(1)}(\mathbf{X}_4/LJ).$$ (B.11)

Introducing these into (B.6), we have

$$\left\langle \exp\left(-\beta \sum_l B_l^{HB}\right)\right\rangle_{LJ} \approx 1 + 4y + 6y^2 + 4y^3 + y^4,$$

(B.12)

where

$$y = \exp(-\beta\varepsilon_{HB})\rho^{(1)}(\mathbf{X}/LJ)2\Delta\mathbf{X},$$

(B.13)

where ε_{HB} is the HB energy and $\Delta\mathbf{X}$ has been estimated in Appendix A.

Thus, with the assumption of weak conditional correlation between water molecules at the distance of 4.5 Å (i.e., simultaneously occupying two small cubes in Fig. 7.2), we factored the lhs of (B.1) into four averages:

$$\left\langle \exp\left(-\beta \sum_{l=1}^{4} B_l^{HB}\right)\right\rangle_{LJ} = (1 + y)^4,$$

(B.14)

which is effectively equivalent to the assumption of independence of the four arms of a water molecule.

Estimate of the Solvent-Induced Interactions between Two, Three, and Four Hϕl Groups

In section 7.7 we estimated the conditional solvation Gibbs energy per arm of water molecule or per arm of any Hϕl group. We extend this calculation to include interactions between 2, 3, and 4 arms belonging to four different molecules (or groups) at some specific configuration.

First, we rewrite the result of section 7.7 in a slightly different notation. We denote by $\Delta G^{*HB}(1/LJ)$ the conditional solvation Gibbs energy of a specific arm, denoted 1, given that the LJ interaction of the solute (or group) carrying this arm is turned on. In the new notation, we have

$$\exp[-\beta \Delta G^*(1/LJ)] = \langle \exp(-\beta B_1^{HB}) \rangle_{LJ} \approx 1$$

$$+ \int d\mathbf{X}_w \{\exp[-\beta U^{HB}(1, \mathbf{X}_w)] - 1\} \rho(\mathbf{X}_w/LJ), \qquad (C.1)$$

where the integration is carried out over all configurations of a water molecule. $\rho(\mathbf{X}_w/LJ)$ is the conditional density of water molecules at $\mathbf{X}_w$, given the LJ of solute 1.

We now consider the conditional solvation Gibbs energy of two arms 1, 2 of two solutes (or groups) at a specific configuration such that they form simultaneous HBs with a water molecule, as in Fig. C.1, i.e.,

$$\exp[-\beta \Delta G^{*HB}(1, 2/LJ)] = \langle \exp(-\beta B_{1,2}^{HB}) \rangle_{LJ}. \qquad (C.2)$$

To compute the quantity on the rhs of (C.2), we repeat essentially the same computation as in section 7.7, the result being

$$\langle \exp(-\beta B_{1,2}^{HB}) \rangle_{LJ} = 1 + 2 \int d\mathbf{X}_w \{\exp[-\beta U^{HB}(1, \mathbf{X}_w)] - 1\} \rho(\mathbf{X}_w/LJ)$$

$$+ \int d\mathbf{X}_w \{\exp[-\beta U^{HB}(1, \mathbf{X}_w) - \beta U^{HB}(2, \mathbf{X}_w)] - 1\} \rho(\mathbf{X}_w/LJ)$$

$$\approx 1 + 2[\exp(-\beta \varepsilon_{HB}) - 1]4\rho_w \frac{2(\Delta \mathbf{X} - \Delta \mathbf{X}')}{8\pi^2}$$

$$+ [\exp(-\beta 2\varepsilon_{HB}) - 1]4\rho_w \frac{2\Delta \mathbf{X}'}{8\pi^2}$$

$$\approx 1 + 251 + 4.33 \times 10^5. \qquad (C.3)$$

Hence,

$$\Delta G^{*HB}(1, 2/LJ) = -7.7 \text{ kcal/mol}. \tag{C.4}$$

In the first term on the rhs of (C.3), we first consider the two possibilities that w will form one HB either with 1 or with 2. The configurational range for doing so is $\Delta X - \Delta X'$, where $\Delta X'$ is the configurational range for w to form two HBs with 1 and 2. The factor 4 is taken to account for the normal pair correlation between the LJ part of w and the LJ part of 1 and 2. [This is roughly $\exp(0.8/0.6) \approx 3.8$; see Eq. (D.4) of Appendix D.] The factor 2 multiplying $\Delta X - \Delta X'$ accounts for the two possible ways w can form HBs with either 1 or 2. The last term on the rhs of (C.3) is the contribution due to the formation of simultaneous HBs between w and 1, 2, from the range of configurations $\Delta X'$. ΔX and $\Delta X'$ are estimated in Appendix A.

We define the solvent-induced HϕI interaction between 1 and 2 at this particular configuration by

$$\delta G^{HB/LJ}(1, 2) = \Delta G^{*HB}(1,2/LJ) - 2\Delta G^{*HB}(1/LJ). \tag{C.5}$$

This is the indirect work required to bring arms 1 and 2 (here assumed to be identical) from an infinite separation to the final configuration as in Fig. C.1a.

In section 7.7 we have estimated that

$$\Delta G^{*HB}(1/LJ) = -2.25 \text{ kcal/mol (experimental)}$$
$$= -2.5 \text{ kcal/mol (theoretical)}. \tag{C.6}$$

The estimate of $\delta G^{HB/LJ}(1,2)$ is either -3.2 kcal/mol or -2.7 kcal/mol, according to whether we choose the experimental or the theoretical estimates in (C.6), respectively. In Chapters 7 and 8 we use the average value of about

$$\delta G^{HB/LJ}(1, 2) \approx -3 \text{ kcal/mol}. \tag{C.7}$$

The qualitative reason for this large solvent-induced effect is the following: The conditional solvation Gibbs energy of one independent arm, say 1, is about -2.25 kcal/ mol. When this arm is brought to a distance of about 4.5 Å and to the correct orientation from a second arm 2, as in Fig. C.1a, the solvation of the first arm is enhanced because the second arm already, with high probability, orients the water molecule in the region

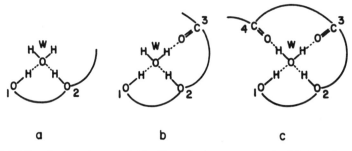

a   b   c

FIGURE C.1. (a) Two functional groups bridged by a single water molecule. (b) Three functional groups bridged by a single water molecule. (c) Four functional groups bridged by a single water molecule.

from which it can form a HB with arm 1. Mathematically, this can be written as

$$\exp[-\beta \delta G^{HB/LJ}(1, 2)] = \frac{\langle \exp(-\beta B_{1,2}^{HB}) \rangle_{LJ}}{\langle \exp(-\beta B_1^{HB}) \rangle_{LJ}^2} = \frac{\langle \exp(-\beta B_1^{HB}) \rangle_{LJ,2}}{\langle \exp(-\beta B_1^{HB}) \rangle_{LJ}}. \tag{C.8}$$

The difference in the two averages on the rhs of (C.8) is that the denominator has only the LJ condition, whereas the numerator has both the LJ and the presence of 2 as conditions.

Next, we estimate the solvent-induced interaction among three HϕI groups forming an equilateral triangle with an edge of 4.5 Å and oriented so that they can simultaneously form three HBs with one water molecule, as in Fig. C.1b. If we neglect quantities of order one, we have approximately, as in (C.3),

$$\exp[-\beta \Delta G^{*HB}(1, 2, 3/LJ)] = \langle \exp(-\beta B_{1,2,3}^{HB}) \rangle_{LJ}$$

$$= 1 + 3 \exp(-\beta \varepsilon_{HB}) 4 \rho_w \frac{2(\Delta \mathbf{X} - \Delta \mathbf{X}')}{8\pi^2}$$

$$+ \exp(-\beta 3 \varepsilon_{HB}) 4 \rho_w \frac{\Delta \mathbf{X}'}{8\pi^2}, \tag{C.9}$$

where the first term accounts for the three possible ways w can form a HB with one arm. The region could be either $\Delta \mathbf{X} - \Delta \mathbf{X}'$ or $\Delta \mathbf{X} - 2\Delta \mathbf{X}'$, depending on the type of arms (donors or acceptors). Once we are in the configurational range to form two HBs, the third arm of w will also be favorably oriented to form the third HB. Therefore we do not account separately for the formation of two HBs. The region $\Delta \mathbf{X}'$ is nearly the same as estimated for two HBs in Appendix A. Note that the three arms cannot be identical; i.e., there must be either two acceptors and one donor, or *vice versa*, to form three HBs with w. The factor 4 was taken as in the previous estimate, though it should be slightly larger in this case.

Thus we estimate that

$$\exp[-\beta \Delta G^{*HB}(1, 2, 3/LJ)] = 1 + 375 + 1.26 \times 10^{10} \tag{C.10}$$

or

$$\Delta G^{*HB}(1, 2, 3/LJ) = -13.8 \text{ kcal/mol}. \tag{C.11}$$

Therefore the triplet HϕI interaction is given by

$$\delta G^{HB/LJ} = \Delta G^{*HB}(1, 2, 3/LJ) - 3\Delta G^{*HB}(1/LJ)$$

$$\sim -13.8 + 3 \times 2.25$$

$$= -7.05 \text{ kcal/mol},$$

or

$$\delta G^{HB/LJ}(1, 2, 3) \approx -13.8 + 3 \times 2.5 = -6.3 \text{ kcal/mol}, \tag{C.12}$$

if we use the experimental and the theoretical results from section 7.7. The average value is −6.67 kcal/mol.

An interesting conclusion can be drawn from the following difference:

$$\Delta G^{*HB}(3/1, 2, LJ) \equiv \Delta G^{*HB}(1, 2, 3/LJ) - \Delta G^{*HB}(1, 2/LJ)$$

$$= -13.8 + 7.7 = -6.1 \text{ kcal/mol}. \tag{C.13}$$

This quantity is the conditional work of turning on arm 3, given that arms 1 and 2 are already turned on and in a favorable configuration, as in Fig. C.1b. Therefore, in this situation the solvation Gibbs energy of arm 3 is almost -6.1 kcal/mol. This means that arm 3 will, with very high probability, form a HB with w. Therefore, its conditional Gibbs energy of solvation is almost equal to the HB energy ε_{HB}.

Clearly, if we already have three arms 1, 2, 3 at the correct configuration and turn on a fourth arm so that all the four arms can form HBs with a single water molecule, we must have, almost with certainty,

$$\Delta G^{*HB}(4/1, 2, 3, LJ) \equiv \Delta G^{*HB}(1, 2, 3, 4/LJ) - \Delta G^{*HB}(1, 2, 3/LJ) = -6.5 \text{ kcal/mol},$$

$$(C.14)$$

from which we estimate

$$\Delta G^{*HB}(1, 2, 3, 4/LJ) = -6.5 - 13.8 = -20.3 \text{ kcal/mol}. \qquad (C.15)$$

Therefore, the HϕI interaction among the four arms at this configuration is

$$\delta G^{HB/LJ}(1, 2, 3, 4) = \Delta G^{*HB}(1, 2, 3, 4/LJ) - 4\Delta G^{*HB}(1/LJ)$$

$$= -20.3 + 4 \times 2.25$$

$$= -11.3 \text{ kcal/mol}$$

or

$$\delta G^{HB/LJ}(1, 2, 3, 4) = -20.3 + 4 \times 2.5 = -10.3 \text{ kcal/mol}, \qquad (C.16)$$

where the two values in (C.16) are for the experimental and the theoretical estimates from section 7.7.

Local Densities of Water Molecules near Hφl Groups Having a Fixed Orientation

The local density of water molecules at $\mathbf{R}_2$, given a water molecule at $\mathbf{R}_1$, is given by

$$\rho(\mathbf{R}_2/\mathbf{R}_1) = \rho_w g(\mathbf{R}_1, \mathbf{R}_2), \tag{D.1}$$

where g is the radial distribution function for water.

If the distance $R = |\mathbf{R}_2 - \mathbf{R}_1|$ is about 2.8 Å, then $g(R)$ is about 2; hence, the average local density at $\mathbf{R}_2$ is about

$$\rho(\mathbf{R}_2/\mathbf{R}_1) \approx 2\rho_w \sim 2 \times 5.55 \times 10^{-2}\,\text{mol/cm}^3, \tag{D.2}$$

where the bulk density of water at room temperature is $\rho_w = 5.55 \times 10^{-2}\,\text{mol/cm}^3$.

From simulation calculations it is found that the average local density of water at contact distance from a nonpolar molecule, say methane, is about $2\rho_w$. We consider the values of $g(R)$ cited above to be normal—normal compared with much larger effects discussed below.

The local density of water molecules at a distance of 2.8 Å and at a relative configuration such that each one can form a HB with either another water molecule or one arm of a functional group is

$$\rho(\mathbf{X}_w/\mathbf{X}_1) = \frac{\rho_w}{8\pi^2} g(\mathbf{X}_w, \mathbf{X}_1) = \frac{\rho_w}{8\pi^2} \exp[-\beta W(\mathbf{X}_w, \mathbf{X}_1)]. \tag{D.3}$$

We estimate the potential of average force at $\mathbf{X}_w$, $\mathbf{X}_1$, as follows:

$$
\begin{aligned}
W(\mathbf{X}_w, \mathbf{X}_1) &= U(\mathbf{X}_w, \mathbf{X}_1) + \delta G(\mathbf{X}_w, \mathbf{X}_1) \\
&= \varepsilon_{\text{HB}} + \varepsilon_{\text{LJ}} + \delta G^{\text{LJ}} + \delta G^{\text{HB/LJ}} \\
&= -6.5 - 0.8 + 2 \times 2.25 \\
&= -2.8\,\text{kcal/mol},
\end{aligned}
\tag{D.4}
$$

where we have chosen $\varepsilon_{\text{HB}} = -6.5\,\text{kcal/mol}$ for the HB energy. $\varepsilon_{\text{LJ}} + \delta G^{\text{LJ}}$ is the potential of average force for two Lennard–Jones particles in water. This value of about $-0.8\,\text{kcal/mol}$ was taken from simulation results of Ravishanker et al. (1982),† and

---

† G. Ravishanker, M. Mezei, and D. L. Beveridge, *Faraday Sym. Chem. Soc.* **17**, 79 (1982).

$\delta G^{HB/LJ}$ is the contribution due to the HBing capability of the two molecules. $\delta G^{HB/LJ}$ was estimated to be twice the loss of the conditional solvation Gibbs energy per arm (see section 7.7). Therefore, at this specific configuration we have

$$\rho(\mathbf{X}_w/\mathbf{X}_1) = \rho(\mathbf{X}_w)g(\mathbf{X}_w, \mathbf{X}_1) = \frac{113\rho_w}{8\pi^2}. \tag{D.5}$$

Thus, the local density at $\mathbf{X}_w$ due to the presence of one arm is about 113 times larger than the bulk density. Note that this value refers to a specific configuration of w with respect to 1. The probability of finding a water molecule in a correct configuration with respect to arm 1, so that it can form a HB with it, is

$$\rho(\mathbf{X}_w/\mathbf{X}_1)\frac{2\Delta\mathbf{X}}{8\pi^2} = \frac{113\rho_w 2\Delta\mathbf{X}}{8\pi^2}. \tag{D.6}$$

The factor 2 is introduced here because a water molecule can offer two different arms to form a HB with arm 1. $\rho_w\Delta\mathbf{X}/8\pi^2$ is the unconditional probability of finding a water molecule within the element of configurations $\Delta\mathbf{X}$. From Appendix A we have an estimate of $\Delta\mathbf{X}/8\pi^2$; hence, the probability in (D.6) is

$$\rho(\mathbf{X}_w/\mathbf{X}_1)\frac{2\Delta\mathbf{X}}{8\pi^2} \approx 113 \times 5.55 \times 10^{-2} \times 2 \times 5.1 \times 10^{-3} = 0.064. \tag{D.7}$$

Next, we compute the local density at $\mathbf{X}_w$ relative to two arms, say 1 and 2, at a distance of 4.5 Å from each other, in such a way that w can simultaneously form two HBs with 1 and 2. The work required to bring w from an infinite position to the final position, as in Fig. C.1a, is

$$W(\mathbf{X}_w|\mathbf{X}_1, \mathbf{X}_2) = -2 \times 6.5 - 2 \times 0.8 + 7.7 + 2.25 = -4.65 \text{ kcal/mol}. \tag{D.8}$$

Here $7.7 + 2.25$ is the loss of conditional solvation Gibbs energy of the pair of arms 1 and 2 and of one arm of w (section 7.7 and Appendix C). Hence the local density at $\mathbf{X}_w$, given 1 and 2, is

$$\rho(\mathbf{X}_w/\mathbf{X}_1, \mathbf{X}_2) - \rho(\mathbf{X}_w)g(\mathbf{X}_w|\mathbf{X}_1, \mathbf{X}_2)$$

$$= \frac{\rho_w}{8\pi^2}\exp\left(\frac{4.65}{kT}\right) = \frac{\rho_w}{8\pi^2}2.58 \times 10^3. \tag{D.9}$$

Thus, the local density at $\mathbf{X}_w$, given 1 and 2, is about 2.58×10^3 times larger than the bulk density at $\mathbf{X}_w$. The probability of finding a water molecule at a range of configurations such that it can simultaneously form two HBs with 1 and 2 is thus

$$\rho(\mathbf{X}_w/\mathbf{X}_1, \mathbf{X}_2)\frac{2\Delta\mathbf{X}'}{8\pi^2} \approx \frac{5.1 \times 10^{-3}}{18}2\rho_w 2.58 \times 10^3 = 0.08. \tag{D.10}$$

Note that again we have a factor of 2 since a water molecule can still form two HBs with 1 and 2 from two different configurations. Although the density at $\mathbf{X}_w$ is much higher [Eq. (D.9)], the range of configuration is now reduced from $\Delta\mathbf{X}$ to $\Delta\mathbf{X}'$ due to the restricted rotation about the O—O line. We estimate in Appendix A that

$$\Delta\mathbf{X}' \approx \Delta\mathbf{X}\frac{20}{360}. \tag{D.11}$$

The third case occurs when w is brought to a configuration such that it can simultaneously form three HBs with 1, 2, 3, as in Fig. C.1b. The corresponding work in this case is

$$W(\mathbf{X}_w|\mathbf{X}_1, \mathbf{X}_2, \mathbf{X}_3) = -3 \times 6.5 - 3 \times 0.8 + 13.8 + 2.25 = -5.85 \text{ kcal/mol}. \quad (\text{D}.12)$$

Hence the local density is now

$$\rho(\mathbf{X}_w/\mathbf{X}_1, \mathbf{X}_2, \mathbf{X}_3) = \rho(\mathbf{X}_w) \exp\left(\frac{5.85}{kT}\right) = 1.96 \times 10^4 \frac{\rho_w}{8\pi^2}. \quad (\text{D}.13)$$

The probability of finding a water molecule in a configuration in which it can simultaneously form three HBs with 1, 2 and 3 is therefore

$$\rho(\mathbf{X}_w/\mathbf{X}_1, \mathbf{X}_2, \mathbf{X}_3) \frac{\Delta\mathbf{X}'}{8\pi^2} \approx 0.31. \quad (\text{D}.14)$$

This is the conditional probability of finding a water molecule at a specific configuration within the range $\Delta\mathbf{X}'$. Here, in contrast to the former case, there is only one possibility that a water molecule can form three HBs with 1, 2 and 3; hence, we need not add a factor of 2. Of course, in order to form three HBs simultaneously, one must have either two donors and one acceptor or *vice versa* [see Fig. C.1b].

The result (D.14) is intuitively clear. With three arms pointing in the correct directions, say as the three next-nearest oxygens on the surface of ice, there is a large probability of finding a water molecule to form a triplet of HBs. Finally, by having four arms properly directed (say, by removing one water molecule from the interior of the ice structure), the local density of water molecules at that point will be about

$$\rho(\mathbf{X}_w/\mathbf{X}_1, \mathbf{X}_2, \mathbf{X}_3, \mathbf{X}_4) = \rho(\mathbf{X}_w) \exp\left(\frac{6.65}{kT}\right) \approx 7.5 \times 10^4 \frac{\rho_w}{8\pi^2}. \quad (\text{D}.15)$$

This value is uncertain. It is based on the estimate of the work required to bring w, given 1, 2, 3, 4, which is roughly

$$W(\mathbf{X}_w|\mathbf{X}_1, \mathbf{X}_2, \mathbf{X}_3, \mathbf{X}_4) = -4 \times 6.5 - 4 \times 0.8 + 20.3 + 2.25 = -6.65 \text{ kcal/mol}. \quad (\text{D}.16)$$

If we calculate the probability of finding a water molecule at $\mathbf{X}_w$ such that it can form four simultaneous HBs with 1, 2, 3, and 4, we obtain from (D.15) the value of 1.18. This calculation is based on the estimate (D.16) and on the same $\Delta\mathbf{X}'$ as in Appendix B. We consider this probability to be practically unity. Having four arms properly directed to form four HBs with a single molecule (Fig. C.1c), we should find, almost with certainty, a water molecule able to form four bonds. This situation can also occur in the interior of proteins.

The Chemical Potential in Various Ensembles

In section 5.9 we derived the statistical mechanical expression for the chemical potential in the T, V, N ensemble. Using the definition of the chemical potential

$$\mu = \left(\frac{\partial A}{\partial N}\right)_{T,V} = A(T, V, N + 1) - A(T, V, N), \qquad (E.1)$$

we obtained the general result

$$\mu = \mu^* + kT \ln \rho \Lambda^3, \qquad (E.2)$$

where $\rho = N/V$ and μ^* is the pseudochemical potential.

Similarly in the T, P, N ensemble we use the equivalent definition of the chemical potential

$$\mu = \left(\frac{\partial G}{\partial N}\right)_{T,P} = G(T, P, N + 1) - G(T, P, N) \qquad (E.3)$$

to obtain the same expression as in (E.2), but with a reinterpretation of the density ρ as

$$\rho = N/\langle V \rangle, \qquad (E.4)$$

where $\langle V \rangle$ is the average volume in the T, P, N ensemble.

From the two equalities (E.1) and (E.2), it follows that if we choose two systems characterized by the variables T, P, N and T, V, N, respectively, such that the average volume $\langle V \rangle$ in the former is equal to the volume in the latter, the two chemical potentials in Eqs. (E.1) and (E.3) will have equal values. (The same is true if we require that the pressure of the T, V, N ensemble be the same as the pressure in the T, P, N ensemble.)

If we require that $V = \langle V \rangle$, then the densities $\rho = N/V$ in Eq. (E.2) and $\rho = N/\langle V \rangle$ in Eq. (E.4) will also be the same. Hence, the liberation Helmholtz energy and the liberation Gibbs energy are equal, and therefore the pseudochemical potentials in these two systems are also equal:

$$\mu^*(T, V, N) = \mu^*(T, P, N), \qquad (E.5)$$

with the requirement that $V = \langle V \rangle$.

It also follows that the solvation Gibbs energy (at T, P, N constant) is equal to the solvation Helmholtz energy (at T, V, N with $V = \langle V \rangle$); i.e.,

$$\Delta G^*(T, P, N) = \Delta A^*(T, V, N) \qquad \text{with } V = \langle V \rangle. \tag{E.6}$$

We note, however, that ΔG^* and ΔA^* in this equation do not pertain to the same process of solvation. If we restrict ourselves to the same process, say at T, P, N constant, then we have the equality

$$\Delta G^*(T, P, N) = \Delta A^*(T, P, N) + P \Delta V^*(T, P, N). \tag{E.7}$$

Thus ΔG^* and ΔA^* for the same process are, in general, different quantities. In all our examples this difference is quite small, as shown in Appendix F.

Similarly, for the T, V, μ ensemble one can obtain the same formal expression for the chemical potential as in (E.2), but now the density ρ is reinterpreted as

$$\rho = \frac{\langle N \rangle}{V}, \tag{E.8}$$

where $\langle N \rangle$ is the average number of particles in the T, V, μ ensemble.[†] Note that μ in this case is one of the independent variables used to characterize the system.

[†] A. Ben-Naim, *Solvation Thermodynamics*, Plenum Press, New York (1987).

Estimates of the $P\Delta V_s^*$ Term for Some Simple Solvation Processes

In several places throughout this book we stated that ΔG_s^* and ΔA_s^* for the same solvation process defined in section 6.16 are almost identical. Likewise, the difference between ΔH_s^* and ΔE_s^* was said to be negligible. The exact differences between these quantities are

$$\Delta G_s^* = \Delta A_s^* + P\Delta V_s^* \tag{F.1}$$

and

$$\Delta H_s^* = \Delta E_s^* + P\Delta V_s^*. \tag{F.2}$$

We now estimate the values of $P\Delta V_s^*$ for three cases.†

   1. Solvation of hexane in pure hexane (at 25°C). The molar volume of hexane is $V_m = 130.3$ cm^3 mol^{-1}; hence, the solvation volume of hexane in pure hexane is

$$\Delta V_s^* = V_m - RT\kappa_T = 130.3 - 3.97 = 126.3 \text{ cm}^3 \text{ mol}^{-1}, \tag{F.3}$$

and $P\Delta V_s^*$ at 1 atm is (1 cm^3 atm $= 0.1013$ J)

$$P\Delta V_s^* = 12.79 \times 10^{-3} \text{ kJ mol}^{-1}. \tag{F.4}$$

The Gibbs energy of solvation of hexane in pure hexane is

$$\Delta G_s^* = 11.00 \text{ kJ mol}^{-1}. \tag{F.5}$$

Therefore, $P\Delta V_s^*$ is quite negligible compared with ΔG_s^*.

   2. Solvation of water in pure water (at 25°C). For pure water the molar volume is $V_m = 18$ cm^3 mol^{-1}; hence,

$$\Delta V_s^* = 18 - 1.13 = 16.87 \text{ cm}^3 \text{ mol}^{-1}, \tag{F.6}$$

and

$$P\Delta V_s^* = 1.71 \times 10^{-3} \text{ kJ mol}^{-1}. \tag{F.7}$$

† A. Ben-Naim, *Solvation Thermodynamics*, Plenum Press, New York (1987).

This should be compared with the solvation Gibbs energy of water in pure water given by (see section 7.7)

$$\Delta G_s^* = -26.5 \text{ kJ mol}^{-1}. \tag{F.8}$$

3. *Solvation of methane in water.* In this case we have

$$\Delta V_s^* = 50 \text{ cm}^3 \text{ mol}^{-1}, \tag{F.9}$$

$$P\Delta V_s^* = 5.06 \times 10^{-3} \text{ kJ mol}^{-1}, \tag{F.10}$$

and

$$\Delta G_s^* = 8.33 \text{ kJ mol}^{-1}. \tag{F.11}$$

Finally, we note that, although ΔG_s^* and ΔA_s^* have almost the same values for the process discussed in this appendix, they are different quantities. On the other hand, in Appendix E we discussed the identity of ΔG_s^* (at given T and P) and ΔA_s^* (at given T and V). These two quantities have the same value provided $\langle V \rangle$ in the T, P system is equal to V in the T, V system.

Transferability of the Conditional Solvation Gibbs Energy

The principle of transferability is commonly used in the construction of the intramolecular potential function of a macromolecule. It has been recently used to construct intermolecular interactions or solute–solvent interaction.[†] The main idea is to transfer the parameters describing the interaction between small molecules, e.g., methane and water, on to larger molecules, say methane–ethanol, or ethane–water. In this book we used a similar idea to extract information from small model compounds and apply it to biopolymers. The information we are interested in is the conditional solvation Gibbs energies of various groups, e.g., methyl, ethyl, hydroxyl, and so on, and intramolecular solvent-induced interactions between such groups. In this appendix we describe the methodology of this transferability principle and examine its adequacy and extent of its reliability.

Consider first an OH group attached to a protein α. The conditional solvation Gibbs energy of OH, given the protein, is obtained from the exact relation between the solvation Gibbs energy of the protein with OH, which we denote by $\alpha + \text{OH}$, and the protein α from which the OH was cut off, i.e.,

$$
\begin{aligned}
\Delta G^*_{\alpha + \text{OH}} &= -kT \ln \langle \exp(-\beta B_\alpha - \beta B_{\text{OH}}) \rangle_0 \\
&= -kT \ln \langle \exp(-\beta B_\alpha) \rangle_0 \langle \exp(-\beta B_{\text{OH}}) \rangle_\alpha \\
&= \Delta G^*_\alpha + \Delta G^{*\text{OH}/\alpha}_\alpha .
\end{aligned}
\tag{G.1}
$$

This is an exact relation. The first approximation we make is to assume that the solvation Gibbs energy of the radical α is nearly the same as of the molecule $\alpha + \text{H}$ ($\alpha + \text{H}$ is the same protein, $\alpha + \text{OH}$, in which the OH has been replaced by H); i.e.,

$$
\Delta G^*_\alpha \approx \Delta G^*_{\alpha + \text{H}} .
\tag{G.2}
$$

This is indeed a good approximation, as shown at the end of this appendix. With this approximation, the required conditional quantity is given by

$$
\Delta G^{*\text{OH}/\alpha}_\alpha = \Delta G^*_{\alpha + \text{OH}} - \Delta G^*_\alpha \approx \Delta G^*_{\alpha + \text{OH}} - \Delta G^*_{\alpha + \text{H}} .
\tag{G.3}
$$

Note that now we have, on the rhs of (G.3), real molecules for which the solvation Gibbs energies are well-defined quantities. Unfortunately, if α is a large protein, these will not be measurable quantities. However, we note that from the definition of $\Delta G^{*\text{OH}/\alpha}_\alpha$ in (G.1),

[†] F. T. Marchese, P. K. Mehrotra, and D. L. Beveridge, *J. Phys. Chem.* **86**, 2592 (1982); and A. Ben-Naim, *J. Phys. Chem.* **63**, 2064 (1975).

it follows that if we cut a small piece of the protein in the neighborhood of the OH group, then the surroundings of the OH should be similar to the surroundings around OH in the real protein. We shall denote this piece of the protein α', for which we have

$$\Delta G_{\alpha'}^{*\text{OH}/\alpha'} = \Delta G_{\alpha'+\text{OH}}^{*} - \Delta G_{\alpha'}^{*}. \tag{G.4}$$

Clearly, if α' is large enough so that it produces the same surroundings for OH as in α, then the quantity defined in (G.4) should be the same as the quantity we need in (G.3). On the other hand, α' must be small enough so that both quantities on the rhs of (G.4) are measurable. A schematic description of the surroundings of a side chain R_i is shown in Fig. 8.11.

How well does the quantity obtained from the model compound α' represent the required quantity for the real protein? Clearly since we do not have the relevant experimental quantities on the rhs of (G.3), the use of $\Delta G_{\alpha'}^{*\text{OH}/\alpha'}$ instead of $\Delta G_{\alpha}^{*\text{OH}/\alpha}$ relies only on the theoretical argument given above. In order to assess the nature of this approximation we shall use the following trick. Instead of a protein we choose a small model compound α. We then use an even smaller model compound to estimate the quantity $\Delta G_{\alpha}^{*\text{OH}/\alpha}$. Thus, in essence, α' is a model compound used to study the conditional solvation of OH in α, where α is itself a model compound for the protein.

Specifically, let cyclohexanol be a model for our protein $\alpha + \text{OH}$, and suppose for the moment that we do not know the solvation quantities for $\alpha + \text{OH}$ and α. As in the case of proteins, we could not have computed $\Delta G_{\alpha}^{*\text{OH}/\alpha}$. Therefore, we try some small model compounds α'. For primary alcohols, we find that $\Delta G_{\alpha'}^{*\text{OH}/\alpha'}$ has values in the range between -6.78 and -7.11 kcal/mol. A better model would be a secondary alcohol for which we get values between -6.71 and -6.66 kcal/mol. How good are these values? Fortunately, we do have the solvation quantities for our protein, which is represented by the cyclohexanol. These are between -6.70 to -6.69 kcal/mol. Therefore, we conclude that measuring the conditional solvation Gibbs energy of an OH near a small model compound of only a few carbon atoms is enough to mimic the surroundings of an OH group in a larger molecule such as a cyclohexanol or cyclopentanol. In other words, if the solvation quantities for cyclohexanol and cyclohexane were not available, we could have safely used the smaller model compounds to obtain the required values relevant to cyclohexane. The pertinent experimental data is shown in Fig. G.1.

The same methodology is used in Fig. G.2 to calculate the conditional solvation Gibbs energy of a methyl group. Here, because of the smaller absolute values of $\Delta G_{\alpha}^{*\text{CH}_3/\alpha}$, we cannot expect to get very accurate results. Nevertheless, the values obtained for a secondary methyl are quite close to the values corresponding to a methyl attached to a ring.

The two demonstrations given above are sufficient to lend confidence to the methodology of using small model compounds to represent the surroundings of a functional group or a side chain in real proteins.

Figure G.3 shows further data for a methyl group attached to an aromatic molecule. Here the conditional solvation Gibbs energy of the methyl is quite small compared with the solvation Gibbs energies of the molecules themselves. We note, however, that attaching a methyl group to a small model compound next to a double bond gives values which are surprisingly of the same order of magnitude as in the aromatic compound. On the other hand, if the double bond is one bond away from the methyl group, then we almost recover the corresponding values of a methyl group attached to an unsaturated hydrocarbon.

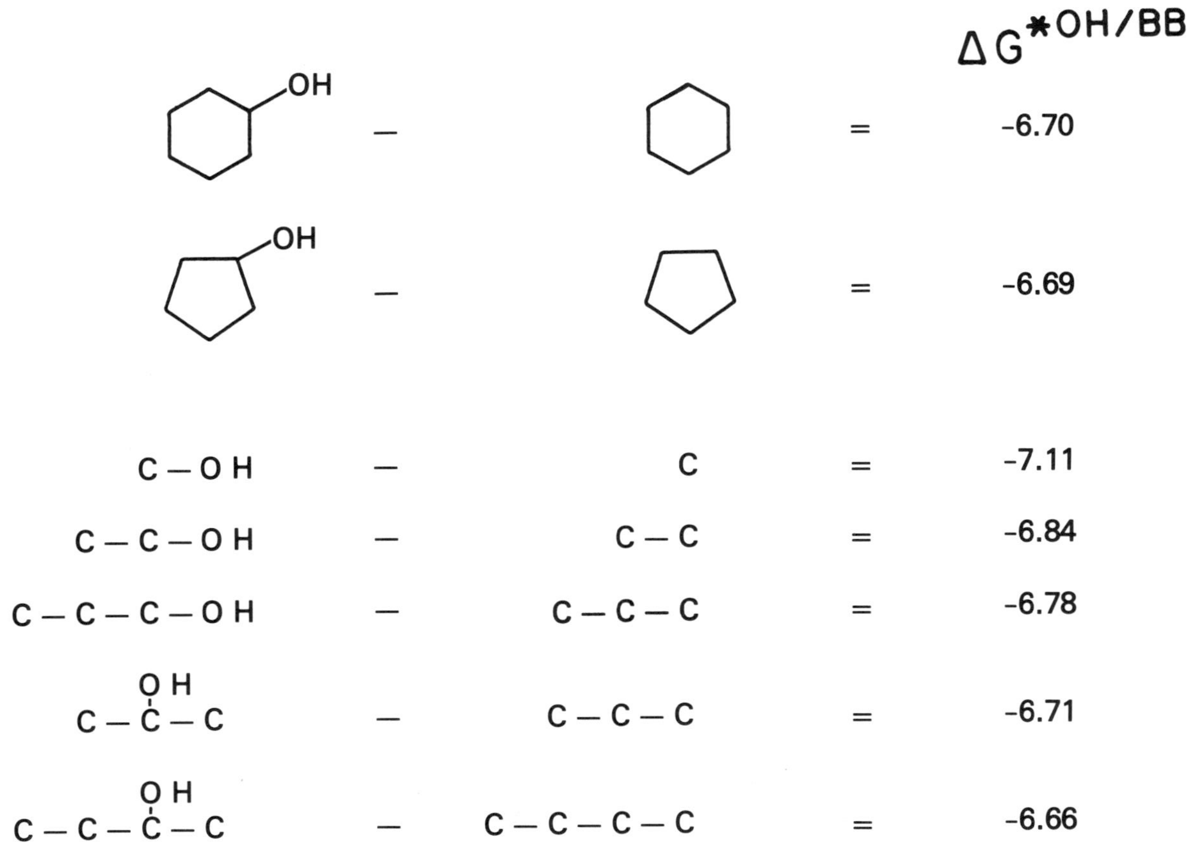

			ΔG^{*}OH/BB

structure		value
cyclohexanol	=	-6.70
cyclopentanol	=	-6.69
C—OH	= C	-7.11
C—C—OH	= C—C	-6.84
C—C—C—OH	= C—C—C	-6.78
$\overset{O\,H}{C-C-C}$	= C—C—C	-6.71
$\overset{O\,H}{C-C-C-C}$	= C—C—C—C	-6.66

FIGURE G.1. Model compounds used to estimate the conditional solvation Gibbs energy of a hydroxyl group. Based on Eq. (G.4) all values are in kcal/mol at 25°C.

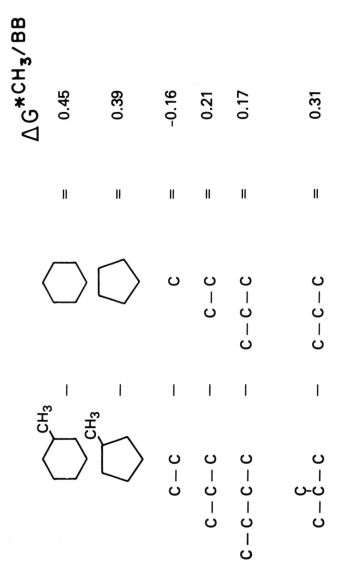

$\Delta G*CH_3/BB$

		$\Delta G*CH_3/BB$

FIGURE G.2. Same as Fig. G.1 for methyl group attached to a hydrocarbon.

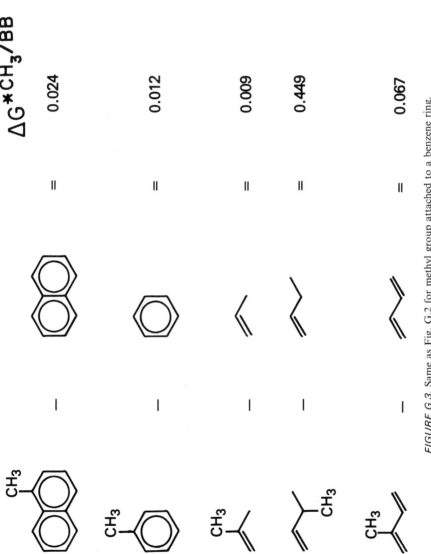

$\Delta G^{*CH_3}/BB$

0.024

0.012

0.009

0.449

0.067

FIGURE G.3. Same as Fig. G.2 for methyl group attached to a benzene ring.

Finally we examine the approximation introduced in (G.2). Clearly, for a large protein, the difference between ΔG_a^* and $\Delta G_{a+\text{H}}^*$ will be negligibly small. For smaller compounds, say normal hydrocarbons, we know that the solvation Gibbs energies of hydrocarbons with n carbon atoms is linear in n, i.e.,

$$\Delta G_n^* = 1.42 + 0.18n. \tag{G.5}$$

The increment per methylene group is about 0.18 kcal/mol. Clearly, the increment per hydrogen atom is smaller than 0.18, for example, one fourth of this value, considering the radius of a hydrogen atom relative to a methylene group. This approximation is introduced in our calculation for each amino-acid unit on the backbone. If we take the solvation free energy of propane [$n = 3$ in (G.5)] we have $\Delta G_3^* \sim 1.96$ kcal/mol. Hence, detaching one hydrogen atom from propane would change its solvation free energy by about 2.3%.

Selected Values of the Conditional Solvation Gibbs Energies of Some Groups Attached to Small Model Compounds

In section 8.7 we discussed two possible ways of dividing the solute–solvent pair interaction [see Eqs. (8.7.11) and (8.7.12)]. Corresponding to these two divisions we have

$$\Delta G_{\alpha+\text{OH}}^{*\text{OH}/\alpha} = \Delta G_{\alpha+\text{OH}}^{*} - \Delta G_{\alpha+\text{H}}^{*} \tag{H.1}$$

$$\Delta G_{\alpha+\text{OH}}^{*\text{HB}/H,S} = \Delta G_{\alpha+\text{OH}}^{*} - \Delta G_{\alpha+\text{OH}}^{*H,S}. \tag{H.2}$$

In the first case, we start with the backbone α (α is the protein from which the OH was cut off; $\alpha+\text{H}$ is obtained from $\alpha+\text{OH}$ by replacing the OH by a hydrogen. We assume that the solvation of α and $\alpha+\text{H}$ are nearly the same; see Appendix G) and "turn on" the entire OH group. In the second case we start with the entire molecule $\alpha+\text{OH}$, but with the HBing capacity turned off, and then turn on the HBing capacity only.

In section 7.7 we estimated the conditional solvation Gibbs energy of the HBing part of water molecules as

$$\Delta G_w^{*\text{HB}/H,S} \approx \Delta G_w^{*} - \Delta G_{\text{Ne}}^{*} = -9.00 \text{ kcal/mol}. \tag{H.3}$$

Since a water molecule has four arms, and assuming that these arms are independently solvated (see section 7.7 and Appendix B), then

$$\Delta G_{\text{one arm}}^{*\text{HB}/H,S} = -2.25 \text{ kcal/mol}. \tag{H.4}$$

An alcohol molecule can be viewed as a water mole, one of the arms of which is replaced by the radical α. Hence, the alcohol has three arms to form HBs. Based on the estimate (H.4), we can write

$$\Delta G_{\alpha+\text{OH}}^{*\text{HB}/H,S} \sim -3 \times 2.25 = -6.75 \text{ kcal/mol}. \tag{H.5}$$

We now use some model compounds to estimate the two quantities on the lhs of (H.1) and (H.2)

$$\Delta G_{\text{CH}_3\text{CH}_2\text{OH}}^{*\text{OH}/\text{CH}_3\text{CH}_2} = \Delta G_{\text{CH}_3\text{CH}_2\text{OH}}^{*} - \Delta G_{\text{CH}_3\text{CH}_3}^{*} = -6.65 \text{ kcal/mol} \tag{H.6}$$

$$\Delta G_{\text{CH}_3\text{CH}_2\text{OH}}^{*\text{HB}/H,S} = \Delta G_{\text{CH}_3\text{CH}_2\text{OH}}^{*} - \Delta G_{\text{CH}_3\text{CH}_2\text{CH}_3}^{*} = -6.78 \text{ kcal/mol}. \tag{H.7}$$

In the first estimate, we attached an OH to ethane, i.e., we replaced a H by an OH. In the second estimate, we replaced a CH_3 by an OH. We see that there is not much difference between the two quantities; the average of these two is -6.71 kcal/mol, which is very close to the estimate based on three solvated arms (H.5).

For a carbonyl group we have similarly

$$\Delta G^{*C=O/H,S}_{\text{butanol}} \approx \Delta G^*_{CH_3CH_2CH_2CHO} - \Delta G^*_{CH_3CH_2CH_2CH_3} = -5.25 \text{ kcal/mol} \quad \text{(H.8)}$$

$$\Delta G^{*C=O/\text{propane}}_{\text{butanol}} \approx \Delta G^*_{CH_3CH_2CH_2CHO} - \Delta G^*_{CH_3CH_2CH_3} = -5.13 \text{ kcal/mol}. \quad \text{(H.9)}$$

Again we see that the values are similar whether we replace a CH_3 by a CHO or a H by a CHO. These values are somewhat larger than the values we would have estimated for two solvated arms, about -4.5 kcal/mol, based on (H.4). This is probably due to the fact that a carbonyl group forms stronger HBs as an acceptor group.

We note also that replacing a methylene group by oxygen to form an ether gives a value of about $-2 \times 2.25 \sim -4.5$ kcal/mol; for instance,

$$\Delta G^*_{CH_3-O-CH_3} - \Delta G^*_{CH_3CH_2CH_3} = -3.85 \text{ kcal/mol} \quad \text{(H.10)}$$

$$\Delta G^*_{CH_3CH_2-O-CH_2CH_3} - \Delta G^*_{CH_3CH_2CH_2CH_2CH_3} = -3.97 \text{ kcal/mol} \quad \text{(H.11)}$$

$$\Delta G^*_{CH_3CH_2CH_2-O-CH_2CH_2CH_3} - \Delta G^*_{CH_3CH_2CH_2CH_2CH_2CH_2CH_3} = -4.3 \text{ kcal/mol}. \quad \text{(H.12)}$$

For attaching a carboxylic group, we have

$$\Delta G^*_{CH_3COOH} - \Delta G^*_{CH_4} = -8.7 \text{ kcal/mol} \quad \text{(H.13)}$$

$$\Delta G^*_{CH_3CH_2COOH} - \Delta G^*_{CH_3CH_3} = -8.3 \text{ kcal/mol} \quad \text{(H.14)}$$

$$\Delta G^*_{CH_3CH_2CH_2COOH} - \Delta G^*_{CH_3CH_2CH_3} = -8.3 \text{ kcal/mol}. \quad \text{(H.15)}$$

This corresponds roughly to the addition of four arms. The reason is that the solvation of the arms of the carbonyl probably interfered with the solvation of the arms of the hydroxyl. Therefore we cannot expect to obtain the sum of the solvation of five independent arms of OH and C=O.

In protein folding, the NH of the backbone is an important hydrogen-bond donor. We have not found a small model compound to mimic the environment of the amide group. We have assumed that the HB conditional Gibbs energy of solvation of NH is about -2.25 kcal/mol.

Correlation Functions in the Canonical and Grand Canonical Ensembles

In Chapter 3 we used correlation functions that were derived from the grand ensembles but interpreted as being defined in the canonical ensemble. Here, we illustrate the relation between the two derivations. We use as an example the model treated in section 3.2.

We start with a closed system of M polymers, as described in section 3.2. The polymers are identical, distinguishable, and independent, each having two equivalent sites. The entire system contains exactly two ligands, i.e., $N = 2$. The corresponding canonical PF is

$$Q(T, M, N = 2) = \sum_{\substack{\text{all states}}} \exp(-\beta E)$$

$$= \sum_{\substack{\text{all states} \\ (1,1)}} + \sum_{\substack{\text{all states} \\ (0,1) \text{ or } (1,0)}}$$

$$= MQ(1,1)Q(0,0)^{M-1} + M(M-1)Q(1,0)^2 Q(0,0)^{M-2}. \qquad \text{(I.1)}$$

There are altogether M^2 states. These are grouped into two groups. There are M states for which the two ligands are on the same polymer [denoted by $(1,1)$], and there are $M(M-1)$ states for which the two ligands are on separated polymers [denoted by $(0,1)$ or $(1,0)$].

Note that $Q(T, M, N = 2)$ pertains to the entire system $(T, M, N = 2)$, whereas $Q(1,1)$ and $Q(0,0)$ pertain to a single polymer.

We write the following probabilities in this system:

$$\text{Pr(one specific polymer doubly occupied)} = \frac{Q(1,1)Q(0,0)^{M-1}}{Q(T, M, N = 2)} \qquad \text{(I.2)}$$

$$\text{Pr(two different specific polymers singly occupied)} = \frac{Q(1,0)^2 Q(0,0)^{M-2}}{Q(T, M, N = 2)} \qquad \text{(I.3)}$$

$$\text{Pr(any polymer doubly occupied)} = \frac{MQ(1,1)Q(0,0)^{M-1}}{Q(T, M, N = 2)} \qquad \text{(I.4)}$$

$$\text{Pr(any two different polymers singly occupied)} = \frac{M(M-1)Q(1,0)^2 Q(0,0)^{M-2}}{Q(T, M, N = 2)}.$$

$$\text{(I.5)}$$

Note that by "singly occupied" we refer to a specific site, say the lhs site of the polymer.

The limiting behavior of these probabilities when $M \to \infty$ (i.e., $N/M \to 0$) is the following:

$$\text{Pr[one specific } (1,1)] \to O\left(\frac{1}{M^2}\right) \tag{I.6}$$

$$\text{Pr[two specific } (1,0)] \to O\left(\frac{1}{M^2}\right) \tag{I.7}$$

$$\text{Pr[any one } (1,1)] \to O\left(\frac{1}{M}\right) \tag{I.8}$$

$$\text{Pr[any two } (1,0)] \to 1, \tag{I.9}$$

where we used a shorthand notation for the same events as in (I.2)–(I.5). For instance, the probability of finding one specific polymer doubly occupied tends to zero as M^{-2}, whereas the probability of finding any two polymers singly occupied tends to 1.

Clearly, if we define the correlation function by

$$\bar{g}(1,1) = \frac{\text{Pr[any one } (1,1)]}{\text{Pr[any two } (1,0)]} \to 0. \tag{I.10}$$

This will tend to zero as M^{-1}. However, if we use the probabilities of the specific events as in (I.2) and (I.3), we have

$$g^0(1,1) = \frac{\text{Pr[one specific } (1,1)]}{\text{Pr[two specific } (1,0)]} = \frac{Q(1,1)Q(0,0)}{[Q(1,0)]^2}. \tag{I.11}$$

This is a constant independent of M. This quantity has been used in Chapter 3 for the ligand–ligand correlation. Clearly, this correlation is related to the work of bringing two ligands from two specific singly occupied polymers to form one doubly occupied specific polymer.

These probabilities can be derived from the grand PF as follows. If we open the system with respect to ligands, the corresponding grand PF is

$$\Xi(T, M, \mu) = \sum_{N=0}^{2M} Q(T, M, N)\lambda^N. \tag{I.12}$$

The probability of finding a system in the grand ensemble with exactly two ligands is

$$\text{Pr}(N=2) = \frac{Q(T, M, N=2)\lambda^2}{\Xi(T, M, \mu)}. \tag{I.13}$$

The probability of finding a system, in the grand ensemble, having precisely two ligands and that these two being situated on a specific polymer is

$$\text{Pr}[N=2, \text{ and } (1,1) \text{ on a specific polymer}] = \frac{Q(1,1)Q(0,0)^{M-1}\lambda^2}{\Xi}, \tag{I.14}$$

and similarly

$$\text{Pr}[N=2, \text{ and } (1,0)^2 \text{ on two specific polymers}] = \frac{Q(1,0)^2 Q(0,0)^{M-2}\lambda^2}{\Xi}. \tag{I.15}$$

The conditional probability of finding (1,1) on a specific polymer, given that there are exactly two ligands in the entire system, is obtained from (I.14) and (I.15):

Pr[(1,1) on a specific polymer, given $N = 2$]

$$= \frac{Pr[N = 2, \text{ and } (1,1) \text{ on a specific polymer}]}{Pr(N = 2)}$$

$$= \frac{Q(1,1)Q(0,0)^{M-1}}{Q(T, M, N = 2)}, \tag{I.16}$$

and similarly

Pr[$(0,1)^2$ on two specific polymers, given $N = 2$]

$$= \frac{Pr[N = 2, \text{ and } (0,1)^2 \text{ on two specific polymers}]}{Pr(N = 2)}$$

$$= \frac{Q(1,0)^2 Q(0,0)^{M-2}}{Q(T, M, N = 2)}. \tag{I.17}$$

Comparing (I.16) with (I.2) and (I.17) with (I.3), we conclude that the probabilities computed in the canonical ensemble with a fixed value of N (here $N = 2$) as in (I.2) and (I.3) are the same as the conditional probabilities computed in the grand ensemble, if the condition specifies the value of N (here $N = 2$).

A Simplified Expression for the Derivative of the Chemical Potential with Respect to the Number of Particles

In several places throughout the book we used the expression for the derivatives

$$\mu_{ij} = \left(\frac{\partial \mu_i}{\partial N_j}\right)_{T,P,N_k(k \neq j)} . \tag{J.1}$$

For a c-component system at a given T and P the general expression for μ_{ij} based on the Kirkwood–Buff theory is (see section 6.7)

$$\mu_{ij} = \frac{kT}{V|\mathbf{B}|} \frac{B^{ij} \sum\limits_{\alpha,\beta} \rho_\alpha \rho_\beta B^{\alpha\beta} - \sum\limits_{\alpha,\beta} \rho_\alpha \rho_\beta B^{\alpha j} B^{i\beta}}{\sum\limits_{\alpha\beta} \rho_\alpha \rho_\beta B^{\alpha\beta}}, \tag{J.2}$$

where $\rho_\alpha = N_\alpha/V$ is the average density of α in the open system. The summations extend over all species in the system $\alpha, \beta = 1, 2, 3, \ldots, c$.

The matrix $\mathbf{B}$ is constructed from the elements

$$B_{ij} = \rho_i \rho_j G_{ij} + \rho_i \delta_{ij}, \tag{J.3}$$

where G_{ij} are defined by

$$G_{ij} = \int_0^\infty [g_{ij}(R) - 1] 4\pi R^2 \, dR \tag{J.4}$$

and $g_{ij}(R)$ is the pair correlation function for the pair of species i and j.

The quantity B^{ij} denotes the cofactor of the element B_{ij} in the determinant $|\mathbf{B}|$. Namely, B^{ij} is a determinant obtained from $|\mathbf{B}|$ by deleting the ith row and the jth column, and the result is multiplied by the sign $(-1)^{i+j}$.

Relation (J.2) is quite simple for $c = 2$ and becomes somewhat cumbersome for $c = 3$. For $c \geq 3$, the application of (J.2) is impractical since it requires handling of a large sum of determinants, each of which when fully expanded is itself a sum of a large number of terms.

Fortunately, there exists a simplification of this expression which we shall quote here. The proof is available elsewhere.†
Define the matrix **G** by

$$\mathbf{G} = \begin{pmatrix} G_{11} + \rho_1^{-1} & G_{12} & G_{13} \cdots \\ G_{21} & G_{22} + \rho_2^{-1} & G_{23} \cdots \\ \vdots & \vdots & \vdots \end{pmatrix}. \tag{J.5}$$

Note that the elements of this matrix are

$$(\mathbf{G})_{ij} = G_{ij} + \delta_{ij}\rho_i^{-1}, \tag{J.6}$$

where G_{ij} are given by (J.4). Thus, care must be exercised to distinguish between G_{ij} and $(\mathbf{G})_{ij}$. Note also that **G** is a symmetric matrix.
By extracting ρ_i from the ith row and ρ_j from the jth column, one can easily get the following relations:

$$|\mathbf{B}| = \boldsymbol{\rho}^2|\mathbf{G}| \tag{J.7}$$

$$B^{ij} = \frac{\boldsymbol{\rho}^2 G^{ij}}{\rho_i \rho_j} \tag{J.8}$$

$$\sum_{\alpha,\beta=1}^{c} \rho_\alpha \rho_\beta B^{\alpha j} B^{i\beta} = \boldsymbol{\rho}^4 \sum_{\alpha,\beta} \frac{G^{\alpha j} G^{i\beta}}{\rho_i \rho_j} \tag{J.9}$$

$$\sum_{\alpha,\beta=1}^{c} \rho_\alpha \rho_\beta B^{\alpha\beta} = \boldsymbol{\rho}^2 \sum_{\alpha,\beta} G^{\alpha\beta}, \tag{J.10}$$

where we have denoted the product of all the densities by

$$\boldsymbol{\rho} = \prod_{i=1}^{c} \rho_i. \tag{J.11}$$

Substituting (J.7)–(J.8) into (J.2), we get‡

$$\mu_{hk} = \frac{kT}{\rho_h \rho_k V |\mathbf{G}|} \frac{\sum_{\alpha,\beta} [G^{hk} G^{\alpha\beta} - G^{h\beta} G^{\alpha k}]}{\sum_{\alpha,\beta} G^{\alpha\beta}}. \tag{J.12}$$

This is a more convenient form than (J.2), since we have eliminated all the ρ_is under the summation signs. (Note, however, that all the G^{ij}s are still dependent on the densities.)
Some of the terms in the numerator of (J.12) vanish. More specifically,

$$G^{hk} G^{\alpha\beta} - G^{h\beta} G^{\alpha k} = 0 \quad \text{for } \alpha = h \text{ (and any } \beta) \quad \text{or} \quad \text{for } \beta = k \text{ (and any } \alpha). \tag{J.13}$$

The nonvanishing terms can be viewed as minors of the adjoint determinant of $|\mathbf{G}|$; i.e.,

$$\begin{vmatrix} G^{hk} & G^{h\beta} \\ G^{\alpha k} & G^{\alpha\beta} \end{vmatrix} \tag{J.14}$$

---

† A. Ben-Naim, *J. Chem. Phys.* **63**, 2064 (1975).
‡ Note that we use k for the Boltzmann constant as well as a subscript indicating a species.

is a minor of order 2 in the adjoint determinant of $|\mathbf{G}|$ provided that the indices h, k, α, B fulfill certain conditions. This observation led to the application of Jacobi's theorem for such minors. The details of the algebraic steps are presented elsewhere.†

The final result for the quantity μ_{hk} in (J.2) is

$$\mu_{hk} = \frac{kT}{\rho_h \rho_k V} \frac{|\mathbf{E}(h, k)|}{|\mathbf{D}|}. \tag{J.15}$$

Here $\mathbf{D}$ is a matrix of order $c + 1$ constructed from $\mathbf{G}$ by appending a row and a column of unities, except for the element D_{11}, which is zero, namely,

$$\mathbf{D} = \begin{pmatrix} 0 & 1 & 1 & 1 & \cdots \\ 1 & & & & \\ 1 & & \mathbf{G} & & \\ \vdots & & & & \end{pmatrix}. \tag{J.16}$$

The general element of $\mathbf{D}$ is

$$D_{nm} = \delta_{n1} + \delta_{m1} - 2\delta_{n1}\delta_{m1} + (\mathbf{G})_{n-1, m-1} \qquad n, m = 1, 2, \ldots, c + 1. \tag{J.17}$$

Note that $(\mathbf{G})_{ij}$ is defined in Eq. 3.6 for $i, j = 1, 2, \ldots, c$, and we put $(\mathbf{G})_{ij} = 0$ for any other indices in (J.17).

The determinant in the numerator of (J.15) is constructed from the matrix $\mathbf{G}$ by replacing the hth row and the kth column of $|\mathbf{G}|$ by unities, except for the element hk, which is replaced by zero. The general element of $\mathbf{E}(h, k)$ is

$$[\mathbf{E}(h, k)]_{\alpha\beta} = \delta_{h\alpha} + \delta_{k\beta} - 2\delta_{h\alpha}\delta_{k\beta} + (\mathbf{G})_{\alpha\beta}(1 - \delta_{h\alpha})(1 - \delta_{k\beta}). \tag{J.18}$$

Relation (J.15) is far easier to apply for a multicomponent system than the original relations (J.2) or (J.12).

† A. Ben-Naim, *J. Chem. Phys.* **63**, 2064 (1975).

List of Abbreviations

BE	Binding energy
CP	Chemical potential
CES	Completely equilibrated system
CN	Coordination number
CMC	Critical micelle concentration
DI	Dilute ideal
FG	Functional group
GPF	Grand partition function
HS	Hard sphere
HB	Hydrogen bond
HϕI	Hydrophilic
HϕO	Hydrophobic
i.g.	Ideal gas
KB	Kirkwood–Buff
lhs	Left-hand side
LJ	Lennard–Jones
MM	McMillan–Mayer
	Mixture model
MDF	Molecular distribution function
1-D	One-dimensional
PES	Partially equilibrated system
PF	Partition function
PS	Preferential solvation
QCDF	Quasi-component distribution function
rhs	Right-hand side
SPT	Scaled particle theory
SC	Side chain
SI	Solvent-induced
	Symmetrical ideal
ST	Statistical thermodynamics
SOW	Structure of water
3-D	Three-dimensional
TSM	Two-structure model
VP	Voronoi polyhedron

Index